Heterostructured Materials

Heterostructured Materials
Novel Materials with Unprecedented Mechanical Properties

edited by

Xiaolei Wu

Yuntian Zhu

Jenny Stanford
Publishing

Published by

Jenny Stanford Publishing Pte. Ltd.
Level 34, Centennial Tower
3 Temasek Avenue
Singapore 039190

Email: editorial@jennystanford.com
Web: www.jennystanford.com

British Library Cataloguing-in-Publication Data
A catalogue record for this book is available from the British Library.

Heterostructured Materials: Novel Materials with Unprecedented Mechanical Properties

Copyright © 2022 by Jenny Stanford Publishing Pte. Ltd.
All rights reserved. This book, or parts thereof, may not be reproduced in any form or by any means, electronic or mechanical, including photocopying, recording or any information storage and retrieval system now known or to be invented, without written permission from the publisher.

For photocopying of material in this volume, please pay a copying fee through the Copyright Clearance Center, Inc., 222 Rosewood Drive, Danvers, MA 01923, USA. In this case permission to photocopy is not required from the publisher.

ISBN 978-981-4877-10-7 (Hardcover)
ISBN 978-1-003-15307-8 (eBook)

Contents

Preface   xxi

Part I   Perspective and Overview

1. Heterogeneous Materials: A New Class of Materials with Unprecedented Mechanical Properties   3
Xiaolei Wu and Yuntian Zhu
- 1.1 Background   4
- 1.2 Definition of Heterostructured Materials   5
- 1.3 Deformation Behavior of Heterostructured Materials   5
- 1.4 HDI Strengthening and HDI Work Hardening   7
- 1.5 Microstructural Requirement for the Optimum Mechanical Properties   10
- 1.6 Future Perspective   11

2. Perspective on Heterogeneous Deformation Induced (HDI) Strengthening and Work Hardening   17
Yuntian Zhu and Xiaolei Wu
- 2.1 Background   18
- 2.2 Brief History of Back Stress   20
- 2.3 Dislocation Models for Back Stress   21
- 2.4 Back Stress and Mechanical Properties   22
- 2.5 Issues with the Back Stress Concept   23
- 2.6 New Definition   25
- 2.7 Outstanding Issues   25

3. Ductility and Plasticity of Nanostructured Metals: Differences and Issues   33
Yuntian Zhu and Xiaolei Wu
- 3.1 Introduction   34
- 3.2 Ductility   35
- 3.3 Plasticity   38

3.4		Relationship between Ductility and Plasticity	39
3.5		Confusions, Misconceptions and Clarifications	39
	3.5.1	Misconception/Confusion 1: Tensile Ductility	39
	3.5.2	Misconception/Confusion 2: Mobile Dislocations Lead to Good Ductility	40
	3.5.3	Misconception/Confusion 3: Low Ductility Equals Low Plasticity	40
	3.5.4	Misconception/Confusion 4: Cross-Area-Reduction as an Indicator of Ductility	41
3.6		Issues for Nanostructured Materials	41
	3.6.1	Sample Size Effect	41
	3.6.2	Approaches to Improve Ductility	42
3.7		Summary	43

Part II Fundamentals of Heterostructured Materials

4. Extraordinary Strain Hardening by Gradient Structure — 53
Xiaolei Wu, Ping Jiang, Liu Chen, Fuping Yuan, and Yuntian Zhu

4.1	Introduction	54
4.2	Microstructural Characterization of Gradient Structure	55
4.3	Unique Mechanical Responses under Uniaxial Tension	56
4.4	Discussion and Summary	61

5. Heterostructured Lamella Structure Unites Ultrafine-Grain Strength with Coarse-Grain Ductility — 73
Xiaolei Wu, Muxin Yang, Fuping Yuan, Guilin Wu, Yujie Wei, Xiaoxu Huang, and Yuntian Zhu

5.1	Introduction	74
5.2	Microstructure of Heterogeneous Lamella Structure	75
5.3	Mechanical Properties and Strain Hardening of HL Structure	76
5.4	Bauschinger Effect and Back Stresses	78
5.5	Strain Partitioning	81

		5.6	Materials and Methods	83
		5.6.1	Materials	83
		5.6.2	Asymmetrical Rolling (AsR) for Heterostructured Lamella (HL) Structured Ti	84
		5.6.3	Tensile Test and Loading-Unloading-Reloading (LUR) Test	84
		5.6.4	EBSD and TEM Observations	85

6. **Synergetic Strengthening by Gradient Structure** — 89

   X. L. Wu, P. Jiang, L. Chen, J. F. Zhang, F. P. Yuan, and Y. T. Zhu

7. **Hetero-Deformation-Induced Strengthening and Strain Hardening in Gradient Structure** — 105

   Muxin Yang, Yue Pan, Fuping Yuan, Yuntian Zhu, and Xiaolei Wu

8. **Residual Stress Provides Significant Strengthening and Ductility in Gradient Structured Materials** — 119

   Muxin Yang, Runguang Li, Ping Jiang, Fuping Yuan, Yandong Wang, Yuntian Zhu, and Xiaolei Wu

8.1	Introduction	120
8.2	Results and Discussion	121
8.3	Conclusion	127

9. **Mechanical Properties of Copper/Bronze Laminates: Role of Boundaries** — 131

   Xiaolong Ma, Chongxiang Huang, Jordan Moering, Mathis Ruppert, Heinz Werner Höppel, Mathias Göken, Jagdish Narayan, and Yuntian Zhu

9.1	Introduction		132
9.2	Experimental Methods		134
9.3	Results		136
	9.3.1	Microstructures	136
	9.3.2	Heterogeneity Across Boundaries	138
	9.3.3	Uniaxial Tensile Tests	139
	9.3.4	Ex-situ EBSD Mapping and GND Characterization	141

9.4	Discussions	146
	9.4.1 Dislocation Pile-Up Model for the GND Density Close to Boundaries	146
	9.4.2 Role of Boundary in Deformation of Nanostructured Bronze	147
	9.4.3 Effect of Boundary Spacing on HDI Hardening	148
9.5	Conclusion	150

10. Hetero-Boundary-Affected Region (HBAR) for Optimal Strength and Ductility in Heterostructured Laminate — 157

C. X. Huang, Y. F. Wang, X. L. Ma, S. Yin, H. W. Höppel, M. Göken, X. L. Wu, H. J. Gao, and Y. T. Zhu

10.1	Introduction	158
10.2	Heterostructured Copper–Bronze Laminates	160
10.3	Hetero-Boundary-Affected Region (HBAR)	161
10.4	Theoretical Modeling of the Critical HBAR Width	163
10.5	Mechanical Behaviors Controlled by Interfacial Spacing	164
10.6	Discussion and Summary	166
10.7	Materials and Methods	168
	10.7.1 Material Preparation	168
	10.7.2 Microstructural Observations	168
	10.7.3 DIC Characterization	168
	10.7.4 Mechanical Testing	169

11. In-situ Observation of Dislocation Dynamics Near Heterostructured Boundary — 177

Hao Zhou, Chongxiang Huang, Xuechao Sha, Lirong Xiao, Xiaolong Ma, Heinz Werner Höppel, Mathias Göken, Xiaolei Wu, Kei Ameyama, Xiaodong Han, and Yuntian Zhu

12. Hetero-Deformation Induced (HDI) Hardening Does Not Increase Linearly with Strain Gradient — 191

Y. F. Wang, C. X. Huang, X. T. Fang, H. W. Höppel, M. Göken, and Y. T. Zhu

13. **Extra Strengthening in a Coarse/Ultrafine Grained Laminate: Role of Gradient Boundaries**   **203**

Y. F. Wang, M. S. Wang, X. T. Fang, Q. He, F. J. Guo, R. O. Scattergood, C. X. Huang, and Y. T. Zhu

13.1	Introduction	204
13.2	Experimental Methods	206
13.3	Results	208
	13.3.1 Microstructural Heterogeneity and Gradient Boundary	208
	13.3.2 Synergistic Strengthening and Strain Hardening	209
	13.3.3 Height Profile and Strain Gradient Across Boundary	210
	13.3.4 DIC and Strain Gradient Across Boundary	212
13.4	Discussion	214
	13.4.1 Formation of Strain Gradient Across Gradient Boundary	214
	13.4.2 GNDs Pile-Up Across Gradient Boundary	215
	13.4.3 Extraordinary Strengthening Effects of Gradient Boundary	217
13.5	Conclusions	220

14. **Ductility by Shear Band Delocalization in the Nano-Layer of Gradient Structure**   **225**

Fuping Yuan, Dingshun Yan, Jiangda Sun, Lingling Zhou, Yuntian Zhu, and Xiaolei Wu

15. **Heterostructure Induced Dispersive Shear Bands in Heterostructured Cu**   **239**

Y. F. Wang, C. X. Huang, Q. He, F. J. Guo, M. S. Wang, L. Y. Song, and Y. T. Zhu

16. **Dense Dispersed Shear Bands in Gradient-Structured Ni**   **251**

Yanfei Wang, Chongxiang Huang, Yusheng Li, Fengjiao Guo, Qiong He, Mingsai Wang, Xiaolei Wu, Ronald O. Scattergood, and Yuntian Zhu

     16.1   Introduction   252

16.2　Experimental Procedures　254
　　16.2.1　Materials and Processing　254
　　16.2.2　Microstructural Characterization and Mechanical Tests　255
　　16.2.3　DIC Strain Characterization　256
16.3　Results　257
　　16.3.1　Surface Roughness of the Gradient Samples　257
　　16.3.2　Gradient Microstructure and Microhardness　257
　　16.3.3　Strength-Ductility Combination　258
　　16.3.4　Dense Dispersed Shear Bands in Nanostructured Layer　259
16.4　Discussion　263
　　16.4.1　Unique Characteristics of Dispersed Shear Bands　263
　　16.4.2　Nucleation of Dispersed Shear Bands　264
　　16.4.3　Stable Evolution of Dispersed Shear Bands　267
　　16.4.4　Microstructure Evolution in Shear Bands　269
　　16.4.5　Effects of Surface Roughness and Strength Heterogeneity on Shear Banding　271
16.5　Conclusions　271

Part III　Gradient Structure

17. Combining Gradient Structure and TRIP Effect to Produce Austenite Stainless Steel with High Strength and Ductility　283

Xiaolei Wu, Muxin Yang, Fuping Yuan, Liu Chen, and Yuntian Zhu

17.1　Introduction　284
17.2　Experimental Procedure　286
　　17.2.1　Materials and SMAT Process　286
　　17.2.2　Mechanical Property Tests　287
　　17.2.3　Microstructural Characterization　287
17.3　Results　288
　　17.3.1　Mechanical Behaviors　288

		17.3.1.1	Tensile property and strain hardening	288
		17.3.1.2	Microhardness evolution	290
		17.3.1.3	Dislocation density evolution	292
	17.3.2	Microstructural Evolution		293
		17.3.2.1	Microstructure by SMAT processing	293
		17.3.2.2	Change of α'-martensite fraction during tensile testing	297
		17.3.2.3	Microstructural evolution during tensile testing	299
17.4	Discussion			300
	17.4.1	Dynamic Strain Partitioning		300
	17.4.2	Deformation-Induced Phase Transformation and TRIP Effect		302
17.5	Conclusions			304

18. Gradient Structure Produces Superior Dynamic Shear Properties — 311

Xiangde Bian, Fuping Yuan, Xiaolei Wu, and Yuntian Zhu

19. On Strain Hardening Mechanism in Gradient Nanostructures — 323

Jianjun Li, G. J. Weng, Shaohua Chen, and Xiaolei Wu

19.1	Introduction		324
19.2	Constitutive Model for Gradient Structure		328
	19.2.1	Flow Stress for Component Homogeneous Layers	330
	19.2.2	Calculation of GNDs Density and HDI Stress	333
	19.2.3	Overall Mechanical Response of Gradient Structure	336
19.3	Results and Discussion		337
	19.3.1	Stress-Strain Curves of Homogeneous IF Steels	341
	19.3.2	Lateral Surface Non-uniform Deformation in Gradient IF Steels	341

		19.3.3	Strain Hardening in Gradient IF Steels	343

	19.4	19.3.4	Strength-Ductility Map for Gradient IF Steels	346
		Conclusions		349

20. Extraordinary Bauschinger Effect in Gradient Structured Copper — 361

Xiaolong Liu, Fuping Yuan, Yuntian Zhu, and Xiaolei Wu

21. Atomistic Tensile Deformation Mechanisms of Fe with Gradient Nano-Grained Structure — 371

Wenbin Li, Fuping Yuan, and Xiaolei Wu

- 21.1 Introduction — 372
- 21.2 Simulation Techniques — 374
- 21.3 Results and Discussions — 376
- 21.4 Concluding Remarks — 385

22. Strain Hardening Behaviors and Strain Rate Sensitivity of Gradient-Grained Fe under Compression over a Wide Range of Strain Rates — 389

Fuping Yuan, Ping Chen, Yanpeng Feng, Ping Jiang, and Xiaolei Wu

- 22.1 Introduction — 390
- 22.2 Experimental Procedures — 392
- 22.3 Experimental Results and Discussions — 395
- 22.4 Conclusions — 412

23. Mechanical Properties and Deformation Mechanism of Mg-Al-Zn Alloy with Gradient Microstructure in Grain Size and Orientation — 417

Liu Chen, Fuping Yuan, Ping Jiang, Jijia Xie, and Xiaolei Wu

- 23.1 Introduction — 418
- 23.2 Experimental Procedures — 421
- 23.3 Results — 424
  - 23.3.1 Gradient Structure in Grain Size and Texture — 424

	23.3.2	Mechanical Properties		426
	23.3.3	Repeated Stress Relaxation Tests		428
	23.3.4	Microstructure and Texture Observation		432
		23.3.4.1	Gradient microstructure after SMAT	432
		23.3.4.2	Texture change during tensile deformation	435
		23.3.4.3	Non-basal dislocation observation	438
23.4	Discussion			439
	23.4.1	Formation of Dual Gradient Microstructure		439
	23.4.2	Influence of Grain Size on Deformation Mechanism		440
	23.4.3	Influence of Orientation and Its Gradient on Deformation Mechanism		441
	23.4.4	Coupling between Size Gradient and Orientation Gradient		443
23.5	Conclusions			444

24. The Evolution of Strain Gradient and Anisotropy in Gradient-Structured Metal — 449

Xiangde Bian, Fuping Yuan, Xiaolei Wu, and Yuntian Zhu

24.1	Introduction		450
24.2	Materials and Experimental Procedures		452
	24.2.1	Materials	452
	24.2.2	Microstructural Characterization	452
	24.2.3	Quasi-Static Uniaxial Tensile Tests Coupled with Digital Image Correlation	453
24.3	Experimental Results and Discussions		455
	24.3.1	Microstructural Characterization and Tensile Properties	455
	24.3.2	Strain Contours and Strain Distributions along the Depth	456
	24.3.3	Evolutions of Strain Gradient and Anisotropy	459

	24.3.4 HDI Hardening	467
24.4	Concluding Remarks	469

25. Influence of Gradient Structure Volume Fraction on the Mechanical Properties of Pure Copper — 473

Xincheng Yang, Xiaolong Ma, Jordan Moering, Hao Zhou, Wei Wang, Yulan Gong, Jingmei Tao, Yuntian Zhu, and Xinkun Zhu

25.1	Introduction	474
25.2	Experimental	474
25.3	Results	476
	25.3.1 Microstructure	476
	25.3.2 Vickers Hardness	476
	25.3.3 Mechanical Behaviors	478
	25.3.4 In-situ SEM Observation	481
25.4	Discussion	484
	25.4.1 Synergetic Strengthening and Extra Strain Hardening	484
	25.4.2 Optimum GS Volume Fraction for Extra Strain Hardening	485
	25.4.3 Strength-Ductility Combinations	486
25.5	Conclusions	487

26. The Role of Shear Strain on Texture and Microstructural Gradients in Low Carbon Steel Processed by Surface Mechanical Attrition Treatment — 489

Jordan Moering, Xiaolong Ma, Guizhen Chen, Pifeng Miao, Guozhong Li, Gang Qian, Suveen Mathaudhu, and Yuntian Zhu

27. Bauschinger Effect and Hetero-Deformation Induced (HDI) Stress in Gradient Cu-Ge Alloy — 499

Xianzhi Hu, Shenbao Jin, Hao Zhou, Zhe Yin, Jian Yang, Yulan Gong, Yuntian Zhu, Gang Sha, and Xinkun Zhu

27.1	Introduction	500
27.2	Experiment	501
27.3	Results and Discussion	501
27.4	Summary	511

28. **Gradient Structured Copper Induced by Rotationally Accelerated Shot Peening**   515

X. Wang, Y. S. Li, Q. Zhang, Y. H. Zhao, and Y. T. Zhu

    28.1   Introduction   516
    28.2   Experimental   517
    28.3   Results and Discussion   518
    28.4   Conclusion   524

29. **Microstructure Evolution and Mechanical Properties of 5052 Alloy with Gradient Structures**   527

Yusheng Li, Lingzhen Li, Jinfeng Nie, Yang Cao, Yonghao Zhao, and Yuntian Zhu

    29.1   Introduction   528
    29.2   Experimental   529
    29.3   Results and Discussion   530
           29.3.1   OM/EBSD Observations   530
           29.3.2   TEM Characterization   531
                   29.3.2.1   Grain refinement via dislocation activities (depth >40 μm)   532
                   29.3.2.2   Grain refinement via DRX (depth <40 μm)   534
           29.3.3   Mechanical Properties   538
    29.4   Conclusions   541

30. **Quantifying the Synergetic Strengthening in Gradient Material**   547

Y. F. Wang, C. X. Huang, M. S. Wang, Y. S. Li, and Y. T. Zhu

31. **Achieving Gradient Martensite Structure and Enhanced Mechanical Properties in a Metastable β Titanium Alloy**   559

Xinkai Ma, Fuguo Li, Zhankun Sun, Junhua Hou, Xiaotian Fang, Yuntian Zhu, and Carl C. Koch

    31.1   Introduction   560
    31.2   Materials and Methods   561
           31.2.1   Sample Preparation   561
           31.2.2   Sample Processing and Mechanical Testing   561

		31.2.3	Microstructure Characterization	562
	31.3	Results		563
		31.3.1	Martensitic Transformation during Torsion Processing and Subsequent Tensile Deformation	563
		31.3.2	Mechanical Behavior of the Torsion-Processed Sample	566
		31.3.3	Gradient Microstructure and Fracture Behavior	568
	31.4	Discussion		571
		31.4.1	Nucleation Mechanism of Gradient α'' Martensite	571
		31.4.2	Evolution of β Phase and α'' Martensite	574
		31.4.3	The Effect of Gradient α'' Martensite Structure on Strain Hardening	576
	31.5	Conclusions		578

Part IV Heterogeneous Grain Structure

32. Dynamically Reinforced Heterogeneous Grain Structure Prolongs Ductility in a Medium-Entropy Alloy with Gigapascal Yield Strength   **585**

Muxin Yang, Dingshun Yan, Fuping Yuan, Ping Jiang, Evan Ma, and Xiaolei Wu

	32.1	Introduction		586
	32.2	Manuscript Text		588
	32.3	Materials and Methods		597
		32.3.1	Material Fabrication and Sample Preparation	597
		32.3.2	Mechanical Property Testing	598
		32.3.3	Microstructural Characterization	598

33. Dynamic Shear Deformation of a CrCoNi Medium-Entropy Alloy with Heterogeneous Grain Structures   **613**

Yan Ma, Fuping Yuan, Muxin Yang, Ping Jiang, Evan Ma, and Xiaolei Wu

	33.1	Introduction	614
	33.2	Materials and Experimental Procedures	616

33.3	Results and Discussions		617
	33.3.1	Microstructural Characterization Before Dynamic Shear Tests	617
	33.3.2	Dynamic Shear Properties	620
	33.3.3	Microstructure Evolution during the Homogeneous Dynamic Shear Deformation	622
	33.3.4	Temperature Rise during the Homogeneous Dynamic Shear Deformation	630
	33.3.5	Characteristics of ASB	632
33.4	Concluding Remarks		634

34. Superior Strength and Ductility of 316L Stainless Steel with Heterostructured Lamella Structure — **641**

Jiansheng Li, Bo Gao, Yang Cao, Yusheng Li, and Yuntian Zhu

34.1	Introduction			642
34.2	Experimental			644
	34.2.1	Characterization of the As-Received Sample		644
	34.2.2	Preparation of HLS		645
	34.2.3	Microstructure Analysis		645
	34.2.4	Mechanical Property Tests		645
34.3	Results			646
	34.3.1	Microstructures		646
		34.3.1.1	XRD analysis	646
		34.3.1.2	Microstructural evolution characterized by EBSD	647
		34.3.1.3	Microstructure observation by TEM	648
	34.3.2	Mechanical Properties		651
		34.3.2.1	Microhardness	651
		34.3.2.2	Tensile behaviors	652
34.4	Discussion			653
	34.4.1	Formation and Evolution Mechanisms of HLS		653
	34.4.2	Enhanced Strength and Ductility		656
34.5	Conclusions			660

Part V Laminate Materials

35. Strain Hardening and Ductility in a Coarse-Grain/Nanostructure Laminate Material — 667
X. L. Ma, C. X. Huang, W. Z. Xu, H. Zhou, X. L. Wu, and Y. T. Zhu

36. Effect of Strain Rate on Mechanical Properties of Cu/Ni Multilayered Composites Processed by Electrodeposition — 679
Zhengrong Fu, Zheng Zhang, Lifang Meng, Baipo Shu, Yuntian Zhu, and Xinkun Zhu
- 36.1 Introduction — 680
- 36.2 Experimental Procedure — 681
- 36.3 Results — 682
- 36.4 Discussions — 687
- 36.5 Conclusions — 690

Part VI Dual-Phase Structure

37. Simultaneous Improvement of Tensile Strength and Ductility in Micro-Duplex Structure Consisting of Austenite and Ferrite — 697
L. Chen, F. P. Yuan, P. Jiang, J. J. Xie, and X. L. Wu
- 37.1 Introduction — 698
- 37.2 Experimental Procedures — 700
- 37.3 Results — 701
  - 37.3.1 Mechanical Property — 701
  - 37.3.2 Microstructure Observation — 702
- 37.4 Discussion — 708
  - 37.4.1 Strengthening Mechanism of Dual Phase Microstructure — 708
  - 37.4.2 Effect of Phase Interaction on Strain Hardening Rate — 713
- 37.5 Conclusion — 716

38. Strain Hardening in Fe–16Mn–10Al–0.86C–5Ni High Specific Strength Steel — 721
M. X. Yang, F. P. Yuan, Q. G. Xie, Y. D. Wang, E. Ma, and X. L. Wu
- 38.1 Introduction — 722

38.2	Materials and Experimental Procedures		724
	38.2.1	Materials	724
	38.2.2	Mechanical Property Tests	725
	38.2.3	Synchrotron Based High Energy X-Ray Diffraction	725
	38.2.4	Microstructural Characterization	726
38.3	Experimental Results		727
	38.3.1	Microstructural Characterization	727
	38.3.2	Tensile Properties	729
	38.3.3	Strain Hardening due to HDI Stress	732
	38.3.4	Load Transfer and Strain Partitioning	734
		38.3.4.1 Load transfer revealed by in situ diffraction measurements	734
		38.3.4.2 Strain partitioning from aspect ratio measurements	736
38.4	Discussion		737
	38.4.1	Plastic Deformation in HSSS	737
	38.4.2	Strain Hardening	739
	38.4.3	Unloading Yield Effect	741
38.5	Conclusions		742

39. Deformation Mechanisms for Superplastic Behaviors in a Dual-Phase High Specific Strength Steel with Ultrafine Grains — **749**

Wei Wang, Muxin Yang, Dingshun Yan, Ping Jiang, Fuping Yuan, and Xiaolei Wu

39.1	Introduction	750
39.2	Materials and Experimental Procedures	751
39.3	Results and Discussions	752
39.4	Conclusions	768

40. Plastic Deformation Mechanisms in a Severely Deformed Fe-Ni-Al-C Alloy with Superior Tensile Properties — **773**

Yan Ma, Muxin Yang, Ping Jiang, Fuping Yuan, and Xiaolei Wu

40.1	Introduction	774
40.2	Results	776

		40.2.1	Microstructures before Tensile Tests and Quasi-Static Uniaxial Tensile Properties	776
		40.2.2	Deformation Mechanisms during Tensile Deformation for Solution Treated and CR Samples	780
	40.3	Discussions		785
	40.4	Materials and Experimental Procedures		789
		40.4.1	Materials	789
		40.4.2	Microstructure Characterizations	790
		40.4.3	Mechanical Testing	790

41. Hetero-Deformation Induced (HDI) Strengthening and Strain Hardening in Dual-Phase Steel — **797**

X. L. Liu, Q. Q. Xue, W. Wang, L. L. Zhou, P. Jiang, H. S. Ma, F. P. Yuan, Y. G. Wei, and X. L. Wu

	41.1	Introduction		798
	41.2	Materials and Experimental Methods		800
	41.3	Experimental Results		801
		41.3.1	Microstructural Characterization	801
		41.3.2	Yield-Point Phenomenon	803
		41.3.3	HDI Stress during Tensile Deformation	805
		41.3.4	Evolution of Schmid Factor and KAM Value	806
	41.4	Modeling HDI Stress in Dual-Phase Microstructure		808
	41.5	Discussions		812
	41.6	Conclusions		814

Index — 821

Preface

It has been a perpetual challenge to make metals and alloys strong and tough. A tough metal needs to possess both high strength and high ductility. However, there exists a well-known banana curve where high strength is usually accompanied by low ductility, and consequently low toughness. According to the Considère criterion for mechanical stability, i.e., for preventing necking, during a tensile test, $d\sigma/d\varepsilon > \sigma$, where σ is the true stress and ε is the true strain, materials with high strength will need a high strain hardening rate ($d\sigma/d\varepsilon$) to maintain the same uniform tensile deformation (ductility) [1, 2]. However, mechanisms that make materials stronger often lower the strain hardening rate, which produces the banana curve. For example, refining grains of metals to ultrafine or nanocrystalline sizes can enhance the strength by several times, but also lower the strain hardening capability to nearly zero, which consequently leads to low ductility.

Recently, heterostructured (HS) materials have been found to be able to possess superior combinations of strength and ductility that are not accessible to their homogeneous counterparts [3]. For example, heterogeneous lamella Ti was found to possess the high strength of ultrafine-grained Ti and at the same time, its ductility is higher than coarse-grained Ti. The high ductility was derived from its high strain hardening rate, which is higher than the coarse-grained Ti [4]. These properties and behaviors can no longer be explained by our conventional textbook knowledge. The concept of back stress strengthening and back stress hardening was proposed to explain the superior properties of the heterostructured materials [3–6]. However, it has been now found that the back-stress concept cannot accurately represent the physical deformation process of heterostructured materials. A new term, hetero-deformation induced (HDI) strengthening and strain hardening has been proposed recently [7].

HS materials are defined as the materials with significant differences in flow stresses among their heterostructured zones, i.e., large mechanical incompatibility among adjacent zones during deformation. The large mechanical incompatibility leads to high heterogeneous (non-uniform) deformation, which produces HDI strengthening and strain hardening, and consequently results in a superior combination of high strength and ductility. It should be noted that conventional homogeneous materials also produce HDI strengthening and HDI hardening due to variations in grain sizes and grain orientations. However, their HDI stress during tensile testing is much lower than the stress caused by dislocation hardening as expressed by the Ashby equation [8, 9]:

$$\tau = \alpha G b \sqrt{\rho_S + \rho_G},$$

where α is a constant, G is the shear modulus, b is the magnitude of Burgers vector, ρ_S is the density of statistically stored dislocations, and ρ_G is the density of geometrically necessary dislocations (GNDs). Consequently, the HDI strengthening and hardening are typically ignored by researchers when discussing the mechanical behavior of conventional homogeneous materials.

For heterostructured materials, the HDI strengthening and work hardening can no longer be ignored since they are much higher than the dislocation hardening. Therefore, HDI stress can be used as a definitive criterion to decide if a material can be called heterostructured material: *A heterostructured material should have HDI stresses higher than its dislocation hardening stresses during tensile deformation.* The HDI stresses can be measured using a loading-unloading-reloading procedure [6]. Materials that meet the definition of heterostructured materials and may satisfy the above criterion include, but are not limited to, gradient materials, heterogeneous lamella materials, laminate materials, dual-phase materials, harmonic (core-shell structured) materials, metal matrix composite, etc. Note that some of these materials, e.g., gradient materials, dual-phase materials, and metal matrix composites, have been studied for over a decade. However, the physical mechanisms that are responsible for their superior mechanical properties are HDI strengthening and hardening, although this was often not well understood or explicitly stated in the earlier literature.

HS materials represent an emerging class of materials that are expected to become a major research field for the communities of materials, mechanics, and physics in the next couple of decades. One of the biggest advantages of HS materials is that they can be produced by large-scale industrial facilities and technologies and therefore can be commercialized without the scaling up and high-cost barriers that are often encountered by other advanced materials.

This collection of papers represents recent progress in the field of HS materials, especially the fundamental physics of HS materials. Since this is a new materials science field, there are still many fundamental issues that need to be studied by experiments, analytical modeling, and computational modeling. Particularly, boundary engineering and boundary-related phenomena such as strain gradient near zone boundaries, GNDs, GND interaction with and transmission through the boundaries, the evolution of back and forward stresses, and shear bands across zone boundaries are critical issues. Papers are arranged in a sequence of chapters that will help new researchers entering the field to have a quick and comprehensive understanding of the HS materials, including the fundamentals and recent progress in their processing, characterization, and properties.

Lastly, the terminology in the field of heterostructured materials has evolved over the last few years as a result of a better understanding of the field and discussions in this growing community. This has posed a problem because the terminology is inconsistent among this collection of papers, which have been published over the past few years. To avoid the confusion of readers, we have attempted to solve this problem by using the most up-to-date terminology in all chapters (papers) in this book.

References

1. Ovid'ko IA, Valiev RZ and Zhu YT. Review on superior strength and enhanced ductility of metallic nanomaterials. *Prog. Mater. Sci.* 2018; **94**:462–540.
2. Valiev RZ, Estrin Y, Horita Z, Langdon TG, Zehetbauer MJ and Zhu YT. Fundamentals of superior properties in bulk nanoSPD materials. *Mater. Res. Lett.* 2016; **4**:1–21.
3. Wu XL and Zhu YT. Heterogeneous materials: a new class of materials with unprecedented mechanical properties. *Mater. Res. Lett.* 2017; 5:527–532.
4. Wu XL, Yang MX, Yuan FP, Wu GL, Wei YJ, Huang XX and Zhu YT. Heterogeneous lamella structure unites ultrafine-grain strength with coarse-grain ductility. *Proc. Natl. Acad. Sci. U.S.A.* 2015; **112**:14501–14505.

5. Xu R, Fan GL, Tan ZQ, Ji G, Chen C, Beausir B, Xiong DB, Guo Q, Guo CP, Li ZQ and Zhang D. Back stress in strain hardening of carbon nanotube/aluminum composites. *Mater. Res. Lett.* 2018; **6**:113–120.
6. Yang MX, Pan Y, Yuan FP, Zhu YT and Wu XL. Back stress strengthening and strain hardening in gradient structure. *Mater. Res. Lett.* 2016; **4**:145–151.
7. Zhu YT and Wu XL. Perspective on heterogeneous deformation induced (HDI) hardening and back stress. *Mater. Res. Lett.* 2019; **7**:393–398.
8. Ashby MF. Deformation of plastically non-homogeneous materials. *Philos. Mag.* 1970; **21**:399.
9. Gao H, Huang Y, Nix WD and Hutchinson JW. Mechanism-based strain gradient plasticity - I. Theory. *J. Mech. Phys. Solids* 1999; **47**:1239–1263.

Part I
Perspective and Overview

Chapter 2

Chapter 1

Heterogeneous Materials: A New Class of Materials with Unprecedented Mechanical Properties

Xiaolei Wu[a,b] and Yuntian Zhu[c,d]

[a]*State Key Laboratory of Nonlinear Mechanics, Institute of Mechanics, Chinese Academy of Science, Beijing 100190, China*
[b]*School of Engineering Science, University of Chinese Academy of Sciences, Beijing 100190, China*
[c]*Department of Materials Science and Engineering, North Carolina State University, Raleigh, NC 27695, USA*
[d]*Nano Structural Materials Center, School of Materials Science and Engineering, Nanjing University of Science and Technology, Nanjing 210094, China*
xlwu@imech.ac.cn, ytzhu@ncsu.edu

Here we present a perspective on heterostructured materials, a new class of materials possessing superior combinations of strength and ductility that are not accessible to their homogeneous counterparts.

Reprinted from *Mater. Res. Lett.*, **5**(8), 527–532, 2017.

Heterostructured Materials: Novel Materials with Unprecedented Mechanical Properties
Edited by Xiaolei Wu and Yuntian Zhu
Text Copyright © 2017 The Author(s)
Layout Copyright © 2022 Jenny Stanford Publishing Pte. Ltd.
ISBN 978-981-4877-10-7 (Hardcover), 978-1-003-15307-8 (eBook)
www.jennystanford.com

Heterostructured materials consist of zones with dramatic strength differences. The zone sizes may vary in the range of micrometers to millimeters. Large strain gradients near zone boundaries are produced during deformation, which produces significant hetero-deformation-induced (HDI) stress to strengthen the material and to produce high HDI-stress work hardening for good ductility. High boundary density is required to maximize the HDI-stress, which is a new strengthening mechanism for improving mechanical properties.

1.1 Background

Materials are either strong or ductile, but rarely both at the same time [1, 2]. Stronger and tougher materials are desired for many structural applications such as transportation vehicles for higher energy efficiency and better performance. For the last three decades, nanostructured (ultrafine-grained) metals have been extensively studied because of their high strength. However, overcoming their low ductility has been a challenge [2–25], which is one of the major reasons why they have not been widely commercialized for industrial applications. Another major obstacle to practical structural applications of nanostructured metals is the challenge in scaling up for industrial production at low cost [26].

After over a century's research, we have almost reached the limit on how much further we can improve the mechanical properties of metals and alloys. Our conventional wisdom from the textbook and literature is to reinforce a weak matrix by a stronger reinforcement such as second phase particles or fibers. A question arises on if there exists yet-to-be-explored new strategies to make the next generation of metals and alloys with a "quantum jump" in strength and ductility instead of the incremental improvements that we have seen for the last several decades.

Recently, there have been several reports on superior combinations of strength and ductility in various metals and alloys that are processed to have widely different microstructures, including the gradient structure [27–32], heterostructured lamella structure [33], bi-modal structure [10, 25, 34, 35], harmonic structure [36–38], laminate structure [39, 40], dual phase steel [41–43], etc [44]. These materials have one common feature: there is a dramatic difference in strength between different zones while the sizes and geometry of the zones may vary widely. In other words, there are huge microstructural heterogeneities in these materials. Therefore, these materials can be considered as heterostructured materials.

In this perspective, we will present the fundamental physics that renders these heterostructured materials superior mechanical properties as well as the microstructures that are required for producing the best mechanical properties.

1.2 Definition of Heterostructured Materials

Heterostructured materials can be defined as materials with dramatic heterogeneity in strength from one zone area to another. This strength heterogeneity can be caused by microstructural heterogeneity, crystal structure heterogeneity or compositional heterogeneity. The zone sizes could be in the range of micrometers to millimeters, and the zone geometry can vary to form very diverse material systems.

1.3 Deformation Behavior of Heterostructured Materials

During deformation, e.g. tensile testing, of the heterostructured materials, with increasing applied strain, the deformation process can be classified into three stages (see Fig. 1.1). **In stage I**, both soft and hard zones deform elastically, which is similar to a conventional homogeneous material.

In stage II, the soft zones will start dislocation slip first to produce plastic strain while the hard zones will remain elastic, which creates a mechanical incompatibility. The soft zones need to deform together with the neighboring hard zones and therefore cannot plastically deform freely. The strain at the zone boundary needs to be continuous, although the softer zones will typically accommodate more strains since they are plastically deforming. Therefore, there will be a plastic strain gradient in the soft zone near the zone boundary. This strain gradient needs to be accommodated by geometrically necessary dislocations, which will make the softer zone appears stronger [33, 45], leading to synergetic strengthening to increase the global measured yield strength of the material [29].

In an extreme/ideal case, the soft zones are completely surrounded by the hard zone matrix so that the soft zone cannot change its shape as required by plastic deformation until the hard zone matrix starts to deform plastically. Geometrically necessary dislocations will pile up at the zone boundaries in the soft zone, but cannot transmit across the zone boundary, building up high HDI-stress (see Fig. 1.2). This can make the soft zone almost as strong as the hard zone matrix, making the global yield strength much higher than what is predicted by the rule of mixtures [33].

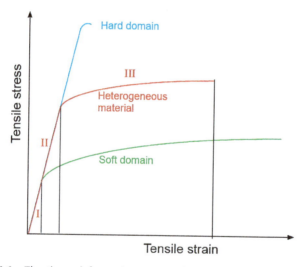

Figure 1.1 The three deformation stages of the heterostructured materials (the red stress-strain curve).

In stage III, both the soft and hard zones deform plastically, but the soft zones sustain much higher strain than the hard zones, producing the so-called strain partitioning [43, 46–50]. When neighboring zones sustain different plastic strains, strain gradients are expected to exist near the zone boundaries in both the soft and hard zones. These strain gradients will become larger with increasing strain partitioning, and consequently produces HDI work hardening. The HDI work hardening will help with prevent necking during tensile testing, thus improving ductility. This is the primary reason why the dual-phase steel has extraordinary work hardening, and consequently high ductility [43, 46, 48, 50].

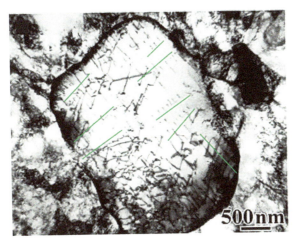

Figure 1.2 A 4-µm soft grain surrounded by hard ultrafine grained matrix in heterostructured lamella Ti. Dislocation pileups are marked by green lines.

1.4 HDI Strengthening and HDI Work Hardening

As discussed above, HDI-stress plays a significant role in the reported extraordinary strength and ductility of heterostructured metals. Two types of dislocations are usually involved in the plastic deformation of metals and alloys: statistically stored dislocations and geometrically necessary dislocations. The flow stress as a function of dislocation density is conventionally calculated as [51–53]

$$\tau = \alpha G b \sqrt{\rho_S + \rho_G} \tag{1.1}$$

where τ is the shear flow stress, α is a constant, G is shear modulus, b is the magnitude of Burgers vector, ρ_S and ρ_G the densities of statistically stored dislocations and geometrically necessary dislocations, respectively. In this equation, the statistically stored dislocation and geometrically necessary dislocations are treated to have the same contribution to the flow stress. Obviously, the HDI-stress caused by geometrically necessary dislocations area is ignored.

For conventional homogeneous metals, Eq. (1.1) has been used to reasonably explain their mechanical behaviors, because the HDI-stress caused by the geometrically necessary dislocations are relatively small. However, for heterostructured materials the HDI-stress can be much higher than the strengthening associated with the statistically stored dislocations [33, 45], and therefore has to be considered. As discussed later, the HDI-stress can be utilized to design heterostructured materials with unprecedented mechanical properties.

What is the physical origin of HDI-stress? To answer this question, let's have a look at the piling up of geometrically dislocations as schematically shown in Fig. 1.3A. Assuming that there is a dislocation source at point X, which emits geometrically dislocations with the same Burgers vector towards the zone boundary on a slip plane. Under an applied shear stress τ_a, there are seven dislocations piled up and the system reached equilibrium. These dislocations collectively produce a long-range stress, τ_b, toward the dislocation source as indicated by the arrow, which counter-balances the applied stress. The effective stress at the dislocation source can be expressed as $\tau_e = \tau_a - \tau_b$. If the critical stress to operate the dislocation source is τ_c, then τ_e has to be higher than τ_c for the dislocation source to emit more dislocations. In other words, higher applied stress is needed for more dislocations to be piled up. Therefore, HDI-stress is a long-range stress created by geometrically necessary dislocations. The above discussion describes how HDI-stress can be produced from an individual dislocation pile. The experimentally measured HDI-stress is the global collective HDI-stress in the whole sample, just like the measured yield stress is a global stress contributed by individual yielding events in the whole sample.

HDI-stress is connected with plastic strain gradient. The plastic strain is produced by the slip of dislocations with each dislocation leaving a displacement of one Burgers vector in its wake. Therefore, in Fig. 1.3A the strain at the boundary is zero at the zone boundary (pile-up head), and the strain is increased to seven Burges vector at the dislocation source. The black curve in Fig. 1.3B shows a smoothed strain curve as a function of distance from the boundary, the other curve in Fig. 1.3B is the corresponding strain gradient curve. Therefore, the pile-up of geometrically necessary dislocations produces strain gradient as well as stress gradient (Fig. 1.3C). In other words, if strain gradient is observed, there will exist the pile up of geometrically necessary dislocations and corresponding HDI-stress.

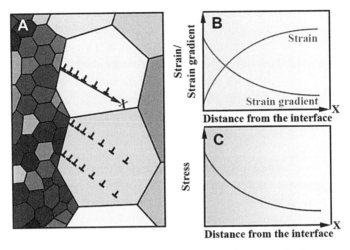

Figure 1.3 (A) Schematics of the piling-up of geometrically necessary dislocations. (B) Plastic strain and strain gradient as a function of distance from the zone boundary. (C) The effective stress as a function of distance from the zone boundary.

Note that another scenario to produce an array of geometrically necessary dislocations and the associated HDI-stress is when dislocations are emitted from a ledge on the zone boundary and or grain boundary, but they form an array on a slip plane near the boundary [54]. This will produce the same HDI-stress to counter balance the stress at the dislocation source although the strain gradient will be different from what is described in Fig. 1.3B. The highest plastic strain always occurs at the dislocation source.

HDI-stress has the same physical origin as the Bauschinger effect [55]. Higher Bauschinger effect corresponds to higher HDI-stress. However, HDI-stress can be used to improve the strength and ductility of metals if appropriate heterostructures can be designed, instead of just phenomenon of mechanical behavior as the Bauschinger effect is often regarded. The HDI-stress and its evolution during a tensile test can be measured experimentally [33, 45].

1.5 Microstructural Requirement for the Optimum Mechanical Properties

After discussing the role of HDI-stress in the mechanical behavior and the physical origin of HDI-stress, it naturally follows that we can design heterostructures to maximize HDI-stress for the best mechanical properties. Since HDI-stress is produced by dislocation pile-up at zone boundaries, we should design the heterostructure with high density of zone boundaries. However, the spacing between the zone boundaries should be large enough to allow effective dislocation pileup in at least the soft zones and ideally in both soft and hard zones. Another design factor is to maximize strain partitioning among heterostructured zones, which will increase the strain gradient, and consequently increase HDI work hardening. This means that the strength between the zones should be large, and the zone geometries should be such that large strain partitioning can be easily realized.

With the above two criteria for high HDI-stress, we can make comments on the effectiveness of various heterostructures. For the gradient structure [27–32], there will be two dynamically migrating boundaries during the tensile tests [28, 29], which allows dislocation accumulation over the whole sample volume. However, the low boundary density also limits its capability of HDI work hardening. For bi-modal structure [10, 25, 34, 35], the boundary density is usually not maximized, which did not make the full use of the HDI hardening potential. For laminate structure [39, 40], the soft and hard laminates are subjected to the same applied strain, which limits their strain partitioning capability and consequently HDI-stress development. For dual phase steel [41–43], the hard martensitic zones typically accounts for 5–30% by volume and are embedded in the soft matrix. Although this allows significant strain partitioning to increase the HDI work hardening and consequently high ductility, the continuous soft matrix also allows the material to yield at low stress, which is why dual-phase materials typically have very high ductility, but limited enhancement in strength. For harmonic structure [36–38], the soft zones are totally surrounded by hard zone layers similar to a cellular structure. It has been observed to enhance the ductility, but the strength improvement is so far limited, which could be improved by reducing the zone boundary spacing and hard zone volume fraction to further constraint the soft zones. For the heterostructured lamella structure [33], the soft lamella zones with a volume fraction of <30% are embedded in a hard matrix (Fig. 1.4), which renders it high strength because the rigid constraint by the hard matrix makes the soft zones almost as strong as the hard matrix during the deformation stage II (see Fig. 1.1). The strong strain partitioning during deformation stage III also render unprecedented high strain hardening, which increases its ductility. Therefore, the heterostructured lamella structure [33] presents a near-ideal heterostructure. This explains why the heterostructured lamella structure has shown the most dramatic improvement of strength and ductility among all reported types of heterostructured materials.

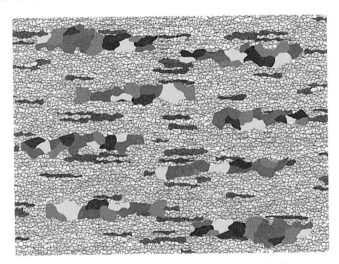

Figure 1.4 Schematics of lamella structure with elongated soft coarse-grained zones embedded in ultrafine-grained matrix.

1.6 Future Perspective

Heterostructured materials is a fast emerging field that is to become a hot research field in the post nanostructured materials era. There is a huge research community in the area of nanostructured materials, which has been extensively studied for over three decades. The maturing of this field and the challenge to the practical applications of nanostructured materials have made it hard to secure research funding in many countries in the world. The heterostructured materials have many similarities to the nanostructured materials because the hard zones could be the nanostructured/ultrafine grained, which makes it easy for researchers in the nanostructured materials community to transit to the heterostructured materials field. In addition, several types of the heterostructured materials can be produced by current industrial processing technology so that their practical applications have a very low barrier.

There are many scientific and engineering issues that need to be addressed by both experimentalists and modelers from the communities of both materials science and mechanics. The heterostructured materials community is quickly growing, with more international conferences and workshops being organized, e.g. the bi-annual TMS symposium on heterogeneous and gradient materials. Sessions on heterostructured materials are also inserted into conventional successful symposia such as the TMS bi-annual meeting on ultrafine-grained materials. These activities help with the fast development in the area of heterostructured materials.

Acknowledgements

X.W. was supported by the National Natural Science Foundation of China (NSFC) under grant Nos. 11472286, 11572328 and 11672313, the Strategic Priority Research Program of the Chinese Academy of Sciences under grant No. XDB22040503, and the National Key Basic Research Program of China under grant Nos. 2012CB932203 and 2012CB937500. Y.Z. was supported by the US Army Research Office, and the Jiangsu Key Laboratory of Advanced Micro & Nano Materials and Technology.

References

1. Valiev RZ, Alexandrov IV, Zhu YT and Lowe TC. Paradox of strength and ductility in metals processed by severe plastic deformation. *J. Mater. Res.*, 2002; 17:5–8.
2. Zhu YT and Liao XZ. Nanostructured metals: retaining ductility. *Nat. Mater.*, 2004; 3:351–352.

3. Jia D, Wang YM, Ramesh KT, Ma E, Zhu YT and Valiev RZ. Deformation behavior and plastic instabilities of ultrafine-grained titanium. *Appl. Phys. Lett.*, 2001; 79:611–613.
4. Zhao YH, Liao XZ, Horita Z, Langdon TG and Zhu YT. Determining the optimal stacking fault energy for achieving high ductility in ultrafine-grained Cu-Zn alloys. *Mater. Sci. Eng. A*, 2008; 493:123–129.
5. Zhao YH, Bingert JF, Zhu YT, Liao XZ, Valiev RZ, Horita Z, Langdon TG, Zhou YZ and Lavernia EJ. Tougher ultrafine grain Cu via high-angle grain boundaries and low dislocation density. *Appl. Phys. Lett.*, 2008; 92:081903.
6. Zhao YH, Zhu YT, Liao XZ, Horita Z and Langdon TG. Tailoring stacking fault energy for high ductility and high strength in ultrafine grained Cu and its alloy. *Appl. Phys. Lett.*, 2006; 89:121906.
7. Wang YM, Ma E, Valiev RZ and Zhu YT. Tough nanostructured metals at cryogenic temperatures. *Adv. Mater.*, 2004; 16:328.
8. Zhao YH, Bingert JE, Liao XZ, Cui BZ, Han K, Sergueeva AV, Mukherjee AK, Valiev RZ, Langdon TG and Zhu YTT. Simultaneously increasing the ductility and strength of ultra-fine-grained pure copper. *Adv. Mater.*, 2006; 18:2949.
9. Zhao YH, Liao XZ, Cheng S, Ma E and Zhu YT. Simultaneously increasing the ductility and strength of nanostructured alloys. *Adv. Mater.*, 2006; 18:2280–2283.
10. Zhao YH, Topping T, Bingert JF, Thornton JJ, Dangelewicz AM, Li Y, Liu W, Zhu YT, Zhou YZ and Lavernia EL. High tensile ductility and strength in bulk nanostructured nickel. *Adv. Mater.*, 2008; 20:3028–3033.
11. Cheng S, Zhao YH, Zhu YT and Ma E. Optimizing the strength and ductility of fine structured 2024 Al alloy by nano-precipitation. *Acta Mater.*, 2007; 55:5822–5832.
12. Estrin Y and Vinogradov A. Extreme grain refinement by severe plastic deformation: a wealth of challenging science. *Acta Mater.*, 2013; 61:782–817.
13. Meyers MA, Mishra A and Benson DJ. Mechanical properties of nanocrystalline materials. *Prog. Mater. Sci.*, 2006; 51:427–556.
14. Lu L, Zhu T, Shen YF, Dao M, Lu K and Suresh S. Stress relaxation and the structure size-dependence of plastic deformation in nanotwinned copper. *Acta Mater.*, 2009; 57:5165–5173.
15. Wang YM and Ma E. Three strategies to achieve uniform tensile deformation in a nanostructured metal. *Acta Mater.*, 2004; 52:1699–1709.

16. Yan FK, Liu GZ, Tao NR and Lu K. Strength and ductility of 316L austenitic stainless steel strengthened by nano-scale twin bundles. *Acta Mater.*, 2012; 60:1059-1071.
17. Youssef K, Sakaliyska M, Bahmanpour H, Scattergood R and Koch C. Effect of stacking fault energy on mechanical behavior of bulk nanocrystalline Cu and Cu alloys. *Acta Mater.*, 2011; 59:5758-5764.
18. An XH, Han WZ, Huang CX, Zhang P, Yang G, Wu SD and Zhang ZF. High strength and utilizable ductility of bulk ultrafine-grained Cu-Al alloys. *Appl. Phys. Lett.*, 2008; 92:201915.
19. Zhang X, Wang H, Scattergood RO, Narayan J, Koch CC, Sergueeva AV and Mukherjee AK. Tensile elongation (110%) observed in ultrafine-grained Zn at room temperature. *Appl. Phys. Lett.*, 2002; 81:823-825.
20. Huang XX, Kamikawa N and Hansen N. Increasing the ductility of nanostructured Al and Fe by deformation. *Mater. Sci. Eng. A*, 2008; 493:184-189.
21. Wang GY, Li GY, Zhao L, Lian JS, Jiang ZH and Jiang Q. The origin of the ultrahigh strength and good ductility in nanotwinned copper. *Mater. Sci. Eng. A*, 2010; 527:4270-4274.
22. Lu L, Shen YF, Chen XH, Qian LH and Lu K. Ultrahigh strength and high electrical conductivity in copper. *Science*, 2004; 304:422-426.
23. An XH, Wu SD, Zhang ZF, Figueiredo RB, Gao N and Langdon TG. Enhanced strength-ductility synergy in nanostructured Cu and Cu-Al alloys processed by high-pressure torsion and subsequent annealing. *Scr. Mater.*, 2012; 66:227-230.
24. Xiao GH, Tao NR and Lu K. Strength-ductility combination of nanostructured Cu-Zn alloy with nanotwin bundles. *Scr. Mater.*, 2011; 65:119-122.
25. Wang YM, Chen MW, Zhou FH and Ma E. High tensile ductility in a nanostructured metal. *Nature*, 2002; 419:912-915.
26. Zhu YT, Lowe TC and Langdon TG. Performance and applications of nanostructured materials produced by severe plastic deformation. *Scr. Mater.*, 2004; 51:825-830.
27. Lu K. Making strong nanomaterials ductile with gradients. *Science*, 2014; 345:1455-1456.
28. Wu XL, Jiang P, Chen L, Yuan FP and Zhu YT. Extraordinary strain hardening by gradient structure. *Proc. Natl. Acad. Sci. USA*, 2014; 111:7197-7201.
29. Wu XL, Jiang P, Chen L, Zhang JF, Yuan FP and Zhu YT. Synergetic strengthening by gradient structure. *Mater. Res. Lett.*, 2014; 2:185-191.

30. Fang TH, Li WL, Tao NR and Lu K. Revealing extraordinary intrinsic tensile plasticity in gradient nano-grained copper. *Science*, 2011; 331:1587–1590.
31. Chen AY, Liu JB, Wang HT, Lu J and Wang YM. Gradient twinned 304 stainless steels for high strength and high ductility. *Mater. Sci. Eng. A*, 2016; 667:179–188.
32. Wei YJ, Li YQ, Zhu LC, Liu Y, Lei XQ, Wang G, Wu YX, Mi ZL, Liu JB, Wang HT and Gao HJ. Evading the strength-ductility trade-off dilemma in steel through gradient hierarchical nanotwins. *Nat. Commun.*, 2014; 5:3580.
33. Wu XL, Yang MX, Yuan FP, Wu GL, Wei YJ, Huang XX and Zhu YT. Heterogeneous lamella structure unites ultrafine-grain strength with coarse-grain ductility. *Proc. Natl. Acad. Sci. USA*, 2015; 112:14501–14505.
34. Han BQ, Huang JY, Zhu YT and Lavernia EJ. Strain rate dependence of properties of cryomilled bimodal 5083 Al alloys. *Acta Mater.*, 2006; 54:3015–3024.
35. Han BQ, Lee Z, Witkin D, Nutt S and Lavernia EJ. Deformation behavior of bimodal nanostructured 5083 Al alloys. *Metall. Mater. Trans. A*, 2005; 36:957–965.
36. Sawangrat C, Kato S, Orlov D and Ameyama K. Harmonic-structured copper: performance and proof of fabrication concept based on severe plastic deformation of powders. *J. Mater. Sci.*, 2014; 49:6579–6585.
37. Zhang Z, Vajpai SK, Orlov D and Ameyama K. Improvement of mechanical properties in SUS304L steel through the control of bimodal microstructure characteristics. *Mater. Sci. Eng. A*, 2014; 598:106–113.
38. Vajpai SK, Ota M, Watanabe T, Maeda R, Sekiguchi T, Kusaka T and Ameyama K. The development of high performance Ti-6Al-4V alloy via a unique microstructural design with bimodal grain size distribution. *Metall. Mater. Trans. A*, 2015; 46:903–914.
39. Ma XL, Huang CX, Moering J, Ruppert M, Höppel HW, Göken M, Narayan J and Zhu YT. Mechanical properties in copper/bronze laminates: role of interfaces. *Acta Mater.*, 2016; 116:43–52.
40. Beyerlein IJ, Mayeur JR, Zheng SJ, Mara NA, Wang J and Misra A. Emergence of stable interfaces under extreme plastic deformation. *Proc. Natl. Acad. Sci. USA*, 2014; 111:4386–4390.
41. Calcagnotto M, Adachi Y, Ponge D and Raabe D. Deformation and fracture mechanisms in fine- and ultrafine-grained ferrite/martensite dual-phase steels and the effect of aging. *Acta Mater.*, 2011; 59:658–670.

42. Li ZM, Pradeep KG, Deng Y, Raabe D and Tasan CC. Metastable high-entropy dual-phase alloys overcome the strength-ductility trade-off. *Nature*, 2016; 534:227.

43. Park K, Nishiyama M, Nakada N, Tsuchiyama T and Takaki S. Effect of the martensite distribution on the strain hardening and ductile fracture behaviors in dual-phase steel. *Mater. Sci. Eng. A*, 2014; 604:135–141.

44. Ma E and Zhu T. Towards strength–ductility synergy through the design of heterogeneous nanostructures in metals. *Mater. Today*, 2017; 20:323–331.

45. Yang MX, Pan Y, Yuan FP, Zhu YT and Wu XL. Back stress strengthening and strain hardening in gradient structure. *Mater. Res. Lett.*, 2016; 4:145–151.

46. Cong ZH, Jia N, Sun X, Ren Y, Almer J and Wang YD. Stress and strain partitioning of ferrite and martensite during deformation. *Metall. Mater. Trans. A*, 2009; 40:1383–1387.

47. Tasan CC, Diehl M, Yan D, Zambaldi C, Shanthraj P, Roters F and Raabe D. Integrated experimental-simulation analysis of stress and strain partitioning in multiphase alloys. *Acta Mater.*, 2014; 81:386–400.

48. Han QH, Asgari A, Hodgson PD and Stanford N. Strain partitioning in dual-phase steels containing tempered martensite. *Mater. Sci. Eng. A*, 2014; 611:90–99.

49. Wang MM, Tasan CC, Ponge D, Dippel AC and Raabe D. Nanolaminate transformation-induced plasticity-twinning-induced plasticity steel with dynamic strain partitioning and enhanced damage resistance. *Acta Mater.*, 2015; 85:216–228.

50. Yang MX, Yuan FP, Xie QG, Wang YD, Ma E and Wu XL. Strain hardening in Fe-16Mn-10Al-0.86C-5Ni high specific strength steel. *Acta Mater.*, 2016; 109:213–222.

51. Gao H, Huang Y, Nix WD and Hutchinson JW. Mechanism-based strain gradient plasticity - I. Theory *J. Mech. Phys. Solids*, 1999; 47:1239–1263.

52. Gao HJ and Huang YG. Geometrically necessary dislocation and size-dependent plasticity. *Scr. Mater.*, 2003; 48:113–118.

53. Ashby MF. Deformation of plastically non-homogeneous materials. *Philos. Mag.*, 1970; 21:399–424.

54. Murr LE. Dislocation ledge sources: dispelling the myth of Frank–Read source importance. *Metall. Mater. Trans. A*, 2016; 47:5811–5826.

55. Kato H, Moat R, Mori T, Sasaki K and Withers P. Back stress work hardening confirmed by Bauschinger effect in a TRIP steel using bending tests. *ISIJ Int.*, 2014; 54:1715–1718.

Chapter 2

Perspective on Heterogeneous Deformation Induced (HDI) Strengthening and Work Hardening

Yuntian Zhu[a,b] and Xiaolei Wu[c,d]

[a]*Nano Structural Materials Center, School of Materials Science and Engineering, Nanjing University of Science and Technology, Nanjing 210094, China*
[b]*Department of Materials Science and Engineering, North Carolina State University, Raleigh, NC 27695, USA*
[c]*State Key Laboratory of Nonlinear Mechanics, Institute of Mechanics, Chinese Academy of Science, Beijing 100190, China*
[d]*School of Engineering Science, University of Chinese Academy of Sciences, Beijing 100190, China*
ytzhu@ncsu.edu, xlwu@imech.ac.cn

Heterostructured materials have been reported as a new class of materials with superior mechanical properties, which has been attributed to the development of back stress. There are

---

Reprinted from *Mater. Res. Lett.*, **7**(10), 393–398, 2019.

Heterostructured Materials: Novel Materials with Unprecedented Mechanical Properties
Edited by Xiaolei Wu and Yuntian Zhu
Text Copyright © 2019 The Author(s)
Layout Copyright © 2022 Jenny Stanford Publishing Pte. Ltd.
ISBN 978-981-4877-10-7 (Hardcover), 978-1-003-15307-8 (eBook)
www.jennystanford.com

numerous reports on back stress theories and measurements with no consensus. Back stress is developed in soft zones to offset the applied stress, making them appear stronger, while forward stress makes hard zones appear weaker. The extra hardening observed in heterostructured materials is resulted from interactions between back stresses and forward stresses, and should be described as heterogeneous deformation induced (HDI) hardening and the measured "back stress" should be renamed HDI stress.

2.1 Background

Heterostructured (HS) materials have recently attracted extensive attention form the materials community, as evidenced by increasing numbers of international conferences and publications in recent years. For example, a symposium entitled "Heterogeneous and Gradient Materials III" was held in the TMS Annual meeting in March 2019. A Gordon Research Conference on Heterostructured materials held in June, 2019. A symposium on Nanostructured, Heterostructured & Gradient Materials held in the Chinese Materials Conference in July 2019. Figure 2.1 shows that the number of papers on HS materials has been increasing exponentially. It is anticipated that HS materials will become a major research field in the post nanostructured materials era.

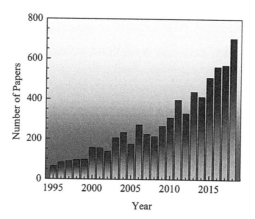

Figure 2.1 The number of papers on heterostructured (HS) materials published in recent years.

HS materials have very diverse microstructures [1], including heterostructured lamella structure [2], gradient structure [3–8], laminate structure [9, 10], dual phase structure [11–13], harmonic structure [14–16], bi-modal structure [17–20], metal matrix composites [21–26], etc. These apparently very diverse structures have a common feature: all of them consist of both soft zones and hard zones with dramatically different flow stresses (or strength) [1].

During tensile deformation of HS materials, the soft zones will start plastic deformation first while the hard zones remain elastic. In this elastic-plastic deformation stage, geometrically necessary dislocations (GNDs) will be blocked by and pile up against zone boundaries, which produces a long-range internal stress, i.e. back stress, in the soft zone. The back stress is directional and offset some applied shear stress, making the soft zone appear stronger to sustain higher applied stress. It is this back stress that is believed to make the heterostructured materials stronger [2, 27]. When both soft and hard zones are deforming plastically, the soft zones will sustain higher plastic strain than the hard zones, producing strain partitioning. Since the plastic strain needs to be the same at the zone boundaries to maintain continuity, there will be strain gradient near the zone boundary to accommodate the strain partitioning. The strain gradient needs to be accommodated by GNDs [28, 29], which in turn produces back-stress induced hardening that helps with retaining ductility [1, 2].

As described above, back stress is believed responsible for the strengthening and extra strain hardening observed in HS materials. Furthermore, various schemes to measure back stress from mechanical testing have been proposed [27, 30, 31], most of which are based on concepts and assumptions instead of fundamental dislocation mechanics. Back stress is also often associated with kinematic hardening, a terminology extensively used in the field of mechanics [32], which describes the mechanical phenomenon without addressing its physical origin. The concept and terminology of back stress are themselves still under debate in the materials community, with some researchers prefer to call them long-range internal stress [33, 34]. Although back stress is usually small in homogeneous metals [27, 35], it becomes critical for HS materials [1, 2]. Therefore, it is of critical importance to understand the relationship between back stress and the mechanical behavior of HS materials.

In this perspective, we will briefly delineate the history of back stress, analyzing the development of back stress and forward stress in HS materials. We will show the inadequacy of using back stress to describe the strengthening and extra strain hardening in HS materials. In addition, back stress cannot be measured from mechanical testing, and the back-stress reported in literature can be more accurately described as heterogeneous deformation induced (HDI) stress. We'll also briefly discuss current fundamental issues that need to be investigated.

2.2 Brief History of Back Stress

In the following we will present a brief history of the back-stress concept, instead of a comprehensive review. Back stress is often related to Bauschinger effect because they have the same physical origin [36–39]. In 1886 Bauschinger found that after an alloy was deformed plastically in tension, it exhibited a compressive yield strength lower than its tensile yield strength, and vice versa [40]. This phenomenon has been referred to as the Bauschinger effect. In 1947 Orowan [41] proposed that dislocations pile-up against second-phase particles produced internal stress to resist the slip of following dislocations. In 1953 Fisher [42] proposed that the yield strength of a precipitation hardened alloy equals the sum of the yield strength of the matrix and the internal stress exerted on the matrix by precipitates. The internal stress proposed by Orowan and Fisher was later named *back stress* in a dislocation model by Ashby [43].

The back stress concept was later used by many groups to explain the strengthening of metal matrix by second-phase particles in the 1970s [44–46]. More recently, back stress has been used to explain the hardening behavior and Bauschinger effect in TRIP steels [47], transient plastic flow during creep [48], the Bauchinger effect in thin film [38], the hysteresis loop under cyclic loading [49], the extra strain hardening of HS materials [1, 2, 27, 50, 51], etc. These reports show that the back stress concept has been widely, if not fully, accepted by the materials community.

2.3 Dislocation Models for Back Stress

Several dislocation models on the formation of back stress have been proposed [1, 33, 43, 52–54]. The back stress is produced by geometrically necessary dislocations (GNDs) [1, 2, 55], which is in turn believed needed to accommodate strain gradient in the gradient plasticity theory [28, 29]. GNDs create lattice bending or misorientation, depending on their geometric pattern of arrangement. As shown in Fig. 2.2, there are two different basic types of GND arrangements. Figure 2.2a illustrates the piling up of GNDs on a slip plane against grain boundary, phase boundary, or zone boundary, which is hereafter referred to as the *Type I* GND arrangement. These GNDs has the same Burgers vector, and can be produced from a Frank-Read dislocation source. The GND pile up produces a stress concentration of $n\tau_a$ at the boundary (the piling up head) [56], where n is the number of GNDs in the pile up and τ_a is the applied shear stress on the slip system. Such a GND pile up will bend slip plane and exert a long range internal stress with a direction that is opposite to the applied stress. This stress is called back stress [1, 2, 43, 57]. It acts to repel more dislocation emissions from the Frank-Read source, which makes soft zone appear stronger in HS materials.

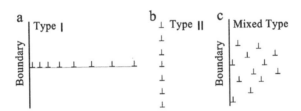

Figure 2.2 Arrangements of GNDs. (a) *Type I*: GNDs piled up against a boundary. Most effective in producing back stress. (b) *Type II*: GNDs aligned up vertically to form a low angle grain boundary. It does not produce back stress. (c) The mixed type. Less effective than *Type I* in producing back stress.

Another basic GND arrangement type is shown in Fig. 2.2b, which is essentially a tilt low-angle grain boundary, and is hereafter referred to *Type II* GND arrangement. It does not produce long-range internal stress, i.e. back stress [57]. However, internal stress exists in short range close to the grain boundary. In a real situation such as bending of a crystal, a mixed type may exist [29], as shown in

Fig. 2.2c, which does produce back stress but is not as effective as the *Type I* GND arrangement.

2.4 Back Stress and Mechanical Properties

Back stress has been considered playing a critical role in the strengthening of metal matrix composites [44–46]. It is also often associated with the Buaschinger effect [36, 38, 47, 50]. However, neither back stress nor Bauschigner effect has been accepted as a major player in strengthening metallic materials by the general materials community. For example, in the literature and textbooks [58], listed mechanisms for strengthening metallic materials include strain (work, dislocation) hardening, grain boundary hardening, solution hardening, and precipitation (dispersion) hardening. It should be noted that the precipitation (dispersion) hardening is explained by the blocking of gliding dislocations instead of the back stress. Back stress is usually not considered for improving mechanical properties. One of the reasons is that the back stress is usually very small in conventional homogeneous polycrystalline materials, and the observed mechanical properties can be adequately explained with the above mechanisms from the textbooks. The back stress in homogenous materials was indeed measured to be very low [35, 59]. This is why back stress has been ignored by the majority of researchers in the materials community.

For HS materials, the enhancement in strength and strain hardening can be so dramatic that it can no longer be explained by the mechanisms listed in the textbooks [1, 2]. This is because the HS zones have big difference in their strength/flow stress, which makes the back stress induced strengthening and work hardening the primary player in the mechanical behavior. For example, an HS lamella Ti was found to have the strength of the ultrafine-grained Ti and uniform elongation higher than that of the coarse-grained Ti [2]. In addition, it also has strain hardening that is higher than the coarse-grained Ti. These mechanical behaviors cannot be explained by the conventional mechanisms in the textbooks. Therefore, back stress induced strengthening and work hardening have been used to explain the extraordinary mechanical properties of HS materials [1, 2, 27, 60].

2.5 Issues with the Back Stress Concept

As discussed above, back stress strengthening and work hardening have been believed responsible for the superior combination of strength and ductility in HS materials [1, 2, 39, 61] as well as the Bauschinger effect [36, 38, 47, 50], although the back stress concept has been largely ignored by many researchers in the materials community. With the fast development of the HS materials, a few types of materials have been reported to possess unprecedented mechanical properties, which have been attributed to the back stress. It is of critical importance to clarify and understand how the back stress affects the mechanical behavior. In addition, methods and equations have been proposed and developed to experimentally measure back stress from mechanical testing. Most of these methods and equations are not well anchored in physics of deformation, i.e. dislocation mechanisms. A question arises on what is actually measured and if the measured "back stress" is the global back stress in the materials.

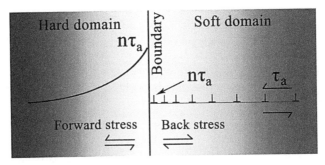

Figure 2.3 Schematics of a GND pile-up, inducing the back stress in the soft zone, which in turn induces the forward stress in the hard zone.

To analyze this issue, let's look at the dislocation model at an HS zone boundary in Fig. 2.3. As shown, there is a GND pileup against the zone boundary in the soft zone under an applied shear stress τ_a. As discussed earlier, this dislocation pileup would produce back stress that is opposite to the applied shear stress in the soft zone to make the soft zone appear stronger. At the head of the pileup, there is a stress concentration of $n\tau_a$, where n is the number of GNDs in the pileup. This stress concentration is applied to the hard zone across the zone boundary, which is in the same direction of the applied stress, and is therefore called the forward stress. In other words, the back stress in the soft zone induced the forward stress in the hard zone, as schematically represented by the curve in the hard zone.

Under the forward stress, the hard zone may behave in three different ways. Note that the forward stress in the hard zone near the zone boundary can be many times higher than the applied stress. First, if the hard zone is not much stronger than the soft zone, the leading dislocation at the head of the GND pileup may be also pushed into and transmit across the zone boundary after a reaction at the boundary. In such a scenario, the buildup of back stress and forward stress is limited, and their influence on the mechanical behavior is also limited, as in the case of GND pileup at grain boundaries in conventional homogeneous materials. Second, if the hard zone is much stronger than the soft zone, the zone boundary will be much more effective in blocking GNDs, and the hard zone will remain elastic until the back stress is very high in the soft zone. This will increase the global yield strength and work hardening when both zones are deforming plastically, especially when the soft zone is fully constrained by the hard zone, as reported in the heterostructured lamella Ti [2]. Third, if the hard zone is not plastically deformable, as in the case of metal matrix composites reinforced by second phase particles. The zone boundary may fail when the stress concentration is too high, and a void may form around the particles, which leads to sample fracture.

As discussed above, the back stress and forward stress are coupled at the zone (or grain) boundary and acted in opposite direction. The back stress makes soft zone appear stronger and the forward strain may make the hard zone appear weaker if it is deformable. Logically both of back stress and forward stress should affect the mechanical behavior and their interaction is not a zero-sum game, as indicated by the unique mechanical behavior and superior mechanical properties of HS materials. This analysis suggests that it is logically not appropriate to attribute the unique mechanical behavior of HS materials to only the back stress, as reported in the literature [2].

In the literature, back stress is measured by the unloading-reloading test during tensile testing [27], or from the unloading curve alone [30]. As shown in Fig. 2.3, if unloading is performed, both back stress and forward stress will be affected because they are a couple. Therefore, the unloading and reloading curves are affected by the interaction of the back stress and forward stress, instead of back stress alone. This effectively makes back stress unmeasurable from the mechanical testing curves, and the measured "back stress" is not the real back stress in a physical sense. It should be noted that since the forward stress is induced by the back stress, we can logically regard the unique mechanical behavior of HS materials as induced by back stress. However, such a statement can be misleading, making people think that the forward stress does not play a role.

2.6 New Definition

As discussed above, a new terminology is needed to describe how the heterostructure affects the mechanical behavior and properties, because the back stress only represents part of what is actually happening. At the fundamental level, the heterostructures leads to heterogeneous deformation among HS zones [1]. First, after the soft zones start yielding, the hard zones will remain elastic, which produces back stress in the soft zone to raise the global yield strength of the sample. After the sample started yielding, the partitioning of plastic strain between the hard and soft zones, i.e. heterogeneous deformation, will continue to produce back stress in the soft zone and forward stress in the hard zone even when the hard zone is also deforming plastically [2]. This produces some extra strain hardening as observed in HS lamella Ti [2]. In other words, it is the heterogeneous deformation that lead to the development of back stress and forward stress, which collectively produced the observed strengthening and extra work hardening. Therefore, the extra hardening in the HS materials can be defined as the heterogeneous deformation induced (HDI) hardening, and the "back stress" measured from the mechanical testing can be defined as the HDI stress. The HDI stress is directional, and associated with GND pileup. It is kinematic in nature, i.e. the kinematic stress as defined in mechanics field.

2.7 Outstanding Issues

The definition of HDI hardening and HDI stress raises some scientific issues for future study. First, the back stress and forward stress are a stress couple with opposite directions. It is not clear how they interact with each other to produce the HDI hardening and HDI stress. Recently, it has been observed that local shear bands occurred across, and hetero-boundary-affected region (HBAR) was formed near the HS zone boundary [62]. This might be a mechanism for the back stress and forward stress to interact with each other, which needs to be further investigated. Second, the HDI stress (measured "back stress" in the literature [2, 27]) appears to increase quickly in the elastic-plastic deformation stage, but slow down during the plastic deformation stage, which need to be investigated. Third, GND pileups lead to the development of back stress, which in turn induces forward stress. GNDs is also believed needed to accommodate the strain gradient. The relationships among back stress, forward stress, strain gradient and HDI stress need to be studied. Understanding these issues will help us with understanding the fundamental physics as well as the heterostructure-mechanical properties of HS materials.

Acknowledgements

This work was funded by National Key R&D Program of China [grant numbers 2017YFA0204402 and 2017YFA0204403], the Natural Science Foundation of China [grant numbers 11572328, 11672313, and 11790293], and the Strategic Priority Research Program of the Chinese Academy of Sciences [grant number XDB22040503]. Y.T.Z. was supported by the US Army Research Office [grant number W911 NF-12-1-0009].

References

1. Wu XL and Zhu YT. Heterogeneous materials: a new class of materials with unprecedented mechanical properties. *Mater. Res. Lett.*, 2017; 5:527–532.
2. Wu XL, Yang MX, Yuan FP, Wu GL, Wei YJ, Huang XX and Zhu YT. Heterogeneous lamella structure unites ultrafine-grain strength with coarse-grain ductility. *Proc. Natl. Acad. Sci. USA*, 2015; 112:14501–14505.
3. Lu K. Making strong nanomaterials ductile with gradients. *Science*, 2014; 345:1455–1456.

4. Wu XL, Jiang P, Chen L, Yuan FP and Zhu YT. Extraordinary strain hardening by gradient structure. *Proc. Natl. Acad. Sci. USA,* 2014; 111:7197–7201.
5. Wu XL, JIang P, Chen L, Zhang JF, Yuan FP and Zhu YT. Synergetic strengthening by gradient structure. *Mater. Res. Lett.,* 2014; 2:185–191.
6. Fang TH, Li WL, Tao NR and Lu K. Revealing extraordinary intrinsic tensile plasticity in gradient nano-grained copper. *Science,* 2011; 331:1587–1590.
7. Chen AY, Liu JB, Wang HT, Lu J and Wang YM. Gradient twinned 304 stainless steels for high strength and high ductility. *Mater. Sci. Eng. A,* 2016; 667:179–188.
8. Wei YJ, Li YQ, Zhu LC, Liu Y, Lei XQ, Wang G, Wu YX, Mi ZL, Liu JB, Wang HT and Gao HJ. Evading the strength-ductility trade-off dilemma in steel through gradient hierarchical nanotwins. *Nat. Commun.,* 2014; 5:3580.
9. Ma XL, Huang CX, Moering J, Ruppert M, Höppel HW, Göken M, Narayan J and Zhu YT. Mechanical properties in copper/bronze laminates: role of interfaces. *Acta Mater.,* 2016; 116:43–52.
10. Beyerlein IJ, Mayeur JR, Zheng SJ, Mara NA, Wang J and Misra A. Emergence of stable interfaces under extreme plastic deformation. *Proc. Natl. Acad. Sci. USA,* 2014; 111:4386–4390.
11. Calcagnotto M, Adachi Y, Ponge D and Raabe D. Deformation and fracture mechanisms in fine- and ultrafine-grained ferrite/martensite dual-phase steels and the effect of aging. *Acta Mater.,* 2011; 59:658–670.
12. Li ZM, Pradeep KG, Deng Y, Raabe D and Tasan CC. Metastable high-entropy dual-phase alloys overcome the strength-ductility trade-off. *Nature,* 2016; 534:227.
13. Park K, Nishiyama M, Nakada N, Tsuchiyama T and Takaki S. Effect of the martensite distribution on the strain hardening and ductile fracture behaviors in dual-phase steel. *Mater. Sci. Eng. A,* 2014; 604:135–141.
14. Sawangrat C, Kato S, Orlov D and Ameyama K. Harmonic-structured copper: performance and proof of fabrication concept based on severe plastic deformation of powders. *J. Mater. Sci.,* 2014; 49:6579–6585.
15. Zhang Z, Vajpai SK, Orlov D and Ameyama K. Improvement of mechanical properties in SUS304L steel through the control of bimodal microstructure characteristics. *Mater. Sci. Eng. A,* 2014; 598:106–113.

16. Vajpai SK, Ota M, Watanabe T, Maeda R, Sekiguchi T, Kusaka T and Ameyama K. The development of high performance Ti-6Al-4V alloy via a unique microstructural design with bimodal grain size distribution. *Metall. Mater. Trans. A*, 2015; 46:903–914.

17. Wang YM, Chen MW, Zhou FH and Ma E. High tensile ductility in a nanostructured metal. *Nature*, 2002; 419:912–915.

18. Han BQ, Huang JY, Zhu YT and Lavernia EJ. Strain rate dependence of properties of cryomilled bimodal 5083 Al alloys. *Acta Mater.*, 2006; 54:3015–3024.

19. Han BQ, Lee Z, Witkin D, Nutt S and Lavernia EJ. Deformation behavior of bimodal nanostructured 5083 Al alloys. *Metall. Mater. Trans. A*, 2005; 36a:957–965.

20. Zhao YH, Topping T, Bingert JF, Thornton JJ, Dangelewicz AM, Li Y, Liu W, Zhu YT, Zhou YZ and Lavernia EL. High tensile ductility and strength in bulk nanostructured nickel. *Adv. Mater.*, 2008; 20:3028–3033.

21. Liu J, Chen Z, Zhang FG, Ji G, Wang ML, Ma Y, Ji V, Zhong SY, Wu Y and Wang HW. Simultaneously increasing strength and ductility of nanoparticles reinforced Al composites via accumulative orthogonal extrusion process. *Mater. Res. Lett.*, 2018; 6:406–412.

22. Zhao L, Guo Q, Li Z, Fan GL, Li ZQ, Xiong DB, Su YS, Tan ZQ, Guo CP and Zhang D. Grain boundary-assisted deformation in graphene-Al nanolaminated composite micro-pillars. *Mater. Res. Lett.*, 2018; 6:41–48.

23. Choudhuri D, Shukla S, Green WB, Gwalani B, Ageh V, Banerjee R and Mishra RS. Crystallographically degenerate B2 precipitation in a plastically deformed fcc-based complex concentrated alloy. *Mater. Res. Lett.*, 2018; 6:171–177.

24. Llorca J, Needleman A and Suresh S. The Bauschinger effect in whisker-reinforced metal-matrix composites. *Scr. Metall. Mater.*, 1990; 24:1203–1208.

25. Thilly L, Van Petegem S, Renault PO, Lecouturier F, Vidal V, Schmitt B and Van Swygenhoven H. A new criterion for elasto-plastic transition in nanomaterials: Application to size and composite effects on Cu-Nb nanocomposite wires. *Acta Mater.*, 2009; 57:3157–3169.

26. Tanaka K and Mori T. Hardening of crystals by non-deforming particles and fibres. *Acta Metall.*, 1970; 18:931.

27. Yang MX, Pan Y, Yuan FP, Zhu YT and Wu XL. Back stress strengthening and strain hardening in gradient structure. *Mater. Res. Lett.*, 2016; 4:145–151.

28. Gao HJ and Huang YG. Geometrically necessary dislocation and size-dependent plasticity. *Scr. Mater.*, 2003; 48:113–118.
29. Ashby MF. Deformation of plastically non-homogeneous materials. *Philos. Mag.*, 1970; 21:399.
30. Feaugas X. On the origin of the tensile flow stress in the stainless steel AISI 316L at 300 K: back stress and effective stress. *Acta Mater.*, 1999; 47:3617–3632.
31. Kuhlmann-Wilsdorf D and Laird C. Dislocation behavior in fatigue. 2. Friction stress and back stress as inferred from an analysis of hysteresis loops. *Mater. Sci. Eng.*, 1979; 37:111–120.
32. Bate PS and Wilson DV. Analysis of the Bauschinger effect. *Acta Metall.*, 1986; 34:1097–1105.
33. Gibeling JC and Nix WD. A numerical study of long-range internal-stresses associated with subgrain boundaries. *Acta Metall.*, 1980; 28:1743–1752.
34. Levine LE, Geantil P, Larson BC, Tischler JZ, Kassner ME, Liu WJ, Stoudt MR and Tavazza F. Disordered long-range internal stresses in deformed copper and the mechanisms underlying plastic deformation. *Acta Mater.*, 2011; 59:5803–5811.
35. Phan TQ, Levine LE, Lee IF, Xu RQ, Tischler JZ, Huang Y, Langdon TG and Kassner ME. Synchrotron X-ray microbeam diffraction measurements of full elastic long range internal strain and stress tensors in commercial-purity aluminum processed by multiple passes of equal-channel angular pressing. *Acta Mater.*, 2016; 112:231–241.
36. Liu XL, Yuan FP, Zhu YT and Wu XL. Extraordinary Bauschinger effect in gradient structured copper. *Scr. Mater.*, 2018; 150:57–60.
37. Mompiou F, Caillard D, Legros M and Mughrabi H. In situ TEM observations of reverse dislocation motion upon unloading in tensile-deformed UFG aluminium. *Acta Mater.*, 2012; 60:3402–3414.
38. Xiang Y and Vlassak JJ. Bauschinger and size effects in thin-film plasticity. *Acta Mater.*, 2006; 54:5449–5460.
39. Taya M, Lulay KE, Wakashima K and Lloyd DJ. Bauschinger effect in particulate Sic-6061 aluminum composites. *Mater. Sci. Eng. A*, 1990; 124:103–111.
40. Bauschinger J. Über die Veränderung der Elasticitätsgrenze und der Festigkeit des Eisens und Stahls durch Strecken und Quetschen, durch Erwärmen und Abkühlen und durch oftmal wiederholte Beanspruchung. Mitteilungen aus dem mechanisch-technischen

Laboratorium der k. polytechnischen Schule, *München*, 1886; 13:1–115.

41. Orowan E. Classification and nomenclature of internal stresses. *J. Inst. Met.*, 1947; 73:47–59.

42. Fisher JC, Hart EW and Pry RH. The hardening of metal crystals by precipitate particles. *Acta Metall.*, 1953; 1:336–339.

43. Ashby MF. Work hardening of dispersion-hardened crystals. *Philos. Mag.*, 1966; 14:1157–1178.

44. Tanaka K and Mori T. Hardening of crystals by non-deforming particles and fibres. *Acta Metall.*, 1970; 18:931.

45. Mori T and Tanaka K. Average stress in matrix and average elastic energy of materials with misfitting inclusions. *Acta Metall.*, 1973; 21:571–574.

46. Brown LM. Back-stresses, image stresses, and work-hardening. *Acta Metall.*, 1973; 21:879–885.

47. Kato H, Moat R, Mori T, Sasaki K and Withers P. Back stress work hardening confirmed by Bauschinger effect in a TRIP steel using bending tests. *ISIJ Int.*, 2014; 54:1715–1718.

48. Gan W, Zhang PH, Wagoner RH and Daehn GS. Effect of load redistribution in transient plastic flow. *Metall. Mater. Trans. A*, 2006; 37a:2097–2106.

49. Fournier B, Sauzay M, Caes C, Mottot M, Noblecourt A and Pineau A. Analysis of the hysteresis loops of a martensitic steel - Part II: Study of the influence of creep and stress relaxation holding times on cyclic behaviour. *Mater. Sci. Eng. A*, 2006; 437:197–211.

50. Hu XZ, Jin SB, Zhou H, Yin Z, Yang J, Gong YL, Zhu YT, Sha G and Zhu XK. Bauschinger effect and back stress in gradient Cu-Ge alloy. *Metall. Mater. Trans. A*, 2017; 48a:3943–3950.

51. Yuan FP, Yan DS, Sun JD, Zhou LL, Zhu YT and Wu XL. Ductility by shear band delocalization in the nano-layer of gradient structure. *Mater. Res. Lett.*, 2019; 7:12–17.

52. Mughrabi H. On the role of strain gradients and long-range internal stresses in the composite model of crystal plasticity. *Mater. Sci. Eng. A*, 2001; 317:171–180.

53. Mughrabi H. The effect of geometrically necessary dislocations on the flow stress of deformed crystals containing a heterogeneous dislocation distribution. *Mater. Sci. Eng. A*, 2001; 319:139–143.

54. Jennings AT, Gross C, Greer F, Aitken ZH, Lee SW, Weinberger CR and Greer JR. Higher compressive strengths and the Bauschinger effect in conformally passivated copper nanopillars. *Acta Mater.*, 2012; 60:3444–3455.
55. Hughes DA, Hansen N and Bammann DJ. Geometrically necessary boundaries, incidental dislocation boundaries and geometrically necessary dislocations. *Scr. Mater.*, 2003; 48:147–153.
56. Hull D and Bacon DJ, *Introduction to Dislocations*, 3rd ed. (Pergamon Press, Oxford, 1984).
57. Hirth JP and Lothe J, *Theory of Dislocations*, 2nd ed. (Krieger Publishing Company, Malabar, FL, 1992).
58. Hertzberg RW, *Deformation and Fracture Mechanics of Engineering Materials*, 3rd ed. (Wiley, New York, 1989).
59. Kassner ME, Geantil P and Levine LE. Long range internal stresses in single-phase crystalline materials. *Int. J. Plast.*, 2013; 45:44–60.
60. Ma XL, Huang CX, Xu WZ, Zhou H, Wu XL and Zhu YT. Strain hardening and ductility in a coarse-grain/nanostructure laminate material. *Scr. Mater.*, 2015; 103:57–60.
61. Xu R, Fan GL, Tan ZQ, Ji G, Chen C, Beausir B, Xiong DB, Guo Q, Guo CP, Li ZQ and Zhang D. Back stress in strain hardening of carbon nanotube/aluminum composites. *Mater. Res. Lett.*, 2018; 6:113–120.
62. Huang CX, Wang YF, Ma XL, Yin S, Höppel HW, Göken M, Wu XL, Gao HJ and Zhu YT. Interface affected zone for optimal strength and ductility in heterogeneous laminate. *Mater. Today*, 2018; 21:713–719.

Chapter 3

Ductility and Plasticity of Nanostructured Metals: Differences and Issues

Yuntian Zhu[a,b] and Xiaolei Wu[c,d]

[a]Nano and Heterostructural Materials Center, Nanjing University of Science and Technology, Nanjing 210094, China
[b]Department of Materials Science and Engineering, North Carolina State University, Raleigh, NC 27695, USA
[c]State Key Laboratory of Nonlinear Mechanics, Institute of Mechanics, Chinese Academy of Sciences, Beijing 100190, China
[d]College of Enginnering Sciences, The University of Chinese Academy of Sciences, Beijing 100049, China
ytzhu@ncsu.edu, xlwu@imech.ac.cn

Ductility is one of the most important mechanical properties for metallic structural materials. It is measured as the elongation to failure of a sample during *standard uniaxial tensile* tests. However, this measurement is problematic and often leads to gross overestimation for nanostructured metals, for which non-standard small samples

---

Reprinted from *Mater. Today Nano*, **2**, 15–20, 2018.

Heterostructured Materials: Novel Materials with Unprecedented Mechanical Properties
Edited by Xiaolei Wu and Yuntian Zhu
Text Copyright © 2018 Elsevier Ltd.
Layout Copyright © 2022 Jenny Stanford Publishing Pte. Ltd.
ISBN 978-981-4877-10-7 (Hardcover), 978-1-003-15307-8 (eBook)
www.jennystanford.com

are usually used. Uniform elongation is a better measure of ductility for small samples because they are less sensitive to sample size. Ductility can be considered as *tensile plasticity*, but is often confused with plasticity. Plasticity is primarily controlled by crystal structure, or the number of available slip systems, while ductility is largely governed by work hardening rate at room temperature, which is significantly affected by microstructure. Plasticity is important for shaping and forming metals into desired shape and geometry to make structural components, while ductility is important for the safety of structural components during service. Nanostructured metals typically have high plasticity, but low ductility, because of their low work hardening capability. Increasing work hardening rate via modifying microstructure is the primary route to improving the ductility of nanostructured metals.

3.1 Introduction

Reasonable ductility (>5%, preferably >10%) is desired to prevent mechanical components or structures from catastrophic failure during service [1]. On the other hand, high strength is also desired so that a metallic structure/component can carry large load at low material weight. This is especially important for future transportation vehicle such as electrical cars, which need to be lightweight to improve their energy efficiency. However, a metallic material is either strong or ductile, but rarely both at the same time [2, 3]. Coarse-grained (CG) metals usually have high ductility but low strength. Refining grains to the nanocrystalline regime in the last few decades has significantly increased the strength, but this is often accompanied with the sacrifice of ductility [4]. The low ductility of nanostructured metals has been a major issue with their potential structural applications.

Ductility of nanostructured metals has been a hot research topic for over a decade [2–7]. Many strategies have been proposed to enhance the ductility of nanostructured metals, including pre-existing and deformation twins [8–36] and second-phase precipitates [37–51]. stacking faults [52, 53], and high-angle grain boundaries [54–56]. However, despite of the extensive research and publications, there still exist wide-spread confusions and misconceptions on the definition and measurement of ductility of nanostructured metals, which have led to the publications of problematic claims and data.

The biggest confusion is on the difference between ductility and plasticity. Plasticity is an important property for metallic materials, which could significantly affect their processing, shaping, and forming ability. Unfortunately, in the academic literature, these two terminologies are often mixed up and inter-changed, which has raised serious issues and sometime led to wrong and/or misleading scientific claims and statements. What is more problematic is that such publications often mislead the community, especially junior researchers and students, as well as the public. To make things worse, the plasticity and ductility are often not well defined in textbooks, e.g. Deformation and Fracture Mechanics of Engineering Materials [57]. These problems become more serious in recent years with the study of nanostructured metals, where very small, non-standard samples are often used to characterize mechanical properties.

In this chapter we will first discuss the differences between ductility and plasticity, clarify some common misconceptions and confusions, and then briefly discuss recent progresses in improving the ductility of nanostructured metals.

3.2 Ductility

In the Wikipedia [https://en.wikipedia.org/wiki/Ductility], ductility is defined as "a solid materials's ability to deform under *tensile stress.*" Quantitatively, ductility is usually measured as the elongation to failure, i.e. the total engineering strain, from the tensile testing of standard large samples. Figure 3.1 schematically shows a typical tensile specimen before and during strain localization (necking) during a tensile test, and Fig. 3.2 shows the corresponding engineering stress-strain curve. The engineering strain is defined as

$$\varepsilon = \frac{\Delta l}{l_0} = \frac{\Delta l_u}{l_0} + \frac{\Delta l_n}{l_0} = \varepsilon_u + \varepsilon_n \qquad (3.1)$$

where l_0 is the initial gage length, Δl is the total gage length change after the tensile test, Δl_u is the uniform gage length change during the tensile test, Δl_n is the local length change in the necking segment, ε_u is the uniform elongation and ε_n is the necking strain.

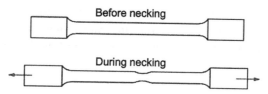

Figure 3.1 Schematic illustration of sample geometry change before and during necking during a tensile test.

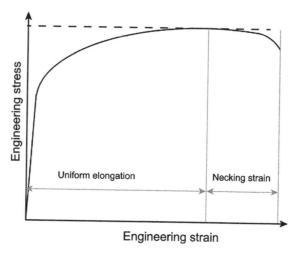

Figure 3.2 Typical tensile engineering stress-strain curve.

As shown in Fig. 3.2, the uniform elongation is determined by the maximum stress in the engineering stress–strain curve. In the true stress–strain curve obtained under a constant strain rate, it is determined by the Considère criterion [4]:

$$\frac{d\sigma_t}{d\varepsilon_t} \geq \sigma_t \qquad (3.2)$$

where σ_t is the true stress and ε_t is the true strain. Note that the Considère criterion can be derived to agree with maximum stress criterion for the engineering stress-strain curve, i.e.

$$\frac{d\sigma_e}{d\varepsilon_e} \geq 0 \qquad (3.3)$$

where σ_e is the engineering stress and ε_e is the engineering strain. In other words, the strain-hardening rate largely determines the uniform deformation, which can be measured as the engineering strain at maximum engineering stress.

It should be noted that the Considère criterion does not consider the influence of strain rate sensitivity, which could also affect the ductility, especially at relatively high homologous temperatures and/or very small grain sizes where the strain rate sensitivity is relatively high [5]. Hart's criterion [58] can take into account both the strain hardening and strain rate sensitivity:

$$\left(\frac{d\sigma_t}{d\varepsilon_t}\right)\frac{1}{\sigma}+m\geq 1 \qquad (3.4)$$

where m is the strain rate sensitivity.

For most nanostructured metals and alloys, the strain rate sensitivity m is very small (<<0.1), and therefor can be ignored, in which case the Considère criterion is applicable. However, there are also cases where m is relatively large at room temperature for nanostructured metals with low melting temperatures [5, 59, 60]. Application of the Hart's criterion is not easy, because the strain rate sensitivity is usually not readily available. But it is clear that the uniform deformation should be higher than what is determined by the Considère criterion if the strain rate sensitivity is significant. In other words, the strain hardening required to maintain uniform elongation is smaller in the Hart's criterion than in the Considère criterion.

To increase the ductility of a metal, one should try to postpone unstable necking by increasing the uniform elongation. The Considère criterion tells us that we need to increase the strain hardening rate to delay localized deformation (necking). Hart's criterion implies that although increasing strain hardening is the primary way to increase ductility, enhancing the strain rate sensitivity also helps by some limited extents for most nanostructured metals.

The ductility is an important property for a metal to undergo tensile forming such as wire drawing or to serve under tensile load such as re-bars in a bridge beam. Without sufficient ductility, a structure or machine part serving under tensile load may fail catastrophically.

3.3 Plasticity

Plasticity is the ability of a solid material to undergo plastic deformation without failure. The plasticity of a metal is mostly determined by its intrinsic crystal structure and available slip systems. According to von Mise's rule [61], five independent slip systems are required to plastically deform a metal without forming discontinuity (crack). Face-centered cubic (fcc) metals have 12 independent slip systems while body-centered cubic (bcc) metals have 48 independent slip systems. Therefore, fcc and bcc metals usually have high plasticity and can be easily shaped and formed by rolling, forging, extrusion, etc. In comparison, hexagonal close packed (hcp) metals have less than five slip systems and usually need deformation twinning to meet the von Mises's requirement, which is why deformation twinning is always activated during their deformation beyond a certain plastic strain. As a result, hcp metals have lower plasticity than fcc and bcc metals. For example, hcp Ti cannot be deformed by equal channel angular pressing using a 90° die at room temperature due to its low plasticity [58]. For the same reason, hcp Mg has low plasticity during cold rolling.

Plasticity can also be affected by microstructures and defect densities. For example, a metal work piece that is cold rolled to a high strain could have high density of entangled immobile dislocations, which reduces its plasticity to make further cold rolling impossible. Annealing the cold-rolled work piece to induce recrystallization is a common practice to remove dislocations and recover its plasticity.

Plasticity is an important property for the shaping and forming of metals such as rolling, forging, extrusion, etc. Plasticity can be usually measured under compressive stress since mechanical stability (necking) normally does not occur under such a loading condition. In cases where local shear-band-induced instability occurs during compressive testing, other deformation mode such as rolling could be used to evaluate the plasticity. It should be noted that the plasticity observed under different deformation mode is different. However, a brittle material will be brittle under any deformation mode and a metal with good plasticity should be able to be deformed to large plastic strains under several, if not all, deformation modes.

3.4 Relationship between Ductility and Plasticity

Ductility can be considered as a special case of plasticity: i.e. plasticity under tensile loading. High plasticity is a prerequisite for high ductility, but does not guarantee high ductility. This is the case for most nanostructrued metals and alloys, which usually have low ductility although they have high plasticity. This is because nanostructured metals have low to zero strain hardening. Ductility is affected more by microstructure. A microstructure that provides for more locations for dislocation accumulation will lead to higher ductility. In contrast, plasticity is primarily determined by the mobility of dislocations. If there are not enough slip systems to meet the von Mise's rule [61], plasticity is largely determined by the activation of other deformation modes such as twinning, grain boundary sliding, grain rotation, etc.

3.5 Confusions, Misconceptions and Clarifications

There has been considerable confusion about ductility and plasticity in our community. These two terminologies are often mixed and interchanged in a large number of *peer-reviewed* academic papers and conference presentations. Therefore, there is an urgent need to clarify these confusions and misconceptions. Below we will list the most common misconceptions/confusions and clarify them.

3.5.1 Misconception/Confusion 1: Tensile Ductility

The terminology "tensile ductility" is often used in the literature. From the first look, there is nothing wrong with it, because ductility is indeed measured from tensile testing. However, this terminology hints there also exist other types of ductility. Indeed, there are papers that reported ductility measured under compressive stress, which is completely wrong. Therefore, one can say ductility or tensile plasticity, but not tensile ductility.

3.5.2 Misconception/Confusion 2: Mobile Dislocations Lead to Good Ductility

If high density of mobile dislocations is observed by transmission electron microscopy (TEM) and/or other characterization techniques, it only means the material has good plasticity, but does not necessarily have good ductility. However, if dislocations are observed to effectively accumulate at many microstructural features such as twins and second phase particles, good ductility can be expected because dislocation accumulation produces work hardening, which promotes ductility [29, 48, 55]. In other words, dislocation gliding produces plasticity while blocking gliding dislocations to force them to accumulate produces ductility.

3.5.3 Misconception/Confusion 3: Low Ductility Equals Low Plasticity

Low plasticity will certainly lead to low ductility, but the reverse may not be true. A material with no plasticity is considered brittle, such as glass and most ceramic materials, which will shatter when deformed under common deformation modes such as tension, compression, bending, etc. at room temperature. Brittle materials typically have no mobile dislocations available to accommodate plastic deformation when deformed. The fracture surfaces of brittle materials often exhibit intra-grain cleavage or grain boundary failure. In contrast, materials with plasticity deforms normally by dislocation slip, and their fracture surfaces exhibit dimples that are generated by dislocation activities at the final failure stage. Such a fracture is called "ductile fracture" in the textbooks, which has helped to create the confusion and mix-up between ductility and plasticity. This problem is not serious for conventional CG metals, because CG metals with ductile fracture typically have good ductility. However, this posts a big problem for nanostructured metals, which typically exhibit dimples on their fracture surface [62], but still have poor ductility. In other words, dimples on the fracture surface are associated with plasticity, not necessarily with ductility. In other words, nanostructured metals typically have dimpled fracture surface, high plasticity, and poor ductility, but are not brittle.

3.5.4 Misconception/Confusion 4: Cross-Area-Reduction as an Indicator of Ductility

There have been suggestions to use the cross-area-reduction at fracture as a measure/indicator of ductility for nanostructured metals. This is reasonable for conventional CG metals because these two material properties are well co-related. However, for nanostructured metals, they are disconnected. Nanostructured metals usually have large cross-area reduction although they have low ductility. Therefore, the cross-area reduction is correlated with plasticity, but not ductility, just like the dimples at the fracture surface.

3.6 Issues for Nanostructured Materials

3.6.1 Sample Size Effect

Most research groups have been using small, non-standard samples to test the mechanical properties, and some groups even used non-tensile loading such as compression of micro-sized pillars to measure ductility. The small samples are due to the difficulty in making nanostructured materials large enough for standard mechanical testing.

It has been found that the sample size and geometry have significant effect on the stress-strain curves and hence the measured ductility [63, 64]. For example, the gage length effect can be demonstrated using Eq. 3.1. The Δl_u, which is the uniform gage length change during the tensile test, should be proportional to the original gage length l_0. Therefore, the uniform elongation should not be affected by the gage length of the sample, if the sample gage length is not so short that it induces a three dimensional stress state [64]. On the other hand, the Δl_n is the local necking length, and is independent of the initial gage length l_0. Therefore, the necking strain $\varepsilon_n = \Delta l_n / l_0$ will increase dramatically with decreasing gage length l_0, especially when the gage length is small. This often leads to a situation where the necking strain is much larger than the uniform strain. When the total elongation to failure is used as the measure of ductility, its value becomes largely decided by the gage length.

Therefore, ductility often becomes artificially high and becomes meaningless and the ductility data from different group cannot be compared. To alleviate this problem, it is advised to use the uniform elongation as a measure of ductility, which is not affected by the gage length, unless the gage length is approaching the necking length, which causes three dimensional stresses [64].

3.6.2 Approaches to Improve Ductility

After understanding the definition of ductility and what factors affect it most, we can develop approaches to improve ductility. As shown in Eqs. 3.2–3.4, to improve the ductility of nanostructured metals, the most effective way is to increase their strain hardening rate. In other words, we need to design nanostructures that can effectively block and accumulate dislocations. For nanostructured metals, grain boundaries are no longer effective for dislocation accumulation, because when the grain sizes are so small that dislocation sources no longer exist in the grain interior, and grain boundaries become dislocation source and sink without much dislocation accumulation [8]. It follows that we should make dislocation barriers inside the grain interior where dislocation can be blocked and accumulated. Some successful strategies along this path include growth twins [30], deformation twins [48, 65], stacking faults [53, 66], and second phase particles/precipitates [48]. Since these approaches can also increase the strength, they often lead to simultaneous increase in both yield strength and ductility.

It should be also noted that ultrafine grained metals processed by severe plastic deformation techniques often contain high density of dislocations. This leaves little room for dislocation accumulation during tensile test before dislocation saturation is reached. Therefore, annealing to lower the dislocation density without increasing the grain size is expected to improve strain hardening and ductility.

Strain rate sensitivity may also play a significant rule in increasing ductility for some metals with low melting temperature. In such cases, room temperature represents a relatively high homologous temperature (T/T_m), which when coupled with fine grain sizes could produce a relatively high strain rate sensitivity to make the nanostructured metal deform in a quasi-superplastic behavior [59].

Indeed, for nanostructured metals and alloys with low melting temperatures such as Al and Zn, grain boundary sliding was observed at room temperature, which is a deformation mechanism with high strain rate sensitivity [60, 67].

For nanostructured fcc metals, decreasing the grain size normally leads to the increase in strain rate sensitivity. For nanostructured bcc metals, the strain rate sensitivity has been observed to decrease with decreasing grain size [68, 69]. These changes of strain rate sensitivity with grain size is relatively small when compared with the change in strain hardening rate. The strain hardening rate of nanostructured metals is dramatically decreased, and often becomes close to zero [3–5, 70], which is why nanostructured metals and alloys usually have very poor ductility. Therefore, enhancing the strain hardening rate is the primary route to improve the ductility of nanostructured metals [3–5].

Recently, it is found that heterostructures can also help with improving the ductility [61, 71–76]. Heterostructures produce significant hetero-deformation-induced (HDI) stress induced work hardening, in addition to the conventional dislocation hardening as expressed by the Ashby equation [77, 78]:

$$\tau = \alpha G b \sqrt{\rho_S + \rho_G} \qquad (3.5)$$

where α is a constant, G is the shear modulus, b is the magnitude of Burgers vector, ρ_S is the density of statistically stored dislocations, and ρ_G is the density of geometrically necessary dislocations (GNDs). In the heterostructured metals, the GNDs produce significant long-range back stress in the soft zones to enhance the yield strength as well as the high HDI work hardening to enhance/retain ductility.

3.7 Summary

Ductility and plasticity are two related but different concepts. High plasticity is a perquisite for high ductility but does not guarantee high ductility. Ductility can be regarded as tensile plasticity and can only be measured using tensile test. When small samples are used, uniform elongation is a better way to measure ductility than elongation to failure because uniform elongation is much less affected by sample size.

Nanostructured metals typically have high plasticity but low ductility. The primary approach to improve their ductility is to improve strain hardening rate. Strain hardening can be enhanced by the accumulation of crystalline defects such as dislocations and twins as well as by HDI strain hardening. For metals with relatively low melting temperature, strain rate sensitivity could play a notable role in improving ductility

Acknowledgements

Financial support from the National Key R&D Program of China (2017YFA0204402 and 2017YFA0204403).

References

1. Davis JR (ed). *Properties and Selection: Nonferrous Alloys and Special-Purpose Materials.* ASM Handbook, Vol. 2 (ASM International, Materials Park, Ohio, 2000).
2. Valiev RZ, Alexandrov IV, Zhu YT and Lowe TC. Paradox of strength and ductility in metals processed by severe plastic deformation. *J. Mater. Res.*, 2002; 17:5–8.
3. Zhu YT and Liao XZ. Nanostructured metals: retaining ductility. *Nat. Mater.*, 2004; 3:351–352.
4. Valiev RZ, Estrin Y, Horita Z, Langdon TG, Zehetbauer MJ and Zhu YT. Fundamentals of superior properties in bulk nanoSPD materials. *Mater. Res. Lett.*, 2016; 4:1–21.
5. Ovid'ko IA, Valiev RZ and Zhu YT. Review on superior strength and enhanced ductility of metallic nanomaterials. *Prog. Mater. Sci.*, 2018; 94:462–540.
6. Wang YM and Ma E. Three strategies to achieve uniform tensile deformation in a nanostructured metal. *Acta Mater.*, 2004; 52:1699–1709.
7. Fang TH, Li WL, Tao NR and Lu K. Revealing extraordinary intrinsic tensile plasticity in gradient nano-grained copper. *Science*, 2011; 331:1587–1590.
8. Zhu YT, Liao XZ and Wu XL. Deformation twinning in nanocrystalline materials. *Prog. Mater. Sci.*, 2012; 57:1–62.
9. Christian JW and Mahajan S. Deformation twinning. *Prog. Mater. Sci.*, 1995; 39:1–157.

10. Anderoglu O, Misra A, Wang J, Hoagland RG, Hirth JP and Zhang X. Plastic flow stability of nanotwinned Cu foils. *Int. J. Plast.*, 2010; 26:875–886.

11. Caceres CH, Lukac P and Blake A. Strain hardening due to {10(1)overbar2} twinning in pure magnesium. *Philos. Mag.*, 2008; 88:991–1003.

12. Capolungo L and Beyerlein IJ. Nucleation and stability of twins in hcp metals. *Phys. Rev. B*, 2008; 78:024117.

13. Chen AY, Liu JB, Wang HT, Lu J and Wang YM. Gradient twinned 304 stainless steels for high strength and high ductility. *Mater. Sci. Eng. A*, 2016; 667:179–188.

14. Dao M, Lu L, Shen YF and Suresh S. Strength, strain-rate sensitivity and ductility of copper with nanoscale twins. *Acta Mater.*, 2006; 54:5421–5432.

15. Demkowicz MJ, Anderoglu O, Zhang XH and Misra A. The influence of Sigma 3 twin boundaries on the formation of radiation-induced defect clusters in nanotwinned Cu. *J. Mater. Res.*, 2011; 26:1666–1675.

16. Kibey S, Liu JB, Johnson DD and Sehitoglu H. Generalized planar fault energies and twinning in Cu-Al alloys. *Appl. Phys. Lett.*, 2006; 89:191911.

17. Kibey S, Liu JB, Johnson DD and Sehitoglu H. Predicting twinning stress in fcc metals: linking twin-energy pathways to twin nucleation. *Acta Mater.*, 2007; 55:6843–6851.

18. Kibey SA, Wang LL, Liu JB, Johnson HT, Sehitoglu H and Johnson DD. Quantitative prediction of twinning stress in fcc alloys: application to Cu-Al. *Phys. Rev. B*, 2009; 79:214202.

19. Li XY, Wei YJ, Lu L, Lu K and Gao HJ. Dislocation nucleation governed softening and maximum strength in nano-twinned metals. *Nature*, 2010; 464:877–880.

20. Lu L, Schwaiger R, Shan ZW, Dao M, Lu K and Suresh S. Nano-sized twins induce high rate sensitivity of flow stress in pure copper. *Acta Mater.*, 2005; 53:2169–2179.

21. Meyers MA, Vohringer O and Lubarda VA. The onset of twinning in metals: a constitutive description. *Acta Mater.*, 2001; 49:4025–4039.

22. Ojha A, Sehitoglu H, Patriarca L and Maier HJ. Twin nucleation in Fe-based bcc alloys—modeling and experiments. *Model. Simul. Mater. Sci. Eng.*, 2014; 22.

23. Ojha A, Sehitoglu H, Patriarca L and Maier HJ. Twin migration in Fe-based bcc crystals: theory and experiments. *Philos. Mag.*, 2014; 94:1816–1840.

24. Shen YF, Lu L, Dao M and Suresh S. Strain rate sensitivity of Cu with nanoscale twins. *Scr. Mater.*, 2006; 55:319–322.

25. Shen YF, Lu L, Lu QH, Jin ZH and Lu K. Tensile properties of copper with nano-scale twins. *Scr. Mater.*, 2005; 52:989–994.

26. Wang J, Anderoglu O, Hirth JP, Misra A and Zhang X. Dislocation structures of Sigma 3 {112} twin boundaries in face centered cubic metals. *Appl. Phys. Lett.*, 2009; 95:021908.

27. Wang J, Li N, Anderoglu O, Zhang X, Misra A, Huang JY and Hirth JP. Detwinning mechanisms for growth twins in face-centered cubic metals. *Acta Mater.*, 2010; 58:2262–2270.

28. Yan FK, Liu GZ, Tao NR and Lu K. Strength and ductility of 316L austenitic stainless steel strengthened by nano-scale twin bundles. *Acta Mater.*, 2012; 60:1059–1071.

29. Lu K, Lu L and Suresh S. Strengthening materials by engineering coherent internal boundaries at the nanoscale. *Science*, 2009; 324:349–352.

30. Lu L, Shen YF, Chen XH, Qian LH and Lu K. Ultrahigh strength and high electrical conductivity in copper. *Science*, 2004; 304:422–426.

31. Liao XZ, Zhao YH, Srinivasan SG, Zhu YT, Valiev RZ and Gunderov DV. Deformation twinning in nanocrystalline copper at room temperature and low strain rate. *Appl. Phys. Lett.*, 2004; 84:592–594.

32. Liao XZ, Zhou F, Lavernia EJ, Srinivasan SG, Baskes MI, He DW and Zhu YT. Deformation mechanism in nanocrystalline Al: partial dislocation slip. *Appl. Phys. Lett.*, 2003; 83:632–634.

33. Ma XL, Xu WZ, Zhou H, Moering JA, Narayan J and Zhu YT. Alloying effect on grain-size dependent deformation twinning in nanocrystalline Cu-Zn alloys. *Philos. Mag.*, 2015; 95:301–310.

34. Wu XL, Liao XZ, Srinivasan SG, Zhou F, Lavernia EJ, Valiev RZ and Zhu YT. New deformation twinning mechanism generates zero macroscopic strain in nanocrystalline metals. *Phys. Rev. Lett.*, 2008; 100:095701.

35. Wu XL and Zhu YT. Inverse grain-size effect on twinning in nanocrystalline Ni. *Phys. Rev. Lett.*, 2008; 101:025503.

36. Zhu YT, Liao XZ, Wu XL and Narayan J. Grain size effect on deformation twinning and detwinning. *J. Mater. Sci.*, 2013; 48:4467–4475.

37. Osetsky YN, Bacon DJ and Mohles V. Atomic modelling of strengthening mechanisms due to voids and copper precipitates in alpha-iron. *Philos. Mag.*, 2003; 83:3623–3641.

38. Xu C, Furukawa M, Horita Z and Langdon TG. Influence of ECAP on precipitate distributions in a spray-east aluminum alloy. *Acta Mater.*, 2005; 53:749–758.
39. Cheng S, Zhao YH, Zhu YT and Ma E. Optimizing the strength and ductility of fine structured 2024 Al alloy by nano-precipitation. *Acta Mater.*, 2007; 55:5822–5832.
40. Roven HJ, Liu MP and Werenskiold JC. Dynamic precipitation during severe plastic deformation of an Al-Mg-Si aluminium alloy. *Mater. Sci. Eng. A*, 2008; 483–484:54–58.
41. Raabe D, Ponge D, Dmitrieva O and Sander B. Nanoprecipitate-hardened 1.5 GPa steels with unexpected high ductility. *Scr. Mater.*, 2009; 60:1141–1144.
42. Sha G, Wang YB, Liao XZ, Duan ZC, Ringer SP and Langdon TG. Influence of equal-channel angular pressing on precipitation in an Al-Zn-Mg-Cu alloy. *Acta Mater.*, 2009; 57:3123–3132.
43. Geng J, Chun YB, Stanford N, Davies CHJ, Nie JF and Barnett MR. Processing and properties of Mg-6Gd-1Zn-0.6Zr Part 2. Mechanical properties and particle twin interactions. *Mater. Sci. Eng. A*, 2011; 528:3659–3665.
44. Zheng BL, Ertorer O, Li Y, Zhou YZ, Mathaudhu SN, Tsao CYA and Lavernia EJ. High strength, nano-structured Mg-Al-Zn alloy. *Mater. Sci. Eng. A*, 2011; 528:2180–2191.
45. Nie JF. Precipitation and hardening in magnesium alloys. *Metall. Mater. Trans. A*, 2012; 43A:3891–3939.
46. Tsai MH, Yuan H, Cheng GM, Xu WZ, Jian WWW, Chuang MH, Juan CC, Yeh AC, Lin SJ and Zhu YT. Significant hardening due to the formation of a sigma phase matrix in a high entropy alloy. *Intermetallics*, 2013; 33:81–86.
47. Li K, Beche A, Song M, Sha G, Lu XX, Zhang K, Du Y, Ringer SP and Schryvers D. Atomistic structure of Cu-containing β'' precipitates in an Al-Mg-Si-Cu alloy. *Scr. Mater.*, 2014; 75:86–89.
48. Zhao YH, Liao XZ, Cheng S, Ma E and Zhu YT. Simultaneously increasing the ductility and strength of nanostructured alloys. *Adv. Mater.*, 2006; 18:2280–2283.
49. Zhao YH, Liao XZ, Jin Z, Valiev RZ and Zhu YT. Microstructures and mechanical properties of ultrafine grained 7075 Al alloy processed by ECAP and their evolutions during annealing. *Acta Mater.*, 2004; 52:4589–4599.

50. Liddicoat PV, Liao XZ, Zhao YH, Zhu YT, Murashkin MY, Lavernia EJ, Valiev RZ and Ringer SP. Nanostructural hierarchy increases the strength of aluminum alloys. *Nat. Commun.*, 2010; 1:63.

51. Wu XL, YUan FP, Yang MX, JIang P, Zhang CX, Chen L, Wei YG and Ma E. Nanodomained nickel unite nanocrystal strength with coarse-grain ductility. *Sci. Rep.*, 2015; 5:11728.

52. Karimpoor AA, Erb U, Aust KT and Palumbo G. High strength nanocrystalline cobalt with high tensile ductility. *Scr. Mater.*, 2003; 49:651–656.

53. Jian WW, Cheng GM, Xu WZ, Yuan H, Tsai MH, Wang QD, Koch CC, Zhu YT and Mathaudhu SN. Ultrastrong Mg alloy via nano-spaced stacking faults. *Mater. Res. Lett.*, 2013; 1:61–66.

54. Zhao YH, Bingert JF, Zhu YT, Liao XZ, Valiev RZ, Horita Z, Langdon TG, Zhou YZ and Lavernia EJ. Tougher ultrafine grain Cu via high-angle grain boundaries and low dislocation density. *Appl. Phys. Lett.*, 2008; 92:081903.

55. Zhao YH, Zhu YT and Lavernia EJ. Strategies for improving tensile ductility of bulk nanostructured materials. *Adv. Eng. Mater.*, 2010; 12:769–778.

56. Huang CX, Hu WP, Wang QY, Wabg C, Yang G and Zhu YT. An ideal ultrafine-grained structure for high strength and high ductility. *Mater. Res. Lett.*, 2015; 3:88–94.

57. ASM Handbook, 7th ed. (ASM International, Materials Park, 1990), p. 1328.

58. DeLo DP and Semiatin SL. Hot working of Ti-6Al-4V via equal channel angular extrusion. *Metall. Trans. A*, 1999; 30:2473–2481.

59. Valiev RZ, Murashkin MY, Kilmametov A, Straumal B, Chinh NQ and Langdon TG. Unusual super-ductility at room temperature in an ultrafine-grained aluminum alloy. *J. Mater. Sci.*, 2010; 45:4718–4724.

60. Zhang X, Wang H, Scattergood RO, Narayan J, Koch CC, Sergueeva AV and Mukherjee AK. Tensile elongation (110%) observed in ultrafine-grained Zn at room temperature. *Appl. Phys. Lett.*, 2002; 81:823–825.

61. Wu XL, Jiang P, Chen L, Yuan FP and Zhu YT. Extraordinary strain hardening by gradient structure. *Proc. Natl. Acad. Sci. USA*, 2014; 111:7197–7201.

62. Zhu YT, Huang JY, Gubicza J, Ungar T, Wang YM, Ma E and Valiev RZ. Nanostructures in Ti processed by severe plastic deformation. *J. Mater. Res.*, 2003; 18:1908–1917.

63. Zhao YH, Guo YZ, Wei Q, Dangelewiez AM, Zhu YT, Langdon TG, Zhou YZ, Lavernia EJ and Xu C. Influence of specimen dimensions on the tensile behavior of ultrafine-grained Cu. *Scr. Mater.*, 2008; 59:627–630.

64. Zhao YH, Guo YZ, Wei Q, Topping TD, Dangelewicz AM, Zhu YT, Langdon TG and Lavernia EJ. Influence of specimen dimensions and strain measurement methods on tensile stress-strain curves. *Mater. Sci. Eng. A*, 2009; 525:68–77.

65. Zhao YH, Zhu YT, Liao XZ, Horita Z and Langdon TG. Tailoring stacking fault energy for high ductility and high strength in ultrafine grained Cu and its alloy. *Appl. Phys. Lett.*, 2006; 89:121906.

66. Jian WW, Cheng GM, Xu WZ, Koch CC, Wang QD, Zhu YT and Mathaudhu SN. Physics and model of strengthening by parallel stacking faults. *Appl. Phys. Lett.*, 2013; 103.

67. Mungole T, Kumar P, Kawasaki M and Langdon TG. The contribution of grain boundary sliding in tensile deformation of an ultrafine-grained aluminum alloy having high strength and high ductility. *J. Mater. Sci.*, 2015; 50:3549–3561.

68. Wei Q, Cheng S, Ramesh KT and Ma E. Effect of nanocrystalline and ultrafine grain sizes on the strain rate sensitivity and activation volume: fcc versus bcc metals. *Mater. Sci. Eng. A*, 2004; 381:71–79.

69. Cheng GM, Jian WW, Xu WZ, Yuan H, MIllett PC and Zhu YT. Grain size effect on deformation mechanisms of nanocrystalline bcc metals. *Mater. Res. Lett.*, 2013; 1:26–31.

70. Jia D, Wang YM, Ramesh KT, Ma E, Zhu YT and Valiev RZ. Deformation behavior and plastic instabilities of ultrafine-grained titanium. *Appl. Phys. Lett.*, 2001; 79:611–613.

71. Wu XL and Zhu YT. Heterogeneous materials: a new class of materials with unprecedented mechanical properties. *Mater. Res. Lett.*, 2017; 5:527–532.

72. Wu XL, Yang MX, Yuan FP, Chen L and Zhu YT. Combining gradient structure and TRIP effect to produce austenite steel with high strength and ductility. *Acta Mater.*, 2016; 112:337–346.

73. Wu XL, Yang MX, Yuan FP, Wu GL, Wei YJ, Huang XX and Zhu YT. Heterogeneous lamella structure unites ultrafine-grain strength with coarse-grain ductility. *Proc. Natl. Acad. Sci. USA*, 2015; 112:14501–14505.

74. Wei YJ, Li YQ, Zhu LC, Liu Y, Lei XQ, Wang G, Wu YX, Mi ZL, Liu JB, Wang HT and Gao HJ. Evading the strength- ductility trade-off dilemma in steel through gradient hierarchical nanotwins. *Nat. Commun.*, 2014; 5.

75. Lu K. Making strong nanomaterials ductile with gradients. *Science*, 2014; 345:1455–1456.

76. Yang MX, Pan Y, Yuan FP, Zhu YT and Wu XL. Back stress strengthening and strain hardening in gradient structure. *Mater. Res. Lett.*, 2016; 4:145–151.

77. Ashby MF. Deformation of plastically non-homogeneous materials. *Philos. Mag.*, 1970; 21:399.

78. Gao H, Huang Y, Nix WD and Hutchinson JW. Mechanism-based strain gradient plasticity - I. Theory. *J. Mech. Phys. Solids*, 1999; 47:1239–1263.

Part II
Fundamentals of Heterostructured Materials

Chapter 4

Extraordinary Strain Hardening by Gradient Structure

Xiaolei Wu,[a] Ping Jiang,[a] Liu Chen,[a] Fuping Yuan,[a] and Yuntian Zhu[b,c]

[a]*State Key Laboratory of Nonlinear Mechanics, Institute of Mechanics, Chinese Academy of Sciences, Beijing 100190, China*
[b]*Department of Materials Science and Engineering, North Carolina State University, Raleigh, NC 27695, USA*
[c]*School of Materials Science and Engineering, Nanjing University of Science and Technology, Nanjing 210094, China*
xlwu@imech.ac.cn, ytzhu@ncsu.edu

Gradient structures have evolved over millions of years through natural selection and optimization in many biological systems such as bones and plant stems, where the structures gradually change from the surface to interior. The advantage of gradient structures is their maximization of physical and mechanical performance while minimizing material cost. Here we report that the gradient structure in engineering materials such as metals renders a unique extra strain

---

Reprinted from *Proc. Natl. Acad. Sci. U.S.A.*, **111**(20), 7197–7201, 2014.

Heterostructured Materials: Novel Materials with Unprecedented Mechanical Properties
Edited by Xiaolei Wu and Yuntian Zhu
Text Copyright © 2014 The Author(s)
Layout Copyright © 2022 Jenny Stanford Publishing Pte. Ltd.
ISBN 978-981-4877-10-7 (Hardcover), 978-1-003-15307-8 (eBook)
www.jennystanford.com

hardening, which leads to high ductility. The grain size gradient under uniaxial tension induces a macroscopic strain gradient and converts the applied uniaxial stress to multi-axial stresses due to the evolution of incompatible deformation along the gradient depth. Thereby accumulation and interaction of dislocations are promoted, resulting in an extra hardening and an obvious strain hardening rate up-turn. Such extraordinary strain hardening, which is inherent to gradient structures and does not exist in homogeneous materials, provides a novel strategy to develop strong and ductile materials by architecting heterostructures.

Significance Statement

Nature creates gradient structures for a purpose: to make biological systems strong and tough to survive severe natural forces. The special mechanism related to various functional gradient materials may be clear. But for structural gradient, i.e. the grain size gradient, the deformation physics is still unclear up to now. One wonders if the grain-size gradient in the nano-micro- scale would also benefit materials engineered by mankind. In this chapter, a universal strain hardening mechanism is revealed in the gradient structure. We uncovered a unique extra strain hardening that is intrinsic to the gradient structure. Its mechanism is the presence of strain gradient together with the stress state change. A superior combination of strength and ductility that is not accessible to conventional homogeneous materials is obtained. As a novel mechanism, extra strain hardening renders high ductility in the gradient structured materials.

4.1 Introduction

Mankind has much to learn from the nature on how to make engineering materials with novel and superior physical and mechanical properties. For examples, clay-polymer multilayers mimicking the naturally grown seashells are found to have exceptional mechanical properties [1–3]. Another example is the gradient structure, which exists in many biological systems such as teeth and bamboos. A typical gradient structure exhibits a systematic change in microstructure along the depth on a macroscopic scale. Gradient structures have been evolved and optimized over millions of years to make the biological systems strong and tough to survive the nature. They are much superior to man-made engineering materials with homogeneous microstructures.

Here we report the discovery of a hitherto unknown strain hardening mechanism, which is intrinsic to the gradient structure in an engineering material. The gradient structure shows a surprising extra strain hardening along with an up-turn and subsequent good retention of strain hardening rate. Strain hardening is critical for increasing the material ductility [4–6]. We also show a superior ductility-strength combination in the gradient structure that is not accessible to conventional homogeneous microstructures.

4.2 Microstructural Characterization of Gradient Structure

We demonstrate these behaviors in a gradient structured sample, i.e. the grain-size gradient structured (GS) surface layers sandwiching a coarse-grained (CG) core, produced by the surface mechanical attrition treatment (SMAT) [7] of a 1-mm thick CG interstitial free (IF) steel sheet (see the Supporting Materials). The GS layers on both sides have a gradual grain size increase along the depth (Fig. 4.1A). In the outermost layer of ~25 μm thick are nearly equi-axial nano-grains with a mean size of 96 nm (Fig. 4.1B). The grain size increases gradually to 0.5 and 1 μm at the depths of ~60 and 90 μm, respectively, with subgrains or dislocation cells smaller

Figure 4.1 Gradient structure by SMAT. (a) Variation of average grain sizes along the depth. The error bars represent the standard deviation of the grain sizes. The gradient structured sample was produced by means of the surface mechanical attrition treatment for 5 minutes on both sides of a coarse-grained IF steel sheet. (b) Cross-sectional TEM bright-field image of the nano-grains with a mean grain size of 96 nm at the depth of 10 μm. (c) Electron back-scatter diffraction image of coarse-grains with a mean grain size of 35 μm.

than 100 nm. For convenience, we define the top 90-mm layer as nanostructured layer [8–10] with a grain-size gradient. The whole gradient layer is 120 µm thick, including the deformed CG layer with either dislocation tangles or dislocation cells of sizes ranging from submicrometers to micrometers. The central strain-free CG core has an average grain size of 35 µm (Fig. 4.1C).

4.3 Unique Mechanical Responses under Uniaxial Tension

Figure 4.2A shows the engineering stress-strain (σ_e–ε_e) curves. The gradient structured (GS/CG) sample exhibits not only a large tensile uniform elongation (E_U), comparable to that of the homogeneous CG sample, but also yield strength that is ~2.6 times as high. In contrast, the free-standing nano-structured (NS) film becomes unstable soon after yielding. Interestingly, the GS/CG sample shows a transient hardening in the regime of small tensile strains on its σ_e–ε_e curve between two inflexion points (inset in Fig. 4.2B). This caused an up-turn in the strain hardening rate Θ (inset in Fig. 4.2B). Meanwhile, the unloading/reloading σ_e–ε_e curves also reveal a similar Θ up-turn upon each re-loading (GS/CG$^+$, red curves in Fig. 4.2B). More importantly, the GS/CG sample shows an even slower Θ reduction than that of the CG sample (Fig. 4.2B), indicating a better Θ retention in the GS/CG sample. In contrast, the free-standing GS layer and CG core do not show any Θ up-turn (see the GS layer and CG curves), suggesting that the unique behavior is produced only when these two type of layers form an integral bulk.

The dramatic hardening behavior raises a critical issue: where in the gradient-structured GS/CG sample is strain hardening generated? To answer this question, we measured the microhardness (H) along the depth of GS/CG samples after testing them to varying strains. As shown in Fig. 4.2C, the H values increase with increasing tensile strain. The border, where H values in the un-tested sample no longer drop, demarcates the gradient layer and CG core. Fig. 4.2D shows H increments, ΔH, along the depth caused by testing at various tensile strains. ΔH is an indicator on the magnitude of hardening retained after unloading. For comparison, the ΔH values are also measured in both the free-standing gradient layer (failure strain of 0.05 in Fig. 4.2A) and homogeneous CG after tensile testing them to the strain of 0.05. Remarkably, *the layer in GS/CG exhibits a unique extra strain hardening, i.e. a much higher ΔH than that of the free-standing gradient layer* (dotted area in Fig. 4.2D). This extra hardening originates in the gradient layer, and its peak moves inwards and finally penetrates into the CG core at higher strains. This indicates again that the gradient layer needs to form an integral bulk with the CG core to be effective in produce strain hardening.

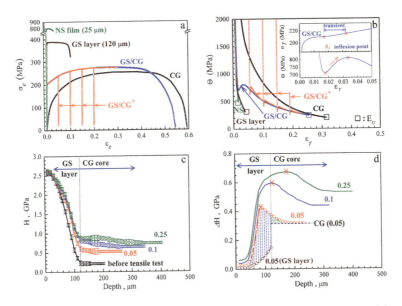

Figure 4.2 Hardening rate up-turn and unique extra strain hardening. (a) Tensile engineering stress–strain (σ_e–ε_e) curves at a quasi-static strain rate of 5×10^{-4} s^{-1}. CG: standalone homogeneous coarse-grained sample; GS layer: grain-size gradient structure layer of 120-μm thick; GS/CG: sandwich sample of 1 mm thick. NS: free-standing, quasi-homogeneous nano-structured film of 20-μm thick peeled from the top surface of GS; GS/CG$^+$: the same sandwich sample subjected to unloading-reloading tensile testing at four separate strains of 0.05, 0.1, 0.15 and 0.2. All tensile samples were dog-bone-shaped, with a gauge dimension of 8 mm × 2.5 mm. (b) Strain hardening rate ($\Theta = d\sigma/d\varepsilon$) vs true strain ($\varepsilon_T$) curves. Inset shows the transient response on the σ_T–ε_T curve of the GS/CG sample between two inflexion points marked by "×", corresponding to the Θ up-turn on its Θ–ε_T curve. GS/CG$^+$ (red curve) shows Θ up-turn upon each re-loading. (c) Vickers microhardness (H) vs depth. The H values were measured on the cross-sectional GS/CG sample after tensile testing them to varying strains (labeled on the curves). The border between GS layer and CG core is located at 120 μm deep. (d) ΔH (H increment) vs depth after varying tensile strains. The dotted area indicates the extra hardening in the GS/CG sample.

The above unique hardening behavior is inherent to the gradient structure and is caused by the gradient-generated multi-axial stresses and strain gradient. Under uniaxial tension (see Fig. 4.3A), necking instability readily occurs in the NS surface layers at very low tensile strains (see NS-film curve in Fig. 4.2A), which is characterized by their fast lateral shrinking. However, lateral instability is constrained and quickly stopped by the neighboring stable layers. Consequently, the strain gradient is produced near the border between the stable central layer and unstable surface layers [11, 12], where strain continuity is required to keep material continuity.

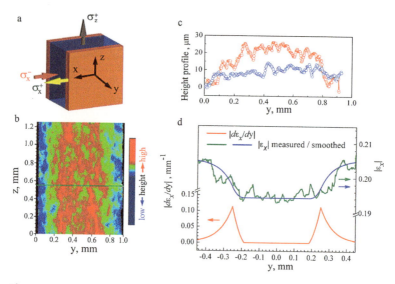

Figure 4.3 Stress state change and strain gradient. (a) Schematic stress state change during uniaxial tension in a GS/CG sample. Outer-layers: the plastically unstable layers. Core: the stable layer. σ_x^+ is the lateral tensile stress in the outer layers, and σ_x^- is the lateral compressive stress in the inner layer in x-direction. σ_z^+ is the applied uniaxial stress. (b) Measured height contour on the side surface, vertical to x-axis, within the gauge section of GS/CG sample at the tensile strain of 0.25. (c) Measured height profiles covering the thickness of both GS/CG sample (along the green line in b) and a standalone homogeneous CG sample after tensile testing to a strain of 0.3. (d) Distribution of lateral strain (ε_x) and strain gradient ($d\varepsilon_x/dy$) across the thickness along the green line in b.

The mutual constraint between the stable central layer and unstable surface layers leads to *stress state changes*, as schematically shown in Fig. 4.3A. When the shrinking surface layers are constrained by the central layer, the constraint is realized in the form of lateral tensile stress in the surface layers, i.e. (σ_x^+) as shown Fig. 4.3A. Since no external lateral stress is applied to the sample, the tensile stresses in the outer surface layers have to be balanced by a lateral compressive stress (σ_x^-) in the inner stable layer. Therefore, the applied uniaxial tensile stress is converted to complex two-dimensional stress states with the outer surface layers under a tension-tension stress state and the central stable layer under a tension-compression stress state.

To evaluate the strain gradient, the height profiles on the lateral surface, i.e. vertical to x-axis in Fig. 4.3A, were measured after suspension of tensile testing exactly prior to necking, as shown in Fig. 4.3B (also Fig. 4.S1A). *The GS/CG sample exhibits marked height difference, i.e. lower on both sides and higher in the middle* (see the red curve in Fig. 4.3C) (also Fig. 4.S1B). This is the direct evidence that non-uniform lateral deformation in GS/CG occurred, with the outer GS layers shrank more than the central zone. In contrast, a homogeneous standalone CG sample only shows surface roughness without a height difference (see the blue curve in Fig. 4.3C). The lateral strain ε_x was calculated as $\varepsilon_x = \delta x / x_0$, where $\delta x = x - x_0$ measured from the contour (Fig. 4.3C) and x_0 is the initial width. This strain is negative due to shrinking and the distribution of its absolute values along the green line in Fig. 4.3B is plotted in Fig. 4.3D. It is also fitted with a smooth curve $\varepsilon_x(y)$ (the blue curve). As shown, the absolute value of ε_x is essentially unchanged in the stable central layer except for the effect of surface roughness. However, the absolute value of ε_x increases gradually towards the surface, which produces strain gradient, $\delta \varepsilon_x / dy$, across the sample thickness as plotted in Fig. 4.3D (the red curve). As shown, *there exists a maximum strain gradient near the boundary*. As discussed later, this maximum strain gradient will promote the accumulation of geometrically necessary dislocations (GNDs) [11, 12] to produce a peak of extra strain hardening.

To probe the physical origin behind the Θ up-turn, the dislocation evolution with strains in gradient-structured sample is studied by stress relaxation tests (see the Supporting Materials), which is complemented by transmission electron microscopy (TEM) observations. Figure 4.4A shows a σ_e-ε_e curve as a function of relaxation time at varying strains (inset in Fig. 4.4A), which were selected carefully to cover the strain range where the Θ upturn occurs (inset in Fig. 4.2B). The ratio Re = ρ_m/ρ_{m0} represents the relative mobile dislocation density evolution [13]. Figure 4.4B shows the evolution of the mobile dislocation density with relaxation time and its inset reveals how the mobile dislocation density after relaxation varies with tensile strain. As shown, with increasing tensile strain, the Re first drops (inset in Fig. 4.4B) and reaches the minimum value at the strain of 0.015, after which Re increases rapidly to reach a near saturated value at the strain of 0.05. Interestingly, the strain value of 0.015 almost coincides with the strain at which minimum Θ is observed in the GS sample (see the inset of Fig. 4.2B). This observation indicates rapid exhaustion of mobile dislocations at low strains, which is consistent with what was reported in nanocrystalline Ni [14]. In addition, the strain of 0.015 is also near the onset of necking instability for NS film (see the green curve in Fig. 4.2A), suggesting that exhaustion of mobile dislocations promoted instability of the nanostructured surface layers [15]. On the other hand, this also creates more space for dislocation accumulation later, setting stage for Θ up-turn (Fig. 4.4E).

TEM observations provide us with information on the evolution of dislocation structures in the nano-grains. At very small strain of 0.008 (soon after yielding), the tangled high-density dislocations are visible either in their interior or at boundaries and sub-boundaries (Fig. 4.4C). At higher strain of 0.015 (prior to the Θ up-turn), the debris of dislocations is visible (Fig. 4.4D), indicating the occurrence of disentanglement and annihilation of the initial dislocation structure [16]. Further increasing strain to 0.035 (soon after Θ up-turn) generates new dislocation structures (Fig. 4.4E). These observations are consistent with and provide insight into the evolution of mobile dislocations and Θ up-turn in the GS sample.

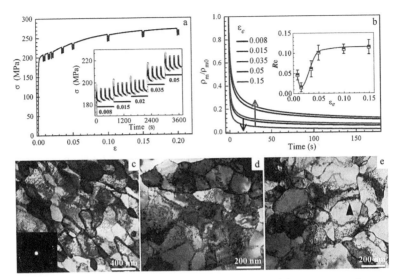

Figure 4.4 Evolution in mobile dislocation density. (a) σ_e–ε_e curve from a stress relaxation test at eight starting strains. Inset is the stress-relaxation time curve. Four relaxation tests are performed at each starting strain. (b) Evolution of the mobile dislocation density, ρ_m/ρ_{m0}, with the first stress relaxation (180 seconds) at varying starting strains. ρ_m is the mobile dislocation density, while ρ_{m0} is the dislocation density at the on set of each relaxation cycle. Two arrows indicate the first drop and later rise in ρ_m/ρ_{m0} with strain. Inset shows $R_e = \rho_m/\rho_{m0}$ after relaxation as a function of starting strain. (c–e) Cross-sectional TEM bright-field images of GS/CG samples after suspension of tenisle testing at varying strains. (c) Dislocation tangled grain boundaries and sub-boundaries at the strain of 0.008 (~20 µm deep). (d) Dislocation debris left inside grain interior and on their boundaries at the strain of 0.015 (~20 µm deep). (e) Newly-formed dislocation sub-boundaries (as pointed by an triangle) at the strain of 0.035 (~25 µm deep).

4.4 Discussion and Summary

The extra strain hardening (Fig. 4.2D) is caused by the strain gradient (Fig. 4.3D), which needs to be accommodated by the generation of the geometrically necessary dislocations (GNDs) [17–20]. The GNDs interact and tangle with mobile dislocations to further promote the dislocation storage [18]. These dislocation activities effectively promotes the dislocation accumulation near the boundary between the unstable and stable layers, which produces the observed extra hardening (ΔH) peak as shown in Fig. 4.2D. With increasing applied strain, stable layers will become unstable, which leads to the migration of the boundary and consequently the ΔH peak toward the CG core. This leaves in its wake of high-densities of dislocations. This explains why the ΔH peak becomes flatter and moves inward as the tensile strain increases. In other words, the extra hardening is accumulative with the dynamically moving boundary. This is the reason that the gradient-structured sample has a slower decrease in Θ than the homogenous CG core with strain (Fig. 4.2B). Furthermore, the multi-axial stress state will activate more slip systems [16, 21], which makes it more likely for dislocations to interact and entangle with each other (Fig. 4.4E), following the initial depletion of dislocations (Fig. 4.4B). The GNDs caused by the strain gradient and the dislocation accumulation caused by the multi-axial stress state are the primary causes for the observed dramatic Θ up-turn and good retention of strain hardening rate.

The nano-grained layers play critical roles in producing high extra strain hardening although no significant extra hardening occurs in the nano-grained layers themselves (Fig. 4.2D). First, the nano-grained layers have a much higher flow stress than the larger-grained inner layer. This ensures high lateral stresses σ_X^+ and σ_X^- (Fig. 4.3A) during the necking of the nano-grained layer, which is constrained by the stable central layer. The high lateral stress will promote the operation of additional slip systems to help with dislocation storage. Second, the early necking by the nano-grained layers activates the multi-axial stresses and strain gradient at early stage of the mechanical testing, which consequently starts the extra strain hardening process in an early stage.

Our preliminary results also suggest that there is a minimum SMAT processing time above which the strain hardening rate up-turn occurs. This minimum processing time is associated with a minimum GS layer thickness. It is our hypothesis that there should be an optimum GS layer thickness that produces the most significant Θ up-turn and the most extra strain hardening. Further systematic investigation is needed to verify this hypothesis. It should be also noted that the mechanism for the good ductility observed here is totally different from that in gradient nano-grained Cu [8], where high ductility was attributed to grain growth due to the low structural stability of the gradient nano-grained Cu. No grain growth is observed in the GS IF-steel in the current study (Fig. 4.S2).

Due to the extra strain hardening, the gradient structure provides for an effective route to a superior combination of good ductility and high strength (Fig. 4.5). When the homogeneous IF-steel is deformed to increase strength, its ductility usually drops dramatically, especially when the strength is above 400 MPa [22–26]. In contrast, the ductility of gradient-structured sample is 5–10 times higher than that of the homogeneous nano-grained structures within the strength range of 450–600 MPa. More importantly, the gradient-structured sample can be easily produced in metallic materials in a cost effective and large-scale way and therefore is expected to be conducive to industrial production.

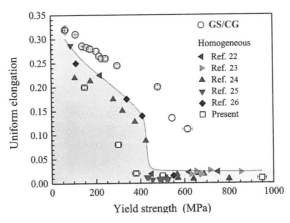

Figure 4.5 Superior mechanical property. Strength and ductility in the GS/CG samples of IF steel as compared with their homogeneous counterparts.

Acknowledgements

X.L.W., P.J., L.C. and F.P.Y. are funded by the NFSC (11002151, 50571110, and 11021262) and 973 Program of China (2012CB92203, 2012CB937500 6138504). Y.T.Z. is funded by the US Army Research Office under the Grant Nos. W911NF-09-1-0427, W911QX-08-C-0083 and by the Nanjing University of Science and Technology.

References

1. Suresh S (2001). Graded materials for resistance to contact deformation and damage. *Science,* **292**:2447–2451.
2. Gao HJ, Ji BH, Jager IL, Arzt E, Fratzl P (2003). Materials become insensitive to flaws at nanoscale: lessons from nature. *Proc. Natl. Acad. Sci. USA,* **100**:5597–5600.
3. Miserez A, Schneberk T, Sun CJ, Zok FW, Waite JH (2008). The transition from stiff to compliant materials in squid beaks. *Science,* **319**:1816–1819.
4. Wang YM, Chen MW, Zhou FH, Ma E (2002). High tensile ductility in a nanostructured metal. *Nature,* **419**:912–915.
5. Lu K, Lu L, Suresh S (2009). Strengthening materials by engineering coherent internal boundaries at the nanoscale. *Science,* **324**:349–352.
6. Ritchie RO (2011). The conflicts between strength and toughness. *Nat. Mater.,* **10**:817–822.
7. Lu K, Lu J (1999). Surface nanocrystallization (SNC) of metallic materials-presentation of the concept behind a new approach. *J. Mater. Sci. Technol.,* **15**:193–197.
8. Fang TH, Li WL, Tao NR, Lu K (2011). Revealing extraordinary intrinsic tensile plasticity in gradient nano-grained copper. *Science,* **331**:1587–1590.
9. Zhu YT, Langdon TG (2004). The fundamentals of nanostructured materials processed by severe plastic deformation. *JOM,* **56**(10):58–63.
10. Huang XX, Hansen N, Tsuji N (2006). Hardening by annealing and softening by deformation in nanostructured metals. *Science,* **312**:249–251.
11. Ashby MF (1970). Deformation of plastically non-homogeneous materials. *Philos. Mag.,* **21**:300–424.
12. Gao H, Huang Y, Nix WD, Hutchinson JW (1999). Mechanism-based strain gradient plasticity - I. Theory. *J. Mech. Phys. Solids,* **47**:1239–1263.
13. Martin JL, Piccolo BLo, Kruml T, Bonneville J (2002). Characterization of thermally activated dislocation mechanisms using transient tests. *Mater. Sci. Eng. A,* **322**:118–125.
14. Wang YM, Hamza AV, Ma E (2005). Activation volume and density of mobile dislocations in plastically deforming nanocrystalline Ni. *Appl. Phys. Lett.,* **86**:241917.

15. Wang YM, Ma E (2004). Three strategies to achieve uniform tensile deformation in a nanostructured metal. *Acta Mater.,* **52**:1699–1709.
16. Wilson DV (1994). Influences of cell-walls and grain-boundaries on transient responses of an IF steel to changes in strain path. *Acta Metall. Mater.,* **42**:1099–1111.
17. Nix WD, Gao HJ (1998). Indentation size effects in crystalline materials: a law for strain gradient plasticity. *J. Mech. Phys. Solids,* **46**:411–425.
18. Gao HJ, Huang YG (2003). Geometrically necessary dislocation and size-dependent plasticity. *Scr. Mater.,* **48**:113–118.
19. Hughes DA, Hansen N, Bammann DJ (2003). Geometrically necessary boundaries, incidental dislocation boundaries and geometrically necessary dislocations. *Scr. Mater.,* **48**:147–153.
20. Hansen N, Huang X (1998). Microstructure and flow stress of polycrystals and single crystals. *Acta Mater.,* **46**:1827–1836.
21. Asaro RJ (1983). Micromechanics of crystals and polycrystals. *Adv. Appl. Mech.,* **23**:1–115.
22. Tsuji N, Ito Y, Saito Y, Minamino Y(2002). Strength and ductility of ultrafine grained aluminum and iron produced by ARB and annealing. *Scr. Mater.,* **47**:893–899.
23. Huang XX, Kamikawa N, Hansen N (2008). Increasing the ductility of nanostructured Al and Fe by deformation. *Mater. Sci. Eng. A,* **493**:184–189.
24. Purcek G, Saray O, Karaman I, Maier HJ (2012). High strength and high ductility of ultrafine-grained interstitial-free steel produced by ECAE and annealing. *Metall. Mater. Trans. A,* **43A**:1884–1894.
25. LapovokR, Orlov D, Timokhina IB, Pougis A, Toth LS, Hodgson PD, Haldar A, Bhattacharjee D (2012). Asymmetric rolling of interstitial-free steel using one idle roll, *Metall. Mater. Trans. A,* **43A**:1328–1340.
26. Hazra SS, Pereloma EV, Gazder AA (2011). Microstructure and mechanical properties after annealing of equal-channel angular pressed interstitial-free steel. *Acta Mater.,* **59**:4015–4029.

Material and Methods

1. Material and Heat Treatment

A 1-mm thick IF steel sheet was used in the present study. Its composition is (wt%) 0.003% C, 0.08% Mn, 0.009% Si, 0.008% S, 0.011% P, 0.037% Al, 0.063% Ti, and 38 ppmN. Sample disks with a diameter of 100 mm were cut from the IF-steel sheet and then pre-annealed at 1173 K for 1 hour to obtain a homogeneous coarse-grained microstructure with a mean grain size of 35 µm.

2. Producing Grain-Size Gradient Structures via SMAT

The techneque of the surface mechanical attrition treatment (SMAT) [7] was used to process the GS/CG sandwich samples with a grain-size gradient structured (GS) layers on both surfaces of the CG core in IF steel. The SMAT technique is based on the impaction of spherical shots (4–6 mm) on the sample disk using high-power ultrasound. Because of the high frequency of the system (20 kHz), the entire surface of the component is peened with a very high number of impacts over a short period of time. The SMAT processing time was the same for both sides of each disk, which varied from 40 seconds to 15 minutes. No crack was observed on the sample surface after SMAT processing.

3. Tensile Test, Unloading-Reloading Test, and Stress-Relaxation Test

Tensile specimens with a gauge length of 8 mm and a width of 2.5 mm were cut from SMAT-processed sample disks. The nano-grained films for Fig. 4.1A were prepared by polishing away the 5-min SMAT-processed samples from one side only, leaving behind a film of desired thickness for tensile testing.

  (1) Quasi-static uniaxial tensile tests were carried out using an Instron 5582 testing machine at strain rate of 5×10^{-4} s^{-1} at room temperature. An extensometer was used to measure strain during the period of uniform tensile deformation.

(2) Loading-unloading tests were conducted using an Instron 5966 testing machine at room temperature. Four loading-unloading cycles were conducted during a uniaixal tensiletest, at tensile strain of 5%, 10%, 15%, and 20%, respectively. Upon straining to certain value (e.g. 5%) at strain rate of 5×10^{-4} s^{-1}, the specimen was unloaded by the stress-control mode to 20 N at unloading rate of 200 N·min^{-1}, followed by reloading at a strain rate of 5×10^{-4} s^{-1} to the same applied stress prior to the next unloading.

(3) Stress-relaxation uniaxial tensile tests were performed using an Instron 5966 testing machine under strain-control mode at room temperature at eight initial applied strains, i.e., 0.8% 1.5%, 2%, 3.5%, 5%, 10%, 15% and 20%. Upon reaching any strain at a strain rate of 5×10^{-4} s^{-1}, the strain was maintained constant while the stress was recorded as a function of time. After the first relaxation over an interval of 180 s, the specimen was reloaded by a strain increment of 0.6% at a strain rate of 1×10^{-4} s^{-1} for the next relaxation. Four stress relaxations were conducted at 0.8%, and then the specimen was strained to next strain at a strain rate of 5×10^{-4} s^{-1}. The relaxation and reloading cycles were then performed with the same testing parameters as used at 0.8%, including the number of cycles, the duration of each relaxation, and the strain increment and strain rate during reloading. In order to obtain reproducible experimental results, the stress relaxations were carried out at least four times. The method to calculate the mobile dislocation density can be found in **7**.

5. Height Profile Measurements

The quantitative and three-dimensional surface height profiles in a large area were measured by means of a non-contact Bruker Contour-I white light interferometry operated in a VSI mode. The maximum resolution in depth was 0.02 µm. The measured data are shown in Fig. 4.S1.

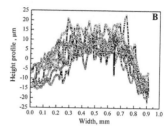

Figure 4.S1 (A) Measured lateral 3D surface topography within the uniform deformation section covering the whole thickness (~0.9 mm) of a GS/CG sample at tensile strain of 0.25. (B) Eight measured height profiles across the lateral surface in (A), showing evident height differences between the outer layers and inner layer. No height differecnce exists in the inner layer except for the surface roughness.

6. *Transmission Electron Microscopy (TEM) Observations*

Cross-sectional TEM observations were conducted to investigate the microstructual evolution and grain size distribution in the gradient structure. TEM samples were cut from gauge section of the tensile sample, and at a depth of ~5 μm from the surface. Dozens of TEM micrographs was taken from each sample, and over 5 hundreds of grains were measured to obtain reliable statistics.

The grain size distribution of nano-grainsin IF steel samples processed by SMAT for 5 minutes prior to and after tensile testing at a strain of 0.25 are shown in Fig. 4.S2, which indicates that there is no observable grain growth during the tensile testing.

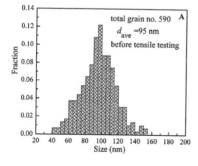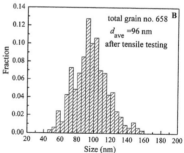

Figure 4.S2 Statistical distribution of grain sizes at the depth of ~5 μm from the treated surface in IF steel sample processed by SMAT for 5 minutes. (A) before tensile testing; (B) after tensile testing at the strain of 0.25.

7. Method to Derive the Mobile Dislocation Density

The applied shear stress during plastic deformation could be generally decomposed into two components [1]:

$$\tau = \tau_\mu(\gamma_p) + \tau^*(T, \dot{\gamma}_p), \quad (4.S1)$$

where τ_μ is an athermal component corresponding to long-range internal stress, the temperature dependence of which is weak and arising mainly due to the temperature effect on the shear modulus. The second component is temperature and strain-rate dependent, corresponding to localized obstacles and energy barriers such as forest dislocations, impurities, solute atoms, etc.

When plastic flow is governed by thermally activated dislocation glide, the dislocation velocity υ depends on activation energy and temperature, which could be described by Arrhenius type function [1, 2]:

$$\upsilon = f_0 \exp\left(-\frac{\Delta G(\tau^*)}{k_B T}\right), \quad (4.S2)$$

where f_0 is a pre-exponential constant, ΔG is the activation energy required to overcome localized obstacles, and k_B is Boltzmann constant. From Orowan equation [3], the plastic deformation rate $\dot{\gamma}_p$ can be expressed as:

$$\dot{\gamma}_p = \rho_m \upsilon b, \quad (4.S3)$$

where ρ_m is mobile dislocation density, b is the magnitude of Burgers vector. The physical activation volume is conventionally defined as:

$$V^* = -\left.\frac{d\Delta G(\tau^*)}{d\tau^*}\right|_T. \quad (4.S4)$$

In repeated stress-relaxation tests, it is reasonable to assume that the variation of mobile dislocation density is negligibly small, as the applied stress jumps occurred in short period from the end of relaxation 1 to the onset of relaxation 2. Accordingly, the variation of the plastic strain rate is dictated by dislocation velocity given by Eq. (4.S2), such that V^* could be determined by:

$$V^* = k_B T \frac{\ln(\dot{\gamma}_2^i/\dot{\gamma}_1^f)}{\Delta \tau^*}, \quad (4.S5)$$

where $\dot{\gamma}_2^i$ and $\dot{\gamma}_1^f$ are the shear strain rates at the onset of relaxation 2 and the end of relaxation 1, respectively.

In practice, the single stress-relaxation transient always exhibits a logarithmic variation of stress drop with time elapse. The apparent activation volume V_a can be determined by fitting the logarithmic stress-relaxation curve [1]:

$$\Delta \tau(t) = -\frac{k_B T}{V_a} \ln\left(1 + \frac{t}{c_r}\right), \tag{4.S6}$$

where $\Delta \tau(t)$ is the stress drop at time t and c_r is a time constant. The physical activation volume reflecting the stress sensitivity of dislocation velocity is generally larger than the apparent activation volume, $\Omega = V_a/V^* \geq 1$.

Combining Eqs. (4.S3) and (4.S4), one can obtain the ratio of dislocation velocity v at time t to the initial dislocation density v_0 at $t = 0$ by:

$$\frac{v}{v_0} = \exp\left(\frac{\Delta \tau^* \cdot V^*}{k_B T}\right), \tag{4.S7}$$

where $\Delta \tau^*$ is the drop of τ^* from time $t = 0$ to time t. It is usually assumed that the mobile dislocation density ρ_m and dislocation velocity v are related by an empirical power law:

$$\frac{\rho_m}{\rho_{m0}} = \left(\frac{v}{v_0}\right)^\beta. \tag{4.S8}$$

Based on Eqs. (4.S1)–(4.S8), Eq. (4.S8) can be transferred to:

$$\frac{\rho_m}{\rho_{m0}} = \left(\frac{c_\gamma}{t + c_\gamma}\right)^{\frac{\beta}{1+\beta}}, \tag{4.S9}$$

where ρ_{m0} is the dislocation density at start of each transient, and β is a dimensionless immobilization parameter:

$$\beta = \frac{\Omega}{1 + K^r/M} - 1, \tag{4.S10}$$

where K^r is the strain hardening coefficient originating from the empirical function of $\frac{d\tau_\mu}{d\gamma_p} = K^r$, which could be replaced by strain hardening rate approximately. M is the stiffness of the specimen-machine system [4, 5].

Accordingly, the density of retained mobile dislocation (*Re* in Fig. 4.4B) could be obtained as $\left.\dfrac{\rho_m}{\rho_{m0}}\right|_{t=180\,s}$.

1. Caillard D and Martin JL (2003). *Thermally Activated Mechanisms in Crystal Plasticity* (Pergamon, Amsterdam).
2. Lu L, et al. (2009). Stress relaxation and the structure size-dependence of plastic deformation in nanotwinned copper. *Acta Mater.*, **57**(17):5165–5173.
3. Orowan E (1940). Problems of plastic gliding. *Proc. Phys. Soc.*, **52**:8–22.
4. Bonneville J, Spatig P, and Martin JL (1995). A new interpretation of stress relaxations in Ni3 (Al, HF) single crystals. *High-Temperature Ordered Intermetallic Alloys Vi, Pts 1 and 2*, Materials Research Society Symposium Proceedings, eds. Horton J, Baker I, Hanada S, Noebe RD, and Schwartz DS), Vol. 364, pp. 369–374.
5. Martin JL, Lo Piccolo B, Kruml T, and Bonneville J (2002). Characterization of thermally activated dislocation mechanisms using transient tests. *Mater. Sci. Eng. A*, **322**(1–2):118–125.

Chapter 5

Heterostructured Lamella Structure Unites Ultrafine-Grain Strength with Coarse-Grain Ductility

Xiaolei Wu,[a] Muxin Yang,[a] Fuping Yuan,[a] Guilin Wu,[b] Yujie Wei,[a] Xiaoxu Huang,[b] and Yuntian Zhu[c,d]

[a]*State Key Laboratory of Nonlinear Mechanics, Institute of Mechanics, Chinese Academy of Sciences, Beijing 100190, China*
[b]*School of Materials Science and Engineering, Chongqing University, Chongqing, China*
[c]*School of Materials Science and Engineering, Nanjing University of Science and Technology, Nanjing 210094, China*
[d]*Department of Materials Science and Engineering, North Carolina State University, Raleigh, NC 27695, USA*
xlwu@imech.ac.cn, ytzhu@ncsu.edu

*Reprinted from *Proc. Natl. Acad. Sci. U.S.A.*, **112**(47), 14501–14505, 2015.

Heterostructured Materials: Novel Materials with Unprecedented Mechanical Properties
Edited by Xiaolei Wu and Yuntian Zhu
Text Copyright © 2015 The Author(s)
Layout Copyright © 2022 Jenny Stanford Publishing Pte. Ltd.
ISBN 978-981-4877-10-7 (Hardcover), 978-1-003-15307-8 (eBook)
www.jennystanford.com

Grain refinement can make conventional metals several times stronger, but this comes at dramatic loss of ductility. Here we report a heterostructured lamella structure in Ti produced by asymmetric rolling and partial recrystallization that can produce an unprecedented property combination: as strong as ultrafine-grained metal and at the same time as ductile as conventional coarse-grained metal. It also has higher strain hardening than coarse-grained Ti, which is hitherto believed impossible. The heterostructured lamella structure is characterized with soft micro-grained lamellae embedded in hard ultrafine-grained lamella matrix. The unusual high strength is obtained with the assistance of high hetero-deformation-induced (HDI) stress developed from heterogeneous yielding, while the high ductility is attributed to HDI hardening and dislocation hardening (defined in Chapter 2). The process discovered here is amenable to large-scale industrial production at low cost, and might be applicable to other metal systems.

5.1 Introduction

Strong or ductile? For centuries engineers have been forced to choose one of them, not both as they would like to do. This is because a material is either strong or ductile but rarely both at the same time. High strength is always desirable, especially under the current challenge of energy crisis and global warming, where stronger materials can help by making transportation vehicles lighter to improve their energy efficiency. However, good ductility is also required to prevent catastrophic failure during service.

Grain refinement has been extensively explored to strengthen metals. Ultrafine-grained (UFG) and nanostructured metals can be many times stronger than their conventional coarse-grained (CG) counterparts [1–5], but low tensile ductility is a roadblock to their practical applications. The low ductility is primarily due to their low strain hardening [6–12], which is caused by their small grain sizes. To further exacerbate the problem, their high strengths require them to have even higher strain hardening than weaker CG metal to maintain the same tensile ductility according to the Considère criterion. This makes it appear hopeless for UFG materials to have high ductility and it has been taken for granted that they are super strong but inevitably much less ductile than their CG counterparts.

5.2 Microstructure of Heterogeneous Lamella Structure

Here we report that a novel heterostructured lamella (HL) structure possesses both the UFG strength and the CG ductility, which has never been realized before. The HL structure was produced by asymmetric rolling [13, 14] and subsequent partial recrystallization (see Method for details). The asymmetric rolling elongated the initial equiaxed grains (Fig. 5.1a) into a lamella structure (Fig. 5.1b), which is heterogeneous with some areas having finer lamella spacing than others. This was caused by the variation of slip systems and plastic strain in grains with different initial orientation [15]. The asymmetric rolling also imparts a higher strain near the sample surface [14], which produces a nanostructured surface layer (Fig. 5.1c) as well as a slight structure gradient before and after recrystallization. During the subsequent partial recrystallization, the lamellae with finer structure recrystallized to form soft microcrystalline lamellae while others underwent recovery to still maintain the hard UFG structure [15] (Fig. 5.1d). Figure 5.1e shows the distribution of recrystallized grains with sizes larger than 1 μm after blacking out UFG areas. Interestingly, most recrystallized grains cluster into long lamellae along the rolling direction, which are distributed within black UFG lamellae. Figure 5.1f is a transmission electron microscopy (TEM) image showing a lamella of recrystallized grains between two UFG lamellae. The recrystallized grains are equiaxed and almost dislocation free. In contrast, the smaller UFG grains still contain a high density of dislocations. Figure 5.1g shows that the volume fraction of recrystallized grains decreases through the depth but their sizes increase slightly. The weak gradients in microstructure before and after recrystallization are reflected in the microhardness (Hv) variation in Fig. 5.1h. Tensile samples with different thickness were prepared by polishing away equal-thickness layers from both sides of the recrystallized sample (Fig. 5.1h).

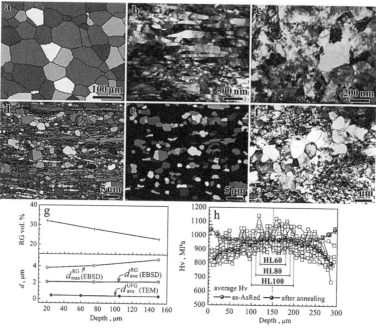

Figure 5.1 Heterostructured lamella (HL) Ti. (a) Electron back-scattered diffraction (EBSD) image of initial coarse-grained Ti. (b, c) Transmission electron microscopy (TEM) images showing non-uniform elongated lamellae in the middle area and nanograins in the surface layer of ~25 μm thick after asymmetric rolling. (d) EBSD image of HL Ti after partial recrystallization. (e) EBCD image of recrystallized grains (RG) larger than 1 μm. (f) Cross-sectional TEM image of RG lamellae with two UFG lamellae on two sides. (g) Weak gradient distribution of RG volume fraction, RG average grain size (d_{ave}^{RG}), maximum size (d_{max}^{RG}), and mean grain size of UFG lamellae (d_{max}^{UFG}) through the depth measured by EBSD and TEM. (h) Microhardness (Hv) gradient in the as-rolled state and after partial recrystallization, together with the location of the tensile samples (e.g. HL60 means 60 mm in thickness). The large scattering in the Hv data after partial recrystallization reflects the heterogeneous nature of HL-Ti.

5.3 Mechanical Properties and Strain Hardening of HL Structure

Figure 5.2a,b shows that HL-Ti, e.g. both HL60 and HL80 from the central 60 mm and 80 mm layer, respectively, is as strong as UFG Ti and as ductile as CG Ti. The squares in Fig. 5.2a mark the start of necking according to the Considère criterion. Both HL60 and HL80 are three times as strong as CG Ti while maintaining the same ductility (the uniform tensile elongation) (Fig. 5.2a). Furthermore, all other HL-Ti samples are also much stronger and more ductile than CG Ti (Fig. 5.2b). In contrast, the UFG Ti became mechanically unstable soon after yielding.

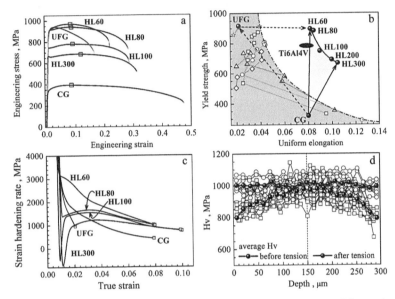

Figure 5.2 Mechanical properties and strain hardening of HL-Ti. (a) Tensile engineering stress-strain curves in HL-Ti at a quasi-static strain rate of 5×10^{-4} s^{-1} in comparison with UFG Ti and CG Ti. UFG sample: 300-μm-thick as-processed by asymmetric rolling. HL-Ti: annealed UFG at 475°C for 5 minutes. The number after HL indicates the sample thickness (μm). The tensile samples were flat and dog-bone shaped, with gauge dimension of 10 mm × 2.5 mm. (b) Yield strength and uniform tensile elongation of HL-Ti. Other data of Ti as well as Ti6Al4V were also shown for comparison. (c) Strain hardening rate ($\Theta = d\sigma/d\varepsilon$) versus true strain of HL-Ti. (d) Change of Vickers microhardness (Hv) before and after tensile testing at tensile strain of 10% in HL300. Note disappearance of a weak Hv gradient across the thickness after tensile testing.

The HL-Ti samples derive their high ductility from their high strain-hardening rate ($\Theta = d\sigma/d\varepsilon$), which becomes unprecedentedly higher than that of CG Ti after some plastic straining (Fig. 5.2c). More remarkably, the HL60 sample has much higher strain hardening than CG Ti for the entire plastic deformation. Thicker HL-Ti samples have a distinctive two-stage strain hardening behavior, with initial low Θ, but higher Θ at larger plastic strains. Specifically, in Stage I, Θ shows a steep drop at first followed by a steep up-turn, typical of discontinuous yielding. This is due to the shortage of dislocations or dislocation sources upon the onset of plastic deformation [8, 16], which makes it necessary for dislocations to glide faster to accommodate the applied constant strain rate. Higher stress is needed to move dislocations faster. Upon yielding, dislocations quickly multiplied, leading to quick Θ increase due to dislocation interaction and entanglement. In Stage II, Θ continues to rise, albeit at a relatively slow rate. This is surprising and has never been observed in either UFG or conventional CG metals, both of which have a typical monotonic drop in Θ.

Surprisingly, the weak microhardness (Hv) gradient disappeared after tensile testing (Fig. 5.2d), in contrast to what was observed in gradient structured steel [11]. This indicates that the recrystallized grains were dramatically hardened by plastic deformation and the weak gradient in microstructure did not play a major role in mechanical properties. Note that the Hv values are still scattered after tensile testing, indicative of the heterogeneous mechanical properties remaining after the tensile testing.

5.4 Bauschinger Effect and Back Stresses

The observed extraordinarily high strength of HL-Ti can be attributed to their composite lamella nature. Under tensile loading, the soft lamellae of recrystallized micro-grains will start plastic deformation first. However, they are constrained by surrounding hard lamellae so that dislocations in such grains are piled up and blocked at lamella interfaces, which are actually also grain boundaries. This produces a long-range back stress [17–19] to make it difficult for dislocations to slip in the micro-grained lamellae until the surrounding UFG lamellae start to yield at a larger global strain. In other words, the back stress has significantly increased the flow stress of the soft lamellae by the time the whole sample is yielding. This is the primary reason for the observed high yield strength of HL-Ti samples, as verified later. This observation is consistent with the high strength caused by back stresses in passivated thin films [20–22] and nanopillars [23, 24].

To probe the origin of the high strain hardening of HL-Ti and the contribution of HDI stress to the observed high yield strength, we conducted loading-unloading-reloading (LUR) testing (Fig. 5.3a) to investigate the Bauschinger effect, from which we can estimate the contributions of the HDI stress and dislocation hardening to the flow stress. Interestingly, HL-Ti shows a very strong Bauschinger effect: during unloading, the reverse plastic flow (σ_{rev}) starts even when the applied stress is still in tension (Fig. 5.3b). A larger hysteresis loop during the unloading-reloading represents stronger Bauschinger effect. As shown in Fig. 5.3b, the hysteresis loop becomes larger with increasing tensile strain for the HL-Ti. Importantly, the hysteresis appeared even during the first unloading-reloading cycle near the yield point. In contrast, the CG Ti has negligible hysteresis (Fig. 5.3b). As schematically shown in Fig. 5.3c, the Bauschinger effect can be described by the reverse plastic strain (ε_{rp}) normalized by the yield strain (ε_y) [21, 24], which increases with increasing plastic strain (Fig. 5.3d). The HDI stress can be calculated as $\sigma_b = \dfrac{(\sigma_f + \sigma_{rev})}{2} - \dfrac{\sigma^*}{2}$ [26], where the σ_f, σ_{rev}, and σ^* are defined in Fig. 5.3c.

As shown in Fig. 5.3d, the HDI stress is about 400 MPa near the yield point. *The soft lamellae in HL-Ti need to overcome this additional high HDI stress to deform plastically, which contributed significantly to the observed high yield strength of the HL samples*, especially the HL60 and HL80 samples. In other words, the observed high yield strength of HL samples was resulted from the high HDI stress.

The HDI stress increased with plastic strain, especially at the early stage, which contributed to the high strain hardening. This is the primary reason why the strain hardening in HL Ti increased with applied global strain and surpassed that of CG Ti, as shown in Fig. 5.2c. The homogeneous CG Ti does not have measurable Bauschinger effect or HDI stress, and its strain hardening decreased monotonically with the global strain. The effective stress, which includes the dislocation hardening and Peierls stress, is much lower than the HDI stress, which is consistent with an earlier report in constrained nanopillars [23]. The increase in the effective stress with tensile strain should be primarily caused by dislocation density, i.e. dislocation hardening. Therefore, the high strain hardening is originated from both HDI stress hardening and dislocation hardening. To the authors' knowledge, the significant HDI hardening has never been reported before.

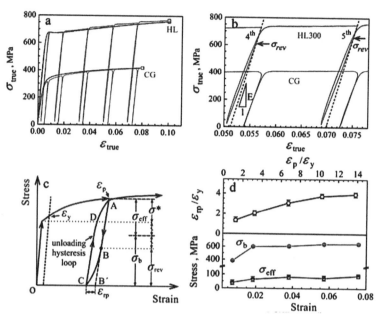

Figure 5.3 Bauschinger effect and HDI stress of HL-Ti. (a) Loading-unloading-reloading stress-strain curves of HL-Ti and CG Ti. (b) Hysteresis loops. The two arrows indicate the deviation from the initial elastic behavior during unloading, i.e. σ_{rev}. (c) Schematic for calculating HDI stress [26]. (d) Normalized reverse plastic strain, HDI stress and effective stress versus applied strain.

The HDI strain hardening is caused by the piling up and accumulation of geometrically necessary dislocations. With increasing tensile strain, dislocation sources in the softer microcrystalline lamellae are activated first. However, the soft lamellae are surrounded and constrained by hard UFG lamellae, which are still deforming elastically. Therefore dislocations in the soft lamellae cannot transmit into the hard UFG lamellae. As shown in Fig. 5.4a, which is a TEM micrograph from an HL Ti sample tested at a tensile train of 2%, dislocations piled up at several locations. All dislocations in an individual pileup are from the same dislocation source and have the same Burgers vector. They consequently produce a long-range back stress to stop the dislocation source from emitting more dislocations. In other words, soft lamellae constrained by hard lamellae matrix appears much stronger than when they are not constrained. This explains why the HL60 and HL80 samples can be as strong as the UFG Ti although they contain over 20% softer microcrystalline lamellae. It appears that the full constraint of the soft lamellae by the hard matrix is a prerequisite for this phenomenon.

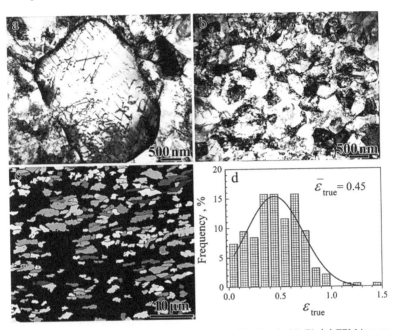

Figure 5.4 Dislocation pile-up and strain partitioning in HL-Ti. (a) TEM image showing pile up of dislocations in a recrystallized micrometer-sized grain at tensile strain of 2%. (b) TEM image showing equiaxed ultrafine grains in a sample tested to a tensile true strain of 9.4%. (c) EBSD image showing elongated recrystallized grains along tensile direction at 9.5% true strain. (d) Distribution of strains in recrystallized grains with an average true strain of 0.45.

5.5 Strain Partitioning

Beyond the yield point, the whole HL Ti sample is deformed plastically. However, the softer lamellae are easier to deform than the hard lamellae. This causes plastic strain partitioning where the soft lamellae carried much higher plastic strain than hard lamellae.

Indeed, after tensile testing for 9% true strain, the ultrafine grains in the hard lamellae remain largely equiaxed (Fig. 5.4b), while most of the recrystallized micrometer-sized grains were deformed from equiaxed shape to elongated shape along the tensile direction (Fig. 5.4c), The true strain in each grain can be calculated from its aspect ratio α as $\varepsilon = (2/3)\ln\alpha$ (see SI). After the HL-Ti was deformed for a global true strain of 9.4%, the average true strain in recrystallized micrometer-sized grains is 45% (Fig. 5.4d), but close to zero in ultrafine grains. It should be noted that the real strain in the UFG lamella cannot be estimated using the grain geometry change because other deformation mechanisms such as coordinated deformation, grain boundary sliding and grain rotation may also contribute to plastics strain when the grain sizes are very small. Nevertheless, the plastic strain in the hard UFG lamellae should be less than 9.4% because the soft lamellae carried much higher plastic strain of 45%. The strain should be continuous at the inter-lamella interfaces, which further leads to strain gradient near these interfaces. Geometrically necessary dislocations will be generated to accommodate the strain gradient [27–29], which will generate long-range HDI stress near the interfaces. In other words, HDI hardening is associated with strain partitioning, i.e. inhomogeneous plastic strain, in the HL-Ti.

In addition to HDI stress, dislocation hardening, which is related to the increase in total dislocation density [29], should also have contributed to the observed high strain hardening. The HL structure promotes the generation and accumulation of two types of the dislocations during the testing. One is the aforementioned geometrically necessary dislocations, and the other is incidental dislocations that do not produce long-range back stress. It should be noted that complex three dimensional stress states may develop from the applied uniaxial stresses due to the plastic incompatibility between the soft and hard lamellae. The stress state change will promote dislocation accumulation and interaction by activating more slips systems [11, 30, 31], similar to what occurs in gradient structures [11, 31]. This will effectively increase the incidental dislocation density.

The HL structure is more effective in producing strain hardening than the reported bi-modal structure [6]. In the HL structure, the recrystallized micrometer-sized grains cluster together to form elongated soft lamellae. It has been reported that the elongated inclusions produce higher strain hardening than spherical ones, especially when its long axis is aligned in the loading direction [32], which is the case in this study. The lamella geometry makes mutual constraint between the soft and hard lamellae more effective, which produces higher HDI stress and higher dislocation density. The second unique feature of the HL structure is that the soft lamellae are embedded in hard lamella matrix. This makes it more effective to constraint the plastic deformation of the soft lamellae to develop higher back stress than the conventional bi-modal structure. The third feature of the HL structure is the high density of inter-lamella interfaces, where dislocation can pile up and accumulate to enhance HDI strain hardening and dislocation hardening.

This work opens a new frontier towards high ductility without sacrificing the high strength of UFG metals. HDI stresses are primarily responsible for the observed high strength. Both HDI strain hardening and dislocation hardening are responsible for the observed extraordinary strain hardening rate and consequent high ductility. The lamella geometry, high constraint of the soft lamellae by hard matrix, and the high density of interfaces make it effective for developing HDI stress and dislocation hardening. The observations here provide a new principle for designing metals with mechanical properties that have not been reachable before. Importantly, the lamella structure is fabricated by asymmetric rolling followed by annealing, which is an industrial process that can be easily scaled up for large-scale production at low cost. The process discovered here might be applicable to other engineering metals and alloys, which needs further study.

5.6 Materials and Methods

5.6.1 Materials

Commercial pure titanium (CP-Ti) sheets of 2.4 mm thick were used in the present study. The compositionwas (wt %) 0.10 C, 0.05 N, 0.015 H, 0.25 O, 0.30 Fe, bal. Ti. The sheets were vacuum-annealed at 700°C for 2 hours to achieve a fully homogeneous coarse-grained (CG) microstructure with a mean grain size of 43 µm.

5.6.2 Asymmetrical Rolling (AsR) for Heterostructured Lamella (HL) Structured Ti

Ti sheets were processed by AsR at room temperature. Rolling was conducted on a rolling mill with 45-mm diameter rolls. The top and bottom rolls were driven at velocities of 1 m/s and 1.3 m/s, respectively, and a rolling reduction was 0.1 mm per pass. The Ti sheet was flipped over, and feeding direction was alternated from one end to the other between AsR passes. The sheets were finally rolled to 300 μm thick after 20 rolling passes with a total rolling reduction of 87.5%. No cracks were observed on the surface of the AsR-processed Ti sheets.

The HL-Ti sheets of 300 μm thick were obtained by subsequent vacuum-annealing of AsR sheets at 475°C for 5 min.

5.6.3 Tensile Test and Loading-Unloading-Reloading (LUR) Test

All tensile specimens were dog-bone shaped, with a gauge length of 10 mm and a width of 2.5 mm. In order to obtain reproducible tensile property, all tensile tests were repeated at least 3–5 times. Tensile samples of central layers with varying thickness were made from 300-μm thick HL samples. The remaining samples were obtained by simultaneously polishing away the surface layers of varying thickness from both sides.

Quasi-static uniaxial tensile tests were carried out using an Instron 5582 testing machine at strain rate of 5×10^{-4} s^{-1} at room temperature. An extensometer was used to measure the strain during the uniform tensile deformation. Loading-unloading tests were conducted using an Instron 5966 testing machine at room temperature. Five loading-unloading cycles were conducted during each uniaixal tensile test at varying tensile strains. Upon straining to a designated strain (e.g. 2%) at strain rate of 5×10^{-4} s^{-1}, the specimen was unloaded by the stress-control mode to 20 N at the unloading rate of 200 N·min^{-1}, followed by reloading at a strain rate of 5×10^{-4} s^{-1} to the same applied stress prior to the next unloading.

5.6.4 EBSD and TEM Observations

The cross-sectional and longitudinal EBSD and TEM observations were conducted to investigate the microstructual evolution in HL-Ti before and after tensile tests. TEM samples were cut from the gauge sections of tensile samples.

Supplementary information is linked to the online version of the chapter.

Acknowledgements

X.L.W., M.X.Y., F.P.Y., M.L.Z., and C.X.Z. are funded by the NFSC (11072243, 11222224, 11472286, 50571110, and 11021262) and MOST (2012CB932203, 2012CB937500, and 6138504). Y.T.Z. is funded by the US Army Research Office (W911 NF-12-1-0009), the US National Science Foundation (DMT-1104667) and by the Nanjing University of Science and Technology. The authors thank Prof. W.D. Nix for many insightful and constructive comments on this chapter.

Author Contributions

X.L.W. managed the project. X.W. and Y.T.Z. conceived the ideas. All analysed the data. M.X.Y. and X.L.W. carried out the experiments. G.L.W. conducted EBSD observations and texture analysis. All contributed to discussions. X.L.W. and Y.T.Z. wrote the chapter.

References

1. Valiev RZ, Islamgaliev RK and Alexandrov IV (2000). Bulk nanostructured materials from severe plastic deformation. *Prog. Mater. Sci.*, **45**:103–189.
2. Langdon TG (2013). Twenty-five years of ultrafine-grained materials: Achieving exceptional properties through grain refinement. *Acta Mater.*, **61**:7035–7059.
3. Valiev RZ, Alexandrov IV, Zhu YT and Lowe TC (2002). Paradox of strength and ductility in metals processed by severe plastic deformation. *J. Mater. Res.*, **17**:5–8.
4. Valiev RZ (2004). Nanostructuring of metals by severe plastic deformation for advanced properties. *Nat. Mater.*, **3**:511–516.

5. Zhu YT and Liao XZ (2004). Nanostructured metals: retaining ductility. *Nat. Mater.*, **3**:351–352.
6. Wang YM, Chen MW, Zhou FH and Ma E (2002). High tensile ductility in a nanostructured metal. *Nature*, **419**:912–915.
7. Lu L, Shen YF, Chen XH, Qian LH and Lu K (2004). Ultrahigh strength and high electrical conductivity in copper. *Science*, **304**:422–426.
8. Huang XX, Hansen N and Tsuji N (2006). Hardening by annealing and softening by deformation in nanostructured metals. *Science*, **312**:249–251.
9. Fang TH, Li WL, Tao NR and Lu K (2011). Revealing extraordinary intrinsic tensile plasticity in gradient nano-grained copper. *Science*, **331**:1587–1590.
10. Wei YJ, Li YQ, Zhu LC, Liu Y, Lei XQ, Wang G, Wu YX, Mi ZL, Liu JB, Wang HT and Gao HJ (2014). Evading the strength-ductility trade-off dilemma in steel through gradient hierarchical nanotwins. *Nat. Commun.*, **5**:3580.
11. Wu XL, Jiang P, Chen L, Yuan FP and Zhu YTT (2014). Extraordinary strain hardening by gradient structure. *Proc. Natl. Acad. Sci. USA*, **111**:7197–7201.
12. Wu XL, Yuan FP, Yang MX, Jiang P, Zhang CX, Chen L, Wei YG, E Ma (2015). Nanodomailed nickle unite nanocrystal strength with coarse-grain ductility. *Sci. Rep.*, **5**:11728.
13. Roumina R and Sinclair CW (2008). Deformation geometry and through-thickness strain gradients in asymmetric rolling. *Metall. Mater. Trans. A*, **39A**:2495–2503.
14. Yu HL, Lu C, Tieu K, Liu XH, Sun Y, Yu QB, Kong C (2012). Asymmetric cryorolling for fabrication of nanostructured aluminum sheets. *Sci. Rep.*, **2**:772.
15. Wu G and Jensen DJ (2005). Recrystallisation kinetics of aluminium AA1200 cold rolled to true strain of 2. *Mater. Sci. Tech.*, **21**:1407–1411.
16. Uchic MD, Dimiduk DM, Florando JN, Nix WD (2004). Sample dimensions influence strength and crystal plasticity. *Science*, **305**:986–989.
17. Gibeling JC, Nix WD (1980). A numerical study of long range internal stresses associated with subgrain boundaries. *Acta Metall.*, **28**(12):1743–1752.
18. Mughrabi H (1983). Dislocation wall and cell structures and long-range internal stresses in deformed metal crystals. *Acta Metall.*, **31**(9):1367–1379.

19. Sinclair CW, Saada G, Embury JD (2006). Role of internal stresses in co-deformed two-phase materials. *Philos. Mag.,* **86**(25–26): 4081–4098.
20. Xiang Y, Vlassak JJ (2006). Bauschinger and size effects in thin-film plasticity. *Acta Mater.,* **54**(20):5449–5460.
21. Xiang Y, Vlassak JJ (2005). Bauschinger effect in thin metal films. *Scr. Mater.,* **53**(2):177–182.
22. Rajagopalan J, Han JH, Saif MTA (2008). Bauschinger effect in unpassivated freestanding nanoscale metal films. *Scr. Mater.,* **59**(7):734–737.
23. Jennings AT, Gross C, Greer F, Aitken ZH, Lee SW, Weinberger CR, Greer JR (2012). Higher compressive strengths and the Bauschinger effect in conformally passivated copper nanopillars. *Acta Mater.,* **60**(8):3444–3455.
24. Lee SW, Jennings AT, Greer JR (2013). Emergence of enhanced strengths and Bauschinger effect in conformally passivated copper nanopillars as revealed by dislocation dynamics. *Acta Mater.,* **61**(6):1872–1885.
25. Li T, Suo, ZG (2006). Deformability of thin metal films on elastomer substrates. *Int. J. Solids Struct.,* **43**:2351–2363.
26. Feaugas X (1999). On the origin of the tensile flow stress in the stainless steel AISI 316L at 300K: back stress and effective stress. *Acta Mater.,* **47**(13):3617–3632.
27. Ashby MF (1970). Deformation of plastically non-homogeneous materials. *Philos Mag.,* **21**:399–411.
28. Gao H, Huang Y, Nix WD and Hutchinson JW (1999). Mechanism-based strain gradient plasticity - I. Theory. *J. Mech. Phys. Solids,* **47**:1239–1263.
29. Gao HJ and Huang YG (2003). Geometrically necessary dislocation and size-dependent plasticity. *Scr. Mater.,* **48**:113–118.
30. Wu XL, Jiang P, Chen L, Zhang JF, Yuan FP and Zhu YT (2014). Synergetic strengthening by gradient structure. *Mater. Res. Lett.,* **2**:185–191.
31. Zandrahimi M, Platias S, Price D, Barrett D, Bate PS, Roberts WT and Wilson DV (1989). Effects of changes in strain path on work-hardening in cubic metals. *Metall. Trans. A,* **20**:2471–2482.
32. Tanaka K, Mori T (1970). The hardening of crystals by non-deforming particles and fibres. *Acta Metall.,* **18**(8):931–941.

Supplementary Information

5.S1 Calculation of True Plastic Strain for Elongated Grains

Assuming after tensile testing, an equiaxed grain of initial size l_0 elongates to an aspect ratio of $\alpha = l/w$, where l and w are the grain length and width, respectively.

During the plastic deformation, the volume is conserved, i.e.

$$l_0^3 = lw^2 = l^3/\alpha^2 \tag{5.S1}$$

Then, the grain length l after tensile testing can be expressed as,

$$l = l_0 \sqrt[3]{\alpha^2} \tag{5.S2}$$

The true plastic strain of the grain can be calculated as

$$\varepsilon = \ln\left(\frac{l}{l_0}\right) = \frac{2}{3}\ln\alpha \tag{5.S3}$$

Chapter 6

Synergetic Strengthening by Gradient Structure

X. L. Wu,[a] P. Jiang,[a] L. Chen,[a] J. F. Zhang,[a] F. P. Yuan,[a] and Y. T. Zhu[b,c]

[a]*State Key Laboratory of Nonlinear Mechanics, Institute of Mechanics, Chinese Academy of Sciences, Beijing 100190, China*
[b]*Department of Materials Science and Engineering, North Carolina State University, Raleigh, NC 27695, USA*
[c]*School of Materials Science and Engineering, Nanjing University of Science and Technology, Nanjing 210094, China*
xlwu@imech.ac.cn, ytzhu@ncsu.edu

Gradient structures are characterized with a systematic change in microstructures on a macroscopic scale. Here we report that gradient structures in engineering materials such as metals produce an intrinsic synergetic strengthening, which is much higher than the sum of separate gradient layers. This is caused by macroscopic

---

Reprinted from *Mater. Res. Lett.*, **2**(4), 185–191, 2014.

stress gradient and the bi-axial stress generated by mechanical incompatibility between different layers. This finding represents a new mechanism for strengthening that exploits the principles of both mechanics and materials science. It may provide for a new strategy for designing material structures with superior properties.

A typical gradient structure is characterized with a microstructural gradient at a macroscopic scale [1]. Gradient structures exist ubiquitously in the nature, because they have superior properties over homogenous structures to weather severe natural environments [1–3]. One wonders if gradient structures would also benefit engineering materials made by mankind. Recently gradient structure was introduced into metals, producing excellent strength and ductility [4–7]. This is significant because strength and ductility are two of the most important, but often mutually exclusive, mechanical properties [8–14]. For example, extensive research for the past three decades has produced ultra-strong nanocrystalline materials [15–17], but they usually have low ductility. Making materials both strong and ductile has been an enduring endeavor for materials scientists [8, 18, 19].

Here we report that gradient structure in metals produces an intrinsic synergetic strengthening effect, with the strength of the gradient-structured sample much higher than the sum of the strength of separate layers, as calculated using the rule of mixtures. The mechanisms for synergetic strengthening are macroscopic stress gradient and complex stress state caused by the gradient structure under uniaxial applied stress.

We produced the gradient structure in an interstitial-free (IF) steel using surface mechanical attrition treatment (SMAT) [20, 21]. The steel sample has a composition of (wt%) 0.003 C, 0.08 Mn, 0.009 Si, 0.008 S, 0.011 P, 0.037 Al, 0.063 Ti, and 38 ppm N. The samples were pre-annealed at 1173 K for 1 hour to obtain a coarse-grained microstructure with a mean grain size around 35 μm. For the purpose of obtaining consistent gradient layers in terms of grain size distribution and layer depth, *etc.*, all samples were processed by SMAT for 5 minutes on both sides, which produced two gradient layers of 120 μm thick each, with the sizes of grains, subgrains and dislocation cells ranging from submicrometers to micrometers. Steel balls with a diameter of 3 mm were used to process the samples. The sample thicknesses varied from the 0.5 mm to 2 mm so that the volume fraction of the gradient layer varied from 0.48 to 0.16. The gradient-structured tensile samples are schematically illustrated in Fig. 6.1. For simplicity, we define such samples as *integrated samples*, which have a coarse-grained layer sandwiched between two gradient-structured layers. The grain size change along the depth is shown in Fig. 6.1.

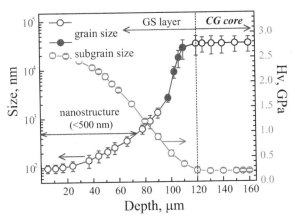

Figure 6.1 Variation of grain sizes and microhardness (Hv) along the depth in samples processed by SMAT for 5 minutes.

Figure 6.2a shows the engineering stress-strain curves of all integrated samples with different volume fractions of the gradient layers. Shown also is the engineering stress-strain curve of the 120-μm-thick gradient layer (GS-layer curve), which was prepared by polishing away the sample from one side until only 120-μm-thick gradient layer was left. The same polishing procedure was used to prepare the 25-μm-thick nanostructured layer (see the NS film curve). As shown, with increasing volume fraction of gradient layer, the yield strength increased dramatically, while the ductility (elongation to failure) decreased slowly, producing a good combination of strength and ductility.

Figure 6.2b demonstrates the synergetic high flow stress (the red curve) in an integrated sample with the GS volume fraction of 24%. The green curve is calculated using the rule-of-mixtures (ROM):

$$\sigma_{ROM} = V_{GS}\sigma_{GS} + (1-V_{GS})\sigma_{CG} \qquad (6.1)$$

where σ_{ROM} is the flow stress of the integrated sample, V_{GS} is the volume fraction of the gradient-structured layers, σ_{GS} is the flow stress of gradient layer, σ_{CG} is the flow stress of coarse-grained sample.

Equation 6.1 can also be used to calculate the yield strength at 0.2% plastic strain (see the blue line in Fig. 6.2c). However, as shown in Fig. 6.2b, the integrated sample yielded at a much large strain than the CG sample. In other words, the CG sample will have a flow stress of σ'_{CG} due to strain hardening, where

$$\sigma'_{CG} = \sigma_{CG} + \Delta\sigma_1 \qquad (6.2)$$

Correspondingly, the ROM described should be modified as

$$\sigma_{ROM}^{mod} = V_{GS}\sigma_{GS} + (1-V_{GS})\sigma'_{CG} \qquad (6.3)$$

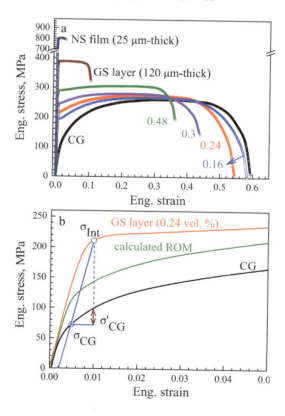

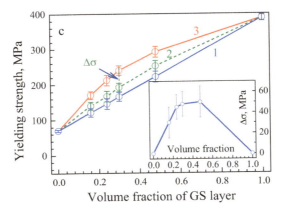

Figure 6.2 (a) Engineering stress-strain curves of integrated samples, the gradient-structured (GS) layers, and the coarse-grained sample. The volume fractions of the gradient layers in the integrated samples are indicated on the curves. (b) The flow stress of the integrated sample (red curve) is much higher than that calculated using ROM (green curve) due to synergetic strengthening. (c) The measured yield strength of the integrated samples (red curve), the calculated yield strength using the rule of mixture (ROM) (blue curve), and the calculated yield strength using the modified rule of mixture (green curve). Inset is the synergetic strengthening $\Delta\sigma$ as a function GS volume fraction.

As shown, the flow stress of the integrated layer (red curve) is much higher than green curve. The difference in the flow stress can be considered as an expression of synergetic strengthening caused by the gradient structure.

Figure 6.2c demonstrates the synergetic strengthening of the gradient structure. The zero percent volume fraction of GS layer represents pure CG sample, while the 100% GS layer represents the data point from the pure gradient-structured layer without a coarse-grained layer. The red line represents the yield strength of the integrated samples that consist of two gradient-structured layers and one coarse-grained central layer. It obviously shows higher yield strength than the blue curve and green curves, which are calculated using equations (6.2) and (6.3), respectively. Therefore, even with consideration of σ'_{CG}, the gradient structure still leads to a synergetic strengthening in the integrated sample.

Assuming we can slice the integrated sample into n thin layers along the thickness direction, and then separately measure their yield strength. The synergetic strengthening by the gradient structure can be calculated as

$$\Delta\sigma = \sigma_{Int} - \sum_{1}^{n} f_i \sigma_i \tag{6.4}$$

where σ_{Int} is the yield strength of the integrated sample, σ_i is the yield strength of the ith layer, and f_i is the thickness fraction of the ith layer. The second term in the equation represents the strength calculated by the ROM. An accurate calculation using Eq. 6.4 requires large a number of layers to be measured, which is experimentally difficult. In Fig. 6.2c, the integrated sample was sliced into three layers (n = 3): two surface layers with a thickness 120 mm and one central layer. Equation 6.3 can also be expressed in a more general form:

$$\Delta\sigma = \sigma_{Int} - \frac{1}{t}\int_0^t \sigma(x)dx \tag{6.5}$$

where t is the thickness of the integrated sample, $\sigma(x)$ is the strength of the sample layer at position x along the thickness.

The synergetic strengthening $\Delta\sigma$ as a function GS volume fraction is shown in the inset of Fig. 6.2c. It reveals that an optimum GS volume fraction exists where the synergetic strengthening is most significant.

The importance of the gradient structured layer in the synergetic strengthening is demonstrated in Figure 6.3 Integrated samples with two 120-mm gradient-structured layers sandwiching a 760-mm coarse-grained layer were used to measure the synergetic strengthening $\Delta\sigma$ as equal-thickness layers were removed from both surfaces. The engineering stress-strain curves of the integrated samples after removing varying thickness of surface layers are shown in Fig. 6.1a. The legend indicates the layer thickness removed from each side of the sample.

Figure 6.3b shows that as the gradient-structured layers were gradually removed, the synergetic strengthening becomes weaker and eventually disappear. This indicates the importance of the gradient-structured layers to the synergetic strengthening. To find out where the synergetic strengthening is from, an integrated sample was tested to a tensile strain of 1%, which corresponds to the 0.2% plastic strain and the change in microhard-ness was measured along the depth. As shown in Fig. 6.3c, there is no change in microhardness for the 70-μm surface layer, indicating that no dislocation accumulation occurred during the tensile testing. This suggests that the high strength surface layer were still deforming largely elastically at 1% tensile strain. Interestingly, there is a peak at the 90 μm depth. This is a very significant feature, which will be discussed later. Figures 6.3b and c indicates that the finer grained gradient layers are essential to activate the synergetic strengthening, but the synergetic strengthening comes mostly from the coarse-grained layer.

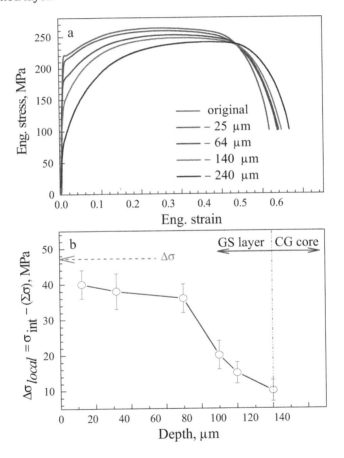

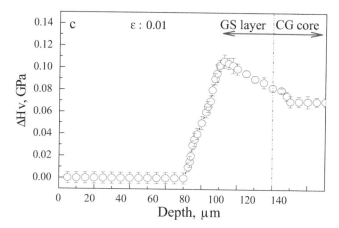

Figure 6.3 (a) Tensile engineering stress–strain curves with surface layers removed from both sides of the integrated sample with initial 120-μm-thick gradient structured layers. The legend indicates the thickness removed from each side. (b) The synergetic strengthening calculated using Eq. 6.4 as a function of surface layer thickness removed. (c) Vickers hardness change along the depth after tensile testing to a strain of 1%.

To probe the mechanism by which the gradient structure strengthens the integrated sample, we analyzed stress, strain and strain gradient distribution across the sample thickness by means of finite element method (FEM) using the commercial ANSYS 14 package. The variation of Poisson's ratio ν with tensile strain was calculated as [22]

$$v = 0.5 - 0.5(v_0) \times E_s/E \qquad (6.6)$$

where E is Young's modulus, $E_s = \sigma/\varepsilon$ is the secant modulus, v_0 (0.3) is Poisson's ratio in the elastic stage.

Figure 6.4 is the schematic FEM model and element grid. The constitutive relationships for both GS-layer and CG-core were taken from the stress-strain curves in Fig. 6.2a, which were obtained from the testing of standalone samples. For higher accuracy, both 3D structured gradient mesh division and twenty nodes element SOLID186 were adopted. The sample was strained under uniaxial tension to a strain of 0.01, which was divided into ten loading-steps.

Figure 6.5 shows the FEM simulation results at the tensile strain of 0.008 in the z-direction. As shown in Fig. 6.5a, the normal stress in the x direction, σ_x, is positive in the central CG layer, but negative in the GS layer. σ_x should be zero in a homogeneous sample because no stress is applied laterally in the x-direction. In other words, the applied uniaxial stress is converted to a bi-axial stress state in the integrated sample. Figure 6.5b shows a large stress gradient near the CG/GS layer boundary. The flow stress in the CG core is slightly higher than in the standalone CG sample, while the flow stress in the GS layer is lower than in the standalone GS sample, which is cause by mechanical incompatibility due to the variation of the Poisson' ratio across the two types of layers. Figure 6.5c shows that the strain in x-direction, ε_x, for the CG core of integrated sample rises evidently as compared with standalone counter-part, while ε_x for GS layer decreases. This is due to the difference of the Poisson's ratio evolution with strain in both GS layer and CG core, i.e. the GS layer is still elastic after yielding of the CG core. Strain compatibility is required across the layer boundary. This is the reason behind stress re-distribution as seen in Fig. 6.5b. Interestingly, as shown in Fig. 6.5d, the peaks of strain gradient $d\varepsilon_x/dy$ and shear gradient $d\gamma_{xz}/dy$ appear near the boundary of the CG/GS layers. In addition, the strain gradients are larger near the sample edge (path 4) than at the center (path 1). The $d\gamma_{xz}/dy$ is much larger than $d\varepsilon_x/dy$ in all paths.

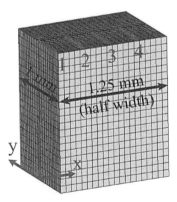

Figure 6.4 Schematic FEM model showing a quarter of the gauge section of tensile sample. $x = 0$ represent the sample centre (path 1) while $x = 1.25$ mm represent the sample edge. The tensile load is applied along the z-direction. The origin of coordinates is located at the sample geometric centre.

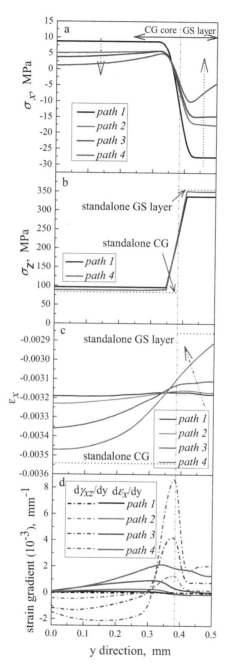

Figure 6.5 FEM simulation results. (a) Stress along x-axis. (b) Stress along z-axis. (c) Strain along x-axis. (d) Normal and shear strain gradients along y-axis.

It should be noted that the above FEM analysis can only be regarded as qualitative. First, it used the constitutive relationships obtained from standalone GS and CG samples. This effectively ignored the effect of the interlayer interaction on the constitutive relationships. Second, it ignored the structure gradient inside the GS layer. Nevertheless, the FEM results in Fig. 6.5 provide insight on the origin of the observed extra strengthening in the integrated sample.

The FEM results in Fig. 6.5 can be understood by the development of mechanical incompatibility between the CG and GS layers and their mutual constraint during the uniaxial tensile testing. At the 0.008 tensile strain, the central CG layer is already deforming plastically while the GS layers still deform elastically. The plastically deforming central layer has an apparent Poisson's ratio close to 0.5 to maintain a constant volume, while the Poisson's ratio in GS layer is close to 0.3. In other words, the central CG layer is trying to shrink more than the outer CG layers in the x direction.

However, the plastic central layer is constrained by the two elastic outer layers to the same lateral strain at their two boundaries. This constraint results in two consequences: First, it causes a strain gradient near the boundaries in the y-direction, as shown in Figure 6.5d, which will lead to the accumulation of geometrically necessary dislocations [23–25]. Second, it leads to tensile normal stress in the plastic central layer and a compressive stress in the outer elastic layer in the x-direction as shown in Fig. 6.5a. Such a bi-axial stress state will activate more slip systems to enhance dislocation interaction and accumulation. These dislocation accumulations consequently contribute to extraordinary strengthening beyond what is predicted by the rule of mixture.

The extra strengthening can be also partially attributed to the stress gradient shown in Fig. 6.5a and 6.5b. It has been reported that stress gradient can significantly increase the yield stress [26–28], as described by

$$\tau' = \frac{\tau_y}{1 - \chi L_{obs}/4} \tag{6.7}$$

where χ is the stress gradient, L_{obs} is the obstacle distance for dislocation slip, and τ_y is the yield strength at $\chi = 0$. This leads to the stress gradient strengthening.

Experimental data in Figs. 6.2c and 6.3c suggest that an optimum gradient layer thickness (volume fraction) exists where the synergetic strengthening is most effective. This can be understood by the bi-axial stress in the integrated sample. It is intuitive to hypothesize that higher magnitude of strain and stress in the x direction will make it more effective to accumulate dislocations since it more effectively promotes dislocation interactions and increase the strain gradient near the boundaries. Since the stress state direction in the plastic central layer is opposite to that in the elastic gradient layers and also because the stresses between these two type of layers need to neutralize each other, there should exist an optimum gradient layer thickness that exert the maximum counter force to the plastically deforming central layer. In other words, there should be an optimum gradient layer thickness that produces the largest synergetic strengthening. This is an issue that need to be further studied to help with designing the gradient structure for maximum synergetic strengthening.

The ΔHv peak in Fig. 6.3c and the strain gradient profile in Fig. 6.5d suggest that the high strain and stress gradient near the boundary indeed promoted the accumulation of dislocations. This ΔHv peak is higher than the hardness increase in the CG layer. This observation is significant because if deformed alone, the CG layer should have highest strain hardening because its large grain sizes make it easier to accumulate dislocation and also because it was subjected to larger plastic deformation.

It should be noted that the concept of gradient structure studied here is very different from the strain gradient plasticity reported in the literature [29–32]. First, the gradient structure spans a macroscopic length of hundreds of grains, which is much larger than the grain size scale typically concerned by the conventional strain gradient plasticity. Second, the gradient structure has grain sizes varying along the depth from the sample surface, which results in a yield strength gradient, but the applied strain is uniform across the entire sample thickness in the z-direction. Thirdly, macroscopic strain gradient is induced in the lateral direction although the applied strain in the z-direction has no macroscopic strain gradient. Lastly, it is should be noted that the strain gradient plasticity can always be applied at the microscopic level, e.g. inside individual grains, for the graded structure.

In summary, gradient structure in IF steel increased the yield strength beyond what is predicted by the rule of mixtures. The synergetic increase in strength is attributed to the mechanical incompatibility caused by the mismatch between the Poisson's ratio in the elastic outer layers and apparent Poisson's ratio in the central plastic layer. The mechanical incompatibility leads to two-dimensional stress states and lateral strain gradient near the plastic-elastic boundaries, which consequently lead to more dislocation interaction and accumulation to quickly increase the strength of the plastically deforming layer. The gradient structure also produces a macroscopic stress gradient across the thickness, especially near the plastic-elastic boundaries. Such stress gradient further contributed to the synergetic strengthening. These mechanisms do not occur in homogenous materials, and are unique deformation characteristics of materials with gradient structures. Our experimental results also indicate an optimum microstructural gradient for the best mechanical properties. These observations provide guidance in designing high-strength metals and alloys with gradient structures.

Acknowledgement

We are grateful for financial support of 973 programs (grants 2012CB932203, 2012CB937500, and 6138504), National Natural Science Foundation of China (grants 11002151, 11072243, 51301187, and 50571110), the Pangu Foundation, and the US Army Research Office (grants W911NF-09-1-0427 and W911QX-08-C-0083).

References

1. Suresh S, Graded materials for resistance to contact deformation and damage. *Science*, 2001; **292**(5526):2447–2451.
2. Gao HJ, Ji BH, Jager IL, Arzt E and Fratzl P, Materials become insensitive to flaws at nanoscale: lessons from nature. *Proc. Natl. Acad. Sci. USA*, 2003; **100**(10):5597–5600.
3. Ray AK, Das SK, Mondal S and Ramachandrarao P, Microstructural characterization of bamboo. *J. Mater. Sci.*, 2004; **39**(3):1055–1060.

4. Fang TH, Li WL, Tao NR and Lu K, Revealing extraordinary intrinsic tensile plasticity in gradient nano-grained copper. *Science*, 2011; **331**(6024):1587–1590.

5. Chen AY, Li DF, Zhang JB, Song HW and Lu J, Make nanostructured metal exceptionally tough by introducing non-localized fracture behaviors. *Scr. Mater.*, 2008; **59**(6):579–582.

6. Hughes DA and Hansen N, Graded nanostructures produced by sliding and exhibiting universal behavior. *Phys. Rev. Lett.*, 2001; **87**(13).

7. Choi IS, Dao M and Suresh S, Mechanics of indentation of plastically graded materials - I: analysis. *J. Mech. Phys. Solids*, 2008; **56**(1):157–171.

8. Zhu YT and Liao XZ, Nanostructured metals: retaining ductility. *Nat. Mater.*, 2004; **3**(6):351–352.

9. Wang YM, Chen MW, Zhou FH and Ma E, High tensile ductility in a nanostructured metal. *Nature*, 2002; **419**(6910):912–915.

10. Huang XX, Hansen N and Tsuji N, Hardening by annealing and softening by deformation in nanostructured metals. *Science*, 2006; **312**(5771):249–251.

11. Zhao YH, Liao XZ, Cheng S, Ma E and Zhu YT, Simultaneously increasing the ductility and strength of nanostructured alloys. *Adv. Mater.*, 2006; **18**(17):2280.

12. Youssef KM, Scattergood RO, Murty KL, Horton JA and Koch CC, Ultrahigh strength and high ductility of bulk nanocrystalline copper. *Appl. Phys. Lett.*, 2005; **87**(9):091904.

13. Liddicoat PV, Liao XZ, Zhao YH, Zhu YT, Murashkin MY, Lavernia EJ, Valiev RZ and Ringer SP, Nanostructural hierarchy increases the strength of aluminum alloys. *Nat. Commun.*, 2010; 163.

14. Lu K, Lu L and Suresh S, Strengthening materials by engineering coherent internal boundaries at the nanoscale. *Science*, 2009; **324**(5925):349–352.

15. Gleiter H, Nanocrystalline materials. *Prog. Mater. Sci.*, 1989; **33**(4):223–315.

16. Meyers MA, Mishra A and Benson DJ, Mechanical properties of nanocrystalline materials. *Prog. Mater. Sci.*, 2006; **51**(4):427–556.

17. Weertman JR, Farkas D, Hemker K, Kung H, Mayo M, Mitra R and Van Swygenhoven H, Structure and mechanical behavior of bulk nanocrystalline materials. *MRS Bull.*, 1999; **24**(2):44–50.

18. Valiev RZ, Alexandrov IV, Zhu YT and Lowe TC, Paradox of strength and ductility in metals processed by severe plastic deformation. *J. Mater. Res.*, 2002; **17**(1):5–8.

19. Jian WW, Cheng GM, Xu WZ, Yuan H, Tsai MH, Wang QD, Koch CC, Zhu YT and Mathaudhu SN, Ultrastrong Mg alloy via nano-spaced stacking faults. *Mater. Res. Lett.*, 2013; **1**(2):61–66.

20. Lu K and Lu J, Surface nanocrystallization (SNC) of metallic materials-presentation of the concept behind a new approach. *J. Mater. Sci. Technol.*, 1999; **15**(3):193–197.

21. Lu K and Lu J, Nanostructured surface layer on metallic materials induced by surface mechanical attrition treatment. *Mater. Sci. Eng. A*, 2004; **375–377**:38–45.

22. Beat CW, Mills EJ and Hyler WS, Effect of variation in poisson's ratio on plastic tensile instability. *J. Fluids. Eng.*, 1967; **89**(1):35–39.

23. Ashby MF, The deformation of plastically non-homogeneous materials. *Philos. Mag.*, 1970; **21**(170): 300–424.

24. Gao HJ, Huang YG, Nix WD and Hutchinson JW, Mechanism-based strain gradient plasticity - I. Theory. *J. Mech. Phys. Solids*, 1999; **47**(6):1239–1263.

25. Gao HJ and Huang YG, Geometrically necessary dislocation and size-dependent plasticity. *Scr. Mater.*, 2003; **48**(2):113–118.

26. Chaudhar P and Scatterg RO, On pile-up model for yielding. *Acta Metall.*, 1966; **14**(5):685.

27. Hirth JP, Dislocation pileups in the presence of stress gradients. *Philos. Mag.*, 2006; **86**(25–26):3959–3963.

28. Chakravarthy SS and Curtin WA, Stress-gradient plasticity. *Proc. Natl. Acad. Sci. USA*, 2011; **108**(38):15716–15720.

29. Abu Al-Rub RK, Interfacial gradient plasticity governs scale-dependent yield strength and strain hardening rates in micro/nano structured metals. *Int. J. Plast.*, 2008; **24**(8):1277–1306.

30. Lapovok R, Dalla Torre FH, Sandlin J, Davies CHJ, Pereloma EV, Thomson PF and Estrin Y, Gradient plasticity constitutive model reflecting the ultrafine micro-structure scale: the case of severely deformed copper. *J. Mech. Phys. Solids*, 2005; **53**(4):729–747.

31. Evans AG and Hutchinson JW, A critical assessment of theories of strain gradient plasticity. *Acta Mater.*, 2009; **57**(5):1675–1688.

32. Mughrabi H, On the current understanding of strain gradient plasticity. *Mater. Sci. Eng. A*, 2004; **387–389**:209–213.

Chapter 7

Hetero-Deformation-Induced Strengthening and Strain Hardening in Gradient Structure

Muxin Yang,[a] Yue Pan,[a] Fuping Yuan,[a] Yuntian Zhu,[b,c] and Xiaolei Wu[a]

[a]*State Key Laboratory of Nonlinear Mechanics, Institute of Mechanics, Chinese Academy of Sciences, Beijing 100190, China*
[b]*Department of Materials Science and Engineering, North Carolina State University, Raleigh, NC 27695, USA*
[c]*School of Materials Science and Engineering, Nanjing University of Science and Technology, Nanjing 210094, China*
xlwu@imech.ac.cn

We report significant hetero-deformation-induce (HDI) strengthening and strain hardening in gradient structured (GS) IF steel. HDI stress is caused by the coupling of back stress in the soft zone

---

Reprinted from *Mater. Res. Lett.*, **4**(3), 145–151, 2016.

and the forward stress in the hard zone. Back-stress is a long-range stress caused by the pileup of geometrically necessary dislocations in the soft zone. We have developed a simple equation and procedure to calculate HDI stress basing on its formation physics from the tensile unloading-reloading hysteresis loop. The gradient structure has mechanical incompatibility due to its grain size gradient. This induces strain gradient, which needs to be accommodated by geometrically necessary dislocations. HDI stress not only raises the yield strength but also significantly enhance strain hardening to increase the ductility.

Gradient structure (GS) in metals represents a new strategy for producing a superior combination of high strength and good ductility [1–6]. The Gradient structure usually consists of a nanostructured (NS) surface layer with increasing grain size along the depth to reach coarse-grained (CG) sizes in the central layer [2, 4].

Gradient structure can promote ductility significantly [2, 4–9], which is measured under tensile loading. The NS layer in a gradient structure may sustain a large amount of tensile strain [2, 4], because they are constrained by the CG layer. It was reported that the gradient structured Cu derives its ductility from the confinement of the soft CG core [2, 10], and from strong grain growth in the NS layer by mechanically-driven grain growth during tensile deformation. Nanostructures in high-purity copper is known to be unstable at room temperature, and mechanical-driven grain growth in nanocrystalline metals have been extensively reported [11–16]. For gradient structured metals with stable nanostructures, however, their high ductility is attributed to extra strain hardening due to the presence of strain gradient and the change of stress states, which generates geometrically necessary dislocations (GNDs) and promotes the generation and interaction of dislocations [3, 4, 17, 18]. Furthermore, the gradient structure is observed to produce an intrinsic synergistic strengthening, with its yield strength much higher than the sum of separate gradient layers [3], which is attributed to the macroscopic stress gradient and plastic incompatibility between layers [3, 4].

The nature of plastic deformation in the gradient structure is still not very clear [1, 2]. In fact, the gradient structure can be approximately regarded as the integration of many thin layers with increasing grain sizes [3, 4]. The gradient structure deforms heterogeneously due to mechanical incompatibilities between neighboring layers with different flow behaviors and stresses under applied strains. As such, it is reasonable to anticipate the development of the strain gradient and internal stresses during plastic deformation, as a result of the mechanical incompatibilities between different layers, similar to what happens in composites [19–21] and dual-phase structures [22].

Back stress has been reported to play a crucial role in strain hardening, strengthening and mechanical properties [21–23]. It is a type of long-range stress exerted by geometrically necessary dislocations that are accumulated and piled up against barriers. It interacts with mobile dislocations to affect their slip [24]. The back stress reduces the effective resolved shear stress for dislocation slip because it always acts in the opposite direction of the applied resolved shear stress. At the same time, the stress concentration at the head of the GND pile-up produces forward stress at the barrier boundary to produce forward stress across the boundary. In a heterogeneous structure, strain will be inhomogeneous but continuous, producing strain gradients, which needs to be accommodated by GNDs [23, 25–27]. This produces back stress in the soft zone and forward stress in the hard zone. The back stress and forward stress are the same in magnitude but opposite in direction at the zone boundary where GNDs are piled up. However, their distributions away from the boundary are different, which produces the hetero-deformation-induced stress, as defined in the Chapter 2.

It has been observed that HDI strengthening and HDI strain hardening are primarily responsible for unprecedented combination of strength and ductility of heterogeneous lamella Ti, which was found as strong as ultrafine grained Ti and as ductile as CG Ti [23]. The gradient structure can be regarded as a type of heterogeneous structure. Therefore, it is reasonable to assume that significant HDI stress will be developed in gradient structure, which should be investigated to have a better understanding on the fundamentals of gradient structure.

Here we report experimental evidences of significant HDI stress and HDI strain hardening in gradient structured IF steel. We will also derive an equation with sound physics to calculate HDI stress from unloading-reloading stress-strain hysteresis loop during a tensile test. A detailed procedure on how to extract useful data from the hysteresis loop for calculating the HDI stress is presented.

A 1-mm thick sheet of IF steel was used as the starting materials with the composition (wt%) 0.003% C, 0.08% Mn, 0.009% Si, 0.008% S, 0.011% P, 0.037% Al, 0.063% Ti, and 38 ppm N. The disk of a 100 mm diameter was cut and annealed at 1173 K for 1 hour to obtain a homogeneous CG microstructure with a mean grain size of 35 μm. Surface mechanical attrition treatment (SMAT) [28] was used to produce the gradient structured sample. The SMAT duration was 5 minutes for both sides of the disk. NS layer of 120 μm thick was formed, which consists of, in sequence, the nanograins (minimum grain size of <100 nm in the top-layer), ultra-fine grains, and deformed coarse grains with dislocation cells towards the central CG core. Microstructural characterization was detailed in our previous papers [3, 4].

Unloading-reloading process during tensile tests was conducted using an Instron 5966 machine at a strain rate of 5×10^{-4} s^{-1} at room temperature. Tensile specimens with a gauge length of 10 mm and a width of 2.5 mm were cut from SMAT-processed disks. An extensometer was used to measure tensile strain during the period of uniform deformation. At a certain unloading strain, the specimen was unloaded in load-control mode to 20 N at an unloading rate of 200 N · min^{-1}, followed by reloading to the same applied load.

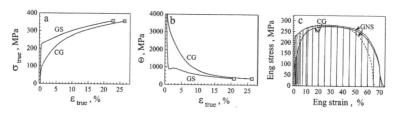

Figure 7.1 (a) Tensile stress-strain curves in gradient structured (GS) and CG IF steel samples. (b) Strain hardening rate ($\Theta = d\sigma/d\varepsilon$) vs. strain. (c) The unloading and reloading test hysteresis loops measured at varying tensile strain for both CG and GS samples.

Figure 7.1a shows the monotonic tensile true stress-true strain (σ-ε) curves in both gradient structured and CG samples. The gradient structured sample shows large ductility comparable to that of CG, but with triple yield strength of CG, which is typical of the excellent combination of strength and ductility in gradient structured metals [2–8]. A transient is visible soon after yielding, characterized by the presence of a short concave segment on the σ-ε curve [4]. During the transient, the strain hardening rate (Θ) sharply drops at first, which is followed by a rapid up-turn, as shown in Fig. 7.1b. Figure 7.1c shows the unloading and reloading test hysteresis loops measured at varying tensile strain for both CG and gradient structured samples.

Unloading-reloading was performed at varying tensile strains to investigate the evolution of HDI stress during tensile test. Figure 7.2a shows schematically the unloading-reloading stress-strain hysteresis loop. As shown, the unloading starts at unloading strains (ε_u) at point A. The segment AB of the unloading curve is quasi-elastic and caused by stress relaxation [29] or viscous flow of the material [30, 31]. The stress drop in this segment is called the thermal component of the flow stress [24, 29], or viscous stress [30, 31]. The segment BC is the linear (elastic) part of the unloading stress with an effective unloading Young's modulus of E_u. The point C is called the unloading yielding point, with a stress of σ_u. Similar segments also exist for the reloading curve with EF as the linear (elastic) part of the reloading stress-strain curve with an effective reloading Young's modulus of E_r, which can be assumed equal to E_u because the microstructure is assumed not changed during the unloading-reloading. The point F is called the reloading yielding point, with a stress of σ_r. Figure 7.2b is measured hysteresis loop from a gradient structured IF steel sample.

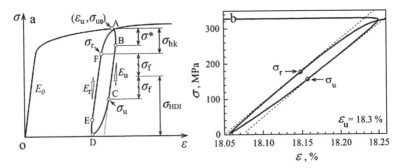

Figure 7.2 (a) The Schematic of the unloading-reloading for defining the unload yielding σ_u, reload yielding σ_r, HDI stress σ_{HDI} and frictional stress σ_f, effective unloading Young's modulus of E_u, effective reloading Young's modulus of E_r. (b) A measured hysteresis loop from a gradient structured IF steel sample with σ_u and σ_r defined.

From the unloading-reloading hysteresis loop we can calculate the HDI stress σ_{HDI}, and the frictional stress σ_f. Note that the stress measured from experimentally measured curves as shown in Fig. 7.2 used to be called back stress, which is not correct, because both back stress and forward stress contributed to the loading-unloading behavior. The frictional stress consists of the Peierls stress as well as other stresses that are needed to overcome the dynamic pinning of dislocations such as solute atoms, second phase, forest dislocations, dislocation debris, dislocation jogs, etc.

To derive the equation for calculating the HDI stress and frictional stress, we first assume that the frictional stress σ_f is a constant during the entire unloading-reloading process. We also assume that the HDI stress does not change with unloading before the unloading yield point C in Fig. 7.2a. This assumption is reasonable because the reverse dislocation motion does not start above this point. In other words, geometrically necessary dislocations that caused the HDI stress do not change their density or configuration before the unloading yield, which keeps the HDI stress approximately constant. This assumption is important and was also adopted by Dickson et al. [29]. During the unloading, the HDI stress is the stress that drives the mobile dislocations to reverse their gliding direction to produce unloading yield. At the unloading yield point C (Fig. 7.2a), the applied stress is low enough that the back stress starts to overcome the applied stress and the frictional stress to make dislocations glide backward, i.e.

$$\sigma_{HDI} = \sigma_u + \sigma_f \qquad (7.1)$$

where σ_u is the unloading yield stress as defined in Fig. 7.2a.

During the reloading, the applied stress needs to overcome the back stress and the frictional stress to drive the dislocation forward at the reloading yield point F, which can be described as

$$\sigma_r = \sigma_{HDI} + \sigma_f \qquad (7.2)$$

where σ_r is the reloading yield stress as defined in Fig. 7.2a.

Here again, we assume that the HDI stress during reloading is the same as the HDI stress during unloading. This is reasonable because during the unloading-reloading process, dislocation configuration can be considered reversible [32]. Solving Eqs. (7.1) and (7.2) yields

$$\sigma_{HDI} = (\sigma_r + \sigma_u) / 2 \qquad (7.3)$$

and

$$\sigma_f = \frac{\sigma_r - \sigma_u}{2} \tag{7.4}$$

Equation (7.3) is similar to an earlier equation proposed for cyclic loading by Cottrell [33] and Kulmann-Wilsdorf and Laird [32], except they used σ_{u0}, the initial flow stress at the beginning of the unloading, in place of the σ_r, i.e.

$$\sigma_b = \frac{\sigma_{u0} + \sigma_u}{2} = \sigma_{HDI} \tag{7.5}$$

where σ_{u0} is the initial unloading stress as defined in Fig. 7.2a, σ_b was called back stress, but should be actually σ_{HDI}.

We argue that Eq. (7.3) is physically sounder than Eq. (7.5) because we are defining unloading yield and reloading yield using the same criterion, i.e. the same deviation of effective Young's modulus as discussed later. It has been recognized that Eq. (7.5) overestimates the HDI stress, and was later modified by Dickson et al. to include the thermal component of the flow stress [24, 29]:

$$\sigma_b = \frac{\sigma_0 + \sigma_u}{2} - \frac{\sigma^*}{2} \tag{7.6}$$

where σ^* is the thermal component of the flow stress as defined in Fig. 7.2a [24, 29], which is also called the viscous stress [31].

The Eq. (7.3) is especially suitable for hysteresis loops with positive unloading yield stresses. If the HDI stress is very small, the unloading yield stress may become negative, in which case σ_u cannot be measured during unloading. As discussed later, the Eq. (7.3) derived here has an important advantage over previously published Eqs. (7.5) and (7.6): it produces consistent HDI stress values with much less scatter. In addition, Eqs. (7.5) and (7.6) are physically problematic because they implicitly used different criteria to define the unloading yield and reloading yield, which is physically unjustifiable.

To extract useful data from the unloading-reloading hysteresis loop, one needs to first determine the unloading yield stress σ_u and reloading yield stress σ_r. However, the real hysteresis loop (e.g. Fig. 7.2b) is not as well defined as in Fig. 7.2a and the practical extraction of the data is not straightforward [31]. The first step is to determine the elastic segments BC as well as its slope (the effective

Young's modulus). The unloading yield point C is usually determined by a plastic strain offset in the range of 5×10^{-6} to 10^{-3}, which have been used by different research groups [24, 31, 34–37]. These offset values are arbitrary and are not well justified. Here we propose to use the deviation of the stress-strain slope from the effective Young's modulus as a physically sound method to determine the yield point. In this study, we choose 5%, 10%, and 15% slope reduction from the effective Young's modulus, E_u. If the strain hardening in the plastically deforming volume is ignored, the slope reduction should be equal to the volume fraction that is plastically deforming. For example, a 10% reduction in E_u means 10% of the sample volume is plastically deforming. We also propose to use $E_r = E_u$, and the same slope reduction values for determining both the unloading yield point and reloading yield point. If the effective Young's modulus, E_u, cannot be easily determined from the loading unloading curve, one can approximately use the initial Young's modulus E_0. However, the error may become larger with increasing tensile strain because microstructure and defect structure may change during the tensile testing.

Figure 7.3 compares the evolution of the unloading yield stress, reloading yield stress, and back stress of the CG and gradient structured IF steel samples with increasing tensile strain at which the unloading was initiated. Several features can be seen from the figure. *First*, the unloading yield stress is affected more than the reloading yield stress by the slope reduction offset value that is used to determine them (Figs. 7.3a and 7.3c). *Second*, using a larger slope reduction leads to lower unloading yield stress σ_u and higher reloading yield stress σ_r. Part of these variations in σ_u and σ_r caused by the choice of slope reduction cancel each other in Eq. (7.3), which leads to smaller scatter in the calculated HDI stress using Eq. 7.3 (Figs. 7.3b and 7.3d). This is an advantage of the Eq. (7.3) for calculating the HDI stress, as compared with the previously reported Eq. (7.6) [23, 24, 29–31]. *Third*, the HDI stresses in both the CG and gradient structured samples increase with the tensile strain. However, the HDI stress is higher in the gradient structured sample than in the CG sample. For example, for the 5% slope reduction, the HDI stress in the gradient structured sample is 10% to 40% higher than those in the CG sample (the red curves in Figs. 7.3b and 7.3d). *Fourth*, Figs. 7.3c and 7.3d shows that if a large slope reduction value is used, the unloading yield stresses for the gradient structured sample at small tensile strains are negative and therefore cannot be measured in the unloading curve. This makes it necessary to use a smaller slope reduction value in determining the HDI stress.

For valid and easy comparison, we propose that the slope reduction value for calculating the HDI stress is marked in the symbol. For example, $\sigma_{HDI,5\%}$ represents HDI stress calculated using 5% slope reduction from the effective Young's modulus, as shown in Figs. 7.3b and 7.3d.

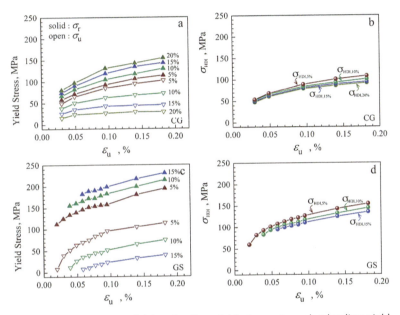

Figure 7.3 Evolution of (a) unloading yield stress σ_u and reloading yield stress σ_r and (b) HDI stress with increasing tensile strain ε_u for CG IF steel, and the evolution of (c) unloading/reloading yield stresses and (d) HDI stress with increasing tensile strain ε_u for gradient structured (GS) IF steel. $\sigma_{HDI,5\%}$ represents HDI stress calculated using 5% slope reduction from the effective Young's modulus.

As shown in Fig. 7.4a, the frictional stress σ_f calculated using Eq. (7.4) is very scattered. A larger slope reduction value leads to significantly higher σ_f. For example, for the CG sample, the σ_f calculated using 20% slope reduction is many times larger than those calculated using 5% slope reduction. This is because Eq. (7.4) adds the absolute values of σ_u and σ_r variations together instead of making them cancel each other as in Eq. (7.3). Therefore, the frictional stress σ_f calculated using Eq. (7.4) is not quantitatively dependable. Nevertheless, Fig. 7.4a consistently shows that for any slope reduction value, the calculated frictional stress is higher in the gradient structured sample than in the CG sample. This is due to the higher dislocation density in the gradient structured sample than in the CG sample [3, 4].

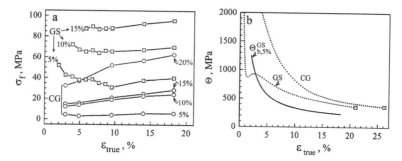

Figure 7.4 The frictional stress σ_f vs. ε_u for gradient structured (GS) and coarse-grained (CG) IF steel samples calculated using Eq. 7.4. (b) The HDI stress strain-hardening difference between the GS and CG structure.

Figure 7.4b shows that the gradient structured sample has much higher HDI stress strain-hardening than the CG sample, especially in the transient range that correlates to Θ up-turn. This indicates that the HDI strain-hardening has significant contribution to the observed Θ up-turn. The rapid HDI stress increase right after the yielding of the gradient structured sample is also obvious in Fig. 7.3d. The observed Θ up-turn has been attributed to fast dislocation accumulation after the initial exhaustion of mobile dislocations [4]. The high HDI stress associated with the observed Θ up-turn observed here suggests that a large quantity of geometrically necessary dislocations is accumulated at this stage. Since the geometrically necessary dislocations are associated with the strain gradient in the sample, this observation also suggests that there was a quick increase in the strain gradient at the beginning of the plastic deformation of the gradient structured IF steel. This is understandable because this is at the deformation stage in which the nanostructured surface layers just started to become unstable and the lateral stresses start to reverse their directions [3, 4]. Specifically, the surface nanostructured layers transit from compressive lateral stress to tensile laterals stress, while the central larger grained layer transits in an opposite way. Such a transition is expected to increase the strain gradient.

In summary, it is found that the gradient-structured IF steel developed strong HDI strengthening and HDI strain-hardening during tensile testing, which arise from the plastic incompatibilities due to its microstructural heterogeneity. The high HDI stress near the beginning of the plastic deformation of the gradient structured IF steel samples should have contributed to the observed synergistic strengthening [3], while the high HDI strain hardening should have contributed to the observed high ductility [4]. The equation derived and the procedure proposed in this work for calculating the HDI stress from the unloading-reloading hysteresis loop produces more consistent HDI stress value than what is previously reported.

Acknowledgements

This work was supported by the NFSC under Grant numbers (11572328, 11072243, 11222224, 11472286, and 11021262); and 973 Projects under Grant numbers (2012CB932203, 2012CB 937500, and 6138504). Y.T.Z. is funded by the US Army Research Office (W911 NF-12-1-0009), the US National Science Foundation (DMT-1104667) and by the Jiangsu Key Laboratory of Advanced Micro&Nano Materials and Technology, Nanjing University of Science and Technology.

References

1. Lu K. Making strong nanomaterials ductile with gradients. *Science*, 2014; **345**:1455–1456.
2. Fang TH, Li WL, Tao NR and Lu K. Revealing extraordinary intrinsic tensile plasticity in gradient nano-grained copper. *Science*, 2011; **331**:1587–1590.
3. Wu XL, JIang P, Chen L, Zhang JF, Yuan FP and Zhu YT. Synergetic Strengthening by gradient structure. *Mater. Res. Lett.*, 2014; **2**:185–191.

4. Wu XL, Jiang P, Chen L, Yuan FP and Zhu YTT. Extraordinary strain hardening by gradient structure. *Proc. Natl. Acad. Sci. USA*, 2014; **111**:7197–7201.

5. Jerusalem A, Dickson W, Perez-Martin MJ, Dao M, Lu J and Galvez F. Grain size gradient length scale in ballistic properties optimization of functionally graded nanocrystalline steel plates. *Scr. Mater.*, 2013; **69**:773–776.

6. Wang HT, Tao NR and Lu K. Architectured surface layer with a gradient nanotwinned structure in a Fe-Mn austenitic steel. *Scr. Mater.*, 2013; **68**:22–27.

7. Kou HN, Lu J and Li Y. High-Strength and high-ductility nanostructured and amorphous metallic materials. *Adv. Mater.*, 2014; **26**:5518–5524.

8. Wei YJ, Li YQ, Zhu LC, Liu Y, Lei XQ, Wang G, Wu YX, Mi ZL, Liu JB, Wang HT and Gao HJ. Evading the strength-ductility trade-off dilemma in steel through gradient hierarchical nanotwins. *Nat. Commun.*, 2014; **5**.

9. Ma XL, Huang CX, Xu WZ, Zhou H, Wu XL and Zhu YT. Strain hardening and ductility in a coarse-grain/nanostructure laminate material. *Scr. Mater.*, 2015; **103**:57–60.

10. Fang TH, Tao NR and Lu K. Tension-induced softening and hardening in gradient nanograined surface layer in copper. *Scr. Mater.*, 2014; **77**:17–20.

11. Weertman JR. Retaining the nano in nanocrystalline alloys. *Science*, 2012; **337**:921–922.

12. Chookajorn T, Murdoch HA and Schuh CA. Design of stable nanocrystalline alloys. *Science*, 2012; **337**:951–954.

13. Zhang K, Weertman JR and Eastman JA. Rapid stress-driven grain coarsening in nanocrystalline Cu at ambient and cryogenic temperatures. *Appl. Phys. Lett.*, 2005; **87**:061921.

14. Zhang K, Weertman JR and Eastman JA. The influence of time, temperature, and grain size on indentation creep in high-purity nanocrystalline and ultrafine grain copper. *Appl. Phys. Lett.*, 2004; **85**:5197–5199.

15. Legros M, Gianola DS and Hemker KJ. In situ TEM observations of fast grain-boundary motion in stressed nanocrystalline aluminum films. *Acta Mater.*, 2008; **56**:3380–3393.

16. Liao XZ, Kilmametov AR, Valiev RZ, Gao HS, Li XD, Mukherjee AK, Bingert JF and Zhu YT. High-pressure torsion-induced grain growth in electrodeposited nanocrystalline Ni. *Appl. Phys. Lett.*, 2006; **88**:021909.

17. Li WB, Yuan FP and Wu XL. Atomistic tensile deformation mechanisms of Fe with gradient nano-grained structure. *AIP Adv.*, 2015; **5**.
18. Li JJ, Chen SH, Wu XL and Soh AK. A physical model revealing strong strain hardening in nano-grained metals induced by grain size gradient structure. *Mater. Sci. Eng. A*, 2015; **620**:16–21.
19. Llorca J, Needleman A and Suresh S. The Bauschinger effect in Whisker-Reinforced metal-matrix composites. *Scr. Metall. Mater.*, 1990; **24**:1203–1208.
20. Sinclair CW, Saada G and Embury JD. Role of internal stresses in co-deformed two-phase materials. *Philos. Mag.*, 2006; **86**:4081–4098.
21. Thilly L, Van Petegem S, Renault PO, Lecouturier F, Vidal V, Schmitt B and Van Swygenhoven H. A new criterion for elasto-plastic transition in nanomaterials: application to size and composite effects on Cu-Nb nanocomposite wires. *Acta Mater.*, 2009; **57**:3157–3169.
22. Calcagnotto M, Adachi Y, Ponge D and Raabe D. Deformation and fracture mechanisms in fine- and ultrafine-grained ferrite/martensite dual-phase steels and the effect of aging. *Acta Mater.*, 2011; **59**:658–670.
23. Wu XL, Yang MX, Yuan FP, Wu GL, Wei YJ, Huang XX and Zhu YT. Heterogeneous lamella structure unites ultrafine-grain strength with coarse-grain ductility. *PNAS*, 2015; **112**:14501–14505.
24. Feaugas X. On the origin of the tensile flow stress in the stainless steel AISI 316L at 300 K: back stress and effective stress. *Acta Mater.*, 1999; **47**:3617–3632.
25. Ashby MF. Deformation of plastically non-homogeneous materials. *Philos. Mag.*, 1970; **21**:399.
26. Gao H, Huang Y, Nix WD and Hutchinson JW. Mechanism-based strain gradient plasticity - I. Theory. *J. Mech. Phys. Solids*, 1999; **47**:1239–1263.
27. Gao HJ and Huang YG. Geometrically necessary dislocation and size-dependent plasticity. *Scr. Mater.*, 2003; **48**:113–118.
28. Lu K and Lu J. Nanostructured surface layer on metallic materials induced by surface mechanical attrition treatment. *Mater. Sci. Eng. A*, 2004; **375–377**:38–45.
29. Dickson JI, Boutin J and Handfield L. A comparison of two simple methods for measuring cyclical internal and effective stresses. *Mater. Sci. Eng.*, 1984; **64**:L7–L11.
30. Fournier B, Sauzay M, Caes C, Mottot M, Noblecourt A and Pineau A. Analysis of the hysteresis loops of a martensitic steel - Part II: study

of the influence of creep and stress relaxation holding times on cyclic behaviour. *Mater. Sci. Eng. A*, 2006; **437**:197–211.

31. Fournier B, Sauzay M, Caes C, Noblecourt M and Mottot M. Analysis of the hysteresis loops of a martensitic steel - Part I: study of the influence of strain amplitude and temperature under pure fatigue loadings using an enhanced stress partitioning method. *Mater. Sci. Eng. A*, 2006; **437**:183–196.

32. Kuhlmann-Wilsdorf D and Laird C. Dislocation behavior in fatigue. 2. Friction stress and back stress as inferred from an analysis of hysteresis loops. *Mater. Sci. Eng.*, 1979; **37**:111–120.

33. Cottrell AH, *Dislocations and Plastic Flow in Crystals* (Oxford University Press, London, 1953).

34. Delobelle P and Oytana C. The study of the laws of behavior at high-temperature, in plasticity-flow, of an austenitic stainless-steel (17-12-Sph). *J. Nucl. Mater.*, 1986; **139**:204–227.

35. Risbet M, Feaugas X and Clavel M. Study of the cyclic softening of an under-aged gamma'-precipitated nickel-base superalloy (Waspaloy). *Journal De Physique IV*, 2001; **11**:293–301.

36. Guillemer-Neel C, Feaugas X and Clavel M. Mechanical behavior and damage kinetics in nodular cast iron: part II. Hardening and damage. *Metall. Trans. A*, 2000; **31**:3075–3085.

37. Morrison DJ, Jia Y and Moosbrugger JC. Cyclic plasticity of nickel at low plastic strain amplitude: hysteresis loop shape analysis. *Mater. Sci. Eng. A*, 2001; **314**:24–30.

Chapter 8

Residual Stress Provides Significant Strengthening and Ductility in Gradient Structured Materials

Muxin Yang,[a,†] Runguang Li,[b,†] Ping Jiang,[a] Fuping Yuan,[a,c] Yandong Wang,[b] Yuntian Zhu,[d,e] and Xiaolei Wu[a,c]

[a]*State Key Laboratory of Nonlinear Mechanics, Institute of Mechanics, Chinese Academy of Sciences, Beijing 100190, China*
[b]*State Key Laboratory for Advanced Metals and Materials, University Science and Technology Beijing, Beijing 100083, China*
[c]*School of Engineering Science, University of Chinese Academy of Sciences, Beijing 100049, China*
[d]*Department of Mater Science and Engineering, North Carolina State University, Raleigh, NC 27695 USA*
[e]*School of Materials Science and Engineering, Nanjing University of Science and Technology, Nanjing 210094, China*
xlwu@imech.ac.cn, ytzhu@ncsu.edu, ydwang@ustb.edu.cn

[†]These authors contributed equally to this work.

Reprinted from *Mater. Res. Lett.*, **7**(11), 433–438, 2019.

Heterostructured Materials: Novel Materials with Unprecedented Mechanical Properties
Edited by Xiaolei Wu and Yuntian Zhu
Text Copyright © 2019 The Author(s)
Layout Copyright © 2022 Jenny Stanford Publishing Pte. Ltd.
ISBN 978-981-4877-10-7 (Hardcover), 978-1-003-15307-8 (eBook)
www.jennystanford.com

Residual stress exists extensively in biological and engineering structures. Here we report that residual stress can be engineered to significantly enhance the strength and ductility of gradient materials. In-situ synchrotron experiments revealed that the strongest strain hardening occurred in the layer with the highest compressive residual stress in a gradient structure. This layer remained elastic longer than adjacent layers during tension, producing high heterodeforamtion induced stress to increase strength and enhancing work hardening even after the disappearance of the compressive stress to increase ductility. This finding provides for a new paradigm for designing gradient structures for superior mechanical properties.

8.1 Introduction

Residual stress exists almost ubiquitously in nature and man-made engineering materials and structures. The advantage of residual stress has long been recognized and well utilized [1–3]. For example, residual stress plays an important role in the function of bio-systems such as blood vessels [4]. Residual stress is engineered into construction structures such as pre-stressed concrete bridge beams to improve their strength and toughness [5]. For metal components, residual stress is often inevitably introduced into their surface layer during forming, machining, and heat treatment [1–3]. The most successful application of residual stress is the introduction of compressive residual stress (CRS) into the surface layer of various components to prolong their fatigue life [1, 6–8]. CRS is also reported to imrpove the fracture toughness of brittle materials such as metallic glass and ceramics [9, 10]. However, residual stress could be detrimental if inadvertently introduced into materials [3], leading to instability and even catastrophic failure during service [11, 12].

Since its inception in 1927, shot peening has been well studied and become a standard industrial practice to impart CRS to improve the fatigue life for metal parts of aircrafts and transportation vehicles [1–3]. However, there has been little knowledge so far as to whether or not such CRS affects the tensile properties such as strength and ductility [13–17]. This is mostly because residual stresses of different types will balance each other internally in a metal component [1, 2], so it is generally believed that residual stresses have little, if any, influence on the tensile mechanical behavior and properties. In addition, residual stress can be released during plastic deformation [2], which reinforces the above belief. For example, it has been recently reported that gradient-structured metals produced by surface mechanical attrition treatment (SMAT) possess superior strength and ductility [14, 15, 18]. SMAT is expected to produce CRS in the sample sub-surface layers [2,6,7,19]. However, due to the above belief the CRS was not considered in explaining the mechanical behavior of these gradient materials.

Here we report that residual stress played a major role in improving the yield strength and ductility of a gradient-structured (GS) interstitial free (IF) steel produced by SMAT. Synchrotron-based X-ray diffraction (XRD) in-situ tension experiment (see Fig. S1 and Note 1 in Supplementary [20, 21]) revealed that the CRS subsurface layer remained elastic longer than other layers, which resulted in direct strain hardening inside the layer itself as well as significant dislocation hardening and hetero-deforamtion induced (HDI) hardening in adjacent tensile layers, which resulted in higher yield strength. The HDI hardening is a more accurate description of the extra hardening in heterostructured materials than back-stress hardening, as discussed in Chapter 2. Surprisingly, the CRS left a legacy of high strain hardening in the CRS layer long after the CRS was eliminated during plastic deformation, which helped with increasing ductility.

8.2 Results and Discussion

Shown in Fig. 8.1a is the gradient structure (GS) from the nanograined (NG) sample surface layer to coarse-grained (CG) central layer after SMAT processing. A microhardness (Hv) gradient was produced along the depth (Fig. 8.1b). The SMAT processing produced a structural gradient [15, 22], as summarized in Fig. 8.1c. Dislocation sub-structures were formed in a sequence of the elongated slip-bands (Fig. 8.1d), dislocation cells (Fig. 8.1e), and subgrains (Fig. 8.1f), which later evolved into ultrafine-grains and nanograins.

XRD in-situ tensile loading (see Fig. S1) was carried out to measure the stress (elastic strain) and dislocation density evolution in discrete layers along the depth with increasing applied strain (see Figs. S2 and S3 in Supplimentary). Figure 8.2a shows the residual stress distribution along the depth in the as-SMAT-processed sample [6, 7]. Figure 8.2b reveals the evolution of stress in the loading direction along the depth with increasing applied strain. The increase in stress can be considered as an indicator of apparent accumulated strain hardening. As shown, a belly-shaped broad peak was quickly established in the depth range of 150–350 μm with increasing applied tensile strain. This indicates that most strain hardening occurred in this depth range. This range corresponds to the shaded ribbon in Fig. 8.1a–c. It is obvious that this layer has relatively large grains (Fig. 8.1a,d–f) and medium-to-low microhardness (Fig. 8.1b). It is puzzling why the strongest strain hardening occurred in this depth range. The conventional wisdom is that the strongest strain hardening should occur in the CG central layer where the grain sizes are largest and the initial dislocation density is low.

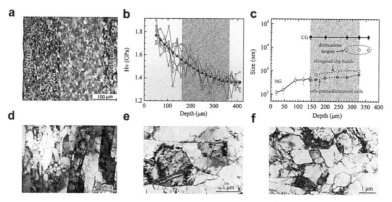

Figure 8.1 Microstructure and microhardness in a gradient structured (GS) IF-steel sample processed by SMAT. (a) Cross-sectional morphology of the GS sample, with nanograined (NG) surface layer and central CG layer. (b) Microhardness (H_V) distribution along the depth. (c) Size distribution of grain, subgrain and defect structures along the depth. The features of dislocation substructures are shown in (d–f) Cross-sectional TEM micrographs showing varying slip-bands, dislocation cells and subgrains at the depth of ~350, 300, and 230 μm, respectively.

Figure 8.2a provides a clue to this puzzle. As shown, CRS existed in the depth range of 150–350 μm, which coincides with the layer of the strongest strain hardening. For simplicity we refer to this layer as the CRS layer, while the adjacent layers with tensile residual stress (TRS) are referred to as the TRS layers. This observation indicates that CRS may be primarily responsible for the observed extraordinary broad strain-hardening peak. This raises an issue:

how does residual stress induces extra strain hardening?

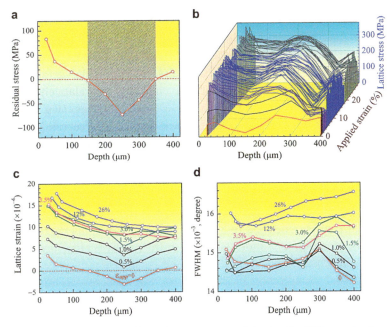

Figure 8.2 Experimental results as measured by in-situ XRD synchrotron tensile loading. (a) Residual stress distribution below the surface prior to tensile loading. (b) Evolution of the strain hardening along the depth at varying applied tensile strain, (c and d) Variations of axial elastic strain and full width at half maximum (FWHM) of diffraction peaks (dislocation density) with applied strain along the depth.

During tensile testing, all layers are subjected to the same applied strain, which are superimposed on the residual elastic strain. With increasing applied strain, all layers in the sample first deform elastically. As shown in Fig. 8.2c, the CRS layer has lower elastic strain than adjacent TRS layers on both sides up to the applied strain of ~3.0%. This is because the applied tensile strain was first offset by the CRS in the CRS layer. This leads to a co-deformation stage in which the TRS layers is already deforming plastically, whereas the CRS layer still remains partially elastic. This produces two elastic-plastic boundaries demarcating the CRS and TRS layers. With increasing applied strain, the thickness of the elastic layer will shrink and the two elastic-plastic boundaries move toward each other and eventually meet at the depth where the highest compressive stress originally existed before tensile testing, after which the elastic layer disappears. This occurred at ~3.0% of strain (Fig. 8.2c).

The elastic-plastic boundaries result in strain gradient near the boundary in the plastic layer, which is accommodated by the geometrically necessary dislocations (GNDs) [15, 22–24]. In other words, dislocations gliding in the adjacent plastically deforming layers may be blocked and pile-up near the two elastic-plastic boundaries. The pile-up of GNDs generates long-range back stress in the plastic layer [15, 22–30]. The back stress needs to be balanced at the boundaries by forward stress in the elastic layer, which contributes to the formation of the observed hardening belly, as shown in Fig. 8.2b. Together, the back stress and forward stress produced the HDI stress [Z1]. Since the sample was SMAT-processed on both sides, there are two CRS layers and consequently four elatic-plastic boundaries to accumulate the GNDs. This makes the HDI hardening much more effective in the gradient structure than previously believed [15].

Careful examination of the data for Fig. 8.2b reveals that the strain haderning belly was formed at 1.5% applied strain, which was before the CRS layer started to yield. This indicates that strain hardening in the CRS layer was significant before yielding. In other words, the CRS directly contributed to enhancing yield strength. In addition, the TRS layers should also have contributed to the enhancement in yield strength by developing strong HDI hardening. Other contributing factors to yield strength include finer grains in the surface layer and high dislocation density produced by SMAT [14–17]. This observation indicates that CRS contributed to the reported synergistic strengthening in gradient IF steel [25].

In the above discussion, dislocation pileup near the elastic-plastic boundary in the plastic layer contributed significantly to enhancing yield strength. This should logically increase the dislocation density in the plastic layers as they propagated into the CRS layer with increasing applied strain. This is verified by the evolution of the full width at half maximum (FWHM) of the synchrotron XRD peaks, which represents the evolution of dislocation density with increasing applied strain. As shown in Fig. 8.2d, the as-processed sample (0% strain) has the highest dislocation density at the depth of 300 µm. This is where the sample was plastically deformed by SMAT to produce dislocation cell and sub-grains (Fig. 8.1e,f) [14].

The dislocation density decreased from this peak point toward the surface because of grain refinement and dislocation recovery at very high plastic strains [15]. The decrease in dislocation density toward the CG center is due to decreasing plastic strain.

With increasing applied strain, a dislocation density valley developed at 250 μm, which corresponds to the location of the highest CRS peak. This is consistent with earlier discussion that the CRS layers remained elastic at the early deformation stage and therefore did not lead to the increase in dislocation density [2]. Figure 8.2d shows that dislocation density quickly increased with increasing tensile strain on both sides at the depths of about 100 μm and above 300 μm. This indicates that dislocation pileups occurred on both sides of the CRS layers near the elastic-plastic boundaries [26, 27].

Interestingly, a dislocation density peak appeared in the original CRS layer at 12% applied strain, which is long after the orginal CRS layer became plastic at ~3.0% strain (see the blue curve in Fig. 8.2c). This indicates that the original CRS layer left a legacy of higher dislocation hardening that lasted for large plastic strains, which is really helpful for preventing necking and maintaining high ductility. In other words, the CRS significantly contributed to maintaining ductility. As shown in Fig. 8.3a–c, high density of dislocations are accumualted at/near the grain boundaries at the initial stage of plastic deformation, which is markedly different from the conventional dislocation sub-structures in the interiors of grains and subgrains. Most of these dislocations are likely GNDs that were generated to accommodate strain gradient at elasto-plastic boundaries. These GNDs interacted and entangled with statistical dislocations, which helped with dislocation accumulation.

CRS contributed to enhancing both the yield strength and ductility, as reflected in Fig. 8.4. Compared with the CG sample, the GS sample has three times of yield strength and slightly higher ductility. This mechanical property is superior over those of other GS samples reported in the literature [14–17].

The initial stage of quick HDI hardening [16, 25, 28] (inset in Fig. 8.4b) coincides with the quick buildup of the dislocation density on both sides of the CRS layer at early stage of tensile deformation (Fig. 8.2b). HDI stress is induced by GND pilingup [26, 31], which indicates significant GND accumulation at the early srtage of deformation. In addition, this process also coincides with the strain hardening uptick [22] (see Fig. 8.4b). As shown in Fig. 8.4b, the strain hardening uptick started at a strain of 1.5%, which corresponds to the establishment of the strain hardening belly. In other words, the strain hardening in the CRS layer is directly related to the starting of the strain hardening uptick. Figure 8.4b also reveals that the strain hardening uptick ended at a strain of ~3%, which is the point the whole CRS layer started plastic deformation. These observations suggest that strain hardening uptick is associated with the direct forward-stress hardening in the shrinking elastic layer and the back-stress buildup in the plastic player as the plastic layers propagated into the two CRS layers.

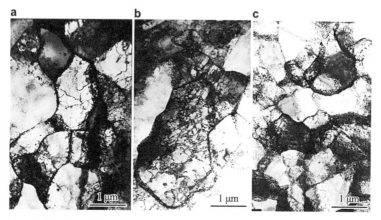

Figure 8.3 TEM observations of dislocation sub-structure evolutions. (a and b) Increase in dislocation entanglement, especially at/near grain boundaries at ~250 and 300 μm deep from top surface and at applied tensile strain of 3%. Also note dislocations in grain interior. (c) Dislocations in grains at ~250 μm deep from top surface and at applied strain of 15%.

The current observations represent a major breakthrough in our understanding of deformation physics and mechanical behavior of gradient materials. It is revealed that the CRS played a significant role in increasing the yield strength, and the HDI hardening as well as the dislocation forest hardening. The effect of CRS on the strength and ductility has never been reported before.

It should be noted that postmortem microhardness measurement is not adequate for studying the strain hardening behavior of gradient materials since it is not very sensitive to effect of residual stress on the mechanical behavior. As shown in Fig. 8.1b, the microhardness did not show obvious deviation in the CRS layer from the monotonic smooth drop in the CRS layer. Lastly, the TRS layer observed here (Fig. 8.2a) usually does not exist in conventional thick metal parts processed by shot peening. It is associated with thin plate samples processed by SMAT. We'd like to stress that the conclusion reached here is not affected by the existence of the TRS layer.

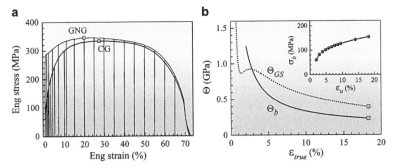

Figure 8.4 Superior synergistic effect between yield strength and ductility in the GS sample. (a) Engineering stress-strain curves of the GS and CG samples. The symbol ⊡ indicates uniform elongation. (b) Strain hardening rate versus applied strain. Inset: back stress versus strain.

8.3 Conclusion

The gradient IF steel samples processed by SMAT have two compressive residual stress layers. Therefore, four dynamic elastic-plastic boundaries existed during tensile testing, where geometrically necessary dislocations are developed and accumulated to develop a strong back stress hardening in the plastic layers. At the same time, a forward stress was produced in the elastic layer to result in higher tensile stress in it. This forward stress is responsible for the establishment of the strain hardening belly in the CRS layer. The back and forward stresses produced the hetero-deforamtion induced hardening. These observations indicates that compressive residual stress played a major role in improving the strength and ductility, which provides some new ideas on the design of gradient materials.

Acknowledgements

X.W. F.Y. and Y.Z. were funded by the National Key R&D Program of China 2017YFA0204402/3, the National Natural Science Foundation of China (Grant Nos. 11572328, 11672313, and 11790293), and the Strategic Priority Research Program of the Chinese Academy of Sciences (Grant No. XDB22040503). Y.Z. was supported by the US Army Research Office (W911 NF-12-1-0009), the US National Science Foundation (DMT-1104667), and the Nanjing University of Science and Technology.

References

1. Withers PJ, Bhadeshia HKDH. Residual stress. Part 2–nature and origins. *Mater. Sci. Technol.*, 2001; **17**:366–375.
2. Lu J. (ed.). *Handbook of Measurement of Residual Stresses* (Fairmont Press, New York, 1996).
3. Webster GA. Role of residual stress in engineering applications. *Mater. Sci. Forum.*, 2000; **347–349**:1–11.
4. Fung YC. What are the residual stresses doing in our blood vessels? *Ann. Biomed. Eng.*, 1991; **19**:237–249.
5. Naaman AE. *Prestressed Concrete Analysis and Design: Fundamentals* (3rd ed.) (Techno Press 3000, Sarasota, 2012).
6. Ya M, Xing Y, Dai F, Lu K, Lu J. Study of residual stress in surface nanostructured AISI 316L stainless steel using two mechanical methods. *Surf. Coat. Technol.*, 2003; **168**:148–155.
7. Roland T, Retraint D, Lu K, Lu J. Fatigue life improvement through surface nanostructuring of stainless steel by means of surface mechanical attrition treatment. *Scr. Mater.*, 2006; **54**:1949–1954.
8. Ortiz AL, Tian JW, Villegas JC, Shaw LL, Liaw PK. Interrogation of the microstructure and residual stress of a nickel-base alloy subjected to surface severe plastic deformation. *Acta Mater.*, 2008; **56**:413–426.
9. Zhang Y, Wang WH, Greer AL. Making metallic glasses plastic by control of residual stress. *Nat. Mater.*, 2006; **5**:857.
10. Green DJ, Tandon RMSV, Sglavo VM. Crack arrest and multiple cracking in glass through the use of designed residual stress profiles. *Science*, 1999; **283**:1295–1297.
11. Withers PJ. Residual stress and its role in failure. *Rep. Prog. Phys.*, 2007; **70**:2211–2264.

12. James MN. Residual stress influences on structural reliability. *Eng. Fail. Anal.*, 2011; **18**:1909–1920.
13. Chen AY, Li DF, Zhang JB, Song HW, Lu J. Make nanostructured metal exceptionally tough by introducing non-localized fracture behaviors. *Scr. Mater.*, 2008; **59**:579–582.
14. Fang TH, Li WL, Tao NR, Lu K. Revealing extraordinary intrinsic tensile plasticity in gradient nano-grained copper. *Science*, 2011; **331**:1587.
15. Wu XL, Jiang P, Chen L, Yuan FP, Zhu YT. Extraordinary strain hardening by gradient structure. *Proc. Natl. Acad. Sci. USA*, 2014; **111**:7197–7201.
16. Kou H, Lu J, Li Y. High-strength and high-ductility nanostructured and amorphous metallic materials. *Adv. Mater.*, 2014; **26**:5518–5524.
17. Moon JH, Baek SM, Lee SG, Seong Y, Amanov A, Lee S, Kim HS. Effects of residual stress on the mechanical properties of copper processed using ultrasonic-nanocrystalline surface modification. *Mater. Res. Lett.*, 2019; **7**:97–102.
18. Yuan FP, Yan DS, Sun JD, Zhou LL, Zhu YT, Wu XL. Ductility by shear band delocalization in the nano-layer of gradient structure. *Mater. Res. Lett.*, 2019; **7**:12–17.
19. Long J, Pan Q, Tao N, Lu L. Residual stress induced tension-compression asymmetry of gradient nanograined copper. *Mater. Res. Lett.*, 2018; **6**:456–461.
20. Wang YD, Peng RL, Almer J, Odén M, Liu YD, Deng JN, He CS, Chen L, Li QL, Zuo L. Grain-to-grain stress interactions in an electrodeposited iron coating. *Adv. Mater.*, 2005; **17**:1221–1226.
21. Jia N, Cong ZH, Sun X, Cheng S, Nie ZH, Ren Y, Liaw PK, Wang YD. An in situ high-energy X-ray diffraction study of micromechanical behavior of multiple phases in advanced high-strength steels. *Acta Mater.*, 2009; **57**:3965–3977.
22. Wu XL, Jiang P, Chen L, Zhang JF, Yuan FP, Zhu YT. Synergetic strengthening by gradient structure. *Mater. Res. Lett.*, 2014; **2**:185–191.
23. Li J, Soh AK. Modeling of the plastic deformation of nanostructured materials with grain size gradient. *Int. J. Plast.*, 2012; **39**:88–102.
24. Zeng Z, Li XY, Xu DS, Lu L, Gao HJ, Zhu T. Gradient plasticity in gradient nano-grained metals. *Extreme Mech. Lett.*, 2016; **8**:213–219.
25. Yang MX, Pan Y, Yuan FP, Zhu YT, Wu XL. Back stress strengthening and strain hardening in gradient structure. *Mater. Res. Lett.*, 2016; **4**:145–151.

26. Wu XL, Zhu YT. Heterogeneous materials: a new class of materials with unprecedented mechanical properties. *Mater. Res. Lett.*, 2017; **5**:527–532.

27. Ma E, Zhu T. Towards strength–ductility synergy through the design of heterogeneous nanostructures in metals. *Mater. Today*, 2017; **20**:323–331.

28. Huang CX, Wang YF, Ma XL, Yin S, Höppel HW, Göken M, Wu XL, Gao HJ, Zhu YT. Interface affected zone for optimal strength and ductility in heterogeneous laminate. *Mater. Today*, 2018; **21**:713–719.

29. Ashby MF. The deformation of plastically non-homogeneous materials. *Philos. Mag.*, 1970; **21**:399–424.

30. Liu XL, Yuan FP, Zhu YT, Wu XL. Extraordinary Bauschinger effect in gradient structured copper. *Scr. Mater.*, 2018; **150**:57–60.

31. Wu XL, Yang MX, Yuan FP, Wu GL, Wei YJ, Huang XX, Zhu YT. Heterogeneous lamella structure unites ultrafine-grain strength with coarse-grain ductility. *Proc. Natl. Acad. Sci. USA*, 2015; **112**:14501–14505.

Chapter 9

Mechanical Properties of Copper/Bronze Laminates: Role of Boundaries

Xiaolong Ma,[a] Chongxiang Huang,[b] Jordan Moering,[a] Mathis Ruppert,[c] Heinz Werner Höppel,[c] Mathias Göken,[c] Jagdish Narayan,[a] and Yuntian Zhu[a,d]

[a]*Department of Materials Science and Engineering, North Carolina State University, Raleigh, NC 27695, USA*
[b]*School of Aeronautics and Astronautics, Sichuan University, Chengdu 610065, China*
[c]*Department of Materials Science and Engineering, Institute I: General Materials Properties, Friedrich-Alexander Universität Erlangen-Nürnberg, Martensstr. 5, 91058 Erlangen, Germany*
[d]*School of Materials Science and Engineering, Nanjing University of Science and Technology, Nanjing 210094, China*
chxhuang@scu.edu.cn, ytzhu@ncsu.edu

Boundaries play a crucial role in mechanical behaviors of both laminated and gradient structured materials. In this work, copper/bronze laminates with varying boundary spacing were fabricated by accumulative roll bonding and subsequent annealing to

---

Reprinted from *Acta Mater.*, **116**, 43–52, 2016.

Heterostructured Materials: Novel Materials with Unprecedented Mechanical Properties
Edited by Xiaolei Wu and Yuntian Zhu
Text Copyright © 2016 Acta Materialia Inc.
Layout Copyright © 2022 Jenny Stanford Publishing Pte. Ltd.
ISBN 978-981-4877-10-7 (Hardcover), 978-1-003-15307-8 (eBook)
www.jennystanford.com

systematically study the boundary effect on mechanical properties. Heterogeneities exist in chemical composition, grain size, hardness and texture across the boundaries. Simultaneous improvement of strength and ductility with decreasing boundary spacing is found in tensile tests. Extra geometrically necessary dislocations (GNDs) are found to accumulate in the vicinity of boundaries, which is due to mechanical incompatibility across the boundaries. Importantly, a hetero-deformation-induced region (HBAR) spanning a few micrometers was found, which is not affected by boundary spacing. These observations suggest the existence of an optimum spacing, which may produce the highest hetero-deformation-induced (HDI) strain hardening and ductility without sacrificing strength.

9.1 Introduction

Laminate and gradient structured metals have recently attracted extensive interests in the materials community for their potential in achieving outstanding mechanical properties [1–6]. Fang et al. and Wu et al. reported that gradient structures with nanocrystalline surface layers and coarse-grained interior produced a superior combination of strength and ductility [1, 5, 7]. Wu et al. reported that heterogeneous lamella structured pure Ti possessed both the high strength of ultrafine grains and the decent ductility of the coarse grains [8]. Both laminate and gradient structures contain boundaries, across which there are differences in chemical compositions and/or microstructures, such as grain sizes and crystallographic orientations (texture) [9–11]. Boundaries were believed to significantly contribute to the observed high strain hardening and ductility in both laminate and gradient materials [3, 12–15]. Kümmel et al. [16] attributed this to an additional grain refinement caused by an increasing number of boundaries, which, however, is not an intrinsic effect from boundaries. Some researchers proposed that gradient distribution of stress near the boundary enhances working hardening in multi-layered and gradient structured metals [3, 17–19]. Wu and Zhu found that strain gradient and the associated HDI strengthening near the boundaries played a critical role in the high strength and high strain hardening rate [5, 8, 20, 21].

Early literatures have linked microstructural heterogeneity with strain gradient evolution and subsequent generation of geometrically necessary dislocations (GNDs) during plastic deformation [22–25]. Applying this general theory to laminates, Ashby et al. developed a reciprocal relationship between the average GND density and boundary spacing in an idealized laminate structure with a single crystal matrix and equally spaced rigid plate-like particles [22]. However, this theory was based on simple assumptions, which do not represent real complex materials very well. In fact, laminate components are mostly engineered from polycrystalline matrix and none of them are absolutely rigid [26, 27]. Additionally, the details of GND density configuration and their dependence on boundary spacing are not well described in the conventional theory and have been rarely investigated experimentally. These issues are critical to understanding the fundamental mechanisms and imperative to practical material design. For example, to what extent and distance does a boundary exert influence during deformation? Does the width of the hetero-boundary-affected region (HBAR) depend on the boundary spacing/layer thickness? Very few systematic studies have been reported to explore these issues.

It has been a grand challenge to investigate the aforementioned issues through direct experimental observations in gradient structures. First, the microstructural boundaries also act as elastic/plastic and necking/stable boundaries during tensile testing. Thus, rather than being stationary, those boundaries migrate dynamically across the samples during deformation due to the strength/ductility gradient, which makes it difficult to identify or track them experimentally [1, 5, 7]. Second, even for laminate structures with stationary boundaries, it is not trivial to fabricate samples with varying boundary spacing and also with similar microstructure across the boundary in terms of the grain size and texture. For example, the majority of laminate metallic structures fabricated by accumulative roll bonding (ARB) always have finer microstructures with increasing rolling cycles [28, 29]. Furthermore, even atomic structure of a boundary might change with decreasing boundary spacing [4, 30]. To effectively probe the effect of boundaries on mechanical performance, identical or very similar boundaries and interfacial structures with varying boundary spacing are needed. Third, it is technically difficult to determine the deformation characteristics, such as dislocation density and their evolutions, near the boundary using conventional approaches such as transmission electron microscopy (TEM) because of the inhomogeneous nature of dislocation slips [31].

In the present study, copper and bronze (Cu-10wt%Zn) laminates with varying interfacial spacing were fabricated using ARB processing and post-annealing. These samples have maintained almost the same level of microstructural difference across their boundaries. Using ex-situ electron back-scattering scattering diffraction (EBSD) technique, the deformation history of boundary regions under tension were successfully recorded, which revealed how and to what extent do these boundaries affect GND activities [32–34]. Unloading-reloading tension tests were performed to further confirm the role of the boundaries in HDI stress evolution.

9.2 Experimental Methods

Commercial pure copper (ASM-C11000) and bronze (ASM-C22000) were selected for this work. The chemical compositions and general mechanical properties of these raw materials are listed in Table 9.1 [35, 36]. The advantage of these two materials is their similar elastic constants so that we don't have to consider the effect of elastic mismatch. 1mm-thick raw copper and bronze sheets were ARB-processed with 2, 3 and 5 cycles to achieve 4, 8 and 32 layers, respectively. Prior to each ARB cycle, the sample surfaces were cleaned by acetone and then wire brushed in order to remove oxide layer and to ensure a well-defined surface roughness and sufficient bonding strength. Subsequently, the two treated surfaces were stacked with an alternate sequence of copper and bronze, and roll-bonded at room temperature using a four high rolling mill (BW 200, CarlWezel, Mühlacker, Germany) at a nominal thickness reduction of 50% per cycle. The bonded sheets were air cooled and halved before performing the next cycle. The edge regions where sheets tend to tear from each other were cut away and the central part with good initial bonding strength was used for following processing. Details of ARB processing can be found in early works [26, 37]. The as-ARB processed samples were annealed together at 250°C for 2 h in a vacuum tube furnace under argon atmosphere and thereafter labeled as N2, N3 and N5, respectively.

Table 9.1 Chemical compositions and general material properties of raw materials [35, 36]

	Chemical compositions (wt.%)					
	Cu	Pb	Fe	O	Zn	Others
Copper	≥99.90	—	—	0.04	—	≤0.06
Bronze	89.0–91.0	≤0.05	≤0.05	—	10.0	—
	Material parameters					
	a (Å)	γ_{SF} (mJ/m^2)	E (GPa)	υ	G (GPa)	
Copper	3.61	45	115	0.324	44	
Bronze	3.64	35	115	0.307	44	

Samples for Ion Channeling Contrast Microscopy (ICCM), micro-hardness testing and EBSD observation were first cut from the annealed samples and then mechanically polished to achieve a mirror-like surface. Electrochemical polishing was then performed for <30 s to remove the strained top-surface layer that may affect following tests. The electrolyte consisted of a phosphoric acid (concentration of 85%), ethanol and deionized water with a volume ratio of 1:1:2. ICCM and EBSD were conducted under an FEI Quanta 3D FEG dual-beam instrument. Texture analysis was based on EBSD maps with a view area of 30 × 100 μm^2 to capture the global characteristic. For local misorientation mapping, each EBSD scan was performed under 30 kV and 16 nA electron beam and with a bin size of 2 × 2 to achieve a decent angular resolution [38, 39]. Scan step size was set at 100 nm to ensure appropriate spatial resolution.

TEM foils were prepared by mechanically polishing specimen to a thickness of ~30 μm, followed by ion milling to perforation. The milling process was performed at −50°C to avoid potential grain growth. TEM observation was performed in a JEM-2010F microscope operating at 200 kV at room temperature. Dog-bone shaped tensile samples with a gauge dimension of 10 × 2 × 1 mm³ were machined from the annealed sheets and tested under uniaxial tension on a Shimadzu AGS machine. Both normal and unloading-reloading tension tests were carried out at room temperature at a strain rate of $9 \times 10^{-4}\,\text{s}^{-1}$ and each test was repeated for at least 3 samples to ensure data reproducibility. Ex-situ EBSD mapping in the interfacial region was carried out on the same specimen at three strain levels: 0%, 3% and uniform elongation strain. The interfacial region of interest was carefully marked by a milling feature.

9.3 Results

9.3.1 Microstructures

Figures 9.1a–c are the optical microscopy images of all sample, which show clearly well defined laminate structure with uniform layer thickness and varying boundary spacing. The red color indicates the copper layers. The layer thickness for N2, N3 and N5 samples are 250, 125 and 31 μm, respectively. ICCM micrographs (Figs. 9.1d–f) reveal similar microstructures in all samples subjected to different ARB cycles. Coarse copper grains with considerable annealing twins are observed in all N2, N3 and N5 samples. The annealing twins were formed by recrystallization during annealing, which is further confirmed by texture analysis later. In contrast, the bronze layer maintained largely the rolling structure in all samples due to its higher thermal stability. Figure 9.1 shows slightly larger grain size in the bronze layer of the N2 sample than in the N3 and N5 samples. This is probably resulted from the relatively low rolling strain in the N2 sample. However, as shown later, the differences in their hardness and texture are not significant. Small quantities of ultrafine and nanocrystalline bronze grains were locally dispersed in the vicinity of the boundaries, which is observed in all samples under ICCM.

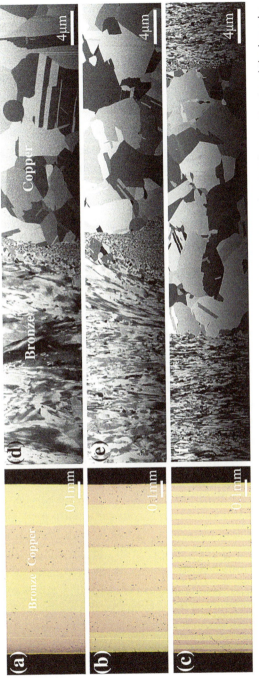

Figure 9.1 Optical microscopy of (a) N2, (b) N3, (c) N5 samples, respectively. It clearly shows the inter-layer contrast and the boundary spacing. Ion channeling contrast microscopy images of microstructures at copper/bronze boundaries in (d) N2, (e) N3 and (f) N5 samples.

9.3.2 Heterogeneity Across Boundaries

A representative copper/bronze boundary in the N5 sample is shown in Fig. 9.2a, which reveals a void-free transition from coarse-grained copper to nanostructured bronze across the boundary. Clearly, the recrystallized coarse-grained Cu grains are much larger and exhibit less dislocation contrast than the bronze layer. The latter still exhibit a deformed microstructure. Figure 9.2b shows more microstructural details in the bronze layer, which reveals elongated grains along the rolling direction as well as dislocation cells and areas with high dislocation density. The statistical distributions of (transverse) grain sizes in the copper and bronze layers are shown in Fig. 9.2c and 9.2d, respectively. As shown, the bronze layer has an average feature size of ~100 nm, while the copper layer has an average grain size of ~5 µm. Such grain size and composition difference are expected to produce a significant mechanical incompatibility across the boundary.

The microstructural and compositional difference across the boundary led to sharp difference in hardness. As shown in Fig. 9.3a, the nanostructured bronze layer has a hardness of ~1500 MPa, which is more than twice that in microcrystalline copper layer (~700 MPa). It is noted that the micro-hardness of both copper and bronze layers are very close in all samples, given the measurement errors, despite varying boundary spacing. Crystallographic orientation (texture) was also found to differ remarkably across the boundaries. As shown in Fig. 9.3b, copper exhibits a strong cube texture component {001}<100>, which is resulted from both recrystallization and grain growth [40]. In contrast, bronze shows {110}<112> texture with slight variation [40, 41], which is consistent with previous reports of brass-type rolling texture in FCC alloys with low stacking-fault energy [42]. Post-annealing after ARB did not change the rolling texture of the bronze layer much, which is consistent with the TEM observation of rolling microstructure in Fig. 9.2. Note that the present pole figures are scanned from the N3 sample, but N2 and N5 exhibit similar textures in the copper and bronze layers.

The above observations indicate that the ARB processing and appropriate annealing produced boundaries with varying spacing but similar microstructure, micro-hardness and texture. This allows us to study the boundary effect on the mechanical properties with complications from other structural factors.

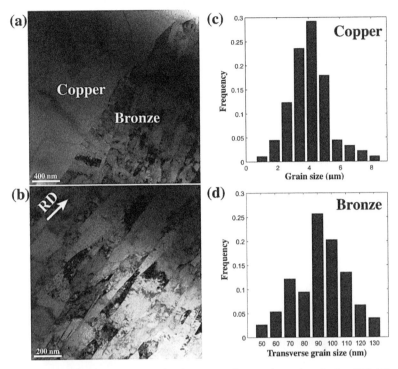

Figure 9.2 (a) TEM micrograph of a copper/bronze boundary in the ARB N5 sample shows void-free boundary and big grain size difference across it. (b) Retained rolling microstructure in the bronze layer. The arrow indicates the rolling direction. (c) Grain size distribution in the copper layers based on ICCM images. (d) Transverse grain size distribution in bronze layers based on TEM micrographs.

9.3.3 Uniaxial Tensile Tests

Figure 9.4a shows the tensile stress-strain curves for N2, N3 and N5 samples. Inset is a photograph of a sample. As shown, both the strength and uniform tensile elongation (ductility) increased with decreasing boundary spacing. Interestingly, the yield strength did not increase as much as the ultimate strength, indicating that decreasing the boundary spacing is more effective in enhancing the strain-hardening rate than yield strength. Figure 9.4b summarizes the variation of ultimate strength and ductility, which confirms their reproducibility. It's worth noting that some earlier ARB studies also reported similar simultaneous improvements in strength and ductility with increasing rolling cycles, but most of these studies attributed this phenomenon to microstructure change instead of the boundary spacing [29, 43, 44].

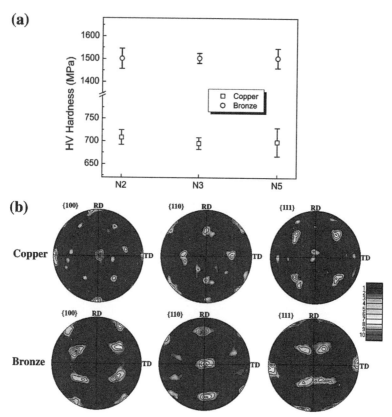

Figure 9.3 (a) Micro-hardness in both the copper and bronze layers in N2, N3, N5 samples after annealing. (b) {100}, {110}, {111} Pole figures in both copper and bronze layers in N3 sample after annealing, showing a strong cube texture and a brass-type texture with slight deviation, respectively. N2 and N5 have similar texture characteristics.

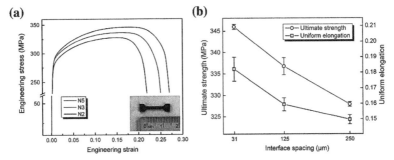

Figure 9.4 (a) Tensile curves of N2, N3 and N5 samples show simultaneous increase of strength and elongation with increasing number of layers. (b) Summary of tensile results. Error bars represent standard deviation from at least 3 data sets.

9.3.4 Ex-situ EBSD Mapping and GND Characterization

Figure 9.5 shows EBSD inverse pole figure mappings of the copper layers along the transverse direction, at zero strain, 3% tensile strain and uniform elongation strain for each sample. Arrows indicate boundaries. Generally, the indexing rate is more than 97% even for copper grains near boundaries, making following interfacial analysis reliable. For the N5 sample, the whole copper layer is captured because it is thin. For each sample, the local crystal orientation in a region near a boundary was measured at the above selected tensile strains to study the GND evolution. In this study, we used the kernel average misorientation (KAM) method to determine the local misorientation from the EBSD orientation data [34]. First, we defined the limit of the general grain boundary misorientation as 3°. Any misorientation greater than this value was excluded in local misorientation calculation since it is caused by a grain boundary, not by GND accumulation. The local misorientation of every single point (100×100 nm^2) was then determined by the 24 surrounding points:

$$\theta_0 = \sum_{i=1}^{24} \theta_i \cdot I_{(\theta_i < \alpha)} \bigg/ \sum_{i=1}^{24} I_{(\theta_i < \alpha)} \tag{9.1}$$

where θ_0 represents the resulted local misorientation for the corresponding point and θ_i is the misorientation between this point and its neighbor point i. i is an indicator function and α is the predefined grain boundary misorientation threshold (3° here). To extrapolate the GND density information, we use a simple method from the strain gradient theory by Gao and Kubin [25, 45]:

$$\rho^{GND} = \frac{2\theta}{ub} \quad (9.2)$$

where ρ^{GND} is the GND density at points of interest, θ represents the local misorientation, b is the Burger's vector (0.255 nm for copper) and u is the unit length (100 nm) of the point. The resulted GND density maps are shown in Fig. 9.6 for all samples under corresponding tensile strains. Clearly, the overall level of GND density is elevated with increasing tensile strain. This is expected from the deformation of polycrystalline materials, which is inhomogeneous [22, 31, 32]. It is also noted that the GND density is not uniform across the whole mapping region, which is not considered in the conventional theory [22, 24]. Figure 9.7 shows the histogram distribution of GND density for each map. The GND distributions in the three samples are very similar at each tensile strain. Here, we need to mention the measurement error in the study. Previous researches on EBSD technique indicated that the measurement error might dominate when the misorientation to measure is very small and the relative error decreases when increasing real misorientation [39, 46]. Taking the extreme case by assuming that the misorientation under zero strain is caused by measurement error, the resulted upper limit of measurement error of GND density is less than 1.73×10^{14} m^{-2}, which is reasonable compared to an early report [47].

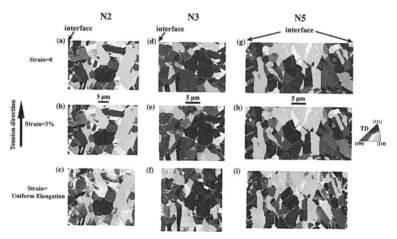

Figure 9.5 Inverse pole figure mapping of the regions around boundaries in (a–c) N2, (d–f) N3 and (g–i) N5 samples under different tensile strain levels: 0%, 3% and maximum uniform elongation.

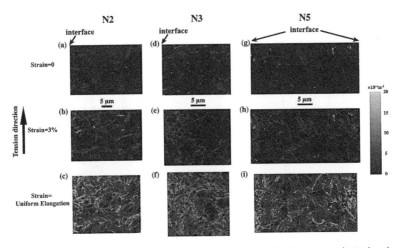

Figure 9.6 GND density mapping based on local misorientation results in (a–c) N2, (d–f) N3 and (g–i) N5 samples under different tensile strains: 0%, 3% and maximum uniform elongation.

Figure 9.7 also shows that the variation of GND density increases with tensile strain. This is definitely a true deformation phenomenon, rather than measurement error, because the latter has an inverse trend [46]. Since the GND density is related to the strain gradient [22, 25], these results indicate that the strain gradient is not uniform and this becomes more severe with increasing plastic strain. One source for the non-uniform strain gradient and GND density is the randomly distributed polycrystalline grain boundaries. They serve as barriers to dislocation motion and therefore locally generate GNDs [32, 34]. The GND density is also affected by the crystallographic orientation, which varies from grain to grain. Another important source would be the macroscopic boundaries, which is the central interest in this study. To extract the effect of boundaries, the mapped regions in Fig. 9.6 were sliced parallel to the boundary and the average GND density in each slice was sampled using the bootstrap method [48] to minimize the interference from grain boundaries, crystallographic orientation and other irrelevant factors. The resulted GND densities versus their equivalent distance from boundaries are shown in Fig. 9.8, which reveals the development of a gradient in GND density near the boundary. In other words, there is a hetero-boundary-affected region (HBAR) where the GNDs were accumulated during the tensile deformation. This zone spans only a few micrometers,

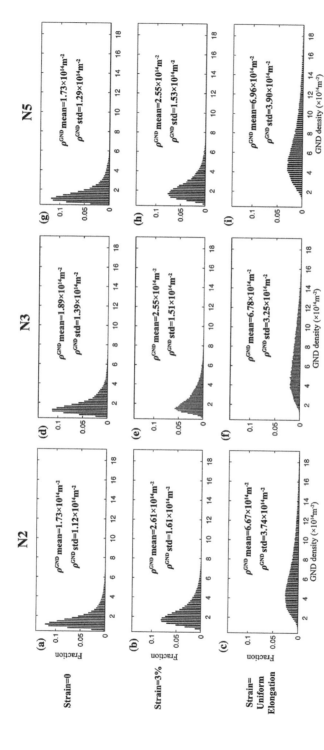

Figure 9.7 Global GND density distribution on corresponding mapping results in Fig. 9.6. The mean value of GND density and the standard deviation are labeled in each histogram.

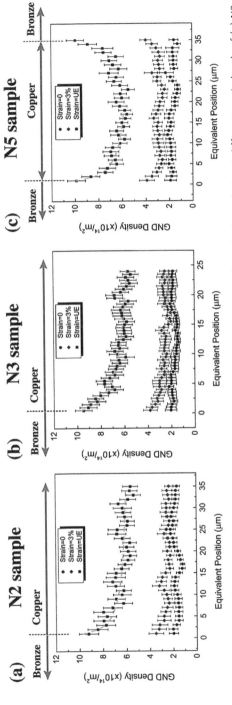

Figure 9.8 Averaged GND density in copper layer versus the equivalent distance from the boundaries at different strain levels of (a) N2 sample, (b) N3 sample, (c) N5 sample. Dashed lines represent pile-up model fitting results.

and is observed in all three samples and becomes more pronounced with increasing applied tensile strain. It is also noted that the GND density in the layer interior becomes rather lower and uniform across all samples, i.e. the boundary effect declined quickly with increasing distance from the boundary.

9.4 Discussions

9.4.1 Dislocation Pile-Up Model for the GND Density Close to Boundaries

The heterogeneity in grain size, strength, stacking fault energy and texture between the copper and bronze layers makes them mechanically incompatible during deformation. However, the two different layers are forced to deform together, which generates strain gradient near the boundary to fit the different strains across the boundary. The strain gradient needs to be accommodated by GND, in the form of dislocation pile-up near the boundary [49–51]. GND distributions were derived based on the conventional pile-up model [50], where the GND density $\rho^{GND}(x)$ is inversely proportional to $k/\sqrt{x(a-x)}$. Here, x is the distance from the boundary, k and a are constants [50]. Defining the averaged GNDs arose from the grain boundary and orientation effects as ρ_0^{GND}, the fitting function can be modified as

$$\rho^{GND}(x) = \frac{k}{\sqrt{x(a-x)}} + \rho_0^{GND} \qquad (9.3)$$

Data sets from the uniform elongation strain level are used to fit the model because they are least affected by measurement noise. The fitting results of GND density are superimposed on the measured data in Fig. 9.8, which demonstrate that the dislocation pile-up model can well describe the GND accumulation at boundaries in laminate structures. The fitting results also reveal that the ρ_0^{GND} values are very close for three samples with different boundary spacing: ρ_0^{GND} = 5.1, 4.8 and 4.9 × 10^{14} m^{-2} in N2, N3 and N5 samples, respectively. Figure 9.8 also shows that GND density close to boundary is ~10 × 10^{14} m^{-2}, indicating that boundaries doubled GND density locally.

Consistent with fitted ρ_0^{GND} in Fig. 9.8, Fig. 9.9 shows that the microhardness values in the copper and bronze layers after tensile testing to uniform elongation are almost the same in three samples. This indicates that the hardening capacities of the layer interior are approximately the same regardless of different boundary density in samples. Therefore, the improved mechanical properties in Fig. 9.4 are largely caused by the boundary effect.

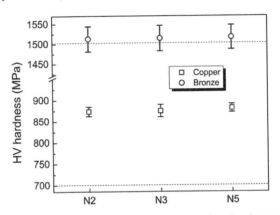

Figure 9.9 Micro-hardness in layer interior at uniformly elongated sample region after tension. Dashed lines represent the base level before tensile tests.

9.4.2 Role of Boundary in Deformation of Nanostructured Bronze

The average transverse grain size in the bronze layers is about 100 nm, which makes it difficult to characterize the deformation history by the current EBSD technique due to its resolution limit. But the extra accumulation of GNDs is still expected to occur in the bronze side as well. Although the hardening in the bronze layer interior during tension is not as high as in the Cu layer (see Fig. 9.9), extra hardening by GND accumulation could be generated in the nanostructured layer close to boundaries, which has been observed in gradient structures [5, 14].

In addition, boundaries are also expected to facilitate other deformation mechanisms in the nanostructured bronze, such as deformation twinning [52], to promote strain hardening and ductility.

Such mechanisms could be activated in grains near the boundary for the following reasons. First, the aforementioned dislocation pile-up will produce a shear stress field near the boundary and this stress field is expected higher than that from ordinary pile-ups at conventional grain boundaries because the plastic incompatibility across such boundary is higher [49, 50]. Second, bi-metal boundaries could act as dislocation sources and sinks to facilitate plastic deformation [30]. Figure 9.10a shows a TEM micrograph of the copper/bronze boundary after tensile deformation. Figure 9.10b is the corresponding high-resolution micrograph. On the right side of the boundary is a large copper grain with an annealing twin and the left is a bronze grain with only 30 nm grain size. A deformation twin in the bronze grain is probably nucleated at the boundary by the transmission of the Shockley partial from the annealing Cu twin [30, 53].

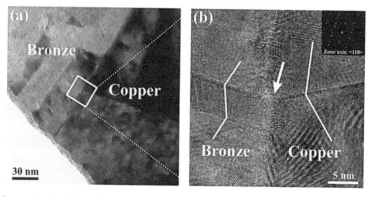

Figure 9.10 (a) A TEM micrograph of copper/bronze boundary in the N5 sample after tensile test. (b) The high-resolution image of the marked square in (a). Inset is a diffraction pattern, which shows the zone axis of copper side is <110>. The symmetric diffraction spots reveal the annealing twin in the copper grain, which is also marked in the image. A deformation twin in bronze side is highlighted as well. Nucleation site for the deformation twin is marked by an arrow.

9.4.3 Effect of Boundary Spacing on HDI Hardening

Dislocation pile-up will produce HDI stress, which is a long-range stress field that prohibits further dislocation emission from the dislocation source [50, 54]. In other words, higher plastic flow stress is needed to overcome this field to sustain further deformation. This mechanism is highlighted in a recent work on Ti lamella microstructures [8]. The macroscopic boundaries here are expected to have similar effect. We calculated the HDI stress from unloading-reloading test curves at different tensile strains (see Fig. 9.11a) using an equation recently proposed by Yang et al. [21]. As shown in Fig. 9.11b, the resulted HDI stress is higher in samples with smaller boundary spacing (higher boundary density). This can be rationalized with the help of Fig. 9.8, which reveals that each boundary produces a HBAR by considerable GND pile-ups. Consequently, higher boundary density amounts to more GND pile-ups in a certain sample volume, leading to a higher observed HDI stress at the present spacing scale.

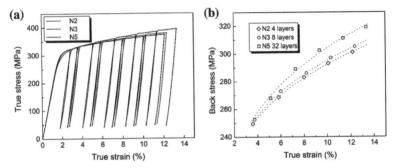

Figure 9.11 (a) Unloading-reloading tensile curves of N2, N3, N5 samples. (b) The calculated HDI stress at corresponding strain levels.

However, further enhancement of HDI work hardening and ductility may be limited when the boundary spacing is smaller than a critical value since the transition of deformation mechanisms may occur [55]. This hypothesis is supported by reports that nanolaminates usually exhibit very high strength but disappointing tensile ductility [56–58]. In other words, there should be an optimum boundary spacing that generates the extra GND pile-ups across the whole layer and yields the highest HDI work hardening and ductility while maintaining high strength. We believe that this optimum spacing is comparable to the observed width of HBAR, i.e. a few micrometers, which is consistent with recent report of aluminum alloy laminate composites [16].

It should be also noted that further reducing boundary spacing via ARB method will probably give rise to the preferred crystallographic orientations (textures) at bi-metal boundaries [2, 4, 59]. Such local and stable texture affects deformation and mechanical properties in its own way. When each layer is much thicker than the grain size dimension, its influence is trivial and the general boundary effect governs the case [60], like in this work. The effect of interfacial texture is expected to interact with boundary effect, which needs to be further studied.

9.5 Conclusion

In summary, we have systematically studied the effect of the copper/bronze boundary on the deformation behavior and mechanical properties of laminate structures. The main conclusions are:

(1) The copper/bronze laminates were produced by means of ARB/annealing with varying boundary spacing but similar heterogeneity across the boundaries including chemical composition, grain size, micro-hardness and texture.

(2) Both ultimate strength and ductility in uniaxial tension are improved with decreasing boundary spacing, which can be primarily attributed to the effect of boundaries since other variables are kept almost the same.

(3) A HBAR during deformation was experimentally observed. It spans a few micrometers regardless of varying boundary spacing. Within this zone, non-uniform strain gradient and GND accumulation were generated via dislocation pile-up. In contrast, hardening capacity in the layer interior is about the same across samples.

(4) The boundary affects adjacent layers during deformation and promotes HDI hardening by producing more GND pile-ups. It's our hypothesis that there exists an optimum boundary spacing, comparable to the width of HBAR, for the highest ductility without sacrifice of strength.

Acknowledgements

This work is supported by the US Army Research Office (W911 NF-12-1-0009), the US National Science Foundation (DMT-1104667) as well as National Natural Science Foundation of China (No. 1117218). Y.T. Zhu also acknowledges the support of Jiangsu Key Laboratory of Advanced Micro & Nano Materials and Technology, Nanjing University of Science and Technology. The authors acknowledge the use of the Analytical Instrumentation Facility (AIF) at North Carolina State University, which is supported by the State of North Carolina and the National Science Foundation.

References

1. T.H. Fang, W.L. Li, N.R. Tao, K. Lu, Revealing extraordinary intrinsic tensile plasticity in gradient nano-grained copper, *Science*, **331** (2011) 1587–1590, doi:10.1126/science.1200177.

2. S. Zheng, I.J. Beyerlein, J.S. Carpenter, K. Kang, J. Wang, W. Han, N.A. Mara, High-strength and thermally stable bulk nanolayered composites due to twin-induced interfaces, *Nat. Commun.*, **4** (2013) 1696, doi:10.1038/ncomms2651.

3. D.K. Yang, P. Cizek, D. Fabijanic, J.T. Wang, P.D. Hodgson, Work hardening in ultrafine-grained titanium: multilayering and grading, *Acta Mater.*, **61** (2013) 2840–2852, doi:http://dx.doi.org/10.1016/j.actamat.2013.01.018.

4. I.J. Beyerlein, J.R. Mayeur, S. Zheng, N.A. Mara, J. Wang, A. Misra, Emergence of stable interfaces under extreme plastic deformation, *Proc. Natl. Acad. Sci.*, **111** (2014) 4386–4390, doi:10.1073/pnas.1319436111.

5. X. Wu, P. Jiang, L. Chen, F. Yuan, Y.T. Zhu, Extraordinary strain hardening by gradient structure, *Proc. Natl. Acad. Sci.*, **111** (2014) 7197–7201, doi:10.1073/pnas.1324069111.

6. J. Lu, H.L. Chan, A.Y. Chen, H.N. Kou, Mechanics of high strength and high ductility materials, *Procedia Eng.*, **10** (2011) 2202–2207, doi:http://dx.doi.org/10.1016/j.proeng.2011.04.364.

7. X.L. Wu, P. Jiang, L. Chen, J.F. Zhang, F.P. Yuan, Y.T. Zhu, Synergetic strengthening by gradient structure, *Mater. Res. Lett.*, (2014) 1–7, doi:10.1080/21663831.2014.935821.

8. X. Wu, M. Yang, F. Yuan, G. Wu, Y. Wei, X. Huang, Y. Zhu, Heterogeneous lamella structure unites ultrafine-grain strength with coarse-grain ductility, *Proc. Natl. Acad. Sci.*, (2015) 201517193, doi:10.1073/pnas.1517193112.

9. S.-B. Lee, J.E. LeDonne, S.C.V. Lim, I.J. Beyerlein, A.D. Rollett, The heterophase interface character distribution of physical vapor-deposited and accumulative roll-bonded Cu–Nb multilayer composites, *Acta Mater.*, **60** (2012) 1747–1761, doi:10.1016/j.actamat.2011.12.007.

10. X.C. Liu, H.W. Zhang, K. Lu, Formation of nano-laminated structure in nickel by means of surface mechanical grinding treatment, *Acta Mater.*, **96** (2015) 24–36, doi:10.1016/j.actamat.2015.06.014.

11. Y. Mishin, M. Asta, J. Li, Atomistic modeling of interfaces and their impact on microstructure and properties, *Acta Mater.*, **58** (2010) 1117–1151, doi:10.1016/j.actamat.2009.10.049.

12. S. Suresh, Graded materials for resistance to contact deformation and damage, *Science*, **292** (2001) 2447–2451, doi:10.1126/science.1059716.

13. L. Jianjun, A. K. Soh, Enhanced ductility of surface nano-crystallized materials by modulating grain size gradient, *Modell. Simul. Mater. Sci. Eng.*, **20** (2012) 085002.

14. J. Li, S. Chen, X. Wu, A.K. Soh, A physical model revealing strong strain hardening in nano-grained metals induced by grain size gradient structure, *Mater. Sci. Eng., A*, **620** (2015) 16–21, doi:10.1016/j.msea.2014.09.117.

15. K. Lu, Making strong nanomaterials ductile with gradients, *Science*, **345** (2014) 1455–1456, doi:10.1126/science.1255940.

16. F. Kümmel, M. Kreuz, T. Hausöl, H. Höppel, M. Göken, Microstructure and mechanical properties of accumulative roll-bonded AA1050A/AA5005 laminated metal composites, *Metals*, **6** (2016) 56, doi:10.3390/met6030056.

17. H.F. Tan, B. Zhang, X.M. Luo, X.D. Sun, G.P. Zhang, Strain rate dependent tensile plasticity of ultrafine-grained Cu/Ni laminated composites, *Mater. Sci. Eng., A*, **609** (2014) 318–322, doi:10.1016/j.msea.2014.04.111.

18. S.S. Chakravarthy, W.A. Curtin, Stress-gradient plasticity, *Proc. Natl. Acad. Sci.*, **108** (2011) 15716–15720, doi:10.1073/pnas.1107035108.

19. Z. Zeng, X. Li, D. Xu, L. Lu, H. Gao, T. Zhu, Gradient plasticity in gradient nano-grained metals, *Extreme Mech. Lett.*, (2015), doi:10.1016/j.eml.2015.12.005.

20. X.L. Ma, C.X. Huang, W.Z. Xu, H. Zhou, X.L. Wu, Y.T. Zhu, Strain hardening and ductility in a coarse-grain/nanostructure laminate material, *Scr. Mater.*, **103** (2015) 57–60, doi:10.1016/j.scriptamat.2015.03.006.

21. M. Yang, Y. Pan, F. Yuan, Y. Zhu, X. Wu, Back stress strengthening and strain hardening in gradient structure, *Mater. Res. Lett.*, (2016) 1–7, doi:10.1080/21663831.2016.1153004.

22. M.F. Ashby, The deformation of plastically non-homogeneous materials, *Philos. Mag.*, **21** (1970) 399–424, doi:10.1080/14786437008238426.

23. N.A. Fleck, G.M. Muller, M.F. Ashby, J.W. Hutchinson, Strain gradient plasticity: theory and experiment, *Acta Metall. Mater.*, **42** (1994) 475–487, doi:10.1016/0956-7151(94)90502-9.

24. W.D. Nix, H. Gao, Indentation size effects in crystalline materials: A law for strain gradient plasticity, *J. Mech. Phys. Solids*, **46** (1998) 411–425, doi:10.1016/S0022-5096(97)00086-0.

25. H. Gao, Mechanism-based strain gradient plasticity I. Theory, *J. Mech. Phys. Solids*, **47** (1999) 1239–1263, doi:10.1016/S0022-5096(98)00103-3.

26. H.P. Ng, T. Przybilla, C. Schmidt, R. Lapovok, D. Orlov, H.-W. Höppel, M. Göken, Asymmetric accumulative roll bonding of aluminium-titanium composite sheets, *Mater. Sci. Eng., A*, **576** (2013) 306–315, doi:10.1016/j.msea.2013.04.027.

27. K. Tanaka, K. Shibata, K. Kurumatani, S. Ikeuchi, S. Kikuchi, R. Kondo, H.T. Takeshita, Formation mechanism of micro/nano-structures through competitive reactions in Mg/Cu super-laminate composites during initial hydrogenation, *J. Alloys Compd.*, **645** (2015) S72–S75, doi:10.1016/j.jallcom.2015.01.196.

28. M. Eizadjou, A. Kazemitalachi, H. Daneshmanesh, H. Shakurshahabi, K. Janghorban, Investigation of structure and mechanical properties of multi-layered Al/Cu composite produced by accumulative roll bonding (ARB) process, *Compos. Sci. Technol.*, **68** (2008) 2003–2009, doi:10.1016/j.compscitech.2008.02.029.

29. M. Ruppert, C. Schunk, D. Hausmann, H.W. Höppel, M. Göken, Global and local strain rate sensitivity of bimodal Al-laminates produced by accumulative roll bonding, *Acta Mater.*, **103** (2016) 643–650, doi:10.1016/j.actamat.2015.11.009.

30. N.A. Mara, I.J. Beyerlein, Review: effect of bimetal interface structure on the mechanical behavior of Cu–Nb fcc–bcc nanolayered composites, *J. Mater. Sci.*, **49** (2014) 6497–6516, doi:10.1007/s10853-014-8342-9.

31. B. Bay, N. Hansen, D.A. Hughes, D. Kuhlmann-Wilsdorf, Overview no. 96 evolution of f.c.c. deformation structures in polyslip, *Acta Metall. Mater.*, **40** (1992) 205–219, doi:10.1016/0956-7151(92)90296-Q.

32. J. Jiang, T.B. Britton, A.J. Wilkinson, Evolution of dislocation density distributions in copper during tensile deformation, *Acta Mater.*, **61** (2013) 7227–7239, doi:10.1016/j.actamat.2013.08.027.

33. P.D. Littlewood, A.J. Wilkinson, Geometrically necessary dislocation density distributions in cyclically deformed Ti–6Al–4V, *Acta Mater.*, **60** (2012) 5516–5525, doi:10.1016/j.actamat.2012.07.003.

34. M. Calcagnotto, D. Ponge, E. Demir, D. Raabe, Orientation gradients and geometrically necessary dislocations in ultrafine grained dual-phase steels studied by 2D and 3D EBSD, *Mater. Sci. Eng., A*, **527** (2010) 2738–2746, doi:10.1016/j.msea.2010.01.004.

35. P.C.J. Gallaghe, Influence of alloying, temperature, and related effects on stacking fault energy, *Metall. Trans.*, **1** (1970) 2429.

36. ASM International, J.R. Davis, eds., *Properties and Selection: Nonferrous Alloys and Special-Purpose Materials*, 10th ed., 6th print (ASM International, Materials Park, Ohio, 2000).

37. C.W. Schmidt, M. Ruppert, H.W. Höppel, F. Nachtrab, A. Dietrich, R. Hanke, M. Göken, Design of graded materials by particle reinforcement during accumulative roll bonding: design of graded materials by particle reinforcement during ARB, *Adv. Eng. Mater.*, **14** (2012) 1009–1017, doi:10.1002/adem.201200046.

38. J. Jiang, T.B. Britton, A.J. Wilkinson, Measurement of geometrically necessary dislocation density with high resolution electron backscatter diffraction: effects of detector binning and step size, *Ultramicroscopy*, **125** (2013) 1–9, doi:10.1016/j.ultramic.2012.11.003.

39. I. Brough, P.S. Bate, F.J. Humphreys, Optimising the angular resolution of EBSD, *Mater. Sci. Technol.*, **22** (2006) 1279–1286, doi:10.1179/174328406X130902.

40. T. Leffers, R.K. Ray, The brass-type texture and its deviation from the copper-type texture, *Prog. Mater. Sci.*, **54** (2009) 351–396, doi:10.1016/j.pmatsci.2008.09.002.

41. J. Hirsch, K. Lücke, Overview no. 76, *Acta Metall.*, **36** (1988) 2863–2882, doi:10.1016/0001-6160(88)90172-1.

42. C. Donadille, R. Valle, P. Dervin, R. Penelle, Development of texture and microstructure during cold-rolling and annealing of F.C.C. alloys: example of an austenitic stainless steel, *Acta Metall.*, **37** (1989) 1547–1571, doi:10.1016/0001-6160(89)90123-5.

43. A. Mozaffari, H. Danesh Manesh, K. Janghorban, Evaluation of mechanical properties and structure of multilayered Al/Ni composites produced by accumulative roll bonding (ARB) process, *J. Alloys Compd.*, **489** (2010) 103–109, doi:10.1016/j.jallcom.2009.09.022.

44. H.-W. Kim, S.-B. Kang, N. Tsuji, Y. Minamino, Elongation increase in ultra-fine grained Al–Fe–Si alloy sheets, *Acta Mater.*, **53** (2005) 1737–1749, doi:10.1016/j.actamat.2004.12.022.

45. L.P. Kubin, A. Mortensen, Geometrically necessary dislocations and strain-gradient plasticity: a few critical issues, *Scr. Mater.*, **48** (2003) 119–125, doi:10.1016/S1359-6462(02)00335-4.

46. A.J. Wilkinson, A new method for determining small misorientations from electron back scatter diffraction patterns, *Scr. Mater.*, **44** (2001) 2379–2385, doi:10.1016/S1359-6462(01)00943-5.

47. A.J. Wilkinson, D. Randman, Determination of elastic strain fields and geometrically necessary dislocation distributions near nanoindents using electron back scatter diffraction, *Philos. Mag.*, **90** (2010) 1159–1177, doi:10.1080/14786430903304145.

48. B. Efron, R. Tibshirani, *An Introduction to the Bootstrap* (Chapman & Hall, New York, 1993).

49. D. Hull, D.J. Bacon, *Introduction to Dislocations*, 5th ed. (Elsevier/Butterworth-Heinemann, Amsterdam, 2011).

50. J.P. Hirth, J. Lothe, *Theory of Dislocations* (Wiley, 1982).

51. J.D. Eshelby, F.C. Frank, F.R.N. Nabarro, XLI. The equilibrium of linear arrays of dislocations, *Lond. Edinb. Dublin Philos. Mag. J. Sci.*, **42** (1951) 351–364, doi:10.1080/14786445108561060.

52. Y.T. Zhu, X.Z. Liao, X.L. Wu, Deformation twinning in nanocrystalline materials, *Prog. Mater. Sci.*, **57** (2012) 1–62, doi:10.1016/j.pmatsci.2011.05.001.

53. W.Z. Han, A. Misra, N.A. Mara, T.C. Germann, J.K. Baldwin, T. Shimada, S.N. Luo, Role of interfaces in shock-induced plasticity in Cu/Nb nanolaminates, *Philos. Mag.*, **91** (2011) 4172–4185, doi:10.1080/14786435.2011.603706.

54. X. Feaugas, On the origin of the tensile flow stress in the stainless steel AISI 316L at 300 K: back stress and effective stress, *Acta Mater.*, **47** (1999) 3617–3632, doi:10.1016/S1359-6454(99)00222-0.

55. A. Misra, J.P. Hirth, R.G. Hoagland, Length-scale-dependent deformation mechanisms in incoherent metallic multilayered composites, *Acta Mater.*, **53** (2005) 4817–4824, doi:10.1016/j.actamat.2005.06.025.

56. N.A. Mara, D. Bhattacharyya, R.G. Hoagland, A. Misra, Tensile behavior of 40nm Cu/Nb nanoscale multilayers, *Scr. Mater.*, **58** (2008) 874–877, doi:10.1016/j.scriptamat.2008.01.005.

57. T.G. Nieh, T.W. Barbee, J. Wadsworth, Tensile properties of a freestanding Cu/Zr nanolaminate (or compositionally-modulated thin film), *Scr. Mater.*, **41** (1999) 929–935, doi:10.1016/S1359-6462(99)00240-7.

58. H.S. Shahabi, M. Eizadjou, H.D. Manesh, Evolution of mechanical properties in SPD processed Cu/Nb nano-layered composites, *Mater. Sci. Eng., A*, **527** (2010) 5790–5795, doi:10.1016/j.msea.2010.05.087.

59. N.A. Mara, I.J. Beyerlein, Interface-dominant multilayers fabricated by severe plastic deformation: Stability under extreme conditions, *Curr. Opin. Solid State Mater. Sci.*, **19** (2015) 265–276, doi:10.1016/j.cossms.2015.04.002.

60. J.S. Carpenter, R.J. McCabe, J.R. Mayeur, N.A. Mara, I.J. Beyerlein, Interface-driven plasticity: the presence of an interface affected zone in metallic lamellar composites: the presence of an interface affected zone in metallic lamellar composites, *Adv. Eng. Mater.*, **17** (2015) 109–114, doi:10.1002/adem.201400210.

Chapter 10

Hetero-Boundary-Affected Region (HBAR) for Optimal Strength and Ductility in Heterostructured Laminate

C. X. Huang,[a,†] Y. F. Wang,[a,†] X. L. Ma,[b] S. Yin,[c] H. W. Höppel,[d] M. Göken,[d] X. L. Wu,[e] H. J. Gao,[c] and Y. T. Zhu[b,f]

[a]*School of Aeronautics and Astronautics, Sichuan University, Chengdu, 610065, China*
[b]*Department of Materials Science and Engineering, North Carolina State University, Raleigh, NC 27695, USA*
[c]*School of Engineering, Brown University, Providence, RI 02912, USA*
[d]*Department of Materials Science and Engineering, Institute I: General Materials Properties, Friedrich-Alexander University of Erlangen-Nürnberg, Martensstr. 5, 91058 Erlangen, Germany*
[e]*State Key Laboratory of Nonlinear Mechanics, Institute of Mechanics, Chinese Academy of Sciences, Beijing 100190, China*
[f]*Jiangsu Key Laboratory of Advanced Micro & Nano Materials and Technology, Nanjing University of Science and Technology, Nanjing 210094, China*
chxhuang@scu.edu.cn, xlwu@imech.ac.cn, huajian_gao@brown.edu, ytzhu@ncsu.edu

---

[†]These two authors contributed equally to this work.

Reprinted from *Mater. Today*, **21**(7), 713–719, 2018.

Heterostructured Materials: Novel Materials with Unprecedented Mechanical Properties
Edited by Xiaolei Wu and Yuntian Zhu
Text Copyright © 2018 Elsevier Ltd.
Layout Copyright © 2022 Jenny Stanford Publishing Pte. Ltd.
ISBN 978-981-4877-10-7 (Hardcover), 978-1-003-15307-8 (eBook)
www.jennystanford.com

Boundaries have been reported to significantly strengthen and toughen metallic materials. However, there has been a long-standing question on whether hetero-boundary-affected region (HBAR) exists, and how it might behave. Here we report in-situ high-resolution strain mapping near boundaries in a copper–bronze heterostructured laminate, which revealed the existence of HBARs. Defined as the region with strain gradient, the HBAR was found to form by the dislocations emitted from the boundary. The HBAR width remained largely constant with a magnitude of a few micrometers with increasing applied strain. Boundaries produced both hetero-deformation-induced (HDI) strengthening and HDI work hardening, which led to both higher strength and higher ductility with decreasing boundary spacing until adjacent HBARs started to overlap, after which a tradeoff between strength and ductility occurred, indicating the existence of an optimum boundary spacing for the best mechanical properties. These findings are expected to help with designing laminates and other heterostructured metals and alloys for superior mechanical properties.

10.1 Introduction

The mechanical properties of materials are largely controlled by their internal boundaries, including grain boundaries, phase boundaries, twin boundaries, lamella boundaries, etc. [1, 2]. By trial and error, mankind has learned to utilize boundaries to produce metals with superior strength and ductility almost two thousand years ago [2]. For example, the Beilian steel was processed by multiple forging and folding to make strong and tough swords in China in the 2nd century [2]. The Damascus steel was used to make blades from 5th century to 19th century [3, 4]. A common feature of these ancient materials is their layered or lamella structures with high density of internal boundaries.

A metallic material is either strong or ductile, but rarely both at the same time [5, 6]. Coarse-grained (CG) metals usually have high ductility but low strength. Refining grains to the nanocrystalline regime in the last few decades has significantly increased the strength, but this is accompanied with the sacrifice of ductility

[7]. In recent years, extensive work has been reported on metals with high density of boundaries [8, 9], including laminated metals with superior mechanical properties [10–15], gradient materials with high strength, extra strain hardening and good ductility [16–22], heterostructured lamella structures with the high strength of ultrafine-grained metal and the high ductility of CG metal [1], metals with growth and deformation twins [23–25], etc. These reports suggest that internal boundaries can be designed to produce superior mechanical properties.

In spite of the above progresses, it remains unclear what mechanisms are activated at boundaries to affect the mechanical properties. The superior mechanical properties observed in lamella and gradient materials have been attributed to HDI strengthening and HDI work hardening caused by the piling up of geometrically necessary dislocations (GNDs) at internal boundaries [1, 11, 26]. These GNDs were generally assumed to be generated by Frank–Read sources and blocked by the boundaries. However, there are also opinions that Frank–Read sources rarely exist in metals [27]. A study on laminated structure reveals possible existence of an optimum layer thickness for the best mechanical properties [11]. Post-mortem examination after tensile testing revealed possible piling-up of GNDs near boundaries [8]. These observations suggest that hetero-boundary-affected region (HBAR) likely exists, but this also raised issues on how the HBARs form and evolve during the plastic deformation.

The strain gradient near the boundaries developed during tensile testing cannot be fully preserved post-mortem because the dislocation pile-up configuration evolves during unloading, which presents a great challenge to quantifying the effect of boundary on the mechanical behaviour. What makes it more challenging is the discrete nature of dislocation pile-up events, which makes it necessary to study the strain gradient statistically along boundaries, rather than local images using conventional transmission electron microscopy (TEM). In this study, we combine the in-situ high-resolution strain mapping, mechanical testing and theoretical modelling to investigate these issues. Copper–bronze laminates were used in this study for their stable boundaries [11].

10.2 Heterostructured Copper–Bronze Laminates

Laminates consisting of CG copper layers and nanostructured (NS) bronze (Cu–10%wt.Zn) layers were fabricated by accumulative roll bonding (Fig. 10.1a). Microhardness was measured using nanoindentation as 0.95 GPa in the CG copper layer, and 2.20 GPa in the NS bronze layer in all laminates (Fig. 10.1b). The layer thickness was systematically varied from 125 μm to 3.7 μm by cold rolling (Fig. 10.S1). All laminate samples were annealed before tensile testing to produce an average grain size of about 4.8 μm in the Cu layer and 100 nm in the bronze layer (Figs. 10.1c,d and 10.S1). Such large differences in grain size and hardness are expected to produce a significant mechanical incompatibility across the boundary during plastic deformation, which consequently generates high HDI stress [8].

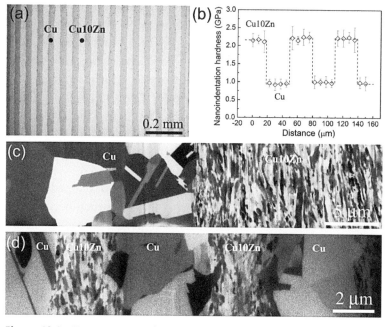

Figure 10.1 Heterostructured copper–bronze laminates. (a) Optical image of the laminate with mean layer thickness of 31 μm. (b) Hardness profile from nanoindentation. (c) Ion Channeling Contrast Microscopy (ICCM) image showing a typical boundary in the laminate with mean layer thickness of 62 μm. (d) ICCM image showing the microstructure in a laminate with mean layer thickness of 3.7 μm.

10.3 Hetero-Boundary-Affected Region (HBAR)

In this study we developed a high-resolution digital image correlation (DIC) analysis technique to map in-situ local strain distribution and evolution near selected boundaries during tensile testing inside a scanning electron microscope (SEM) (see the Materials and Methods section for more details). Figure 10.2a,b shows the strain distributions in the tensile direction ε_{yy} and thickness direction ε_{xx} with increasing applied strain. As shown, many discrete shear bands were developed at about 45° to the loading direction in both the CG Cu and NS bronze layers. The average ε_{yy} strain at and near the boundary is about the same as that away from the boundary (Figs. 10.2c and 10.S2). In other words, the average ε_{yy} strain is uniformly distributed along the thickness. This is because all layers in the laminate are continuous and subjected to the same applied strain in the tensile direction. However, the strain in the thickness direction, ε_{xx}, concentrates on each side of the boundaries to form an obvious region with strain gradient, as shown in Fig. 10.2b,c. This region with strain gradient is here defined as the hetero-boundary-affected region (HBAR).

Plastic strain in a metal is typically dominated by the nucleation and propagation of dislocations. The higher ε_{xx} strain at the boundary indicates that dislocations are emitted from the boundaries as proposed by Li and Murr [27, 28]. This is opposite to the conventional belief that dislocations are emitted from Frank–Read sources and piled up at boundaries [8, 29], which will produce the lowest plastic strain at the boundary.

To quantify the width of the HBARs, the ε_{xx} strain peaks across the boundaries (Figs. 10.2c and 10.S2) can be fitted by a Gaussian distribution function, from which two important parameters, the width W at half maximum of the strain peak and strain peak intensity H can be extracted, as illustrated in Fig. 10.2d. The width at the half strain peak can be assumed equal to half of the peak width at its base. In other words, W can be considered as the HBAR width. Figure 10.2e shows that the HBAR width remains largely constant with increasing applied tensile strain. *This indicates the existence of a characteristic HBAR width that does not change with the applied* tensile strain. However, the intensity of the plastic strain peak in the HBAR increases with increasing tensile strain (Fig. 10.2e), which suggests that the strain gradient becomes larger with increasing tensile strain.

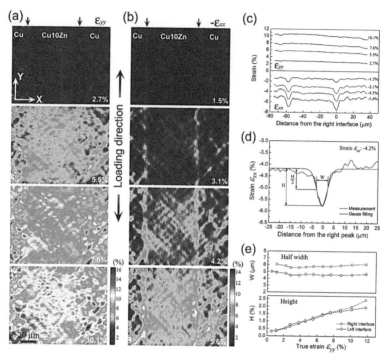

Figure 10.2 Hetero-boundary-affected region (HBAR) measured by in-situ high-resolution DIC technique in a Cu-bronze laminate. (a) The distribution of strain $\varepsilon_{yy} = du_y/dy$ with increasing applied tensile strain, where u_y is the displacement in tensile direction. (b) The distribution of strain $\varepsilon_{xx} = du_x/dx$ with increasing applied tensile strain, where Y is the tensile direction and X is the sample thickness direction. The top black arrows indicate the position of boundaries. (c) The evolution of statistical average strain, ε_{yy} and ε_{xx}, as a function of distance from the right boundary. The ε_{xx} strain peaks at the two boundaries indicate a high strain gradient near them. (d) The definition and calculation of HBAR based on strain peak. Here, H denotes the intensity of strain peak and W is the width at half of the peak maximum ($H/2$), referred to as half width for simplicity. W also equals the width of the HBAR. (e) The evolution of the half width (W) and strain intensity (H) with increasing applied tensile strain. The half widths of the two ε_{xx} strain peaks near boundaries remain largely constant, while their intensities increase with increasing applied tensile strain.

10.4 Theoretical Modeling of the Critical HBAR Width

According to the dislocation ledge source model [27, 28], the ledges on the boundary between the CG copper and NS bronze layers could act as sources to emit dislocations, as observed here (Fig. 10.2). Upon loading, the ledge sources on the boundary will be activated and produce arrays of dislocations as shown in Fig. 10.3, which forms the HBAR.

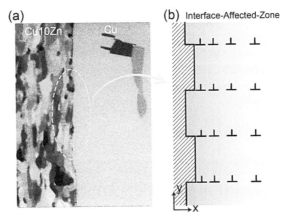

Figure 10.3 Hetero-boundary-affected region (HBAR) by dislocation ledge source model. (a) ICCM micrograph of Cu–bronze boundary. (b) Schematic illustrations of HBAR by dislocation source model.

Li and co-workers [27, 28] have shown that the dislocation emission due to ledge sources gives rise to similar length scale as the classical dislocation pile up model [29] near a boundary. In this model, the length of the emitted dislocation array l, corresponding to the width of the HBAR, can be expressed as

$$l = \frac{\mu n b}{\pi (1-v) \sigma} \quad (10.1)$$

where μ is the shear modulus, n the number of dislocations in the array, b the burgers vector, v the Poisson's ratio, and σ the applied stress.

The local stress field at the root of the dislocation array is:

$$\sigma_{xy} \cong \sigma\sqrt{\frac{l}{x}}, \text{ for } x \ll \frac{l}{2} \qquad (10.2)$$

To estimate the critical width of the HBAR, we assume the strength of the boundary to be on the order of μ and the ledge sources are located at a few Burgers vectors away from the boundary. Plugging these parameters into equation (2) yields:

$$l_{IAZ} \approx \left(\frac{\mu}{\sigma_y}\right)^2 b \qquad (10.3)$$

It can be estimated that the derived length scale of HBAR is on the order of a few micrometers, which is consistent with our experiment observation (Fig. 10.2b,e). This length scale is also consistent with the characteristic length associated with strain gradient plasticity

$$l_{GND} \sim \left(\frac{\mu}{\sigma_y}\right)^2 b \text{ [30, 31]}.$$

10.5 Mechanical Behaviors Controlled by Interfacial Spacing

Figure 10.4a shows that the yield strength and flow stress increased with decreasing boundary spacing, which is expected since the boundaries enhance strength and flow stress [1, 17]. It is also found that the work hardening capability increased with decreasing boundary spacing (Fig. 10.4b), which should help with retaining ductility [6, 32]. Figure 10.4c demonstrates that the ductility first increased with decreasing boundary spacing, reaching its maximum when the boundary spacing is about 15 µm, and then decreased with further reduction of the boundary spacing. In other words, there exists an optimum boundary spacing for the best combination of strength and ductility. It should be noted that this optimum boundary spacing is about twice of the critical HBAR width. In other words, the HBARs from adjacent boundaries start to overlap with each other below this optimum spacing.

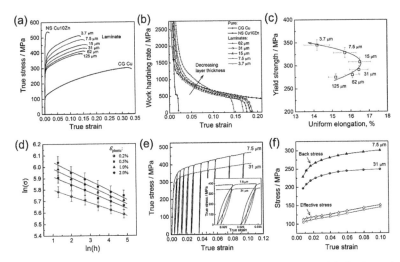

Figure 10.4 The effect of boundary spacing on the mechanical behaviors of the Cu–bronze laminates. (a) True stress-strain curves vs. boundary spacing. (b) Smaller boundary spacing produces higher work hardening. (c) Strength-ductility relationship vs boundary spacing h. The given numbers reflect for the boundary spacing. (d) Linear relationship between ln(σ) and ln(h). (e) Unloading-reloading curves for HDI stress and effective-stress measurements. (f) The effect of boundary spacing on the evolution of HDI stress and effective stress with plastic strain.

Figure 10.4d indicates that the relationship between the boundary spacing h and the flow stress σ can be described as $\sigma = \sigma_0 h^n$, where σ_0 and n are constants. The boundaries are expected to produce strong HDI stress, which can be measured by unloading-reloading experiments (Figs. 10.4e and 10.S3), using a recently proposed procedure [26]. Figure 10.4f shows that the HDI strengthening in the sample is much higher than the effective stress strengthening caused by pure dislocation density strengthening. The HDI stress is primarily caused by GNDs that were emitted from and piled-up near the boundaries. More GNDs are expected with higher boundary density, i.e. decreasing boundary spacing. It is also shown in Fig. 10.4f that the HDI stress increased quickly at the beginning of plastic deformation (<2%) and then slowed down as plastic strain increases, suggesting that the HDI work hardening is most effective in the early stage of plastic deformation. In contrast, the effective stress increased almost linearly with applied strain, indicating a largely constant contribution to work hardening rate. The effective stress, τ_e, is associated with dislocation density [31]: $\tau_e = \alpha \mu b \sqrt{\rho}$, where α is a constant between 0.2–0.5, and ρ is dislocation density. Since $d\tau_e/d\varepsilon$ = constant, as indicated in Figs. 10.4f and 10.S3, it can be derived that $\sqrt{\rho} \propto \varepsilon$.

10.6 Discussion and Summary

It is found that a characteristic HBAR was formed near boundaries in heterostructured laminates. The HBAR was defined as the region with strain gradient on one side of a boundary. There is consensus that the piling up of GNDs near the boundary produces strain gradient [30, 31, 33, 34]. We have found in an earlier work that GND density gradient was developed near the boundary in the CG-Cu/NS-bronze sample with increasing applied strain [11], which suggests that the GND density gradient is related to the strain gradient observed here in the DIC measurement.

As shown in Fig. 10.2d,e, although the critical HBAR width remained largely constant with increasing tensile strain, the strain intensity in the HBAR increased, leading to higher strain gradient and HDI work hardening. During plastic deformation, the emitted dislocations from dislocation sources at the boundary form the HBAR with a characteristic length scale on the order of several micrometers, which corresponds to an inherent internal material length associated with the storage of GNDs. This is consistent with the fact that the critical width of HBAR is also the characteristic length in strain gradient plasticity [30, 31]. Therefore, it can be expected that the effect of GNDs is maximized when the boundary spacing become comparable to twice of the HBAR width at which the HBARs from adjacent boundaries touch each other, as shown in our experiment.

The critical width of the HBAR is about 5–6 μm, which is very close to half of the optimum boundary spacing for the highest ductility of laminate (Fig. 10.4a,c). This suggests that when the HBARs from adjacent boundaries approach each other with decreasing boundary spacing, their HBARs may start to overlap, which limits the effectiveness of HDI hardening, and consequently leads to weaker ability to retain ductility.

According to the experimental observation [27], when the boundary ledge sources were activated, the average number of dislocations in each emission profile become saturated around 1% of plastic strain. This is also consistent with our observation that the width of the critical HBAR does not change with the applied tensile strain (Fig. 10.2e).

Figure 10.5 compares the strength and ductility of the Cu–bronze laminate with those of conventional homogeneous Cu and bronze, which further demonstrates that the laminate spacing can be optimized to produce superior combinations of strength and ductility that are not accessible to their homogeneous counterparts. The optimum/critical boundary spacing revealed in Figs. 10.4c and 10.5 has significant implications in the design of the heterostructures [8, 9] for superior mechanical properties. The key principle is to maximize the HDI strengthening and HDI work hardening. When the boundary spacing is too large, not enough HDI stress is produced. However, when the boundary spacing is too small, the HBARs overlap with each other, leaving insufficient space for dislocation to pile-up, which then limits the work hardening capability for retaining ductility. Therefore, heterostructures should be designed with an optimum interfacial spacing comparable to twice of the HBAR width.

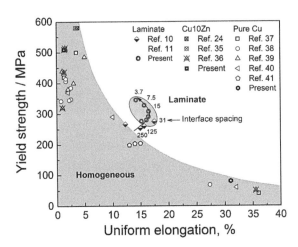

Figure 10.5 Comparison of strength and ductility (uniform elongation) between the laminated CG-Cu/NS-bronze samples and the conventional homogenous materials. An optimum layer thickness exists for the best combination of strength and ductility.

10.7 Materials and Methods

10.7.1 Material Preparation

Cu–bronze laminates were processed by accumulative rolling bonding (ARB) using commercial Cu (99.9%) and bronze (Cu–10 wt.%Zn) sheets [11]. The layer thickness was systematically varied from 125 μm to 3.7 μm by cold rolling. All laminate samples were annealed at 220 °C for 2 hours so that recrystallization occurs in Cu layers but not in bronze layers (Fig. 10.S1).

10.7.2 Microstructural Observations

The laminate microstructure was characterized by optical microscopy, ion channeling contrast microscopy (ICCM) in scanning electron microscope (SEM) and transmission electron microscopy (TEM). ICCM was performed under an FEI Quanta 3D FEG dual-beam instrument, and TEM was conducted in a JEM-2010F microscope. Sample preparation for microstructural observations can be found in our previous works [10, 11].

10.7.3 DIC Characterization

An in-situ quasi-static micro-tensile stage was made in house and set up in a JSM-6510LV microscope. Taking reference images from 10 equidistant points in the lateral surface of the gauge section (see the coordinate system in Fig. 10.S4), a 0.2% engineering strain was applied for the first loading and then a large strain increment of 1.1% was applied until sample failure. Digital images were taken after each strain increment. 2D DIC calculations were performed with a subset size of 25×25 pixel^2, and a step spacing of 3 pixel, as shown in Fig. 10.S5.

Before performing DIC imaging, random pattern with appropriate scale and resolution is needed to cover the sample surface as strain markers. In this work, electrochemical etching was used as an effective patterning method [42], using a solution comprising 10 g FeCl$_3$, 100 ml H$_2$O, and 2.5 ml HCl dilute solution, under a DC voltage of 0.7 V for 15 s. Samples before and after etching were tested in tension and the resulting stress-strain curves were identical, indicating that the etching had negligible effect on the measured mechanical behavior. Figure 10.S5 shows the typical digital speckle image taken under SEM, covering 715×550 pixel^2 effective area with a spatial resolution of 171 nm/pixel. The mean intensity gradient is equal to 51.2, which is of high quality for the speckle patterns used in DIC [43].

10.7.4 Mechanical Testing

Uniaxial tensile and loading-unloading-reloading tests were performed using dog-bone shaped tensile samples with a gauge length of 12 mm and a width of 2 mm at a strain rate of 5×10^{-4} s^{-1}. An extensometer was used to calibrate strain during uniform elongation and each test was repeated for at least 3 times. Nanoindentation experiments were conducted using an MTS Nanoindenter XP equipped with a Berkovich pyramid indenter. Neighboring indents were separated by distances larger than 10 μm to avoid the influence of the plastic zone around the indent and each datum given in the text was averaged from 5 indentations.

Acknowledgements

Financial support from the National Key R&D Program of China (2017YFA0204402 and 2017YFA0204403) and National Natural Science Foundation of China (11672195 and 11572328) are acknowledged. C.X.H. is funded by Sichuan Youth Science and Technology Foundation (2016JQ0047). Y.T.Z. is funded by the US Army Research Office. H.J.G is funded by National Science Foundation through Grant DMR-1709318. H.W.H and M.G. gratefully acknowledge the funding of the German Research Council (DFG) which, within the framework of its 'Excellence Initiative' supports the Cluster of Excellence 'Engineering of Advanced Materials' at the University of Erlangen-Nürnberg.

Author Contributions

C.X.H. and Y.F.W. developed the high-resolution DIC technique and designed experiments. C.X.H., Y.F.W., X.L.M., X.L.W. and Y.T.Z. conceived the initial research issues. C.X.H., Y.F.W. X.L.W. and Y.T.Z. analysed the data. S.Y. and H.J.G. performed the theoretical analysis. H.W.H. and M.G. processed the laminated sample using ARB. All contributed to discussions. C.X.H., Y.F.W., X.L.W. H.J.G. and Y.T.Z. wrote the chapter.

References

1. X. L. Wu, et al. *Proc. Natl. Acad. Sci. U. S. A.*, **112** (2015) 14501–14505.
2. J. T. Wang, *Mater. Sci. Forum*, **503–504** (2006) 363–370.
3. J. D. Verhoeven, A. H. Pendray, W. E. Dauksch, *JOM*, **50** (1998) 58–64.
4. J. Wadsworth, O. D. Sherby, *Prog. Mater. Sci.*, **25** (1980) 35–68.
5. R. Z. Valiev, et al. *J. Mater. Res.*, **17** (2002) 5–8.
6. Y. T. Zhu, X. Z. Liao, *Nat. Mater.*, **3** (2004) 351–352.
7. R. Z. Valiev, et al. *Mater. Res. Lett.*, **4** (2016) 1–21.
8. X. L. Wu, Y. T. Zhu, *Mater. Res. Lett.*, **5** (2017) 527–532.
9. E. Ma, T. Zhu, *Mater. Today*, **20** (2017) 323–331.
10. X. L. Ma, et al. *Scr. Mater.*, **103** (2015) 57–60.
11. X. L. Ma, et al. *Acta Mater.*, **116** (2016) 43–52.
12. A. Misra, J. P. Hirth, R. G. Hoagland, *Acta Mater.*, **53** (2005) 4817–4824.
13. I. N. Mastorakos, et al. *J. Mater. Res.*, **26** (2011) 1179–1187.
14. M. M. Wang, et al. *Acta Mater.*, **85** (2015) 216–228.
15. J. S. Carpenter, et al. *Mater. Res. Lett.*, **3** (2015) 50–57.
16. K. Lu, *Science*, **345** (2014) 1455–1456.
17. X. L. Wu, et al. *Proc. Natl. Acad. Sci. U. S. A.*, **111** (2024) 7197–7201.
18. X. L. Wu, et al. *Mater. Res. Lett.*, **2** (2014) 185–191.
19. T. H. Fang, et al. *Science*, **331** (2011) 1587–1590.
20. A. Y. Chen, et al. *Mater. Sci. Eng. A*, **667** (2016) 179–188.
21. Y. J. Wei, et al. *Nat. Commun.*, **5** (2014) 3580.
22. X. L. Wu, et al. *Acta Mater.*, **112** (2016) 337–346.
23. K. Lu, L. Lu, S. Suresh, *Science* **324** (2009) 349–352.
24. Y. H. Zhao, et al. *Appl. Phys. Lett.*, **89** (2006) 121906.
25. C. X. Huang, et al. *Mater. Res. Lett.*, **3** (2015) 88–94.
26. M. X. Yang, et al. *Mater. Res. Lett.*, **4** (2016) 145–151.
27. L. E. Murr, *Meta. Mater. Tran. A*, **47** (2016) 5811–5826.
28. J. C. M. Li, *Trans. Metal. Soc. AIM*, **227** (1963) 239–247.

29. J. P. Hirth, J. Lothe, *Theory of Dislocations*, 2nd edn, John Wiley Sons, 1982.
30. W. D. Nix, H. Gao, *J. Mech. Phys. Solids*, **46** (1998) 411–425.
31. H. Gao, et al. *J. Mech. Phys. Solids*, **47** (1999) 1239–1263.
32. Y. T. Zhu, X. Z. Liao, X. L. Wu, *Prog. Mater. Sci.*, **57** (2012) 1–62.
33. M. F. Ashby, *Philos. Mag.*, **21** (1970) 399–424.
34. H. Gao, Y. G. Huang, *Scr. Mater.*, **48** (2003) 113–118.
35. X. Y. San, et al. *Mater. Des.*, **35** (2012) 480–483.
36. P. Zhang, et al. *Mater. Sci. Eng. A*, **594** (2014) 309–320.
37. Y. H. Zhao, et al. *Adv. Mater.*, **18** (2006) 2949–2953.
38. F. Dalla Torre, et al. *Acta Mater.*, **52** (2004) 4819–4832.
39. Y. Zhang, N. R. Tao, K. Lu, *Acta Mater.*, **56** (2008) 2429–2440.
40. P. Xue, B. L. Xiao, Z. Y. Ma, *Mater. Sci. Eng. A*, **532** (2012) 106–110.
41. Sh. Ranjbar Bahadori, K. Dehghani, F. Bakhshandeh, *Mater. Sci. Eng. A*, **583** (2013) 36–42.
42. J. C. Stinville, *Acta Mater.*, **98** (2015) 29–42.
43. B. Pan, Z. X. Lu, H. M. Xie, *Optic. Lasers Eng.*, **48** (2010) 469–477.

Supplementary Information

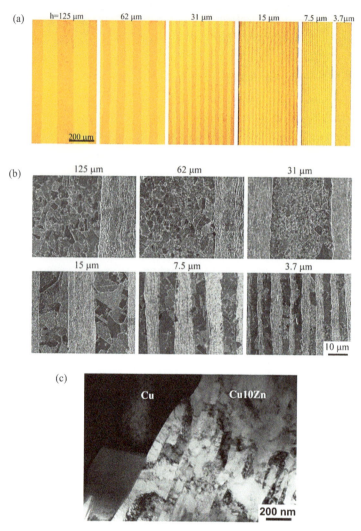

Figure 10.S1 (a) Optical microscopy showing the Cu–bronze laminates. The red color indicates Cu layers, while the yellow color is bronze layers. (b) SEM microstructures of Cu–bronze laminates. After light corrosion, Cu layers with large grain size are clear presented, while bronze layers maintain fine grain structures that cannot be observed in SEM. (c) TEM micrograph showing a typical boundary.

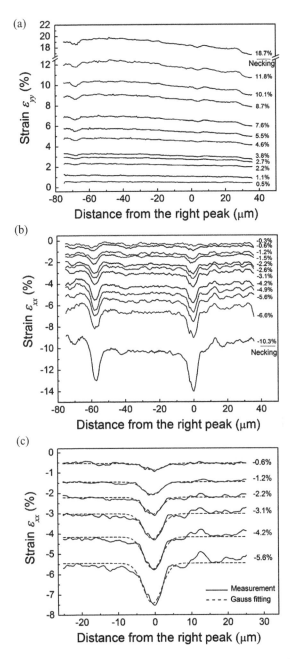

Figure 10.S2 (a) The evolution of statistical average strain ε_{yy} as a function of distance from the right boundary. (b) The evolution of statistical average strain ε_{xy} as a function of distance from right boundary. (c) The Gauss fitting of the strain peaks.

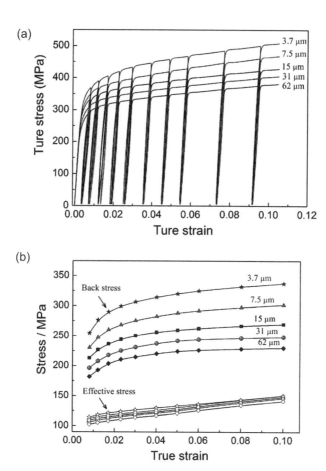

Figure 10.S3 (a) Unloading-reloading curves for measuring the HDI stress. (b) The effect of boundary spacing on the evolution of HDI stress and effective stress with plastic strain.

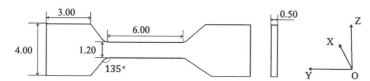

Figure 10.S4 Geometry of the tensile specimen for DIC and reference coordinate system, X, Y and Z represent thickness, length (tensile) and width direction, respectively. DIC measurements are performed on XOY plane. Unit: mm.

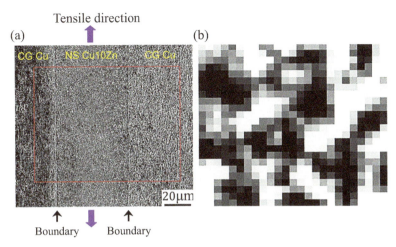

Figure 10.S5 (a) Typical speckle images with spatial resolution of 171 nm/piexl. The red frame marks 715 × 550 piexl² effective correlation area. (b) The gray-scale distribution of a singer correlation window (25 × 25 piexl²).

Chapter 11

In-situ Observation of Dislocation Dynamics Near Heterostructured Boundary

Hao Zhou,[a,b] Chongxiang Huang,[c] Xuechao Sha,[b] Lirong Xiao,[a] Xiaolong Ma,[d] Heinz Werner Höppel,[e] Mathias Göken,[e] Xiaolei Wu,[f] Kei Ameyama,[g] Xiaodong Han,[b] and Yuntian Zhu[a,d]

[a]*Nano and Heterogeneous Material Center, Nanjing University of Science and Technology, Nanjing, China*
[b]*Beijing Key Lab of Microstructure and Property of Advanced Materials, Beijing University of Technology, Beijing, China*
[c]*School of Aeronautics and Astronautics, Sichuan University, Chengdu, China*
[d]*Department of Materials Science and Engineering, North Carolina State University, Raleigh, NC, USA*
[e]*Institute I: General Materials Properties, Friedrich-Alexander Universität Erlangen-Nürnberg, Erlangen, Germany*
[f]*Institute of Mechanics, Chinese Academy of Sciences, Beijing, China*
[g]*Department of Mechanical Engineering, Faculty of Science and Engineering, Ritsumeikan University, Kusatsu, Japan*
ytzhu@ncsu.edu, xdhan@bjut.edu.cn

Reprinted from *Mater. Res. Lett.*, **7**(9), 376–382, 2019.

Heterostructured Materials: Novel Materials with Unprecedented Mechanical Properties
Edited by Xiaolei Wu and Yuntian Zhu
Text Copyright © 2019 The Author(s)
Layout Copyright © 2022 Jenny Stanford Publishing Pte. Ltd.
ISBN 978-981-4877-10-7 (Hardcover), 978-1-003-15307-8 (eBook)
www.jennystanford.com

There has been a long-standing controversy on how dislocations interact with boundaries. Here we report in-situ observations that in a Cu–brass heterostructured TEM film Frank–Read sources are the primary dislocation sources. They were dynamically formed and deactivated throughout the deformation in grain interior, which has never been reported before. This observation indicates that strain gradient near boundaries cannot be quantitatively related to the density gradient of geometrically necessary dislocations, and it was primarily produced by Frank–Read source gradient instead of dislocation pile-ups. These findings provide new insights on how to design heterostructured boundaries to enhance mechanical properties.

Impact Statement: In-situ TEM reveals the dynamic formation and destruction of Frank–Read sources, whose number gradient produced strain gradient near heterostructured boundaries. GND gradient is not proportional to strain gradient.

It has been a perpetual challenge to produce strong and tough materials for engineering applications. Many approaches have been reported to produce strong materials, but these often come at the sacrifice of ductility [1–3]. Good ductility is required for a strong material to be simultaneously tough. Boundary engineering such as introducing twins [4, 5], second-phases [6, 7], and grain refinement [1] have been extensively reported for improving the strength and/or ductility of metals. Their general principle is to utilize dislocation-boundary interactions to modify mechanical behavior.

Recently, heterostructured (HS) materials were found to be able to avoid the trade-off between strength and ductility [3, 8]. This makes it essential to understand how dislocations interact with HS boundaries so as to pave the scientific foundation for this new type of materials. The superior mechanical properties were attributed to hetero-deformation-induced (HDI) synergistic strengthening and work hardening, which are produced by the piling-up of geometrically necessary dislocations (GNDs) at HS zone boundaries [8]. Electron backscatter diffraction (EBSD) indeed revealed GND gradient near boundaries in a Cu–brass heterostructure after tensile testing, which seems to agree with the popular dislocation pile-up theory [9–12]. However, a negative strain gradient was recently found near the Cu–brass boundary, i.e. the higher plastic strain the nearer the boundary [13]. This is contrary to the strain gradient produced by

GND pileup from Frank–Read sources. Therefore, it was concluded that the strain gradient was produced by dislocation emissions from the boundary [13], which is consistent with a minority view that the Frank–Read sources rarely exist in grain interior and boundaries such as grain boundaries are the primary dislocation sources [14, 15]. This minority view has been controversial for over half a century [16]. These controversies cannot be solved by post-mortem transmission electron microscopy (TEM) observations, which cannot differentiate dislocation sources or the dynamic process of dislocation processes.

Here we report in-situ TEM observations of dislocation dynamics in a soft Cu grain near a boundary with brass (Cu-10 wt.% Zn), which were produced by accumulative rolling bonding (For more information, see Supplemental Materials). It is surprisingly found that the strain gradient near the boundary was primarily caused by the dynamic formation and deactivation of Frank–Read sources near the boundary.

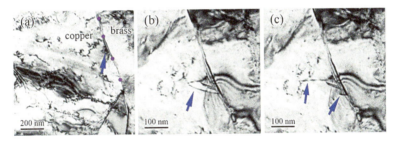

Figure 11.1 Perfect dislocation emissions from a Cu/brass boundary. (a) A Cu grain-brass boundary delineated by four purple dots. The blue arrow indicates a stress concentration spot. (b) A perfect dislocation (marked by the blue arrow) emitted from the boundary at the stress concentration spot. (c) The dislocation reached the sample surface and broke to two segments as indicated by two blue arrows.

Figure 11.1 shows an example of dislocation emission into a Cu grain interior from the Cu/brass boundary observed under in-situ TEM (Supplementary Movie 1). The Cu grain is about 2 μm long and 0.8 μm wide. In Fig. 11.1a the four purple dots delineate the boundary while the blue arrow marks a high-stress location on the boundary, as indicated by the dark strain contrast. Figures 11.1b,c are higher magnification of the same location upon further in-situ straining. As shown in Fig. 11.1b, a perfect dislocation is emitted at the stress concentration spot on the boundary (blue arrow). This dislocation semi-loop is gliding and growing on a slip plane inclined to the sample surface. When a segment of dislocation reaches and glides out of the surface, the dislocation semi-loop is broken into two segments, as indicated by the two blue arrows in Fig. 11.1c. A total of two perfect dislocation emissions from the boundary were observed in this experiment.

Figure 11.2 shows the nucleation of twins through partial dislocation emissions from the boundary and dislocation interactions with the twins (Supplementary Movie 2). Figure 11.2a shows a staking fault growing from the boundary by a slipping partial. The stacking fault is on an inclined {111} slip plane, which produced parallel Moiré fringes. It intersects both the top and bottom surfaces, making it look like a ribbon. It appeared narrow because it has a large inclination angle to the surface. Upon further straining, two more partials are emitted in adjacent planes, but at varying distances from the boundary (Fig. 11.2b), which formed a twin with a two-layer segment as marked by the blue arrow and a three-layer segment as indicated by the green arrow. Further straining moved the three partials close to the twinning front (Fig. 11.2c), while a dislocation is interacting with the twin (see the orange arrow). The dislocation caused local detwinning (Fig. 11.2d, blue arrow). The detwinning mechanism by a perfect dislocation can be found in our earlier work [17].

Further straining leads to the nucleation and growth of another twin, which is shorter and parallel to the first twin (Fig. 11.2e). The second twin has three layers spaced at a distance from each other, suggesting that they repel each other. In other words, they may have the same Burgers vector. Therefore, this twin is likely formed by the monotonic emission of partials from the boundary, and will produce a twinning strain [18]. Interestingly, there is a Frank–Read source between the two twins, which emitted perfect dislocations toward the boundary. We propose a possible mechanism for the Frank–Read source to operate between the two twins (Supplemental Fig. S6). Figure 11.2f indicates extensive interactions between dislocations and twins near the boundary, which produced a high intensity and complex strain field.

In-situ Observation of Dislocation Dynamics Near Heterostructured Boundary | 181

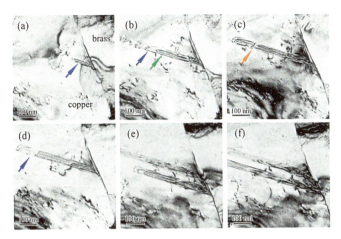

Figure 11.2 Twin nucleation and growth from the zone boundary and twin-dislocation interactions. (a) A stacking fault nucleated from the boundary. (b) A three-layer deformation twin (green arrow) changes to two layers near the growth end (blue arrow). (c) A perfect dislocation interacting with the twin (orange arrow). (d) Detwinning caused by a dislocation (blue arrow). (e) Second twin from the boundary and dislocations emitted from a Frank–Read source between the two twins. (f) Strain field near the boundary.

Frank–Read dislocation sources were found dynamically formed, and deactivated after emitting some dislocations (Supplementary Movie 3). A total of ten Frank–Read sources were observed at different stage of straining (Supplemental Figs. S7–S14). These dislocations typically glide toward and piled up against the boundary. When a Frank–Read source was deactivated, the dislocation pileup associated with it often disappears. Figure 11.3a shows five dislocations that were emitted from a Frank–Read source and piled up against the boundary. Upon further straining, the Frank–Read source was deactivated with only two remnant dislocations left (Fig. 11.3b). Note that these dislocations slipped for a relatively long distance without encountering the sample surface, indicating that they are on a {111} slip plane that is almost parallel to the sample surface. Figure 11.3c shows another Frank–Read source that was formed after the one in Fig. 11.3a was deactivated. As shown, a half dislocation loop was formed. However, with further straining the dislocation loop reached the surface, and was broken into two segments as indicated by two black arrows. This indicates that the slip plane of this Frank–Read source is different from those discussed earlier.

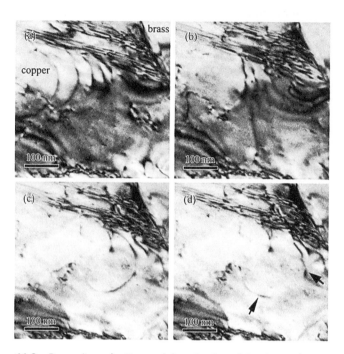

Figure 11.3 Dynamic nucleation and deactivation of Frank–Read sources. (a) A Frank–Read source produced five dislocations piling-up against the boundary, (b) The deactivation of the dislocation source; (c) Another Frank–Read source was activated and emitted a dislocation loop. (d) The dislocation loop was broken at the sample surface.

The in-situ TEM observation here solved three critical issues in dislocation-boundary interactions. The first is the mechanism for the formation of the negative strain gradient near the boundary. GNDs are responsible for producing the HDI stress observed in heterostructured materials [8, 19, 20]. It is generally believed that in coarse-grained metals Frank–Read sources in grain interior are main dislocation sources [21–23]. If the dislocations emitted from a Frank–Read source piles up against a boundary, they become the type of GNDs that produce long-range back stress [20]. The other type of GNDs are those that form low-angle grain boundaries, which do not produce HDI stress. For dislocation pile-up, the plastic strain increases from the boundary to the Frank–Read source, which hereafter we refer to as the positive strain gradient. The plastic strain at the boundary is zero if no dislocation is pushed into the boundary. It was recently observed that a negative strain gradient exists near the Cu–brass boundary [13], which is opposite to the above Frank–Read source scenario. Therefore, it was concluded that boundaries were the dislocation sources [13], which is consistent with the decades-old minority view that dislocation ledges at grain boundaries are the primary dislocation sources [14, 16].

In the current in-situ TEM observation, only two perfect dislocations and two thin twins were observed emitted from the boundary, as compared to about ten Frank–Read sources that were dynamically generated, each emitting a number of dislocations toward the boundary before being eventually deactivated. The dislocation pileups mostly disappeared after the deactivation of the dislocation sources. These observations indicate that the Frank–Read sources are primarily responsible for the observed negative strain gradient. Logically, since most Frank–Read sources did not produce GND pile-up, only the density gradient of the Frank–Read sources can produce the observed negative strain gradient, since the dislocation pileup will produce a positive strain gradient. Since both dislocation density and stresses are higher as it gets closer to the boundary, it is logically reasonable to hypothesize that the closer to the boundary, the higher probability to generate Frank–Read sources. This will produce the observed density gradient of the Frank–Read sources to yield a negative strain gradient. This hypothesis is verified by analyzing the distances of Frank–Read sources from the boundary (Supplemental Fig. S15).

The second long-standing issue possibly solved here is whether Frank–Read source plays a primary role in the plastic deformation of coarse-grained metals. The majority view in textbooks is that Frank–Read sources are primarily responsible in coarse-grained metals [21, 23, 24], while grain boundaries may become the dominant dislocation source in nanocrystalline metals [4, 25–28]. However, Murr [14] argued that Frank–Read sources rarely exist in the grain interior and grain boundary ledges are the primary dislocation sources, as proposed by Li in 1961 [16]. Murr used postmortem TEM observations [29] to support his hypothesis. The in-situ TEM observations here clearly indicate that dynamic Frank–Read sources are the primary dislocation sources in coarse-grained Cu, and postmortem TEM observations are not reliable since the Frank–Read sources are dynamically generated and deactivated during the deformation process, which leaves behind little evidence for postmortem TEM observation.

The third issue solved here is the relationship between the strain gradient and GND density gradient. It is believed that the strain gradient has to be accommodated by GNDs [30], which hints at a quantitative relationship between the magnitude of strain gradient and GND density gradient. The dynamic nature of Frank–Read sources and the disappearance of dislocation pileups make it necessary to revise the current strain plasticity theory [30] as well as our understanding of the relationship between the strain gradient and HDI stress evolution [20]. Specifically, the strain gradient near the boundary is related to the density gradient of Frank–Read sources, not directly to the GND density gradient, although a GND density gradient was indeed developed [9]. Since most GNDs are destroyed instead of being accumulated, with increasing applied strain it can be logically expected that the increase in GND pileups will slow down and eventually reach a dynamic saturation, which will lead to a corresponding slowing down and saturation of back stress. This was exactly what has been observed in heterostructured materials [8, 9].

The dynamic nature of the Frank–Read sources is surprising, but can be understood with the concept of dislocation intersections [21, 24]. When two dislocations gliding on different slip planes intersect each other, they will produce jogs in the other dislocation line. The magnitude and orientation of the jog in one dislocation line equal those of the Burgers vector of the other dislocation (see Fig. 11.4a). Since the jog is not on the original slip plane of the dislocation, it acts to pin the dislocation line, a scenario that has been observed in the current in-situ TEM experiment (Supplemental Fig. S4). If two jogs are formed along a gliding dislocation and their spatial distance are large enough for the dislocation segment between them to bow out to emit dislocations under the local stress, a Frank–Read source will be generated (Fig. 11.4b,c). Considering the large number of slip systems on the four {111} planes in fcc Cu, the probability of forming such Frank–Read sources should be very high, which is why so many Frank–Read sources were observed during the in-situ TEM experiments.

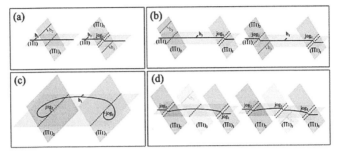

Figure 11.4 Schematic diagram of dynamic nature of the Frank–Read source. (a) Intersection of two dislocations gliding on different slip planes, (b) Two pinning jogs formed along a gliding dislocation, (c) A Frank–Read source is generated to emit dislocations under the local stress, and (d) The Frank–Read source is deactivated by a new jog formed between the two pinning jogs.

The Frank–Read source could be deactivated if a new jog forms on the dislocation segment between the two pinning jogs, in which the local stress may no longer be able to bow out the shorter remaining free dislocation segments. The critical shear stress needed to activate a Frank–Read source is [21]

$$\tau_c = \frac{2Gb}{d} \tag{11.1}$$

where G is the shear modulus, b is the Burgers vector magnitude and d is the distance between two pinning points. When a new pinning point is formed on the dislocation line between the two original pinning points, d will decrease and the critical stress needed to bow out the shorter dislocation segment will increase, leading to the deactivation of the Frank–Read source (Fig. 11.4d). We have indeed observed many dislocation segments that cannot emit dislocations because they are too short.

It should be noted that TEM observations can only be performed on thin films [31]. This casts some uncertainty on our in-situ TEM results since the thin film may be under different stress state from the bulk, and the free surfaces may affect dislocation slip. We note that the current observations can be logically explained using current textbook knowledge without invoking thin film, and believe that the results at least qualitatively reflect the physics of dislocation interaction with zone boundaries. Nevertheless, more study is needed, possibly by computer modeling.

The probability of observing dislocation pile-ups is affected by the tendency for planar slip, which is in turn affected by stacking fault energy (SFE), and lattice friction caused by solute atoms [32]. Pure Cu is a wavy slip material [32], meaning easy dislocation cross-slip, which is largely responsible for the observed disappearance of dislocation pile-ups.

Low SFE makes it difficult for a perfect dislocation to cross-slip, meaning stable dislocation pile-ups. Indeed, we have observed extensive dislocation pile-ups in 304 L stainless steel (Supplemental Fig. S5), which has lower SFE (18 mJ/m^2) [33] than Cu (54.6 mJ/m^2) [4]. Since GND piling-up can effectively produce back stress [20], it can be logically deduced that low SFE should promote HDI hardening to produce superior mechanical properties. Lastly, HDI hardening has been recently renamed hetero-deformation induced (HDI) hardening for more precise description of physics [34].

Acknowledgement

The authors thanks the support of the Materials Characterization and Research Center of Nanjing University of Science and Technology and the Jiangsu Key Laboratory of Advanced Micro & Nano Materials and Technology. YTZ acknowledge the support of the US Army Research Office.

Disclosure statement

No potential conflict of interest was reported by the authors.

Funding

This work was supported by the National Key R&D Program of China [2017YFB0204403], National Natural Science Foundation of China [51601003 and 11672195]. C.X.H. was funded by Sichuan Youth Science and Technology Foundation [2016JQ0047]. Y.T.Z. was funded by US Army Research Office.

References

1. Ovid'ko IA, Valiev RZ, Zhu YT. Review on superior strength and enhanced ductility of metallic nanomaterials. *Prog Mater Sci*, 2018; **94**:462–540.
2. Wang YM, Chen MW, Zhou FH, et al. High tensile ductility in a nanostructured metal. *Nature*, 2002; **419**:912–915.
3. Wei YJ, Li YQ, Zhu L, et al. Evading the strength-ductility trade-off dilemma in steel through gradient hierarchical nanotwins. *Nat Commun.*, 2014; **5**:3580.
4. Zhu YT, Liao XZ, Wu XL. Deformation twinning in nanocrystalline materials. *Prog Mater Sci*, 2012; **57**: 1–62.
5. Lu K, Lu L, S. Suresh S. Strengthening materials by engineering coherent internal boundaries at the nanoscale. *Science*, 2009; **324**(5925):349–352.
6. Araki K, Takada Y, Nakaoka K. Work-hardening of continuously annealed dual phase steels. *Trans Iron Steel Inst Jpn*, 1977; **17**:710–717.
7. Kundu A, Field DP. Influence of plastic deformation heterogeneity on development of geometrically necessary dislocation density in dual phase steel. *Mater Sci Eng A*, 2016; **667**:435–443.
8. Wu XL, Yang MX, Yuan FP, et al. Heterogeneous lamella structure unites ultrafine-grain strength with coarse-grain ductility. *Proc Natl Acad Sci USA*, 2015; **112**:14501–14505.
9. Ma XL, Huang CX, Moering J, et al. Mechanical properties in copper/bronze laminates: role of interfaces. *Acta Mater*, 2016; **116**:43–52.
10. Chassagne M, Legros M, Rodney D. Atomic-scale simulation of screw dislocation/coherent twin boundary interaction in Al, Au, Cu and Ni. *Acta Mater*, 2011; **59**:1456–1463.
11. Kondo S, Mitsuma T, Shibata N, et al. Direct observation of individual dislocation interaction processes with grain boundaries. *Sci Adv*, 2016; **2**(11):1501926.
12. Kacher J, Eftink BP, Cui B, et al. Dislocation interactions with grain boundaries. *Curr Opin Solid State Mater Sci*, 2014; **18**:227–243.
13. Huang CX, Wang YF, Ma XL, et al. Interface affected zone for optimal strength and ductility in heterogeneous laminate. *Mater Today*, 2018; **21**(7):713–719.
14. Murr LE. Dislocation ledge sources: dispelling the myth of Frank–Read source importance. *Metall Mater Trans A*, 2016; **47**(12):5811–5826.

15. Mompiou F, Legros M, Boé A, et al. Inter- and intragranular plasticity mechanisms in ultrafine-grained Al thin films: an in situ TEM study. *Acta Mater*, 2013; **61**:205–216.
16. Li JCM. High-angle tilt boundary-a dislocation core model. *J Appl Phys*, 1961; **32**(3):525–530.
17. Zhu YT, Wu XL, Liao XZ,et al. Dislocation-twin interactions in nanocrystalline fcc metals. *Acta Mater*, 2011; **59**:812–821.
18. Wu XL, Liao XZ, Srinivasan SG, et al. New deformation twinning mechanism generates zero macroscopic strain in nanocrystalline metals. *Phys Rev Lett*, 2008; **100**:095701.
19. Ma E, Zhu T. Towards strength-ductility synergy through the design of heterogeneous nanostructures in metals. *Mater Today*, 2017; **20**(6):323–331.
20. Wu XL, Zhu YT. Heterogeneous materials: a new class of materials with unprecedented mechanical properties. *Mater Res Lett*, 2017; **5**(8):527–532.
21. Hirth JP, Lothe J. *Theory of Dislocations*. Malabar, FL: Krieger Publishing Company; 1992.
22. Hertzberg RW. *Deformation and Fracture Mechanics of Engineering Materials*. New York: Wiley;1989.
23. Frank FC, Read WT. Multiplication processes for slow moving dislocations. *Phys Rev*, 1950; **79**:722–723.
24. D. Hull and D. J. Bacon, *Introduction to Dislocations*. Oxford: Pergamon Press;1984.
25. Yamakov V, Wolf D, Phillpot SR, et al. Deformation mechanism crossover and mechanical behaviour in nanocrystalline materials. *Philos Mag Lett*, 2003; **83**:385–393.
26. Yamakov V, Wolf D, Phillpot SR, et al. Deformation-mechanism map for nanocrystalline metals by molecular-dynamics simulation. *Nat Mater*, 2004; **3**:43–47.
27. Swygenhoven HV, Weertman JR. Deformation in nanocrystalline metals. *Mater Today*, 2006; **9**(5):24–31.
28. Swygenhoven HV, Derlet PM, Froseth AG. Stacking fault energies and slip in nanocrystalline metals. *Nat Mater*, 2004; **3**:399–403.
29. Murr LE, Wang SH. Comparison of microstructural evolution associated with the stress-strain diagrams for nickel and 304 stainless steel:an electron microscope study of micro-yielding and plastic-flow. *Res Mech*, 1982; **4**:237–274.

30. Gao H, Huang Y, Nix WD, et al. Mechanism-based strain gradient plasticity-I. Theory. *J Mech Phys Solids*, 1999; **47**(6):1239–1263.
31. Mompiou F, Caillard D, Lergos M, et al. In situ TEM observations of reverse dislocation motion upon unloading in tensile-deformed UFG aluminum. *Acta Mater*, 2016; **60**:3402–3414.
32. Hong SI, Laird C. Mechanisms of slip mode modification in F.C.C. solid-solutions. *Acta Metall Mater*, 1990; **38**(8):1581–1594.
33. Lee TH, Shin E, Oh CS, Ha HY, Kim SJ. Correlation between stacking fault energy and deformation microstructure in high-interstitial-alloyed austenitic steels. *Acta Mater*, 2010; **58**: 3173–3186.
34. Zhu, YT, Wu, XL. Perspective on hetero-deformation induced (HDI) hardening and back stress. *Mater Res Lett*, 2019; **7**(10):393–398.

Chapter 12

Hetero-Deformation Induced (HDI) Hardening Does Not Increase Linearly with Strain Gradient

Y. F. Wang,[a] C. X. Huang,[a] X. T. Fang,[b] H. W. Höppel,[c] M. Göken,[c] and Y. T. Zhu[b,d]

[a]*School of Aeronautics and Astronautics, Sichuan University, Chengdu 610065, China*
[b]*Department of Materials Science and Engineering, North Carolina State University, Raleigh, NC 27695, USA*
[c]*Department of Materials Science and Engineering, Institute I: General Materials Properties, Friedrich-Alexander University of Erlangen-Nürnberg, Martensstr. 5, 91058 Erlangen, Germany*
[d]*Nano and Heterogeneous Materials Center, School of Materials Science and Engineering, Nanjing University of Science and Technology, Nanjing 210094, China*
chxhuang@scu.edu.cn, ytzhu@ncsu.edu

Hetero-deformation induced (HDI) hardening has been attributed to geometrically necessary dislocations (GNDs) that are

Reprinted from *Scr. Mater.*, **174**, 19–23, 2020.

Heterostructured Materials: Novel Materials with Unprecedented Mechanical Properties
Edited by Xiaolei Wu and Yuntian Zhu
Text Copyright © 2019 Acta Materialia Inc.
Layout Copyright © 2022 Jenny Stanford Publishing Pte. Ltd.
ISBN 978-981-4877-10-7 (Hardcover), 978-1-003-15307-8 (eBook)
www.jennystanford.com

needed to accommodate strain gradient near the boundaries of heterostructured zones. Here we report that HDI hardening does not increase linearly with increasing strain gradient in the hetero-boundary-affected region. This is because some GND pileups may be absorbed by the boundary and consequently does not contribute to HDI hardening with increasing strain gradient. Higher mechanical incompatibility across boundary produces higher strain gradient. The strain gradient-dependent strengthening effect of heterostructured boundary mainly originates from the development of HDI stress.

Heterostructured metals and alloys have been reported to possess exciting potential of overcoming the strength-ductility tradeoff by introducing high-density of heterostructured zone boundaries [1–13]. These zone boundaries are different from grain boundaries and twin boundaries in conventional homogeneous materials. There are dramatic differences in strength and strain hardening capability across these boundaries [1–3, 7–9]. Hetero-deformation caused by this mechanical incompatibility can introduce strain gradient near the boundaries [2–4, 8, 11]. As suggested by the strain gradient plasticity theory in micromechanics, pileup of geometrically necessary dislocations (GNDs) is needed to accommodate the plastic strain gradient [14, 15]. These GNDs are expected to produce HDI strengthening and hardening, in addition to the isotropic strengthening caused by the increase of total dislocation density [1, 2, 16, 17]. It was recently realized that the terms of back-stress strengthening and hardening cannot accurately describe the real physics of what occurred across the boundary, because GND pileups produces not only back stress in the soft zones, but also forward stress in the hard zones [18]. Back stress and forward stress are conjugated long-range internal stresses with opposite directions, which collectively affect flow stress during the unloading and reloading. Therefore, "back stress" measured from the mechanical testing (e.g. unloading-reloading [9, 19]) was renamed hetero-deformation induced (HDI) stress by Zhu and Wu [18] to accurately describe the extra strain hardening in heterostructured materials [2–4, 19–21].

It was assumed in the gradient plasticity theory [14, 15] that the GND density and gradient are quantitatively related to the strain gradient near the boundary. GND gradient was found to indeed exist near zone boundaries [12]. Strain gradient was measured and an hetero-boundary-affected region (HBAR) was found near the zone boundary [11]. These earlier works led to the belief that the HDI hardening increases with strain gradient and a quantitative relationship exists between them. However, a recent in-situ transmission electron microscopy (TEM) examination revealed that Frank–Read sources and dislocation pileups were dynamically formed and disappeared during tensile test [22], which suggested that the GND density and gradient may not be quantitatively related to the strain gradient. Consequently, the HDI hardening may not be quantitatively related to the strain gradient. To clarify this issue, it is essential to compare the developments of the strain gradient around zone boundary and the associated HDI stress.

In this study, microscale digital image correlation (μ-DIC) was used to characterize the evolution of strain gradient in the hetero-boundary-affected region (HBAR) during tensile testing. The result was compared with the evolution of HDI hardening, which was measured by the unloading-reloading approach [2, 9, 19]. The comparison revealed that the HDI hardening does not have a linear relationship with increasing strain gradient in the HBAR.

Three types of laminate samples stacked with alternate sequences of copper and brass (Cu-10 wt.% Zn), copper and brass (Cu-30 wt.% Zn), copper and copper were accumulatively roll-bonded (ARB) to a layer thickness of ~62 μm at room temperature. The as-received copper-brass (Cu-10 wt.% Zn) laminates were further rolled to a layer thickness of 31 μm, 15 μm, or 7.5 μm. Thereafter, all laminates were annealed at 220°C for 2 h so that recrystallization occurs in Cu layers but not in Cu–Zn alloy layers. The Cu layers in as-annealed laminates are characterized with fully recrystallized coarse grains (CG), while the Cu–Zn alloy layers exhibit severely deformed nanostructures (NS) [11, 12]. For simplicity, three types of laminates were labeled as CG/NS$_{10Zn}$, CG/NS$_{30Zn,}$ and CG/CG, respectively, in which CG represents coarse-grained Cu layer.

Hardness was measured using an MTS Nanoindenter XP equipped with a Berkovich pyramid indenter. As shown in Fig. 12.1a, sharp boundaries with significant hardness incompatibility in adjacent zones exist in heterostructured CG/NS$_{10Zn}$ and CG/NS$_{30Zn}$ laminates, while no hardness difference across the boundaries in CG/CG sample.

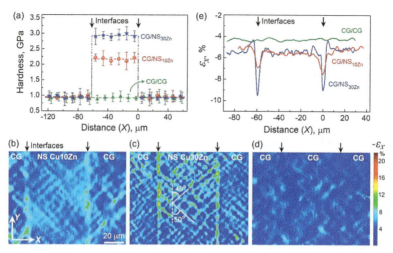

Figure 12.1 (a) Hardness profiles of three types of laminates. Every data point was averaged from 5 nanoindentations. The $-\varepsilon_x$ strain maps at the tensile strain of ~10%: (b) CG/NS$_{10Zn}$ [11], (c) CG/NS$_{30Zn}$ and (d) CG/CG. In the coordinate, Y is the tensile direction, and X is the sample thickness direction perpendicular to boundary. (e) The distribution of statistical average strain ε_x as a function of distance from the right boundary. The white lines in (c) mark the shear banding direction.

Dog-bone-shaped small sample with the gauge length parallel to the rolling direction was prepared for μ-DIC strain measurement. Refined speckle pattern prepared by electrochemical etching was recorded by secondary electron imaging. The detailed test procedures are reported in our previous work [11]. Figures 12.1b–d present the distribution of strain $-\varepsilon_x$ across boundaries. The strain distribution is averaged along the boundary and plotted as a function of distance from a boundary (Fig. 12.1e), which provides insights on the statistical nature of strain distribution near the boundary. Obviously, hetero-boundary-affected regions (HBARs) spanning ~10 μm with a negative strain gradient are formed at heterostructured boundaries in CG/NS$_{10Zn}$ and CG/NS$_{30Zn}$ laminates. In contrast, strain ε_x in CG/CG sample evolves uniformly across the homogeneous boundary (the green curve in Fig. 12.1e). Comparing Fig. 12.1a with Fig. 12.1e reveals that the magnitude of strain gradient increases with higher mechanical incompatibility.

The negative gradient of strain ε_x was caused by the gradient distribution of dynamically generated Frank–Read sources near the heterostructured boundary [22]. The mechanical incompatibility leads to synergistic constraint between heterostructured zones during tension, which changes the local stress state and causes stress concentration near the boundary [4, 23, 24]. The closer to the boundary, the higher probability to activate Frank–Read sources because of the higher local stresses. The Frank–Read sources are essentially dislocation segments with various lengths between dislocation jogs, which are formed by the interaction of intersecting dislocations gliding on different slip planes [22, 25]. Therefore, it is logical to expect that sites closer to boundary should experience more dislocation activities, and thus have higher measurable strain. Higher mechanical incompatibility induces larger stress gradient near boundary, which will cause more intense Frank–Read source gradient. This is the reason why the strain gradient at the CG/NS$_{30Zn}$ boundary is much steeper than that at the CG/NS$_{10Zn}$ boundary (Fig. 12.1a and e).

Figure 12.2a shows the statistical evolution of strain ε_x in CG/NS$_{10Zn}$ laminate with increasing applied strain ε_y. The ratio of the height (H) and the width at half height (W) of strain concentration peak, H/W, can be used to quantify the mean intensity of strain gradient $|d\varepsilon_x/dx|$ in HBAR. As shown in Fig. 12.2b, the mean $|d\varepsilon_x/dx|$ in the HBAR increases almost linearly with applied strain. In other words, the strain gradient in the HBAR increases linearly with applied strain. As discussed latter, this can be attributed to the formation and deactivation of dislocation pileups near the heterostructured boundary.

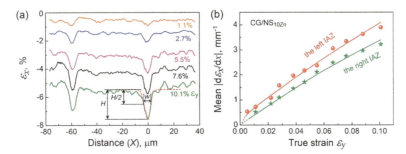

Figure 12.2 Evolution of strain ε_x in CG/NS$_{10Zn}$ with increasing applied strain ε_y: (a) statistical averaged distribution [11], (b) the mean strain gradient $|d\varepsilon_x/dx|$ in HBAR. The intensity (H) and the width at half intensity (W) of strain concentration peak at boundary are extracted from the Gaussian fitting (red dotted line). Mean $|d\varepsilon_x/dx|$ is approximately equal to the ratio H/W.

Since the width of the HBAR is not affected by layer thickness until adjacent HBARs start to overlap [11, 12], the volume fraction of HBAR (V_{HBAR}) is primarily determined by the boundary density. The V_{HBAR} in the four groups of CG/NS$_{10Zn}$ samples with layer thicknesses of 62 μm, 31 μm, 15 μm and 7.5 μm are approximately 16.8%, 33.6%, 69.5% and 100%, respectively. These samples are used to deduce the strain gradient-related strengthening effects of heterostructured boundaries. Tensile samples with a gauge length of 12 mm and a width of 2 mm were machined for both uniaxial and loading–unloading–reloading (LUR) tensile tests. HDI stress is deduced from hysteresis loops (Fig. 12.3a) using the procedure that was used to measure the "back stress" [2, 9]. As shown in Fig. 12.3b, the measured HDI stress (σ_{HDI}) increases with applied strain, and the sample with higher V_{HBAR} has higher HDI stress.

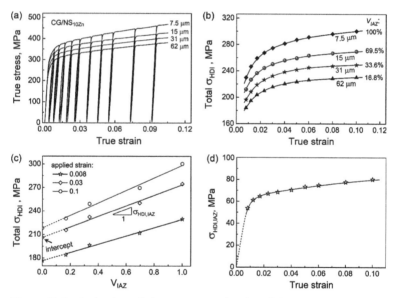

Figure 12.3 Deduction of the HDI stress developed from per unit volume fraction of HBAR ($\sigma_{HDI,HBAR}$). (a) LUR curves and hysteresis loops of CG/NS$_{10Zn}$ laminates with different layer thickness, i.e., CG/NS$_{10Zn}$ samples with varying volume fraction of HBAR (V_{HBAR}). (b) Total σ_{HDI} of laminates. (c) Linear fitting of the total σ_{HDI} as a function of V_{HBAR}, showing the deduction of $\sigma_{HDI,HBAR}$ at a certain strain. (d) Evolution of $\sigma_{HDI,HBAR}$ with applied strain.

Note that microscopically plastic strain produced by dislocations is intrinsically heterogeneous in polycrystalline materials [26]. The back stress and forward stress can also be built up in homogeneous materials due to the non-uniform dislocation pileups against grain boundary or dislocation walls [16]. For example, pronounced hysteresis effects were observed in work-hardened polycrystalline and freestanding NS samples [9, 16, 25]. Therefore, the above measured HDI stress at a certain applied strain should be the total σ_{HDI} of laminate sample which can be expressed as

$$\sigma_{HDI} = (V_{NS}\sigma_{HDI,NS} + V_{CG}\sigma_{HDI,CG}) + V_{IAZ}\sigma_{HDI,IAZ}, \quad (12.1)$$

where $\sigma_{HDI,NS}$ and $\sigma_{HDI,CG}$ are the HDI stress intrinsic to NS and CG components, respectively. $\sigma_{HDI,IAZ}$ represents the HDI stress developed from per unite HBAR volume fraction, i.e., the HDI stress induced by the hetero-deformation in the HBAR. The contribution of these HDI stress components in a laminate can be evaluated basing on the volume fraction of their origin.

Figure 12.3c shows the fitting of the total σ_{HDI} as a function of V_{HBAR} according to the linear relationship described in Eq. (12.1). As shown, the slop and intercept of the fitting line represent the magnitude of $\sigma_{HDI,IAZ}$ and $V_{NS}\sigma_{HDI,NS} + V_{CG}\sigma_{HDI,CG}$ at a certain applied strain, respectively. Figure 12.3d shows the evolution of the $\sigma_{HDI,IAZ}$ with increasing applied strain. Surprisingly, the $\sigma_{HDI,IAZ}$ increased quickly at the early strain stage, and then slowed down at a tensile strain of ~2%. Note that the derived $\sigma_{HDI,IAZ}$ is the upper limit of HDI stress in a laminate with non-overlapping HBARs. The evolution of $\sigma_{HDI,IAZ}$ with increasing applied strain (Fig. 12.3d) is different from the linear evolution of strain gradient intensity (Fig. 12.2b), which indicates that the HDI stress in the HBAR does not increase linearly with increasing strain gradient in the HBAR. This observation contradicts earlier theory and assumption that the development of HDI stress depends proportionally on the evolution strain gradient [1, 11, 12, 21].

The development of HDI stress from HBAR is caused by the piling up and accumulation of GNDs [18]. Specifically, the spatial gradient of GND density ($\nabla\rho_{GND}$) contributes to the net long-range internal stress. The HDI stress at a specific position can be in theory calculated by integrating the stress field of each individual GND as $\sigma_{HDI} \propto \nabla\rho_{GND}$ [21, 25, 27–29]. This expression can be further rephrased as $\sigma_{HDI} \propto \nabla\eta$ if one assumes that ρ_{GND} is proportional to the strain gradient η [14, 15]. Accordingly, during straining, the HDI stress caused by GND pileups in local strain gradient zone should develop proportionally with the evolvement of strain gradient intensity. In other words, a certain increase in strain gradient should lead to a constant increase of HDI stress. This is why HDI stress was assumed or hinted to be linearly increase with strain gradient intensity in earlier works [1, 11, 12, 21].

The fast increase of HDI hardening before 2% applied strain (Fig. 12.3d) indicates that a quick GND piling-up process at the elastic-plastic transition and low-plastic strain stages [18]. With further tensile straining, the dramatic slowdown of HDI hardening implies that the piling-up slowed down. Logically, this could be caused by the slowdown of strain gradient increase. However, Fig. 12.2b shows that strain gradient in the HBAR increased linearly with applied strain during the entire tensile testing. In other words, the GND accumulation slowed down after 2% applied strain in the HBAR despite the continuous increase in the strain gradient. This can be attributed to the dynamic formation and disappearance of GND pileups in the HBAR [22].

As revealed by in-situ TEM observation [22], the piling up of GNDs near boundary is accompanied with the formation and activation of Frank–Read source, which may be absorbed by the heterostructured boundary as the associated source becomes deactivated in subsequent straining process. Disappeared GND pileups take away their contribution on HDI stress but leave measurable strain in the slipping path. At the large strain stage, the new formation and disappearance of pileups tend to reach a dynamic saturation, leading to a significant slowing down of HDI stress (Fig. 12.3d) but a continuous increase of strain accumulation near boundary (Fig. 12.2b). In short, the primary physics behind the non-proportional relationship between the developments of HDI stress and strain gradient is that the GND pileups and strain gradient in HBAR are accumulated asynchronously.

The true stress-strain curves of CG/NS_{10Zn} samples are plotted in Fig. 12.4a. As shown, the flow stress increases with decreasing layer thickness. The measured flow stress (σ_f) of laminates at a certain applied strain can be expressed as

$$\sigma_f = (V_{NS}\sigma_{f,NS} + V_{CG}\sigma_{f,CG}) + V_{IAZ}\sigma_{f,IAZ}, \qquad (12.2)$$

where $\sigma_{f,NS}$ and $\sigma_{f,CG}$ are the flow stress intrinsic to NS and CG components, respectively. $\sigma_{f,IAZ}$ is the flow stress developed from per unit HBAR volume fraction, i.e., the extra strengthening caused by GND pileups in an HBAR. Similar to the fitting process shown in Fig. 12.3c, the $\sigma_{f,IAZ}$ at certain strain can be extracted by extracting the slope of the linear relationship between σ_f and V_{IAZ}. The green curve in Fig. 12.4b shows the evolution of $\sigma_{f,IAZ}$ with applied strain.

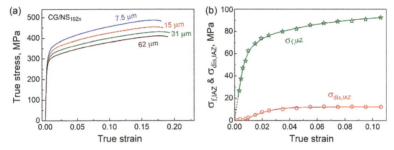

Figure 12.4 (a) Uniaxial tensile response of CG/NS$_{10Zn}$ laminates [11]. (b) The total strengthening effect ($\sigma_{f,HBAR}$) and the dislocation strengthening ($\sigma_{dis,HBAR}$) developed from per unit volume fraction of HBAR.

Since the piling up of GNDs in HBAR can also produce extra dislocation strengthening ($\sigma_{dis,IAZ}$), i.e., Taylor strengthening due to the increase of total dislocation density [14], the $\sigma_{f,IAZ}$ is the sum of $\sigma_{HDI,IAZ}$ and $\sigma_{dis,IAZ}$,

$$\sigma_{f,IAZ} = \sigma_{HDI,IAZ} + \sigma_{dis,IAZ}. \tag{12.3}$$

$\sigma_{dis,IAZ}$ can be deduced by subtracting the $\sigma_{HDI,IAZ}$ from the $\sigma_{f,IAZ}$. Interestingly, $\sigma_{dis,IAZ}$ (the red curve in Fig. 12.4b) is much smaller than $\sigma_{f,IAZ}$, especially at the early strain stage. This indicates that the extra strain hardening in the HBAR primarily comes from the HDI hardening, instead of the isotropic dislocation hardening, especially at the early elastic-plastic stage. This result provides direct evidence to the earlier claim that long-range internal stress dominated extraordinary strengthening and work hardening in heterostructured materials [2, 3, 9, 19].

Note that the parameters $\sigma_{HDI,NS}$ and $\sigma_{f,NS}$ used in Eq. (12.1) and Eq. (12.3) may not be equal to the corresponding those in a freestanding NS layer, because the stress state of NS layers in a laminate is modified by the synergistic constraint from neighboring

CG layers. As shown in Figs. 12.1b and c, stable shear bands are dispersed in the NS layer, which helps with accommodating applied strain and retaining uniform elongation [30]. This unique deformation mode did not occur in a freestanding NS counterpart. This observation indicates that the boundary effects, $\sigma_{HDI,IAZ}$ and $\sigma_{f,IAZ}$, cannot be simply deduced from the mechanical responses of freestanding components because of the synergistic interactions among different layers [6, 23].

In summary, it is found that the mechanical incompatibility induces the formation of HBAR, where negative strain gradient is developed in the direction perpendicular to the boundary. The intensity of strain gradient in HBAR increases linearly with applied strain, while the HDI stress developed from the GNDs pileup in HBAR increases quickly at the early strain stage and then slows down to approximate saturation. These are due to the dynamical formation and disappearance of dislocation pileups near boundary during straining, and indicate that there is no quantitative relationship between them. Importantly, the strain gradient-related strengthening effect of heterostructured boundary is mainly originated from the development of HDI stress, instead of the increase of total dislocation density. This work sheds light on the possibility of enhancing the hetero-deformation induced hardening by architecting heterostructured boundary.

Acknowledgements

This work was supported by the National Key R&D Program of China (2017YFA0204403), the National Natural Science Foundation of China (Nos.11672195) and Sichuan Youth Science and Technology Foundation (2016JQ0047). Y.F.W. would like to acknowledge the support from China Scholar Council. Y.T.Z. was funded by the US Army Research Office.

References

1. X. L. Wu, Y. T. Zhu, *Mater. Res. Lett.*, **5** (2017) 527–532.
2. X. L. Wu, M. X. Yang, F. P. Yuan, G. L. Wu, Y. J. Wei, X. X. Huang, Y. T. Zhu, *Proc. Natl. Acad. Sci. U.S.A.*, **112** (2015) 14501–14505.

3. H. K. Park, K. Ameyama, J. Yoo, H. Hwang, H. S. Kim, *Mater. Res. Lett.*, **6** (2018) 261–267.
4. X. L. Wu, P. Jiang, L. Chen, F. P. Yuan, Y. T. Zhu, *Proc. Natl. Acad. Sci. U.S.A.*, **111** (2014) 7197–7201.
5. S. W. Wu, G. Wang, Q. Wang, Y. D. Jia, J. Yi, Q. J. Zhai, J. B. Liu, B. A. Sun, H. J. Chu, J. Shen, P. K. Liaw, C. T. Liu, T. Y. Zhang, *Acta Mater.*, **165** (2019) 444–458.
6. Y. F. Wang, C. X. Huang, M. S. Wang, Y. S. Li, Y. T. Zhu, *Scr. Mater.*, **150** (2018) 22–25.
7. Z. Cheng, H. F. Zhou, Q. H. Lu, H. J. Gao, L. Lu, *Science*, **362** (2018) eaau1925.
8. Z. Cheng, L. Lu, *Scr. Mater.*, **164** (2019) 130–134.
9. Y. F. Wang, M. X. Yang, X. L. Ma, M. S. Wang, K. Yin, A. H. Huang, C. X. Huang, *Mater. Sci. Eng. A*, **727** (2018) 113–118.
10. M. Huang, C. Xu, G. H. Fan, E. Maawad, W. M. Gan, L. Geng, F. X. Lin, G. Z. Tang, H. Wu, Y. Du, D. Y. Li, K. S. Miao, T. T. Zhang, X. S. Yang, Y. P. Xia, G. J. Cao, H. J. Kang, T. M. Wang, T. Q. Xiao, H. L. Xie, *Acta Mater.*, **153** (2018) 235–249.
11. C. X. Huang, Y. F. Wang, X. L. Ma, S. Yin, H. W. Höppel, M. Göken, X. L. Wu, H. J. Gao, Y. T. Zhu, *Mater. Today*, **21** (2018) 713–719.
12. X. L. Ma, C. X. Huang, J. Moering, M. Ruppert, H. W. Höppel, M. Göken, J. Narayan, Y. T. Zhu, *Acta Mater.*, **116** (2016) 43–52.
13. J. Wang, A. Misra, *Curr. Opin. Solid State Mater. Sci.*, **15** (2011) 20–28.
14. H. Gao, Y. Huang, W. D. Nix, J. W. Hutchinson, *J. Mech. Phys. Solids*, **47** (1999) 1239–1263.
15. J. W. Hutchinson, *Int. J. Solid. Struct.*, **37** (2000) 225–238.
16. H. Mughrabi, *Acta Metall.*, **31** (1983) 1367–1379.
17. X. Feaugas, *Acta Mater.*, **47** (1999) 3617–3632.
18. Y. T. Zhu, X. L. Wu, *Mater. Res. Lett.*, **7** (2019) 393–398.
19. M. X. Yang, Y. Pan, F. P. Yuan, Y. T. Zhu, X. L. Wu, *Mater. Res. Lett.*, **4** (2016) 145–151.
20. H. H. Lee, J. I. Yoon, H. K. Park, H. S. Kim, *Acta Mater.*, **166** (2019) 638–649.
21. J. J. Li, G. J. Weng, S. H. Chen, X. L. Wu, *Int. J. Plast.*, **88** (2017) 89–107.
22. H. Zhou, C. X. Huang, X. C. Sha, L. R. Xiao, X. L. Ma, H. W. Höppel, M. Göken, X. L. Wu, K. Ameyama, X. D. Han, Y. T. Zhu, *Mater. Res. Lett.*, **7** (2019) 376–382.

23. X. L. Wu, P. Jiang, L. Chen, J. F. Zhang, F. P. Yuan, Y. T. Zhu, *Mater. Res. Lett.*, **2** (2014) 185–191.
24. Y. F. Wang, M. S. Wang, K. Yin, A. H. Huang, Y. S. Li, C. X. Huang, *T. Nonferr. Metal. Soc.*, **29** (2019) 588–594.
25. F. Mompiou, D. Caillard, M. Legros, H. Mughrabi, *Acta Mater.*, **60** (2012) 3402–3414.
26. B. Bay, N. Hansen, D. A. Hughes, D. Kuhlmann-Wilsdorf, *Acta Metall. Mater.*, **40** (1992) 205–219.
27. C. J. Bayley, W. A. M. Brekelmans, M. G. D. Geers, *Int. J. Solid. Struct.*, **43** (2006) 7268–7286.
28. L. P. Evers, W. A. M. Brekelmans, M. G. D. Geers, *Int. J. Solid. Struct.*, **41** (2004) 5209–5230.
29. M. Kuroda, V. Tvergaard, *J. Mech. Phys. Solids*, **56** (2008) 1591–1608.
30. Y. F. Wang, C. X. Huang, Q. He, F. J. Guo, M. S. Wang, L. Y. Song, Y. T. Zhu, *Scr. Mater.*, **170** (2019) 76–80.

Chapter 13

Extra Strengthening in a Coarse/Ultrafine Grained Laminate: Role of Gradient Boundaries

Y. F. Wang,[a,b] M. S. Wang,[a] X. T. Fang,[b] Q. He,[a] F. J. Guo,[a] R. O. Scattergood,[b] C. X. Huang,[a] and Y. T. Zhu[b,c]

[a]*School of Aeronautics and Astronautics, Sichuan University, Chengdu 610065, China*
[b]*Department of Materials Science and Engineering, North Carolina State University, Raleigh, NC 27695, USA*
[c]*Nano and Heterogeneous Structural Materials Center, School of Materials Science and Engineering, Nanjing University of Science and Technology, Nanjing 210094, China*
chxhuang@scu.edu.cn

The boundaries introduced in metals play crucial roles in mechanical behaviors. Here the effect of gradient boundaries on mechanical behavior was investigated in a laminated Cu-30Zn sample composed of coarse-grained and ultrafine-grained layers. Tensile

---

Reprinted from *Int. J. Plast.*, **123**, 196–207, 2019.

Heterostructured Materials: Novel Materials with Unprecedented Mechanical Properties
Edited by Xiaolei Wu and Yuntian Zhu
Text Copyright © 2019 Elsevier Ltd.
Layout Copyright © 2022 Jenny Stanford Publishing Pte. Ltd.
ISBN 978-981-4877-10-7 (Hardcover), 978-1-003-15307-8 (eBook)
www.jennystanford.com

tests revealed a superior strength-ductility combination with extraordinary strengthening and work hardening. By combining the measurements of height contour and strain distribution using digital image correlation, the accumulation of strain gradient is detected in the hetero-boundary-affected region (HBAR) during tension, which is caused by the mechanical incompatibilities across boundaries and the synergetic constraint between layers. The intensity of strain gradient in the HBAR increased with tensile strain, which was accommodated by the accumulation of geometrically necessary dislocations in the HBAR, thereby resulting in extra dislocations and hetero-deformation-induced (HDI) stress for synergistic strengthening.

13.1 Introduction

Heterostructured metal is a new class of materials with unique structural and mechanical properties (PNAS Ti, Wu and Zhu, 2017). Their microstructures are characterized with multiple zones with dramatically different flow stresses. The inter-zone boundaries can be gradient or sharp (Huang et al., 2018; Park et al., 2018; Wu et al., 2015, 2014a). Due to the synergetic interaction among mechanically incompatible zones, the mechanical properties are usually superior to what is predicted by the rule of mixture (Cheng et al., 2018; Wang et al., 2018c; Wu et al., 2015, 2014b) [3–5, 8, 11]. In recent years, tailored strength-ductility synergy and enhanced work hardening have been achieved in several types of heterostructures, including the laminate (Huang et al., 2018; Ma et al., 2016; Wu et al., 2017), gradient (Cheng et al., 2018; Fang et al., 2011; Wang et al., 2018b; Wei et al., 2014; Wu et al., 2014a), hierarchical lamellar (Li et al., 2018; Wu et al., 2015) and harmonic (Park et al., 2018) structures.

A critical question arises on where in such heterostructures is the improved strength and strain hardening originated. Although there are big differences in the detailed microstructures of these heterostructured materials, the fundamental deformation principles, such as inhomogeneous deformation across boundaries and hierarchical strain/stress partitioning among zones, might be universal (PNAS Ti, Wu and Zhu, 2017). The strain partitioning and accumulation are associated with the formation of strain gradient across zone boundaries in order to maintain strain continuity (Ashby, 1970; Huang et al., 2018). As proposed and simulated in gradient structure, plastic strain gradient could be accumulated with the migration of elastic/plastic boundary during straining (PNAS Fe, Zeng et al., 2016). It is also recognized that the plastic strain gradient needs to be accommodated by the accumulation of geometrically necessary dislocations (GNDs), which is effective in developing internal stresses (Ashby, 1970; Kassner et al., 2013; Mughrabi, 2006). This unusual dislocation activity provides an opportunity to promote extraordinary strengthening and strain hardening, and change the stress-strain behavior of component zones (Cheng et al., 2018; Li et al., 2017b; Yang et al., 2016). Therefore, the boundaries between zones are believed to significantly affect the strain accommodation and mechanical responses of heterostructures. On the other hand, the heterostructured boundaries themselves may have direct effects on the nucleation and pile-up of dislocations as well (Hirth and Lothe, 1982; Murr, 2016).

Our previous works on the sharp boundaries in the NS/CG laminates revealed a strain gradient-dominated hetero-boundary-affected region (HBAR) where extra GNDs were accumulated (Huang et al., 2018; Ma et al., 2016). However, for some other heterostructures, such as gradient materials and the engineering composites synthesized by welding, the boundary between heterostructured zones generally has transitional microstructure, i.e., a gradient boundary (Wu et al., 2014a; Zhu et al., 2017). Such gradient boundary avoids the early formation of serious stress concentration and local cracking problems that are typical in a sharp boundary, and is expected to produce wider affected zone and be more efficient in enhancing mechanical performances. In spite of the above progresses, direct visualization and quantitative analysis for the plastic behaviors of gradient boundary are still lacking. These are also essential for understanding the fundamental deformation mechanism and optimizing the structural design of heterostructured materials.

In this chapter we present a CG/ultrafine-grained (UFG) laminate structure with two discrete gradient boundaries. The configuration and evolution of 3D strain gradient near gradient boundary were quantitatively visualized using height contour measurement and in-situ digital image correlation (DIC). The formation of strain gradient and its effect on strengthening and strain hardening behaviors will be discussed.

13.2 Experimental Methods

Friction stir processing (FSP) was used to fabricate the laminate samples with gradient boundaries. An annealed Cu-30 wt.%Zn plate with a thickness of 5.6 mm was selected as CG base metal (BM). The advantage of this material is its low stacking fault energy and proper melting temperature, which are conducive to the formation of recrystallized UFG in string zone (SZ) during FSP (Chang et al., 2004; Wang et al., 2018a). The FSP tool has a shoulder in diameter of 12 mm and an unthreaded cylindrical pin in length of 1.8 mm and diameter of 3 mm. Two FSP procedures were symmetrically conducted on both sides of the plate under flowing water at a constant rotational speed of 1400 rpm and a traverse speed of 200 mm/min. The as-processed plate was annealed at 250°C for 48 h in order to relieve residual stress. Figure 13.1a shows the cross-sectional morphology of the as-prepared sample observed by scanning electron microscope (SEM). There is clear metallurgic contrast between BM and SZ zones, suggesting great difference in their microstructures. As verified later, homogeneous UFG was formed in SZ. The heterostructures were characterized using transmission electron microscopy (TEM) and electron back-scattered diffraction (EBSD).

Dog-bone shaped tensile specimens with a gauge length of 18 mm were machined parallel to the processing direction in as-FSP processed plate. The colored frames in Fig. 13.1a show the position and dimension of the gauge cross-section of tensile samples. The cross-sectional dimension of heterostructured laminate (red), freestanding CG (light blue) and UFG (yellow) samples are 1.5×3.7 mm^2, 1×1.7 mm^2 and 1.7×1 mm^2, respectively. Figure 13.1b shows the optical morphology of the lateral surface of laminate sample. Clearly, it is composed of a CG core layer and two UFG surface layers with a thickness ratio of 1:1.7:1. The thickness and width of all samples were precisely controlled by polishing off redundant surface. Tests for each type of sample were performed three times at a nominal strain rate of $5 \times 10^{-4}\,s^{-1}$. In order to evaluate the mechanical incompatibility and the deformation-induced hardening in hetero-boundary-affected region (HBAR), microhardness measurement was carried out across the boundary between SZ and BM on both samples before and after tensile deformation, using a load of 25 g. To confirm reproducibility, the microhardness distribution for each type of sample was repeatedly measured at four independent positions.

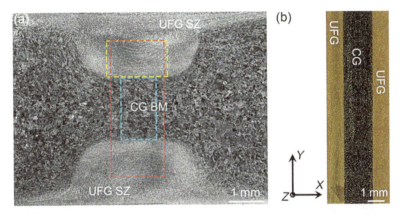

Figure 13.1 (a) Scanning electron morphology of the cross-section of FSP processed heterostructure, showing a CG layer sandwiched between two UFG zones. Colored frames illustrate the cross-section of heterostructured laminate (red), freestanding CG (light blue) and UFG (yellow) samples used to tensile testing, respectively. (b) An optical overview on the lateral surface (18 × 3.7 mm² surface) of heterostructured laminate sample, showing straight boundaries between layers. In the coordinate, Y is the tensile loading direction; X and Z are the thickness and the width directions of laminate sample, respectively.

The height contours on the lateral surface (Fig. 13.1b) of the laminate samples before and after tensile test were measured using a white light interferometry in a vertical scanning mode. The height resolution is 5 nm. In-situ 2D DIC was also conducted on the lateral surface of laminate sample, using a short-focus optical lens. Before performing DIC imaging, a random pattern was prepared by spraying black paints on white background.

13.3 Results

13.3.1 Microstructural Heterogeneity and Gradient Boundary

As the cross-sectional morphology shown in Fig. 13.1a, there is a well-defined straight outline on the bottom of SZ. Figure 13.2a is a representative TEM image of SZ, showing equiaxed UFG structure with clear and sharp grain boundaries. FSP is a severe thermal-plastic deformation process, during which subdivision and fragmentation of parent large grains and dynamic recrystallization to form new small grains occurred simultaneously (Wang et al., 2018a). In this study, the formation of UFG microstructure can be primarily attributed to the low stacking fault energy of material and the fast cooling rate under flowing water during processing (Chang et al., 2004).

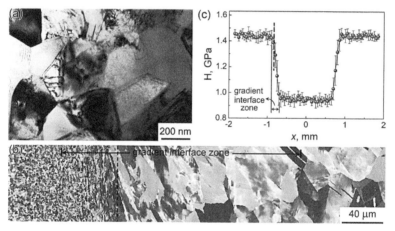

Figure 13.2 The microstructural and mechanical heterogeneities of FSP processed sample. (a) Typical TEM image in the SZ. (b) An EBSD morphology showing the gradient boundary between CG and UFG layers. (c) The hardness (*H*) distribution measured along the longer side of the red frame in Fig. 13.1. Every datum was averaged from 4 indentations. The black dotted lines in (b) and (c) denote the layer boundary.

Figure 13.2b shows the microstructural details between SZ and BM. A continuous transition in grain size from recrystallized UFG to plastically elongated CG and then equiaxed CG is produced, forming a gradient transition zone between SZ and BM. The hardness distribution measured along the long side of the red frame in Fig. 13.1a is shown in Fig. 13.2c. Clearly, the hardness within both UFG and CG layers is rather homogeneous, while transitions with a discrepancy of ~0.48 GPa appears between them. The black dotted line in Fig. 13.2c marks the starting point of the precipitous drop of hardness, which corresponds strictly to the position in microstructure indicated by black dotted line in Fig. 13.2b where the grain size increases sharply. This geometric position can be easily located by metallographic observation, which is referred to as the boundary of UFG and CG layers. Note that the widths of freestanding UFG and CG tensile samples were precisely polished by referring to this layer boundary. The region with hardness and microstructure transition is defined as the gradient boundary zone between CG and UFG layer.

Such heterostructured laminate structure with discrete gradient boundaries avoids the tracking problem of transferable boundary in gradient structure (Zeng et al., 2016). Therefore, the CG/UFG laminate specimen cut as the red cross-section in Fig. 13.1 can be viewed as a model sample for studying the plastic behavior around a single gradient boundary.

13.3.2 Synergistic Strengthening and Strain Hardening

Figure 13.3a shows the tensile stress-strain curves for CG/UFG laminate, freestanding CG and UFG samples. The ultimate strength of CG/UFG laminate is measured as 467 MPa, which is comparable to that of pure UFG sample (481 MPa), while the uniform elongation of laminate sample (34.3%) is much larger than that of freestanding UFG (18.4%). Such a surprising strength–ductility combination is superior to what is predicted using the volume fraction-based rule of mixture (green curve in Fig. 13.3a) (Wu et al., 2014b). Accordingly, a shadow region could be drawn between the prediction and experimentally measured curves, as seen in Fig. 13.3a. This indicates an extra flow stress achieved in CG/UFG laminate sample, demonstrating a considerable synergistic strengthening in such heterostructure (Wang et al., 2018b; Wu et al., 2014b). The extra flow stress at the strain of 10% is measured as 71 MPa, which accounts for 16.7% of the flow strength.

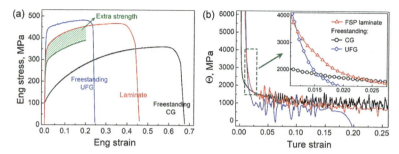

Figure 13.3 Uniaxial tensile behaviors of CG/UFG laminate, freestanding CG and UFG samples: (a) engineering stress-strain curves; (b) strain hardening rate (Θ) vs. true strain. The green curve in (a) is the predicted response for CG/UFG laminate sample basing on rule of mixture. The inset in (b) is the magnified curve at low strains where the CG/UFG laminate sample exhibits extremely high Θ.

The comparison of strain hardening behaviors is shown in Fig. 13.3b. Surprisingly, the strain hardening rate (Θ) of laminate sample is higher than that of both freestanding UFG and CG samples at a low plastic strain stage (~1%–2.5%). Interestingly, the strain regime for the high strain hardening rate is consistent with that (1.8%–3.2%) of Θ-up-turn in gradient IF steel (Wu et al., 2014a). This implies that some underlying mechanisms such as complex stress/strain states induced by mutual constraint between layers should be the same in these two structures.

13.3.3 Height Profile and Strain Gradient Across Boundary

It is interesting to find that the heterostructured layers in CG/UFG laminate (Fig. 13.1b) after tension exhibits remarkable difference in lateral height contour. Figure 13.4a and b show the 3D height profiles measured on the lateral surface before and after tensile deformation, respectively. Obviously, the initial electrochemically polished surface has no height difference between layers except minor surface roughness. However, the central CG layer in the deformed sample is much higher than the UFG layers on both sides, as seen in Fig. 13.4b. These height contours were statistically averaged along the tensile direction (Y) and plotted versus the sample thickness (X) in Fig. 13.4c. The black dotted lines in Fig. 13.4c denote the position of layer boundaries, which were accurately located by polishing off the surface roughness firstly and then metallographic examining using SEM. It is seen that the height difference across layer boundary is as high as ~6 μm.

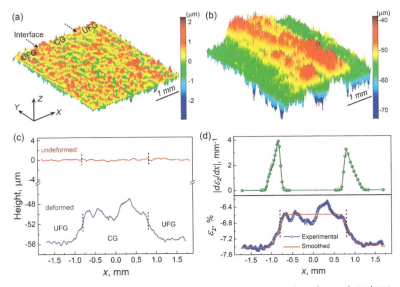

Figure 13.4 3D height contour measured on the lateral surface of CG/UFG laminate sample: (a) before tension, (b) after tensile testing to a strain of 14%. (c) The distribution of statistical average height as a function of sample thickness (X). (d) The distribution of strain ε_z and strain gradient $|d\varepsilon_z/dx|$ deduced from the height profile in (c). The dotted arrows and lines mark the position of layer boundary.

The shrinking strain in sample width direction ε_z, i.e. the lateral strain parallel to the boundary, can be deduced from the height profiles (Wu et al., 2014a). As the blue curve plotted in Fig. 13.4d, the surface UFG layers have higher $|\varepsilon_z|$ than the central CG layer, and ε_z changes gradually in HBAR. The green curve in Fig. 13.4d represents the distribution of strain gradient $|d\varepsilon_z/dx|$ across boundaries, which was deduced from the smoothed ε_z (red curve). It is obvious that the HBARs present pronounced strain gradient, and the average width of strain gradient zones (~500 μm) is much larger than that (~240 μm) of the gradient boundary (Fig. 13.2b and c). It is technically difficult to locate the exact position of strain gradient peak because the derivative of smoothed strain distribution is varied with the applied mathematical function. But it is certain that the peak is at or adjacent to the layer boundary, where a remarkable change in hardness and grain size occurs (Fig. 13.2). Moreover, the zone with strain gradient in UFG side is wider than that in CG side, suggesting a higher efficiency in accommodating strain gradient for the latter.

13.3.4 DIC and Strain Gradient Across Boundary

2D DIC experiments were conducted on the lateral surface to extract the in-plane strains, i.e. the normal strains in the tensile direction ε_y and the sample thickness direction ε_x. The digital speckle image covers the effective correlation area of 1630 × 382 pixel² with a spatial resolution of 9.7 mm/pixel (Fig. 13.5a). Figure 13.5b shows the detailed distribution of gray scale with clear contrast in a single correlation window. This high-quality speckle pattern is an essential prerequisite for accurate strain calculation.

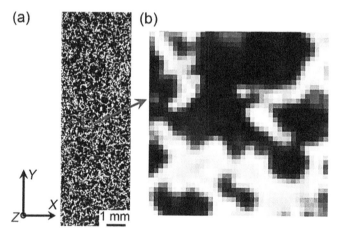

Figure 13.5 DIC speckle image taken from the lateral surface of CG/UFG laminate sample: (a) typical speckle pattern; (b) detailed distribution of gray scale in a single correlation window (40 × 40 pixel²).

Figures 13.6a and c are the typical contours of ε_y and corresponding ε_x mapped at various tensile strains. Figures 13.6b and d show the evolution of ε_y and ε_x with increasing tensile strain, respectively, which were statistically averaged along the tensile direction and plotted as a function of thickness. Several general features can be observed from these strain distributions. First, ε_y distributes homogeneously along both the tensile direction Y and the thickness direction X, confirming that the UFG layers elongated concurrently with CG core in laminate structure (Wu et al., 2017). Second, the CG layer exhibits higher ε_x than the UFG layers, as seen in Fig. 13.6d. Third, ε_x exhibits obvious gradient distribution across the boundaries, i.e. high $|d\varepsilon_x/dx|$ in near boundary zones (Figs. 13.6c and d). Furthermore, the position and width of the gradient zone of strain ε_x are consistent well with that of ε_z revealed in Fig. 13.4d.

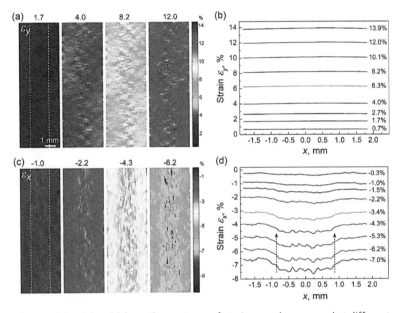

Figure 13.6 (a) and (c) are the contours of strain ε_y and ε_x mapped at different tensile strains, respectively. (b) and (d) are the distribution of statistical average strain ε_y and ε_x plotted as a function of distance from the center of thickness, respectively. The dotted arrows and lines mark the position of layer boundaries.

Figure 13.7a shows the plot of strain gradient ($|d\varepsilon_x/dx|$) against the sample thickness, which is extracted from the distribution of ε_x. The evolution of the width (W) and intensity (I) for two strain gradient peaks are plotted in Fig. 13.7b. As shown, the intensity I increases with increasing applied strain. Since the strain gradient is resulted from the mutual constraint between incompatible zones, the persistent increase of I suggests that the interaction between CG and UFG layers was never interrupted or attenuated during uniform elongation. Surprisingly, the width W (~500 μm) remains largely constant during straining. One possible reason for this puzzle is that the zone for the accumulation of GNDs induced by the mutual interaction across the gradient boundary is limited within ~500 μm. This is similar to the situation in CG/NS laminates that the strain gradient zone around sharp boundaries maintained a constant width during tension (Huang et al., 2018; Ma et al., 2016).

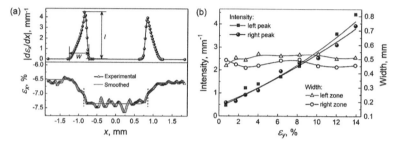

Figure 13.7 (a) The distribution of strain ε_x and strain gradient $|d\varepsilon_x/dx|$ at a tensile strain of 13.9%. (b) The evolution of the intensity (I) and width (W) of strain gradient peaks.

13.4 Discussion

The inhomogeneous distribution of strains across the gradient boundary measured by height contour measurement and DIC strain characterization can verify the validity of each other, according to the criterion that metals keep a constant volume during plastic deformation, i.e. $\varepsilon_y = -(\varepsilon_x + \varepsilon_z)$. Note that the strain gradient of two lateral shrinking strains only exists in the X direction with structural heterogeneity. This demonstrates that such strain gradients are inherent to the heterostructurres.

13.4.1 Formation of Strain Gradient Across Gradient Boundary

The mechanical incompatibilities induced by structure heterogeneity, such as the discrepancies in elastic limit, strain hardening capability and propagation of plastic strain, lead to synergistic constraint between CG and UFG layers during deformation (Wang et al., 2018c;

Wu et al., 2017, 2014a). This is capable of reshaping the strain path of component layers by strain partitioning between them (Huang et al., 2016). In order to maintain the strain continuity and avoid the formation of stress singularity in the HBAR, the formation of strain gradient was inevitable accordingly (Ashby, 1970; Mughrabi, 2006).

In the yielding stage, the big difference in elastic limit across boundary led to a long elastic-plastic transition process, during which the fast shrinking of yielded CG in HBAR can be suppressed by a lateral tensile constraint from elastic UFG layer, i.e. elastic/plastic interaction (PNAS Fe, Wang et al., 2018c; Yang et al., 2019; Zeng et al., 2016). At the end of yielding, the discrepancy of lateral elastic strain between CG and UFG layer is calculated to be about 0.05%.

After yielding, the CG and UFG layers deformed plastically together. However, the mechanical incompatibilities are still prevalent across the boundaries. Here, several reasons associated with structure heterogeneity are summarized for the prevalence of strain discrepancy and the accumulation of strain gradient during the uniform plastic deformation stage. Firstly, the UFG layers have earlier strain localization tendency due to their relatively low strain hardening capability. Such unstable tendency is incompatible with the stable plastic deformation of CG core, and can be constrained by the latter (Wu et al., 2014a). This interaction process contributes to the accumulation of strain gradient near the CG/UFG boundaries.

Secondly, the propagation of early damages in UFG layers, such as micro strain concentration bands, can be effectively passivated and/or impeded by the gradient boundary (Chen et al., 2008; Huang et al., 2016). This procedure can make the propagation of damages preferentially along the direction parallel to boundaries, i.e., reshaping the strain path of UFG layer by promoting the shrinking strain along direction Z (Fig. 13.4). It is for this reason that the ε_z in UFG layer is higher than that in CG layer (Fig. 13.4d). As a result, the strain discrepancy across boundary is enhanced, and the intensity of strain gradient is further increased.

13.4.2 GNDs Pile-Up Across Gradient Boundary

Following the plastic model of compatible heterogeneous deformation proposed by Ashby (Ashby, 1970), the density and arrangement of GNDs in strain gradient zone can be derived by inserting an array of dislocations to compensate the extraordinary strain and/or displacement (Cheng et al., 2018; Li et al., 2017b, 2017a). For the zone with known inhomogeneous distribution of plastic strain, the density of GNDs, ρ_G, can be calculated by (Gao et al., 1999)

$$\rho_G = \eta/b, \tag{13.1}$$

where b is the Burgers vector. η is the equivalent strain gradient which is expressed as

$$\eta = \frac{1}{2}\sqrt{\eta_{ijk}\eta_{ijk}}, \tag{13.2}$$

and

$$\eta_{ijk} = u_{k,ij}, \tag{13.3}$$

where η_{ijk} and u_k are the strain gradient and the displacement tensors, respectively. In the current study, the statistically averaged ε_y is uniformly distributed in the three axial directions, and ε_x and ε_z exhibit gradient in the X direction only (Figs. 13.4 and 13.6), i.e. $\varepsilon_y = C$, $\varepsilon_x = h(x)$ and $\varepsilon_z = -C - h(x)$. Therefore, the first-order approximation of displacement functions u in X, Y and Z directions can be written as

$$u_x = \int h(x)\, d_x, \tag{13.4a}$$

$$u_y = Cy, \tag{13.4b}$$

and

$$u_z = -Cz - h(x)z, \tag{13.4c}$$

where C is a constant. The non-zero components of strain gradient η_{ijk} derived from Eqs. (13.3) and (13.4) are

$$\eta_{xxx} = -\eta_{xzz} = -\eta_{zxz} = h'(x), \tag{13.5a}$$

and

$$\eta_{xxz} = -h''(x)\, z, \tag{13.5b}$$

Note that the hydrostatic part of above η_{ijk} components is also equal to 0, which is another necessary condition for plastic incompressibility. Submitting Eqs. (13.5) and (13.2) into Eq. (13.1) leads to the detailed function of ρ_G as follow

$$\rho_G = \sqrt{3h'^2(x) + h''^2(x)z^2}/2b. \tag{13.6}$$

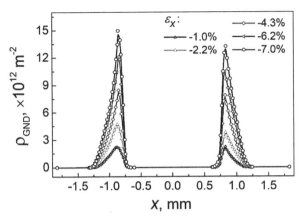

Figure 13.8 The distribution of GNDs density, showing the accumulation in HBAR with increasing strain.

By slicing the strain gradient zone into n unit layers, one can assume that for each layer GNDs are uniformly arranged in the plane paralleled with the boundary (Li et al., 2017a, 2017b). This assumption can be greatly approximated if there is no in-layer structure inhomogeneity. Therefore, the $h''^2(x)z^2$ in Eq. (13.6) can be interpreted as a second-order small component that is negligible. The $h'(x)$ in Eq. (13.6) was calculated numerically by differentiating the smoothed data matrix of ε_x (Fig. 13.6d and 13.7a) along X direction. The calculated ρ_G is plotted as the distance from thickness center in Fig. 13.8, which shows pronounced increase of GNDs pile-up in HBARs. This is similar to the GNDs pile-up around the sharp boundary between CG and NS layers (Huang et al., 2018; Ma et al., 2016). These results confirm the roles of gradient boundary in undertaking strain gradient and accumulating GNDs, i.e., accommodating strain inhomogeneity. It should be noted that the above discussion is based on the assumption that all strain gradient is accommodated by the GNDs, a concept that has been well accepted, but needs to be verified experimentally.

13.4.3 Extraordinary Strengthening Effects of Gradient Boundary

The hardness distributions of both laminate and freestanding samples measured at the tensile strain of ~14% are compared with that of as-prepared samples in Fig. 13.9a. It is clear that two hardness peaks appear at the boundaries after tensile deformation. In Fig. 13.9b, the blue data represents the hardness increment (ΔH) of laminate sample, and the red dotted line is a fitted ΔH distribution basing on the ΔH of freestanding componential layers. It can be seen that big ΔH gaps (the green shadow regions) between the measured and fitted results appear in the HBARs, suggesting extra strain hardening retained in strain gradient regions. This confirms that the enhanced flow stress and strain hardening rate found in laminate sample (Fig. 13.3) are largely caused by the strain gradient-related strengthening effects of gradient boundary (Huang et al., 2018).

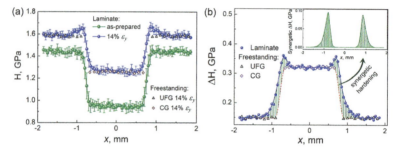

Figure 13.9 (a) Hardness distribution of laminate, freestanding CG and UFG samples measured before and after tension. (b) The hardness increment (ΔH) of samples. The insert and the shadow region in (b) are the extraordinary hardness increment in the HBAR of laminate sample.

The GND accumulation in strain gradient zone affects dislocation activities in two ways. Firstly, GNDs can act as local obstacles to mobile dislocations, i.e. playing a similar role as statistically stored dislocations (SSDs) via. Taylor-type strengthening (σ_T) (Gao et al., 1999; Mughrabi, 2006)

$$\sigma_T = M\alpha\mu b\sqrt{\rho_s + \rho_G}, \quad (13.7)$$

where M, α, μ and ρ_s are the Taylor factor, Taylor constant, shear modulus and density of SSDs, respectively. Another is that the gradient variation in spatial distribution of ρ_G is capable of producing a directional back stress, which can reduce the resolved effective stress of dislocation slip in a long-range manner (Ashby, 1970; Kassner et al., 2013; Mughrabi, 2006), leading to hetero-deformation-induced (HDI) strengthening (Wu and Zhu, 2019). For the present laminate sample, the von-Mises equivalent back stress (σ_b) produced by local edge GNDs can be expressed as (Bayley et al., 2006; Li et al., 2017b)

$$\sigma_b = \sqrt{16v^2 - 16v + 7}\,\frac{\mu b R^2}{8(1-v)}\frac{\partial \rho_G}{\partial x}, \quad (13.8)$$

where v is the Poisson's ratio. R is the integral circular for GNDs to contribute to the back stress, which should be on the order of the characteristic material length associated with strain gradient plasticity (Gao et al., 1999; Huang et al., 2018).

Importantly, the improved multiaxial stresses induced by synergetic constraint between layers in strain gradient zone provide an opportunity to activate more slip systems (Asaro, 1983). This can promote the interaction and entanglement of dislocations, thereby contributing to the improvement of total dislocation density rather than only GNDs. As verified in gradient nanotwinned Cu, the total density of dislocations in constituent layers was several times higher than that in their standalone counterparts (Cheng et al., 2018).

Considering the above strengthening mechanisms in strain gradient zone, the extra flow stress generated from the gradient boundary in CG/UFG laminate ($\Delta \sigma$) can be approximately calculated by superposing the contribution of each unite layer

$$\Delta \sigma = \sum_{i=1}^{n} f_i (\Delta \sigma_{Ti} + \sigma_{bi}), \quad (13.9a)$$

and

$$\Delta \sigma_{Ti} = M\alpha \mu b \left(\sqrt{\rho_{si} + \rho_{Gi}} - \sqrt{\rho_{sialone}} \right), \quad (13.9b)$$

$$\sigma_{bi} = \sqrt{16v^2 - 16v + 7}\,\frac{\mu b R^2}{8(1-v)}\frac{\Delta \rho_{Gi}}{\Delta x_i}, \quad (13.9c)$$

where f_i, ρ_{si}, ρ_{Gi} and $\Delta \rho_{Gi}/\Delta x_i$ represent the volume fraction, density of SSDs, density of GNDs and its average gradient of the ith layer, respectively. $\rho_{sialone}$ is the density of SSDs of the freestanding ith layer. These extraordinary strengthening effects around gradient boundary are responsible for the extra flow stress in Fig. 13.3a. In addition, the gradual increase of GND pile-up also helps with maintaining high strain hardening rate (Fig. 13.3b), i.e. producing synergistic work hardening (Park et al., 2018; Wu et al., 2015).

Note that the extraordinary mechanical responses detected in present UFG/CG laminate originated from the synergistic effects around only two discrete boundaries. This may be attributed to the large affected zone (~500 mm) of gradient boundary which is much wider than that (~10 mm) of the sharp NS/CG boundary observed in roll-bonded laminate (Huang et al., 2018; Ma et al., 2016). These results suggest that the gradient boundary is more efficient in producing synergetic strengthening and work hardening. It is anticipated that more pronounced synergistic effects could be obtained if denser gradient boundaries are introduced in the heterostructure (Wu and Zhu, 2017). Since the synergistic interaction across boundary is predominated by the mechanical incompatibility between zones, the macroscopic boundary discussed here can also be viewed as an broaden model boundary for understanding the effects of microscopic boundary in the structure composed of heterostructured zones.

13.5 Conclusions

In summary, the effects of gradient boundary on the strain behavior and mechanical responses of a CG/UFG laminate were quantitatively studied by tensile test, height profile and DIC strain measurements. The main conclusions are summarized as below:

1. An enhanced strength-ductility synergy was found in the heterostructured laminate, which can be primarily attributed to the extraordinary strengthening and work hardening effects originated from gradient boundaries.
2. The gradient boundary plays an important role in accommodating the strain incompatibility between layers via forming gradient strain distribution and accumulating dense GNDs. The width of the zone with strain gradient remains a constant, whereas the intensity of strain gradient increases with increasing applied strain.
3. The gradient boundary is more effective than sharp boundary in strengthening. The accumulation of GNDs near gradient boundaries leads to extra dislocation and HDI strengthening as well as work hardening for laminate.

Acknowledgements

This work was supported by the National Key R&D Program of China (2017YFA0204403), National Natural Science Foundation of China (No. 11672195) and Sichuan Youth Science and Technology Foundation (2016JQ0047). Yanfei Wang would like to acknowledge the support from Chinese Scholar Council (CSC) in China for his visit to NCSU.

References

Asaro, R. J., 1983. Micromechanics of crystals and polycrystals, *Adv. Appl. Mech.*, **23**, 1–115. https://doi.org/10.1016/S0065-2156(08)70242-4

Ashby, M. F., 1970. The deformation of plastically non-homogeneous materials. *Philos. Mag.*, **21**, 399–424. https://doi.org/10.1080/14786437008238426

Bayley, C. J., Brekelmans, W. A. M., Geers, M. G. D., 2006. A comparison of dislocation induced back stress formulations in strain gradient crystal plasticity. *Int. J. Solids Struct.*, **43**, 7268–7286. https://doi.org/10.1016/j.ijsolstr.2006.05.011

Chang, C. I., Lee, C. J., Huang, J. C., 2004. Relationship between grain size and Zener–Holloman parameter during friction stir processing in AZ31 Mg alloys. *Scr. Mater.*, **51**, 509–514. https://doi.org/10.1016/j.scriptamat.2004.05.043

Chen, A., Li, D., Zhang, J., Song, H., Lu, J., 2008. Make nanostructured metal exceptionally tough by introducing non-localized fracture behaviors. *Scr. Mater.*, **59**, 579–582. https://doi.org/10.1016/j.scriptamat.2008.04.048

Cheng, Z., Zhou, H. F., Lu, Q. H., Gao, H. J., Lu, L., 2018. Extra strengthening and work hardening in gradient nanotwinned metals. *Science*, **362**, eaau1925. https://doi.org/10.1126/science.aau1925

Fang, T. H., Li, W. L., Tao, N. R., Lu, K., 2011. Revealing extraordinary intrinsic tensile plasticity in gradient nano-grained copper. *Science*, **331**, 1587–1590. https://doi.org/10.1126/science.1200177

Gao, H., Huang, Y., Nix, W. D., Hutchinson, J. W., 1999. Mechanism-based strain gradient plasticity-I. Theory. *J. Mech. Phys. Solids*, **47**, 1239–1263. https://doi.org/10.1016/S0022-5096(98)00103-3

Hirth, J. P., Lothe, J., 1982. *Theory of Dislocations*. 2nd ed., John Wiley Sons.

Huang, C. X., Wang, Y. F., Ma, X. L., Yin, S., Höppel, H. W., Göken, M., Wu, X. L., Gao, H. J., Zhu, Y. T., 2018. Interface affected zone for optimal strength and ductility in heterogeneous laminate. *Mater. Today*, **21**, 713–719. https://doi.org/10.1016/j.mattod.2018.03.006

Huang, M., Fan, G. H., Geng, L., Cao, G. J., Du, Y., Wu, H., Zhang, T. T., Kang, H. J., Wang, T. M., Du, G. H., Xie, H. L., 2016. Revealing extraordinary tensile plasticity in layered Ti-Al metal composite. *Sci. Rep.*, **6**, 38461. https://doi.org/10.1038/srep38461

Kassner, M. E., Geantil, P., Levine, L. E., 2013. Long range internal stresses in single-phase crystalline materials. *Int. J. Plast.*, **45**, 44–60. https://doi.org/10.1016/j.ijplas.2012.10.003

Li, J. J., Lu, W. J., Zhang, S. Y., Raabe, D., 2017a. Large strain synergetic material deformation enabled by hybrid nanolayer architectures. *Sci. Rep.*, **7**, 11371. https://doi.org/10.1038/s41598-017-11001-w

Li, J. J., Weng, G. J., Chen, S. H., Wu, X. L., 2017b. On strain hardening mechanism in gradient nanostructures. *Int. J. Plast.*, **88**, 89–107. https://doi.org/10.1016/j.ijplas.2016.10.003

Li, J. S., Cao, Y., Gao, B., Li, Y. S., Zhu, Y. T., 2018. Superior strength and ductility of 316L stainless steel with heterogeneous lamella structure. *J. Mater. Sci.*, **53**, 10442–10456. https://doi.org/10.1007/s10853-018-2322-4

Ma, E., Zhu, T., 2017. Towards strength–ductility synergy through the design of heterogeneous nanostructures in metals. *Mater. Today*, **20**, 323–331. https://doi.org/10.1016/j.mattod.2017.02.003

Ma, X. L., Huang, C. X., Moering, J., Ruppert, M., Höppel, H. W., Göken, M., Narayan, J., Zhu, Y., 2016. Mechanical properties of copper/bronze laminates: Role of interfaces. *Acta Mater.*, **116**, 43–52. https://doi.org/10.1016/j.actamat.2016.06.023

Mughrabi, H., 2006. Dual role of deformation-induced geometrically necessary dislocations with respect to lattice plane misorientations and/or long-range internal stresses☆. *Acta Mater.*, **54**, 3417–3427. https://doi.org/10.1016/j.actamat.2006.03.047

Murr, L. E., 2016. Dislocation ledge sources: dispelling the myth of Frank–read source importance. *Metall. Mater. Trans. A*, **47**, 5811–5826. https://doi.org/10.1007/s11661-015-3286-5

Park, H. K., Ameyama, K., Yoo, J., Hwang, H., Kim, H. S., 2018. Additional hardening in harmonic structured materials by strain partitioning and back stress. *Mater. Res. Lett.*, **6**, 261–267. https://doi.org/10.1080/21663831.2018.1439115

Wang, Y. F., An, J., Yin, K., Wang, M. S., Li, Y. S., Huang, C. X., 2018a. Ultrafine-grained microstructure and improved mechanical behaviors of friction stir welded Cu and Cu–30Zn joints. *Acta Metall. Sin. (Engl. Lett.)*, **31**, 878–886. https://doi.org/10.1007/s40195-018-0719-3

Wang, Y. F., Huang, C. X., Wang, M. S., Li, Y. S., Zhu, Y. T., 2018b. Quantifying the synergetic strengthening in gradient material. *Scr. Mater.*, **150**, 22–25. https://doi.org/10.1016/j.scriptamat.2018.02.039

Wang, Y. F., Yang, M. X., Ma, X. L., Wang, M. S., Yin, K., Huang, A. H., Huang, C. X., 2018c. Improved back stress and synergetic strain hardening in coarse-grain/nanostructure laminates. *Mater. Sci. Eng. A*, **727**, 113–118. https://doi.org/10.1016/j.msea.2018.04.107

Wei, Y. J., Li, Y. Q., Zhu, L. C., Liu, Y., Lei, X., Wang, G., Wu, Y. X., Mi, Z. L., Liu, J. B., Wang, H. T., Gao, H. J., 2014. Evading the strength–ductility trade-off dilemma in steel through gradient hierarchical nanotwins. *Nat. Commun.*, **5**, 1–8. https://doi.org/10.1038/ncomms4580

Wu, H., Fan, G. H., Huang, M., Geng, L., Cui, X. P., Xie, H. L., 2017. Deformation behavior of brittle/ductile multilayered composites under interface constraint effect. *Int. J. Plast.*, **89**, 96–109. https://doi.org/10.1016/j.ijplas.2016.11.005

Wu, X. L., Jiang, P., Chen, L., Yuan, F. P., Zhu, Y. T., 2014a. Extraordinary strain hardening by gradient structure. *Proc. Natl. Acad. Sci. U.S.A.*, **111**, 7197–7201. https://doi.org/10.1073/pnas.1324069111

Wu, X. L., Jiang, P., Chen, L., Zhang, J. F., Yuan, F. P., Zhu, Y. T., 2014b. Synergetic strengthening by gradient structure. *Mater. Res. Lett.*, **2**, 185–191. https://doi.org/10.1080/21663831.2014.935821

Wu, X. L., Yang, M. X., Yuan, F. P., Wu, G. L., Wei, Y. J., Huang, X. X., Zhu, Y. T., 2015. Heterogeneous lamella structure unites ultrafine-grain strength with coarse-grain ductility. *Proc. Natl. Acad. Sci. U.S.A.*, **112**, 14501–14505. https://doi.org/10.1073/pnas.1517193112

Wu, X. L., Zhu, Y. T., 2017. Heterogeneous materials: a new class of materials with unprecedented mechanical properties. *Mater. Res. Lett.*, **5**, 527–532. https://doi.org/10.1080/21663831.2017.1343208

Yang, M. X., Pan, Y., Yuan, F. P., Zhu, Y. T., Wu, X. L., 2016. Back stress strengthening and strain hardening in gradient structure. *Mater. Res. Lett.*, **4**, 145–151. https://doi.org/10.1080/21663831.2016.1153004

Yang, M. X., Li, R. G., Jiang, P., Yuan, F. P., Wang, Y. D., Zhu, Y. T., Wu, X. L., 2019. Residual stress provides significant strengthening and ductility in gradient structured materials. *Mater. Res. Lett.*, **7**, 433–438.

Zeng, Z., Li, X. Y., Xu, D. S., Lu, L., Gao, H. J., Zhu, T., 2016. Gradient plasticity in gradient nano-grained metals. *Extreme Mech. Lett.*, **8**, 213–219. https://doi.org/10.1016/j.eml.2015.12.005

Zhu, L. L., Ruan, H. H., Chen, A. Y., Guo, X., Lu, J., 2017. Microstructures-based constitutive analysis for mechanical properties of gradient-nanostructured 304 stainless steels. *Acta Mater.*, **128**, 375–390. https://doi.org/10.1016/j.actamat.2017.02.035

Zhu, Y. T., Wu, X. L., 2019. Perspective on heter-deformation-induced (HDI) hardening and back stress. *Mater. Res. Lett.*, **7**, 393–398.

Chapter 14

Ductility by Shear Band Delocalization in the Nano-Layer of Gradient Structure

Fuping Yuan,[a,b] Dingshun Yan,[a] Jiangda Sun,[a] Lingling Zhou,[a] Yuntian Zhu,[c,d] and Xiaolei Wu[a,b]

[a]*State Key Laboratory of Nonlinear Mechanics, Institute of Mechanics, Chinese Academy of Science, 15 Beisihuan West Road, Beijing 100190, China*
[b]*School of Engineering Science, University of Chinese Academy of Sciences, 19A Yuquan Road, Beijing 100049, China*
[c]*Nano Structural Materials Center, School of Materials Science and Engineering, Nanjing University of Science and Technology, 200 Xiaolingwei, Nanjing 210094, China*
[d]*Department of Materials Science and Engineering, North Carolina State University, Campus Box 7919, Raleigh, NC 27695, USA*
ytzhu@ncsu.edu, xlwu@imech.ac.cn

Nanostructured metals typically fail soon after yielding, starting with the formation of narrow shear band. Here we report the

---

Reprinted from *Mater. Res. Lett.*, 7(1), 12–17, 2018.

Heterostructured Materials: Novel Materials with Unprecedented Mechanical Properties
Edited by Xiaolei Wu and Yuntian Zhu
Text Copyright © 2018 The Author(s)
Layout Copyright © 2022 Jenny Stanford Publishing Pte. Ltd.
ISBN 978-981-4877-10-7 (Hardcover), 978-1-003-15307-8 (eBook)
www.jennystanford.com

observation of shear band delocalization in gradient metals. Shear bands were nucleated and delocalized in the nanostructured layers by propagating along the gage length soon after yielding, converting the shear band into localized strain zone (LSZ). Synergistic work hardening was developed in the LSZ by regaining dislocation hardening capability, and by hetero-deformation-induced (HDI) hardening from the strain gradients in the axial and depth directions, which helped with enhancing global ductility.

Nanostructured (NS) metals feature high strength [1–8], but typically have low ductility [1, 9–12]. When stretched alone, NS metals intrinsically tend to develop shear bands soon after yielding, which propagates quickly through the cross-section [9–12], ending uniform elongation. This is because NS metals have low strain hardening capability [1, 9–12]. Strategies for improving ductility [9–19] are mostly related to recovering dislocation accumulation capability [5, 10–14], which is limited by either the small grain sizes or initial near-saturation dislocation density [2, 3, 9, 10], both of which limit dislocation accumulation during tensile testing [9–11].

Ductility is generally defined as uniform elongation in the specimen gage length during tensile testing [20]. It can be determined by the Considère criterion [21], $(d\sigma/d\varepsilon)/\sigma \geq 1$, where σ is true stress and ε true strain, or by the Hart's criterion [20], $(d\sigma/d\varepsilon)/\sigma + m \geq 1$, where m is the strain rate sensitivity. Neither criterion literally specify whether homogeneous strain in the gage section is a must for ductility, which raises a question: is it possible to maintain high ductility if strain is not uniform?

We investigated this issue using a gradient structured (GS) interstitial-free (IF) steel, consisting of NS surface layers with a continuous increase in grain sizes along the depth to the central coarse-grained (CG) layer [11, 12]. It is found that shear bands were formed in the NS layers at very early stage during tensile testing, as NS metals typically do. Unexpectedly, the shear bands became stabilized due to strain gradient and propagated slowly along the gage length to become a localized strain zone (LSZ), which produced synergistic strain hardening to help with retaining ductility. In other words, the shear bands helped with retaining ductility, contrary to our conventional understanding of strain localization [22–24].

Coarse-grained (CG) interstitial-free (IF) steel plate of 1 mm thick with a mean grain size of 26 μm was used as the initial material.

Gradient structure (GS) was prepared using surface mechanical attrition treatment (SMAT) technique [25]. The nanostructured layer is ~40 μm thick, with a mean grain size of 200 nm.

The tensile samples were dog-bone shaped with the gauge section of 1 mm × 2 mm × 10 mm. Uniaxial tensile tests were conducted at a strain rate ($\dot{\varepsilon}_{app}$) of 8×10^{-4} s^{-1} under room temperature. Digital image correlation (DIC) imaging was performed on the top NS surface layer (see Supplementary Material). Microstructures, texture, and Vickers microhardness (H_v) were characterized on samples subjected to various tensile strains. H_v was measured with load of 25 g and dwell period of 15 seconds on NS surface after grinding off surface roughness by ~10 μm deep. Focused ion beam (FIB) was used to cut transmission electron microscopy (TEM) samples precisely in the shear band and LSZ in the NS layer according to the DIC image.

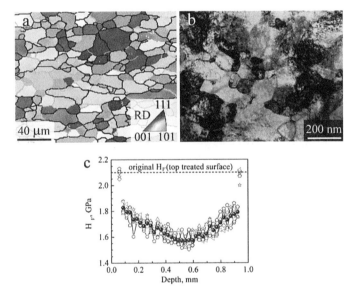

Figure 14.1 Microstructure and microhardness (H_V) in GS IF steel. (a) Electron back-scatter diffraction image showing coarse-grains in the central CG layer. (b) TEM image showing nano-grains with high density of dislocations at ~20 μm depth in NS layer. (c) H_V gradient along the depth.

Figure 14.1a shows the microstructure of the central CG layer with an average grain size of 26 μm, and Fig. 14.1b shows the nanostructure in the NS surface layer at ~50 μm depth, which reveals entangled dislocations in grain interior and the average grain size is 200 nm, typical of severely deformed metals [3, 4, 26]. A microhardness (H_V) gradient is found from the surface to the center (Fig. 14.1c).

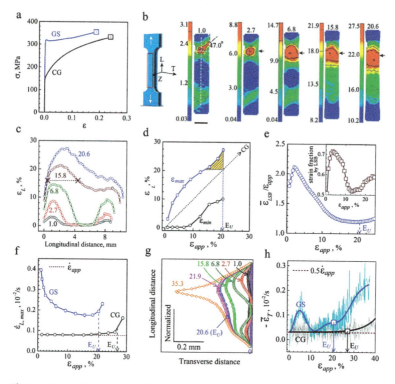

Figure 14.2 Strain localization and delocalization in LSZ. (a) Tensile true stress–strain (σ–ε) curves measured by DIC testing. GS: gradient-structured sample. CG: coarse-grained sample. Square on each curve: uniform elongation (E_U). (b) Contour maps of axial strain (ε_L) on NS surface versus applied tensile strain (ε_{app}). Surface area of the gauge section: 2 × 10 mm². Number above each map: ε_{app} (%). Scale bar (color): range of ε_L in whole contour, with maximum (defined as ε_{max}) (top number) and minimum ε_L. Scale bar (horizontal): 2 mm. (c) Distribution of ε_L at varying ε_{app} along axial position, e.g. white vertical line in the first contour in (d). (d) Maximum and minimum ε_L (ε_{max} and ε_{min}) vs ε_{app}. Diagonal dashed line: uniform ε_L of CG sample. (e) $\bar{\varepsilon}_{LSB}/\varepsilon_{app}$ versus ε_{app}. $\bar{\varepsilon}_{LSB}$: axial average strain in LSZ. Inset: Strain fraction supplied by the LSZ at varying ε_L. (f) Axial maximum strain rate ($\dot{\varepsilon}_L = d\varepsilon_L/dL$) at the propagating front of the LSZ vs ε_{app}. (g) Profiles of lateral shrinkages in (b) at varying ε_{app} (number, %). (h) Average lateral shrinkage rate at the upper LSZ vs ε_{app}.

Figure 14.2a shows the true stress–strain (σ–ε) curves of GS and CG samples. The GS sample shows a uniform strain (E_U) of 20.6%, which retained about 80% of that (26%) of the CG sample, while doubled yield strength. Figure 14.2b shows a set of typical contour maps showing longitudinal (axial) strain (ε_L, %) distribution at varying applied tensile strains (ε_{app}, %). At ε_{app} = 1%, two shear bands crossing each other were formed in the upper part of the sample. The shear bands are measured at ~45° to tensile axis, with the orientation of maximum resolved shear stress. Thus, the plastic response in NS layer begins with the nucleation of shear bands. These two shear bands propagated downward with increasing ε_{app} along the gauge length and continually broadens to form a LSZ, see maps with increasing ε_{app} from 2.7% to 15.8% until necking (E_U) at ε_{app} = 20.6% (Fig. 14.2a). A weaker LSZ is also visible in the lower part but fail to propagate much. In every sample tested, there is only one dominant LSZ, which led to the failure of whole sample.

The LSZ accumulated plastic strains continuously. Figure 14.2c shows heterogeneous ε_L at varying ε_{app} as a result of the propagating LSZ. ε_L was measured along a longitudinal line which goes through the maximum ε_L (ε_{max}) in each contour, e.g. white line at ε_{app} = 1% in Fig. 14.2b. The LSZ is defined, here, as that with $\varepsilon_L > \varepsilon_{app}$, e.g. the segment bounded by two × marks in the curve at ε_{app} of 15.8%. Figure 14.2d shows the evolution of ε_{max} and minimum ε_L (ε_{min}) in NS layer. Note that ε_{max} is always in the center of the LSZ. Figure 14.2e shows the evolution of average ε_L in LSZ, $\bar{\varepsilon}_{LSB}$. Figure 14.2f shows the axial maximum strain rate $\dot{\varepsilon}_L$ ($\dot{\varepsilon}_{max}$) in NS layer calculated by $\dot{\varepsilon}_L = \partial\varepsilon/\partial t$. $\dot{\varepsilon}_{max}$ is found always at the propagating front of the LSZ and can be used as an indicator for the propagating rate of LSZ.

From Figs. 14.2b–f, several features of the plastic deformation can be drawn. Firstly, the shear band/LSZ sustained more strain in its interior than outside with increasing ε_{app} (Figs. 14.2c and d), typical of strain localization. The left and right peaks of ε_L, e.g. at $\varepsilon_{app} \leq 6.8\%$ (Fig. 14.2c), represent the upper and the lower shear bands, while the latter disappeared at ε_{app} = 15.8%. Secondly, ε_L is not uniform in the NS layer during the whole testing (Figs. 14.2b–d). Moreover, ε_{max} in shear band/LSZ is larger than ε_{app}, even than E_U (shadowed area in Fig. 14.2d). In contrast, ε_L is equal to ε_{app} in CG (see Fig. S1 in the Supplementary Material), as represented by the diagonal dotted line, due to uniform deformation in the stable CG layer before necking.

Thirdly, the ratio of $\bar{\varepsilon}_{LSB}/\varepsilon_{app}$ in the LSZ (Fig. 14.2e) can be seen as an indicator on the severity of strain localization. As seen, strain localization started at low ε_{app} of ~1%, reached the maximum at ε_{app} = 2.0%, and then decreased until the end of uniform elongation (E_U). Fourthly, $\dot{\varepsilon}_{max}$ started about one order of magnitude larger than $\dot{\varepsilon}_{app}$ = 8 × 10^{-4} s^{-1} (Fig. 14.2f) and dropped monotonously. In other words, the shear band/LSZ propagated fast initially, but slowed down later. In contrast, the CG sample shows a nearly constant $\dot{\varepsilon}_L$ until necking (Fig. 14.2f and Fig. S1 in SI). Most importantly, the applied strain is mostly sustained in the LSZ. The deformation fraction supplied by LSZ is calculated by ($\bar{\varepsilon}_{LSB}$ × area of LSB)/(ε_{app} × gauge area). As seen in the inset of Fig. 14.2e, the LSZ accommodated the majority of the applied tensile strains.

Figure 14.2g shows the changing profiles of lateral shrinkage at varying ε_{app} by subtracting local width from the largest real-time width. The location of localized shrinkage coincides with that of the LSZ, see the arrows in contour maps at varying ε_{app} in Fig. 14.2b. In other words, the LSZ triggered localized lateral shrinkage at the very early stage. This is unexpected because the sample was still globally stable with strong strain hardening. Fig. 14.2h shows the average lateral shrinkage rate ($\dot{\varepsilon}_T$). It reached the peak at ε_{app} ~ 5%, decreased subsequently, and increased again until global necking (E_U). In contrast, the CG sample has constant $\dot{\varepsilon}_T$, equal to half of $\dot{\varepsilon}_{app}$, typical of uniform plastic deformation.

The heterogeneous ε_L (Fig. 14.2c) caused axial strain gradient, $d\varepsilon_L/dL$, near the LSZ boundaries in the NS layer. The maximum strain gradient lies always at the front of propagating LSZ, and increased with ε_{app} (Fig. 14.3a). Strain softening occurred in the center of the LSZ at the early stage of shear band formation (Fig. 14.3b), showing dramatic drop of H_V. H_V increased later with increasing ε_{app} from 6.8% to 15.8%, indicating that it recovered some strain hardening capability (Fig. 14.3c).

Strain gradient arises from mechanical incompatibility. The propagating front of the LSZ demarcates its boundary (Fig. 14.3b). As shown, there is a steep strength gradient at the LSZ boundary, which will lead to strain gradient during tensile deformation. Geometrically necessary dislocations (GNDs) need to be produced to facilitate the strain gradient [27–29], which will produce strong

HDI hardening [14, 30–36]. The HDI hardening will impede the axial rapid propagation of the LSZ, which helps with the stabilization and delocalization of the shear band/the LSZ, leading to the drop of $\dot{\varepsilon}_L$ with ε_{app} (Fig. 14.2g).

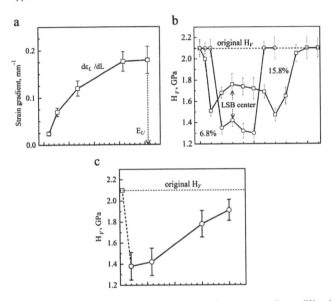

Figure 14.3 Evolution of both maximum axial strain gradient (Ψ_{max}) and microhardness (H_V). (a) Ψ_{max} at the front of the propagating LSZ vs ε_{app}. (b) H_V variation along axial LSZ at ε_{app} of 6.8% and 15.8%. (c) Change of H_V at the center of the LSZ vs ε_{app}.

The initial drop and later rise of H_V in the center of the LSZ (Figs. 14.3b and c) indicate the recovery of dislocation strain hardening. Figure 14.4a shows a compression texture (a-1) due to SMAT before tensile testing (weak compressive shear texture with (110)//compressive axis), which evolved later into tensile textures at ε_{app} of 15.8%. The tensile texture strength is especially strong at the center of the LSZ (strong tensile texture with (110)//tensile axis, as indicated in Fig. 14.4a-3), as compared with locations outside of the LSZ (weak tensile texture with (110)//tensile axis, as indicated in Fig. 14.4a-2). This indicates strong dislocation activities in nanograins in the LSZ during tensile deformation.

The dislocation activity is corroborated by TEM observations. At ε_{app} of 4% (Fig. 14.4b), the dislocations are hardly seen inside most grains in the center of the LSZ, in contrast to high density of dislocations before tensile testing (Fig. 14.1b). Inset at the top left corner of Fig. 14.4b reveals dislocation debris in a few grains. This indicates dismantlement of original dislocation sub-structure due to the change of strain path and stress state [12, 37]. This is the reason of the observed strain softening in the shear band/LSZ (Fig. 14.3b). Further straining led to the formation of new dislocation networks near grain boundaries, as shown in Fig. 14.4c (ε_{app} = 20.6%). This is caused by the complex stress state [12, 37] in the LSZ, where multiple slip systems are activated, which in turn forms new dislocation entanglements and accumulation (see the inset). The change in dislocation density coincides with that of H_V (Fig. 14.3c), suggesting their close relationship. Furthermore, the mean grain size is maintained close to ~200 nm in the LSZ during the tensile testing, indicating no grain growth in the LSZ.

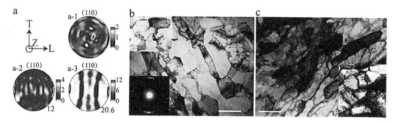

Figure 14.4 Evolution of texture and dislocation density. (a) a-1: initial compressive texture. a-2 and a-3 at ε_{app} = 15.8%: tensile texture outside and at the center of LSZ. Number: local axial strain. (b) and (c) TEM micrographs showing dislocation density in the LSZ at ε_{app} of 4% and 15.8%, respectively. Inset in (b) (lower left): selected area electron diffraction pattern in the NC layer. Inset in (b) (upper left): dislocation debris. Inset in (c): newly-generated dislocations at grain boundaries. Scale bars: 200 nm.

It should be noted that initial softening and recovered hardening observed in the LSZ has some similarity and difference from the reported hardness fluctuation observed during the severe plastic deformation of a nanocrystalline Ni–Fe alloy [38]. Although both cases were linked to dislocation density change, the initial softening and recovered hardening in the LSZ was caused by the change of strain path, while the latter was observed during severe straining in the strain direction.

The NS layer and central CG layer in the GS sample were subjected to the same amount applied tensile strain. The NS layer formed propagating shear bands (Figs. 14.2b and c), while the CG layer deformed uniformly. Strain delocalization in the shear bands/LSZ helped with retaining ductility in NS layer. In contrast, shear bands would have failed homogeneous NS metals quickly [2, 3, 11, 12]. The LSZ regained strain hardening capability after the initial softening (Fig. 14.4c), a phenomenon that would rarely occur in homogeneous NS metals [38]. Both forest dislocation hardening and HDI hardening acted to stabilize the shear band [14, 33], making it possible for it delocalize gradually. The propagation of shear band into the sample depth is deterred because the underneath CG layer is stable. This induces strain gradient in the depth direction and HDI hardening to prevent LSB from propagating into the depth [12, 31, 34].

The strategy of utilizing the shear band delocalization to develop synergistic work hardening for improving the ductility is expected also applicable to other heterostructured metals consisting of NS and CG zones. Another way to delocalize strain caused by shear bands is to develop high-density of them all over the NS zones so that no individual shear bands will fail the specimen. Similar synergistic work hardening as discussed above should also work in this situation to improve ductility. In fact, such types of LSBs have been observed in layered heterostructures although their effect on work hardening was not discussed [39].

In conclusion, strain localization by shear bands seems unavoidable in nanostructures. However, in gradient structured materials the detrimental shear bands could be harnessed to benefit ductility. Specifically, in a gradient structured specimen shear bands nucleated early in the nanostructured and propagated along the gage length, instead of across the specimen cross-section as normally observed in homogeneous materials. This delocalized the shear bands to form a LSZ. Strain gradient was produced in the propagating front of the LSZ, which produced HDI hardening [14, 34] to stabilize the propagating shear bands. In addition, strain gradient were also produced near the boundaries between the LSZ and coarse-grained central layer, which produces more HDI hardening. Dislocation hardening capability in the LSZ was recovered after initial strain softening, which, along with HDI hardening, induces synergistic strain hardening to help with ductility in NS layer.

Acknowledgements

This work was supported by the National Key Basic Research Program of China under Grant Nos. (2017YFA0204402, 2017YFA0204403), the National Natural Science Foundation of China (NSFC) under Grant Nos. (11472286, 11572328, 11672313 and 11790293), and the Strategic Priority Research Program of the Chinese Academy of Sciences under Grant No. (XDB22040503). Y.Z. was also supported by the US Army Research Office (W911 NF-12-1-0009).

Declaration of interest statement: We declare no conflicts of interest.

Supplemental data for this article can be accessed here https://doi.org/10.1080/21663831.2018.1546238.

References

1. Lu K. Stabilizing nanostructures in metals using grain and twin boundary architectures. *Nat Rev Mater*, 2016; **1**:16019.
2. Meyers MA, Mishra A, Benson DJ. Mechanical properties of nanocrystalline materials. *Prog Mater Sci*, 2006; **51**:427–556.
3. Ovid'ko IA, Valiev RZ, Zhu YT. Review on superior strength and enhanced ductility of metallic nanomaterilas. *Prog Matet Sci*, 2018; **94**:462–540.
4. Valiev RZ, Estrin Y, Horita Z, Langdon TG, Zehetbauer MJ, Zhu YT. Fundamentals of superior properties in bulk nanoSPD materials. *Mater Res Lett*, 2016; **4**:1–21.
5. Lu L, Chen X, Huang X, Lu K. Revealing the maximum strength in nanotwinned copper. *Science*, 2009; **323**:607–610.
6. Wu G, Chan K-C, Zhu LL, Sun LG, Lu J. Dual-phase nanostructuring as a route to high-strength magnesium alloys. *Nature*, 2017; **545**:80–83.
7. Liu XC, Zhang HW, Lu K. Strain-induced ultrahard and ultrastable nanolaminated structure in nickel. *Science*, 2013; **342**:337–340.
8. Zhou X, Li XY, Lu K. Enhanced thermal stability of nanograined metals below a critical grain size. *Science*, 2018; **360**:526–530.

9. Zhu YT, Liao XZ. Nanostructured metals-retaining ductility. *Nat Mater*, 2004; **3**:351–352.
10. Wang YM, Chen MW, Zhou FH, Ma E. High tensile ductility in a nanostructured metal. *Nature*, 2002; **419**:912–915.
11. Fang TH, Li WL, Tao NR, Lu K. Revealing extraordinary intrinsic tensile plasticity in gradient nano-grained copper. *Science*, 2011; **331**:1587–1590.
12. Wu XL, Jiang P, Chen L, Yuan FP, Zhu YT. Extraordinary strain hardening by gradient structure. *Proc Natl Acad Sci USA*, 2014; **111**:7197–7201.
13. He BB, Hu B, Yen HW, Cheng GJ, Wang ZK, Luo HW, Huang MX. High dislocation density-induced large ductility in deformed and partitioned steels. *Science*, 2017; **357**:1029–1032.
14. Wu XL, Yang MX, Yuan FP, Wu GL, Wei YJ, Huang XX, Zhu YT. Heterogeneous lamellar structure unites ultrafine-grain strength with coarse-grain ductility. *Proc Natl Acad Sci USA*, 2015; **112**:14501–14505.
15. Yang MX, Yan DS, Yuan FP, Jiang P, Ma E, Wu XL. Dynamically reinforced heterogeneous grain structure prolongs ductility in a medium-entropy alloy with gigapascal yield strength. *Proc Natl Acad Sci USA*, 2018; **115**:7224–7229.
16. Kou HN, Lu J, Li Y. High-strength and high-ductility nanostructured and amorphous metallic materials. *Adv Mater*, 2014; **26**:5518–5524.
17. Wu XL, Yuan FP, Yang MX, Jiang P, Zhang CX, Chen L, Wei YG, Ma E. Nanodomained nickel unite nanocrystal strength with coarse-grain ductility. *Sci Rep*, 2015; **5**:11728.
18. Jia D, Wang YM, Ramesh KT, Ma E, Zhu YT, Valiev RZ. Deformation behavior and plastic instabilities of ultrafine-grained titanium. *Appl Phys Lett*, 2001; **79**:611–613.
19. Chen AY, Li DF, Zhang JB, Song HW, Lu J. Make nanostructured metal exceptionally tough by introducing non-localized fracture behaviors. *Scr Mater*, 2008; **59**:579–582.
20. Hart EW. A theory for flow of polycrystals. *Acta Metall*, 1967; **15**:1548–1549.
21. Weber L, Kouzeli M, San Marchi C, Mortensen A. On the use of Considere's criterion in tensile testing of materials which accumulate internal damage. *Scr Mater*, 1999; **41**:549–551.

22. Wei Q, Kecskes L, Jiao T, Hartwig KT, Ramesh KT, Ma E. Adiabatic shear banding in ultrafine-grained Fe processed by severe plastic deformation. *Acta Mater*, 2004; **52**:1859–1869.
23. Bian XD, Yuan FP, Zhu YT, Wu XL. Gradient structure produces superior dynamic shear properties. *Mater Res Lett*, 2017; **5**:501–507.
24. Ma Y, Yuan FP, Yang MX, Jiang, P, Ma E, Wu XL. Dynamic shear deformation of a CrCoNi medium-entropy alloy with heterogeneous grain structures. *Acta Mater*, 2018; **148**:407–418.
25. Lu K, Lu J. Surface nanocrystallization (SNC) of metallic materials: presentation of the concept behind a new approach. *J Mater Sci Technol*, 1999; **15**:193–197.
26. Cao Y, Ni S, Liao XZ, Song M, Zhu YT. Structural evolutions of metallic materials processed by severe plastic deformation. *Mater Sci Eng R*, 2018; **133**:1–59.
27. Ashby MF. The deformation of plastically non-homogeneous materials. *Philos Mag*, 1970; **21**:399–424.
28. Gao HJ, Huang Y, Nix WD, Hutchinson JW. Mechanism-based strain gradient plasticity - I. Theory. *J Mech Phys Solids*, 1999; **47**:1239–1263.
29. Gao HJ, Huang YG. Geometrically necessary dislocation and size-dependent plasticity. *Scr Mater*, 2003; **48**:113–118.
30. Zeng Z, Li XY, Xu DS, Lu L, Gao HJ, Zhu T. Gradient plasticity in gradient nano-grained metals. *Extreme Mech Lett*, 2016; **8**:213–219.
31. Wu XL, Jiang P, Chen L, Zhang JF, Yuan FP, Zhu YT. Synergetic strengthening by gradient structure. *Mater Res Lett*, 2014; **2**:185–191.
32. Li JJ, Weng GJ, Chen SH, Wu XL. On strain hardening mechanism in gradient nanostructure. *Int J Plast*, 2017; **88**:89–107.
33. Wu XL, Zhu YT, Heterogeneous materials: a new class of materials with unprecedented mechanical properties. *Mater Res Lett*, 2017; **5**:527–532.
34. Yang MX, Pan Y, Yuan FP, Zhu YT, Wu XL. Back stress strengthening and strain hardening in gradient structure. *Mater Res Lett*, 2016; **4**:145–151.
35. Liu XL, Yuan FP, Zhu YT, Wu XL. Extraordinary Bauschinger effect in gradient structured copper. *Scr Mater*, 2018; **150**:57–60.
36. Ma E, Zhu T. Towards strength-ductility synergy through the design of heterogeneous nanostructures in metals. *Mater Today*, 2017; **20**:323–331.

37. Wilson DV. Influences of cell-walls and grain-boundaries on transient responses of an IF steel to changes in strain path. *Acta Metall Mater*, 1994; **42**:1099–1111.

38. Ni, S, Wang YB, Liao XZ, Alhajeri SN, Li HQ, Zhao YH, Lavernia EJ, Ringer SP, Langdon TG, Zhu YT. Strain hardening and softening in a nanocrystalline Ni–Fe alloy induced by severe plastic deformation. *Mater Sci Eng A*, 2011; **528**:3398–3403.

39. Huang CX, Wang YF, Ma XL, Yin S, Höppel HW, Göken M, Wu XL, Gao HJ, Zhu YT. Interface affected zone for optimal strength and ductility in heterogeneous laminate. *Mater Today*, 2018; **21**:713–719.

Chapter 15

Heterostructure Induced Dispersive Shear Bands in Heterostructured Cu

Y. F. Wang,[a] C. X. Huang,[a] Q. He,[a] F. J. Guo,[a] M. S. Wang,[a] L. Y. Song,[a] and Y. T. Zhu[b,c]

[a]*School of Aeronautics and Astronautics, Sichuan University, Chengdu 610065, China*
[b]*Nano and Heterogeneous Structural Materials Center, Nanjing University of Science and Technology, Nanjing 210094, China*
[c]*Department of Materials Science and Engineering, North Carolina State University, Raleigh, NC 27695, USA*
chxhuang@scu.edu.cn, ytzhu@ncsu.edu

Here we report the formation of dispersive shear bands in a heterostructured Cu composed of coarse-grained (CG) and ultrafine-grained (UFG) zones. Microscale digital image correlation revealed that dense shear bands evolved in a stable manner over the whole gauge section. Our observation suggests that the limited strain hardening of UFG zones and deformation heterogeneity promoted

Reprinted from *Scr. Mater.*, **170**, 76–80, 2019.

Heterostructured Materials: Novel Materials with Unprecedented Mechanical Properties
Edited by Xiaolei Wu and Yuntian Zhu
Text Copyright © 2019 Acta Materialia Inc.
Layout Copyright © 2022 Jenny Stanford Publishing Pte. Ltd.
ISBN 978-981-4877-10-7 (Hardcover), 978-1-003-15307-8 (eBook)
www.jennystanford.com

shear banding, which is a major deformation mechanism in heterostructured materials. The dispersive shear banding helped with ductility retention.

Due to their low strain hardening capability, ultrafine-grained (UFG) and nanostructured (NS) metals under tensile loading often become unstable soon after yielding by forming individual catastrophic shear bands (SBs) [1]. Efforts for regaining the ductility of these materials by annealing are usually accompanied by a significant sacrifice of strength, i.e., a tradeoff between strength and ductility [1, 2]. As a common deformation phenomenon in high-strength NS metals under tension, shear banding can sustain severe local plastic strain [3]. However, for conventional homogeneous UFG or NS samples under quasi-static tensile straining, well-developed SBs can generally form only in the region near fracture, which are unstable due to the lack of strain hardening to deter their propagation [4–6]. Therefore, shear banding is often associated with failure for UFG/NS metals.

It has been recently reported that heterostructures with coarse-grained (CG) soft zones embedded in UFG or NS matrix can produce high strength while retaining reasonable ductility [7–13]. For example, the lamella Ti composed of UFG matrix and CG lamellae was as strong as UFG Ti, and at the same time its ductility was even larger than homogenous CG Ti [8]. Heterostructures synthesized by abnormal grain growth in cryogenically deformed matrix retained excellent work hardening at limited strength loss [11–13].

A critical question arisen here is how the heterostructure accommodates the large uniform applied strain. Strain partitioning where the softer zones sustain higher plastic strain has been observed [8, 10]. It has been reported that the synergistic deformation of heterostructured zones induces pile-up of geometrically necessary dislocations near zone boundaries, which contributes to the development of hetero-deformation induced (HDI) stress and extra strain hardening [8–11, 14–16]. However, strain accommodation mechanism in different zones has not been well studied. Our recent experiments in NS/CG laminates revealed that the synergistic constraint between layers induces highly-dispersed stable SBs in the NS layers [14]. This suggests a new synergistic deformation mechanism. But it is not clear whether this also occurs in heterostructures composed of randomly arranged heterostructured zones.

To study the above issue, the plastic behaviors of UFG/CG bimodal-grained heterostructures are probed here using microscale digital image correlation (μ-DIC). Commercial Cu (99.9 wt.%) rod after equal channel angular pressing for eight passes via route B_c was cut into plates with a thickness of 4 mm, followed by rolling to a final thickness of 0.5 mm. Thereafter, three groups of as-rolled samples were annealed at 140°C for 2 h, 140°C for 6 h, and 230°C for 2 h, respectively.

Microstructure of the as-processed samples was characterized using electron back-scattered diffraction (EBSD) under an FEI Quanta 3D FEG instrument. Dog-bone shaped tension specimens with gauge dimension of $12 \times 2 \times 0.5$ mm^3 were cut along the rolling direction and uniaxially tested at a quasi-static rate of 5×10^{-4}. μ-DIC was conducted in a JSM-6510LV electron microscope. Detailed speckle preparation, imaging and calculation procedures can be found in [14].

Figure 15.1 shows the microstructure and grain size distribution of as-processed samples. The microstructure of as-rolled sample is characterized by homogeneous UFG with an average grain size of ~0.89 μm (Fig. 15.1a1 and a2). During annealing treatment at 140°C, some recrystallized grains grew up to form coarse-grains, while the matrix maintained largely the UFG structure (Fig. 15.1b1 and c1), i.e., partial recrystallization occurred to form a heterostructure with bimodal grain size distribution (Fig. 15.1b2 and c2). The area fraction of recrystallized CG in samples annealed at 140°C for 2 h and 6 h were statistically measured as ~30% and ~70%, respectively. In contrast, complete recrystallization occurred in the sample annealed at 230°C for 2 h, forming homogeneous CG structure with the grain sizes ranging from 5 μm to 15 μm (Fig. 15.1d1 and d2).

Note that the size ranges of grains in the UFG and CG zones are approximately the same to the grain sizes of homogeneous UFG and CG samples, respectively. For simplicity, the two types of heterostructures are labeled as $Cu_{30\%}$ and $Cu_{70\%}$, respectively, where the subscripts $x\%$ represent the area fraction of CG zones.

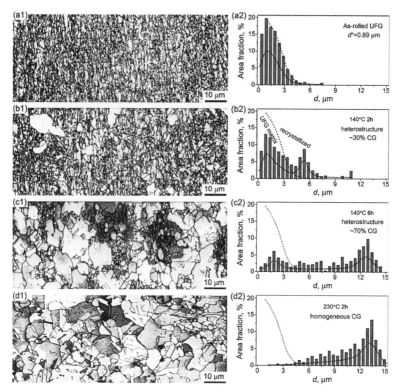

Figure 15.1 EBSD micrograph (the left column) and statistical distribution of grain size (the right column) of as-processed samples: (a1–a2) as-rolled UFG, (b1–b2) heterostructured sample with ~30% CG ($Cu_{30\%}$), (c1–c2) heterostructured sample with ~70% CG ($Cu_{70\%}$), and (d1–d2) homogeneous CG. The dotted line is a fitting curve of the grain size distribution of as-rolled UFG sample.

The tensile properties of the sample in varying processing states are compared in Fig. 15.2. The great differences in both strength and ductility between UFG and CG samples suggest significant mechanical incompatibility between the constituent zones in heterostructured samples. Remarkably, $Cu_{30\%}$ and $Cu_{70\%}$ samples displayed not only relatively high yield strengths, but also excellent uniform elongations. This implies that the microstructural heterogeneity induced an unusual deformation process for heterostructured samples.

The strain configurations across multiple CG and UFG zones in heterostructured samples are compared with that of homogeneous materials in Fig. 15.3. The μ-DIC characterization in homogeneous

UFG sample was focused in an area in the necking zone but still far away from fracture, whereas in other samples it was performed in the uniform section.

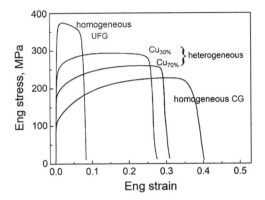

Figure 15.2 Tensile engineering stress-strain curves for as-rolled UFG, heterostructured $Cu_{30\%}$ and $Cu_{70\%}$, and homogeneous CG samples.

As shown, dense SBs in widths of several micrometers, oriented at the ~±50° direction with respect to the loading axis, are dispersed in heterostructured samples (Fig. 15.3a1–a2 and b). These SBs have significant strain accumulation but none of them carry enough strain localization to fail the sample. Therefore, these SBs can be referred to as stable SBs. However, such shear band configuration has never been reported in the gauge section of homogeneous FCC metals under tension.

Although embryonic SBs with weak strain accumulation are also formed in homogeneous UFG at the position that experienced an ultimate strain of ~4.4% (Fig. 15.3c), they have no chance to accumulate more plastic strain due to the stress relief caused by individual catastrophic SB at the most serious necking position [4–6].

No SB was observed in homogeneous CG (Fig. 15.3d) which displays high work hardening due to the uniform dislocation activity. This implies that the low strain hardening capability in UFG zone is a prerequisite for nucleating SBs. These results demonstrate that the UFG and CG zones need to co-exist in an integral bulk heterostructure to be effective in activating dispersed intense SBs and maintaining their stable evolution.

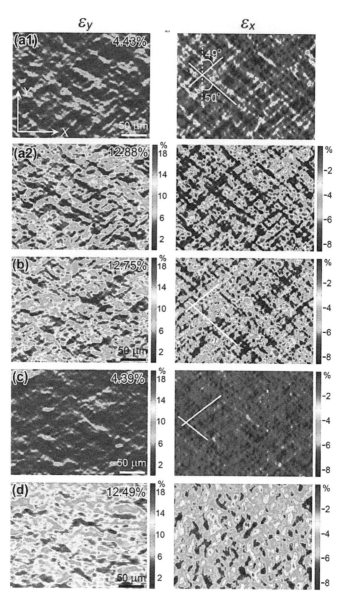

Figure 15.3 Strain maps at a low magnification in (a1–a2) Cu$_{30\%}$, (b) Cu$_{70\%}$, (c) as-rolled UFG, and (d) homogeneous CG samples, showing the shear banding deformation in heterostructures. ε_y (the left column) is the strain in tensile loading direction and ε_x (the right column) is the corresponding strain in sample width direction. The shear bands are warm-colored in ε_y contour and cold-colored in ε_x contour. The number in each ε_y subgraph represents the average true tensile strain applied to the sample.

Owing to the dispersed distribution and stable evolution, SBs evolved stably during the uniform elongation of heterostructured samples, which contributed to plastic strain. Taking the $Cu_{30\%}$ sample as an example, at the tensile strain of 12.75%, the shear banding region accounts for ~64% of gauge area, and the strain in shear banding region accounts for ~82% of total applied strain. These results indicate that dispersive shear banding is the main strain accommodation mechanism in such heterostructured material. Comparing Fig. 15.3a2 with Fig. 15.3b reveals that both the density and strain intensity of SBs in the $Cu_{30\%}$ sample are higher than those in the $Cu_{70\%}$ sample. This might be the reason that the elongation of $Cu_{30\%}$ sample is comparable to that experienced by the $Cu_{70\%}$ sample although its volume fraction of CG zone is far less than that in the $Cu_{70\%}$ sample (Fig. 15.2).

Figures 15.4a and b present the strain maps measured at a relatively high magnification in the $Cu_{30\%}$ sample, which provide insight into the interaction between SBs and heterostructured zone boundary. The CG zones are outlined by solid lines. Many strain concentration sites are frequently observed on the CG/UFG zone boundaries as indicated by the red arrows in Fig. 15.4a and b. As denoted by the white arrows, some of the SBs are stopped in front of the zone boundary or obstructed by the CG zone, whereas some other propagating SBs bypass or cut through the CG zone, but with a significant decrease in strain intensity. These observations confirm the roles of zone boundary and soft CG zone in arresting the propagation and stabilizing the strain accumulation of SBs [17–19].

The nucleation of shear banding is a plastic response of local instability, which is preferentially activated at sites with stress concentration, which may be built high enough to surpass the rate of local strain hardening [19–21]. Comparing Fig. 15.3c with a1 reveals that at similar applied strain (~4.4%) the SBs in $Cu_{30\%}$ sample is much stronger and denser than those in homogeneous UFG. Since the partially recrystallized $Cu_{30\%}$ sample displayed higher strain hardening rate, this may imply that the $Cu_{30\%}$ sample sustained even higher local stress than homogeneous UFG according to the Considère criterion. Such high stress concentration should be a mechanical response induced by structural heterogeneity.

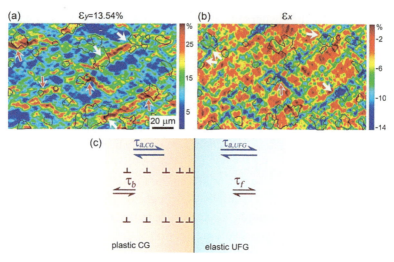

Figure 15.4 Strain maps of $Cu_{30\%}$ sample at a high magnification: (a) ε_y and (b) ε_x. The solid lines outlined the boundary of CG zones. The white arrows denote the arresting and buffering effects of CG zone on shear banding, and the red arrows mark the high strain concentration in near-boundary zone. (c) Schematic illustration of the dislocation pile-up-induced long-range back stress (τ_b) and forward stress (τ_f) around zone boundary.

As suggested by the strain concentration sites in Fig. 15.4a and b, stress concentration sites are formed preferentially on or near zone boundaries. The large difference in yield strength between CG and UFG zones induces a long elastic-plastic transition stage for heterostructured samples, during which the CG zones yield plastically, while the UFG zones continue to deform elastically. The elastic/plastic zone boundary is not penetrable to the glide dislocations in CG zone, which leads to dislocation pile-ups against zone boundaries [8, 11], as shown schematically in Fig. 15.4c. Such dislocation pile-ups produce long-range internal stresses, i.e., HDI stress (σ_b) in the CG zone and forward stress (τ_f) in the hard UFG zone (Fig. 15.4c) [22, 23]. The HDI stress acts in the opposite direction of the applied shear stress ($\tau_{a,CG}$), which strengthens CG zone by reducing the resolved shear stress ($\tau_{a,CG} - \tau_b$) and permits it to sustain higher external stress. The forward stress acts in the direction of applied stress, which leads to a higher total stress ($\sigma_{a,UFG} + \sigma_f$) in the UFG zone. Before completely yielding, the long-range internal stresses increase with increasing the number of pile-up dislocations, which contributes to extremely high stress concentration in the boundary zone near the head of dislocation pile-up [23–25].

The zone boundary will become penetrable for gliding dislocations once the total internal stress in the UFG side reaches its elastic limit [26]. This process can be promoted by the forward stress. Under such high stress concentration, local shear instability on the boundary or in the UFG side near boundary can be nucleated once a local softening is trigged by the penetration of plastic events across the heterostructured boundary [19, 20]. The UFG zone does not have enough strain hardening capability to arrest the local shear instability, which permits the propagation of local shear instability along the preferred shear orientation. This leads to the nucleation of early SBs near zone boundaries.

In contrast to the development of catastrophic SB in homogeneous UFG materials, the propagating SBs in heterostructured $Cu_{30\%}$ and $Cu_{70\%}$ samples can be effectively arrested by the neighboring CG zone after cut through the UFG zone [17], as observed above in Fig. 15.4. The plastic deformation of the CG zone ahead of a propagating shear band lowers the stress intensity at the propagating front, while simultaneously producing strain hardening to stabilize the SBs [18, 21]. On the other hand, the confined early SBs can only relieve the stress concentration locally, which provides opportunity to nucleate more SBs in other regions until they are dispersed over the whole gauge section. This appears to be the primary mechanism for the formation of dense and dispersed stable SBs in the heterostructured samples (Fig. 15.3).

Note that dispersed SBs were also experimentally observed in NS/dendrite composited Ti [17] and gradient Cu [27], stimulated in the NS surface layer of gradient Cu [28] and ferrite/martensite dual phase steel [29, 30]. The common features of these materials are that all of them are composed of multiple zones with significant mechanical incompatibility and low strain hardening capability in the harder zone. This suggests that dispersed shear banding might be a universal strain accommodation mechanism for heterostructures. More experimental and computational investigations are needed to verify this theory.

In summary, heterostructure induced the formation of dense and dispersive stable SBs in UFG/CG heterostructured Cu sample during tensile tests. The low strain hardening capability of UFG zone is a prerequisite for the nucleation of SBs. The long-range internal stress induced by elastic/plastic interaction between UFG and CG zones led to high stress concentration around zone boundaries, which promoted the nucleation of SBs from zone boundaries. The CG zone helped with stabilizing the SBs and arresting their propagation, which allowed SBs to be dispersed throughout the gauge section. This work demonstrates the possibility of achieving excellent ductility in high-strength UFG and NS materials by architecting structural heterogeneity.

Acknowledgements

This work was supported by the National Key R&D Program of China (2017YFA0204403), the National Natural Science Foundation of China (No.11672195) and Sichuan Youth Science and Technology Foundation (2016JQ0047). Y.F.W. would like to acknowledge the support from Chinese Scholar Council. Y.T.Z. was funded by the US Army Research Office.

References

1. M. A. Meyers, A. Mishra, D. J. Benson, *Prog. Mater. Sci.*, **51** (2006) 427–556.
2. I. A. Ovid'ko, R. Z. Valiev, Y. T. Zhu, *Prog. Mater. Sci.*, **94** (2018) 462–540.
3. J. E. Carsley, A. Fisher, W. W. Milligan, E. C. Aifantis, *Metall. Mater. Trans. A*, **29** (1998) 2261–2271.
4. S. Cheng, E. Ma, Y. Wang, L. Kecskes, K. Youssef, C. Koch, U. Trociewitz, K. Han, *Acta Mater.*, **53** (2005) 1521–1533.
5. K. Yang, Y. Ivanisenko, A. Caron, A. Chuvilin, L. Kurmanaeva, T. Scherer, R. Z. Valiev, H. J. Fecht, *Acta Mater.*, **58** (2010) 967–978.
6. A. M. Hodge, T. A. Furnish, A. A. Navid, T. W. Barbee, *Scr. Mater.*, **65** (2011) 1006–1009.
7. X. L. Wu, Y. T. Zhu, *Mater. Res. Lett.*, **5** (2017) 527–532.
8. X. L. Wu, M. X. Yang, F. P. Yuan, G. L. Wu, Y. J. Wei, X. X. Huang, Y. T. Zhu, *Proc. Natl. Acad. Sci. U.S.A.*, **112** (2015) 14501–14505.
9. J. S. Li, Y. Cao, B. Gao, Y. S. Li, Y. T. Zhu, *J. Mater. Sci.*, **53** (2018) 10442–10456.

10. H. K. Park, K. Ameyama, J. Yoo, H. Hwang, H. S. Kim, *Mater. Res. Lett.*, **6** (2018) 261–267.
11. S. W. Wu, G. Wang, Q. Wang, Y. D. Jia, J. Yi, Q. J. Zhai, J. B. Liu, B. A. Sun, H. J. Chu, J. Shen, P. K. Liaw, C. T. Liu, T. Y. Zhang, *Acta Mater.*, **165** (2019) 444–458.
12. Y. M. Wang, M. W. Chen, F. H. Zhou, E. Ma, *Nature*, **419** (2002) 912–915.
13. Y. H. Zhao, T. Topping, J. F. Bingert, J. J. Thornton, A. M. Dangelewicz, Y. Li, W. Liu, Y. Zhu, Y. Zhou, E. J. Lavernia, *Adv. Mater.*, **20** (2008) 3028–3033.
14. C. X. Huang, Y. F. Wang, X. L. Ma, S. Yin, H. W. Höppel, M. Göken, X. L. Wu, H. J. Gao, Y. T. Zhu, *Mater. Today*, **21** (2018) 713–719.
15. X. L. Ma, C. X. Huang, J. Moering, M. Ruppert, H. W. Höppel, M. Göken, J. Narayan, Y. T. Zhu, *Acta Mater.*, **116** (2016) 43–52.
16. Y. F. Wang, M. X. Yang, X. L. Ma, M. S. Wang, K. Yin, A. H. Huang, C. X. Huang, *Mater. Sci. Eng. A*, **727** (2018) 113–118.
17. G. He, J. Eckert, W. Löser, L. Schultz, *Nat. Mater.*, **2** (2003) 33–37.
18. M. Huang, C. Xu, G. H. Fan, E. Maawad, W. M. Gan, L. Geng, F. X. Lin, G. Z. Tang, H. Wu, Y. Du, D. Y. Li, K. S. Miao, T. T. Zhang, X. S. Yang, Y. P. Xia, G. J. Cao, H. J. Kang, T. M. Wang, T. Q. Xiao, H. L. Xie, *Acta Mater.*, **153** (2018) 235–249.
19. N. Jia, F. Roters, P. Eisenlohr, D. Raabe, X. Zhao, *Acta Mater.*, **61** (2013) 4591–4606.
20. Y. P. Li, G. P. Zhang, *Acta Mater.*, **58** (2010) 3877–3887.
21. C. C. Hays, C. P. Kim, W. L. Johnson, *Phys. Rev. Lett.*, **84** (2000) 2901–2904.
22. H. Mughrabi, *Acta Metall.*, **31** (1983) 1367–1379.
23. H. Mughrabi, T. Ungár, W. Kienle, M. Wilkens, *Philos. Mag.*, **53** (1986) 793–813.
24. F. Mompiou, D. Caillard, M. Legros, H. Mughrabi, *Acta Mater.*, **60** (2012) 3402–3414.
25. R. Schouwenaars, M. Seefeldt, P. V. Houtte, *Acta Mater.*, **58** (2010) 4344–4353.
26. Y. T. Zhu, X. L. Wu, *Mater. Res. Lett.*, **10** (2019) 393–398.
27. Y. F. Wang, F. J. Guo, Q. He, L. Y. Song, M. S. Wang, A. H. Huang, Y. S. Li, C. X. Huang, *Mater. Sci. Eng. A*, **752** (2019) 217–222.
28. Y. Lin, J. Pan, H. F. Zhou, H. J. Gao, Y. Li, *Acta Mater.*, **153** (2018) 279–289.

29. M. I. Latypov, S. Shin, B. C. De Cooman, H. S. Kim, *Acta Mater.*, **108** (2016) 219–228.
30. M. Jafari, S. Ziaei-Rad, N. Saeidi, M. Jamshidian, *Mater. Sci. Eng. A*, **670** (2016) 57–67.

Chapter 16

Dense Dispersed Shear Bands in Gradient-Structured Ni

Yanfei Wang,[a,b] Chongxiang Huang,[a] Yusheng Li,[c] Fengjiao Guo,[a] Qiong He,[a] Mingsai Wang,[a] Xiaolei Wu,[d] Ronald O. Scattergood,[b] and Yuntian Zhu[b,c]

[a]*School of Aeronautics and Astronautics, Sichuan University, Chengdu 610065,China*
[b]*Department of Materials Science and Engineering, North Carolina State University, Raleigh, NC 27695, USA*
[c]*Nano and Heterogeneous Structural Materials Center, School of Materials Science and Engineering, Nanjing University of Science and Technology, Nanjing 210094, China*
[d]*State Key Laboratory of Nonlinear Mechanics, Institute of Mechanics, Chinese Academy of Sciences, Beijing 100190, China*
chxhuang@scu.edu.cn, ytzhu@ncsu.edu

During tensile deformation, nanostructured (NS) metals often fail soon after yielding by forming a localized shear band. Here we report the observation of high density of shear bands that are homogeneously dispersed in the NS layer of a gradient Ni sample. These shear bands were nucleated at early elastic/plastic strain

---

Reprinted from *Int. J. Plast.*, **124**, 186–198, 2020.

Heterostructured Materials: Novel Materials with Unprecedented Mechanical Properties
Edited by Xiaolei Wu and Yuntian Zhu
Text Copyright © 2019 Elsevier Ltd.
Layout Copyright © 2022 Jenny Stanford Publishing Pte. Ltd.
ISBN 978-981-4877-10-7 (Hardcover), 978-1-003-15307-8 (eBook)
www.jennystanford.com

stage, reached number saturation at ~3% strain, and remained arrested by the central CG matrix during the entire plastic deformation, resulting in a uniform tensile plasticity comparable to that of CG matrix. The formation of dispersed shear bands was promoted by the elastic/plastic interaction between NS surface layer and CG matrix, and affected by the surface roughness and the hardness variation in the NS surface layer. The width of shear bands remained constant, but the intensity of strain accumulation increased almost linearly with applied tensile strain, suggesting a stable shear banding process. Microstructure examination revealed that the strain in shear bands was accommodated by mechanically driven grain boundary migration and grain coarsening. These results not only provide insight into the deformation of gradient structures, but also demonstrate the possibility of achieving large uniform elongation in NS materials by activating dispersed stable shear bands.

16.1 Introduction

Gradient-structured materials with increasing grain sizes from nanostructured (NS) surface layers to coarse-grained (CG) central layer have attracted intensive interests due to their superior combination of strength and ductility (Wu et al., 2014a; Cheng et al., 2018; Wu et al., 2016, 2014b; Wu and Zhu, 2017; Lu, 2014; Fang et al., 2011; Wei et al., 2014; Lin et al., 2018; Zhu et al., 2019; Wang et al., 2018a; Lu et al., 2019). They were found to have synergistic strengthening to produce yield strength higher than what is predicted by the rule of mixture (Cheng et al., 2018; Wang et al., 2018a; Wu et al., 2014b) and extra strain hardening (Li et al., 2017; Wu et al., 2014a) to retain good ductility.

The mechanical incompatibility between different layers is believed responsible for the observed mechanical behavior and superior properties (Wu et al., 2014a; Wu and Zhu, 2017). Specifically, during tensile testing the softer CG layer yields first to start plastic deformation, creating plastic/elastic boundaries where geometrically necessary dislocations are piled-up to create hetero-deformation induced (HDI) strengthening and extra work-hardening (Cheng et al., 2018; Kassner et al., 2013; Ming et al., 2019; Wang et al., 2018b; Yang et al., 2016; Zhu and Wu, 2019). When the whole sample is plastically deforming, the NS surface layers start unstable necking first, causing larger lateral shrinking in the NS layers than in the CG layer. This converts the applied uniaxial strain into multiaxial strains, which further helps with activating multiple slip systems and promoting dislocation hardening (Asaro, 1983; Wu et al., 2014a). In addition, residual stress produced during the processing of the gradient structure was also found to enhance mechanical properties (Moon et al., 2019; Yang et al., 2019).

A question arises on how the NS layer deforms in a gradient-structured sample. This question is critical for understanding the deformation mechanism and mechanical behavior of gradient materials. When deformed alone, NS metals often quickly start strain localization after yielding without much uniform elongation due to their low stain hardening capability (Ovid'ko et al., 2018; Valiev et al., 2016; Zhu and Wu, 2018). In gradient materials produced by surface mechanical attrition treatment, the NS layers have been reported to sustain very high plastic strains without apparent failure (Wu et al., 2014a; Cheng et al., 2018; Wu et al., 2016; Lu, 2014; Fang et al., 2011; Wei et al., 2014; Lin et al., 2018; Zhu et al., 2019). For a long time it was believed that the NS layers were deformed uniformly due to the constraint and support by the CG central layer (Fang et al., 2011; Lu, 2014; Wei et al., 2014; Wu et al., 2014a). However, it was recently reported that large shear bands were formed in the NS layer of gradient IF steel, which was delocalized along the gauge length to develop into a large strain accumulation zone with increasing tensile strain (Yuan et al., 2019). The shear bands accommodated the majority of applied tensile strain and provided some strain hardening to help with improving the overall strain hardening and ductility. This finding is surprising, because in both homogeneous NS and work-hardened polycrystalline materials the formation of such macroscopic SBs generally indicates the development of catastrophic strain localization, which will induce early fracture (Cheng et al., 2005; Ovid'ko et al., 2018; Yang et al., 2010). But it is not clear whether such stable shear banding is a universal behavior of the NS layers in gradient materials.

Shear banding is shear strain localization in a narrow zone, which often runs across multiple grain boundaries and twin boundaries during plastic deformation of polycrystalline metals. It is caused by local strain instability and accompanied with dramatic local orientation and texture change (Hong et al., 2010; Jia et al., 2013, 2012). Strain hardening was believed necessary to prevent shear banding (Borg, 2007; Mahesh, 2006). Nanostructured materials are especially prone to shear banding due to their low strain hardening capability. In the gradient structured metals, shear bands may be initiated in the NS layers. However, it may be difficult for the shear bands to propagate through the thickness/cross-section of the sample because the central CG layer typically has much higher strain hardening capability, which may act to stabilize the shear bands (Yuan et al., 2019). This needs to be further studied.

In this study we systematically studied shear band formation in the NS layer of gradient structured Ni plate with a large thickness of 3.6 mm using in-situ digital image correlation (DIC) technique. Nanostructured layers and microstructure gradients were produced with different processing techniques and parameters to study the mechanism of shear band nucleation and growth. It is found that high density of uniformly dispersed shear bands was formed instead of individual catastrophic shear bands that are typically reported in conventional homogeneous polycrystalline samples. Shear banding appears to be a primary mechanism for the NS layers to accommodate large applied plastic strain instead of being a pre-failure phenomenon.

16.2 Experimental Procedures

16.2.1 Materials and Processing

Commercial-pure Ni (99.60 wt.%) plates with a dimension of 100 mm × 90 mm × 3.6 mm were used for this study. The Ni plates were firstly annealed in vacuum at 750°C for 4 h, forming a fully recrystallized CG structure. Gradient plates with different surface roughness were symmetrically processed by means of rotationally accelerated shot peening (RASP) and piezoelectric surface nanocrystallization (PSNC) on both sides of as-annealed plates. RASP is a new surface mechanical attrition technique, which can vary both the speed and diameter of impacting balls (Wang et al., 2017). PSNC has relatively low processing efficiency but can produce extremely smooth NS surfaces (Li et al., 2016). Table 16.1 lists the main processing parameters for three different types of gradient samples $Ni_{RASP\text{-}\phi2}$, $Ni_{RASP\text{-}\phi1}$ and Ni_{PSNC}. In the RASP process, the samples peened by big balls are further treated using small balls for a longer time in order to reduce surface roughness.

Table 16.1 The processing parameters and referential label of gradient samples, where ϕ is the ball diameter, v is velocity, t is the processing time in RASP, and d is the penetration depth of the indenter in PSNC

Sample	RASP step I			RASP step II			PSNC
	ϕ, mm	v, m/s	t, min	ϕ, mm	v, m/s	t, min	d, mm
$Ni_{RASP\text{-}\phi2}$	2	40	5	0.5	40	10	—
$Ni_{RASP\text{-}\phi1}$	1	40	5	0.5	40	10	—
Ni_{PSNC}	—	—	—	—	—	—	0.3

16.2.2 Microstructural Characterization and Mechanical Tests

The microstructures of gradient samples were characterized using scanning electron microscopy (SEM) and transmission electron microscopy (TEM). Focused ion beam was used to extract TEM foils precisely from selected positions. A Ni coating was deposited on the penned surface to protect the microstructure of topmost layer before ion beam cutting. TEM observation was performed in a FEI Tecnai G2 T20 microscope at 200 KV.

The variation of Vickers hardness along the depth was measured using a standard pyramid indenter at a load of 25 g for 15 s. Tests for each sample were repeated at four independent locations. Indentations were arranged along a zigzag line, and the space between neighboring indentations was three times longer than the diagonal of impression. Dog-bone shaped tensile specimens with a gauge dimension of $18 \times 4.9 \times 3.6$ mm³ were machined from as-processed gradient plates. Figure 16.1A illustrates the geometry of gradient tensile specimen. Tensile specimens with only the topmost 400-μm-thick layer or the central 2.0-mm-thick core layer of $Ni_{RASP\text{-}\phi1}$ material were prepared by polishing away the other layers of gradient tensile specimen. All tensile tests were performed at a strain rate of 5×10^{-4} s^{-1}.

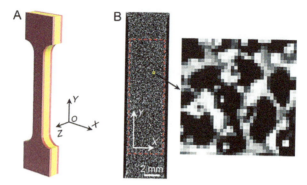

Figure 16.1 (A) Illustration of the geometry of gradient tensile specimen. In the coordinate, Y is the tensile loading direction, X is the sample width direction, and Z is the sample thickness direction with microstructure gradient. The XOY plane parallels to the NS surface. (B) Typical speckle pattern (the left subgraph) and detailed distribution of gray scale (the right subgraph) on the NS surface on gauge section. The red frame marks the 1452 × 450 pixel² effective calculation area.

16.2.3 DIC Strain Characterization

The strain distribution and evolution on the nanostructured surface and the lateral surface (the surface parallel to the YOZ plane) of gradient tensile specimens during tension were in-situly recorded by DIC, using a short-focus optical lens. A random speckle pattern was prepared by spraying black paints on white background before performing DIC imaging. Figure 16.1B presents a typical speckle image taken from the nanostructured surface. As indicated by the dotted red frame, an effective area of 1452 × 450 pixel² with a resolution of 9.7 µm/pixel was used for DIC calculation. The right subgraph shows the distribution of gray scale in a representative correlation window (36 × 36 pixel²). The mean intensity gradient of the gray profile of present speckle pattern is calculated as 48.6, meaning a high correlation coefficient for strain calculation (Pan et al., 2010).

16.3 Results

16.3.1 Surface Roughness of the Gradient Samples

Figure 16.2 shows the linear distribution of the surface roughness of the gradient samples, which were measured using a white light interferometry. The height resolution is ~20 nm. The maximum height difference between the convex peak and concave valley of roughness contour for $Ni_{RASP\text{-}\phi2}$, $Ni_{RASP\text{-}\phi1}$ and Ni_{PSNC} samples are measured as 80.8 μm, 37.1 μm and 2.1 μm, respectively. As shown, the sample processed with larger balls has higher roughness. The PSNC produced a very smooth surface due to its high striking frequency.

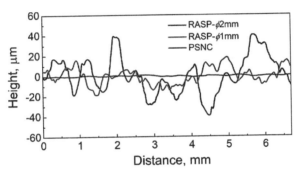

Figure 16.2 The linear distribution of surface roughness measured in the three types of gradient samples.

16.3.2 Gradient Microstructure and Microhardness

The $Ni_{RASP\text{-}\phi1}$ sample was used to show typical gradient microstructure. Figure 16.3A is a representative SEM image from the top NS layer to the CG central layer, which reveals a severely deformed sub-surface layer with obliterated initial grain boundary. Figure 16.3B is a typical TEM image taken at the depth range of ~2–3 μm below the penned surface, showing a mixture of elongated and equiaxed nanostructures with high density of dislocations. The solid symbols in Fig. 16.3C mark the microhardness profiles of the three types of gradient samples before tensile testing. As shown, the topmost layers of all samples have a hardness of ~270 Hv, which is about twice that of the CG matrix. The thicknesses of the gradient layers of the $Ni_{RASP-\phi2}$, $Ni_{RASP-\phi1}$ and Ni_{PSNC} samples measured from these hardness profiles are ~900 μm, ~780 μm and ~450 μm, respectively. The detailed gradient microstructure and formation mechanism of such gradient samples were reported in previous works (Liu et al., 2015; Wang et al., 2006).

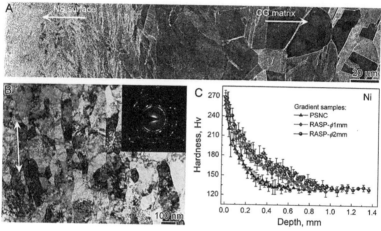

Figure 16.3 (A) A gradient microstructure from nanostructured surface layer to coarse-grained central layer; (B) Bright-field TEM image showing the NS at the depth range of ~2–3 μm. (C) Micro-hardness profiles measured in the mechanical gradient layer of as-received gradient samples. The double-arrowed white line in (B) indicates the direction parallel to surface.

16.3.3 Strength-Ductility Combination

Tensile specimens with only the topmost 400-μm-thick layer and the central 2.0-mm-thick core layer of $Ni_{RASP-\phi1}$ material are referred to as freestanding nanostructured gradient surface layer (NGSL) and homogeneous CG core, respectively. Figure 16.4 compares the tensile behaviors of gradient samples, the freestanding NGSL and homogeneous CG layer. The gradient samples exhibit improved yield strength and excellent ductility (uniform elongation), which indicates a superior strength-ductility combination. Interestingly, the yield strength of the freestanding NGSL reaches 504 ± 6 MPa, while its ductility is lowered to ~6.8%. These results indicate that the NS surface layers were well supported by the CG layer and did not fail prematurely.

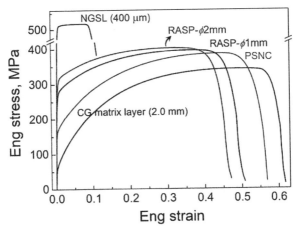

Figure 16.4 Tensile engineering stress-strain curves for gradient samples and the freestanding NGSL and homogeneous CG matrix layer.

16.3.4 Dense Dispersed Shear Bands in Nanostructured Layer

Figures 16.5A–C present the distribution and evolution of strains in the tensile direction ε_y (the left five columns) and width direction ε_x (the right five columns) on the nanostructured surface of gradient samples during uniform elongation. Surprisingly, dense macroscopic shear bands (SBs), orientated at 49° ~ 55° with respect to the tensile axis, are uniformly distributed over the whole surface of all gradient samples. In contrast, such SBs are not found in the homogeneous CG sample (Fig. 16.5D), which implies that the low strain hardening capability of the NS layer is a precondition for the formation of SBs. It is also clear that the three types of gradient samples exhibit obvious difference in the morphology of SBs and the extent of strain inhomogeneity, which indicates significant effect of surface roughness and gradient layer thickness on the SB dimension, density and strain intensity.

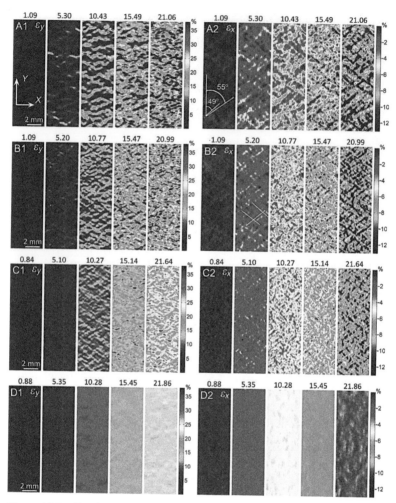

Figure 16.5 The distributions of strain ε_y (the left column) and corresponding ε_x (the right column) on the surface of $Ni_{RASP\text{-}\phi2}$ (A1–A2), $Ni_{RASP\text{-}\phi1}$ (B1–B2) and Ni_{PSNC} (C1–C2) gradient samples and homogeneous CG (D1–D2) sample. They reveal the distribution and evolution of dispersive shear bands (SBs) in gradient sample. The shear bands are warm-colored in ε_y contour and cold-colored in ε_x contour. In the coordinate, Y is the tensile loading direction and X is the sample width direction. The number above each subgraph represents the average true tensile strain applied to the sample.

For every type of gradient sample, the distribution and evolution of strains in 8 randomly selected SBs were analyzed. Taking a representative SB in the $Ni_{RASP\text{-}\phi2}$ sample as an example, the distributions of ε_y and ε_x across the band at different tensile strains were statistically averaged along the shear banding direction and plotted in Figures 16.6A and B, respectively. As shown, the peak strain in SB is nearly twice that in the background (i.e. non-shear banding zone). It should be noted that the strains in both SBs and background increase with applied strain.

Figure 16.6C demonstrates that the strain peak profile across a SB can be fitted with a Gauss function to extract two important parameters: the strain intensity (I) and width at half maximum (W). Figure 16.6D compares the evolutions of the average strain intensity $\left(\overline{I}\right)$ and width at half maximum $\left(\overline{W}\right)$ extracted from the ε_x profile of the selected SBs, with increasing applied strain. As shown, the $|\overline{I}|$ increased linearly with increasing applied strain, suggesting a stable plastic deformation in SBs. In addition, the $|\overline{I}|$ increased faster in the $Ni_{RASP-\phi 2}$ sample than in the Ni_{PSNC} sample. The $\overline{W}$ remains largely constant during the whole plastic straining process, indicating constant SB widths, which contrasts the width growth in gradient IF steel (Yuan et al., 2019). The $\overline{W}$ of ε_x strain peaks in the $Ni_{RASP-\phi 2}$, $Ni_{RASP-\phi 1}$ and Ni_{PSNC} samples are ~0.33 mm, ~0.27 mm and ~0.24 mm, respectively. These values are much larger than those in homogeneous NS/UFG bulks (Jia et al., 2013; Hong et al., 2010; Jia et al., 2003; Carsley et al., 1995, 1998).

Figure 16.7 shows SB nucleation at early strain stages in a $Ni_{RASP-\phi 2}$ sample. SBs started to nucleate at the elastic-plastic transition stage (indicated by the white arrows). The density evolution of SBs, total length of SBs per unit area, as a function of applied strain is shown in Fig. 16.8. It is revealed that for all samples the SB density increased quickly at the elastic-plastic transition and low plastic-strain stages, and then reached saturation at ~3% strain. It is also clear that the saturated density of SBs in the Ni_{PSNC} sample with smoother surface is nearly double that in the $Ni_{RASP-\phi 2}$ sample.

Figure 16.9 shows the SB evolution in a stand-alone NGSL. As indicate by the white arrow, a dominant SB with extremely high strain concentration quickly developed, penetrated through the cross-section and then caused fracture, leading to low ductility. Meanwhile, the SBs in uniform gauge section exhibit a lower density than in the integrated $Ni_{RASP-\phi 1}$ sample (Fig. 16.5B). These results suggest that dense dispersive SBs can be activated only when NS surface layers and CG matrix are deformed together.

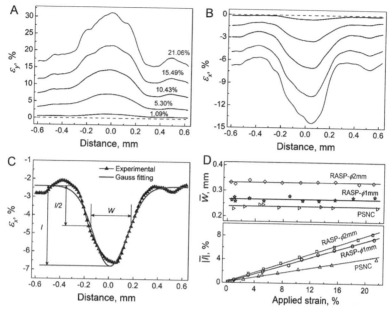

Figure 16.6 Strain distribution and evolution in SBs. (A) and (B) are the statistically averaged distribution of ε_y and ε_x across a representative SBs in $\text{Ni}_{\text{RASP-}\phi 2}$. Curves in (A) and (B) with the same color are strain distributions at the same strain state. (C) Gauss fitting of the strain peak of a SB, from which the strain intensity (I) and the full width at half maximum (W) are measured. (D) The evolution of $|\bar{I}|$ and $\bar{W}$ in the ε_x peak of SBs in three types of gradient samples, with increasing applied strain. Each data point in (D) was averaged from the values of 8 randomly selected SBs.

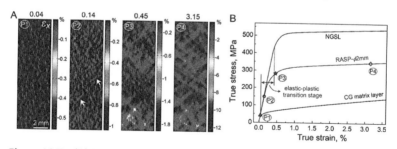

Figure 16.7 (A) SBs on the surface of $\text{Ni}_{\text{RASP-}\phi 2}$ sample at different strain stages, revealing the fast increase of SBs density in elastic-plastic transition stage (P2–P3) and low plastic-strain stage (P3–P4). The white arrows indicate the initial embryos of inclined strain bands. (B) True stress-strain curves, showing the strain stage of P1–P4.

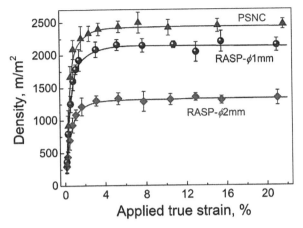

Figure 16.8 The density evolution of SBs in gradient samples.

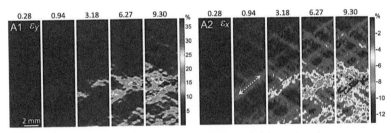

Figure 16.9 SBs on the surface of freestanding NGSL that peeled from $Ni_{RASP-\phi1}$ material: (A1) ε_y; (A2) ε_x. The double-headed arrow indicates the run-away SB.

16.4 Discussion

The experimental results show that the enhanced plasticity of NS surface layer in gradient structure can be primarily attributed to the formation of dense macroscopic SBs and their stable evolution. Such dispersed and stable shear banding deformation was never observed in homogeneous metals under tension, which suggests that this mechanism is unique to the NS-CG gradient structure.

16.4.1 Unique Characteristics of Dispersed Shear Bands

As a non-crystallographic deformation mode, SBs are also often observed in materials without enough strain hardening to maintain uniform plastic flow, such as NS/UFG bulk metals and metallic glasses (Yang et al., 2010; Cheng et al., 2005; Hong et al., 2010; Jia et al., 2003; Carsley et al., 1995, 1998; Lu et al., 2013; Hays et al., 2000). It is interesting that both the configuration and evolution of the SBs observed here are very different from those in conventional homogeneous materials.

First, the SBs in gradient Ni are observed to distribute uniformly over the whole gauge section, and the plastic deformation in the SBs evolved stably with increasing applied strain (Fig. 16.6D). In contrast, in NS/UFG metals, SBs propagate quickly to fail the sample due to the low strain hardening capacity and lack of constraint, and the gauge region far away from neck or fracture zone generally exhibits no SBs (Cheng et al., 2005; Yang et al., 2010). Second, the SBs in the gradient sample have macroscopic sizes in the order of millimeters (Fig. 16.6 and Fig. 16.5A–C). In contrast, except for the predominant unstable SB, the height of most SBs in the NS/UFG bulks is in the range of several tens or hundreds of nanometers, and their width generally ranges from submicron to several microns, i.e., micro SBs (Jia et al., 2003; Lu et al., 2013; Yang et al., 2010). Third, the density of SBs in gradient Ni reached saturation at ~3% strain (Fig. 16.8), while new SBs and branches in NS materials are nucleated through the whole plastic straining process (Jia et al., 2003). Fourth, the constant width of SBs in the gradient Ni differs greatly from the continuous thickening behavior of micro SBs in NS metals (Carsley et al., 1995; Hong et al., 2010; Jia et al., 2003). This is also different from the earlier report of SB widening in gradient IF steel (Yuan et al., 2019).

16.4.2 Nucleation of Dispersed Shear Bands

SBs may be nucleated when the local shear stress (τ_θ) on the preferential shear banding plane (θ) reaches the critical value for shear instability ($\tau_{c\theta}$), which is preferentially activated at stress concentration sites under load (Ardeljan et al., 2015; He et al., 2003; Jia et al., 2013). In heterostructured materials, the boundaries between soft and hard zones are likely sites for stress concentration due to the mutual constraint-induced complex stress state and dislocation pile-up (C. X. Huang et al., 2018; M. Huang et al., 2018; Wang et al., 2019c; Wu et al., 2015).

For the gradient structure, during the elastic-plastic transition stage (see Fig. 16.7B), the elastic NS surface layer and the plastic inner layers form an elastic-plastic zone boundary. Due to the incompatibility in lateral shrinking rate between the inner plastic CG (apparent Poisson ratio $v \approx 0.5$) and the elastic NS surface layer ($v \approx 0.33$), a lateral compression stress (σ_x^-) is added to the NS surface layer (see Fig. 16.10) (Wang et al., 2018b; Wu et al., 2014b). This increases the shear stress τ_θ from $\sigma_y \sin\theta \cos\theta$ to $(\sigma_y + |\sigma_x^-|)\sin\theta \cos\theta$. In addition, the dislocation pile-up near this elastic/plastic interface also contributes to stress concentration (Wu et al., 2015; Zeng et al., 2016). As verified by the Vickers hardness map in Fig. 16.11, the surface layers themselves exhibit significant hardness variation, which indicates the existence of in-layer soft/hard zone boundary. This is expected to introduce stress/strain inhomogeneity and concentration sites as well (Sun et al., 2009; Wu and Zhu, 2017). Hence, the NS surface layers in gradient materials are believed to have abundant dispersed potential sites for nucleating SBs.

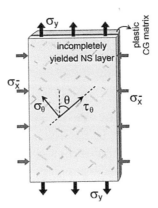

Figure 16.10 Schematic illustration of the load and constraint applied in the NS surface layer of gradient sample at the elastic-plastic transition stage. σ_y is external load, and σ_x^- is the compressive constraint from plastic CG matrix. θ is the preferential shear banding plane. The tilted short bands indicate the formation of early SBs.

During deformation, once the stress concentration is built high enough (reaches $\tau_{c\theta}$) to induce plastic deformation across a zone boundary, the local plastic deformation may become unstable in the

NS surface layer because the harder zone likely has finer grain size and low strain hardening capability (Ardeljan et al., 2015; Wang et al., 2019b). This will nucleate a local SB, whose propagation in the length direction can only be stopped by another hard zone with even higher strength in the NS layer. This effect is similar to the activation of SBs in NS Ti composite (He et al., 2003), bimodal structured Cu (Wang et al., 2019b), and heterophase laminate (Jia et al., 2013), where the SB nucleation was promoted by the interaction-induced high internal stress at zone boundaries.

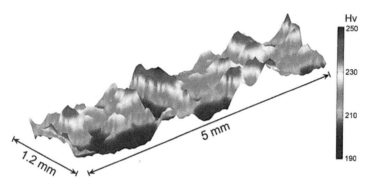

Figure 16.11 Vickers hardness contour measured on the sub-surface layer (at the depth of 130 μm) of as-received Ni$_{RASP-\phi2}$ sample. The data map contains the value of 2400 homogeneously distributed indentations. The indent depth is ~2.75–3.15 μm.

As discussed above, SBs can be effectively arrested by the neighboring harder zones in the NS layer due to the strength inhomogeneity. In the depth direction, a propagating SB will meet layers with larger grains whose higher plasticity will lower the stress at the band front, while their stronger work hardening helps with arresting the propagation. The arrest of propagating SBs provides opportunity to nucleate more SBs in less optimal regions, which eventually leads to the nucleation of high density of dispersed SBs that are homogeneously distributed over the whole NS layer. This is the primary mechanism for the fast increase of SB density at the yielding and low plastic-strain stages (Fig. 16.7 and Fig. 16.8). This process is similar to the arrestment of propagating SBs and the activation of multiple SBs in inhomogeneous metallic glasses composited with ductile phase or layers (Hays et al., 2000; Kosiba et al., 2019; Lu et al., 2013; Sha et al., 2017; Zhou et al., 2013).

The orientation of SBs observed here is not along the plane with maximum shear stress, i.e., 45° with respect to the loading axis. This is a glasslike local shear instability response (Carsley et al., 1995; Donovan, 1989; Zhang et al., 2003), which has also been observed in high-strength NS and UFG metals. For example, micro SBs in UFG Pd–Ag alloy oriented ~50° to the tensile direction (Yang et al., 2010). Asymmetric shear banding orientations under tension and compression were observed in NS Fe–Cu alloy (Carsley et al., 1998, 1995). It has been proposed that such behaviors can be attributed to the yielding-related shear banding process and the normal stress-dependent yielding mechanism (Carsley et al., 1998). As described in the Mohr-Coulomb criterion ($\tau_{y\theta} = \tau_0 - \mu\sigma_\theta$), the shear yielding stress ($\tau_{y\theta}$) on inclined θ plane of NS metals is sensitive to the in-plane normal stress (σ_θ) (Donovan, 1989; Zhang et al., 2003). For the NS surface layer of gradient Ni sample under tension, the $\tau_{y\theta}$ is met more easily on planes 49° ~ 55° to the tensile axis, which induces plastic shear banding in this angle range.

16.4.3 Stable Evolution of Dispersed Shear Bands

As demonstrated by the linear evolution of strain intensity in SBs (Fig. 16.6D), SBs in NS layer remain stable to very large applied tensile strains without serious strain localization to fracture the sample.

Without the CG central layer the freestanding NGSL formed a dominant SB that caused the early fracture (Fig. 16.9), which verifies the critical role of the CG central layer in stabilizing SB propagation. Figure 16.12 presents the DIC strain mappings characterized on the lateral surface of gradient sample. As shown, SBs were originated from NS surface layers and weakened along depth with a gradual decrease in strain intensity. They partially intersected with each other, and were terminated by CG interior at the depth of ~650 μm. On the other hand, the in-layer mutual intersection of SBs constrained their in-layer propagation (Fig. 16.5A–C). These processes blunted bands sharpness, and hindered their catastrophic development, leading to a stable propagation process. Stabilization of strain localization by the neighboring softer zone or incompatible interface has also been reported in other heterostructures (Azizi et al., 2018; Chen et al., 2008; Wu et al., 2017; Zhu and Lu, 2012).

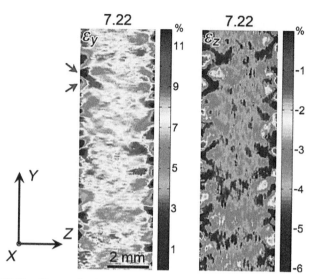

Figure 16.12 SBs on the lateral surface of $Ni_{RASP-\phi 1}$ sample, measured at the applied strain of 7.22%. Z represents the thickness direction with microstructure gradient.

The CG interior of present gradient Ni sample may also play a role in constraining the widening of dispersed SBs. In this study, the SB width remained constant with increasing strain (Fig. 16.6D), which is very different from the sustained widening of SB observed in the NS layer of gradient IF steel (Yuan et al., 2019). A closer examination of the gradient IF steel sample found that the gradient layer extended all the way to the center due to thin whole sample thickness (1 mm). In contrast, the Ni sample used here is much thicker (3.6 mm), which resulted in a much thicker CG central layer. Therefore, the likely reason for this difference is that the CG core in the gradient IF steel is too thin to effectively constrain the widening of SBs.

Moreover, the strain intensity of SBs increased gradually with applied strain (Fig. 16.6), which indicates that large strain gradient was formed in the shear banding zone during straining. As reported in gradient structured IF steel and CG/NS laminates, such strain gradient increase is known to lead to the accumulation of geometrically necessary dislocations and the development of HDI work hardening (C. X. Huang et al., 2018; Li et al., 2017; Yang et al., 2016). The enhanced strain hardening will help to stabilize the plastic flow in SBs.

16.4.4 Microstructure Evolution in Shear Bands

Dispersed SBs stabilizes the uniform elongation of NS surface layer, which verifies a long-term conjecture that NS metals may serve as ductile materials as long as catastrophic strain localization is effectively suppressed (Fang et al., 2011; Yuan et al., 2019). A crucial curiosity arising here is how the microstructure in SBs evolves to accommodate such large tensile strain.

We examined the microstructure of the topmost NS layers of $Ni_{RASP-\phi 1}$ sample before and after tension (Fig. 16.13). The microstructure of as-processed surface layer is characterized by largely elongated grains with an average transversal grain size of 68 nm and an aspect ratio of 5.17 (Fig. 16.13A1 and A2). After deformation, surprisingly, equiaxed grains with even larger size and lower dislocation density are developed in shear banding zone (Fig. 16.13B1 and B2). These observations suggest that the plastic deformation of shear banding zone is dominated by grain coarsening, which has been interpreted as a mechanically-driven grain boundary migration mechanism in NS metals (Chen et al., 2017; Fang et al., 2011; Wang et al., 2019a). The increase of grain size and change of aspect ratio are also detected in the zones outside SBs (Fig. 16.13C1 and C2). However, comparing to that in shear banding zone, the extent of grain coarsening is less significant and the initial NS remains in local regions (marked by dotted cycles), suggesting that the rate and extent of grain boundary migration are positively correlated with stress/strain concentrations (Rupert et al., 2009).

As marked by red arrows in Fig. 16.13C1, under stress circumstance grain boundary migration occurred at the expense of neighboring highly-defective microstructures. This procedure results in dismantlement and annihilation of initial defects, thereby accommodating applied strain. The migration of grain boundary can be interpreted as a releasing process of distortion energy (Legros et al., 2008; Wang et al., 2012). Accompanying grain boundary migration in SBs, the formation of equiaxed grains with relatively low dislocation density provides substantial available room for defects storage (Fig. 16.13B1), which implies the regaining of strain hardening capability for shear banding zone (Huang et al., 2015;

Yuan et al., 2019). With the further increase of applied strain, the dominant plastic mechanism in shear banding zone will shift from grain boundary migration to conventional dislocation slip as the size of grown-up grains tends to saturation (Fang et al., 2011). This is the reason why some equiaxed coarse grains in shear banding zone exhibit high dislocation density (Fig. 16.13B1). More importantly, the regained strain hardening capability may also play crucial roles in stabilizing shear banding deformation.

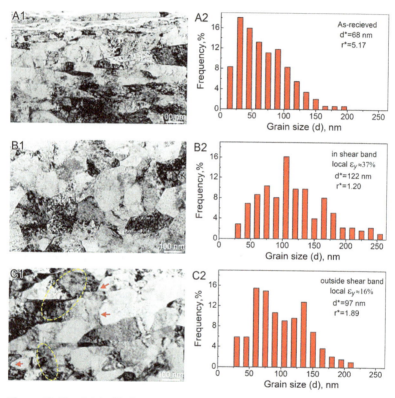

Figure 16.13 Bright-filed TEM micrographs and the statistical distribution of grain size in the topmost layer of $Ni_{RASP-\phi 1}$: (A1, A2) As-received; (B1, B2) shear banding zone with a local ε_y of ~37%; (C1, C2) non-shear banding matrix with a local ε_y of ~16%. TEM foils of (B) and (C) were extracted from the SB and the neighboring matrix indicated by red arrows in Fig. 16.12, at the applied strain (whole-field average) of 21.66%. Dotted cycles and red arrows in (C1) mark the residue initial NS and the migrating grain boundary, respectively.

16.4.5 Effects of Surface Roughness and Strength Heterogeneity on Shear Banding

Surface roughness could significantly affect the morphology of SBs, but is not the primary reason for SB nucleation. As shown in Fig. 16.2, the PSNC sample had a very smooth surface, but dispersed SBs appeared in its NS layer (Fig. 16.5C) with higher density than in the RASP samples which had much higher surface roughness (Fig. 16.8).

The statistical structural parameters of gradient samples, and the geometrical parameters and strain intensity of SBs are summarized in Table 16.2. It can be seen that SBs in samples with bigger surface roughness exhibit greater width, higher strain intensity but lower spatial density. This can be attributed to the effect of concave-convex roughness pattern on the development of stress concentration. Bigger peening balls produced the wider and deeper valleys on the surface layer (Table 16.2 and Fig. 16.2), which consequently led to larger stress concentration sites with longer inter-site distance and higher stress concentration under load.

Since the NS surface layers in all of the gradient samples were produced by the impact of high-speed balls or indenter, it was inevitable for the deformation to be heterogeneous (Panin et al., 2015). This led to inhomogeneous distribution of hardness (strength) in surface layers (Fig. 16.11), and thus provided extra soft/hard incompatible boundaries for stress concentration and strength heterogeneity to arrest the propagation of early SBs during deformation.

16.5 Conclusions

In summary, a unique plastic strain accommodation mechanism, formation of dense dispersed SBs, was experimentally revealed in gradient-structured Ni using DIC strain measurements. The main conclusions can be drawn as follow.

(1) Dense dispersed SBs were uniformly distributed over the NS surface layer, and evolved stably during the entire plastic deformation, resulting in excellent uniform elongations (larger than 30.6%) which were much larger than that of a freestanding NS surface layer (~6.8%).

Table 16.2 The structural parameters of gradient samples and the geometrical parameters of SBs, including the maximum height difference between the convex peak and concave valley (H) and the average space between neighboring concave valleys ($\overline{d_v}$) of surface roughness contour, the thickness of gradient layer (t_G), the thickness ratio of gradient layer to sample (t_G/t_S), the saturated SB density (ρ_b), the average space between neighboring SBs ($\overline{d_b}$), the average strain intensity ($|\overline{I}|$) and width at half intensity ($\overline{W}$) of ε_x peaks at the tensile strain of ~15.5%

Material	Roughness		Gradient layer			SBs		
	H, μm	$\overline{d_v}$, mm	t_G, μm	t_G/t_t	ρ_b, m/m²	$\overline{d_b}$, mm	$\overline{W}$, mm	$|\overline{I}|$, %
Ni$_{RASP-\phi2}$	80.8	0.82 ± 0.46	900	0.5	1330 ± 100	0.91 ± 0.67	0.33 ± 0.12	6.20 ± 3.8
Ni$_{RASP-\phi1}$	37.1	0.38 ± 0.23	780	0.43	2140 ± 150	0.56 ± 0.44	0.27 ± 0.09	5.62 ± 4.0
Ni$_{PSNC}$	2.1	—	450	0.25	2430 ± 120	0.49 ± 0.31	0.24 ± 0.11	3.13 ± 2.1

(2) The nucleation of dispersed SBs started at an early elastic-plastic transition stage, and reached saturation at ~3% strain. The gradient microstructure, surface roughness and hardness variation introduced abundant soft/hard zone boundaries in the NS surface layer. The stress concentration induced by elastic/plastic interaction at zone boundaries promoted the early nucleation of SBs. The CG matrix and neighboring hard zones arrested the propagation of early SBs, which provided an opportunity to nucleate more SBs at dispersed stress concentration sites.

(3) Stable shear banding at the large strain stage was maintained by the stabilization and mechanical constraints from the central CG matrix and the mutual intersection of SBs.

(4) The large applied strain in shear bands was accommodated by mechanically-driven grain boundary migration. Grain coarsening in shear bands led to the regaining of strain hardening capability, which helps to stabilize shear banding deformation.

(5) Surface roughness affected the morphology of SBs by affecting the stress concentration in the surface layer.

Acknowledgements

This work was supported by the National Key R&D Program of China (2017YFA0204403), the National Natural Science Foundation of China (Nos.11672195 and 51741106) and Sichuan Youth Science and Technology Foundation (2016JQ0047). Yanfei Wang would like to acknowledge the support from Chinese Scholar Council.

References

Ardeljan, M., Knezevic, M., Nizolek, T., Beyerlein, I. J., Mara, N. A., Pollock, T. M., 2015. A study of microstructure-driven strain localizations in two-phase polycrystalline HCP/BCC composites using a multi-scale model. *Int. J. Plast.*, **74**, 35–57. https://doi.org/10.1016/j.ijplas.2015.06.003.

Asaro, R. J., 1983. Micromechanics of crystals and polycrystals, in: *Adv. Appl. Mech.*, Elsevier, pp. 1–115. https://doi.org/10.1016/S0065-2156(08)70242-4.

Azizi, H., Zurob, H. S., Embury, D., Wang, X., Wang, K., Bose, B., 2018. Using architectured materials to control localized shear fracture. *Acta Mater.*, **143**, 298–305. https://doi.org/10.1016/j.actamat.2017.10.027.

Borg, U., 2007. Strain gradient crystal plasticity effects on flow localization. *Int. J. Plast.*, 23, 1400–1416. https://doi.org/10.1016/j.ijplas.2007.01.003.

Carsley, J. E., Fisher, A., Milligan, W. W., Aifantis, E. C., 1998. Mechanical behavior of a bulk nanostructured iron alloy. *Metall. Mater. Trans. A*, **29**, 2261–2271. https://doi.org/10.1007/s11661-998-0104-3.

Carsley, J. E., Milligan, W. W., Hackney, S. A., Aifantis, E. C., 1995. Glasslike behavior in a nanostructured Fe/Cu alloy. *Metall. Mater. Trans. A*, **26**, 2479–2481. https://doi.org/10.1007/BF02671262.

Chen, A., Li, D., Zhang, J., Song, H., Lu, J., 2008. Make nanostructured metal exceptionally tough by introducing non-localized fracture behaviors. *Scr. Mater.*, **59**, 579–582. https://doi.org/10.1016/j.scriptamat.2008.04.048.

Chen, W., You, Z. S., Tao, N. R., Jin, Z. H., Lu, L., 2017. Mechanically-induced grain coarsening in gradient nano-grained copper. *Acta Mater.*, **125**, 255–264. https://doi.org/10.1016/j.actamat.2016.12.006.

Cheng, S., Ma, E., Wang, Y., Kecskes, L., Youssef, K., Koch, C., Trociewitz, U., Han, K., 2005. Tensile properties of in situ consolidated nanocrystalline Cu. *Acta Mater.*, **53**, 1521–1533. https://doi.org/10.1016/j.actamat.2004.12.005.

Cheng, Z., Zhou, H. F., Lu, Q. H., Gao, H. J., Lu, L., 2018. Extra strengthening and work hardening in gradient nanotwinned metals. *Science*, **362**, eaau1925. https://doi.org/10.1126/science.aau1925.

Donovan, P. E., 1989. A yield criterion for Pd40Ni40P20 metallic glass. *Acta Metall.*, **37**, 445–456. https://doi.org/10.1016/0001-6160(89)90228-9.

Fang, T. H., Li, W. L., Tao, N. R., Lu, K., 2011. Revealing extraordinary intrinsic tensile plasticity in gradient nano-grained copper. *Science*, **331**, 1587–1590. https://doi.org/10.1126/science.1200177.

Hays, C. C., Kim, C. P., Johnson, W. L., 2000. Microstructure controlled shear band pattern formation and enhanced plasticity of bulk metallic glasses containing *in situ* formed ductile phase dendrite dispersions. *Phys. Rev. Lett.*, **84**, 2901–2904. https://doi.org/10.1103/PhysRevLett.84.2901.

He, G., Eckert, J., Löser, W., Schultz, L., 2003. Novel Ti-base nanostructure-dendrite composite with enhanced plasticity. *Nat. Mater.*, **2**, 33–37. https://doi.org/10.1038/nmat792.

Hong, C. S., Tao, N. R., Huang, X., Lu, K., 2010. Nucleation and thickening of shear bands in nano-scale twin/matrix lamellae of a Cu–Al alloy processed by dynamic plastic deformation. *Acta Mater.*, **58**, 3103–3116. https://doi.org/10.1016/j.actamat.2010.01.049.

Huang, C. X., Hu, W. P., Wang, Q. Y., Wang, C., Yang, G., Zhu, Y. T., 2015. An ideal ultrafine-grained structure for high strength and high ductility. *Mater. Res. Lett.*, **3**, 88–94. https://doi.org/10.1080/21663831.2014.968680.

Huang, C. X., Wang, Y. F., Ma, X. L., Yin, S., Höppel, H. W., Göken, M., Wu, X. L., Gao, H. J., Zhu, Y. T., 2018. Interface affected zone for optimal strength and ductility in heterogeneous laminate. *Mater. Today*, **21**, 713–719. https://doi.org/10.1016/j.mattod.2018.03.006.

Huang, M., Xu, C., Fan, G. H., Maawad, E., Gan, W. M., Geng, L., Lin, F. X., Tang, G. Z., Wu, H., Du, Y., Li, D. Y., Miao, K. S., Zhang, T. T., Yang, X. S., Xia, Y. P., Cao, G. J., Kang, H. J., Wang, T. M., Xiao, T. Q., Xie, H. L., 2018. Role of layered structure in ductility improvement of layered Ti-Al metal composite. *Acta Mater.*, **153**, 235–249. https://doi.org/10.1016/j.actamat.2018.05.005.

Jia, D., Ramesh, K. T., Ma, E., 2003. Effects of nanocrystalline and ultrafine grain sizes on constitutive behavior and shear bands in iron. *Acta Mater.*, **51**, 3495–3509. https://doi.org/10.1016/S1359-6454(03)00169-1.

Jia, N., Roters, F., Eisenlohr, P., Kords, C., Raabe, D., 2012. Non-crystallographic shear banding in crystal plasticity FEM simulations: example of texture evolution in α-brass. *Acta Mater.*, **60**, 1099–1115. https://doi.org/10.1016/j.actamat.2011.10.047.

Jia, N., Roters, F., Eisenlohr, P., Raabe, D., Zhao, X., 2013. Simulation of shear banding in heterophase co-deformation: example of plane strain compressed Cu–Ag and Cu–Nb metal matrix composites. *Acta Mater.*, **61**, 4591–4606. https://doi.org/10.1016/j.actamat.2013.04.029.

Kassner, M. E., Geantil, P., Levine, L. E., 2013. Long range internal stresses in single-phase crystalline materials. *Int. J. Plast.*, **45**, 44–60. https://doi.org/10.1016/j.ijplas.2012.10.003.

Kosiba, K., Şopu, D., Scudino, S., Zhang, L., Bednarcik, J., Pauly, S., 2019. Modulating heterogeneity and plasticity in bulk metallic glasses: role of interfaces on shear banding. *Int. J. Plast.*, **119**, 156–170. https://doi.org/10.1016/j.ijplas.2019.03.007.

Legros, M., Gianola, D. S., Hemker, K. J., 2008. In situ TEM observations of fast grain-boundary motion in stressed nanocrystalline aluminum films. *Acta Mater.*, **56**, 3380–3393. https://doi.org/10.1016/j.actamat.2008.03.032.

Li, J. J., Weng, G. J., Chen, S. H., Wu, X. L., 2017. On strain hardening mechanism in gradient nanostructures. *Int. J. Plast.*, **88**, 89–107. https://doi.org/10.1016/j.ijplas.2016.10.003.

Li, Y. S., Li, Y. L., Zhu, Y. T., Liu, J., 2016. A surface nanocrystallization device of metallic materials driven by piezoelectricity. Chinese Patent, CN 201621380875.X.

Lin, Y., Pan, J., Zhou, H. F., Gao, H. J., Li, Y., 2018. Mechanical properties and optimal grain size distribution profile of gradient grained nickel. *Acta Mater.*, **153**, 279–289. https://doi.org/10.1016/j.actamat.2018.04.065.

Liu, X. C., Zhang, H. W., Lu, K., 2015. Formation of nano-laminated structure in nickel by means of surface mechanical grinding treatment. *Acta Mater.*, **96**, 24–36. https://doi.org/10.1016/j.actamat.2015.06.014.

Lu, K., 2014. Making strong nanomaterials ductile with gradients. *Science*, **345**, 1455–1456.

Lu, X. C., Zhang, X., Shi, M. X., Roters, F., Kang, G. Z., Raabe, D., 2019. Dislocation mechanism based size-dependent crystal plasticity modeling and simulation of gradient nano-grained copper. *Int. J. Plast.*, **113**, 52–73. https://doi.org/10.1016/j.ijplas.2018.09.007.

Lu, X. L., Lu, Q. H., Li, Y., Lu, L., 2013. Gradient confinement induced uniform tensile ductility in metallic glass. *Sci. Rep.*, **3**, 3319. https://doi.org/10.1038/srep03319.

Mahesh, S., 2006. Deformation banding and shear banding in single crystals. *Acta Mater.*, **54**, 4565–4574. https://doi.org/10.1016/j.actamat.2006.05.043.

Ming, K. S., Bi, X. F., Wang, J., 2019. Strength and ductility of CrFeCoNiMo alloy with hierarchical microstructures. *Int. J. Plast.*, **113**, 255–268. https://doi.org/10.1016/j.ijplas.2018.10.005.

Moon, J. H., Baek, S. M., Lee, S. G., Seong, Y., Amanov, A., Lee, S., Kim, H. S., 2019. Effects of residual stress on the mechanical properties of copper processed using ultrasonic-nanocrystalline surface modification. *Mater. Res. Lett.*, **7**, 97–102. https://doi.org/10.1080/21663831.2018.1560370.

Ovid'ko, I. A., Valiev, R. Z., Zhu, Y. T., 2018. Review on superior strength and enhanced ductility of metallic nanomaterials. *Prog. Mater. Sci.*, **94**, 462–540. https://doi.org/10.1016/j.pmatsci.2018.02.002.

Pan, B., Lu, Z. X., Xie, H. M., 2010. Mean intensity gradient: An effective global parameter for quality assessment of the speckle patterns used in digital image correlation. *Opt. Lasers Eng.*, **48**, 469–477. https://doi.org/10.1016/j.optlaseng.2009.08.010.

Panin, A. V., Kazachenok, M. S., Kozelskaya, A. I., Hairullin, R. R., Sinyakova, E. A., 2015. Mechanisms of surface roughening of commercial purity titanium during ultrasonic impact treatment. *Mater. Sci. Eng. A*, **647**, 43–50. https://doi.org/10.1016/j.msea.2015.08.086.

Rupert, T. J., Gianola, D. S., Gan, Y., Hemker, K. J., 2009. Experimental observations of stress-driven grain boundary migration. *Science*, **326**, 1686–1690. https://doi.org/10.1126/science.1178226.

Sha, Z. D., Branicio, P. S., Lee, H. P., Tay, T. E., 2017. Strong and ductile nanolaminate composites combining metallic glasses and nanoglasses. *Int. J. Plast.*, **90**, 231–241. https://doi.org/10.1016/j.ijplas.2017.01.010.

Sun, X., Choi, K. S., Liu, W. N., Khaleel, M. A., 2009. Predicting failure modes and ductility of dual phase steels using plastic strain localization. *Int. J. Plast.*, **25**, 1888–1909. https://doi.org/10.1016/j.ijplas.2008.12.012.

Valiev, R. Z., Estrin, Y., Horita, Z., Langdon, T. G., Zehetbauer, M. J., Zhu, Y. T., 2016. Fundamentals of superior properties in bulk nanoSPD materials. *Mater. Res. Lett.*, **4**, 1–21. https://doi.org/10.1080/21663831.2015.1060543.

Wang, B., Idrissi, H., Galceran, M., Colla, M. S., Turner, S., Hui, S., Raskin, J. P., Pardoen, T., Godet, S., Schryvers, D., 2012. Advanced TEM investigation of the plasticity mechanisms in nanocrystalline freestanding palladium films with nanoscale twins. *Int. J. Plast.*, **37**, 140–156. https://doi.org/10.1016/j.ijplas.2012.04.003.

Wang, K., Tao, N. R., Liu, G., Lu, J., Lu, K., 2006. Plastic strain-induced grain refinement at the nanometer scale in copper. *Acta Mater.*, **54**, 5281–5291. https://doi.org/10.1016/j.actamat.2006.07.013.

Wang, X., Li, Y. S., Zhang, Q., Zhao, Y. H., Zhu, Y. T., 2017. Gradient structured copper by rotationally accelerated shot peening. *J. Mater. Sci. Technol.*, **33**, 758–761. https://doi.org/10.1016/j.jmst.2016.11.006.

Wang, Y. F., Guo, F. J., He, Q., Song, L. Y., Wang, M. S., Huang, A. H., Li, Y. S., Huang, C. X., 2019a. Synergetic deformation-induced extraordinary softening and hardening in gradient copper. *Mater. Sci. Eng. A*, **752**, 217–222. https://doi.org/10.1016/j.msea.2019.03.020.

Wang, Y. F., Huang, C. X., He, Q., Guo, F. J., Wang, M. S., Song, L. Y., Zhu, Y. T., 2019b. Heterostructure induced dispersive shear bands in heterostructured Cu. *Scr. Mater.*, **170**, 76–80. https://doi.org/10.1016/j.scriptamat.2019.05.036.

Wang, Y. F., Huang, C. X., Wang, M. S., Li, Y. S., Zhu, Y. T., 2018a. Quantifying the synergetic strengthening in gradient material. *Scr. Mater.*, **150**, 22–25. https://doi.org/10.1016/j.scriptamat.2018.02.039.

Wang, Y. F., Wang, M. S., Yin, K., Huang, A. H., Li, Y. S., Huang, C. X., 2019c. Yielding and fracture behaviors of coarse-grain/ultrafine-grain heterogeneous-structured copper with transitional interface. *Trans. Nonferrous Metals Soc. China*, **29**, 588–594. https://doi.org/10.1016/S1003-6326(19)64967-8.

Wang, Y. F., Yang, M. X., Ma, X. L., Wang, M. S., Yin, K., Huang, A. H., Huang, C. X., 2018b. Improved back stress and synergetic strain hardening in coarse-grain/nanostructure laminates. *Mater. Sci. Eng. A*, **727**, 113–118. https://doi.org/10.1016/j.msea.2018.04.107.

Wei, Y. J., Li, Y. Q., Zhu, L. C., Liu, Y., Lei, X., Wang, G., Wu, Y. X., Mi, Z. L., Liu, J. B., Wang, H. T., Gao, H. J., 2014. Evading the strength–ductility trade-off dilemma in steel through gradient hierarchical nanotwins. *Nat. Commun.*, **5**, 1–8. https://doi.org/10.1038/ncomms4580.

Wu, H., Fan, G. H., Huang, M., Geng, L., Cui, X. P., Xie, H. L., 2017. Deformation behavior of brittle/ductile multilayered composites under interface constraint effect. *Int. J. Plast.*, **89**, 96–109. https://doi.org/10.1016/j.ijplas.2016.11.005.

Wu, X. L., Jiang, P., Chen, L., Yuan, F. P., Zhu, Y. T., 2014a. Extraordinary strain hardening by gradient structure. *Proc. Natl. Acad. Sci. U.S.A.*, **111**, 7197–7201. https://doi.org/10.1073/pnas.1324069111.

Wu, X. L., Jiang, P., Chen, L., Zhang, J. F., Yuan, F. P., Zhu, Y. T., 2014b. Synergetic strengthening by gradient structure. *Mater. Res. Lett.*, **2**, 185–191. https://doi.org/10.1080/21663831.2014.935821.

Wu, X. L., Yang, M. X., Yuan, F. P., Chen, L., Zhu, Y. T., 2016. Combining gradient structure and TRIP effect to produce austenite stainless steel with high strength and ductility. *Acta Mater.*, **112**, 337–346. https://doi.org/10.1016/j.actamat.2016.04.045.

Wu, X. L., Yang, M. X., Yuan, F. P., Wu, G. L., Wei, Y. J., Huang, X. X., Zhu, Y. T., 2015. Heterogeneous lamella structure unites ultrafine-grain strength with coarse-grain ductility. *Proc. Natl. Acad. Sci. U.S.A.*, **112**, 14501–14505. https://doi.org/10.1073/pnas.1517193112.

Wu, X. L., Zhu, Y. T., 2017. Heterogeneous materials: a new class of materials with unprecedented mechanical properties. *Mater. Res. Lett.*, **5**, 527–532. https://doi.org/10.1080/21663831.2017.1343208.

Yang, K., Ivanisenko, Y., Caron, A., Chuvilin, A., Kurmanaeva, L., Scherer, T., Valiev, R. Z., Fecht, H. J., 2010. Mechanical behaviour and in situ observation of shear bands in ultrafine grained Pd and Pd-Ag alloys. *Acta Mater.*, **58**, 967–978. https://doi.org/10.1016/j.actamat.2009.10.013.

Yang, M. X., Li, R. G., Jiang, P., Yuan, F. P., Wang, Y. D., Zhu, Y. T., Wu, X. L., 2019. Residual stress provides significant strengthening and ductility in gradient structured materials. *Mater. Res. Lett.*, **7**, 433–438. https://doi.org/10.1080/21663831.2019.1635537.

Yang, M. X., Pan, Y., Yuan, F. P., Zhu, Y. T., Wu, X. L., 2016. Back stress strengthening and strain hardening in gradient structure. *Mater. Res. Lett.*, **4**, 145–151. https://doi.org/10.1080/21663831.2016.1153004.

Yuan, F. P., Yan, D. S., Sun, J. D., Zhou, L. L., Zhu, Y. T., Wu, X. L., 2019. Ductility by shear band delocalization in the nano-layer of gradient structure. *Mater. Res. Lett.*, **7**, 12–17. https://doi.org/10.1080/21663831.2018.1546238.

Zeng, Z., Li, X. Y., Xu, D. S., Lu, L., Gao, H. J., Zhu, T., 2016. Gradient plasticity in gradient nano-grained metals. *Extreme Mech. Lett.*, **8**, 213–219. https://doi.org/10.1016/j.eml.2015.12.005.

Zhang, Z. F., Eckert, J., Schultz, L., 2003. Difference in compressive and tensile fracture mechanisms of Zr59Cu20Al10Ni8Ti3 bulk metallic glass. *Acta Mater.*, **51**, 1167–1179. https://doi.org/10.1016/S1359-6454(02)00521-9.

Zhou, H. F., Qu, S. X., Yang, W., 2013. An atomistic investigation of structural evolution in metallic glass matrix composites. *Int. J. Plast.*, **44**, 147–160. https://doi.org/10.1016/j.ijplas.2013.01.002.

Zhu, L. L., Lu, J., 2012. Modelling the plastic deformation of nanostructured metals with bimodal grain size distribution. *Int. J. Plast.*, **30–31**, 166–184. https://doi.org/10.1016/j.ijplas.2011.10.003.

Zhu, L. L., Wen, C. S., Gao, C. Y., Guo, X., Chen, Z., Lu, J., 2019. Static and dynamic mechanical behaviors of gradient-nanotwinned stainless steel with a composite structure: Experiments and modeling. *Int. J. Plast.*, **114**, 272–288. https://doi.org/10.1016/j.ijplas.2018.11.005.

Zhu, Y. T., Wu, X. L., 2019. Perspective on hetero-deformation induced (HDI) hardening and back stress. *Mater. Res. Lett.*, **7**, 393–398. https://doi.org/10.1080/21663831.2019.1616331.

Zhu, Y. T., Wu, X. L., 2018. Ductility and plasticity of nanostructured metals: differences and issues. *Mater. Today Nano*, **2**, 15–20. https://doi.org/10.1016/j.mtnano.2018.09.004.

Part III
Gradient Structure

Chapter 17

Combining Gradient Structure and TRIP Effect to Produce Austenite Stainless Steel with High Strength and Ductility

Xiaolei Wu,[a] Muxin Yang,[a] Fuping Yuan,[a] Liu Chen,[a] and Yuntian Zhu[b,c]

[a]*State Key Laboratory of Nonlinear Mechanics, Institute of Mechanics, Chinese Academy of Sciences, Beijing 100190, China*
[b]*Department of Materials Science and Engineering, North Carolina State University, Raleigh, NC 27695, USA*
[c]*Nano Structural Materials Center, School of Materials Science and Engineering, Nanjing University of Science and Technology, Nanjing, Jiangsu 210094, China*
xlwu@imech.ac.cn, mxyang@lnm.imech.ac.cn, fpyuan@lnm.imech.ac.cn, liuchen@imech.ac.cn, ytzhu@ncsu.edu

We report a design strategy to combine the benefits from both gradient structure and transformation-induced plasticity (TRIP). The resultant TRIP-gradient steel takes advantage of both mechanisms, allowing strain hardening to last to a larger plastic strain. 304

Reprinted from *Acta Mater.*, **112**, 337–346, 2016.

Heterostructured Materials: Novel Materials with Unprecedented Mechanical Properties
Edited by Xiaolei Wu and Yuntian Zhu
Text Copyright © 2016 Acta Materialia Inc.
Layout Copyright © 2022 Jenny Stanford Publishing Pte. Ltd.
ISBN 978-981-4877-10-7 (Hardcover), 978-1-003-15307-8 (eBook)
www.jennystanford.com

stainless steel sheets were treated by surface mechanical attrition to synthesize gradient structure with a central coarse-grained layer sandwiched between two grain-size gradient layers. The gradient layer is composed of submicron-sized parallelepiped austenite zones separated by intersecting ε-martensite plates, with increasing zone size along the depth. Significant microhardness heterogeneity exists not only macroscopically between the soft coarse-grained core and the hard gradient layers, but also microscopically between the austenite zone and ε-martensite walls. During tensile testing, the gradient structure caused strain partitioning, which evolved with applied strain, and lasted to large strains. The $\gamma \to \alpha'$ martensitic transformation is triggered successively with an increase of the applied strain and flow stress. Importantly, the gradient structure prolonged the TRIP effect to large plastic strains. As a result, the gradient structure in the 304 stainless steel provides a new route towards a good combination of high strength and ductility, via the co-operation of both the dynamic strain partitioning and TRIP effect.

17.1 Introduction

It has been a challenge to produce metals and alloys with both high strength and high ductility [1–5]. High strength can be achieved by well-known strategies such as grain refinement, deformation twinning, nano-twins, second-phase particle strengthening, solution hardening, and work hardening. However, this usually comes with sacrifice of ductility [4–9]. Ductility is measured under tensile loading either as total elongation to failure or as uniform elongation. The decrease in ductility with increasing strength is often observed because (i) high strength metals often have low strain hardening rate, as in the case of cold worked metals and nanostructured metals [1, 3–7]; ii) according to the Considère criterion, a high-strength metal would need a higher strain hardening rate to reach the same uniform elongation as that of a low-strength metal [8, 10].

A promising strategy for simultaneously improving the strength and ductility is proposed recently by combining several plastic deformation and strain hardening mechanisms in steels [11–15]. Each hardening mechanism may be active over a certain strain range, but several of them together could collectively produce a high strain-hardening rate to improve ductility, when at least one of them persists over a wide range of the imposed tensile strains [11, 12]. One of the most effective approaches to increase both strain hardening and ductility is the transformation-induced plasticity, i.e. the TRIP effect, which is observed in dual-/multi-phase steels [11–13, 16–18], TRIP steels [11, 12, 19–23], bainite steels [15, 24], etc. In these steels, the phase-specific deformation mechanism takes effect at different stress or strain levels and therefore, the global plastic response is composite-like, which critically depends on the evolution of stress transfer and dynamic strain partitioning among different phases [11, 12]. In addition, martensitic transformation is activated at varying local stress or strain levels during the tensile tests, which provides not only higher strain hardening but also excellent plasticity, when the TRIP effect is designed to sustain over a wide strain range [11, 12]. In addition, to produce good TRIP effect, the phase transformation needs to persist up to large strains. High strain hardening rate is reported to originate from the harder phase produced by phase transformation as well as mechanical incompatibility, i.e. strain partitioning, between different phases with different flow stresses [11–13, 16]. Strain partitioning causes strain gradient at phase boundaries, which is accommodated by geometrically necessary dislocations [25, 26] and consequently enhances strain hardening [27]. Recently, a microstructural design concept, referred to the TRIP-maraging steel, is proposed [11]. The strength and ductility were improved simultaneously from three effects, i.e. the TRIP effect by nano-scale austenite, maraging effect by nano-precipitation, and composite effect due to strain partitioning.

Recently, another approach employing gradient structure (GS) has been reported to produce a superior combination of yield strength and ductility [28–34]. The GS often consists of the grain-size gradient layers (GLs) with increasing grain sizes along the depth towards the central coarse-grained (CG) core layer. For gradient structured metals with stable gradient structures, high ductility is attributed to an extra strain hardening due to the presence of strain gradient and the change of stress states, which generates geometrically necessary dislocations and promotes the generation and interaction of forest dislocations [29]. The GS may produce an intrinsic synergetic strengthening, with its yield strength much higher than the sum of separate gradient layers [34]. This is due to the macroscopic strength gradient and mechanical incompatibility between layers [29–34]. Such a mechanical incompatibility was also found to produce high hetero-deformation induced (HDI) stress, which may contribute to both strength and ductility [3, 35].

The objective of this investigation is to propose a design strategy, referred to as the TRIP-gradient steel, to produce high strength and high ductility. The 304 austenite stainless steel (304ss) is chosen as the model material due to its TRIP effect during tensile testing [36–44]. The mechanical properties of 304ss have been extensively studied [45–48]. In early studies, gradient structure in 304ss was reported to produce a high yield strength of 950 MPa with a ductility of 30% [49, 50], but its plastic deformation and strain hardening mechanisms were not explored. In this work, we combined the gradient structure with the TRIP effect in 304ss. In particular, the TRIP effect was designed to operate in a wider strain range, which depends on dynamic strain partitioning associated with the composite-like deformation in the GS. Microstructural observations and measurements were performed on tensile deformed samples to correlate the excellent strength and ductility to the strain partitioning and TRIP effect.

17.2 Experimental Procedure

17.2.1 Materials and SMAT Process

In the present investigation, as-annealed AISI 304ss disks were used, with a diameter of 100 mm and thicknesses of 0.5, 1 and 2 mm. The chemical composition was 0.04 C, 0.49 Si, 1.65 Mn, 7.8 Ni, 16.8 Cr, 0.37 Mo and the balance Fe (all in wt%).

The surface mechanical attrition treatment (SMAT) technique was used to produce the gradient structured 304ss specimens, with a central CG layer sandwiched between two gradient layers (GLs). During the SMAT process, spherical steel shots of 3-mm in diameter were accelerated to high speeds using high-power ultra-sound to impact on the sample disk [51, 52]. Because of the high frequency of the system (20 kHz), the entire surface of the component is peened with high density of impacts over a short period of time. The SMAT processing time was the same for both sides of each disk, which was

1 to 15 minutes. No crack was observed on the sample surface after SMAT processing.

17.2.2 Mechanical Property Tests

Tensile specimens were cut from the SMAT-processed disks. They are dog-bone shaped with a gauge length of 15 mm and width of 3 mm. Specimens of varying thicknesses in the gradient layers were also prepared for both tensile testing and X-ray diffraction (XRD) measuements of phase transitions at different depths and applied tensile strains. These specimens were prepared by polishing away the whole SMAT-processed sample from one side via mechanical polishing at first and later electro-polishing of at least 150 μm thick at room temperature, leaving behind the specimen of desired thickness.

Uniaxial tensile tests were carried out at a strain rate of 5×10^{-4} s^{-1} at room temperature using an MTS Landmark testing machine. An extensometer was used to measure the strain during tensile testing. Uniaxial tensile stress-relaxation tests were also performed under strain-control mode at room temperature. Upon reaching a designated relaxation strain, the strain was maintained constant while the stress was recorded as a function of time. After the first relaxation over an interval of 60 s, the specimen was reloaded by a strain increment of 0.6% at a strain rate of 5×10^{-4} s^{-1} for next relaxation. Three stress relaxations were conducted for each designated strain. All tensile tests were performed three times on average for each condition to verify the reproducibility of the monotonic and cyclic tensile stress-strain curves.

Vickers microhardness (Hv) indentations were made in the gauge section to measure the Hv evolution, prior to and after tensile testing, under a 25 g load across the cross-section of the gradient structured specimen.

17.2.3 Microstructural Characterization

Transmission electron microscopy (TEM) observations were conducted to investigate microstructual evolution in the GS samples during tensile tests in a JEM 2010 microscope with an operating voltage of 200 kV. TEM disks were cut from the gauge section of tensile samples after suspension of tensile testing at designated strains.

Electron back-scattered diffraction (EBSD) observations were conducted using a high-resolution field emission Cambridge S-360 SEM equipped with a fully automatic Oxford Instruments Aztec 2.0 EBSD system (Channel 5 Software). A scanning step of 0.03–0.08 μm was performed during the EBSD acquisition. Due to spatial resolution of the EBSD system, the collected Kikuchi patterns can be obtained automatically at a step resolution of 0.02 μm and correspondingly misorientations less than 2° cannot be identified. The samples for EBSD examinations were mechanically polished carefully followed by electro-polishing using an electrolyte of 90 vol.% acetic acid and 10 vol.% perchloric acid with a voltage of 40–45 V at about −40°C in a Struers LectroPol-5 facility.

X-ray diffraction (XRD) measuements were performed to obtain the phase transformation information during tensile tests using a Philips Xpert X-ray diffractometer with Cu Kα radiation. The samples were carefully prepared by mechanical polishing plus electropolishing from the surface to reach the designated depths after suspending tensile testing at varying designated strains. The phase volume fraction was estimated from the peak integrated intensities I_{hkl} after background subtraction. The volume fraction of α'-martensite, $V_{\alpha'}$, was calculated using the equation,

$$V_{\alpha'} = \frac{I_{bcc(110)}}{I_{bcc(110)} + I_{fcc(111)}}.$$

17.3 Results

17.3.1 Mechanical Behaviors

17.3.1.1 Tensile property and strain hardening

Figure 17.1 shows the tensile behaviors in various gradient structured specimens as well as the CG specimens for comparison. Figure 17.1a displays the quasi-static engineering stress-strain curves. A two-stage flow behavior is visible in the stress-strain (σ – ε) curves of both the GS and CG samples. Taking CG as an example, the σ – ε curve consists of a parabolic segment at the initial stage, followed by a sigmoidal or S-shape segment at the second stage. This leads to two inflection points on the strain hardening rate $\left(\Theta = \dfrac{\partial \sigma}{\partial \varepsilon}\right)$ versus true strain curve as shown in Fig. 17.1d. Two inflection points are marked by I_1 and I_2, respectively, corresponding to the onset of Θ up-turn and down-turn. The initial deformation stage is characterized by a low flow stress and a decreasing strain hardening rate with strain. This stage is associated with planar slip of partial and dissociated dislocations and the formation of ε-martensite plates [38, 39]. The second stage begins with a rapid rise in Θ, i.e. Θ rise, due to the deformation-induced $\gamma \rightarrow \alpha'$-martensite transformation [41–48], which is one of the major contributors to the TRIP effect. The final drop in Θ is caused by dynamic recovery.

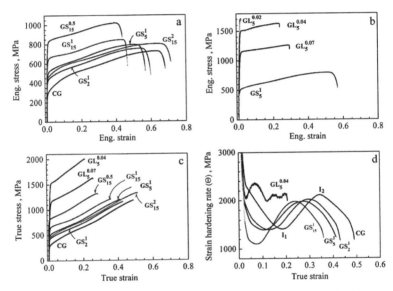

Figure 17.1 Tensile mechanical behavior of gradient structured (GS) 304ss. (a) Engineering tensile stress-strain ($\sigma - \varepsilon$) curves at the strain rate of 5×10^{-4} s^{-1}. CG: coarse grain. Subscript and superscript of GS represent SMAT period (minutes) and sample thickness (mm), respectively. (b) Engineering $\sigma - \varepsilon$ curves. GL: gradient layer. Superscript: GL thickness (mm). (c) True $\sigma - \varepsilon$ curves. (d) Strain hardening rate (Θ) versus true strain curves. I_1 and I_2 on the $\sigma - \varepsilon$ curve of CG: inflection points. Note that the strain is applied global strain.

Figure 17.1a shows the increase of yield strength ($\sigma_{0.2}$) of all gradient-structured specimens of 1-mm thick (superscript represents specimen thickness in mm) with increasing SMAT duration (subscript, minutes), with a drop of uniform tensile elongation ε_u by a much lower percentage. For example, the processing by SMAT for 15 minutes, i.e. GS^1_{15}, increased $\sigma_{0.2}$ by 132% while ε_u dropped by only 30%, which is consistent with an early report on gradient structured IF steel [29]. Interestingly, the combination of $\sigma_{0.2}$ and ε_u may be optimized by adjusting either the SMAT processing period or sample thickness [34]. Taking the GS^2_{15} sample (2-mm thick) as an example, which is twice thick as the GS^1_{15}, both its $\sigma_{0.2}$ and ε_u increased over those of the CG sample. As the original thickness was reduced from 1 to 0.5 mm, $\sigma_{0.2}$ increased significantly, but uniform tensile elongation also decreased.

Stand-alone gradient layers of various thicknesses from GS^1_5 were tested further to investigate their mechanical behaviors. The $\sigma - \varepsilon$ curves are shown in Fig. 17.1b. The top layer of 20 μm thick, i.e. $GL^{0.02}_5$ film, has $\sigma_{0.2}$ as high as 1700 MPa but necking occurred soon after yielding. Increasing the thickness to 40 and 70 μm ($GL^{0.04}_5$ and $GL^{0.07}_5$ films) led to typical two-stage flow behavior similar to that in GS^1_5, but with uniform elongation even above 20%.

Figure 17.1c shows the true stress-strain curves of all GS samples and GL films, which show high strain hardening during the uniform deformation. The flow stress reached as high as 2 GPa in $GL^{0.04}_5$. Figure 17.1d shows global strain hardening responses of all GS samples. For the CG sample, Θ decreases first, which is followed by a large increase, and then, drops again all the way to final necking. All GS samples show a similar trend of Θ evolutions. The only difference is the shifting of Θ rise to low starting stain. For example, the strain is 0.1 for Θ rise in GS^1_{15}, which is much less than 0.2 for CG. It is noteworthy, however, that the Θ rise in the GL films, e.g. $GL^{0.04}_5$, begins at a very small strain as compared to that in GS samples.

17.3.1.2 Microhardness evolution

The microhardness (Hv) evolution may reveal strain hardening behaviors locally in the GS samples during tensile deformation. Based on the $\Theta - \varepsilon$ curve of GS^1_5 sample shown in Fig. 17.1d, several tensile strains were selected for conducting Hv measurements to correlate the global strain hardening with local strain hardening.

Figure 17.2a shows the Hv profile at varying applied global tensile strains in GS_5^1. Before tensile testing, Hv gradient appeared along the depth, clearly demarcating the central CG core and GL of ~250 μm deep on both sides. The Hv value near the treated surface is 3.5 GPa, much larger than 2 GPa in the central CG core. At the tensile strains of 0.05 and 0.1 prior to the onset of Θ rise (Fig. 17.1d), Hv increased uniformly in the central CG layer, but hardly changed in most of the gradient layer. Even when the applied global tensile strain increased to 0.22, and the Θ rise has already started (Fig. 17.1d), the Hv in the top surface hardly changed at all, in contrast to the significant Hv increase in the CG layer. An obvious Hv rise in the top surface is not observed until tensile strain reached 0.35. Note that this strain is far beyond the Θ peak and already in the strain range of final Θ drop. Finally, by the end of uniform elongation at the strain of 0.4, the top layer showed an even higher Hv increase while the initial Hv gradient almost disappeared, which is very different from the reported Hv evolution in IF steel [29]. The Hv evolution was also measured in the free-standing $GL_5^{0.07}$ film during tensile deformation as shown in Fig. 17.2b. Hv rises instantly even at true strain of 0.1, in sharp contrast to what is observed in GS_5^1 where Hv remains unchanged as shown in Fig. 17.2a.

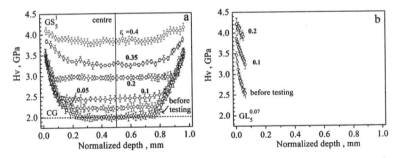

Figure 17.2 (a) Hv evolution with tensile strains along the depth in GS_5^1 sample. Number indicates true strain for Hv measurements after suspension of tensile testing. Dotted line: Hv of as-annealed CG sample. (b) Hv evolution along the depth in $GL_5^{0.07}$ sample.

17.3.1.3 Dislocation density evolution

Figure 17.1d reveals a Θ drop soon after yielding in GS samples much faster than that in CG. The stress relaxation tests were then conducted to correlate the Θ drop with the mobile dislocation behavior during this initial strain stage. Figure 17.3a shows the $\sigma - \varepsilon$ curve in both GS_5^1 and CG samples by stress relaxation testing. A number of relaxation strains were selected to cover the initial strain stage. Figure 17.3b shows the decay curves of the mobile dislocation density, ρ_m, with time at varying relaxation strains. The details for calculating ρ_m can be found in Refs. [29] and [53]. ρ_m is featured by an initial rapid drop followed by leveling-off as relaxation time extends. The final mobile dislocation density, ρ_{equi}, is shown in Fig. 17.3c as a function of relaxation strains.

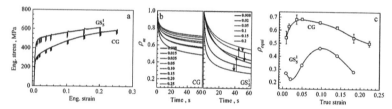

Figure 17.3 Stress relaxation tensile testing. (a) True stress-strain curves. (b) Evolution of mobile dislocation density (ρ_m) vs. time at various relaxation strains in the CG (left) and GS_5^1 sample (right). Arrows in right figure indicate ρ_{equi} change with strains. (c) ρ_{equi} vs. true strain in GS_5^1 and CG samples. Note that the strain is applied global strain.

Interestingly, ρ_{equi} dropped initially in GS_5^1 sample until a strain of 0.03 and then increased until a strain of ~0.1, followed by a final drop. In contrast, the ρ_{equi} in the CG sample rose first up to a strain of 0.04 and then dropped continuously. The rise in ρ_{equi}, which relates to the total dislocation density, is usually related to the increase in strain hardening [29, 53]. Comparing Figs. 17.1d and 17.3c reveals that the increase in ρ_{equi} coincides with the initial drop of Θ up to the strain of 0.1. Hence, even with a rapid Θ drop, the uniform deformation may still continue mainly due to the accumulation of dislocations. This is well in line with the observation in the gradient structured IF steel [29].

17.3.2 Microstructural Evolution

We studied both the microstructure and the $\gamma \rightarrow \alpha'$-martensitic transformation before and after tensile testing in GS_5^1 samples to correlate the tensile property and strain hardening with the microstructural evolution.

17.3.2.1 Microstructure by SMAT processing

Figure 17.4 shows EBSD images in GS_5^1 sample after SMAT processing. Figure 17.4a shows equi-axial grains of austenite with an average grain size of 25 µm in the central CG layer. Figure 17.4a-1 is an overlay of the corresponding EBSD image map (in grayscale) and the martensite phase map (in green), which indicates the presence of α'-martensite with a volume fraction of ~6% in original austenite of CG samples. The phases become indiscernible in the gradient layer mainly due to the small microstructural sizes in the severely deformed state. As seen in Figs. 17.4b/b-1 and 17.4c/c-1, the high density of band-like stripes are visible in most grains. This indicates significant refinement of original coarse grains by these stripes, even though the contours of most of original grain boundaries are still visible.

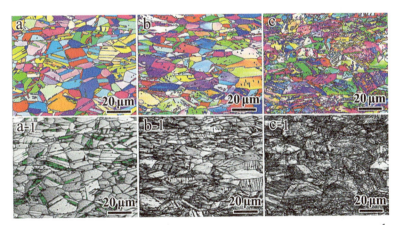

Figure 17.4 EBSD images at varying locations in gradient structured GS_5^1 sample before tensile testing. (a)–(c) Microstructures at central CG layer and at depths of ~180 and 60 µm deep, respectively, in the gradient layer. (a-1) Overlay of the EBSD image quality map (in grayscale) and martensite phase map (in green). (b-1) and (c-1) EBSD image quality map showing deformation fringes.

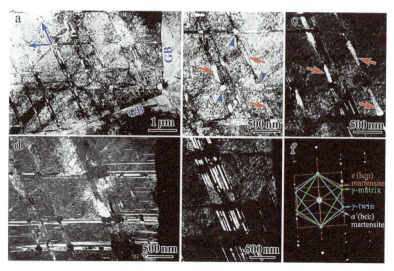

Figure 17.5 TEM micrographs in the gradient layer in GS_5^1 sample at the depth of ~180 μm below treated surface. a. Bright-field image showing parallelepiped zones separated by ε-martensite plates in two orientations as marked by two blue arrows. b and c. Bright- and dark-field images, respectively, showing α'-martensite formed either at intersections of ε-martensite plates (triangles in blue) or inside ε-martensite plates (arrows in red). d and e. Dark-field TEM images showing deformation twins in $(11\bar{1})_\gamma$ and $(1\bar{1}1)_\gamma$ planes respectively. Note deformation twins parallel to ε-martensite bands and inside each parallelepiped zone. f. Selected area electron diffraction (SAED) pattern covering several parallelepiped zones. Note the co-existence of fcc-γ, γ-twin, ε- and α'-martensite.

Detailed TEM observations were then conducted to investigate the microstructure and constituent phases in the gradient layer of GS_5^1 sample. Figure 17.5 is a series of TEM images showing typical microstructure in the original coarse-grained interior at the depth of ~180 μm. Figure 17.5a shows the intersecting bands of two $\{111\}_\gamma$ orientations which are originated from grain boundaries and extend into the grain interior. The initial constituent phase in these bands is ε-martensite, which was formed at low strains by the planar slip of partial dislocations due to low stacking fault energy of 304ss [37, 38]. These ε-martensite bands subdivide the original coarse grains into a large number of submicron-sized parallelepiped zones with austenite interior [37, 38, 44, 50]. Figures 17.5b and 17.5c are enlarged bright- and dark-field images and its corresponding selected area electron diffraction (SAED) pattern is given in Fig. 17.5f. Two types of α'-martensite are observed in the shear bands. One is granular, as indicated by blue triangles, which is nucleated at the intersections of ε-martensite bands [37, 44]. The other is plate-shaped, which is located inside and along ε-martensite bands, as indicated by red arrows shown in Figs. 17.5b and 17.5c. Both were formed via the $\varepsilon \rightarrow \alpha'$-martensite transformation [42–44]. Moreover, deformation twins of two different orientations, parallel to ε-martensite bands, are revealed by dark-field TEM observations, as shown in Figs. 17.5d and 17.5e. These twins are close to ε-martensite bands, but their growth is blocked by ε-martensite bands.

The parallelepiped zone is typical of the microstructural units in the gradient layer during SMAT processing. Figures 17.6a and 17.6b show the presence of the submicron-sized zones by subdivision of ε-martensite bands at the depth of ~100 and 50 µm, respectively. Parallel zone walls are clearly seen in Fig. 17.6b, oriented along two <110> zone axes so that these two zone walls are edge on. The third zone wall can be seen in Fig. 17.6a because it is not edge on. As shown in Fig. 17.6b, the zones are elongated with one side much longer than the other. Figure 17.7 shows that the sizes of both zone sides decrease when approaching the treated surface in the gradient layer.

Shear bands are occasionally observed in the gradient layers. The SMAT may produce a strain rate as high as 10^3 s^{-1} near the treated surface [51]. Two types of microstructures are observed in shear bands. One, as shown in Fig. 17.8a, consists of ε-martensite bands with high density of stacking faults, with the formation of granular α'-martensite by the $\varepsilon \rightarrow \alpha'$ transformation. The other, as shown in Fig. 17.8b, consists of ultrafine-grained austenite, which is likely formed by the $\alpha' \rightarrow \gamma$ reverse transformation due to heat effect in the shear bands [54–56]. The upper-right inset is an enlarged image showing dislocation-free, recrystallized austenite grains. The bottom-left inset is an SAED pattern, indicating low angle grain boundaries of these austenite grains.

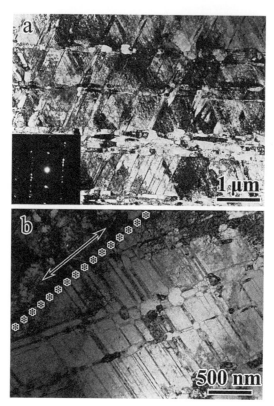

Figure 17.6 TEM micrographs showing parallelepiped austenite zones separated by ε-martensite bands in the gradient layer at the depth of (a) ~100 μm and (b) 50 μm.

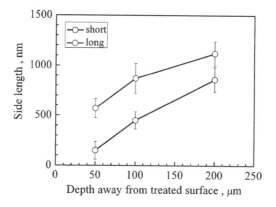

Figure 17.7 Statistic side lengths of parallelepiped zones along the depth.

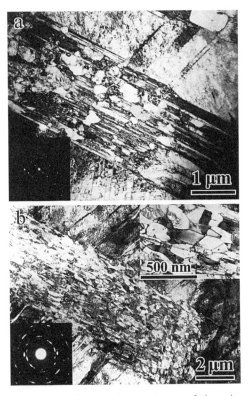

Figure 17.8 TEM micrographs showing two types of shear bands in gradient layer. (a) Stacking faults in a shear band that contains granular α'-martensite at ~80 μm deep from treated surface. Lower left inset shows the SAED pattern indicating the ε-martensite with a $[\bar{1}100]$ zone axis. (b) Recrystallized austenite grains in a shear band. The upper right inset is an enlarged image of dislocation-free austenite grains in the shear band and the lower left inset is an SAED pattern showing low-angle misorientations between the austenite grains.

17.3.2.2 Change of α'-martensite fraction during tensile testing

The $\gamma \rightarrow \alpha'$-martensitic transformation is expected to be affected by both the austenite zone size in the gradient layer and the local applied strains and stresses during the tensile test. The zone size gradient observed in the gradient layer is expected to affect the $\gamma \rightarrow \alpha'$-martensitic transformation and consequently the TRIP effect during the tensile testing. A series of XRD measurements was carried out to measure $V_{\alpha'}$ at varying depths and tensile strains during tensile testing of GS_5^1 sample. The strains were chosen based on the $\Theta - \varepsilon$ curve in Fig. 17.1d and Hv change in Fig. 17.2. The objective is to correlate $V_{\alpha'}$ with the global and local strain hardening and to clarify the effect of gradient structure on TRIP effect.

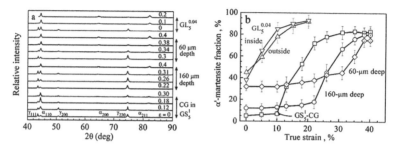

Figure 17.9 Quantitative α'-martensite transformation during tensile testing in GS_5^1 and $GL_5^{0.04}$. (a) XRD spectra measured at several elected tensile strains for free-standing $GL_5^{0.04}$ film and three locations in gradient structured GS_5^1. The number on spectrum represents the true tensile strain at which XRD measurement was carried out. (b) Volume fraction of α'-martensite as a function of tensile strains.

Figure 17.9a shows several typical XRD spectra at varying depths and applied strains in the GS_5^1 sample. Figure 17.9b plots the calculated $V_{\alpha'}$ as a function of applied tensile strain at varying depths. The results clearly indicate the dependence of $V_{\alpha'}$ on both depths and global tensile strains. For the central CG core, $V_{\alpha'}$ remains around 5% at low strains. It does not rise rapidly until the strain of 0.1 and lasts up to 0.22. Finally, $V_{\alpha'}$ levels off at the strain of 0.25 and keeps unchanged until the end of the uniform deformation. It is noted that the strain range for the rapid rise of $V_{\alpha'}$ coincide with the rise of the global Θ as shown in Fig. 17.1d. The level-off of $V_{\alpha'}$ indicates the absence of $\gamma \rightarrow \alpha'$-martensite transformation at large strains. In other words, for the CG core, the $\gamma \rightarrow \alpha'$-martensitic deformation and the related TRIP effect occur only in the global strain range of 0.1 to 0.22. This is consistent with what was observed in the uniform CG 304ss [38, 39]. Figure 17.9b shows the high initial α'-martensite content produced during SMAT processing in the gradient layer. Remarkably, a substantial rise of $V_{\alpha'}$ in the gradient layers starts at larger strains. For example, the $\gamma \rightarrow \alpha'$-martensite transformation occurs in the global strain range of 0.22 to 0.38 at the location of

160 µm deep from the surface. In other words, the $\gamma \to \alpha'$-martensite transformation in the gradient layers lagged behind the CG core.

Comparing Fig. 17.9b with Fig. 17.1d reveals that the $\gamma \to \alpha'$-martensite transformation occurred during the whole range of uniform deformation. In other words, the gradient structure enabled the $\gamma \to \alpha'$-martensite transformation to last much longer than CG to help improving the ductility. In addition, the evolution of $V_{\alpha'}$ correlates well with the Hv changes in Fig. 17.2.

Figure 17.9b also shows that $V_{\alpha'}$ in the free-standing $GL_5^{0.04}$ film rises rapidly even at the low small strain. By the end of the uniform elongation, $V_{\alpha'}$ at both sides is close to 97%, which is larger than those at varying depths in the gradient layer. This is probably due to their much higher flow stresses.

17.3.2.3 Microstructural evolution during tensile testing

The parallelepiped zones shown in Figs. 17.5 and 17.6 is the basic microstructural unit for accommodating plastic deformation in the gradient layer. The plastic response determines the mechanical properties and strain hardening behavior of the gradient structured samples during tensile deformation. TEM observations were used to study the microstructural evolution in the parallelepiped zones with tensile strains in GS_5^1 sample. The applied tensile strain for TEM studies was determined based on the $V_{\alpha'}$ evolution in Fig. 17.9.

Figure 17.10a shows the microstructure at ~160 µm deep at an applied tensile strain of 0.25. A row of elongated zones was analyzed in detail. Several long plates (marked by red arrows) are determined to be α'-martensite according to SAED patterns. Each zone interior is still austenite. A large number of granular α'-martensite grains, indicated by blue triangles, are also seen, which are formed via $\varepsilon \to \alpha'$ phase transformation. Both the volume fraction and grain sizes of α'-martensite are increased as compared to those shown in Figs. 17.5 and 17.6. Figure 17.10b shows a zone taken at ~120 µm deep at an applied tensile strain of 0.3. Four-side frames have been transformed completely to α'-martensite according to the SAED pattern. The α'-martensite deformed plastically and dislocation entanglement was formed in their interiors. The zone interior is still austenite but with dislocation cells. This type of microstructure should have the most differences in flow stresses between the zone frames and interior, as the frames have been transformed completely to α'-martensite. Therefore, local strain partitioning is strongest. As the applied tensile strain further increased to 0.35 at ~50 μm depth as seen in Fig. 17.10c, all zone interiors were transformed into α'-martensite. The SAED pattern shows a discontinuous ring of the bcc structure. Dislocations are observed in α'-martensite. Figure 17.10d shows a large number of granular and plate-shaped α'-martensite with a little ε-martensite left.

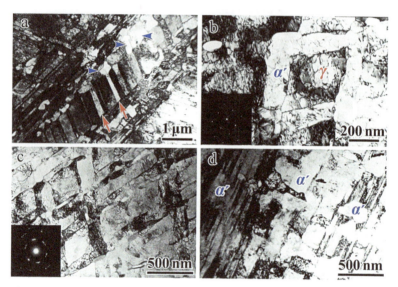

Figure 17.10 TEM micrographs showing α'-martensite transformation in the gradient layer of the GS_5^1 sample during tensile testing. (a) α'-martensite plates as indicated by red arrows. Blue triangles mark granular α'-martensite at intersecting ε-martensite. Interior: austenite. (b) α'-martensite plates on four sides with an austenitic interior. Inset is the SAED pattern showing the bcc lattice with the [001] zone axis taken from the upper side. (c) α'-martensite on both the sides and interiors. Inset shows the ring-like SAED pattern indicating a bcc lattice. (d) α'-martensite.

17.4 Discussion

17.4.1 Dynamic Strain Partitioning

The gradient structure reported here will cause strain partitioning during tensile tests. The strain partitioning has two origins at two different scales. One is macroscopic across the whole sample thickness, while the other is microscopic between the zone interior and walls of each parallelepiped zone. Both types of strain partitioning evolve during tensile testing due to the dependence of the $\gamma \rightarrow \alpha'$-martensite transformation with applied tensile strains and the austenite stability of zone sizes as well. In this sense, the dynamic strain partitioning is expected, similar to that in TRIP-maraging steel [12].

Macroscopically, a gradient Hv exists across the gradient structured sample as shown in Fig. 17.2. The Hv is lower in the CG core than that in gradient layers on both sides. The former yields first, while the latter is still deforming elastically. Hence, the gradient structure will begin with a composite-like deformation. As a result, the CG core bears more plastic deformation. This is evidenced by Hv evolutions with tensile strains as shown in Fig. 17.2. The Hv keeps nearly unchanged in most of the GL at tensile strains of 0.05 and 0.1, respectively, in contrast to a gradual rise of Hv in the CG core due to strain hardening. This indicates the onset of strain partitioning between the central CG core and gradient layers. Furthermore, the whole gradient structured sample can be approximately regarded as the integration of many thin layers with increasing Hv toward the surface [34]. The plastic deformation occurs in sequence of neighboring layers with low to high flow stresses. As a result, plastic strain partitioning continues as the GL begins to deform plastically and lasts up to the end of the uniform elongation. The strain partitioning at the macroscale is, actually, an intrinsic property of the gradient structure due to the presence of the initial Hv gradient.

Microscopically, large Hv difference exists between the parallelepiped zones and martensitic zone walls. The austenite zones are softer than ε-martensite walls that surround zones from all sides as shown in Fig. 17.5. In addition, this Hv difference will be increased after the onset of the $\gamma \rightarrow \alpha'$-martensite transformation in the zone walls upon further straining as seen in Figs. 17.10a and 17.10b. The α'-martensite may deform plastically and store entangled dislocations. Besides, the $\gamma \rightarrow \alpha'$-martensite transformation in the gradient layers begins at large strains and finishes by the end of uniform elongation based on the XRD measurements shown in Fig. 17.9b. These observations suggest that the deformation is not uniform between the zone interiors and walls at the microscopic scale. In other words, dynamic strain partitioning occurs microscopically in each parallelepiped zone and continues until the final stage of uniform deformation. This microscopic strain partitioning can be mapped semi-in-situ using EBSD and image analysis [57].

The dynamic strain partitioning is associated with the mechanical incompatibility at both the macroscale and microscale. Firstly, two boundaries exist, demarcating the elastic and plastic layers in the gradient structure during tensile deformation. These boundaries will migrate toward the surface in the gradient layer with increasing applied strain as shown in Fig. 17.2. Mechanical incompatibility is produced at elastic/plastic boundaries. Secondly, mechanical incompatibility still exists between soft zones and their hard walls. As a result of these two types of mechanical incompatibility, strain gradients are generated near both elastic/plastic boundaries [34] and austenite/martensite boundaries. These strain gradients produce geometrically necessary dislocations and increase both the yield strength and strain hardening [29, 34, 58]. It has been reported that mechanical incompatibility will also cause high HDI stress and HDI hardening [59, 60]. Detailed investigations are being carried out to study this issue.

Both the soft CG layer and soft austenite zones will bear more strain than the hard ones during tensile deformation of the gradient structure. As a result, dynamic strain partitioning causes an enhanced uniform ductility, especially when strain partitioning lasts up to the end of uniform deformation.

17.4.2 Deformation-Induced Phase Transformation and TRIP Effect

The deformation-induced $\gamma \rightarrow \alpha'$-martensite transformation plays a crucial role in strain hardening in homogeneous CG 304ss [38–43]. It induces a Θ rise as shown in Fig. 17.1d. However, it is difficult for the phase transformation to last until the final strain stage. XRD measurements shown in Fig. 17.9b reveal the onset of the $\gamma \rightarrow \alpha'$ transformation at first in the CG core in the gradient structured GS_5^1. The $V_{\alpha'}$ rise in the CG core begins at true strain of ~0.1, slows down after the strain of 0.22 and finally stops at the strain of 0.3. In other words, the TRIP effect occurs only in the strain range from 0.1 to 0.3 for the CG core. It is noted that the strain of 0.1 happens to match that for the onset of Θ rise as shown in Fig. 17.1d. This indicates that the $\gamma \rightarrow \alpha'$ transformation in the CG core is responsible for Θ rise.

The starting strain for the onset of $\gamma \rightarrow \alpha'$ transformation shifts to 0.1 in GS_5^1, much smaller than 0.18 in homogeneous CG 304ss. This is likely caused by the complex evolution of the stress-state in the gradient structured sample. Plastically deforming metals should have an apparent Poisson's ratio of 0.5, while elastic deformations usually have a Poisson's ratio of ~0.3 [34, 61]. As revealed in Fig. 17.2a, the CG core has already started plastic deformation at the strain larger than 0.05%, while gradient layers are still largely elastic. Under uniaxial tension, the CG core should laterally shrink more than elastic gradient layers. This will lead to tensile lateral stress in the CG layer and compressive lateral stress in the gradient layers. Since the $\gamma \rightarrow \alpha'$ transformation is accompanied with volume expansion, the lateral tensile stress should promote the phase transformation [62, 63], which is the reason why the onset of the $\gamma \rightarrow \alpha'$ martensitic transformation for the CG layer is shifted to lower tensile strains in gradient structured samples.

Interestingly, a strong delay of the $\gamma \rightarrow \alpha'$ transformation is visible in the gradient layers as seen in Fig. 17.9b. This delay is caused mainly by three reasons. First, the austenite zones become smaller and smaller with decreasing distance to the surface. The small grain sizes deter the onset of phase transformation [11]. Second, the gradient layer only bears a lower plastic strain than the CG core due to the strain partitioning. Third, the lateral compressive stress in the gradient layer should also have played a significant role in delaying the martensitic transformation. It is further deduced from Figs. 17.1 and 17.9b that the $\gamma \rightarrow \alpha'$ martensite transformation is controlled not only by the strain level and local stress state as discussed above, but also by the applied stress level. There exists probably a minimum stress level to trigger the phase transformation. The free-standing $GL_5^{0.04}$ film started the plastic deformation at very high stresses, which is why the $\gamma \rightarrow \alpha'$-martensitic transformation started almost immediately upon plastic deformation.

In the gradient structure the α'-martensite transformation takes place in the whole strain range from 0.1 to the sample failure, as seen in Fig. 17.9b. In other words, the gradient structure enabled the α'-martensite transformation, and consequently the TRIP effect, to last over a large strain rage, which effectively increases the uniform elongation. Taking the GS_5^1 sample as an example, the α'-martensite transformation is prolonged to the strain of more than 0.3, as compared to ~0.25 beyond which the α'-martensite transformation largely stopped in the CG sample (see Fig. 17.9b).

The combination of the gradient structure and the TRIP effect produced a combination of strength and ductility that is superior to what was reported earlier, as shown in Fig. 17.11. This is especially true in the high strength regime, where the TRIP-gradient 304ss has much higher ductility than those reported for other homogeneous samples.

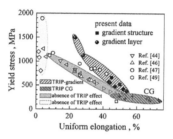

Figure 17.11 Yield strength and uniform elongation produced by combination of gradient structure and TRIP effect in 304s, as compared with those reported in 304ss with other types of microstructures.

17.5 Conclusions

SMAT processing of 304 stainless steel produced a gradient structure with two gradient structured layers sandwiching a coarse-grained core layer. The gradient layers consist of parallelepiped austenite zones separated by hard ε martensite walls, and a zone size gradient along the depth.

During tensile testing of the gradient structured 304ss samples, the CG core started plastic deformation first, while the gradient layer was still deforming elastically, causing macroscopic strain partitioning and mechanical incompatibility, which consequently produced a tensile lateral stress in the CG core and compressive lateral stress in the gradient layers. The tensile lateral stress in the CG core promoted early $\gamma \rightarrow \alpha'$-martensite transformation, while the compressive stress in the gradient layers delayed TRIP to a larger applied tensile strain. In addition, smaller zone size in the gradient layer also delayed the α'-martensite transformation. As a result, the TRIP effect is prolonged to the end of the tensile test, in contrast to the TRIP effect in homogeneous CG samples, where TRIP effect usually ends early. In other words, the gradient structure can spread the TRIP effect over a larger strain range to enhance ductility.

The TRIP-gradient 304ss shows an excellent combination of yield strength and uniform elongation. The multiple plastic responses, including the dynamic strain partitioning and phase transformation intrinsically were inter-linked with each other and co-contributed to the observed superior mechanical properties.

Acknowledgements

This work was financially supported by the National Natural Science Foundation of China (NSFC) under grant Nos. 11572328, 11222224, 11472286, the 973 Programs under grant Nos. 2012CB932203, 2012CB937500, and 6138504. Y.Z. was supported by the US Army Research Office (W911 NF-12-1-0009), the US National Science Foundation (DMT-1104667), and the Nanjing University of Science and Technology.

References

1. R.Z. Valiev, Y. Estrin, Z. Horita, T.G. Langdon, M.J. Zehetbauer, Y.T. Zhu, Fundamentals of superior properties in bulk nanoSPD materials, *Mater. Res. Lett.* **4** (2016) 1–21.
2. L. Lu, Y.F. Shen, X.H. Chen, L.H. Qian, K. Lu, Ultrahigh strength and high electrical conductivity in copper, *Science* **304** (2004) 422–426.
3. X.L. Wu, M.X. Yang, F.P. Yuan, G.L. Wu, Y.J. Wei, X.X. Huang, Y.T. Zhu, Heterogeneous lamella structure unites ultrafine-grain strength with coarse-grain ductility, *Proc. Natl. Acad. Sci. U.S.A.* **112** (2015) 14501–14505.
4. Y.M. Wang, M.W. Chen, F.H. Zhou, E. Ma, High tensile ductility in a nanostructured metal, *Nature* **419** (2002) 912–915.
5. X.X. Huang, N. Hansen, N. Tsuji, Hardening by annealing and softening by deformation in nanostructured metals, *Science* **312** (2006) 249–251.
6. R.Z. Valiev, Nanostructuring of metals by severe plastic deformation for advanced properties, *Nat. Mater.* **3** (2004) 511–516.

7. Y.T. Zhu, X.Z. Liao, Nanostructured metals: retaining ductility, *Nat. Mater.* **3** (2004) 351–352.

8. X.L. Wu, F.P. Yuan, M.X. Yang, C.X. Zhang, P. Jiang, L. Chen, Y.G. Wei, E. Ma, Nanodomained nickel unite nanocrystal strength with coarse-grain ductility, *Sci. Rep.* **5** (2015) 11728.

9. R.Z. Valiev, Y. Estrin, Z. Horita, T.G. Langdon, M.J. Zehetbauer, Y.T. Zhu, Producing bulk ultrafine-grained materials by severe plastic deformation: ten years later, *JOM* **68** (2016) 1216–1226.

10. Y.M. Wang, E. Ma, Three strategies to achieve uniform tensile deformation in a nanostructured metal, *Acta Mater.* **52** (2004) 1699–1709.

11. M.M. Wang, C.C. Tasan, D. Ponge, A. Kostka, D. Raabe, Smaller is less stable: size effects on twinning vs. transformation of reverted austenite in TRIP-maraging steels, *Acta Mater.* **79** (2014) 268–281.

12. M.M. Wang, C.C. Tasan, D. Ponge, A.C. Dippel, D. Raabe, Nanolaminate transformation-induced plasticity-twinning-induced plasticity steel with dynamic strain partitioning and enhanced damage resistance, *Acta Mater.* **85** (2015) 216–228.

13. M. Calcagnotto, Y. Adachi, D. Ponge, D. Raabe, Deformation and fracture mechanisms in fine- and ultrafine-grained ferrite/martensite dual-phase steels and the effect of aging, *Acta Mater.* **59** (2011) 658–670.

14. D. Raabe, D. Ponge, O. Dmitrieva, B. Sander, Nanoprecipitate-hardened 1.5 GPa steels with unexpected high ductility, *Scr. Mater.* **60** (2009) 1141–1144.

15. H.K.D.H. Bhadeshia, Nanostructured bainite, *Proc. Royal Soc. A* **466** (2010) 3–18.

16. J.H. Ryu, D.-I. Kim, H.S. Kim, H.K.D.H. Bhadeshia, D.-W. Suh, Strain partitioning and mechanical stability of retained austenite, *Scr. Mater.* **63** (2010) 297–299.

17. C.C. Tasan, M. Diehl, D. Yan, C. Zambaldi, P. Shanthraj, F. Roters, D. Raabe, Integrated experimental-simulation analysis of stress and strain partitioning in multiphase alloys, *Acta Mater.* **81** (2014) 386–400.

18. K. Sugimoto, M. Kobayashi, S. Hashimoto, Ductility and strain-induced transformation in a high-strength transformation-induced plasticity-aided dual-phase steel, *Metall. Mater. Trans. A* **23** (1992) 3085–3091.

19. C. Herrera, D. Ponge, D. Raabe, Design of a novel Mn-based 1 GPa duplex stainless TRIP steel with 60% ductility by a reduction of austenite stability, *Acta Mater.* **59** (2011) 4653–4664.

20. H.L. Yi, K.Y. Lee, H.K.D.H. Bhadeshia, Extraordinary ductility in Al-bearing delta-TRIP steel, *Proc. Royal Soc. A* **467** (2011) 234–243.
21. S. Zaefferer, J. Ohlert, W. Bleck, A study of microstructure, transformation mechanisms and correlation between microstructure and mechanical properties of a low alloyed TRIP steel, *Acta Mater.* **52** (2004) 2765–2778.
22. K.X. Tao, H. Choo, H.Q. Li, B. Clausen, J.E. Jin, Y.K. Lee, Transformation-induced plasticity in an ultrafine-grained steel: an in situ neutron diffraction study, *Appl. Phys. Lett.* **90** (2007) 101911.
23. Y.S. Jung, Y.K. Lee, Effect of pre-deformation on the tensile properties of a metastable austenitic steel, *Scr. Mater.* **59** (2008) 47–50.
24. L. Morales-Rivas, H.W. Yen, B.M. Huang, M. Kuntz, F.G. Caballero, J.R. Yang, C. Garcia-Mateo, Tensile response of two nanoscale bainite composite-like structures, *JOM* **67** (2015) 2223–2235.
25. M.F. Ashby MF, Deformation of plastically non-homogeneous materials, *Philos. Mag.* **21** (1970) 399–411.
26. H.J. Gao, Y.G. Huang, W.D. Nix, J.W. Hutchinson, Mechanism-based strain gradient plasticity - I. Theory, *J. Mech. Phys. Solids* **47** (1999) 1239–1263.
27. H.J. Gao, Y.G. Huang, Geometrically necessary dislocation and size-dependent plasticity, *Scr. Mater.* **48** (2003) 113–118.
28. T.H. Fang, W.L. Li, N.R. Tao, K. Lu, Revealing extraordinary intrinsic tensile plasticity in gradient nano-grained copper, *Science* **331** (2011) 1587–1590.
29. X.L. Wu, M.X. Yang, F.P. Yuan, G.L. Wu, Y.J. Wei, X.X. Huang, Y.T. Zhu, Heterogeneous lamella structure unites ultrafine-grain strength with coarse-grain ductility, *Proc. Natl. Acad. Sci. U.S.A.* **112** (2015) 14501–14505.
30. Y.J. Wei, Y.Q. Li, L.C. Zhu, Y. Liu, X.Q. Lei, G. Wang, Y.X. Wu, Z.L. Mi, J.B. Liu, H.T. Wang, H.J. Gao, Evading the strength- ductility trade-off dilemma in steel through gradient hierarchical nanotwins, *Nat. Commun.* **5** (2014) 3580.
31. H.N. Kou, J. Lu, Y. Li, High-strength and high-ductility nanostructured and amorphous metallic materials, *Adv. Mater.* **26** (2014) 5518–5524
32. K. Lu, Making strong nanomaterials ductile with gradients, *Science* **345** (2014) 1455–1456.
33. W.B. Li, F.P. Yuan, X.L. Wu, Atomistic tensile deformation mechanisms of Fe with gradient nano-grained structure, *AIP Adv.* **5** (2015) 087120.

34. X.L. Wu, P. Jiang, L. Chen, J.F. Zhang, F.P. Yuan, Y.T. Zhu, Synergetic strengthening by gradient structure, *Mater. Res. Lett.* **2** (2014) 185–191.

35. M.X. Yang, Y. Pan, F.P. Yuan, Y.T. Zhu and X.L. Wu, Back stress strengthening and strain hardening in gradient structure, *Mater. Res. Lett.* **4** (2016) 145–151.

36. G.L. Huang, D.K. Matlock, G. Krauss, Martensite formation, strain rate sensitivity, and deformation-behavior of type-304 stainless-steel sheet, *Metall. Trans. A* **20** (1989) 1239–1246.

37. H.W. Zhang, Z.K. Hei, G. Liu, J. Lu, K. Lu, Formation of nanostructured surface layer on AISI 304 stainless steel by means of surface mechanical attrition treatment, *Acta Mater.* **51** (2003) 1871–1881.

38. L.E. Murr, K.P. Staudhammer, S.S. Hecker, Effects of strain rate and strain rate on deformation- induced transformation in 304 stainless-steel: Part II. Microstructure study, *Metall. Trans. A* **13** (1982) 627–635.

39. A.K. De, J.G. Speer, D.K. Matlock, D.C. Murdock, M.C. Mataya, R.J. Comstock, Deformation-induced phase transformation and strain hardening in type 304 austenitic stainless steel, *Metall. Mater. Trans. A* **37A** (2006) 1875–1886.

40. Z.H. Cong, N. Jia, X. Sun, Y. Ren, J. Almer, Y.D. Wang, Stress and strain partitioning of ferrite and martensite during deformation, *Metall. Mater. Trans. A* **40A** (2009) 1383–1387.

41. H.C. Shin, T.K. Ta, Y.W. Chang, Kinetics of deformation induced martensitc transformation in a 304 stainless steel, *Scr. Mater.* **45** (2001) 823–829.

42. S. Cheng, X.L. Wang, Z.L. Feng, B. Clausen, H. Choo, P.K. Liaw, Probing the characteristic deformation behaviors of transformation-induced plasticity steels, *Metall. Mater. Trans. A* **39A** (2008) 3105–3112.

43. J. Talonen, P. Nenonen, G. Pape, H. Hanninen, Effect of strain rate on the strain-induced $\gamma \rightarrow \alpha'$- martensite transformation and mechanical properties of austenitic stainless steels, *Metall. Mater. Trans. A* **36A** (2005) 421–432.

44. X.S. Yang, S. Sun, X.L. Wu, E. Ma, T.Y. Zhang, Dissecting the mechanism of martensitic transformation via atomic-scale observations, *Sci. Rep.* **4** (2014) 6141.

45. H.Y. Yi, F.K. Yan, N.R. Tao, K. Lu, Comparison of strength-ductility combinations between nanotwinned austenite and martensite-austenite stainless steels, *Mater. Sci. Eng. A* **647** (2015) 152–156.

46. H.Y. Yi, F.K. Yan, N.R. Tao, K. Lu, Work hardening behavior of nanotwinned austenitic grains in a metastable austenitic stainless steel, *Scr. Mater.* **114** (2016) 133–136.

47. C.X. Huang, W.P. Hu, Q.Y. Wang, C. Wabg, G. Yang, Y.T. Zhu, An ideal ultrafine-grained structure for high strength and high ductility, *Mater. Res. Lett.* **3** (2015) 88–94.

48. Y.F. Shen, N. Jia, Y.D. Wang, X. Sun, L. Zuo, D. Raabe, Suppression of twinning and phase transformation in an ultrafine grained 2 GPa strong metastable austenitic steel: Experiment and simulation, *Acta Mater.* **97** (2015) 305–315.

49. A.Y. Chen, D.F. Li, J.B. Zhang, H.W. Song, J. Lu, Make nanostructured metal exceptionally tough by introducing non-localized fracture behaviors, *Scr. Mater.* **59** (2008) 579–582.

50. A.Y. Chen, H.H. Ruan, J. Wang, H.L. Chan, Q. Wang, Q. Li, J. Lu, The influence of strain rate on the microstructure transition of 304 stainless steel, *Acta Mater.* **59** (2011) 3697–3709.

51. K. Lu, J. Lu, Nanostructured surface layer on metallic materials induced by surface mechanical attrition treatment, *Mater. Sci. Eng. A* **375** (2004) 38–45.

52. X. Wu, N. Tao, Y. Hong, B. Xu, J. Lu, K. Lu, Microstructure and evolution of mechanically-induced ultrafine grain in surface layer of Al-alloy subjected to USSP, *Acta Mater.* **50** (2002) 2075–2084.

53. J.L. Martin, B. Lo Piccolo, J. Bonneville, The role of thermal activation in the strength anomaly of Ni_3Al, *Intermetallics* **8** (2000) 1013–1018.

54. A. Mishra, M. Martin, N.N. Thadhani, B.K. Kad, E.A. Kenik, M.A. Meyers, High-strain-rate response of ultra-fine-grained copper, *Acta Mater.* **56** (2008) 2770–2783.

55. F.P. Yuan, P. Jiang, X.L. Wu, Annealing effect on the evolution of adiabatic shear band under dynamic shear loading in ultra-fine-grained iron, *Int. J. Imp. Eng.* **50** (2012) 1–8.

56. F.P. Yuan, X.D. Bian, P. Jiang, M.X. Yang, X.L. Wu, Dynamic shear response and evolution mechanisms of adiabatic shear band in ultrafine-grained austenite-ferrite duplex steel, *Mech. Mater.* **89** (2015) 47–58.

57. J.H. Ryu, D.I. Kim, H.S. Kim, H.K.D.H. Bhadeshia and D.W. Suh. Strain partitioning and mechanical stability of retained austenite. *Scr. Mater.* **63** (2010) 297–299.

58. J.J. Li, S.H. Chen, X.L. Wu, A.K. Soh, A physical model revealing strong strain hardening in nano-grained metals induced by grain size gradient structure, *Mater. Sci. Eng. A* **620** (2014) 16–21.

59. M.X. Yang, Y. Pan, F.P. Yuan, Y.T. Zhu, X.L. Wu, Back stress strengthening and strain hardening in gradient structure, *Mater. Res. Lett.*, on line (http://dx.doi.org/10.1080/21663831.2016.1153004).

60. M.X. Yang, F.P. Yuan, Q.G. Xie, Y.D. Wang, E. Ma, X.L. Wu, Strain hardening in Fe-16Mn-10Al-0.86C-5Ni high specific strength steel, *Acta Mater.* **109** (2016) 213–222.

61. C.W. Beat, E.J. Mills, W.S. Hyler, Effect of variation in Poisson's ratio on plastic tensile instability, *J. Fluids Eng.* **89** (1967) 35–39.

62. A.A. Lebedev, V.V. Kosarchuk, Influence of phase transformations on the mechanical properties of austenitic stainless steels, *Int. J. Plast.* **16** (2000) 749–767.

63. P.J. Jacques, Q. Furnémont, F. Lani, T. Pardoen, F. Delannay, Multiscale mechanics of TRIP-assisted multiphase steels: I. Characterization and mechanical testing, *Acta Mater.* **55** (2007) 3681–3693.

Chapter 18

Gradient Structure Produces Superior Dynamic Shear Properties

Xiangde Bian,[a,b] Fuping Yuan,[a,b] Xiaolei Wu,[a,b] and Yuntian Zhu[c,d]

[a]*State Key Laboratory of Nonlinear Mechanics, Institute of Mechanics, Chinese Academy of Science, Beijing 100190, China*
[b]*School of Engineering Science, University of Chinese Academy of Sciences, Beijing 100190, China*
[c]*Department of Materials Science and Engineering, North Carolina State University, Raleigh, NC 27695, USA*
[d]*Nano Structural Materials Center, School of Materials Science and Engineering, Nanjing University of Science and Technology, Nanjing 210094, China*
fpyuan@lnm.imech.ac.cn, xlwu@imech.ac.cn

We report that gradient-structured metals have much superior dynamic shear properties over homogeneous nanostructured (NS) metals. The gradient structure was found to delay the nucleation of adiabatic shear bands (ASB) at the NS surface layers and to reduce

---

Reprinted from *Mater. Res. Lett.*, **5**(7), 501–507, 2017.

the propagation speed of ASB by an order of magnitude as compared with homogeneous materials. The conventional maximum stress criterion on ASB initiation for homogeneous is found not valid for gradient material. These findings may provide for a novel strategy for designing impact-tolerant structures and materials with high impact toughness.

Metals and alloys with both high strength and ductility have been pursued by scientists and engineers for several decades [1–11]. Gradient structure, where the grain/subgrain-structure size changes gradually from the surface to interior [12–19] shows superior ductility-strength combination under quasi-static tensile loading, which was attributed to either mechanically driven grain growth for unstable nanostructured surface layer [15] or extra strain hardening by the presence of strain gradient together with the stress state change for the mechanically stable gradient structure [13, 17]. Hetero-deformation induced (HDI) hardening has been reported to play a significant role in the mechanical behavior and properties of gradient structures [12], similar to what occurs in other heterostructures [10, 20, 21].

The flow behaviors of structural materials are strongly affected by the applied strain rate [22–25]. At high strain rates the adiabatic heating results in substantial temperature rise, thermal softening and formation of adiabatic shear band (ASB), which triggers fracture [25–29]. The dynamic shear response and evolution mechanisms of ASB in homogeneous materials with coarse grains (CG), ultra-fine grains (UFG) and nanograins (NG) have been extensively investigated [25–35]. Typically, CG materials have low yield strength, high critical strain for onset of ASB and high impact toughness, while UFG and NG materials possess high yield strength, low uniform strain and low impact toughness. In other words, there exists a strength-toughness trade-off for homogeneous metals under dynamic shear loading. A question arises on if the superior strength-ductility combination of gradient structures under quasi-static strain rates can translate into superior strength-toughness under high strain rates.

Here, we report that the gradient structured metals can produce extraordinary strength-toughness combination that is not accessible to homogeneous materials. The underlying mechanisms were revealed and will be discussed.

Hot-rolled TWIP steel with (wt%) 0.6C, 23Mn, 0.035Nb, 0.035Ti, and the balance Fe was annealed at 700°C for 1 h, followed by immediate water quenching, to obtain CG austenite samples with average grain size of 20 μm (Fig. 18.1a). 3-mm thick disks were sliced from the CG samples and processed by surface mechanical attrition treatment (SMAT) [13, 36] to produce gradient structures for 10–20 min on both surfaces of the disks. For comparison, homogeneous samples were processed by cold rolling the CG samples with thickness reduction of 20%, 30% and 70%, which produces different yield strengths.

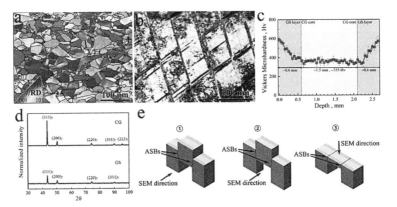

Figure 18.1 (a) EBSD image for the untreated CG sample; (b) TEM micrograph at the depth of 50 μm; (c) Vickers microhardness along the depth in a sample processed by SMAT for 20 min; (d) X-ray diffraction patterns; (e) schematics of sample cross-sections for microstructural examination.

Microhardness along the depth for a sample processed by SMAT for 20 min is shown in Fig. 18.1c. The sample consists of a 1.5 mm CG core, sandwiched between two 0.6 mm gradient layers. The microhardness is about 355 Hv in the core and 600 Hv near the surface. The hardness decreases almost linearly along the depth, which is different from gradient structure in IS steel [13]. Figure 18.1b shows the TEM micrograph at the depth of 50 μm, which reveals parallel deformation twins along two {111} planes with inter-twin spacing of 100–300 nm, and intersecting each other High density of dislocations is also observed in the interior of grains, especially near the twin boundaries. The high microhardness in the surface layer can be attributed to the reduction of the dislocation mean free path by intersecting twins [7] and dislocation entanglements.

XRD analysis did not detect any phase transformation from the SMAT process (Fig. 18.1d).

The set-up of Hopkinson bar experiments, the geometry and dimensions of the hat-shaped specimens and the specimen holders can be found in [34]. The extraction of samples for microstructure observations are illustrated in Fig. 18.1e. Hat-shaped Hopkinson bar experiments have been widely used to study ASB in metals and alloys [25, 26, 30–32, 34, 35]. Strong cylindrical maraging steel specimen holders were used to ensure an approximate pure shear deformation by constraining the lateral expansion of two legs for the hat-shaped specimens and to freeze the microstructure at specific shear displacements by varying the height of the specimen holders. Other experimental details can be found in [25, 26, 34].

The dynamic shear properties for the gradient, CG and deformed samples (one sample was annealed at 500°C for 1 h) are shown in Fig. 18.2. The shear stress-shear displacement curves are shown in Fig. 18.2a, and the corresponding shear stress-nominal shear strain curves are shown in Fig. 18.2b. The nominal shear strain was calculated from the shear displacement and the shear zone thickness with assumption of homogeneous deformation in the concentrated shear zone. The deformed sample with 70% thickness reduction has a dynamic shear yield strength of 1300 MPa, but stress drops quickly after yielding. The maximum-stress point is considered as the initiation point of the strain localization according to the conventional maximum stress criterion for homogeneous materials [25–30], after which ASB begins to form and propagate. Thus, the uniform dynamic plastic shear strain after yielding for this sample is almost zero, indicating a very low strain hardening capability under dynamic shear loading. In contrast, a two-stage dynamic shear flow behavior is visible for the CG sample before the stress drop: a strong linear strain hardening stage followed by a short plateau stage. The CG structure has a low dynamic shear yield strength (~320 MPa), with a very high uniform shear strain of ~4.5.

The dynamic shear toughness can be calculated from the area under the shear stress-shear strain curves in Fig. 18.2b, which is an important parameter for structural components serving under dynamic loading. As shown in Fig. 18.2c, the strength and dynamic shear toughness of homogeneous materials shows a typical trade-off. In contrast, the gradient structures show an excellent combination of strength and impact dynamic toughness that is out of reach of homogeneous structure (CG sample and deformed samples with various CR thickness reductions. The dynamic properties might be further optimized by optimizing the gradient structure [19].

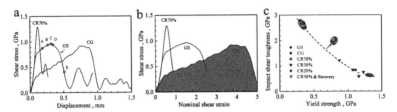

Figure 18.2 (a) Shear stress vs. shear displacement curves of various samples; (b) shear stress vs. nominal shear strain curves; (c) impact shear toughness vs. dynamic shear yield strength of the TWIP steel samples with different microstructures.

In order to understand the corresponding deformation mechanisms under the dynamic shear loading for the gradient structures, dynamic shear experiments with controlled shear deformation at five different interrupted displacements (0.18, 0.26, 0.31, 0.35, 0.53 mm), as marked as "A, B, C, D, E" in the curve of Fig. 18.2a, were conducted on samples processed by SMAT for 20 min to investigate the microstructural evolution. Figure 18.3 shows the SEM micrographs taken under BSE mode for the microstructural evolution of the treated surface at the five different interrupted shear displacements. The hat is on the top, and the base ends of hat-shaped specimens are at the bottom. As shown, although the grains are severely sheared, shear band did not form at the shear displacements smaller than 0.26 mm. ASB started to form at shear displacement of 0.26 mm and propagated across the whole shear zone at the shear displacement of 0.31 mm.

The microhardness of the homogeneously deformed sample with 70% CR thickness reduction is about 600 Hv, which is close to the surface microhardness of the gradient structured sample. However, as shown in Fig. 18.2a, ASB formed in this sample basing on the maximum stress criterion for homogeneous materials [25–30]. This is much lower than the ASB onset shear displacement (0.26 mm) at the surface of the gradient structured sample. The delayed initiation of ASB in the nanostructured surface layer under dynamic shear loading could also be attributed to the strain gradients and the strain partitioning [10, 12, 13, 17], the interaction between different layers and the strong strain hardening for the CG center.

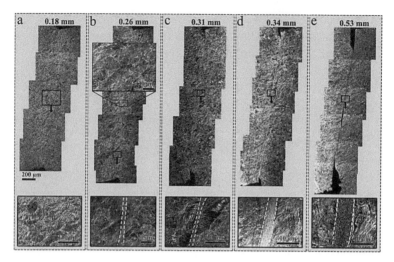

Figure 18.3 SEM micrographs under BSE mode for the microstructural evolution of the nanostructured surface of a gradient structured sample at shear displacements of (a) 0.18 mm, (b) 0.26 mm, (c) 0.31 mm, (d) 0.34 mm, and (f) 0.53 mm.

Figure 18.4 shows the SEM micrographs of the microstructural evolutions of the CG center at two different shear displacements (0.34, 0.53 mm). The shear band is not formed at the shear displacement of 0.34 mm, while a thin shear band is clearly observed at the shear displacement of 0.53 mm. Thus, the critical shear displacement for the initiation of ASB at the CG center is larger than that of the nanostructured surface layer. This is due to the fact that the CG center has much higher strain hardening capability for preserving longer uniform shear deformation than the nanostructured surface layer. This also indicates that ASB first nucleated in the nanostructured surface and then propagated to the CG center.

In order to probe the ASB propagation from the nanostructured surface to the center, the cross-section along the depth in the concentrated shear zone was examined, and the images at three shear displacements (0.31, 0.34 and 0.53 mm) are displayed in Fig. 18.5. It is well known that once an ASB is initiated in a specific location in homogeneous materials, it can propagate with a speed determined by a number of material parameters and the applied shear impact velocity [38, 39]. The propagation speed was estimated to be about 600 m/s for a CRS 1018 steel when the applied shear impact velocity is about 25 m/s (similar to the shear impact velocity applied in the present study) [39]. Two ASB nucleated at the shear displacement of

0.26 on the two nanostructured surfaces. At the shear displacement of 0.31 mm, two ASBs propagated for a distance about 0.2 mm from the two surfaces towards the center. The ASB propagated another 0.5 mm when the shear displacement increased to 0.34 mm. The two ASB tips joined coalesced when the shear displacement increases to 0.53 mm, forming a large ASB across the whole sample thickness. The propagation speed of the ASBs along the thickness in the gradient structures can be estimated to be about 50 m/s from the Fig. 18.5, which is an order of magnitude slower than that in the homogeneous materials [38, 39]. This could be attributed to the strain gradients and the strain partitioning [10, 12, 13, 17], the interaction between different layers and the strong strain hardening for the CG center.

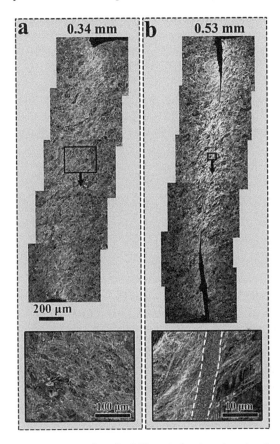

Figure 18.4 SEM micrograph under BSE mode for the microstructure evolution of the CG center at two different interrupted shear displacements: (a) 0.34 mm and (b) 0.53 mm.

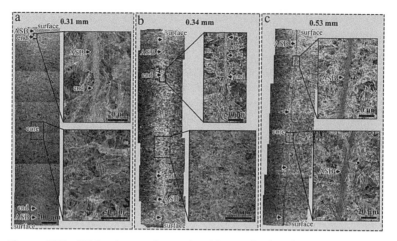

Figure 18.5 SEM micrographs under BSE mode for the microstructural evolution of the cross section at three different interrupted shear displacements: (a) 0.31 mm, (b) 0.34 mm, and (c) 0.53 mm.

Moreover, the maximum stress criterion in which the shear stress should start to drop once ASB is initiated in the homogeneous materials [25–30] *is no longer applicable* to gradient structures. As indicated from Figs. 18.2 and 18.3, the ASB nucleated on the nanostructured surface at the shear displacement of 0.26 mm, while the shear stress did not start to drop until a larger shear displacement of 0.31 mm. The gradient structures can be considered as consisting of numerous thin layers with systematically varying grain sizes/substructure sizes. Therefore, their global average dynamic shear flow stress should be at least the sum of the shear flow stress of different layers, as calculated using the rule of mixture (ROM), and might be even higher than what is calculated by ROM due to synergetic extra strengthening caused by the strain gradient associated with the gradient structure, which need to be accommodated by geometrically necessary dislocations (GNDs) and consequently produces HDI hardening [12, 13, 17].

When the shear displacement is in between 0.26 and 0.31 mm, the nanostructured surface layer is under strain softening due to the formation of ASB, while the CG center is still under strain hardening due to the absence of ASB and the high strain hardening capability of CG structure. The competition between the softening of surface layers and the hardening of center layers would determine the trend of the flow stress of the overall gradient structures with increasing applied shear displacement. At the early stage of ASB propagation from the surface to the center, the overall flow stress of the gradient structures can still increase with increasing shear displacement. However, when ASB propagates toward the center layers, the fraction of strain softening part would increase. The overall flow stress would start to drop with increasing shear displacement at a critical shear displacement. Apparently, this critical shear displacement should be larger than that for onset of ASB at the nanostructured surface layer, making the well-known maximum stress criterion invalid for gradient structured.

In summary, gradient materials behave very differently from homogeneous materials under dynamic loading. It has a combination of strength and dynamic shear toughness that is not accessible to its homogeneous counterparts. The gradient structure delayed the initiation of adiabatic shear bands (ASB) on the nanostructured surface and dramatically slowed down the ASB propagation by an order of magnitude, which renders the gradient materials higher dynamic shear toughness. The superior dynamic properties of the gradient materials are derived from its strain gradient during dynamic shear deformation, which produces a geometrically necessary dislocations and strong HDI hardening. Our findings here provide a new strategy for designing strong and tough metals and alloys for dynamic applications.

Acknowledgements

This work was supported by the National Natural Science Foundation of China (NSFC) under grant Nos. 11472286, 11572328 and 11672313, the Strategic Priority Research Program of the Chinese Academy of Sciences under grant No. XDB22040503, and the National Key Basic Research Program of China under grant Nos. 2012CB932203 and 2012CB937500. Y.Z. was supported by the US Army Research Office (W911 NF-12-1-0009), and the Nanjing University of Science and Technology.

References

1. Youssef KM, Scattergood RO, Murty KL, Horton JA, Koch CC. Ultrahigh strength and high ductility of bulk nanocrystalline copper. *Appl Phys Lett.* 2005; **87**(9):091904.
2. Zhu YT, Liao XZ. Nanostructured metals-retaining ductility. *Nat Mater.* 2004; **3**(6):351–352.
3. Meyers MA, Mishra A, Benson DJ. Mechanical properties of nanocrystalline materials. *Prog Mater Sci.* 2006; **51**(4):427–556.
4. Wang YM, Chen MW, Zhou FH, Ma E. High tensile ductility in a nanostructured metal. *Nature.* 2002; **419**(6910):912–915.
5. Zhao YH, Liao XZ, Cheng S, Ma E, Zhu YT. Simultaneously increasing the ductility and strength of nanostructured alloys. *Adv Mater.* 2006; **18**(17):2280–2283.
6. Lu L, Chen X, Huang X, Lu K. Revealing the maximum strength in nanotwinned copper. *Science.* 2009; **323**(5914):607–610.
7. Kou HN, Lu J, Li Y. High-strength and high-ductility nanostructured and amorphous metallic materials. *Adv Mater.* 2014; **26**:5518–5524.
8. Liddicoat PV, Liao XZ, Zhao YH, Zhu YT, Murashkin MY, Lavernia EJ, Valiev RZ, Ringer SP. Nanostructural hierarchy increases the strength of aluminum alloys. *Nat Commun.* 2010; **1**:63.
9. Liu G, Zhang GJ, Jiang F, Ding XD, Sun YJ, Sun J, Ma E. Nanostructured high-strength molybdenum alloys with unprecedented tensile ductility. *Nat Mater.* 2013; **12**:344–350.
10. Wu XL, Yang MX, Yuan FP, Wu GL, Wei YJ, Huang XX, Zhu YT. Heterogeneous lamella structure unites ultrafine grain strength with coarse-grain ductility. *Proc Natl Acad Sci USA.* 2015; **112**(47):14501–14505.
11. Valiev RZ. Nanostructuring of metals by severe plastic deformation for advanced properties. *Nat Mater.* 2004; **3**:511–516.
12. Yang MX, Pan Y, Yuan, FP, Zhu, YT, Wu XL. Back stress strengthening and strain hardening in gradient structure. *Mater Res Lett.* 2016; **4**(3):141–151.
13. Wu XL, Jiang P, Chen L, Yuan FP, Zhu YT. Extraordinary strain hardening by gradient structure. *Proc Natl Acad Sci USA.* 2014; **111**(20):7197–7201.
14. Suresh S. Graded materials for resistance to contact deformation and damage. *Science.* 2001; **292**(5526):2447–2451.

15. Fang TH, Li WL, Tao NR, Lu K. Revealing extraordinary intrinsic tensile plasticity in gradient nano-grained copper. *Science*. 2011; **331**(6024):1587–1590.
16. Wei YJ, Li YQ, Zhu LC, Liu Y, Lei XQ, Wang G, Wu YX, Mi ZL, Liu JB, Wang HT, Gao HJ. Evading the strength-ductility trade-off dilemma in steel through gradient hierarchical nanotwins. *Nat Commun*. 2014; **5**:3580.
17. Wu XL, Jiang P, Chen L, Zhang JF, Yuan FP, Zhu, YT. Synergetic strengthening by gradient structure. *Mater Res Lett*. 2014; **2**(4):185–191.
18. Hughes DA, Hansen N. Graded nano-structures produced by sliding and exhibiting universal behavior. *Phys Rev Lett*. 2001; **87**(13):135503.
19. Zeng Z, Li XY, Xu DS, Lu L, Gao HJ, Zhu T. Gradient plasticity in gradient nano-grained metals. *Extreme Mech Lett*. 2015; **8**:213–219.
20. Feaugas X. On the origin of the tensile flow stress in the stainless steel AISI 316L at 300 K: back stress and effective stress. *Acta Mater*. 1999; **47**(13):3617–3632.
21. Ashby MF. The deformation of plastically nonhomogeneous materials. *Philos Mag*. 1970; **21**(170):399–424.
22. Zener C, Hollomon JH. Effect of strain rate upon plastic flow of steel. *J Appl Phys*. 1944; **15**:22–32.
23. Nemat-Nasser S, Guo WG. Thermomechanical response of DH-36 structural steel over a wide range of strain rate and temperatures. *Mech Mater*. 2003; **35**:1023–1047.
24. Suo T, Li YL, Zhao F, Fan XL, Guo WG. Compressive behavior and rate-controlling mechanisms of ultrafine grained copper over wide temperature and strain rate ranges. *Mech Mater*. 2013; **61**:1–10.
25. Mishra A, Martin M, Thadhani NN, Kad BK, Kenik EA, Meyers MA. High-strain rate response of ultra-fine-grained copper. *Acta Mater*. 2008; **56**:2770–2783.
26. Yuan FP, Bian XD, Jiang P, Yang MX, Wu XL. Dynamic shear response and evolution mechanisms of adiabatic shear band in an ultrafine-grained austenite-ferrite duplex steel. *Mech Mater*. 2015; **89**:47–58.
27. Meyers MA, Subhash G, Kad BK, Prasad L. Evolution of microstructure and shear-band formation in α-hcp titanium. *Mech Mater*. 1994; **17**:175–193.
28. Bai Y, Dodd B. *Adiabatic Shear Localization*. Pergamon Press, New York. 1992.
29. Wright TW. *The Physics and Mathematics of Adiabatic Shear Bands*. Cambridge University Press, Cambridge. 2002.

30. Xue Q, Gray III GT, Henrie BL, Maloy SA, Chen SR. Influence of shock prestraining on the formation of shear localization in 304 stainless steel. *Metall Mater Trans A*. 2005; **36A**:1471–1486.

31. Yang Y, Jiang F, Zhou BM, Li XM, Zheng HG, Zhang QM. Influence of shock prestraining on the formation of shear localization in 304 stainless steel. *Mater Sci Eng A*. 2011; **528**:2787–2794.

32. Yuan FP, Jiang P, Wu XL. Annealing effect on the evolution of adiabatic shear band under dynamic shear loading in ultra-finegrained iron. *Int J Impact Eng*. 2012; **50**:1–8.

33. Wei Q, Kecskes L, Jiao T, Hartwig KT, Ramesh KT, Ma E. Adiabatic shear banding in ultrafine-grained Fe processed by severe plastic deformation. *Acta Mater*. 2004; **52**:1859–1869.

34. Xing JX, Yuan FP, Wu XL. Enhanced quasi-static and dynamic shear properties by heterogeneous gradient and lamella structures in 301 stainless steels. *Mater Sci Eng A*. 2017; **680**:305–316.

35. Johansson J, Persson C, Lai HP, Colliander MH. Microstructural examination of shear localisation during high strain rate deformation of alloy 718. *Mater Sci Eng A*. 2016; **662**:363–372.

36. Lu K, Lu J. Surface nanocrystallization (SNC) of metallic materials-presentation of the concept behind a new approach. *J Mater Sci Technol*. 1999; **15**:193–197.

37. Sun, JL, Trimby, PW, Yan, FK, Liao, XZ, Tao, NR, Wang, JT. Shear banding in commercial pure titanium deformed by dynamic compression. *Acta Mater*. 2014; **79**:47–58.

38. Xue, Q, Meyers, MA. Self-organization of shear bands in titanium and Ti-6Al-4V alloy. *Acta Mater*. 2002; **50**:575–596.

39. Bonnet-Lebouvier, AS, Molinari, A, Lipinski, P. Analysis of the dynamic propagation of adiabatic shear bands. *Inter J Solids Struc*. 2002; **39**:4249–4269.

Chapter 19

On Strain Hardening Mechanism in Gradient Nanostructures

Jianjun Li,[a,b] G. J. Weng,[c] Shaohua Chen,[d] and Xiaolei Wu[e]

[a]*Department of Engineering Mechanics, School of Mechanics,*
Civil Engineering and Architecture, Northwestern Polytechnical University,
Xi'an 710129, Shaanxi, China
[b]*Department of Microstructure Physics and Alloy Design,*
Max-Planck-Institut für Eisenforschung GmbH, Düsseldorf 40237, Germany
[c]*Department of Mechanical & Aerospace Engineering, Rutgers University,*
New Brunswick, NJ 08903, USA
[d]*Institute of Advanced Structure Technology, Beijing Institute of Technology,*
Beijing 100081, China
[e]*State Key Laboratory of Nonlinear Mechanics, Institute of Mechanics,*
Chinese Academy of Sciences, Beijing 100190, China
jianjunli.mech@hotmail.com, j.li@mpie.de, mejjli@nwpu.edu.cn

Experiments have shown that a gradient design, in which grain size spans over four orders of magnitude, can make strong nanomaterials ductile. The enhanced ductility is attributed to the considerable

---

Reprinted from *Int. J. Plast.*, **88**, 89–107, 2017.

Heterostructured Materials: Novel Materials with Unprecedented Mechanical Properties
Edited by Xiaolei Wu and Yuntian Zhu
Text Copyright © 2016 The Author(s).
Layout Copyright © 2022 Jenny Stanford Publishing Pte. Ltd.
ISBN 978-981-4877-10-7 (Hardcover), 978-1-003-15307-8 (eBook)
www.jennystanford.com

strain hardening capability obtained in the gradient metals. A non-uniform deformation on the lateral sample surface is also observed. This might generate geometrically necessary dislocations (GNDs) into the sample. However, no direct evidence has been provided. Therefore the issues remain: why can the gradient structure generate high strain hardening, and how does it reconcile the good strength-ductility combination of gradient nanostructures? Here for the first time we quantitatively investigate the strain hardening of a gradient interstitial-free steel by developing a dislocation density-based continuum plasticity model, in which the interaction of the component layers in the gradient structure is represented by incorporating GNDs and hetero-deformation induced (HDI) stress. It is demonstrated that both the surface non-uniform deformation and the strain-hardening rate up-turn can be quantitatively predicted. The results also show that the strain hardening rate of the gradient sample can reach as high as that of its coarse-grained counterpart. A strength-ductility map is then plotted, which clearly show that the gradient samples perform much more superior to their homogeneous counterparts. The predicted map has been verified by a series of experimental data. A detailed analysis on GNDs distribution and HDI stress evolution at the end further substantiates our view that the good strain hardening capability results from the generation of abundant GNDs by the surface non-uniform deformation into the nano-grained layers of the gradient sample.

19.1 Introduction

Refining grains down to nanoscale renders metals ultra-high strength but low ductility (Beyerlein et al., 2015; Farrokh and Khan, 2009; Gleiter, 1989; Khan et al., 2000; Khan and Liu, 2016; Li et al., 2014; Meyers et al., 2006; Ovid'ko, 2002; Rodríguez-Galán et al., 2015; Rupert, 2016; Weng, 2011; Zhu and Li, 2010). The ultra-high strength opens a new challenge of reconciling both strength and ductility in nanostructured metallic metals and alloys. This strength-ductility trade-off dilemma has led to a severe setback to their engineering applications. Therefore, over the last two decades it has been constantly pursued to have the unity of nano-grain's ultra-high strength and coarse-grain (CG)'s excellent ductility in the same sample (Kou et al., 2014; Li et al., 2010b; Li et al., 2016; Liu et al., 2015; Lu et al., 2009; Ma, 2006; Ovid'ko and Langdon, 2012; Wang et al., 2002; Wei et al., 2014; Youssef et al., 2005; Zhao et al., 2008). Recently, gradient nanostructured metals have attracted much interests due to their high strength and high ductility (Fang et al., 2011; Kou et al., 2014; Li and Soh, 2012a; Lu, 2014; Moering et al., 2015; Moering et al., 2016; Wu et al., 2014a; Yang et al., 2016; Yin et al., 2016; Yuan et al., 2016; Zeng et al., 2016). For example, a gradient copper made by surface mechanical grinding treatment possesses almost the same uniform tensile elongation and a doubled yield strength as compared with its CG counterpart (Fang et al., 2011).

This class of materials is marked by a gradual grain-size variation from the sample surface into its interior by four orders of magnitude. As shown in Figs. 19.1a and b, the grain size in the gradient structure is typically tens of nanometers on the surface layer, but deep in the sample it can increase to the CG size (typically tens of micrometers). This is in sharp contrast to those homogeneous samples with uniform grain size distribution (Fig. 19.1c). Generally, the gradient structure has two gradient surface layers (one on the top and the other at the bottom), and in-between a sandwiched CG core (Fig. 19.2a). This type of microstructures can be produced by surface nano-crystallization technique, which has been introduced in the form of many process variations that are both cost-effective and amenable to large-scale industrial production, such as surface mechanical attrition treatment (Lu and Lu, 2004) and surface mechanical grinding treatment (Li et al., 2008).

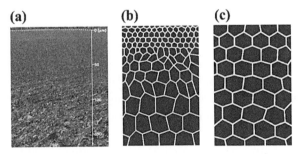

Figure 19.1 (a) Microstructure of gradient sample (Lu, 2014) and (b) its schematic diagram. (c) Schematic of homogeneous structure.

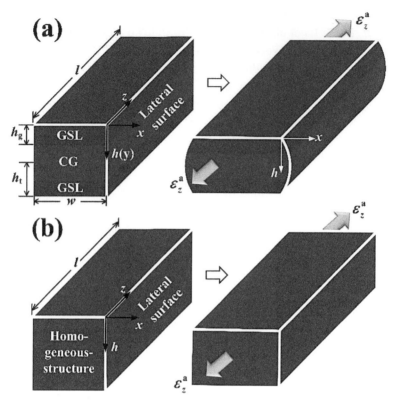

Figure 19.2 Schematic diagram of the deformation of IF steel samples subjected to uniform uniaxial tensile strain ε_z^a: (a) gradient sample consisting of two grain size gradient surface layers (GSL) sandwiching a coarse-grained (CG) core, in which h_t is the half thickness of the whole sample; (b) homogeneous sample. Note the non-uniform deformation on the lateral surface of the gradient sample in (a).

The good balance of strength and ductility of gradient sample is found to be resulted from the considerable strain hardening capability obtained in the gradient structure (Wu et al., 2014a; Yang et al., 2015). For example, a strain hardening up-turn is observed in the gradient interstitial free (IF) steel sample fabricated by Wu et al. (2014a) using surface mechanical attrition treatment. The up-turn, also found in copper (Yang et al., 2015), is believed to be responsible for the high strength and ductility in the gradient IF steel sample, i.e., 2.5 times higher yield strength at a loss of only 3.3% uniform elongation as compared with the CG counterpart. They also observed a non-uniform deformation on the lateral surface of the gradient sample with a height profile up to 30 μm that is absent in ordinary homogeneous structures (Fig. 19.2). The non-uniform deformation might generate geometrically necessary dislocations (GNDs) into the gradient sample but no evidence has been provided.

As compared with so many experimental investigations, corresponding simulation and theoretical works are scarce due to the huge grain size variation in the gradient structure. All the existing crystal plasticity and molecular dynamics studies are limited to 2D structures, in which the grain size varies only across two length scales (~10 nm to ~100 nm) (Li et al., 2015b; Liu and Mishnaevsky Jr, 2014; Zeng et al., 2016), because of the high computational cost, whereas the theoretical investigations tend to focus on 1D stress state without incorporating the boundary effect of different layers (Li et al., 2010a; Li and Soh, 2012b). Although one could get some insights from the above investigations, they cannot describe the 3D nature of the problem with the experimentally observed lateral surface non-uniform deformation. Therefore, there are several critical questions remain to be answered: (1) what leads to the strong strain hardening in the gradient structure? (2) how does the non-uniform deformation form on the lateral surface? (3) are the GNDs that are associated with the surface non-uniform deformation responsible for the strong strain hardening capability? (4) if the answer of (3) is yes, what is their quantitative relation? and (5) how does the gradient microstructure modulate the strain hardening? In order to answer the above critical questions, we embark on the development of a dislocation density-based theoretical model to quantitatively describe the strain hardening in gradient structure. Since non-uniform deformation is generally associated with GNDs (Ashby, 1970; Demir et al., 2009; Fleck and Hutchinson, 1993; Gao et al., 1999; Lyu et al., 2015; Martínez-Pañeda and Niordson, 2016), it is important to incorporate the effect of GNDs in the present model.

The chapter is organized as follows. The developed model will be presented in Section 19.2 to identify the constitutive behavior of all the component layers by considering GNDs and HDI stress. The overall mechanical response is then calculated based on the properties of these individual layers through volume averaging. The results and discussion will be given in Section 19.3, which is followed by conclusions (Section 19.4).

19.2 Constitutive Model for Gradient Structure

A gradient IF steel sample is assumed to be a perfectly-bonded multilayered structure. The grain size is assumed to be uniformly distributed in each layer (called homogeneous layer). The constitutive behavior of the constituent homogeneous layers with different grain size will be established first by using two sub-models, i.e., the nanocrystalline (NC) sub-model for layers with grain size varying from tens of nanometers to several hundred nanometers (called NC layers), while the microcrystalline (MC) sub-model is for those with grain size ranging from submicrometer to tens of micrometers (called MC layers). We assume that there are n layers in the gradient region, and the CG core can be viewed as the $(n + 1)$-th layer, where the sample is subjected to uniaxial tension in z direction (Figs. 19.2a and 19.3a). The lateral and thickness directions are set as x and h (or y), respectively. Therefore, the z-h plane is the lateral surface (Fig. 19.2a).

In order to describe how the surface non-uniform deformation forms, we first consider the deformation of free-standing layers with different grain sizes ($I = 1, 2, ..., n + 1$, and grain size increases as i increases) subjected to uniaxial tensile strain ε_z^a, as shown in Fig. 19.3a. Generally, the free-standing layers with smaller grain sizes (outer layers) are more prone to plastic instability (or necking/shrinking in the lateral (x) direction) than those with larger grain sizes (inner layers). Hence, the shrinking magnitude in the necked region, i.e., $-\varepsilon_{xi}w$, or $\varepsilon_{zi}w/2$ for incompressible material, varies in different layers (Fig. 19.3b). The subscripts x and z denote directions and i represents the i-th layer. In the gradient structure, the interaction between different layers will modify the deformation behavior of each layer, that is, the faster lateral necking of the outer layers is suppressed by the higher capability of the inner layers to resist plastic instability, which produces the surface non-uniform deformation, as schematically shown by the curved boundary in Fig. 19.3c. According to Ashby (Ashby, 1970), the non-uniform deformation is able to inject GNDs into the gradient sample (Fig. 19.3c). We assume that all GNDs in the gradient sample are induced by the surface non-uniform deformation. As a result, the effect of layer boundaries can be considered by incorporating GNDs in the present constitutive model. In the following, we will first provide the constitutive framework for gradient structure based on the conventional strain gradient theory of plasticity (Huang et al., 2004) and Kocks-Mecking-Estrin (KME) model (Estrin and Mecking, 1984) with flow stress modified (Section 19.2.1). The surface non-uniform deformation and the GNDs density are then calculated by obtaining the shrinking strain of each layer (ε_{xi} or ε_{zi}) using a Marciniak-Kuczyński (M-K) type analysis (Marciniak et al., 1973) and by transforming the gradient problem into a plane strain thermal stress one for the sake of easy solution using finite element simulations (Section 19.2.2).

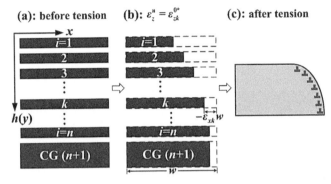

Figure 19.3 The formation of non-uniform deformation on the lateral surface (z-h plane) of a gradient sample. (a) free-standing layers with different grain sizes before tension; (b) free-standing layers during deformation as layer k starts necking in the lateral (x) direction; (c) gradient structure with injections of GNDs after tension.

The total strain rate tensor $\dot{\varepsilon}$ for each constituent layer can be expressed as (the layer index i is omitted for clarity):

$$\dot{\varepsilon} = \frac{1}{2\mu}\dot{\sigma}' + \frac{1-2\nu}{3E}\delta\delta:\dot{\sigma} + \frac{3\dot{\varepsilon}^p}{2\sigma}\sigma', \qquad (19.1)$$

where σ is the stress tensor and a dot denotes differentiation with respect to time t; and δ, $\sigma' = \sigma - \delta\delta:\sigma/3$, $\sigma = \sqrt{3\sigma':\sigma'/2}$, $\dot{\varepsilon}^p$, $\dot{\varepsilon}^p = \sqrt{2\dot{\varepsilon}^p:\dot{\varepsilon}^p/3}$, E, μ and ν are Kronecker's delta, deviatoric stress, equivalent von Mises stress, plastic strain rate, equivalent von Mises plastic strain rate, Young's modulus, shear modulus and Poisson's ratio, respectively. A kinetic equation is then applied to correlate $\dot{\varepsilon}^p$ and σ as (Huang et al., 2004):

$$\dot{\varepsilon}^p = \dot{\varepsilon}(\sigma/\sigma_f)^m, \qquad (19.2)$$

where $\dot{\varepsilon} = \sqrt{2\dot{\varepsilon}':\dot{\varepsilon}'/3}$ ($\dot{\varepsilon}' = \dot{\varepsilon} - \delta\delta:\dot{\varepsilon}/3$ is the deviatoric strain rate), σ_f and m are the equivalent strain rate, flow stress and rate-sensitivity exponent, respectively. The flow stress σ_f for the component homogeneous NC and MC layers in gradient structure will be addressed in the following subsection 19.2.1.

Specifically, for incompressible material ($v = 0.5$) as adopted in the present analysis, the constitutive equations of each layer (Eq. (19.1)) in the gradient structure subjected to uniform uniaxial tension $\dot{\varepsilon}_z^a$ can be specified as (i.e., Eq. (19.A5) in Appendix A), see Fig. 19.2a for the coordinate system used:

$$\dot{\varepsilon}_x = \frac{1}{2}\left(\frac{1}{2\mu}\dot{\sigma}_x + \eta\sigma_x\right) - \frac{1}{2}\dot{\varepsilon}_z$$

$$\dot{\varepsilon}_y = -\frac{1}{2}\left(\frac{1}{2\mu}\dot{\sigma}_x + \eta\sigma_x\right) - \frac{1}{2}\dot{\varepsilon}_z, \qquad (19.3)$$

$$\dot{\gamma}_{xy} = \frac{1}{\mu}\dot{\tau}_{xy} + 2\eta\tau_{xy}$$

where $\eta = \dfrac{3\dot{\varepsilon}^p}{2\sigma}$. Here, considering the geometry of each layer with thickness much less than the other two dimensions, we adopted the assumption of $\sigma_y = 0$, as used in the optimal, shear-lag model for multi-layered elastic structure (Nairn and Mendels, 2001). The above assumption has been validated by the following finite element calculations (Supplementary Note 1), which shows that the magnitude of σ_y is ignorable as compared with those of σ_x and σ_z except in only a very small region near the right (lateral) surface (Supplementary Fig. 1b). Equations (19.2)–(19.3) describe the triaxial constitutive relations for all component layers of different grain sizes in a gradient structure subjected to uniaxial tension.

19.2.1 Flow Stress for Component Homogeneous Layers

The flow stress in the NC layers can be approximated as:

$$\sigma_f = \sigma_0 + M\alpha\mu b\sqrt{\rho_s + \rho_G} + \sigma_b + \sigma_{GB}, \qquad (19.4)$$

where M, α, b, ρ_s, ρ_G, σ_b and σ_0 are the Taylor factor, Taylor constant, magnitude of the Burgers vector, statistically stored dislocation density, GNDs density, HDI stress and lattice friction stress, respectively. The GNDs density in gradient structure has been given by Li et al (2015a). The essential expressions are summarized here for completeness. All GNDs are assumed to be edge type and positioned uniformly along the lateral (x) direction (Fig. 19.3c). The dislocation line of each GND is idealized as a straight line running through the entire sample along the tensile (z) direction (Figs. 19.2a and 19.3c). Therefore, the total length of the GNDs ζ injected by the non-uniform lateral deformation in the unit layer with thickness Δh can be expressed as

$$\zeta = \frac{l\Delta h}{s} = \frac{la}{b}, \qquad (19.5)$$

where l is the total length of the gradient sample (Fig. 19.2a), $s = b(\Delta h/a)$ is the spacing between individual slip steps near the lateral surface, a (>0) is the displacement difference along the x-axis between two adjacent deformed layers, which characterizes the lateral surface non-uniform deformation. By assuming that all GNDs are uniformly distributed in each layer, the volume occupied by GNDs can be approximated as $V = l(\Delta h)^2$. Thus, the GNDs density can be given as

$$\rho_G = \frac{\zeta}{V} = \frac{a}{b(\Delta h)^2}. \qquad (19.6)$$

On the other hand, the predicted GNDs density (Eq. (19.6)) can be evaluated by the measured hardness along the sample depth since it is difficult to measure GNDs density directly. By using Tabor's factor λ to convert the flow stress σ_f (Eq. (19.4)) to the measured hardness $H = \lambda \sigma_f = \lambda \left(\sigma_0 + \sigma_{GB} + \sigma_b + M\alpha\mu b \sqrt{\rho_s + \rho_G} \right)$, we have

$$H - H_0 = \lambda M\alpha\mu b \sqrt{\rho_s + \rho_G}, \qquad (19.7)$$

where $H_0 = \lambda(\sigma_0 + \sigma_{GB} + \sigma_b)$ represents the hardness along the thickness on the lateral surface of gradient sample before deformation. As for free standing component layers after deformation without incorporating GNDs (i.e., $\rho_G = 0$), their hardness H_s can be obtained from Eq. (19.7)

$$H_s - H_0 = \lambda M\alpha\mu b \sqrt{\rho_s}. \qquad (19.8)$$

By combining Eqs. (19.7) and (19.8), ρ_G can be expressed as

$$\rho_G = \frac{(H-H_0)^2 - (H_s - H_0)^2}{(\lambda M \alpha \mu b)^2}. \quad (19.9)$$

The hardness parameters H, H_0 and H_s can be measured through experiments (Fig. 19.2c of Wu et a., 2014a).

As observed in experiments (Yang et al., 2016), HDI stress plays a role in enhancing strain hardening of gradient structure. The HDI stress in the gradient structure is a long range internal stress caused by the interaction and pile-up of GNDs near the layer boundaries. Since it is the spatial variation of GNDs density that contributes to the long-range internal stress (Bayley et al., 2006; Evers et al., 2004; Groma et al., 2003; Yefimov et al., 2004), the internal stress resulted from single edge GND can be given as (Bayley et al., 2006)

$$\sigma_x = \frac{3\mu b R^2}{8(1-v)} \frac{\partial \rho_G}{\partial y}, \quad \sigma_{xy} = -\frac{\mu b R^2}{8(1-v)} \frac{\partial \rho_G}{\partial x}$$

$$\sigma_y = \frac{\mu b R^2}{8(1-v)} \frac{\partial \rho_G}{\partial y}, \quad \sigma_{xz} = 0, \quad (19.10)$$

$$\sigma_z = \frac{\mu b v R^2}{2(1-v)} \frac{\partial \rho_G}{\partial y}, \quad \sigma_{yz} = 0$$

where v, R are Poisson's ratio and the integral circular domain for GNDs to contribute to the HDI stress, respectively. Based on the assumption that GNDs are positioned uniformly along the x direction in each unit layer, we have $\frac{\partial \rho_G}{\partial x} = 0$, and $\frac{\partial \rho_G}{\partial y}$ can be approximated as $\frac{\Delta \rho_G}{\Delta h}$, in which $\Delta \rho_G$ is the difference of GNDs density between neighboring layers. Then the non-zero internal stress components are σ_x, σ_y and σ_z. As a result, the GNDs-induced HDI stress can be approximated as the von-Mises equivalence of the above stress components, i.e.,

$$\sigma_b = \sqrt{16v^2 - 16v + 7} \frac{\mu b R^2}{8(1-v)} \frac{\Delta \rho_G}{\Delta h}. \quad (19.11)$$

The parameter R should be in the order of layer thickness Δh in order to model the interaction of adjacent layers. The last term $\sigma_{GB} = k_{HP}/\sqrt{d}$ in flow stress arises from a Hall-Petch type strengthening due to the refined grains (Capolungo et al., 2007; Carvalho Resende et al., 2013; Delince et al., 2007), where k_{HP} is a constant Hall-Petch slope, and d denotes grain size.

The evolution of statistically stored dislocation density ρ with respect to plastic strain can be expressed as:

$$\frac{d\rho_s}{d\varepsilon^p} = M(k + k_1\sqrt{\rho_s} - k_2\rho_s - k_e\rho_s), \qquad (19.12)$$

where $k = k_3/(bd)$, $k_1 = \psi/b$, $k_2 = k_{20}(\dot{\varepsilon}^p/\dot{\varepsilon}_0)^{-1/n_0}$, ψ is a proportionality factor, k_{20} and $\dot{\varepsilon}_0$ are material constants, k_3 is a geometric factor related to the grain shape and proportion of dislocations arriving at the grain boundaries, n_0 is inversely proportional to temperature (Estrin and Mecking, 1984), and $k_e = (d_e/d)^2$ is an additional dynamic recovery factor, where d_e is a reference grain size corresponding to the critical grain size at which enhanced dynamic recovery occurs (Li and Soh, 2012b).

The expressions of the flow stress for MC layers are set to be the same as those of NC layers, but with different values for parameters according to experimental observations (Wu et al., 2014a), such as Hall-Petch slope k_{HP}, reference grain size d_e and dynamics recovery factor k_2. This makes the two sub-models consistent.

19.2.2 Calculation of GNDs Density and HDI Stress

The objective of this section is to calculate the GNDs density presented in Eq. (19.6) and HDI stress in Eq. (19.11) by solving the 3D plastic deformation problem of a gradient structure subjected to uniaxial tension, i.e., Eqs. (19.2) and (19.3). As for the layers without considering GNDs (i.e., $\rho_G = 0$, $\sigma_b = 0$), these equations can be solved numerically independently layer by layer. However, for the layers with GNDs injected by the lateral surface non-uniform deformation, the above equations should be solved by combining all the component layers. The key is to calculate the magnitude of shrinking of each layer (Fig. 19.3b), that is, ε_{zi} (= $-2\varepsilon_{xi}$) as appeared in Eq. (19.3). Figure 19.3b shows that as the applied strain of the gradient structure approaches the limit strain of the k-th layer, i.e., $\varepsilon_z^a = \varepsilon_{zk}^{0*}$

(the strain in the section of uniform deformation as a neck sets in), it starts shrinking/necking. The strain in the necked region along z direction, i.e., ε_z, for layer k is $\varepsilon_{zk} = -2\varepsilon_{xk} = \varepsilon_{zk}^* = \chi_k \varepsilon_{zk}^{0*} = \chi_k \varepsilon_z^a$, where ε_{zk}^* is the strain in the necked region and $\chi_k = \varepsilon_{zk}^* / \varepsilon_{zk}^{0*}$ is the limit strain ratio. The values of ε_{zk}^{0*} and ε_{zk}^* can be calculated using the following M-K analysis. Moreover, the layers 1-$(k-1)$ have already necked since their limit strains are smaller than that of the layer k. However, in order to model the surface non-uniform deformation, we assume that they are able to afford to deform even after necking. This assumption is conceivable because in a gradient structure the necking of the layers 1-$(k-1)$ could be suppressed by the inner layers with better resistance to plastic instability, as schematically shown in Fig. 19.3b. The ε_z of these layers are set to be similar in form to that of the k-th layer, i.e., $\varepsilon_{zi} = \chi_i \varepsilon_z^a$, $i = 1, 2, ..., (k-1)$, where χ_i has the same definition as the above χ_k. Finally the ε_z in the stable layers ($i = (k+1)$, $(k+2), ..., (n+1)$) are identical to the applied strain ε_z^a, i.e., $\varepsilon_{zi} = \varepsilon_z^a$. As a result, the z-component strains for all component layers in a gradient structure subjected to uniaxial tensile strain ε_z^a ($\varepsilon_z^a = \varepsilon_{zk}^{0*}$) can be summarized as

$$\varepsilon_{zi} = \begin{cases} \chi_i \varepsilon_z^a, & \text{if } \varepsilon_z^a \geq \varepsilon_{zi}^{0*} \quad \text{(for necked layers, } i = 1, 2..., k) \\ \varepsilon_z^a, & \text{if } \varepsilon_z^a < \varepsilon_{zi}^{0*} \quad \text{(for stable layers, } i = k+1, k+2,..., n+1) \end{cases}.$$

(19.13)

The values of χ_i for the i-th layer will be obtained by implementing M-K analysis based on the constitutive behavior of layer i without incorporating GNDs (i.e., $\rho_G = 0$).

The original M-K method is aimed at analyzing sheet necking behavior and calculating the forming limit diagrams of the sheets by introducing an initial non-uniformity (Hutchinson and Neale, 1978; Marciniak et al., 1973). Therefore, M-K analysis is an appropriate approach to find the limit strain ratio χ_i of layer i as defined above because the free-standing component layers in a gradient sample could be viewed as a sheet. The M-K analysis used here is along the line of that by Hutchinson and Neale (1978) but using the material behavior developed in the present model. For brevity, the subscript 'i' is omitted whenever there is no danger of confusion. Assume the material is rigid-plastic and incompressible, the law of plastic flow (Eq. (19.1)) becomes

$$d\varepsilon = \frac{3}{2}\frac{d\bar{\varepsilon}}{\bar{\sigma}}\sigma', \qquad (19.14)$$

where $d\bar{\varepsilon} = \sqrt{2d\varepsilon : d\varepsilon/3}$. The yield condition is $\bar{\sigma} = \sigma_f$ for rigid-plastic material (see Eq. (19.2)), and for plane stress condition $\bar{\sigma} = \sqrt{\sigma_z^2 - \sigma_z\sigma_x + \sigma_x^2}$. The detailed M-K analysis is given in Appendix C, which yields two differential equations for solving the equivalent strains ε^0 in the uniform section and ε in the necked region, i.e., Eqs. (19.C9) and (19.C11). The two equations can be rewritten as

$$\frac{d\varepsilon^0}{d\varepsilon} = \pm\sqrt{\frac{1}{B} - \frac{1}{(1-\xi)^2}\frac{1-B-G}{B}\left(\frac{\bar{\sigma}^0}{\bar{\sigma}}\right)^2 \exp\left(-2C\varepsilon^0 - 2\varepsilon_y\right)},$$

$$\frac{d\varepsilon_y}{d\varepsilon} = -\frac{A\sqrt{1-B-G}}{(1-\xi)}\frac{\bar{\sigma}^0}{\bar{\sigma}}\exp\left(-C\varepsilon^0 - \varepsilon_y\right) - D\frac{d\varepsilon^0}{d\varepsilon}$$

(19.15)

where ξ is the initial geometric non-uniformity, and the expressions of A, B, C, D and G are given in Appendix C. Physically, the value of $d\varepsilon^0/d\varepsilon$ should be selected as positive. Now we can use the fourth-order Runge-Kutta method to solve the two differential equations in (19.15), which gives the variation of ε^0 with respect to ε, then the critical strain ε_z^*, ε_z^{0*}, and the ratio $\chi = \varepsilon_z^*/\varepsilon_z^{0*}$ as well can be obtained by using the criterion $d\varepsilon^0/d\varepsilon = 0$.

After getting the value of ε_z in Eq. (19.3) for each layer, it is ready to solve the plastic deformation problem of gradient structure subjected to uniaxial tension. However, since the value of η differs in different layers and it evolves during deformation, it is difficult, even not impossible, to solve them numerically. In this study, the above difficulty has been circumvented by transforming the above 3D plastic deformation problem into a plane strain problem of the same structure subjected to thermal load $\dot{T}$. In the following, we will prove this transformation.

As for a gradient structure in plane strain condition subjected to thermal load $\dot{T}$, the constitutive relations of each layer are given as follows (Appendix B).

$$\dot{\varepsilon}_x = \frac{1}{2}\left(\frac{1}{2\mu}\dot{\sigma}_x + \eta\sigma_x\right) + \frac{3}{2}\alpha\dot{T}$$

$$\dot{\varepsilon}_y = -\frac{1}{2}\left(\frac{1}{2\mu}\dot{\sigma}_x + \eta\sigma_x\right) + \frac{3}{2}\alpha\dot{T}, \qquad (19.16)$$

$$\dot{\gamma}_{xy} = \frac{1}{\mu}\dot{\tau}_{xy} + 2\eta\tau_{xy}$$

where α is thermal expansion coefficient. Compare Eqs. (19.3) and (19.16), if we set $\frac{3}{2}\alpha\dot{T} = -\frac{1}{2}\dot{\varepsilon}_z$, and by integration we get $\frac{3}{2}\alpha\Delta T = -\frac{1}{2}\varepsilon_z$. Then by setting $\Delta T = -1°C$, we have

$$\alpha = \frac{1}{3}\varepsilon_z. \qquad (19.17)$$

Now the two sets of equations turn out to be identical. That is, the solution of the 3D plastic deformation problem of a gradient sample subjected to uniaxial tension, i.e., Eqs. (19.2) and (19.3), is equivalent to that of a plane strain thermal-stress problem, i.e., Eqs. (19.2) and (19.16). The latter can be easily solved by finite element simulations (Appendix D), which yields the values of the non-uniform deformation parameter a, and the GNDs density and HDI stress, see Eqs. (19.6) and (19.11).

19.2.3 Overall Mechanical Response of Gradient Structure

Since the focus of the present study is on uniaxial tensioning of gradient structure in which the sample is subjected to uniaxial uniform strain ε_z^a through its thickness in experiments, the overall stress-strain response σ_z^G in the gradient sample can be approximately obtained through volume averaging over all the component layers as:

$$\sigma_z^G = \frac{\sum_{i=1}^{n^*}\sigma_{zi}^{GND}h_i}{h_t}, \qquad (19.18)$$

where σ_{zi}^{GND} denotes the z-component stress of layer i in the gradient sample considering both GNDs and HDI stress; h_i and h_t are the thickness of layer i and that the half thickness of the entire gradient sample, respectively; and n^* is the layer number of the whole sample adopted in the finite element model. According to the experimental measurements (Wu et al., 2014a), the grain size d along the sample thickness h in the gradient region can be approximated as $\log_{10}(d) = a_g h^2 + b_g$, where a_g and b_g are constants depending on the grain size of the topmost phase d_1 and CG core d_c. In order to calculate the uniform elongation of the homogeneous or gradient IF steel samples, Considère criterion is implemented, i.e., $d\sigma_z^G / d\varepsilon_z^a \leq \sigma_z^G$. As for yield strength, it is commonly defined as 0.2% offset stress.

19.3 Results and Discussion

Since the MC model is proposed for homogeneous IF steels with grain size ranging from submicrometer to tens of micrometers, while the NC model is for those with grain size varying from tens of nanometers to submicrometer, and considering that (i) the existing modified versions of KME models for coarse- and fine-grained metals can be applied down to 1 μm (Delince et al., 2007) and those for nano- and ultrafine-grained metals can be applied up to 583 nm (Liu and Mishnaevsky Jr, 2014); and (ii) the enhanced dynamic recovery usually occurs in submicrometers, e.g., 700 nm for copper (Duhamel et al., 2010), the boundary should lie in the range of submicrometers. The boundary value is determined to be 680nm by fitting the model predictions with the experimental data so that the proposed model could be able to describe accurately the grain size dependence of the constitutive behavior of homogeneous IF steels with grain sizes ranging from nanoscale to microscale (Figs. 19.4 and 19.5). The main difference between the two sub-models lies in the specific values of the parameters in Hall-Petch constants σ_0 and k_{HP} as well as the dynamic recovery factors k_2 and k_e (or d_e) due to the grain size effect. The Hall-Petch constants should be set as different for the two sub-models according to the experimental measurements that the Hall-Petch slope (k_{HP}) in the submicrometer to micrometer range is smaller than that in the nanometer to submicrometer range, whereas the corresponding lattice frictional stress (σ_0) is larger for the former than the latter (Fig. 19.5a). The dynamic recovery factor k_2 of NC sub-model should be larger than that of MC sub-model because the dynamic recovery is much larger in refined grains than that in larger grains due to the sharply increased grain boundary region volume fraction as grain size is reduced down into nanoscale (Meyers et al., 2006). For the same reason, the reference grain size d_e in $k_e = d_e/d)^2$ corresponding to the initiation of enhanced dynamic recovery should also be smaller in NC model than that in the MC model. There are totally seven adjustable parameters in the proposed model; and all can be determined by fitting with experimental data while the material and geometric parameters are directly taken from experiments or existing KME models. The number of the layers in the gradient region has been selected to be large enough to enable each layer to have homogeneous grain structure. The effect of the number of the layers, i.e., n_g, in the gradient region has been presented in Supplementary Note 2. The Tabor's factor is determined by dividing the measured hardness of the topmost layer (with grain size of 96nm) of the gradient IF steel sample with $h_g = 120$ μm by its yield strength. The parameter values used in the analysis are listed in Tables 19.1 and 19.2. In addition, except otherwise stated, the experimental data used in the present analysis for comparison with the model predictions have all been adopted from the experiments of Wu et al. (2014a). For brevity no more statement on them will be given further.

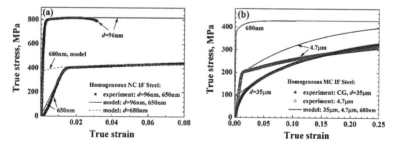

Figure 19.4 True stress-true strain relations of homogeneous IF steel samples predicted by the proposed NC (a) and MC (b) sub-models. The experimental data for d = 96 nm, 650 nm, 4.7 μm and CG IF steel samples are included for comparison.

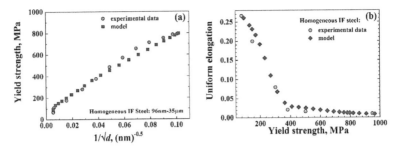

Figure 19.5 The predicted Hall-Petch relation (a) and strength-ductility relation (b) for homogeneous IF steel samples with grain sizes ranging from tens of nanometers to tens of micrometers. The experimental data for the two relations are also included for comparison.

Table 19.1 Material parameters for NC and MC IF steel samples

Parameter	Symbol	NC IF steel	MC IF steel
Grain size	d	10 nm–680 nm	680 nm–35 μm
Elastic modulus (GPa)	E	200	200
Shear modulus (GPa)	μ	77	77
Poisson's ratio	ν	0.28	0.28
Magnitude of Burgers vector (nm)	b	0.26	0.26
Taylor factor	M	3.06	3.06
Taylor constant	α	0.3	0.3
Hall-Petch slope (MPa·m$^{1/2}$)	k_{HP}	0.194	0.252
Lattice frictional stress (MPa)	σ_0	110.3	40
Tabor's factor	λ	3.23	3.23
Rate sensitivity exponent	m	20	20
Proportionality factor	ψ	0.019	0.019
Dynamic recovery factor	k_2	29.05	4.5
Dynamic recovery constant 1	k_{20}	21	3.25

(Continued)

Table 19.1 (Continued)

Parameter	Symbol	NC IF steel	MC IF steel
Dynamic recovery constant 2	n_0	21.25	21.25
Geometric factor	k_3	0.05	0.05
Reference strain rate (s^{-1})	$\dot{\varepsilon}_0$	1	1
Applied total strain rate (s^{-1})	$\dot{\varepsilon}_z^a$	1×10^{-3}	1×10^{-3}
Reference grain size	d_e	950 nm	3.5 μm
Initial dislocation density (m^{-2})	ρ_0	3×10^{12}	3×10^{12} (2×10^{11} for CG)
Original orientation of the groove	β	25.3°	25.3°
Initial geometric non-uniformity	ξ	0.015	0.015
Proportional loading ratio (uniaxial tension)	r	-0.5	-0.5

Table 19.2 Material and geometric parameters for gradient IF steel

Parameter	Symbol	Magnitude
Grain size of coarse grained core (μm)	d_c	35
Grain size of topmost phase (nm)	d_1	200, 96, 30 or 10
Phase number in the gradient region	n	20, 24, 49, 50, or 98
Thickness of gradient surface layer (μm)	h_g	50, 120, 150, 200, 250, 300, 350, 400, or 490
Half thickness of the total sample (μm)	h_t	500
Half width of the total sample (mm)	w	1.25
Layer thickness in finite element simulations (μm)	Δh	9
Integral domain for HDI stress (μm)	R	3

19.3.1 Stress-Strain Curves of Homogeneous IF Steels

The constitutive behavior of homogeneous NC and MC IF steel samples without incorporating GNDs ($\rho_G = 0$) and HDI stress ($\sigma_b = 0$) predicted by the proposed model will be validated first before investigating the overall mechanical response of gradient structure. Figure 19.4 presents the predicted true stress-true strain curves for several NC and MC IF steel samples with grain size $d = 35$ μm, 4.7 μm, 680 nm, 650 nm and 96 nm; and the corresponding experimental data for $d = 35$ μm, 4.7 μm, 650 nm and 96 nm are included for comparison. The elastic moduli of the predicted curves are tuned to coincide the experimental measurements. The two 680 nm curves in (a) and (b) are almost identical but they are respectively predicted by the NC and MC sub-models. It can be clearly shown that the predicted curves are in good agreement with the experimental data in terms of both strength and strain hardening behavior except the case of $d = 4.7$ μm. In the latter case the yield strength and uniform elongation measured in experiments (i.e., 188.7 MPa and 21.7%) are also well reproduced by the proposed model (i.e., 192.85 MPa and 21.5%) in spite of predicting higher flow stress. The lower measured flow stress is might due to the specific situation of the tested sample such as annealing technique and microstructure. In order to obtain an overview of the predictive capability of the proposed model for homogeneous IF steel samples, we also plotted the Hall-Petch relation and the yield strength-uniform elongation map predicted by the model, as presented in Figs. 19.5a and b, respectively. The overall quantitative agreement between the model prediction and the experimental data demonstrated the validity of the proposed model to identify the constitutive behavior of homogeneous NC and MC IF steels with different grain sizes ranging from nanoscale to microscale. The above results also show that the proposed sub-models could be applied up to a strain of the uniform elongation of the CG sample, i.e., 26% (Fig. 19.5b).

19.3.2 Lateral Surface Non-uniform Deformation in Gradient IF Steels

The lateral surface non-uniform deformation can be represented by the height profile that can be obtained from the difference between the lateral (x) displacements along the sample depth and that of the topmost layer on the lateral surface. Figure 19.6 presents the variation of the calculated height profile with respect to the sample depth h for a gradient IF steel sample (d_1 = 96 nm, d_c = 35 µm, h_g = 120 µm, h_t = 500 µm) subjected to various uniaxial tensile engineering strains, i.e., ε_z^e = 0.05, 0.1 and 0.25. Six measured data for height profile at ε_z^e = 0.25 are also included, see the symbols in Fig. 19.6a. The comparison shows that the predicted height profile agrees quantitatively very well with the experimental data for the case of 0.25, which validates the proposed model's capability to predict the non-uniform deformation on the lateral surface of a gradient structure. With height profile known, the predicted GNDs density can be calculated using Eq. (19.6), as shown in Fig. 19.6b. The results show that ρ_G first increases and then decreases with respect to the sample depth, which generates a maximum near the interface between the gradient region and the CG core and that the maximum move towards to the CG core as the applied engineering strain increases from 0.05 to 0.25, i.e., the maximum ρ_G increases from 1.66 to 5.94 × 10^{13} m^{-2}, while the corresponding critical h increases from 53.64 µm to 80.42 µm, see the black balls in Fig. 19.6b. Figure 19.6 also pointed out that the magnitudes of the height profile and the GNDs density are increased considerably as the applied strain rises, which leads to more GNDs and stronger strain hardening in the gradient structure. The GNDs density distributions that are derived from the measured hardness along the sample depth of the gradient sample are also shown in Fig. 19.6b for comparison (Eq. (19.9)). Note that it is difficult, even not impossible, to measure the value of hardness of each free standing component layer after deformation, i.e., H_s, in the gradient surface layer ($h \in$ [0, 120 µm]). Here they are approximated as those of the hardness along the sample depth of the free-standing gradient surface layer with thickness of 120 µm. However, all the values of the hardness parameters in Eq. (19.9) in the CG core ($h \geq$ 120 µm) has been given in Fig. 19.2c of Wu et al., 2014. The comparison for the CG region shows that, although there are some quantitative discrepancies between the predictions and the experimental data, they agree well with each other overall for all the three applied strains. And it is important to note that no parameter adjustment has been made. As for the experimental derivation of GNDs density in the gradient region, there exist also maximums near the interface between the gradient layer and the CG core for all the three applied strains, i.e., the maximum (ρ_G, h) increases from (98.2 μm, 5.41 × 10^{13} m^{-2}) to (107.23 μm, 11.57 × 10^{13} m^{-2}). The comparison in the gradient layer show that the model predictions are in qualitatively agreement with the experimental data. The quantitative difference might result from the underestimated H_s in Eq. (19.9) in the gradient layer for the experimental calculations.

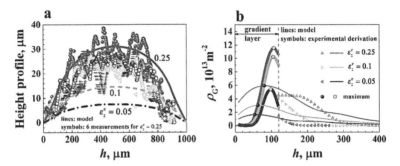

Figure 19.6 (a) The predicted height profile on the lateral surface of a gradient IF steel sample subjected to various uniaxial tensile engineering strains; and (b) corresponding distribution of GNDs density along the sample depth h (Eq. (19.6)). The GNDs densities derived from the measured hardness along the sample depth using Eq. (19.9) are included in (b) for comparison.

19.3.3 Strain Hardening in Gradient IF Steels

The overall true stress-true strain curves of gradient IF steels (d_1 = 96 nm, d_c = 35 μm, h_g = 120 μm, h_t = 500 μm) with both GNDs and HDI stress incorporated are shown in Fig. 19.7a. The results for the sample with GNDs only and those without GNDs and HDI stress are also included for comparison. Their corresponding strain hardening rates are presented in Fig. 19.7b, while Fig. 19.7c compares the predicted strain hardening rates with the experimental measurements for both CG and gradient samples. Two phenomena are found intriguing. First, there is an upturn appearing in the plot of the strain hardening rate of the gradient sample with GNDs at the strain of around 1.3% (green solid line in Fig. 19.7b), and the incorporation of HDI stress further enhances the magnitude (red dashed line). The predicted appearance of the upturn is in quantitatively good accord with the experimental observation, in

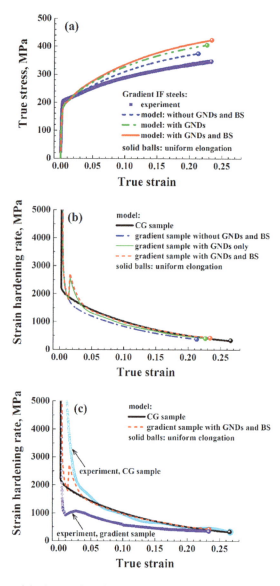

Figure 19.7 (a) The predicted true stress-true strain curves of gradient IF steel samples incorporating GNDs (and HDI stress (BS)). The predicted curve of gradient IF steel sample without incorporating GNDs (and BS) and the corresponding experimental data are also included for comparison. (b) The predicted strain hardening rate corresponding to (a). The predicted data for CG sample are also included in (b) for comparison. (c) Comparison of the predictions and the experimental data for both the gradient and CG samples.

which the upturn occurs at the strain of around 1.1% (Fig. 19.7c). This upturn is resulted from the initiation of the injection of the GNDs into all the component layers as the topmost layer starts to neck that is nevertheless suppressed by the other layers with higher plasticity in the gradient structure during uniaxial tension. Second, the strain hardening rate after upturn is even higher than that in CG IF steel sample. The upturn is followed by a decrease of strain hardening that eventually reaches as high as the value of CG sample, see the red dashed line for gradient sample and the black solid line for CG sample. As a result, the upturn enhances considerably the strain hardening capability in the gradient structure, see the red dashed line versus the blue dot-dashed one. Furthermore, Fig. 19.7c shows that the predicted strain hardening behavior and the experimental observation coincide overall although the predicted value is higher than the measurement for the gradient sample, the reasons of which will be discussed later. Finally the uniform elongation of the gradient IF steel sample with GNDs and HDI stress is predicted as 0.2336 (red ball in Fig. 19.7c), which is in good agreement with the experimental value, i.e., 0.2329 (violet ball in Fig. 19.7c). The predicted yield strength is 177.35 MPa, is also very close to the experimental datum (205 MPa).

It should be noted from Figs. 19.7a and c that the predicted flow stress and strain hardening rate after yielding of the gradient IF steel sample are larger than the experimental data. There are possibly three reasons accounting for the over-predictions. The first one could be the assumption of zero σ_y. The existence of non-zero σ_y in the very small region near the lateral surface (Supplementary Fig. 1b) may influence the model predictions. The second one is the simplified assumption of the perfectly bonded multi-layered structure. The idealized layer interface in the present model is actually grain boundary in the real structure. The former is sharp and clear whereas the latter is curved and has many triple junctions. The damage of the actual grain boundary in real gradient structure during deformation could reduce the overall strain hardening capability as compared with the predicted behavior (Figs. 19.7a and c). The last one might come from the simplified assumption of multilayered structure with two gradient surface layers and a CG core. Actually, there exists a narrow deformed CG region (around 10–20 μm) observed adjacent to the undeformed CG core, in which the grain boundary is not actually high-angle boundary but just sub-grain boundary. However, in the present analysis the deformed CG region with different sub-grain sizes are viewed as different layers with different grain sizes. Therefore, the flow stress of the layers in the deformed CG region is overestimated. That is one of the possible reasons why the predicted flow stress is higher than the experimental data. Considering the complicated gradient microstructure, the predicted curves are in reasonably agreement with the experimental data.

19.3.4 Strength-Ductility Map for Gradient IF Steels

In order to comprehensively investigate the effect of GNDs and HDI stress on the strength-ductility performance, sixteen case studies are conducted to obtain a yield strength-uniform elongation (or strength-ductility) map. The CG core maintains a constant grain size of 35 µm, whereas the grain size of the topmost layer d_1 is tuned to be 10, 30, 96, and 200 nm. For each d_1 the gradient surface layer thickness varies from 50 µm to 490 µm (with corresponding gradient layer thickness/volume fraction ranging from 10% to 98%) as the half thickness of the total sample keeps constant at 500 µm. The strength-ductility map of gradient IF steel without incorporating GNDs and HDI stress is plotted in Fig. 19.8a (Region II). The predicted and experimental data for homogeneous IF steel samples are also included for comparison (Region I). The results clearly show that the gradient sample possess better synergy of strength and ductility as compared with the homogeneous samples except small and high strength regions, even though no GNDs are considered.

In addition to the results presented in Fig. 19.8a, the predicted strength-ductility data of the gradient IF steel samples with GNDs and HDI stress incorporated (Region IV) are plotted in Fig. 19.8b. The data of homogeneous samples (Region I), gradient samples without GNDs and HDI stress (Region II) and gradient samples with GNDs only (Region III) are all included from comparison. The results show that the GNDs and HDI stress can further considerably enhance the strength-ductility synergy (Regions III and IV) as compared with the gradient samples without them (Region II). Figure 19.8b also shows that GNDs play a dominant role (Region III) in enhancing the strength-ductility balance of gradient structure, whereas the HDI stress plays a minor role (Region IV). The results suggest that

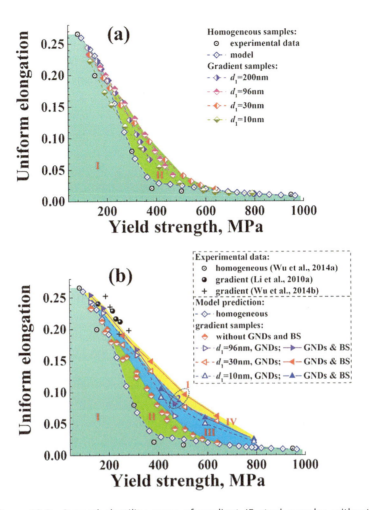

Figure 19.8 Strength-ductility maps of gradient IF steel samples without (a) (Region II) and with (b) GNDs and HDI stress (BS) (Regions III and IV). The experimental data for homogeneous (Wu et al., 2014a) and gradient (Li et al., 2010a; Wu et al., 2014b) IF steel samples as well as the model predictions for homogeneous samples are included in (a) and (b) for comparison. In (b), the predicted data without GNDs and HDI stress for d_1 = 96 and 30nm as presented in (a) (Region II) and those with GNDs only (Region III) are also included for comparison.

in order to gain a good balance of strength-ductility in gradient IF steel sample, the topmost layer grain size should not be too larger or too small and should be less than 100 nm and larger than 10 nm.

For example, the d_1 = 30 nm, h_g = 400 µm sample possesses a good strength-ductility balance at the strength of 508 MPa and ductility of 9.67%, see the numeral 1 in the oval of Fig. 19.8b, which are increased by 227 MPa and 7.26% (or 81% and 301% in percent), respectively, as compared with the homogeneous IF steels. By using Considère criterion, a series of experimental strength-ductility data are obtained from the published stress-strain relations for gradient IF steel samples (Li et al., 2010a; Wu et al., 2014b) with d_c = 35 µm, h_t = 500 µm but with different surface mechanical attrition treatment times, which generates different topmost layer grain size d_1 and gradient layer thickness h_g. The above experimental data are also shown in Fig. 19.8b for comparison. Note that nearly all the experimental data for gradient samples are located along the outer boundary of the yellow region predicted by the proposed model, which clearly demonstrated the proposed model's capability in predicting the strength and ductility of gradient IF steel samples. Figure 19.8b also show that the proposed model could be applied up to a strain of at least 25.44% depending on the thickness of the gradient region h_g and the grain size of the topmost layer d_1 (Fig. 19.8b).

Finally, a detailed analysis on GNDs distribution and HDI stress evolution is made (Fig. 19.9). Two cases are selected, i.e., d_1 = 30 nm, h_g = 400 µm sample (case 1), and d_1 = 96 nm, h_g = 490 µm (case 2) with a strength-ductility synergy of (465.39 MPa, 8.18%) is analyzed (Fig. 19.9). Note case 1 possesses both higher strength and higher ductility than case 2, see the two symbols in the oval of Fig. 19.8b. Figure 19.9 show that the GNDs density and HDI stress in case 1 are much higher than case 2. It is noteworthy that the maximum ρ_G of case 1 occurs in small sample depth (161–170 µm) (Fig. 19.9a), i.e., in nano-grained layers (grain size ranging from 93.9 nm–106.9 nm for applied strains larger than 0.04), whereas that of case 2 appears in much larger sample depth (205–321 µm) (Fig. 19.9b), i.e., in micro-grained layers (grain size varying from 0.924 µm to 1.22 µm for applied strains larger than 0.05), which indicates that the nano-grained layers of case 1 has received much more GNDs and thus gained much higher strain hardening capability than case 2. Therefore, the plastic instability of nano-grained layers in case 1 is much more suppressed than case 2. Moreover, the stress-strain relations and the corresponding strain hardening rates for the two cases (Supplementary Fig. 4) show that the strength and enhancements of strain hardening capability thanks to the existence of GNDs and HDI stress in case 1 are much larger than those in case 2, which undoubtedly leads to a better strength-ductility synergy in the former than the latter.

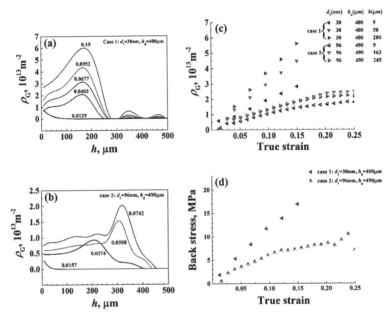

Figure 19.9 GNDs density distribution in case 1 (d_1 = 30 nm, h_g = 400 μm) (a) and case 2 (d_1 = 96 nm, h_g = 490 μm) (b) at different applied strains, see the numerals above the lines; and the evolution of GNDs density at different sample depth (c) and HDI stress (d) with respect to applied strain for the two cases.

19.4 Conclusions

In this chapter a dislocation density-based theoretical model has been proposed to describe the strain hardening of gradient IF steels by incorporating GNDs that are associated with the non-uniform deformation as experimentally observed on the lateral surface. A principle assumption has been made that all GNDs in the gradient structure are injected by the observed surface non-uniform deformation. The main findings are summarized as follows:

(1) The predicted non-uniform deformation on the lateral surface is in quantitatively good agreement with the measurements for a gradient IF steel sample.
(2) The strain hardening rate up-turn as observed in experiments is well predicted. The strain hardening rate is able to reach as high as that of coarse-grained sample.
(3) The developed strength-ductility map for gradient IF steel samples show that the strength-ductility performance of gradient samples is remarkably enhanced as compared with their homogeneous counterparts. The predicted map has been verified by a series of experimental data.
(4) By analyzing the GNDs density distribution and HDI stress evolution in detail, we can conclude that the high strain hardening capability results from the generation of abundant GNDs into the nano-grained layers of the gradient structure.

As a concluding remark, the proposed theoretical framework for gradient nanostructures could be implemented into a finite element scheme as done by (Martínez-Pañeda et al., 2016; Nielsen and Niordson, 2014) to improve the accuracy without imposing some assumptions such as zero stress component along the thickness direction.

Acknowledgements

This work was supported by the Alexander von Humboldt Foundation, the National Natural Science Foundation of China (NSFC) (Grant No. 11402203), and the Fundamental Research Funds for the Central Universities (Grant No. 3102015BJ(II)JGZ025). G.J.W. thanks the support of NSF Mechanics of Materials and Structures Program, under CMMI-1162431. S.H.C. acknowledges the support of NSFC through Grants #11372317, # 11532013 and the 973 Nano-project (2012CB937500). X.L.W. thanks the support of NSFC under grant No. 11572328 and the 973 Program under grant No. 2012CB932203. The authors are also grateful to Prof. Dierk Raabe for his fruitful discussion and insightful comments.

Appendix A: Constitutive Equations for a Gradient Structure Subjected to Uniaxial Tension

As for a gradient structure subjected to uniform uniaxial tension $\dot{\varepsilon}_z$, the constitutive relations (Eq. (19.1)) for each component layer under incompressible material assumption become

$$\dot{\varepsilon} = \frac{1}{2\mu}\dot{\sigma}' + \eta\sigma', \qquad (19.A1)$$

The non-zero stresses components are σ_x, σ_z and τ_{xy}, and the non-zero strain components are ε_x, ε_y, ε_z and γ_{xy}. By adopting $\sigma_y = 0$, the constitutive relations can be specified as

$$\dot{\varepsilon}_x = \frac{1}{2\mu}\left(\frac{2}{3}\dot{\sigma}_x - \frac{1}{3}\dot{\sigma}_z\right) + \eta\left(\frac{2}{3}\sigma_x - \frac{1}{3}\sigma_z\right)$$

$$\dot{\varepsilon}_y = \frac{1}{2\mu}\left(-\frac{1}{3}\dot{\sigma}_x - \frac{1}{3}\dot{\sigma}_z\right) + \eta\left(-\frac{1}{3}\sigma_x - \frac{1}{3}\sigma_z\right) \qquad (19.A2)$$

$$\dot{\varepsilon}_z = \frac{1}{2\mu}\left(-\frac{1}{3}\dot{\sigma}_x + \frac{2}{3}\dot{\sigma}_z\right) + \eta\left(-\frac{1}{3}\sigma_x + \frac{2}{3}\sigma_z\right)$$

$$\dot{\gamma}_{xy} = \frac{1}{\mu}\dot{\tau}_{xy} + 2\eta\tau_{xy}$$

The incompressible condition is:

$$\dot{\varepsilon}_x + \dot{\varepsilon}_y + \dot{\varepsilon}_z = 0. \qquad (19.A3)$$

From the first and second equations in (19.A2), we have

$$\dot{\varepsilon}_x - \dot{\varepsilon}_y = \frac{1}{2\mu}\dot{\sigma}_x + \eta\sigma_x. \qquad (19.A4)$$

The combination of Eqs. (19.A3) and (19.A4) yields

$$\dot{\varepsilon}_x = \frac{1}{2}\left(\frac{1}{2\mu}\dot{\sigma}_x + \eta\sigma_x\right) - \frac{1}{2}\dot{\varepsilon}_z$$

$$\dot{\varepsilon}_y = -\frac{1}{2}\left(\frac{1}{2\mu}\dot{\sigma}_x + \eta\sigma_x\right) - \frac{1}{2}\dot{\varepsilon}_z. \qquad (19.A5)$$

$$\dot{\gamma}_{xy} = \frac{1}{\mu}\dot{\tau}_{xy} + 2\eta\tau_{xy}$$

Appendix B: Constitutive Relations for a Gradient Structure Subjected to Thermal Load

As for a gradient structure (in plane strain condition) subjected to thermal load $\dot{T}$, the constitutive relation is

$$\dot{\varepsilon} = \frac{1}{2\mu}\dot{\sigma}' + \eta\sigma' + \alpha\dot{T}\delta. \qquad (19.\text{B1})$$

Non-zero stress components are σ_x, σ_z, τ_{xy} (the assumption of $\sigma_y = 0$ is also adopted here); and non-zero strain components are ε_x, ε_y, and γ_{xy}. The constitutive relations are given as

$$\dot{\varepsilon}_x = \frac{1}{2\mu}\left(\frac{2}{3}\dot{\sigma}_x - \frac{1}{3}\dot{\sigma}_z\right) + \eta\left(\frac{2}{3}\sigma_x - \frac{1}{3}\sigma_z\right) + \alpha\dot{T}$$

$$\dot{\varepsilon}_y = \frac{1}{2\mu}\left(-\frac{1}{3}\dot{\sigma}_x - \frac{1}{3}\dot{\sigma}_z\right) + \eta\left(-\frac{1}{3}\sigma_x - \frac{1}{3}\sigma_z\right) + \alpha\dot{T} \qquad (19.\text{B2})$$

$$\dot{\varepsilon}_z = \frac{1}{2\mu}\left(-\frac{1}{3}\dot{\sigma}_x + \frac{2}{3}\dot{\sigma}_z\right) + \eta\left(-\frac{1}{3}\sigma_x + \frac{2}{3}\sigma_z\right) + \alpha\dot{T}$$

$$\dot{\gamma}_{xy} = \frac{1}{\mu}\dot{\tau}_{xy} + 2\eta\tau_{xy}$$

The plane strain condition yields

$$\dot{\varepsilon}_z = 0. \qquad (19.\text{B3})$$

From (19.B3) and the third equation of (19.B2), we can derive

$$\frac{1}{2\mu}\left(-\frac{1}{3}\dot{\sigma}_z\right) + \eta\left(-\frac{1}{3}\sigma_z\right) = \frac{1}{2}\left[\frac{1}{2\mu}\left(-\frac{1}{3}\dot{\sigma}_x\right) + \eta\left(-\frac{1}{3}\sigma_x\right) + \alpha\dot{T}\right].$$

$$(19.\text{B4})$$

Substitute (19.B4) into the first and second equations of (19.B2), we have

$$\dot{\varepsilon}_x = \frac{1}{2}\left(\frac{1}{2\mu}\dot{\sigma}_x + \eta\sigma_x\right) + \frac{3}{2}\alpha\dot{T}$$

$$\dot{\varepsilon}_y = -\frac{1}{2}\left(\frac{1}{2\mu}\dot{\sigma}_x + \eta\sigma_x\right) + \frac{3}{2}\alpha\dot{T}. \qquad (19.\text{B5})$$

$$\dot{\gamma}_{xy} = \frac{1}{\mu}\dot{\tau}_{xy} + 2\eta\tau_{xy}$$

Appendix C: M-K Type Analysis for Component Layers

Let us consider a free-standing layer with a geometric non-uniformity in the form of a groove or band in the center with an initial thickness $\bar{h}(0)$, while the thickness of the outside region (uniform section) is $\bar{h}_0(0)$, which gives the initial geometric non-uniformity $\xi = \left[\bar{h}_0(0) - \bar{h}(0)\right]/\bar{h}_0(0) \geq 0$ (Fig. 19.C1). Throughout the manuscript, the subscript or superscript '0' denotes the quantities in the uniform section while its absence refer to those in the groove or band. The band width-to-thickness ratio is assumed to be very large. The dashed and solid bands in Fig. 19.C1 are those before and after deformation, respectively, the orientations of which are β and γ, respectively.

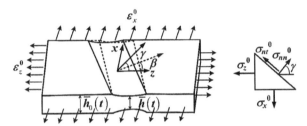

Figure 19.C1 Schematic and conventions for M-K analysis for single free-standing phase/layer in a gradient sample. The dashed and solid bands denote those before and after deformation, respectively.

Proportional loading is imposed on the edges of the layer, i.e.,

$$\frac{\varepsilon_x^0}{\varepsilon_z^0} = r = \text{const}, \quad \frac{\sigma_z^0}{\sigma^0} = \frac{2+r}{\sqrt{3(1+r+r^2)}}, \quad \frac{\sigma_x^0}{\sigma^0} = \frac{1+2r}{\sqrt{3(1+r+r^2)}}. \tag{19.C1}$$

The equilibrium conditions across the band yield

$$\sigma_{nn}^0 \bar{h}_0(t) = \sigma_{nn} \bar{h}(t), \quad \sigma_{nt}^0 \bar{h}_0(t) = \sigma_{nt} \bar{h}(t), \text{ or } \frac{\sigma_{nt}}{\sigma_{nn}} = \frac{\sigma_{nt}^0}{\sigma_{nn}^0}, \tag{19.C2}$$

where σ_{nn}^0 and σ_{nt}^0 are the normal and shear stress components in the band. By combining the equilibrium condition and the definition of logarithmic strain, we have

$$\frac{\sigma_{nn}^0/\sigma^0}{\sigma_{nn}/\sigma} = (1-\xi)\frac{\sigma}{\sigma^0}\exp\left(\varepsilon_y - \varepsilon_y^0\right). \tag{19.C3}$$

From the relation between the normal stress and the principle stress in the uniform section (Fig. 19.C1), we can obtain

$$\frac{\sigma_{nn}^0}{\sigma^0} = \frac{(2+r)\cos^2\gamma + (1+2r)\sin^2\gamma}{\sqrt{3(1+r+r^2)}}. \tag{19.C4}$$

By applying the incompressibility condition ($\varepsilon_x^0 + \varepsilon_y^0 + \varepsilon_z^0 = 0$) and the definition of equivalent strain in the uniform section, the strain in the thickness dimension is

$$\varepsilon_y^0 = -\frac{\sqrt{3}(1+r)}{2\sqrt{1+r+r^2}}\varepsilon^0 = -C\varepsilon^0. \tag{19.C5}$$

Assume

$$d\varepsilon_{tt} = d\varepsilon_{tt}^0, \tag{19.C6}$$

and consider the law of plastic flow Eq. (19.14) as well as the equilibrium condition Eq. (19.C2), we get

$$\frac{\sigma_{nn}}{\sigma} = \frac{1}{AH}\sqrt{1-B\left(\frac{d\varepsilon^0}{d\varepsilon}\right)^2}, \tag{19.C7}$$

$$A = \frac{\sqrt{3}}{2}, \quad B = \frac{3(\sin^2\gamma + r\cos^2\gamma)^2}{4(1+r+r^2)},$$

$$H = \sqrt{1 + \left[\frac{2(r-1)\sin\gamma\cos\gamma}{(2+r)\cos^2\gamma + (1+2r)\sin^2\gamma}\right]^2}.$$

$$\tag{19.C8}$$

Substitute (19.C4), (19.C5) and (19.C7) into the equilibrium condition (19.C3), we have

$$H\sqrt{(1-B-G)}\left[1 - B\left(\frac{d\varepsilon^0}{d\varepsilon}\right)^2\right]^{-1/2} = (1-\xi)\frac{\sigma}{\sigma^0}\exp(C\varepsilon^0 + \varepsilon_y), \tag{19.C9}$$

$$G = \frac{(r-1)^2\sin^2\gamma\cos^2\gamma}{(1+r+r^2)}. \tag{19.C10}$$

From Eqs. (19.14), (19.C6) and (19.C7), we get

$$\frac{d\varepsilon_y}{d\varepsilon} = -\frac{A}{H}\left[1-B\left(\frac{d\varepsilon^0}{d\varepsilon}\right)^2\right]^{1/2} - D\frac{d\varepsilon^0}{d\varepsilon}, \qquad (19.C11)$$

$$D = \frac{\sqrt{3}(\sin^2\gamma + r\cos^2\gamma)}{4\sqrt{(1+r+r^2)}}. \qquad (19.C12)$$

Equations (19.C9) and (19.C11) can be combined to calculate the equivalent strains ε^0 and ε. Note that if we set $\sigma = K\varepsilon^N$ (for power law hardening material), the above equations are identical to those given by Hutchinson and Neale (1978).

Appendix D: Finite Element Simulations for Gradient Structure

In order to calculate the lateral surface non-uniform deformation as well as GNDs density and HDI stress, we use ABAQUS (2012) by constructing a 1/4 symmetric finite element model considering the geometric symmetry and the uniaxial load condition. The element is chosen to be four-node bilinear plane strain element CPE4. The detailed procedure is given as follows. (1) the limit strain ε_{zi}^{0*} or the limit strain ration χ_i of free-standing layer i (i = 1, 2, ..., n + 1) in the uniform section is calculated using the M-K type analysis. (2) select a series of applied strain ε_z^a. (3) when the applied strain ε_z^a is equal to the limit strain of layer k, ε_{zk}^{0*}, the z-component strains ε_{zi} for each constituent layers of the gradient structure can be obtained using Eq. (19.13). (4) calculate the value of thermal expansion coefficient α_i for each layer by $\alpha_i = \varepsilon_{zi}/3$ (Eq. (19.17)). (5) obtain from the finite element simulations the lateral (x) displacement of the j-th node on the lateral surface along the sample depth u_j, j = 2, 3, ..., n^* + 1 (n^* is the layer number of the whole sample in the finite element model), for the given applied strain ε_z^a. (6) calculate the lateral surface non-uniform deformation parameter a for layer j-1 by using $a_{j-1} = (u_j - u_{j-1})$. (7) obtain parameter a for another given applied strain by repeating procedures (3)–(6). (8) calculate the GNDs density ρ_G (Eq. (19.6)) and the HDI stress σ_h (Eq. (19.11)) as well as the flow stress of each layer subjected to uniaxial tension (Eq. (19.4)).

References

ABAQUS, 2012. ABAQUS/Standard User's Manual Version 6.12-2.

Ashby, M., 1970. The deformation of plastically non-homogeneous materials. *Philos. Mag.* **21**, 399–424.

Bayley, C. J., Brekelmans, W. A. M., and Geers, M. G. D., 2006. A comparison of dislocation induced back stress formulations in strain gradient crystal plasticity. *Int. J. Solids Struct.* **43**, 7268–7286.

Beyerlein, I. J., Demkowicz, M. J., Misra, A., and Uberuaga, B. P., 2015. Defect-interface interactions. *Prog. Mater. Sci.* **74**, 125–210.

Capolungo, L., Spearot, D. E., Cherkaoui, M., McDowell, D. L., Qu, J., and Jacob, K. I., 2007. Dislocation nucleation from bicrystal interfaces and grain boundary ledges: Relationship to nanocrystalline deformation. *J. Mech. Phys. Solids* **55**, 2300–2327.

Carvalho Resende, T., Bouvier, S., Abed-Meraim, F., Balan, T., and Sablin, S. S., 2013. Dislocation-based model for the prediction of the behavior of b.c.c. materials – Grain size and strain path effects. *Int. J. Plast.* **47**, 29–48.

Delince, M., Brechet, Y., Embury, J. D., Geers, M. G. D., Jacques, P. J., and Pardoen, T., 2007. Structure-property optimization of ultrafine-grained dual-phase steels using a micro structure-based strain hardening model. *Acta Mater.* **55**, 2337–2350.

Demir, E., Raabe, D., Zaafarani, N., and Zaefferer, S., 2009. Investigation of the indentation size effect through the measurement of the geometrically necessary dislocations beneath small indents of different depths using EBSD tomography. *Acta Mater.* **57**, 559–569.

Duhamel, C., Brechet, Y., and Champion, Y., 2010. Activation volume and deviation from Cottrell-Stokes law at small grain size. *Int. J. Plast.* **26**, 747–757.

Estrin, Y., and Mecking, H., 1984. A unified phenomenological description of work-hardening and creep based on one-parameter models. *Acta Metall.* **32**, 57–70.

Evers, L. P., Brekelmans, W. A. M., and Geers, M. G. D., 2004. Scale dependent crystal plasticity framework with dislocation density and grain boundary effects. *Int. J. Solids Struct.* **41**, 5209–5230.

Fang, T. H., Li, W. L., Tao, N. R., and Lu, K., 2011. Revealing extraordinary intrinsic tensile plasticity in gradient nano-grained copper. *Science* **331**, 1587–1590.

Farrokh, B., and Khan, A. S., 2009. Grain size, strain rate, and temperature dependence of flow stress in ultra-fine grained and nanocrystalline Cu and Al: Synthesis, experiment, and constitutive modeling. *Int. J. Plast.* **25**, 715–732.

Fleck, N. A., and Hutchinson, J. W., 1993. A phenomenological theory for strain gradient effects in plasticity. *J. Mech. Phys. Solids* **41**, 1825–1857.

Gao, H., Huang, Y., Nix, W. D., and Hutchinson, J. W., 1999. Mechanism-based strain gradient plasticity— I. Theory. *J. Mech. Phys. Solids* **47**, 1239–1263.

Gleiter, H., 1989. Nanocrystalline materials. *Prog. Mater. Sci.* 33, 223-315.

Groma, I., Csikor, F. F., and Zaiser, M., 2003. Spatial correlations and higher-order gradient terms in a continuum description of dislocation dynamics. *Acta Mater.* **51**, 1271–1281.

Huang, Y., Qu, S., Hwang, K. C., Li, M., and Gao, H., 2004. A conventional theory of mechanism-based strain gradient plasticity. *Int. J. Plast.* **20**, 753–782.

Hutchinson, J., and Neale, K., 1978. Sheet necking-II. Time-independent behavior, in: Koistinen, D. P., (Eds.), *Mechanics of Sheet Metal Forming*. Plenum Press, New York, pp. 127–153.

Khan, A. S., Zhang, H. Y., and Takacs, L., 2000. Mechanical response and modeling of fully compacted nanocrystalline iron and copper. *Int. J. Plast.* 16, 1459-1476.

Khan, A. S., and Liu, J., 2016. A deformation mechanism based crystal plasticity model of ultrafine-grained/nanocrystalline FCC polycrystals. *Int. J. Plast.*, http://dx.doi.org/10.1016/j.ijplas.2016.1008.1001.

Kou, H., Lu, J., and Li, Y., 2014. High-strength and high-ductility nanostructured and amorphous metallic materials. *Adv. Mater.* **26**, 5518–5524.

Li, J., Chen, S., Wu, X., Soh, A. K., and Lu, J., 2010a. The main factor influencing the tensile properties of surface nano-crystallized graded materials. *Mater. Sci. Eng. A* **527**, 7040–7044.

Li, J., and Soh, A. K., 2012a. Enhanced ductility of surface nano-crystallized materials by modulating grain size gradient. *Modell. Simul. Mater. Sci. Eng.* **20**, 085002.

Li, J., and Soh, A. K., 2012b. Modeling of the plastic deformation of nanostructured materials with grain size gradient. *Int. J. Plast.* **39**, 88–102.

Li, J., Chen, S., Wu, X., and Soh, A. K., 2015a. A physical model revealing strong strain hardening in nano-grained metals induced by grain size gradient structure. *Mater. Sci. Eng. A* **620**, 16–21.

Li, W., Yuan, F., and Wu, X., 2015b. Atomistic tensile deformation mechanisms of Fe with gradient nano-grained structure. *AIP Adv.* **5**, 087120.

Li, W. L., Tao, N. R., and Lu, K., 2008. Fabrication of a gradient nano-microstructured surface layer on bulk copper by means of a surface mechanical grinding treatment. *Scr. Mater.* **59**, 546–549.

Li, X. Y., Wei, Y. J., Lu, L., Lu, K., and Gao, H. J., 2010b. Dislocation nucleation governed softening and maximum strength in nano-twinned metals. *Nature* **464**, 877–880.

Li, Y. J., Raabe, D., Herbig, M., Choi, P. P., Goto, S., Kostka, A., Yarita, H., Borchers, C., and Kirchheim, R., 2014. Segregation stabilizes nanocrystalline bulk steel with near theoretical strength. *Phys. Rev. Lett.* **113**, 106104.

Li, Z., Pradeep, K. G., Deng, Y., Raabe, D., and Tasan, C. C., 2016. Metastable high entropy dual phase alloys overcome the strength ductility trade off. *Nature*, 1–4.

Liu, H., and Mishnaevsky Jr, L., 2014. Gradient ultrafine-grained titanium: Computational study of mechanical and damage behavior. *Acta Mater.* **71**, 220–233.

Liu, J., Khan, A. S., Takacs, L., and Meredith, C. S., 2015. Mechanical behavior of ultrafine-grained/nanocrystalline titanium synthesized by mechanical milling plus consolidation: Experiments, modeling and simulation. *Int. J. Plast.* **64**, 151–163.

Lu, K., and Lu, J., 2004. Nanostructured surface layer on metallic materials induced by surface mechanical attrition treatment. *Mater. Sci. Eng. A* **375–377**, 38–45.

Lu, K., 2014. Making strong nanomaterials ductile with gradients. *Science* **345**, 1455–1456.

Lu, L., Chen, X., Huang, X., and Lu, K., 2009. Revealing the maximum strength in nanotwinned copper. *Science* **323**, 607–610.

Lyu, H., Ruimi, A., and Zbib, H. M., 2015. A dislocation-based model for deformation and size effect in multi-phase steels. *Int. J. Plast.* **72**, 44–59.

Ma, E., 2006. Eight routes to improve the tensile ductility of bulk nanostructured metals and alloys. *JOM* **58**, 49–53.

Marciniak, Z., Kuczyński, K., and Pokora, T., 1973. Influence of the plastic properties of a material on the forming limit diagram for sheet metal in tension. *Int. J. Mech. Sci.* **15**, 789–800.

Martínez-Pañeda, E., and Niordson, C. F., 2016. On fracture in finite strain gradient plasticity. *Int. J. Plast.* **80**, 154–167.

Martínez-Pañeda, E., Niordson, C. F., and Bardella, L., 2016. A finite element framework for distortion gradient plasticity with applications to bending of thin foils. *Int. J. Solids Struct.* **96**, 288–299.

Meyers, M. A., Mishra, A., and Benson, D. J., 2006. Mechanical properties of nanocrystalline materials. *Prog. Mater. Sci.* **51**, 427–556.

Moering, J., Ma, X., Chen, G., Miao, P., Li, G., Qian, G., Mathaudhu, S., and Zhu, Y., 2015. The role of shear strain on texture and microstructural gradients in low carbon steel processed by Surface Mechanical Attrition Treatment. *Scr. Mater.* **108**, 100–103.

Moering, J., Ma, X., Malkin, J., Yang, M., Zhu, Y., and Mathaudhu, S., 2016. Synergetic strengthening far beyond rule of mixtures in gradient structured aluminum rod. *Scr. Mater.* **122**, 106–109.

Nairn, J. A., and Mendels, D.-A., 2001. On the use of planar shear-lag methods for stress-transfer analysis of multilayered composites. *Mech. Mater.* **33**, 335–362.

Nielsen, K. L., and Niordson, C. F., 2014. A numerical basis for strain-gradient plasticity theory: Rate-independent and rate-dependent formulations. *J. Mech. Phys. Solids* **63**, 113–127.

Ovid'ko, I., and Langdon, T., 2012. Enhanced ductility of nanocrystalline and ultrafine-grained metals. *Rev. Adv. Mater. Sci.* **30**, 103–111.

Ovid'ko, I. A., 2002. Materials science - Deformation of nanostructures. *Science* **295**, 2386–2386.

Rodríguez-Galán, D., Sabirov, I., and Segurado, J., 2015. Temperature and stain rate effect on the deformation of nanostructured pure titanium. *Int. J. Plast.* **70**, 191–205.

Rupert, T. J., 2016. The role of complexions in metallic nano-grain stability and deformation. *Curr. Opin. Solid State Mater. Sci.*, **20**, 257–267.

Wang, Y. M., Chen, M. W., Zhou, F. H., and Ma, E., 2002. High tensile ductility in a nanostructured metal. *Nature* **419**, 912–915.

Wei, Y., Li, Y., Zhu, L., Liu, Y., Lei, X., Wang, G., Wu, Y., Mi, Z., Liu, J., and Wang, H., 2014. Evading the strength–ductility trade-off dilemma in steel through gradient hierarchical nanotwins. *Nat. Commun.* **5**, 3580.

Weng, G. J., 2011. A composite model of nanocrystalline materials, in: Li, J. C. M., (Eds.), *Mechanical Properties of Nanocrystalline Materials*. Pan Stanford Publishing, Hackensack, NJ, pp. 93–135.

Wu, X., Jiang, P., Chen, L., Yuan, F., and Zhu, Y. T., 2014a. Extraordinary strain hardening by gradient structure. *Proc. Natl. Acad. Sci. U. S. A.* **111**, 7197–7201.

Wu, X., Jiang, P., Chen, L., Zhang, J., Yuan, F., and Zhu, Y., 2014b. Synergetic strengthening by gradient structure. *Mater. Res. Lett.* **2**, 185–191.

Yang, M., Pan, Y., Yuan, F., Zhu, Y., and Wu, X., 2016. Back stress strengthening and strain hardening in gradient structure. *Mater. Res. Lett.*, DOI:10.10 80/21663831.21662016.21153004.

Yang, X., Ma, X., Moering, J., Zhou, H., Wang, W., Gong, Y., Tao, J., Zhu, Y., and Zhu, X., 2015. Influence of gradient structure volume fraction on the mechanical properties of pure copper. *Mater. Sci. Eng. A* **645**, 280–285.

Yefimov, S., Groma, I., and van der Giessen, E., 2004. A comparison of a statistical-mechanics based plasticity model with discrete dislocation plasticity calculations. *J. Mech. Phys. Solids* **52**, 279–300.

Yin, Z., Yang, X., Ma, X., Moering, J., Yang, J., Gong, Y., Zhu, Y., and Zhu, X., 2016. Strength and ductility of gradient structured copper obtained by surface mechanical attrition treatment. *Mater. Des.* **105**, 89–95.

Youssef, K. M., Scattergood, R. O., Murty, K. L., Horton, J. A., and Koch, C. C., 2005. Ultrahigh strength and high ductility of bulk nanocrystalline copper. *Appl. Phys. Lett.* **87**, 091904.

Yuan, F., Chen, P., Feng, Y., Jiang, P., and Wu, X., 2016. Strain hardening behaviors and strain rate sensitivity of gradient-grained Fe under compression over a wide range of strain rates. *Mech. Mater.* **95**, 71–82.

Zeng, Z., Li, X., Xu, D., Lu, L., Gao, H., and Zhu, T., 2016. Gradient plasticity in gradient nano-grained metals. *Extr. Mech. Lett.* **8**, 213–219.

Zhao, Y. H., Topping, T., Bingert, J. F., Thornton, J. J., Dangelewicz, A. M., Li, Y., Liu, W., Zhu, Y. T., Zhou, Y. Z., and Lavernia, E. L., 2008. High tensile ductility and strength in bulk nanostructured nickel. *Adv. Mater.* **20**, 3028–3033.

Zhu, T., and Li, J., 2010. Ultra-strength materials. *Prog. Mater. Sci.* **55**, 710–757.

Chapter 20

Extraordinary Bauschinger Effect in Gradient Structured Copper

Xiaolong Liu,[a] Fuping Yuan,[a] Yuntian Zhu,[c,d] and Xiaolei Wu[a,b]
[a]*State Key Laboratory of Nonlinear Mechanics, Institute of Mechanics, Chinese Academy of Sciences, 15 Beisihuan West Road, Beijing 100190, China*
[b]*College of Engineering Sciences, University of Chinese Academy of Sciences, Yuquan Road, Beijing 100049, China*
[c]*Department of Materials Science and Engineering, North Carolina State University, Raleigh, NC 27695, USA*
[d]*School of Materials Science and Engineering, Nanjing University of Science and Technology, Nanjing 210094, China*
xlwu@imech.ac.cn

Bauschinger effect is a well-known phenomenon, in which the tensile stress is higher than the reverse compressive stresses. Here we report that gradient structured copper exhibits an extraordinarily large Bauschinger effect. We propose to use the reverse yield softening, $\Delta\sigma_b$, as a quantitative parameter to represent the Bauschinger

---

Reprinted from *Scr. Mater.*, **150**, 57–60, 2018.

Heterostructured Materials: Novel Materials with Unprecedented Mechanical Properties
Edited by Xiaolei Wu and Yuntian Zhu
Text Copyright © 2018 Acta Materialia Inc.
Layout Copyright © 2022 Jenny Stanford Publishing Pte. Ltd.
ISBN 978-981-4877-10-7 (Hardcover), 978-1-003-15307-8 (eBook)
www.jennystanford.com

effect. $\Delta\sigma_b$ evolves in the same trend as the hetero-deformation induced (HDI) stress with pre-strain, and can be used to evaluate the effectiveness of a heterostructure in producing HDI stress for superior mechanical properties.

Bauschinger effect has been extensively reported in conventional homogenous materials [1–3]. However, the Bauschinger effect in homogenous materials is usually not very strong, and has not been related with mechanical properties such as strength and ductility. Recently, heterostructured materials have been found to have both strong Bauschinger effect and high hetero-deformation induced (HDI) stress [4]. Gradient structure can be considered a type of heterostructure [5], and was also reported to have high HDI stress [6]. In addition, HDI work hardening was found, i.e. the HDI stress increases with pre-applied tensile strain (pre-strain). These reports raise two interesting issues: (1) How does the Bauschinger effect evolve with pre-strain in heterostructured materials? (2) Can the Bauschinger effect be quantified and related to the mechanical properties such as strength and ductility?

Gradient structure is characterized with increasing grain size along the depth from NS surface layers to coarse-grained (CG) central layer. It has been found to suppress the strain localization and represents a new strategy for producing a superior combination of high strength and good ductility [7, 8]. The superior properties of gradient structured (GS) interstitial-free (IF) steel were attributed to extraordinary high strain hardening [9, 10], a significant part of which is attributed to the HDI work hardening [4–6]. For example, it was found that the GS IF steel developed high HDI stress to increase its strength and high HDI work hardening to maintain good ductility [6]. In other words, HDI stress plays a major role in the mechanical behavior of gradient materials. It will be of interest to probe if the Bauschinger effect is related to superior mechanical properties in gradient materials.

Strain hardening is the primary approach for improving ductility [11]. There are two mechanisms to produce strain hardening in metals, i.e., forest dislocation hardening and HDI hardening [12–15]. The forest dislocation hardening is attributed to the accumulation of dislocations. The flow stress of a metal is described by [16, 17]:

$$\tau = \alpha G b \sqrt{\rho_s + \rho_G} \qquad (20.1)$$

where τ is the shear flow stress, α is a constant, G is the shear modulus, b is the magnitude of Burgers vector, ρ_S is the density of statistically stored dislocations, ρ_G is the density of geometrically necessary dislocations. Therefore, the strain hardening caused by dislocation accumulation can be derived as

$$\frac{d\tau}{d\varepsilon} = \alpha G b \frac{d\sqrt{\rho_s + \rho_G}}{d\varepsilon} \qquad (20.2)$$

where ε is the applied strain. Eq. 20.2 indicates the strain hardening is caused by the increase of total dislocation density with strain.

For conventional homogeneous metals, dislocation accumulation is often the primary mechanism for strain hardening [18], for which Eqs. (20.1) and (20.2) can reasonably explain their mechanical behaviors. HDI stress is rarely included in the explanation of the mechanical properties, because it is much weaker than forest dislocation hardening [5]. Bauschinger effect is usually not related to the mechanical properties either. However, for heterostructured metals, HDI stress contributes much more to work hardening than the forest dislocations do, and the Bauschinger effect is also much stronger [4]. HDI hardening is caused by the pileups of geometrically necessary dislocations [16, 17], which are needed to accommodate strain gradient near the boundaries when strain partitioning occurs between the hard and soft zones [19–21]. For example, the excellent strain hardening capacity of the heterostructured materials, such as nano-composites [22, 23], precipitation alloys [24–26], and passivated films [27] is attributed to heterogeneous deformation between the soft and hard phases, which trigger the HDI hardening.

It is the objective of this work to investigate the evolution of Bauschinger effect with pre-strain, to quantify the Bauschinger effect and to relate the Bauschinger effect with the mechanical properties of heterostructured materials. Gradient structured (GS) Cu is used as for the experimental study, which was found to have an extraordinarily high Bauschinger effect due to its high HDI stress, because both Bauschinger effect and the HDI stress have the same physical origin.

Oxygen-free copper with the composition (wt%): P, 0.002%, Fe, 0.004%, Ni, 0.002%, Sn, 0.002%, S, 0.004%, Zn, 0.003%, O, 0.03% was used in this study. A copper rod of 12 mm in diameter was cut and annealed in vacuum at 873 K for 2 h to obtain homogeneous CG microstructure (Fig. 20.1a) with a mean grain size of 78 μm. Test specimens are machined into a dog-bone shape with a gauge diameter of 3 mm and gauge length of 15 mm from the annealed bar. Some specimens were processed by means of surface mechanical attrition treatment (SMAT) to form a GS surface layer in the gauge section.

The SMAT duration was 30 minutes for each specimen. GS layer of 250 μm thick was formed, in which the grain size increases gradually with an increasing depth (Fig. 20.1b). In topmost 20 μm thick layer the transversal grain sizes increase from 200 nm to 400 nm along the depth as shown in Fig. 20.1c and d (the location is framed by a box in Fig. 20.1b). Figure 20.1e presents the variation of average transversal grain sizes along depth from the surface. The original coarse grains are below the 250-μm depth.

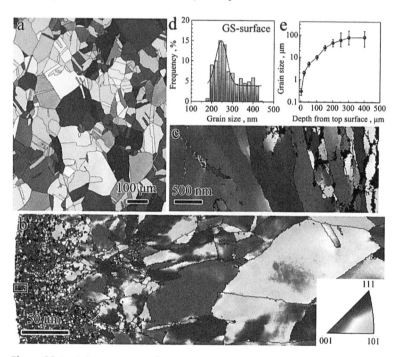

Figure 20.1 Microstructure of CG-Cu and GS-Cu, (a) CG-Cu, (b) GS-Cu, (c) EBSD image and (d) grain size distribution for the topmost 3 μm thick layer of GS-Cu, and (e) variation of average transversal grain sizes along depth.

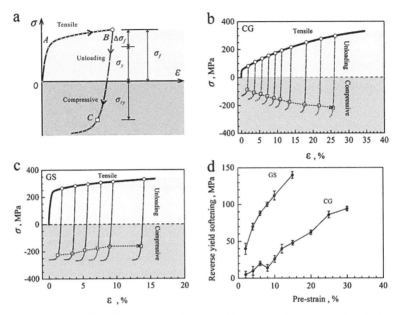

Figure 20.2 (a) Schematic of the Bauschinger tensile-compressive tests, tensile curves and tensile-compressive curves for CG-Cu (b) and GS-Cu (c), and (d) reverse yield softening vs. pre-strain.

Mechanical tests including the tensile tests and Bauschinger tensile-compressive tests were conducted using a MTS Landmark machine at strain rate of 5×10^{-4} s^{-1} under strain control mode. The strain is measured with an extensometer. Full Bauschinger tensile-compressive tests were performed as schematically presented in Fig. 20.2a. Test programs were developed to conduct the Bauschinger tensile-compressive tests, i.e. a specimen is stretched to the certain strain (O-A-B in Fig. 20.2a) under strain control mode, and then unloaded and compressed beyond reverse yielding (B-C in Fig. 20.2a). Ten samples were used For CG-Cu, including one for monotonous tensile loading and nine for tensile-compressive tests (forward loading pre-strains are 2%, 4%, 6%, 8%, 10%, 12%, 15%, 20% and 25%). Seven samples of GS-Cu were used in the tests, including one for monotonous tensile loading and six for tensile-compressive loading (pre-strains are 2%, 4%, 6%, 8%, 10% and 15%). It is worth noting that no buckling occurred in our tests, partially because the compressive strain is only about 2% in our tests, and the reverse yielding strength was measured at a strain offset of 0.2%. The initial yield strength σ_y, flow stress σ_f (marked by circle in Fig. 20.2a), reverse yield strength σ_{ry} (marked by square in Fig. 20.2a) can be obtained from tensile-compressive curves. It was proposed that HDI stress σ_b can be determined by the following equations [8]:

$$\sigma_b = \frac{\sigma_f + \sigma_{ry}}{2} \qquad (20.3)$$

However, configurations of geometrically necessary dislocations at the 0.2% reverse compressive strain are expected to be different from those at the beginning of unloading. In other words, the configurations of geometrically necessary dislocations would have been at least partially changed at such a large reverse strain. Consequently, the HDI stress measured using Eq. 20.3 would be lower than the real HDI stress at the beginning of the unloading. Therefore, the HDI stress calculated using Eq. 20.3 can be considered the lower bound of the HDI stress. More accurate measurement of HDI stress can be measured using an unloading-reloading procedure [6].

Bauschinger effect is a phenomenon that the reverse compressive yield stress is lower than the initial forward flow stress at the beginning of unloading. Therefore, it is reasonable to use the difference between the forward flow stress at the beginning of the unloading and the reverse compressive yield stress at 0.2% offset strain, i.e.

$$\Delta\sigma_b = \sigma_f - |\sigma_{ry}| = \sigma_f + \sigma_{ry} \qquad (20.4)$$

Here we define $\Delta\sigma_b$ as the Bauschinger reverse yield softening and propose to use $\Delta\sigma_b$ *as a quantitative measure of the Bauschinger effect*. Obviously, the value of $\Delta\sigma_b$ is twice the value of the lower-bound HDI stress calculated using Eq. 20.3. In other words, the Bauschinger reverse yield softening $\Delta\sigma_b$ can be related to the magnitude of the HDI stress. Since the HDI stress has been used as an indicator of the effectiveness of a heterostructure in producing a superior combination of strength and ductility, the $\Delta\sigma_b$ can be also used to evaluate and optimize the heterostructure.

The fundamental basis for the above co-relationship between the Bauschinger effect and HDI stress is that they are both produced by the same physical phenomenon: the piling up of geometrically necessary dislocations. Several other parameters have also been used to represent the Bauschinger effect, including Bauschinger strain and stress parameter [3], permanent softening [28, 29]. The Bauschinger reverse yield softening $\Delta\sigma_b$ proposed here has the advantage of clear physical definition, easy to measure, and can effectively relates the Bauschinger effect with mechanical behavior and properties of heterostructured materials. In other words, higher Bauschinger effect, as quantitatively represented by $\Delta\sigma_b$, indicates a more effective heterostructure for superior mechanical properties.

Figure 20.2b and c presents the monotonic tensile true stress-strain curves and the tensile-compressive curves for CG-Cu and GS-Cu. The GS-Cu has excellent combination of high strength and good ductility. It is worth noting that no transient response in tensile curve of GS-Cu was observed, which was reported in other round bar GS specimens [7, 30]. The differences in the tensile parts of the tensile-compressive curves among different specimens in each group are very small and negligible. As shown in Fig. 20.2b, for the homogeneous CG copper both the forward stress and reverse yield stress (marked by circle and square in Fig. 20.2b) increase with increasing applied strain, which is typical of isotropic hardening or forest dislocation hardening. This is because the dislocation density typically increases with increasing plastic strain. In contrast, Fig. 20.2c shows that for GS-Cu with the increasing applied tensile strain, the forward yield strength (marked by circles) increases, while the magnitude of the reverse yield strength (marked by squares) decreases. This is extraordinary and very different from behavior of the homogeneous CG sample. This is because the GS sample has very strong directional HDI stress, which caused extraordinarily high Bauschinger effect.

Figure 20.2d shows the evolution of the reverse yield softening $\Delta\sigma_b$ with increasing applied forward tensile pre-strain. As shown, the GS Cu has a much stronger Bauschinger effect than its homogeneous CG counterpart. This means that the GS Cu has much higher HDI work hardening to help it with enhancing the strength and retaining the ductility.

The microstructure of CG-Cu and GS-Cu is totally different as shown in Fig. 20.1. The CG-Cu has a nearly homogeneous microstructure within coarse grains. As a pure metal the plastic deformation of CG-Cu is homogeneous. Statistically-stored dislocations (SSDs) accumulate during homogeneous plastic straining. Their accumulation is mostly result of chance encounters, which led to mutual trapping and dislocation accumulation [16]. The strain hardening caused by such dislocation density increase can be described by Eq. 20.2, and is non-directional. It should affect the forward flow stress and reverse yield stress the same way and therefore does not cause Bauchinger effect. However, even homogenous CG copper usually has some heterogeneity, which produces some Bauschinger effect with lower magnitude residual HDI stress as compared to the GS-Cu, as shown in Fig. 20.2d.

The microstructure of GS-Cu is heterogeneous and the grain sizes vary from nanoscale to microscale. Layers with different grain sizes have different flow stress, which first leads to the development of two dynamically migrating plastic/elastic boundaries, and later two migrating necking/stable boundaries [9]. High strain gradient will be developed near the boundaries, which need to be accommodated by GNDs. The accumulation of GNDs will cause a directional long-range HDI stress, which impedes dislocation slip in the tensile direction, and promote dislocation slip in the reverse direction. Correspondingly, the tensile-compressive curves show a high forward flow stress and a reduced reverse yield stress as shown in Fig. 20.2c. In other words, this leads to both large Bauschinger effect and high HDI stress.

In summary, the gradient structured (GS) copper shows extraordinary Bauschinger effect in which the reverse yield stress increases with increasing pre-strain. This was caused by the large Bauschinger effect, as represented quantitatively by the large Bauschinger reverse yield softening, $\Delta\sigma_b$, which is closely related with the HDI stress because they both have the same physical origin: piling up of geometrically necessary dislocations. The $\Delta\sigma_b$ can be used as an effective parameter to quantitatively represent the magnitude of the Bauschigner effect. It can also be used as a parameter to evaluate the effectiveness of heterostructures in producing superior mechanical properties. With the larger pre-strain, the Bauschinger effect increases faster in GS-Cu than in CG-Cu, indicating high HDI hardening in the GS Cu.

Acknowledgements

This work was supported by National Key R&D Program of China (2017YFA0204403), and the National Natural Science Foundation of China under Grant Nos. 11572328, 11472286, 51601204, and 51471039. The Strategic Priority Research Program of the Chinese Academy of Sciences under Grant No. XDB22040503. YTZ acknowledges the support of the US Army Research Office (W911NF-17-1-0350).

References

1. G.D. Moan, J.D. Embury, *Acta Metall.* **27** (1979) 903.
2. T. Kishi, I. Gokyu, *Metall. Mater. Trans. B* **4** (1973) 390.
3. A. Abel, H. Muir, *Philos. Mag.* **26** (1972) 489.
4. X.L. Wu, M.X. Yang, F.P. Yuan, G.L. Wu, Y.J. Wei, X.X. Huang, Y.T. Zhu, *Proc. Natl. Acad. Sci. U.S.A.* **112** (2015) 14501.
5. X.L. Wu, Y.T. Zhu, *Mater. Res. Lett.* **5** (2017) 527.
6. M.X. Yang, Y. Pan, F.P. Yuan, Y.T. Zhu, X.L. Wu, *Mater. Res. Lett.* **1** (2016) 1.
7. T.H. Fang, W.L. Li, N.R. Tao, K. Lu, *Science* **331** (2011) 1587.
8. X.C. Liu, H.W. Zhang, K. Lu, *Science* **342** (2013) 337.
9. X.L. Wu, P. Jiang, L. Chen, F.P. Yuan, Y.T. Zhu, *Proc. Natl. Acad. Sci. U.S.A.* **111** (2014) 7197.
10. X.L. Wu, P. Jiang, L. Chen, Y.T. Zhu, *Mater. Res. Lett.* **2** (2014) 185.
11. R.Z. Valiev, Y. Estrin, Z. Horita, T.G. Langdon, M.J. Zehetbauer, Y.T. Zhu, *Mater. Res. Lett.* **4** (2016) 1.
12. K.M.J. Akhtar S. Khan, *Int. J Plast.* **15** (1999) 1265.
13. X. Feaugas, *Acta Mater.* **47** (1999) 3617.
14. G. Fribourg, Y. Bréchet, A. Deschamps, A. Simar, *Acta Mater.* **59** (2011) 3621.
15. K. Maciejewski, H. Ghonem, *Int. J Fatigue* **68** (2014) 123.
16. M.F. Ashby, *Philos. Mag.* **21** (1970) 399.
17. H. Gao, Y. Huang, W.D. Nix, J.W. Hutchinson, *J. Mech. Phys. Sol.* **47** (1999) 1239.
18. Y.T. Zhu, X.Z. Liao, X.L. Wu, *Prog. Mater. Sci.* **57** (2012) 1.

19. Z. Sun, G. Song, T.A. Sisneros, B. Clausen, C. Pu, L. Li, Y. Gao, P.K. Liaw, *Sci. Rep.* **6** (2016) 1.
20. J. Gilsevillano, *Scr. Mater.* **60** (2009) 336.
21. D.S. Yan, C.C. Tasan, D. Raabe, *Acta Mater.* **96** (2015) 399.
22. G. Badinier, C.W. Sinclair, S. Allain, O. Bouaziz, *Mater. Sci. Eng. A* **597** (2014) 10.
23. L. Thilly, S.V. Petegem, P.-O. Renault, F. Lecouturier, V. Vidal, B. Schmitt, H.V. Swygenhoven, *Acta Mater.* **57** (2009) 3157.
24. A. Reynolds, J. Lyons, *Metal. Mater. Trans. A* **28** (1997) 1205.
25. S.H. Kim, H. Kim, N.J. Kim, *Nature* **518** (2015) 77.
26. M.X. Yang, F.P. Yuan, Q.G. Xie, Y.D. Wang, E. Ma, X.L. Wu, *Acta Mater.* **109** (2016) 213.
27. Y. Xiang, J.J. Vlassak, *Scr. Mater.* **53** (2005) 177.
28. B.K. Chun, J.T. Jinn, J.K. Lee, *Int. J. Plast.* **18** (2002) 571.
29. M.P. Miller, E.J. Harley, D.J. Bammann, *Int. J. Plast.* **15** (1999) 93.
30. H.W. Huang, Z.B. Wang, J. Lu, K. Lu, *Acta Mater.* **87** (2015) 150.

Chapter 21

Atomistic Tensile Deformation Mechanisms of Fe with Gradient Nano-Grained Structure

Wenbin Li, Fuping Yuan, and Xiaolei Wu
State Key Laboratory of Nonlinear Mechanics, Institute of Mechanics, Chinese Academy of Science, No. 15, North 4th Ring, West Road, Beijing 100190, China
liwenbin0115@163.com, fpyuan@lnm.imech.ac.cn, xlwu@imech.ac.cn

Large-scale molecular dynamics (MD) simulations have been performed to investigate the tensile properties and the related atomistic deformation mechanisms of the gradient nano-grained (GNG) structure of bcc Fe (gradient grains with d from 25 nm to 105 nm), and comparisons were made with the uniform nano-grained (NG) structure of bcc Fe (grains with d = 25 nm). The grain size gradient in the nano-scale converts the applied uniaxial stress to multi-axial stresses and affects the dislocation behaviors in the

---

Reprinted from *AIP Adv.*, **5**, 087120, 2015.

Heterostructured Materials: Novel Materials with Unprecedented Mechanical Properties
Edited by Xiaolei Wu and Yuntian Zhu
Text Copyright © 2015 Author(s).
Layout Copyright © 2022 Jenny Stanford Publishing Pte. Ltd.
ISBN 978-981-4877-10-7 (Hardcover), 978-1-003-15307-8 (eBook)
www.jennystanford.com

GNG structure, which results in extra hardening and flow strength. Thus, the GNG structure shows slightly higher flow stress at the early plastic deformation stage when compared to the uniform NG structure (even with smaller grain size). In the GNG structure, the dominant deformation mechanisms are closely related to the grain sizes. For grains with $d = 25$ nm, the deformation mechanisms are dominated by GB migration, grain rotation and grain coalescence although a few dislocations are observed. For grains with $d = 54$ nm, dislocation nucleation, propagation and formation of dislocation walls near GBs are observed. Moreover, formation of dislocation walls and dislocation pile-ups near GBs are observed for grains with $d = 105$ nm, which is the first observation by MD simulations to our best knowledge. The strain compatibility among different layers with various grain sizes in the GNG structure should promote the dislocation behaviors and the flow stress of the whole structure, and the present results should provide insights for designing the microstructures for developing strong-and-ductile metals.

21.1 Introduction

Nano-grained (NG) metals usually have ultra-high strength, but show reduced strain hardening rate and limited ductility compared to their coarse grained (CG) counterparts, due to the incapability of effectively accumulating dislocations inside the nano-grains [1–3]. The structural applications in modern industry always demand stronger and tougher metals and alloys. Such expectations have been realized by several strategies developed recently through tailoring nano-scale microstructures, such as, pre-existing growth nano-twins, nano-precipitates, bimodal grain size distribution, and gradient nano-grained (GNG) structure [4–16].

The GNG/CG architecture, consisting of the GNG surface layers and the CG core, can produce high strength and ductility [11–16]. Based on surface nanocrystallization and warm co-rolling technologies, high strength and exceptional ductility could be obtained in the periodically layered GNG/CG structures by introducing non-localized fracture behaviors [12]. High tensile plasticity can also be achieved in the NG Cu film when confined by a CG substrate with a gradient grain size transition, which attributes largely to a mechanically-driven grain boundary migration process and a strain-induced growth of nano-grains [13] where strain localization is suppressed [17, 18]. Evidences have been indicated that large plastic strains could be obtained in NG metals without apparent grain growth by other deformation modes, such as compression and rolling [19–21], in which complex stress states exist in the specimens instead of uniaxial stress state. These experimental observations hint that a stable GNG layer may also intrinsically improve ductility without grain growth, and the strain compatibility between different layers may suppress the strain localization of the NG surface layers for improved ductility, as indicated in our previous work [14]. The strain compatibility between different layers usually induces stress partitioning, stress state change and hetero-deformation induced (HDI) stress in the GNG structure, and the HDI stress usually refers to the stress associated with a strain process providing long-range interactions with mobile dislocations (such as geometrically necessary dislocations, GNDs) for HDI hardening [22]. The generation of GNDs increases the dislocation density in the materials and makes it more likely for dislocations to interact and entangle with each other, which increases the flow stress [14, 23, 24]. Moreover, the strain gradient should be introduced when the constraint exists, which produces the extra hardening and the higher strength compared to the rule-of-mixture due to the grain size gradient [14, 24, 25].

Nanocrystalline (NC) bcc metals have been found to behave very differently from NC metals with other crystal structures [26–28]. With decreasing grain size, the density of mixed and edge dislocations increases while the screw dislocations become less dominated for bcc metals [27]. Moreover, the overall dislocation density decreases with decreasing grain size when the grain sizes are very small [27]. The strain rate sensitivity was found to decrease with decreasing grain size for bcc metals due to these observations [26–28]. MD simulations have been shown to be particularly helpful for investigating the atomistic deformation mechanisms of nanostructured metals, in which the real-time microstructural responses of the system, the atomistic and macroscopic stress and strain, and the stress state can be examined in detail [7, 29–41]. For example, previous investigations [29, 30, 36, 39] have elucidated a transition in the dominant deformation mechanisms with decreasing grain size from dislocation-mediated plasticity to GB-associated plasticity in NC metals. Previous MD simulations on bcc Fe have also indicated that the type of dislocations is full lattice dislocation even when the grain size is below 20 nm [33, 39–41]. The fracture resistance of bcc NC Fe has been shown to increase with decreasing grain size below a critical grain size, and the "most brittle grain size" appears to be in accordance with the "strongest grain size" [33]. MD simulations have also been used to characterize the dislocation-core structure in the framework of the Peierls-Nabarro model and study the mechanism of the dislocation motion at different temperatures in bcc Fe [40]. The dislocations were found to move by nucleation and propagation of kink-pairs along its line at low temperature.

MD simulations may be helpful and provide hints for the atomistic tensile deformation mechanisms in the GNG structure. In this regard, large-scale MD simulations were utilized in this work to investigate and compare the tensile properties and the related atomistic deformation mechanisms of two microstructures, i.e., the GNG structure of bcc Fe and the uniform NG structure of bcc Fe.

21.2 Simulation Techniques

The MD simulations were carried out using the large-scale atomic/molecular massively parallel simulator (LAMMPS) code and the force interactions between atoms were described by an Fe EAM potential developed by Mendelev et al. [42]. This potential was calibrated according to the ab initio values or the experimental data of lattice constant, elastic constants, point-defect energies, bcc-fcc transformation energy, and relaxed core structure of the dislocations. In order to explore deformation mechanisms of GNG structure, similar to the configuration used by Yamakov et al. [29], a columnar grain structure was considered in which grains larger than those possible in fully 3-dimensional simulations could be simulated. The <111> direction was chosen for the column axis (z direction) in the present study, which ensures that dislocations can glide on either of three {110} slip systems in each grain following their nucleation. In this study, two configurations were considered, i.e., the GNG structure of bcc Fe and the uniform NG structure of bcc Fe. The relaxed structures for these two configurations are shown in Fig. 21.1, and the color coding is based on the common neighbor analysis (CNA) values. Green color stands for perfect bcc atoms, red color stands for grain boundaries (GBs), dislocation core, and free surface atoms, blue color stands for fcc atoms, pink color stands for hcp atoms. The same CNA color coding is used in all figures of present study. In Fig. 21.1a, a GNG structure of 140 grains with various orientations was constructed by the Voronoi method, and the grain size increases gradually from 25 nm at the surface to 105 nm at the center. In Fig. 21.1b, a uniform NG structure of 464 grains with various orientations was also constructed by the Voronoi method. In Fig. 21.1, the x direction is along $[11\bar{2}]$ and the y direction is along $[1\bar{1}0]$ for the grain with 0 angle, and the grains with other angles were rotated about the z axis from this reference grain. Both configurations are symmetric about the center line along y direction. The surface layers have the same grain sizes, structures and orientations for both configurations, as shown in Fig. 21.1. Both samples have dimensions of $400 \times 6500 \times .99$ nm^3, and contain approximately 24,000,000 atoms. The thickness (0.99 nm) of the columnar grains was chosen to be larger than the cutoff distance (0.53 nm) of the EAM interatomic potential in order to ensure the periodicity along the z direction. Unlike the bowed-out dislocations with curvature from GBs of real nanocrystals, vertical GBs of an ideal columnar structure normally produce infinite straight dislocations, however previous research [7,29] have indicated that such simulation structure can also provide insights for atomistic deformation mechanisms of nanostructured metals with proper orientation selection. Periodic boundary conditions were imposed along x and z directions, while free boundary conditions were imposed along y directions. Both as-created samples were first subjected to energy minimization using the conjugate gradient method before tensile loading, then gradually heated up to the desired temperature and finally relaxed in the Nose/Hoover isobaric-isothermal ensemble (NPT) under both the pressure 0 bar and the desired temperature (1 K) for enough time (100 ps). After relaxation, an 8% strain was applied to both samples with a constant engineering strain rate of $2 \times 10^8 s^{-1}$ along x direction. During the tensile loading, the overall pressures in the y and z directions were kept to zero.

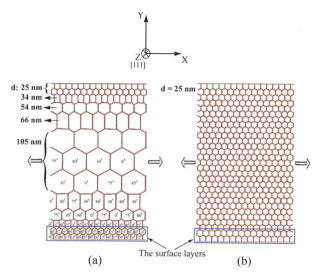

Figure 21.1 The relaxed tensile configurations for (a) GNG structure of bcc Fe; (b) uniform NG structure of bcc Fe. Perfect bcc atoms are not shown in this figure.

21.3 Results and Discussions

In the GNG/CG sandwich samples produced by the surface mechanical attrition treatment (SMAT) or the surface mechanical grinding treatment (SMGT) [11–14], the gradient layer generally has gradient grain sizes along the depth from several tens of nm to several tens of μm. Under uniaxial tension, the outer NG layers will first become plastically unstable at lower tensile strains, resulting in fast lateral shrinking of unstable layers. However, the lateral shrinking is constrained and stopped by the neighboring stable large-grained layer since the strain compatibility is required to maintain material continuity. Moreover, the constraint induces the lateral tensile stress in the NG layers and the lateral compressive stress in the inner layers [14].

In our simulation model of the GNG structure, the gradient grain size transition is only from 25 nm at the surface to 105 nm at the center due to the computation limitations of MD simulations. However, the stress states for both the surface layers and the center layer in the GNG sample are still observed to change from uniaxial tension to multi-axial states during the deformation, as indicated in Fig. 21.2. Figure 21.2a shows simulated stress (σ_x) as a function of tensile strain for both the GNG structure and the uniform NG structure. Figures 21.2b and 21.2c show simulated lateral stresses (σ_y and σ_z) as a function of tensile strain for the surface layers in both configurations. In Figs. 21.2b and 21.2c, the lateral stresses represent the average lateral stresses based on all atoms from the surface layers in both configuration. As indicated, the constraint between the surface layers and the center layer induces lateral tensile stresses in the surface layers for the GNG structure, which is contrast to the uniform NG structure where nearly zero lateral stresses are observed in both the surface layers and the center layer. The overall lateral stresses are still zero in the GNG structure during the tensile loading (the overall lateral stresses are monitored during loading, and the overall sample is still under uniaxial loading). Correspondingly, lateral compressive stresses should be induced in the center layer since the overall pressures in the y and z directions were kept to zero for the whole GNG structure during the loading. It should be noted that free boundary condition is intended to impose on y direction like the real GNG sample, and the strain compatibility between the surface layers and the center layer may suppress the strain localization of the surface layers, which results in the stress-state change.

Figure 21.3a shows the number of dislocations as a function of tensile strain for the surface layers in both configurations. Figures 21.3b and 21.3c show the simulated deformation patterns of selected grains in the surface layers for the uniform NG structure and the GNG structure, respectively. The strongest grain size in bcc Fe was found to be approximately 15 nm by previous MD simulations [39]. This indicates that the 25 nm grains should be stronger than the 105 nm grains in our simulation model. However, as indicated in Fig. 21.2a, the flow stress of the GNG structure (with a high volume fraction for grains with d = 105 nm) at early plastic deformation stage is comparable to and even slightly higher than that of the uniform NG structure (d = 25 nm). The underlying mechanisms should be twofold. First, the stress state change is expected to increase the dislocation density in the materials and make it more likely for dislocations to interact and entangle with each other, which increases the flow stress [23]. As indicated in Fig. 21.3a, higher dislocation density is observed in the GNG structure for the same surface layer due to the stress state change and the constraint, when compared to the uniform NG structure. Second, the strain gradient occurs when the constraint exists, which produces the GNDs and the extra hardening due to the grain size gradient [14, 24]. Several dislocations can be observed as red dots marked by arrows within grain interiors of the surface layers for the uniform NG structure (Fig. 21.3b). However, as indicated in Fig. 21.3c, more dislocations can be found in the same grains for the GNG structures, and these dislocations seem to nucleate at GBs since there are no initial dislocation sources inside the grains.

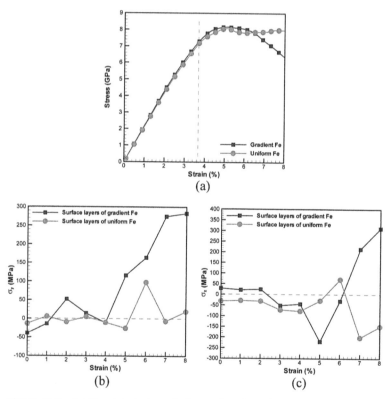

Figure 21.2 (a) Simulated stress (σ_x) as a function of tensile strain for both configurations; (b) Simulated lateral stress σ_y as a function of tensile strain for the surface layers in both configurations; (c) Simulated lateral stress σ_z as a function of tensile strain for the surface layers in both configurations.

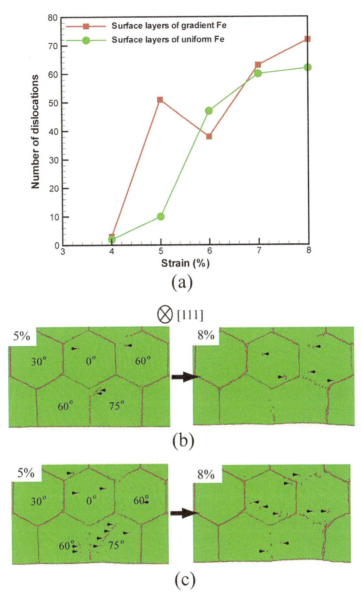

Figure 21.3 (a) Number of dislocations as a function of tensile strain for the surface layers in both configurations. Simulated deformation patterns at 5% and 8% tensile strains of selected grains in the surface layers (d = 25 nm) for (b) the uniform NG structure of bcc Fe; (c) the GNG structure of bcc Fe.

Figure 21.4 shows the overall simulated deformation patterns for the GNG structure of bcc Fe at tensile strains of 4% and 6%. A small crack is observed to nucleate at the GB of two large grains at a tensile strain of 6%, which produces a abrupt drop in the flow stress for the GNG structure as shown in the Fig. 21.2a. A few dislocations are observed in the small grains (d = 25 nm). Larger number of dislocations are observed in the interiors of large grains (d = 105 nm). Moreover, as indicated later, the dominant deformation mechanisms are closely related to the grain sizes in the GNG structure.

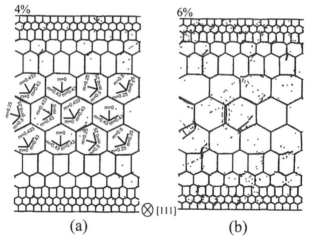

Figure 21.4 The overall simulated deformation patterns for GNG structure of bcc Fe at a tensile strain of (a) 4%; (b) 6%. A small crack is observed to nucleate at GB of large grains, indicated by a dash ellipse. The Schmid Factors for three {110} slip systems in each large grain (d = 105 nm) are also indicated in Fig. 4(a).

Figures 21.5a and 21.5b show nucleation of a dislocation from GBs and propagation of the dislocation in the interior of grains, respectively. This confirms that GBs are the sources of dislocations, and the dislocation type is full lattice dislocation. Figures 21.5c and 21.5d show a typical edge dislocation with edge component of $1/3[\bar{1}\bar{1}2]$ projected onto [111] plane and a typical screw dislocation in the interior of grains, respectively. Crystallographic analysis and image simulations reveal that the best way to study dislocations with edge components in bcc systems is to take images along <110> zone axis, from which it is possible to identify 1/2 <111> pure edge dislocations, and edge components of 1/2 <111> and 1/2 <001> mixed dislocations [27, 28]. However, in order to activate more slip systems in the 2D columnar structure, the zone axis was chosen as [111] in the present study. Both dislocations with Burges vectors of $1/2[\bar{1}\bar{1}1]$ and [100] have the same edge components of $1/3[\bar{1}\bar{1}2]$ when projected onto [111] plane. So, it is hard to differentiate a $1/2[\bar{1}\bar{1}1]$ dislocation from a [100] dislocation from the projection along [111] zone axis, and it is needed to examine 3D structure of dislocations and indentify the dislocation type. The dislocation types were then investigated and can be identified by Burgers circuits and DDA (dislocation detection algorithm) tools in the present study [43]. As shown in Figs. 21.5c–d, the Burges vector of a screw dislocation is parallel to the dislocation line, while the Burges vector of an edge dislocation is perpendicular to the dislocation line.

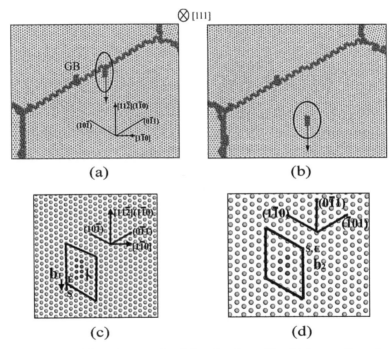

Figure 21.5 (a) Nucleation of a dislocation from GBs; (b) Propagation of a dislocation in the interior of grains; (c) A typical edge dislocation ($\vec{b}_1 = 1/2[\bar{1}\bar{1}1]$) with edge component of $1/3[\bar{1}\bar{1}2]$ projected onto [111] plane in the interior of grains; (d) A typical screw dislocation ($\vec{b}_2 = 1/2[111]$) in the interior of grains.

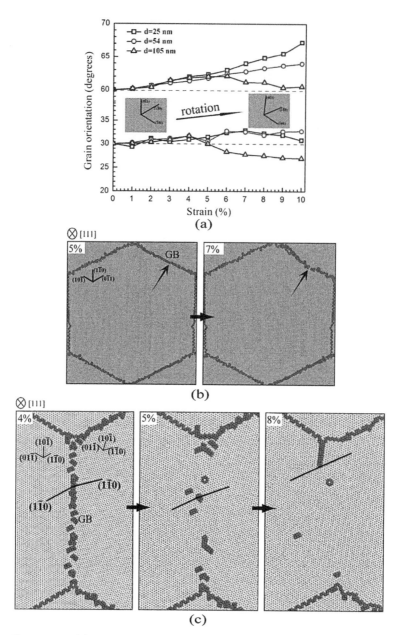

Figure 21.6 (a) The grain orientations as a function of tensile strain for selected grains with various grain sizes in the GNG structure; (b) Simulated deformation patterns at 5% and 7% tensile strains in selected 25 nm grains showing GB migration; (d) Simulated deformation patterns at 4%, 5% and 7% tensile strains in selected 25 nm grains showing grain rotation and grain coalescence.

Figure 21.6a shows the grain orientations as a function of tensile strain for selected grains with various grain sizes in the GNG structure. Figure 21.6b shows simulated deformation patterns at 5% and 7% tensile strains in selected 25 nm grains showing GB migration. Figure 21.6c shows simulated deformation patterns at 4%, 5% and 7% tensile strains in selected 25 nm grains showing grain rotation and grain coalescence. Based on observations from Fig. 21.6a, three conclusions could be drawn as follows: (a) The smaller the grain, the easier for grain rotation when the initial orientation is 60°; (b) The grain rotations are similar for various grain sizes when the initial orientation is 30°, which indicates the grain rotations are not only related to grain sizes but also the initial grain orientations; (c) The grain rotations could be clockwise or counter-clockwise. Previous investigations [29, 44] also suggest that the GB activities increase with decreasing grain sizes for tens of nm grain size range. A unified approach to four fundamental phenomena associated with GBs and grain growth has recently been formulated: (a) Normal GB motion; (b) Relative translation of the grains parallel to the GB plane coupled to normal GB motion; (c) GB sliding; (d) Grain rotation [45]. This indicates that GB motions are generally coupled and curvature driven, which is similar to our observations as indicated in Fig. 21.6b. The grain growth mechanisms include GB migration and grain-rotation-induced grain coalescence [45, 46], which are also observed in our simulation for a GB with initial misorientation of 15° (Fig. 21.6c). For grains with d = 25 nm in the GNG structure, the deformation mechanisms are dominated by GB migration, grain rotation and grain coalescence although a few dislocations are observed.

Figure 21.7 shows simulated deformation patterns at 3%, 5% and 8% tensile strains in selected 54 nm grains showing dislocation nucleation, propagation and formation of dislocation wall near GBs. Based on MD simulations, a transition have been elucidated in deformation mechanisms with decreasing grain size from dislocation-mediated plasticity to GB-associated plasticity in nanocrystalline metals [29, 30, 36, 39, 47]. Above "the strongest grain size", the prevailing mechanisms for nanocrystalline metals depend on the stacking-fault energy (SFE), the elastic properties of metals, and the magnitude of the applied stress [47]. The metals with high SFE exhibit a series of loops of perfect dislocations propagating through the grain interiors. By contrast, in the metals with low SFE, complete dislocations could not be nucleated and only partial dislocations could be nucleated from GBs and absorbed by opposite GBs, leaving stacking-fault (SF) behind in the grain interior. In the grains with d = 50 nm in the GNG structure of bcc Fe, perfect dislocations are observed to be nucleated from GBs and propagate to the grain interior, and finally a dislocation wall is formed near GBs (Fig. 21.7). This dislocation wall can impede further dislocation nucleation and propagation, resulting in strain hardening for the whole structure.

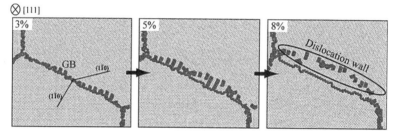

Figure 21.7 Simulated deformation patterns at 3%, 5% and 8% tensile strains in selected 54 nm grains showing dislocation nucleation, propagation and formation of dislocation wall near GBs.

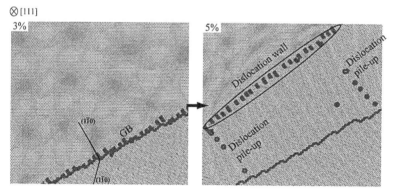

Figure 21.8 Simulated deformation patterns at 3% and 5% tensile strains in selected 105 nm grains showing formation of dislocation wall and dislocation pile-up near GBs.

Figure 21.8 shows simulated deformation patterns at 3% and 5% tensile strains in selected 105 nm grains showing formation of dislocation wall and dislocation pile-up near GBs. Besides the formation of dislocation wall near GBs, dislocation pile-up is also observed between the initial GB and the newly formed dislocation wall (Fig. 21.8), which is the first observation by MD simulations to our best knowledge. This dislocation pile-up should contribute significantly to the strain hardening of the whole structure.

21.4 Concluding Remarks

Large-scale MD simulations have been used to elucidate the atomistic deformation mechanisms of the GNG structure of bcc Fe and the uniform NG structure of bcc Fe in the present study. The grain size gradient in the GNG structure converts the applied uniaxial stress to multi-axial stresses and promotes the dislocation behaviors, thus resulting in extra hardening and slightly higher flow strength when compared to the uniform NG structure even with the stronger grain size. The dominant deformation mechanisms are found to be closely related to the grain sizes in the GNG structure. The deformation mechanisms are dominated by GB activities although few dislocations are observed in the grain interior with $d = 25$ nm. However, dislocation nucleation from GBs, propagation in the grain interior and formation of dislocation wall near GBs are observed for larger grains ($d = 54, 105$ nm). Moreover, dislocation pile-up between the newly formed dislocation wall and the initial GBs is observed for grains with $d = 105$ nm, which is the first observation by MD simulations to our best knowledge. The formation of dislocation wall and dislocation pile-up should contribute significantly to the strain hardening during the plastic deformation. The present findings should provide insights for developing the metals and alloys with superior mechanical properties.

Acknowledgements

The financial supports of this work were provided by the National Key Basic Research Program of China (2012CB932203 and 2012CB937500) and NSFC (11222224, 11472286, and 11021262). The simulations reported here were performed at Supercomputing Center of Chinese Academy of Sciences.

References

1. H. Gleiter, *Prog. Mater. Sci.* **33**, 223 (1989).
2. M. A. Meyers, A. Mishra, and D. J. Benson, *Prog. Mater. Sci.* **51**, 427 (2006).
3. R. O. Ritchie, *Nat. Mater.* **10**, 817 (2011).
4. L. Lu, Y. Shen, X. Chen, L. Qian, and K. Lu, *Science* **304**, 422 (2004).
5. L. Lu, X. Chen, X. Huang, and K. Lu, *Science* **323**, 607 (2009).
6. K. Lu, L. Lu, and S. Suresh, *Science* **324**, 349 (2009).
7. X. Y. Li, Y. J. Wei, L. Lu, K. Lu, and H. J. Gao, *Nature* **464**, 877 (2010).
8. Y. H. Zhao, X. Z. Liao, S. Cheng, E. Ma, Y. T. Zhu, *Adv. Mater.* **18**, 2280 (2006).
9. P. V. Liddicoat, X. Z. Liao, Y. H. Zhao, Y. T. Zhu, M. Y. Murashkin, E. J. Lavernia, R. Z. Valiev, and S. P. Ringer, *Nat. Commun.* **1**, 63 (2010).
10. Y. M. Wang, M. W. Chen, F. H. Zhou, and E. Ma, *Nature* **419**, 912 (2002).
11. K. Lu and J. Lu, *J. Mater. Sci. Technol.* **15**, 193 (1999).
12. A. Y. Chen, D. F. Li, J. B. Zhang, H. W. Song, and J. Lu, *Scr. Mater.* **59**, 579 (2008).
13. T. H. Fang, W. L. Li, N. R. Tao, and K. Lu, *Science* **331**, 1587 (2011).
14. X. L. Wu, P. Jiang, L. Chen, F. P. Yuan, and Y. T. Zhu, *Proc. Natl. Acad. Sci. U.S.A.* **111**, 7197 (2014).
15. Y. J. Wei, Y. Q. Li, L. C. Zhu, Y. Liu, X. Q. Lei, G. Wang, Y. X. Wu, Z. L. Mi, J. B. Liu, H. T. Wang, and H. J. Gao, *Nat. Commun.* **5**, 3580 (2014).
16. K. Lu, *Science* **345**, 1455 (2014).
17. Y. Xiang, T. Li, Z. Suo, and J. Vlassak, *Appl. Phys. Lett.* **87**, 161910 (2005).
18. T. Li and Z. Suo, *Inter. J Solids Struct.* **43**, 2351 (2006).
19. Y. M. Wang, E. Ma, and M. W. Chen, *Appl. Phys. Lett.* **80**, 2395 (2002).
20. L. Lu, M. L. Sui, and K. Lu, *Science* **287**, 1463 (2000).
21. X. L. Wu, Y. T. Zhu, Y. G. Wei, and Q. M. Wei, *Phys. Rev. Lett.* **103**, 205504 (2009).
22. X. Feaugas, *Acta Mater.* **47**, 3617 (1999).
23. R. J. Asaro, *Adv. Appl. Mech.* **23**, 1 (1983).
24. H. Gao, Y. Huang, W. D. Nix, and J. W. Hutchinson, *J. Mech. Phys. Solids* **47**, 1239 (1999).
25. X. L. Wu, P. Jiang, L. Chen, J. F. Zhang, F. P. Yuan, and Y. T. Zhu, *Mater. Res. Lett.* **2**, 185 (2014).

26. Q. Wei, *J. Mater. Sci.* **42**, 1709 (2007).
27. G. M. Cheng, W. W. Jian, W. Z. Xu, H. Yuan, P. C. Millett, and Y. T. Zhu, *Mater. Res. Lett.* **1**, 26 (2013).
28. G. M. Cheng, W. Z. Xu, W. W. Jian, H. Yuan, M. H. Tsai, Y. T. Zhu, Y. F. Zhang, P. C. Millett, *J. Mater. Res.* **28**, 1820 (2013).
29. V. Yamakov, D. Wolf, S. R. Phillpot, A. K. Mukherjee, and H. Gleiter, *Nat. Mater.* **1**, 45 (2002).
30. J. Schiotz and K. W. Jacobsen, *Science* **301**, 1357 (2003).
31. H. Van Swygenhoven, P. M. Derlet, and A. G. Froseth, *Nat. Mater.* **3**, 399 (2004).
32. J. Wang and H. C. Huang, *Appl. Phys. Lett.* **85**, 5983 (2004).
33. D. Farkas and B. Hyde, *Nano Lett.* **5**, 2403 (2005).
34. A. C. Lund and C. A. Schuh, *Acta Mater.* **53**, 3193 (2005).
35. D. Wolf, V. Yamakov, S. R. Phillpot, A. K. Mukherjee, and H. Gleiter, *Acta Mater.* **53**, 1 (2005).
36. Z. L. Pan, Y. L. Li, and Q. Wei, *Acta Mater.* **56**, 3470 (2008).
37. M. F. Horstemeyer, D. Farkas, S. Kim, T. Tang, and G. Potirniche, *Inter. J. Fatigue* **32**, 1473 (2010).
38. T Zhu and J Li, *Prog. Mater. Sci.* **55**, 710 (2010).
39. J. B. Jeon, B. Lee and Y. W. Chang, *Scr. Mater.* **64**, 494 (2011).
40. G. Monnet and D. Terentyev, *Acta Mater.* **57**, 1416 (2009).
41. A. Spielmannová, A. Machová, and P. Hora, *Acta Mater.* **57**, 4065 (2009).
42. M. I. Mendelev, S. Han S, D. J. Srolovitz, G. J. Ackland, D. Y. Sun, and M. Asta, *Philos. Mag.* **83**, 3977 (2003).
43. A. Stukowski and K. Albe, *Modell. Simul. Maters. Sci. Eng.* **18**, 025016 (2010).
44. V. Yamakov, D. Moldovan, K. Rastogi, and D. Wolf, *Acta Mater.* **54**, 4053 (2006).
45. J. W. Cahn, Y. Mishin, A. Suzuki, *Acta Mater.* **54**, 4953 (2006).
46. D. Moldovan, V. Yamakov, D. Wolf, and S. P. Phillpot, *Phys. Rev. Lett.* **89**, 206101 (2002).
47. V. Yamakov, D. Wolf, S. R. Phillpot, A. K. Mukherjee, and H. Gleiter, *Nat. Mater.* **3**, 43 (2004).

Chapter 22

Strain Hardening Behaviors and Strain Rate Sensitivity of Gradient-Grained Fe under Compression over a Wide Range of Strain Rates

Fuping Yuan, Ping Chen, Yanpeng Feng, Ping Jiang, and Xiaolei Wu

State Key Laboratory of Nonlinear Mechanics, Institute of Mechanics, Chinese Academy of Science, No. 15, North 4th Ring, West Road, Beijing 100190, China
fpyuan@lnm.imech.ac.cn

In the present work, gradient-grained Fe was synthesized by means of surface mechanical grinding treatment, and the compression behaviors of the coarse-grained Fe and the gradient-grained Fe were investigated under both quasi-static and dynamic loading conditions over a wide range of strain rates (from 5×10^{-4} to 10^4 s^{-1}). After

---

Reprinted from *Mech. Mater.*, **95**, 71–82, 2016.

Heterostructured Materials: Novel Materials with Unprecedented Mechanical Properties
Edited by Xiaolei Wu and Yuntian Zhu
Text Copyright © 2016 Elsevier Ltd.
Layout Copyright © 2022 Jenny Stanford Publishing Pte. Ltd.
ISBN 978-981-4877-10-7 (Hardcover), 978-1-003-15307-8 (eBook)
www.jennystanford.com

surface mechanical grinding treatment, equiaxed ultrafine grains, elongated lamellar ultrafine grains, fully-developed sub-grains with dense dislocations walls, non-fully-developed dislocation cells, and deformed coarse grains are sequentially observed along the depth from the treated surface. The grain/cell size increases while the measured micro-hardness decreases along the depth for the gradient-grained Fe. The gradient-grained structure shows apparent strain hardening behaviors at all strain rates up to 10^4 s^{-1} although the strain hardening exponent (n) for the gradient-grained Fe is smaller than that of the coarse-grained Fe at the same strain rate. This apparent hardening behavior is attributed to the hardening from both the coarse-grained center and the surface gradient layers when the strain localization trend for the ultrafine-grained surface layers are suppressed by the coarse-grained center. The extra hardening might be due to the hetero-deformation induced (HDI) hardening associated with the constraint and mechanical incompatibility between different layers in the gradient-grained structure. The dynamic strain rate sensitivity of the gradient-grained Fe is observed to be slightly larger than that of the coarse-grained Fe, which contradicts the general observation that strain rate sensitivity should decrease with reduction of grain size for BCC metals. The geometrically necessary dislocations associated with the HDI hardening and the grain size gradient result in additional increase in dislocation density, which may be the reason for the enhanced dynamic strain rate sensitivity in the gradient-grained Fe although it has smaller average grain size compared to the coarse-grained Fe. The present results should provide insights for the applications of gradient-grained structure under dynamic conditions.

22.1 Introduction

In recent years, bulk ultrafine-grained (UFG) and nano-grained (NG) metals, which are commonly produced by means of severe plastic deformation (SPD) techniques, have drawn increasing interests due to their ultra-high strength (Meyers et al., 2006; Dao et al., 2007). However, they show limited ductility compared to their coarse-grained (CG) counterparts due to the reduced strain hardening rate (Valiev, 2004; Zhu and Liao, 2004). Stronger and tougher metals and alloys are always desired for the structural applications in modern industry, and such demands have been realized recently by several novel strategies through tailoring microstructures at nano-scale, such as, bimodal grain size design, nano-precipitate dispersion, pre-existing growth nano-twins, and gradient nano-grained (GNG) structure (Wang et al., 2002; Liddicoat et al., 2010; Liu et al., 2013; Lu et al., 2009; Suresh, 2011; Fang et al., 2011; Wu et al., 2014a; Wei et al., 2014).

The GNG/CG hierarchical architecture, consisting of the CG core sandwiched by two GNG surface layers, can produce excellent synergy of strength and ductility (Lu and Lu, 2004; Chen et al., 2008; Fang et al., 2011; Wu et al., 2014a; Wei et al., 2014). Strain localization can be suppressed in the NG Cu film when confined by a CG core with a gradient grain size transition due to a mechanically-driven and strain-induced growth of nano-grains (Fang et al., 2011). Our previous work showed that large plastic strain could also be achieved in NG surface layer of Fe without apparent grain growth when confined by the CG core (Wu et al., 2014a). These evidences suggested that the variant trends for strain instability between different layers in GNG/CG architecture usually induce stress state change to suppress the strain localization of the NG surface layers. The mechanical incompatibility should also produce HDI hardening associated with a strain process providing long-range interactions with mobile dislocations (such as geometrically necessary dislocations, GNDs) (Feaugas, 1999; Elliot et al., 2004). Moreover, the constraint and mechanical incompatibility between different layers should also induce the strain gradient, which produces the extra hardening and the synergic strengthening due to the grain size gradient (Gao et al., 1999; Wu et al., 2014a; Wu et al., 2014b).

The strain hardening behaviors and the observed flow stresses for resisting plastic deformation of metals and alloys also highly depend on the rate associated with loading (Zener and Hollomon, 1944; Subhash, 1995; Nemat-Nasser and Guo, 2003; Song et al., 2007; Mishra et al., 2008). Previous research have indicated that the grain size (especially down to the UFG or NG regimes) has strong influences on the strain rate dependent plastic deformation, such as the strain rate sensitivity (SRS) and the rate-controlling deformation mechanisms (Wei et al., 2004; Wei et al., 2006a; Wei et al., 2006b; Wei, 2007; Mishra et al., 2008; Suo et al., 2011; Suo et al., 2013a; Yu et al., 2015). The effect of grain size on SRS has been observed to be strongly dependent on the lattice structures (Wei, 2007). The SRS of FCC metal increases with the reduction of grain size, while the SRS of BCC metals shows opposite behaviors. To date, numerous efforts have been undertaken to investigate the tensile properties and the strain hardening behaviors of nanostructured metals under quasi-static conditions (Malow and Koch, 1998; Tsuji et al., 2002; Han et al., 2003; Conrad, 2004; Han et al., 2004; Valiev, 2004; Giga et al., 2006; Meyers et al., 2006; Hazra et al., 2011). There also have been a number of investigations on the effect of high strain rate on the mechanical behaviors of UFG or NG metals (Jia et al., 1999; Wei et al., 2004; Wei et al., 2006a; Mishra et al., 2008; Suo et al., 2011; Suo et al., 2013a; Yu et al., 2015). However, most of reported research work on dynamic deformation of nanostructured metals were based on UFG or NG metals with relatively homogeneous grain size. There is clearly a lack of understanding of the effect of strain rate on the plastic deformation mechanisms for gradient-grained structure. In this regard, strain hardening behaviors and strain rate sensitivity of gradient-grained Fe under compression are systematically examined over a wide range of strain rates, and the corresponding plastic deformation behaviors are also compared with the CG counterpart.

22.2 Experimental Procedures

The commercial pure iron used in the present study was received in the form of rods of 10 mm in diameter. The as-received materials have a composition, in weight percentage, of 0.008 C, 0.02 Si, 0.08 Mn, 0.009 P, 0.008 S, 0.01 Cr, 0.01 Ni, 0.01 Cu 0.06 Al, and the balance of Fe. The as-received rods were first annealed at 900°C for 2 h to obtain a CG polycrystalline structure with a single ferrite phase. After annealing, the rods were processed by surface mechanical grinding treatment (SMGT) to synthesize a gradient-grained structure. SMGT is a method adopted from machining at high strain rates (up to 10^3–10^6 s^{-1}). As shown in Fig. 22.1a, a cylindrical WC/Co tool tip repetitively slides at a velocity of v_1 along the gauge section of the dog-bone shaped sample (with a diameter of 10 mm for the ends and a diameter of 3 mm for the gauge section), which rotates at a velocity of v_2 with respect to the tool tip. A high-rate plastic shear deformation zone with high strain gradient is induced underneath the tip and is applied to the top surface layer of the sample when a preset penetration depth (20 μm for each time) into the sample is applied to the tool tip. The magnitude of induced plastic strain and the depth of the deformed layer are closely related to the tip diameter (3 mm) and the total penetration depth (180 μm), while the shear deformation rate is determined by the velocities of v_1 (3 mm/s) and v_2 (600 rpm). Effective cooling system with liquid nitrogen is applied to the processed material and the tool tip during SMGT in order to refine the started coarse grains into nanoscale regime.

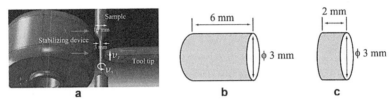

Figure 22.1 (a) Schematic of SMGT; (b) dimensions of the specimens for quasi-static compression experiments; (c) dimensions of the specimens for dynamic compression experiments.

The microstructure after SMGT was examined by optical microscope (OM), electron backscattered diffraction (EBSD), transmission electron microscope (TEM) and micro-hardness measurements. The sample surfaces for OM were polished to 2000 grit and finally polished with 0.25 μm diamond paste. This was followed by etching with 5% Nital. The cross-sectional surfaces of the SMGT samples for EBSD were first polished to 2000 grit and finally polished with 0.25 μm diamond paste, and then were electro-polished with 5% perchloric acid at 37 V voltage and −20°C for 10~15 s to reveal the microstructure. Using a field emission gun and the low accelerating voltage, the spatial resolution of EBSD can be significantly improved, making it possible to successfully explore the microstructure even in severely deformed state (Chen et al., 2011a; Chen et al., 2011b; Sun et al., 2014; Yuan et al., 2015). The cross-sectional disks for TEM were cut with a thickness of 500 μm and polished down to 30 μm using 2000 grid SiC papers. Final thinning to electron transparency was achieved by ion milling at various depths from the treated surface. Micro-hardness measurements were also made on the cross-sectional surface of the SMGT samples before and after dynamic compression tests using a Vickers diamond indenter at a load of 10 g for 10 s dwell time. The light load is especially suitable for measurements along the depth for the gradient structure, in which the indentation size is ~7 μm and the vertical spacing for indentations is ~22 μm (the real spacing for indentations is larger since sawtooth pattern was used) at the top surface layers.

All SMGT samples for quasi-static and dynamic compression testing were machined from the gauge section of the processed samples by wire saw with loading direction parallel to the axis of rods. The dimensions of samples for quasi-static and dynamic compression testing are given in Fig. 22.1b and Fig. 22.1c, respectively. Quasi-static uniaxial compression tests were carried out using a MTS 810 testing machine with a maximum load capacity of 25 kN at a strain rate of 0.0005 s^{-1}. An extensometer was used to measure the strain during the compression loading.

Dynamic compression tests were performed using Hopkinson-bar techniques. Details of the Hopkinson-bar technique and the data analysis for true stress, true strain and true strain rate can be found elsewhere (Subhash et al., 1997; Song et al., 2007, Sunny et al., 2009). Grease was used between the bars and the specimens to ensure low frictions. A pulse shaper was also placed between the striker and the input bar to control the shape of the incident pulse, and thus promote equilibrium conditions within the specimens. Semiconductor strain gages were used on both the incident and transmitted bars to obtain a very high signal-to-noise ratio for the strain measurements on both bars. Then, the recorded reflected and transmitted signals can be used to calculate the engineering stress, the engineering strain and the engineering strain rate:

$$\sigma_s = E\left(\frac{A}{A_s}\right)\varepsilon_T \qquad (22.1)$$

$$\varepsilon_s = -\frac{2C_0}{l_s}\int_0^t \varepsilon_R d\tau \qquad (22.2)$$

$$\dot{\varepsilon}_s = -\frac{2C_0}{l_s}\varepsilon_R \qquad (22.3)$$

where ε_T and ε_R are the transmitted and reflected strain pulses for the input and output bars, respectively; C_0, E and A are the longitudinal elastic wave velocity, Young's modulus and the cross-sectional area of the loading bars, respectively; l_s and A_s are the length and the cross-sectional area of the specimens, respectively. Finally, the true stress-true strain curves can be obtained from these engineering stress-engineering strain curves. For the dynamic compression tests, the strain rates were controlled at ~1500, ~6000, ~10000 s^{-1} depending on the striker bar velocity. At least two experiments for each strain rate were performed.

22.3 Experimental Results and Discussions

The as-annealed structure and the gradient-grained structure after SMGT by OM are shown in Fig. 22.2a and Fig. 22.2b, respectively. The grain size for the as-annealed sample is varied from 10 μm to 100 μm, and the average grain size is about 50 μm. While the SMGT sample shows apparent deformed band structure at the top surface of about 150 μm. The treated surface is on the right side. The flow direction of the band structure is consistent with the shear deformation direction.

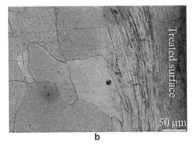

Figure 22.2 (a) OM of the as-annealed Fe; (b) OM of the gradient-grained Fe.

The deformed microstructures at the areas close to the treated surface (up to 70 μm from the treated surface) were characterized using EBSD. Figure 22.3a shows the image quality map by EBSD, in which blue color (>15°) represents high-angle boundaries, green color (5°–15°) and red color (2°–5°) represent low-angle boundaries. Figure 22.3b shows the EBSD orientation map. The orientation map shown in the present study is based on the inverse pole figure coloring scheme relative to the direction of cross section. The color

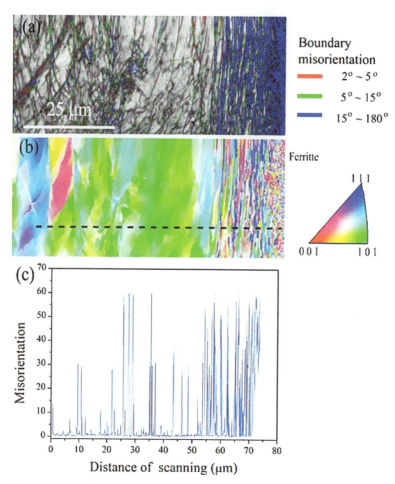

Figure 22.3 (a) EBSD micrograph with misorientation angles for the gradient-grained Fe; (b) EBSD orientation map for the gradient-grained Fe; (c) misorientation angle distribution along the dash line in Fig. 22.3b.

codes follow the inserted triangle in Fig. 22.3b, in which red, green and blue colors represent the grains having [001], [101] and [111] directions parallel to the direction of cross section, respectively. The misorientation angle distribution along the dash line of Fig. 22.3b is shown in Fig. 22.3c, in which the distance of scanning is from the left to the right. As observed, the outermost layer of ~5 μm thickness are nearly equiaxed ultrafine grains with high-angle boundaries, while the sub-surface layer of ~20 μm thickness are elongated lamellar ultrafine grains with high-angle boundaries. At the depths of 25–50 μm, the microstructure exhibits strong <110> texture, which may be due to <110> texture trend by cold work for BCC metals. The misorientation angles for various boundaries decrease with increasing depth from the treated surface due to the strain gradients induced by SMGT, as shown in Fig. 22.3c.

The plastic strain deformation by SMGT can induce high strain gradients at the surface layers of the samples. Thus the surface layers of the samples after SMGT also have characteristic of high gradient microstructures (Li et al., 2008; Fang et al., 2011; Wu et al., 2014a). Figures 22.4a–f show bright-field TEM micrographs at various depths from the treated surface for the gradient-grained Fe. The corresponding selected area diffraction (SAD) patterns for Figs. 22.4a and b are also shown in the insets. The outermost layer of the gradient-grained Fe (Fig. 22.4a) are nearly equiaxed ultrafine grains with a mean size of ~100 nm, and the SAD patterns show random orientations for all grains. At the depth of 20 μm (Fig. 22.4b), the microstructure shows elongated lamellar ultrafine grains with a mean lamellar thickness of ~200 nm, and high density dislocations are also observed inside the lamellar grains. At the depths of 80 and 130 μm (Figs. 22.4c–d), the initial coarse grains are divided into sub-grains with a mean size of 700~800 nm by dense dislocation walls (DDWs) or fully-developed dislocation cells. High density of dislocations and dislocation tangles are also observed in the sub-grains. In Fig. 22.4d, the boundaries of dislocation cells are marked by the red dash lines, and DDWs are marked by the red arrows. At the depths of 180 and 230 μm (Figs. 22.4e–f), the deformed microstructures are the non-fully-developed dislocation cells and dislocation tangles are mainly concentrated at the boundaries of cells. In Fig. 22.4f, the areas with high density dislocation tangles are marked by the red dash lines and the red arrow. Based on observations from TEM and the statistical analysis, the variation of average grain/cell size along the depth from the treated surface is plotted in Fig. 22.4g. Up to the depth of 30 μm, equiaxed ultrafine grains or elongated lamellar ultrafine grains are observed. Between the depths of 30 μm and 175 μm, sub-grains and full-developed dislocations walls are observed. Between the depths of 175 μm and 290 μm, non-fully-developed dislocation cells are observed. With further increasing depths, the deformed coarse grains are observed since the dislocation density further

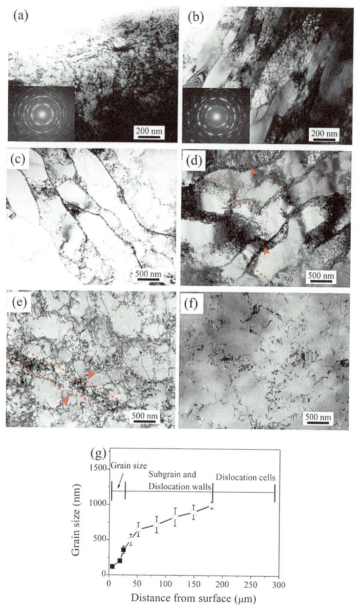

Figure 22.4 Bright-field TEM micrographs at various depths from the treated surface for the gradient-grained Fe: (a) close to the surface area; (b) ~20 μm; (c) ~80 μm; (d) ~130 μm; (e) ~180 μm; (f) ~230 μm. (g) Variation of average grain/cell size along the depth from the treated surface.

decreases. As we know, Fe is a typical BCC metal with a high stacking faults energy (SFE) of about 200 mJ/m². Thus, the grain refinement process during SMGT can be summarized as the following four steps and mechanisms: (i) increasing of dislocation density in the coarse grains; (ii) development of DDWs or dislocation cells; (iii) transformation of DDWs and dislocation cells into sub-grains with small misorientations; (iv) evolution of sub-grains to highly-misoriented grains with equiaxed shape or lamellar shape.

Figure 22.5 shows Vickers micro-hardness distributions along the depth from the treated surface for the gradient-grained Fe. Seven groups of measurements along the depths were made, and the average value was taken for reducing the physical errors. The standard deviations for seven group measurements are plotted as error bars in Fig. 22.5b. The hardness for the as-annealed Fe is also plotted as a straight line in Fig. 22.5b for comparison. As shown, the micro-hardness decreases from 180 Hv at the top surface (10 μm from the treated surface) to 110 Hv at the center (slightly higher than the 105 Hv for un-deformed coarse grains) for SMGT Fe. The continuously decreasing hardness along the depth is consistent with the increasing grain/cell size, as shown in Fig. 22.4g.

The mechanical responses of the as-annealed Fe and the gradient-grained Fe under quasi-static compression (strain rate of 0.0005 s⁻¹) are presented here. Figure 22.6a shows the engineering stress-strain curves, Fig. 22.6b shows the true stress-strain curves, while Fig. 22.6c shows the hardening rate $\left(\Theta = \dfrac{d\sigma}{d\varepsilon}\right)$ curves as a function of the true strain. The yield strength of the gradient-grained Fe is estimated to be ~300 MPa, which is almost two times of that for the as-annealed Fe. Although both the as-annealed Fe and the gradient-grained Fe show apparent strain hardening behaviors up to strain of 60%, the strain hardening ability of the gradient-grained Fe is observed to be smaller compared to that of the as-annealed Fe. As shown in Fig. 22.6b, the difference of the flow stress between the as-annealed Fe and the gradient-grained Fe decreases with increasing strain. The as-annealed Fe also shows a larger strain hardening rate compared to the gradient-grained Fe, as shown in Fig. 22.6c. This enhanced strength and the reduced strain hardening rate should be attributed to the refined strain size and the increased dislocation density at the surface layer of the gradient-grained Fe.

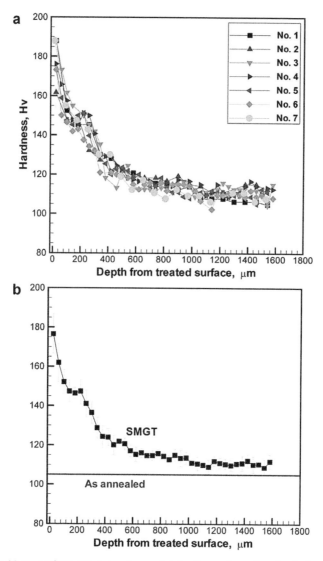

Figure 22.5 Vickers micro-hardness distributions along the depth from the treated surface for SMGT Fe: (a) seven groups of measurements; (b) average hardness.

Figures 22.7a and b show typical true stress-true strain curves under dynamic compression at various strain rates (~1500, ~6000, ~10000 s^{-1}) for the as-annealed Fe and the gradient-grained Fe, respectively. In order to further show the strain hardening behaviors of the as-annealed Fe and the gradient-grained Fe under various strain rates, Ludwik–Hollomon equation was used to fit the true stress–true strain curves as follows:

$$\sigma = \sigma_0 + K\varepsilon^n \tag{22.4}$$

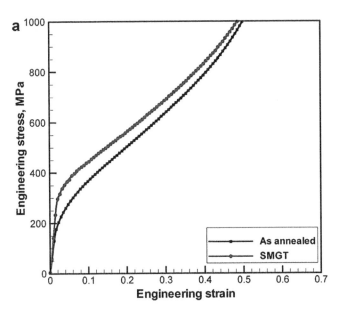

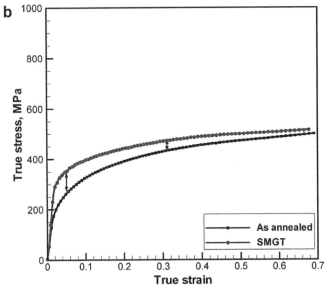

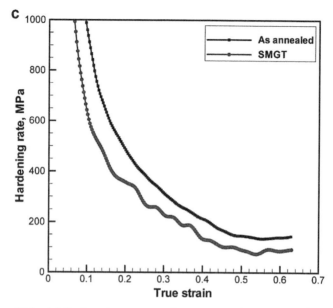

Figure 22.6 (a) Engineering compressive stress-strain curves under quasi-static condition for as-annealed Fe and SMGT Fe; (b) true compressive stress-strain curves under quasi-static condition for as-annealed Fe and SMGT Fe; (c) strain hardening rate vs. true strain curves for as-annealed Fe and SMGT Fe.

where, σ_0 is the yield strength, K is the strain hardening factor and n is the strain hardening exponent which typically reflects the stain hardening ability. According to the Ludwik–Hollomon equation, double logarithm true stress- true strain curves at various strain rates are plotted in Figs. 22.8a and b in order to obtain the strain hardening exponents for the as-annealed Fe and the gradient-grained Fe. Typically, the CG metals exhibit relatively small yield stresses and apparent strain hardening behaviors even under dynamic compression, while much higher yield stresses but reduced strain hardening behaviors, sometimes even strain softening behaviors due to the thermal effect can be observed in UFG or NG metals under dynamic compression (Wei et al., 2004; Wei et al., 2006a; Mishra et al., 2008; Suo et al., 2011; Suo et al., 2013a). For the as-annealed Fe (Fig. 22.7a), apparent hardening behaviors are observed at all strain rates like other CG metals and the strain hardening ability (strain hardening exponent, n) decreases with increasing strain rate (Fig. 22.8a). Although the strain hardening exponent (n) for the gradient-grained Fe is smaller than that of the CG Fe at the same strain rate, it is interesting to note that the gradient-grained Fe still shows apparent strain hardening behaviors at all strain rates even when the thermal softening effect exists under dynamic compression.

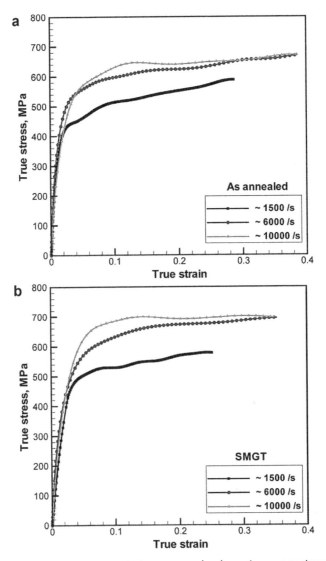

Figure 22.7 True stress-true strain curves under dynamic compression: (a) as-annealed Fe; (b) SMGT Fe.

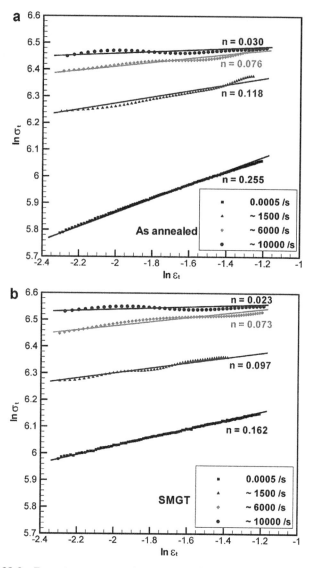

Figure 22.8 True stress-true strain curves with double logarithm forms at various strain rates under compression: (a) as-annealed Fe; (b) SMGT Fe. The solid lines are the linear fit for the data.

The apparent hardening behavior under dynamic compression for the gradient-grained Fe raises a critical issue: where is the strain hardening generated? To answer this question, we measured the micro-hardness (Hv) along the depth from the treated surface for the gradient-grained Fe before and after dynamic compression. As shown in Fig. 22.9a, the micro-hardness values increase at all depths after dynamic compression. The hardness increment (ΔH) along the depth after dynamic compression is also shown in Fig. 22.9a, and ΔH is an indicator on the magnitude of hardening retained after unloading. It should be noted that the first indentation is at the depth of 10 μm from the treated surface and the indentation size is about 7 μm, thus the hardness contribution for the first indentation should come from both the equiaxed ultrafine grains (with grain size of ~100 nm) and the elongated lamellar ultrafine grains (with grain size of ~200 nm). It is not surprised that strong hardness increment (about 60 Hv) occurs in the CG center since the coarse grains have the abilities for strain hardening. However, it is interesting to note that the surface layer with equiaxed/lamellar ultrafine grains (100–200 nm) also shows large hardness increment (about 60 Hv) after dynamic compression, which is completely different to the general observations that disappeared strain hardening behaviors, sometimes even strain softening behaviors should be observed in UFG or NG metals under dynamic compression (Wei et al., 2004; Wei et al., 2006a; Mishra et al., 2008; Suo et al., 2011; Suo et al., 2013a). The strain softening behaviors in UFG or NG metals under dynamic compression were explained due to the thermal softening and the formation of adiabatic shear band (Wei et al., 2004; Wei et al., 2006a; Mishra et al., 2008), however this strain localization trend for the surface layer of the gradient-grained Fe would be suppressed by the CG center and the mechanical constraint would be generated between different layers in the gradient-grained Fe. Our optical micrograph after dynamic compression confirmed this suggestion, and no adiabatic shear band was observed for the surface layer of the gradient-grained Fe (as shown in Fig. 22.9b). Thus, based on these observations mentioned above, the apparent strain hardening behaviors under dynamic compression for the gradient-grained Fe could be understood as three-fold: (1) first, the CG center provides the important part of the strain hardening ability for the gradient-grained Fe; (2) second, the constraint and mechanical incompatibility between different layers should produce extra hardening for the surface gradient layers due to HDI hardening (Feaugas, 1999; Elliot et al., 2004; Wu et al., 2015) and the GNDs induced by the grain size gradient (Gao et al., 1999; Wu et al., 2014a; Wu et al., 2014b); (3) the restrain of dynamic recovery of dislocations at high strain rates

(Suo et al., 2011) may also contribute to the strain hardening under dynamic conditions.

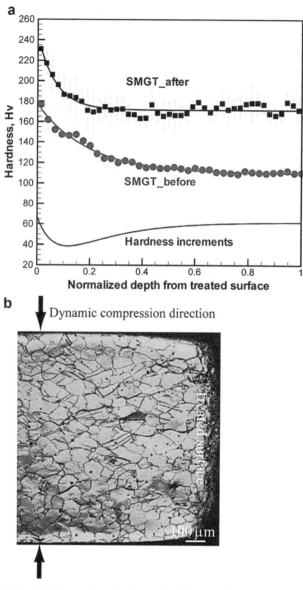

Figure 22.9 (a) Vickers micro-hardness distributions for SMGT Fe before and after dynamic compression (strain rate of ~1500 s^{-1}), the symbols are the measured data points and the solid lines are the fitting curves; (b) OM of the gradient-grained Fe after dynamic compression.

Due to the suppression of adiabatic shear band by the CG center, the dynamic homogeneous compression strain can be much improved in the gradient-grained Fe. The energy absorption under dynamic compression up to a specific strain can be calculated as the areas under the true stress-strain curves in Fig. 22.7a. Driven by the need to retain impact toughness while reaping the strengthening benefits from the gradient-grained structure, the gradient-grained structure reported here can be used to design components of energy absorbers for the automotive industry, which is illustrated in Fig. 22.10. In Fig. 22.10, the energy absorption under dynamic compression is plotted against the strain rate for the as-annealed Fe and the gradient-grained Fe. As indicated, besides with higher yield strength, the gradient-grained Fe also has higher energy absorption under dynamic compression at all strain rates, when compared to the as-annealed Fe.

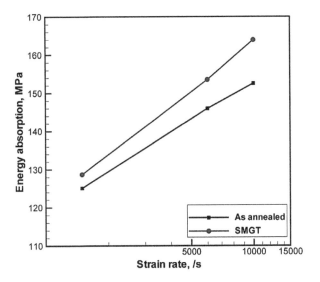

Figure 22.10 Energy absorption up to a true strain of 0.25 vs. strain rate: (a) as-annealed Fe; (b) SMGT Fe.

Figures 22.11a–c show the flow stress as a function of the strain rate with double logarithmic coordinates for both the as-annealed Fe and the gradient-grained Fe at fixed true strains of 10%, 15% and 20%, respectively. At all fixed true strains, the flow stress of the gradient-grained Fe is clearly observed to increase slightly faster than that of the as-annealed Fe with increasing strain rate. Both the as-annealed Fe and the gradient-grained Fe show positive SRS under dynamic compression.

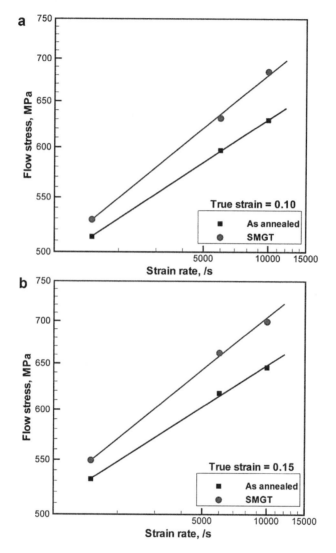

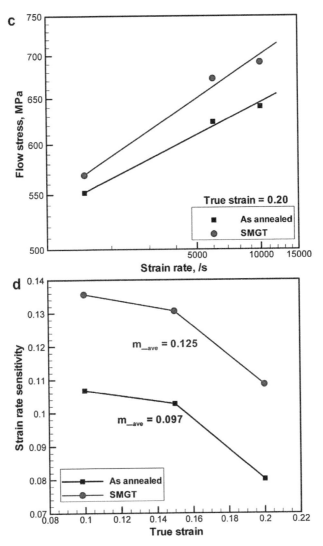

Figure 22.11 Flow stress vs. strain rate for as-annealed Fe and SMGT Fe: (a) at 10% true strain; (b) at 15% true strain; (c) at 20% true strain. Double logarithm coordinates are used in Figs. 22.11a–c, and the solid lines are the linear fit for the data. (d) Strain rate sensitivity vs. true strain for as-annealed Fe and SMGT Fe.

In general, the plastic deformation of metals and alloys depends on the strain, the strain rate and the deformation temperature, thus the flow stress can be written as:

$$\sigma = f(\varepsilon, \dot{\varepsilon}, T) \tag{22.5}$$

where ε is the true strain, $\dot{\varepsilon}$ is the true strain rate, and T is the environment temperature associated with the loading. Based on Eq. (22.5), the SRS of flow stress can be defined as:

$$m = \left(\frac{\partial \ln \sigma}{\partial \ln \dot{\varepsilon}}\right)_{\varepsilon, T} \tag{22.6}$$

In practice, the SRS is often calculated at certain fixed temperature and at several fixed strains. Under quasi-static conditions, the SRS is generally determined using the strain rate jump tests or stress relaxation tests. However, these two experiments are hard to employ under dynamic conditions. Alternatively, the SRS can also be obtained approximately by using the true stress-strain curves with double logarithmic form at a fixed true strain, where the SRS is derived as the slope of a linear regression fit for the curves. By using this method, the dynamic SRS as a function of the true strain is plotted in Fig. 22.11d for both the as-annealed Fe and the gradient-grained Fe. The dynamic SRS is observed to decrease from 0.107 to 0.08 when the true strain is changed from 10% to 20% for the as-annealed Fe, and the average dynamic SRS is about 0.097. While, the dynamic SRS decreases from 0.136 to 0.109 when the true strain is changed from 10% to 20% for the gradient-grained Fe, and the average dynamic SRS is about 0.125. It is interesting to note that the dynamic SRS in our experiments are much higher than those from the literature (<0.04) (Wei, 2007). The SRS values from previous research for UFG or NG metals (Wei, 2007) are mostly obtained from quasi-static loading (nanoindentation, strain rate jump tests or stress relaxation tests), and the SRS values under quasi-static conditions are typically lower than those under dynamic conditions due to possibly different deformation mechanisms (Meyers, 1994; Suo et al., 2013b). According to the previous research (Wei, 2007), the SRS of BCC metal is observed generally to decrease with the reduction of grain size. The gradient-grained Fe has smaller average grain size and should have smaller SRS according to previous work (Wei, 2007), when compared to the as-annealed Fe. However, the results in the present study show different trends for SRS of BCC metals with reduction of grain size compared to previous work (Wei, 2007), and the possible reasons will be two-fold: (1) first, the SRS values are obtained from different strain rate regimes for our data and the data from previous work (Wei, 2007); (2) more importantly, the other possible reason is the reduced activation volume by the gradient structure, and will be presented in details in the following discussion.

Base on the concept of thermal-activation theory, the apparent activation volume (V^*) can be estimated:

$$V^* = \frac{\sqrt{3}kT}{\sigma m} \tag{22.7}$$

where k is Boltzmann's constant, T is the absolute temperature associated with loading, m is the SRS, and σ is the flow stress.

Thus, the SRS can be written as

$$m = \frac{\sqrt{3}kT}{\sigma V^*} \tag{22.8}$$

Typically, the physical activation volume is highly dependent on the dislocation density. The higher the dislocation density, the lower the activation volume (Wei, 2007).

According to the previous research (Wei, 2007), the activation volume decreases to nearly a constant when the stress is increased to a moderate level for BCC metals, however the stress is still following the Hall-Petch relation when the grain size is refined. Thus, the SRS should decrease with reduced grain size for BCC metals based on Eq. (22.8).

However, the dislocation density could be further increased by the additional dislocations from the GNDs associated with the HDI hardening (Feaugas, 1999; Elliot et al., 2004; Wu et al., 2015) and the grain size gradient (Gao et al., 1999; Wu et al., 2014a; Wu et al., 2014b) when the constraint and mechanical incompatibility between different layers are produced in the gradient-grained structure. This additional increase in dislocation density could possibly result in lower activation volume, thus larger SRS (m), which may be the reason why the gradient-grained Fe has slightly higher SRS even it has smaller average grain size compared to the as-annealed Fe.

22.4 Conclusions

In the present work, gradient-grained Fe was synthesized using SMGT, and then the microstructure after SMGT was examined by OM, EBSD, TEM and micro-hardness measurements. The strain hardening behaviors and strain rate sensitivity of the as-annealed Fe and the gradient-grained Fe under compression were then studied over a wide range of strain rates (from 5×10^{-4} to 10^4 s^{-1}). The main findings are summarized as follows:

(1) From the treated surface to the center, equiaxed ultrafine grains, elongated lamellar ultrafine grains, sub-grains and full-developed dislocations walls, non-fully-developed dislocation cells, and deformed coarse grains are sequentially observed for the gradient-grained Fe. The grain/cell size increases while the measured micro-hardness decreases along the depth for the gradient-grained Fe.

(2) Although the strain hardening exponent (n) for the gradient-grained Fe is smaller than that of the as-annealed Fe at the same strain rate, the gradient-grained structure still shows apparent strain hardening behaviors at all strain rates up to 10^4 s^{-1}. Based on the micro-hardness measurements before and after dynamic compression, this apparent hardening behavior could be attributed to both hardening from the CG center and the surface gradient layers. The constraint and mechanical incompatibility between different layers in gradient-grained structure should be produced when the strain localization trend for the surface layer of the gradient-grained Fe is suppressed by the CG center. The constraint and mechanical incompatibility between different layers in gradient-grained structure should produce extra hardening for the surface gradient layers due to HDI hardening (Feaugas, 1999; Elliot et al., 2004; Wu et al., 2015).

(3) The dynamic SRS of the gradient-grained Fe is slightly larger than that of the as-annealed Fe, which is controversial to the general observations from previous work (Wei, 2007) that SRS should decrease with reduction of grain size for BCC metals. The GNDs associated with the HDI hardening and the grain size gradient can result in additional increase of dislocation

density, which may be the reason why the gradient-grained Fe has slightly higher SRS even it has smaller average grain size compared to the as-annealed Fe.

The present results advance our understanding of compression behaviors for the gradient-grained structure under extreme conditions. The gradient structure should also have promising applications for structural materials and energy absorption due to the high strength, the apparent strain hardening behaviors at all strain rates up to 10^4 s^{-1} and the enhanced dynamic strain rate sensitivity.

Acknowledgments

The authors would like to acknowledge the financial support from the National Key Basic Research Program of China under Grants No. 2012CB932203 and No. 2012CB937500; and NSFC under Grants No. 11222224, No. 11472286, 11072243, and No. 11021262. The authors would like to thank Dr. Husheng Zhang for helping to conduct the dynamic compression experiments.

References

Chen, A.Y., Li, D.F., Zhang, J.B., Song, H.W., Lu, J., 2008. Make nanostructured metal exceptionally tough by introducing non-localized fracture behaviors. *Scr. Mater.* **59**, 579–581.

Chen, D., Kuo, J.C., Wu, W.T., 2011a. Effect of microscopic parameters on EBSD spatial resolution. *Ultramicroscopy* **111**, 1488–1494.

Chen, Y.J., Li, Y.J., Walmsley, J.C., Dumoulin, S., Gireesh, S.S., Armada, S., 2011b. Quantitative analysis of grain refinement in titanium during equal channel angular pressing. *Scr. Mater.* **64**, 904–907.

Conrad, H., 2004. Grain-size dependence of the flow stress of Cu from millimeters to nanometers. *Metall. Mater. Trans. A* **35**, 2681–2695.

Dao, M., Lu, L, Asaro, R.J., De, Hosson, J.T.M., Ma, E., 2007. Toward a quantitative understanding of mechanical behavior of nanocrystalline metals. *Acta Mater.* **55**, 4041–4065.

Elliot, R.A., Orowan, E., Udoguchi, T., Argon, A.S., 2004. Absence of yield points in iron on strain reversal after aging, and the Bauschinger overshoot. *Mech. Mater.* **36**, 1143–1153.

Fang, T.H., Li, W.L., Tao, N.R., Lu, K., 2011. Revealing extraordinary intrinsic tensile plasticity in gradient nano-grained copper. *Science* **331**, 1587–1590.

Feaugas, X., 1999. On the origin of the tensile flow stress in the stainless steel AISI 316 L at 300 K: back stress and effective stress. *Acta Mater.* **47**, 3617–3632.

Gao, H., Huang, Y., Nix, W.D., Hutchinson, J.W., 1999. Mechanisms-based strain gradient plasticity - I. Theory. *J. Mech. Phys. Solids* **47**, 1239–1263.

Giga, A., Kimoto, Y., Takigawa, Y., Higashi, K., 2006. Demonstration of an inverse Hall-Petch relationship in electrodeposited nanocrystalline Ni-W alloys through tensile testing. *Scr. Mater.* **55**, 143–146.

Han B.Q., Lavernia E.J., Mohamed A., 2003. Mechanical properties of iron processed by severe plastic deformation. *Metall. Mater. Trans. A* **34A**, 71–83.

Han B.Q., Lavernia E.J., Mohamed A., 2004. Dislocation structure and deformation in iron processed by Equal-Channel-Angular pressing. *Metall. Mater. Trans. A* **35A**, 1343–1350.

Hazra, S.S., Pereloma, E.V., Gazder, A.A., 2011. Microstructure and mechanical properties after annealing of equal-channel angular pressed interstitial-free steel. *Acta Mater.* **59**, 4015–4029.

Jia, D., Ramesh, K.T., Ma, E., 1999. Failure mode and dynamic of nanophase iron under compression. *Scr. Mater.* **42**, 73–78.

Li, W.L., Tao, N.R., Lu, K., 2008. Fabrication of a gradient nano-micro-structured surface layer on bulk copper by means of a surface mechanical grinding treatment. *Scr. Mater.* **59**, 546–549.

Liddicoat, P.V., Liao, X.Z., Zhao, Y.H., Zhu, Y.T., Murashkin, M.Y., Lavernia, E.J., Valiev, R.Z., Ringer, S.P., 2010. Nanostructural hierarchy increases the strength of aluminium alloys. *Nat. Commun.* **1**, 63.

Liu, G., Zhang, G.J., Jiang, F., Ding, X.D., Sun, Y.J., Sun, J., Ma, E., 2013. Nanostructured high-strength molybdenum alloys with unprecedented tensile ductility. *Nat. Mater.* **12**, 344–350.

Lu K., Lu, J., 2004. Nanostructured surface layer on metallic materials induced by surface mechanical attrition treatment. *Mater. Sci. Eng. A* **375–377**, 38–45.

Lu, K., Lu, L., Suresh, S., 2009. Strengthening materials by engineering coherent internal boundaries at the nanoscale. *Science* **324**, 349–352.

Malow, T.R., Koch, C.C., 1998. Mechanical properties, ductility, and grain size of nanocrystalline iron produced by mechanical attrition. *Metall. Mater. Trans. A* **29A**, 2285–2295.

Meyers, M.A., 1994. *Dynamic Behavior of Materials*. Wiley-Interscience, New York, USA, pp. 323–326.

Meyers, M.A., Mishra, A., Benson, D.J., 2006. Mechanical properties of nanocrystalline materials. *Prog. Mater. Sci.* **51**, 427–556.

Mishra A., Martin M., Thadhani N.N., Kad B.K., Kenik E.A., Meyers M.A., 2008. High-strain rate response of ultra-fine-grained copper. *Acta Mater.* **56**, 2770–2783.

Nemat-Nasser, S., Guo, W.G., 2003. Thermomechanical response of DH-36 structural steel over a wide range of strain rate and temperatures. *Mech. Mater.* **35**, 1023–1047.

Song, B., Chen, W., Antoun, B.R., 2007. Determination of early flow stress for ductile specimens at high strain rates by using a SHPB. *Exp. Mech.* **47**, 671–679.

Subhash, G., 1995. The constitutive behavior of refractory-metals as a function of strain-rate. *JOM* **5**, 55–58.

Subhash, G., Ravichandran, G., Pletka, B.J., 1997. Plastic deformation of hafnium under uniaxial compression. *Metall. Mater. Trans. A* **28A**, 1479–1487.

Sun, J.L., Trimby, P.W., Yan, F.K., Liao, X.Z., Tao, N.R., Wang, J.T., 2014. Shear banding in commercial pure titanium deformed by dynamic compression. *Acta Mater.* **79**, 47–58.

Sunny, G., Yuan, F.P., Lewandowski, J.J., Prakash, V., 2009. Design of inserts for split-Hopkinson pressure bar testing of low strain-to-failure materials. *Exp. Mech.* **49**, 479–490.

Suo, T., Li, Y.L., Xie, K., Zhao, F., Zhang, K.S., Deng, Q., 2011. Experimental investigation on strain rate sensitivity of ultra-fine grained copper at elevated temperatures. *Mech. Mater.* **43**, 111–118.

Suo, T., Li, Y.L., Zhao F., Fan, X.L., Guo, W.G., 2013a. Compressive behavior and rate-controlling mechanisms of ultrafine grained copper over wide temperature and strain rate ranges. *Mech. Mater.* **61**, 1–10.

Suo, T., Chen, Y.Z., Li, Y.L., Wang, C.X., Fan, X.L., 2013b. Strain rate sensitivity and deformation kinetics of ECAPed aluminum over a wide range of strain rates. *Mater. Sci. Eng. A* **560**, 545–551.

Suresh, S., 2011. Graded materials for resistance to contact deformation and damage. *Science* **292**, 2447–2451.

Tsuji, N., Ito, Y., Saito, Y., Minamino, Y., 2002. Strength and ductility of ultrafine grained aluminum and iron produced by ARB and annealing. *Scr. Mater.* **47**, 893–899.

Valiev R., 2004. Nanostructuring of metals by severe plastic deformation for advanced properties. *Nat. Mater.* **3**, 511–516.

Wang, Y.M., Chen, M.W., Zhou, F.H., Ma, E., 2002. High tensile ductility in a nanostructured metal. *Nature* **419**, 912–915.

Wei Q., Kecskes L., Jiao T., Hartwig K.T., Ramesh K.T., Ma E., 2004. Adiabatic shear banding in ultrafine-grained Fe processed by severe plastic deformation. *Acta Mater.* **52**, 1859–1869.

Wei. Q., Jiao, T., Ramesh, K.T., Ma, E., Kecskes, L.J., Magness, L., Dowding, R., Kazykhanov, V.U., Valiev, R.Z., 2006a. Mechanical behavior and dynamic failure of high-strength ultrafine grained tungsten under uniaxial compression. *Acta Mater.* **54**, 77–87.

Wei, Q., Zhang, H.T., Schuster, B.E., Ramesh, K.T., Valiev, R.Z., Kecskes, L.J., Dowding, R.J., Magness, L., Cho, K., 2006b. Microstructure and mechanical properties of super-strong nanocrystalline tungsten processed by high-pressure torsion. *Acta Mater.* **54**, 4079–4089.

Wei Q., 2007. Strain rate effects in the ultrafine grain and nanocrystalline regimes e influence on some constitutive responses. *J. Mater. Sci.* **42**, 1709–1727.

Wei, Y.J., Li, Y.Q., Zhu, L.C., Liu, Y., Lei, X.Q., Wang, G., Wu, Y.X., Mi, Z.L., Liu, J.B., Wang, H.T., Gao, H.J., 2014. Evading the strength–ductility trade-off dilemma in steel through gradient hierarchical nanotwins. *Nat. Commun.* **5**, 3580.

Wu, X.L., Jiang, P., Chen, L., Yuan, F.P., Zhu, Y.T., 2014a. Extraordinary strain hardening by gradient structure. *Proc. Natl. Acad. Sci. U.S.A.* **111**, 7197–7201.

Wu, X.L., Jiang, P., Chen, L., Zhang, J.F., Yuan, F.P., Zhu, Y.T., 2014b. Synergetic strengthening by gradient structure. *Mater. Res. Lett.* **2**, 185–191.

Wu, X.L., Yang, M.X., Yuan, F.P., Wu, G.L., Wei, Y.J., Huang, X.X., Zhu, Y.T., 2015. Heterogeneous lamellar structure unites ultrafine-grain strength with coarse-grain ductility. *Proc. Natl. Acad. Sci. U.S.A.* **112**, 14501–14505.

Yu, X., Li, Y.L., Wei, Q.M., Guo, Y.Z., Suo, T., Zhao F., 2015. Microstructure and mechanical behavior of ECAP processed AZ31B over a wide range of loading rates under compression and tension. *Mech. Mater.* **86**, 55–70.

Yuan, F.P., Bian, X.D., Jiang, P., Yang, M.X., Wu, X.L., 2015. Dynamic shear response and evolution mechanisms of adiabatic shear band in an ultrafine-grained austenite-ferrite duplex steel. *Mech. Mater.* **89**, 47–58.

Zener C., Hollomon J.H., 1944. Effect of strain rate upon plastic flow of steel. *J. Appl. Phys.* **15**, 22–32.

Zhu, Y.T., Liao, X.Z., 2004. Nanostructured metals - retaining ductility. *Nat. Mater.* **3**, 351–352.

Chapter 23

Mechanical Properties and Deformation Mechanism of Mg-Al-Zn Alloy with Gradient Microstructure in Grain Size and Orientation

Liu Chen,* Fuping Yuan, Ping Jiang, Jijia Xie, and Xiaolei Wu

State Key Laboratory of Nonlinear Mechanics, Institute of Mechanics, Chinese Academy of Science, Beijing 100190, China
xlwu@imech.ac.cn

The surface mechanical attrition treatment was taken to fabricate the gradient structure in AZ31 magnesium alloy sheet. Microstructural investigations demonstrate the formation of dual gradients with

---

*Present address: Beijing Institute of Aeronautical Materials, Beijing 100095, China

Reprinted from *Mater. Sci. Eng. A*, **694**, 98–109, 2017.

Heterostructured Materials: Novel Materials with Unprecedented Mechanical Properties
Edited by Xiaolei Wu and Yuntian Zhu
Text Copyright © 2017 Elsevier B.V.
Layout Copyright © 2022 Jenny Stanford Publishing Pte. Ltd.
ISBN 978-981-4877-10-7 (Hardcover), 978-1-003-15307-8 (eBook)
www.jennystanford.com

respect to grain size and orientation, where the microstructural sizes decreased from several micrometers to about 200 nm from center area to treated surface, while the c-axis gradually inclined from being vertical to treated plane towards parallel with it. According to tensile results, the gradient structured sample has yield strength of 305 MPa in average, which is increased by about 4 times when compared with its coarse-grained counterpart. Meanwhile, contrary to quick failure after necking in most traditional magnesium alloys, the failure process of gradient structure appears more gently, which gives it 6.5% uniform elongation but 11.5% total elongation. Further comparative tensile tests on separated gradient layers and corresponding cores demonstrate that the gradient structured sample has higher elongation both in uniform or in post-uniform stages. In order to elucidate the relationship between mechanical properties and deformation mechanisms for this dual gradient structure, repeated stress relaxation tests and pole figure examinations via X-ray diffraction were conducted in constituent gradient layer and corresponding core, as well as gradient structured sample. The results show that pyramidal dislocations in dual gradient structure are activated through the whole thickness of sample. Together with the contribution of grain-size gradient, more dislocations are activated in dual gradient structure under tensile loading, which results in stronger strain hardening and hence higher ductility.

23.1 Introduction

Exploring lightweight materials with high strength and high ductility is clearly attractive but challenging for scientists [1]. Lightweight alloys with high strength have great benefits in energy saving, especially in aeronautical and vehicle fields. However, the strength and ductility are always exclusive for most traditional metals and alloys. For instance, nanostructured metals have strength 10 times higher than their coarse-grained (CG) counterparts due to grain boundary strengthening, but negligible ductility [2, 3].

Magnesium alloys are one of the most important lightweight materials and have attracted large amount of interest. One of the roadblocks for their massive application is the low ductility at room temperature. In general, magnesium alloys are of hexagonal close-packed (hcp) crystallographic structure, resulting in quite different deformation mechanisms from high-symmetry materials, e.g., face-centered cubic (fcc) and body-centered cubic (bcc) alloys. There are three main dislocation slip modes for magnesium alloys during plastic deformation, i.e., basal $\langle a \rangle$, prismatic $\langle a \rangle$ and pyramidal $\langle c + a \rangle$. Since the basal $\langle a \rangle$ slip has the lowest critical resolved shear stress (CRSS), it is the easiest one to be activated during deformation. The CRSS of pyramidal $\langle c + a \rangle$ slips are believed significantly higher than that of basal $\langle a \rangle$ slip [4, 5]. Both basal and prismatic $\langle a \rangle$ slips have no components along $\langle c \rangle$ direction, and only four independent slip systems could be provided by them together, which make it difficult to meet the von-Mises criterion for arbitrary deformation [6]. Twinning is another deformation mode for magnesium alloys to accommodate the deformation along $\langle c \rangle$ direction, which has been investigated intensively during past two decades [7–10]. Additionally, stimulating the activation of pyramidal $\langle c + a \rangle$ slip is another effective approach to enhance the dislocation behavior and thus improve the mechanical properties.

The activation of pyramidal slip in magnesium alloys has been observed in many investigations under tensile deformation, e.g., directly changing the crystal structure (i.e., c/a ratio) through alloying with lithium [11, 12], increasing the deformation temperature to reduce the CRSS of pyramidal $\langle c + a \rangle$ slip [6], or decreasing the grain size to reduce the yield stress anisotropy between basal and pyramidal slips [4, 13]. Since grain refining is a more universal strategy which is suitable for different magnesium alloys with a variety of chemical components, it has attracted increasing attention in recent years. As observed in traditional fcc and bcc metals where yield stress increases with decreasing of grain size, the yield stress of magnesium alloys also increases if grain size is reduced, no matter which slip mode dominates their yielding. However, different slip mode displays different sensitivity to grain size change, which apparently leads to different slopes in Hall-Petch equations. Generally speaking, the pyramidal slip has a less slope than basal one, which therefore results in a decreasing anisotropy in yield stress (as characterized by $\tau_{CRSS}^{basal} / \tau_{CRSS}^{pyramidal}$) with grain refinement. Yu et al. have shown the pyramidal slip in micro magnesium samples with several hundred nanometers in geometry size, demonstrating the enhancement of ductility due to reduction of yield stress anisotropy between basal and pyramidal slips [13]. However, as predicted by trade-off between strength and ductility in traditional metals, simple grain refinement down to nanoscale will significantly lose their ductility since the tiny grains are hardly to store dislocations [14]. Therefore, to retain the favorable contribution due to propensity to trigger pyramidal slip after grain refinement, it is necessary to take strategy to store dislocations inside small grains. Additionally, the activation of pyramidal slip could also be influenced by stress state, where external hydrostatic stress [15] and compatible stress near triple junctions [4] have been observed contributing pyramidal slips.

Recently, it has been evidenced that superior combination between strength and ductility could be achieved through gradient microstructure (GS), where the grain sizes are gradually increased from nanoscale at surface to microscale at center core along thickness direction [16–19]. This microstructure will generate geometrically necessary dislocations (GNDs) during uniaxial stretching due to strain gradient and the change in stress states, which therefore produces an extra strain hardening greater than that given by the mixture law, and results in the enhancement of mechanical properties [18–20]. Meanwhile, the deformation behavior in grains at different depths is quite different due to different yield stresses and strain hardening rates, which leads to strain partitioning and stress transfer between layers at different depths, and thus generates high hetero-deformation induced (HDI) stress to improve both strength and ductility [21].

In current investigation, the GS generated through surface mechanical attrition treatment (SMAT) has been employed to improve the mechanical properties of magnesium alloys. This strategy is designed to take advantage of the reduction in yield stress anisotropy due to grain refinement, together with the dislocation accumulation in small grains due to strain gradient and compatible deformation. An AZ31 magnesium alloy was taken as a model material, which has been extensively investigated with respect to its mechanical property and deformation mechanism. The gradient microstructure and mechanical properties, as well as the change of texture resulting from SMAT and its subsequent influence on plastic deformation have been investigated and analyzed in detail.

23.2 Experimental Procedures

The material for this study has a chemical compositions (wt%) of 3.05Al, 1.04Zn, 0.2Mn and Mg in balance. The raw materials were melted in electric furnace under protection gas consisting of CO_2 and SF_6 (99:1 in vol.), then casted into the mold pre-heated to 300°C. The cast ingot was homogenized at 400°C for 10 hours, and then hot forged to bar with 90 mm × 50 mm in cross section and 200 mm in length. Afterward, the forged bar was annealed at 300°C for 30 min and quenched in water. The plates with thickness of 2 mm thereafter were sliced out crosswise using electro discharging machining, which are denoted as as-received samples throughout this study.

The SMAT technique was used to produce the gradient microstructure. The surfaces of as-received plates were mechanically polished through silicon carbide abrasive paper to 2000 grit. During SMAT process, 100 number of stainless steel balls with 3 mm in diameter were accelerated in chamber by vibrator with frequency of 20 kHz, flying toward the sample surface along different directions. Each side of plates had been treated for 15 min and the final thickness of treated sample is about 1.7 mm.

Tensile tests were conducted using an Instron 5966 test machine at room temperature with strain rate of $5 \times 10^{-4} s^{-1}$. Since the as-received plates have strong in-plane (0001) texture as shown in Fig. 23.1, the dog-bone shaped tensile specimen in as-received and SMATed states were cut along two directions with length parallel with and vertical to c-axis, which are designated respectively as CG(0°) and CG(90°) for as-received sample, or GS(0°) and GS(90°) in SMATed state. Tensile samples have gauge sizes of 25 × 18 mm^2 in width and length, while three different thicknesses (1.7, 0.2, and 1.3 mm) were designed for bulk, gradient layer (GL) and corresponding center core. All GLs and cores were obtained from CG(0°) samples. The engineering strain was measured using extensometer with 5 mm in gauge length.

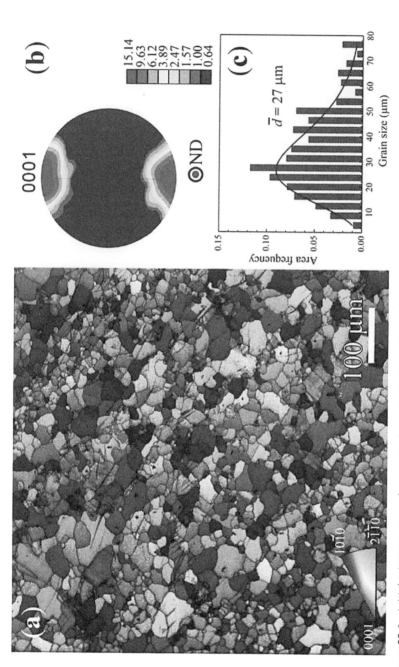

Figure 23.1 Initial microstructure of as-received sample in planar view. (a) Map of IPF plus IQ. (b) Corresponding PF exhibiting strong texture with c-axis parallel with plane. (c) Grain size distribution showing an average diameter of 27 μm.

The repeated stress relaxation tests were performed using the same Instron machine and extensometer under strain-controlled mode at room temperature with a serial of preset strains. Once attaining any strain with strain rate of $5 \times 10^{-4} s^{-1}$, the extensometer strain was maintained constant whereas the stress was recorded successively against time. After the first relaxation over an interval of 60 s, the sample was re-stretched by a strain increment of 0.5% with strain rate of $5 \times 10^{-4} s^{-1}$ for next relaxation. Four relaxation cycles were conducted at initial preset strain, and then the sample was strained to next strain at strain rate of $5 \times 10^{-4} s^{-1}$. Three times repeat were carried for stress relaxation test, and the calculation methods are referred to references.

The FM100 micro-hardness tester was employed to characterize the distribution of hardness along thickness at cross-section after SMAT. The indenting process was automatically carried under a load of 5 g with holding time of 15 s, and the tests were repeated for more than ten times at the same depth.

The X-ray diffraction (XRD) was used to investigate the texture at different depth through planar scanning, where the surface layer with proper thickness was removed by mechanical polishing to reach to the prescribed depth. The Rigaku SmartLab diffractometer was employed in this study, using Cu K_α radiation at the voltage of 40 kV and current of 30 mA.

Electron back-scattered diffraction (EBSD) observations were conducted using a field emission JSM 7100F scanning electron microscope (SEM), which equipped with a backscattered electron detector. The samples for EBSD and backscattered image (BSE) examinations were mechanically polished followed by electro-polishing using an electrolyte of 97% ethylalcohol and 3% perchloricacid (HClO4) with a voltage of 38 V at about −40°C. The step size (l_s) of EBSD scanning was determined by $l_s = d_a/N$, where d_a is the approximate value of grain diameter, while N is a number ranging from 10 to 20. Accordingly, the step size for as-received sample was set as about 1.5 μm, while equals about 100 nm for gradient layers. Additionally, the minimum value of misorientation angle was limited to 2° during the counting of grain sizes.

Transmission electron microscopy (TEM) observations were conducted through JEM 2010 microscope with an operating voltage of 200 kV, and the foil samples for TEM observation were prepared by twin-jet polishing under voltage of 30 V with the same solution as electro-polishing.

23.3 Results

23.3.1 Gradient Structure in Grain Size and Texture

Since magnesium alloys always exhibit strong texture which will then exert strong influence on mechanical properties, the initial texture of as-received samples was characterized by EBSD, and the results are shown in Fig. 23.1. Figure 23.1a displays the map overlapped with inversed pole figure (IPF) and image quality (IQ). Figure 23.1b shows the (0001) pole figure (PF), which manifests the strong in-plane texture with c-axis parallel with scanning plane. Therefore, two special directions parallel and vertical to c-axes will be designated as $0°$ and $90°$ in suffixes. Figure 23.1c shows the distribution of grain size, and the average value is measured as 27 μm.

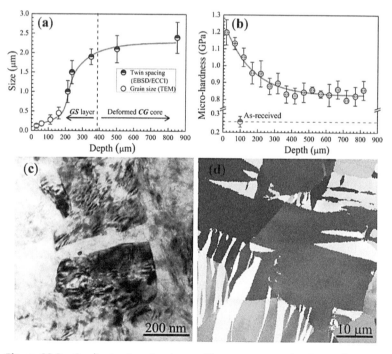

Figure 23.2 Gradient microstructure with respect to grain size and micro-hardness. (a) Grain size gradient and (b) micro-hardness gradient. (c) Bright-field TEM micrograph at the depth of about 50 μm away from treated surface showing grains with about 300 nm in diameter. (d) IPF map in cross-sectional view at center section of gradient structured sample exhibiting deformation twins and larger grains with diameters of about 30 μm.

The severely plastic deformation during SMAT will efficiently generate a gradient microstructure with respect to grain size attributed from gradient magnitudes of strain and stress along thickness. Figure 23.2 displays the gradient microstructure in grain size and micro-hardness. The size gradient includes the grain size as well as the twin spacing when approaching the center area. As shown in Fig. 23.2a, the gradient layer with approximate thickness of 400 μm has a sharp change of grain size with depth, and furthermore, the deformation twinning have occurred in center area which make the microstructural size much less than its original grain diameter. Figure 23.2b exhibits the gradient of micro-hardness, which is in line with the tendency of size gradient. The typical microstructures close to treated surface and center area are shown in Fig. 23.2c and d, which are demonstrated by the TEM micrograph and EBSD IPF map, respectively. Figure 23.2c shows a near equiaxed grain of about 300 nm at depth of 50 μm, while much larger size of about several microns after partitioning by twin boundaries is exhibited in Fig. 23.2d.

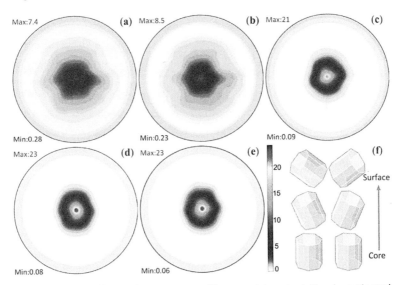

Figure 23.3 Gradient microstructure with respect to orientation investigated by XRD. (0001) PFs at the depth of (a) 0 μm (the treated surface), (b) 110 μm, (c) 280 μm, (d) 530 μm, and (e) 850 μm, respectively. (f) Schematic illustration of orientation gradient where (0001) axes gradually inclined towards treated plane from center to surface layers. (All PFs plotted with TD in East and LD in North directions).

In respect to texture after SMAT, Fig. 23.3 displays an important feature different from other fcc and bcc alloys treated by the same technique, where the orientation gradient from center to surface is clearly shown with c-axes gradually changing from vertical to parallel directions (schematically illustrated in Fig. 23.3f). The (0001) PFs at different depth were examined by XRD and plotted in Figs. 23.3a–e, which were scanned in planes parallel with surface. The relatively stronger (0001) out-of-plane texture with many c-axes nearly parallel with normal directions are shown in Fig. 23.3d and e, revealing a 90° turning of c-axis during SMAT. With decreasing of depth towards surface, the c-axes become inclined gradually and the corresponding maximum intensities are also decreased, demonstrating the formation of orientation gradient along with grain-size gradient during SMAT. Accordingly, a dual gradient microstructure was generated in magnesium alloys.

23.3.2 Mechanical Properties

Figure 23.4 shows the mechanical properties according to uniaxial tensile tests. Figure 23.4a shows engineering stresses plotted against engineering strains. The as-received samples show different mechanical response when stretch along different directions, which therefore shows lower yield stress (70 MPa) and higher tensile elongation along c-axes ($GS^{0°}$), but higher yield stress (138 MPa) and lower tensile elongation vertical to c-axes ($GS^{90°}$). After SMAT, the difference in respect to tensile curve along two directions becomes disappearing, which makes the tensile curves seem nearly overlapped with each other, manifesting the formation of similar microstructure viewed from two directions. In addition, the post-uniform elongation in GS sample is about 5%, which is nearly negligible for as-received sample. Since the voids within twins are prone to formation in coarse grains, it makes the magnesium alloys deformed localized and ending up with failure prior to diffuse necking [22]. However, the GS samples have shown significant elongation after maximum load, revealing higher deformability attributed from gradient microstructure. Actually, the total elongation of GS sample is even higher than that of CG specimen which was loaded vertical to c-axis ($GS^{90°}$).

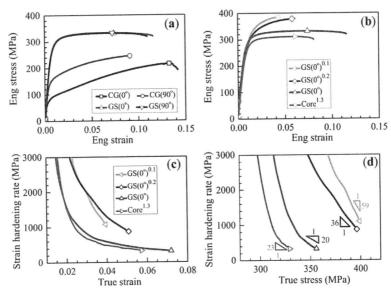

Figure 23.4 Mechanical properties and unique tensile elongation of gradient structure as compared with its gradient layer and corresponding center core. (a) Tensile engineering stress-strain curves of gradient and as-received samples. (b) Tensile engineering stresses plotted against strains for gradient structure, gradient layers and center core. Corresponding strain hardening rates plotted against (c) true strain and (d) true stress, respectively.

The gradient structured sample is always viewed as an integration of different layers with increasing grain sizes along thickness, and the interaction between layers will be generated during deformation due to the requirement of compatibilities across different layers [18, 19, 23]. In order to reveal the interaction between layers, the GL with thickness of 200 μm and corresponding core were prepared and tested, which are shown in Fig. 23.4b. For comparison, the tensile curves of GS sample and 100-μm-thick GL are also plotted here. According to the yield stresses of GL and core (320 MPa in GL and 280 MPa in corresponding core), the rule of mixture (ROM) gives an average yield stress of 289 ± 15 MPa, while the GS sample exhibits a real yield stress of 305 ± 10 MPa, revealing a slightly higher value than ROM prediction. This result is in alignment with the previous investigations for extra hardening of gradient microstructure [18, 19]. Furthermore, it should be noted that the uniform elongation of GS samples is amazingly greater than those of both GL and corresponding core, and this phenomenon will be more significant when it comes to the total elongation. The mechanism attributed to this phenomenon will be analyzed later in detail based on the unique dual gradient microstructure in GS samples.

Figure 23.4c and d display the strain hardening rate corresponding to Fig. 23.4b, which are plotted against true strain and true stress, respectively. In general, the uniform deformation during tensile tests requires a consistent high strain hardening to balance the geometrical softening due to reduction of cross-sectional area. However, as shown in Fig. 23.4c, the strain hardening rate of GLs decreases sharply with the increasing of strain, which hence triggers the Considere criterion at small strain, leading to low uniform elongation [24]. The GS sample has a moderate strain hardening rate, gently decreasing with the increase of tensile strain, and thus generates the highest uniform elongation. Therefore, not only the level of strain hardening rate itself, but also the decreasing trend with tensile strain, is important to determine the magnitude of uniform elongation. The decreasing of strain hardening rate is always resulted from dynamic recovery, and could be described by the slope of strain hardening rate curve plotted against true stress [25]. Figure 23.4d shows the rates of dynamic recovery corresponding to Fig. 23.4b, where GS specimen exhibits the lowest rate. The dynamic recovery is generally related to annihilation of dislocations due to cross-slip in CG metals or exhaustion of dislocation due to grain boundary trap in nanostructured metals [25, 26], and could be suppressed by the generation of GNDs due to interaction between different phases [27]. The lowest rate for GS sample manifests that the multiplication of dislocations was not drastically lowered during tensile deformation, thus enable the GS sample to be stretched to higher strain.

23.3.3 Repeated Stress Relaxation Tests

As just mentioned above, the GS samples might have unique ability to enhance the multiplication and interaction of dislocations. Meanwhile, twinning mechanism seems hard to be activated during tensile test in current gradient microstructure due to small grain size and unfavorable orientation, which will be discussed later. Therefore, the evolution of mobile dislocations was examined through repeated stress relaxation test to reveal the strain hardening mechanism of gradient structure.

Figure 23.5 displays the measured data from relaxation tests. Figure 23.5a shows the engineering stresses plotted against strain for GS, 200-μm-thick GL and corresponding center core. Their shear stress decays for each first relaxation at different strains are exhibited in Fig. 23.5b, c, and d, respectively. The more the shear stress drops, the larger the plastic deformation occurs in gauge section during relaxation. The shear stress decays are hence related to mobile dislocation behavior, which could be further calculated to obtain the evolution of mobile dislocation density for intuitive viewing.

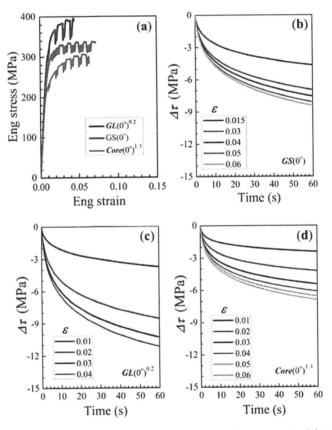

Figure 23.5 Stress decays during repeated stress relaxation tests. (a) Tensile engineering stress-strain curves. Shear stress decays with the elapse of time at different engineering strains for (b) gradient structure, corresponding (c) gradient layer and (d) center core.

Figure 23.6 exhibits the calculated data of repeated stress relaxation tests according to measured data in Fig. 23.5. Figure 23.6a–c displays the exhaustion curves of mobile dislocation for GS, GL and core samples, where the vertical axes denote the retained fraction of mobile dislocations. With regards to GS sample, the exhaustion curves are rising up in general with increase of tensile strain, demonstrating increasing retained mobile dislocations within gradient microstructure. This is in alignment with previous observations for gradient structured IF steel, which were mainly attributed from the generation of dislocation due to stress state changes and strain gradients [18]. However, the difference between exhaustion curves at different tensile strains for GL specimen is nearly negligible, while a slight increase is observed for center core as shown in Fig. 23.6c. The retained density of mobile dislocation at the end of relaxation is further plotted in Fig. 23.6d against tensile strains, which is more intuitive to exhibit the more significant increase of mobile dislocation for gradient structure than those for both GL and corresponding core.

Figure 23.6e shows the physical activation volumes of GS, GL and core samples. When tensile strain increases, the physical activation volumes in three structures begin to decrease. In order to understand the physical meaning reflected in this figure, it is necessary to look at what the physical activation volume depends on. According to textbook [28], the equation $V^* = bdl$ reveals that the physical activation volume depends on the magnitude of Burgers vector (b), the activation length or size of barrier (d), and the mean free path or distance between barriers (l). Therefore, with increasing of strain, the mean free path decreases due to multiplication of dislocations, and thus leads to reduction of physical activation volume. Accordingly, as evidenced in Fig. 23.6d, the GS sample which has great ability to produce mobile dislocation should display a lower physical activation volume at least than the ROM predictions. However, as shown in Fig. 23.6e, the GS sample demonstrates high V^* values, which are even higher than the physical activation volumes of center core at strains larger than 2%. Since the activation length is nearly constant for dislocation-dislocation interaction, it is more possible that the magnitude of Burgers vector was changed because different type of dislocation with greater magnitude of Burgers vector was activated in GS sample. It is well-known the pyramidal ⟨c + a⟩ dislocations have Burgers vectors significantly longer than basal or prismatic ⟨a⟩ dislocations, and hence are energetically less favored [22, 29]. For instance, in magnesium with c/a ratio of 1.6, the magnitude of pyramidal dislocation $1/3\langle 11\bar{2}3\rangle$ is about 1.9 times higher than basal or prismatic dislocation $1/3\langle 11\bar{2}0\rangle$. Accordingly, the formation of pyramidal dislocations seems the primary reason to make the physical activation volume in GS sample higher than those in GL and core samples.

Figure 23.6f displays the physical activation volumes plotted against true stress on log-log axes. Since the mean free path of dislocation-dislocation interaction has a relationship of [30]:

$$l \approx \frac{1}{\sqrt{\rho}}, \qquad (23.1)$$

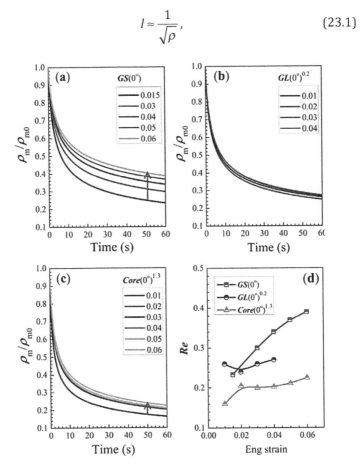

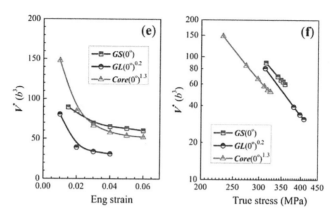

Figure 23.6 Higher mobile dislocation density and physical activation volume exceeding its constituent portions in gradient structured sample. Mobile dislocation decays at different engineering strains with time as compared with the initial values for (a) gradient structure, corresponding (b) gradient layer and (c) center core. Contrast of (d) mobile dislocation densities at the end of each relaxation and (e) physical activation volume for three structures at different engineering strains. (f) Physical activation volumes plotted against true stresses in log coordinate.

where ρ denotes dislocation density, and the stress depends on dislocation density according to:

$$\sigma = \alpha M \mu b \sqrt{\rho}, \quad (23.2)$$

where α, M and μ are constant, Taylor factor and shear modulus, the physical activation volume hence could be expressed as:

$$V^* \approx \alpha M \mu d b^2 / \sigma. \quad (23.3)$$

Therefore, for dislocation-dislocation interaction, the curves of V^* versus stress on log-log axes will exhibit linear relationship. Figure 23.6f shows the linear relationship for three different microstructures, manifests that corresponding plastic deformation was dominated by dislocation mechanisms.

23.3.4 Microstructure and Texture Observation

23.3.4.1 Gradient microstructure after SMAT

Figure 23.7 shows the microstructure at center area after SMAT. The IPF map in planar view is displayed in Fig. 23.7a, while the IPF plus IQ map is exhibited in Fig. 23.7b, where plenty of deformation bands are visible, manifesting the plastic deformation in even center area. It is different from previous investigations in respect to gradient microstructure for fcc and bcc metals in terms of resultant microstructure, which exhibited a deformation-free core sandwiched by GLs with thickness of about 200 μm [18, 23, 31]. Figure 23.7c is the corresponding PF with transversal direction (TD) in east and loading direction (LD) in north, demonstrating a strong (0001) texture, which is in line with the XRD result shown in Fig. 23.3e. Figure 23.7d and e display the BSE micrograph and IPF map in cross-sectional view with higher magnification than Fig. 23.7a and b. High density of extensive twins is clearly shown, and some of them are marked in Fig. 23.7d and e with symbols of **T**1, **T**2 and **T**3. Given the low stress level to activate extensive twinning (at the same level to basal slip [32]), the center core could be easy to deform plastically through deformation twinning mechanism during SMAT process.

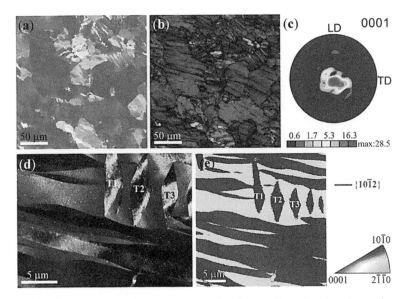

Figure 23.7 Micrographs at center section for gradient structured sample. (a) IPF and (b) IPF plus IQ maps in planar view. (c) Corresponding PF showing strong texture with c-axis parallel to normal direction. Cross-sectional views of deformed microstructures obtained through (d) BSE and (e) EBSD showing $\{10\bar{1}2\}$ deformation twins [with the same TD as that in (c)].

Figure 23.8 displays the microstructure beneath the treated surface of about 200 μm. Figure 23.8a–c shows three typical microstructures in cross-sectional view obtained by BSE scanning, where high density of deformation twins on different orientations are exhibited. Since different grains have different crystallographic orientations as compared with the macroscopic coordinate, the twinning propensity hence would be varied from grain to grain, then resulting in different twined microstructure as shown in Fig. 23.8a–c. The deformed microstructure was further examined by EBSD, as shown in Fig. 23.8–e, which exhibit more severely deformed microstructure than that in center area. Figure 23.8f shows the corresponding PF, which in alignment with the XRD investigations, exhibits a clear divergence of orientations from the center point.

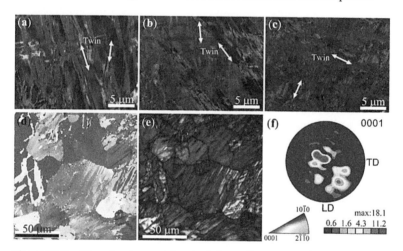

Figure 23.8 Microstructures at depth of about 200 μm away from the treated surface. (a–c) BSE micrographs in cross-sectional view showing different twin spacing and directions in different grains. EBSD maps of (d) IPF and (e) IPF plus IQ in planar view, and (f) corresponding (0001) PF.

Figure 23.9 displays the microstructure beneath the surface of about 50 μm. Figure 23.9a is a BSE micrograph, showing a dramatically deformed microstructure. With TEM observation, as shown in Fig. 23.9b, the grains of about 200 nm are formed. Meanwhile, some twins with thickness less than 100 nm are observable. Within these grains, high density of dislocation was produced due to severely plastic deformation during SMAT.

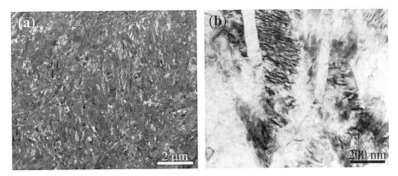

Figure 23.9 Microstructure at the depth of about 50 μm away from the treated surface. (a) The micrograph obtained via BSE in cross-sectional view. (b) Bright-field TEM micrograph in planar view.

23.3.4.2 Texture change during tensile deformation

Since the interaction between different layers along thickness is a collective behavior, the texture observation together with previously mentioned stress relaxation was used to analyze the collective change in microstructure and dislocation behavior. According to the relationship between crystal rotation and dislocation slip, the rotation axis can be determined by Taylor axis through the equation of:

$$T_s = n_s \times d_s, \qquad (23.4)$$

where T_s, n_s and d_s are Taylor axis, slip plane normal and slip direction, respectively. The Taylor axes have been calculated in many investigations and used to analyze which slip mode was activated primarily [19–21]. Based on this theory, the prismatic $\langle a \rangle$ slip produces a rotation axis of $\langle 0001 \rangle$, while basal $\langle a \rangle$ and pyramidal $\langle c + a \rangle$ have the same as their Taylor axes. Figure 23.10 shows the pole figures of gradient layer and center core before and after tensile deformation. A contraction of contour along LD is observed in (0001) pole figure for gradient layer, revealing a rotation history around TD. Given the inclined orientation in this layer as shown in Fig. 23.10a, only Taylor axis of could be parallel with TD, which therefore manifests the activation of basal $\langle a \rangle$ or pyramidal $\langle c + a \rangle$ slips in gradient layer. In comparison, as shown in Fig. 23.10c and d, the pole figure of core sample shows a rotation of contour around ND, which has been evidenced nearly parallel with $\langle 0001 \rangle$

direction for core sample (Fig. 23.3). Hence, this rotation around ND demonstrates the activation of prismatic $\langle a \rangle$ slip in the core sample during tensile deformation.

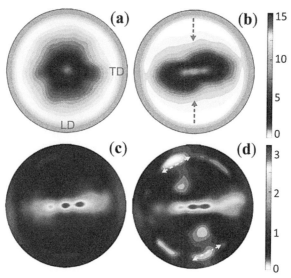

Figure 23.10 Deformation mechanisms of the standalone gradient layer and center core. (a) and (b) displaying the (0001) PFs on treated surface for standalone gradient layer before after tensile deformation. (c) and (d) showing ($10\bar{1}0$) PFs on its surface for corresponding core prior to and after tensile deformation. (All PFs plotted with TD in East and LD in North directions).

Figure 23.11 shows the PFs at different depths before and after tensile deformation. The left column displays the PFs before stretching, while the corresponding figures after deformation are exhibited in right column for comparison. A vertical contraction along LD is clearly seen in each figure on right column, demonstrating the occurrence of basal $\langle a \rangle$ or pyramidal $\langle c + a \rangle$ slips through the entirety of thickness during tensile deformation. Since the texture near center area is initially favorable to prismatic $\langle a \rangle$ slip, the occurrence of strong basal or pyramidal slip features manifests new slip mode was activated in dual gradient microstructure. According to the analysis before, considering the relaxation results and the favorable orientation, the mode should be the pyramidal slip. In addition, the maximum value of intensity at all depths has been increased after deformation, which is alignment with the vertical contraction due to a rotation towards the center point.

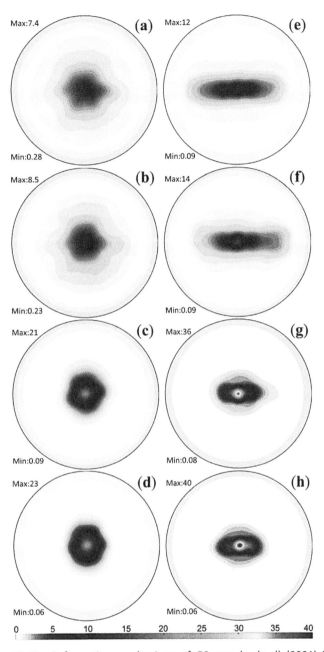

Figure 23.11 Deformation mechanisms of GS sample. (a–d) (0001) PFs at depth of 0 μm, 110 μm, 280 μm, 530 μm and 850 μm. Corresponding PFs after 5% elongation are shown in (e–h). (All PFs plotted with TD in East and LD in North directions).

23.3.4.3 Non-basal dislocation observation

Figure 23.12 displays the TEM micrographs for investigation of ⟨c + a⟩ dislocations in gradient microstructure. Figure 23.12a shows a bright-field micrograph in planer view at the depth of about 30 μm after 5% elongation. The areas enclosed by symbols of solid circle and asterisk in Fig. 23.12a are correspondingly shown in Fig. 23.12b and c, which were taken under weak-beam dark field conditions with diffraction vector of g = 0002 and, respectively.

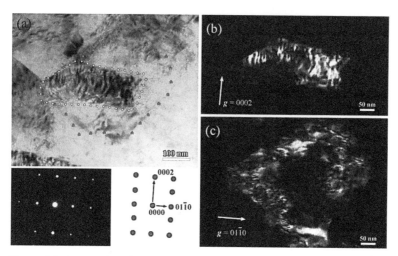

Figure 23.12 Non-basal slip in gradient structured sample after 5% tensile elongation. (a) TEM bright-field micrograph in normal plane at the depth of about 30 μm. Weak-beam dark field micrographs using diffraction vector of (b) g = 0002 and (c) g = $01\bar{1}0$.

When the diffraction vector is g = 0002, both basal and prismatic ⟨a⟩ dislocations are invisible or just exhibit extremely weak contrast, since the condition of $g \cdot b = 0$ will right set in for ⟨a⟩-type (basal or prismatic ⟨a⟩) dislocations having Burgers vector of $1/3\langle11\bar{2}0\rangle$ Accordingly, the dislocations shown in Fig. 23.12b are pyramidal ⟨c + a⟩ dislocations. If the diffraction vector $g = 01\bar{1}0$ is as shown in Fig. 23.12c, either the ⟨a⟩-type or the pyramidal ⟨c + a⟩ dislocation could be visible if the Burgers vector not vertical to $\langle01\bar{1}0\rangle$, such as basal $1/3\langle11\bar{2}0\rangle$, prismatic $1/3\langle1\bar{2}10\rangle$ and pyramidal $1/3\langle1\bar{2}13\rangle$ Therefore, TEM observation demonstrates the activation of pyramidal dislocations in gradient structure along with the activation of basal dislocations.

23.4 Discussion

23.4.1 Formation of Dual Gradient Microstructure

With respect to microstructure, the most important feature in GS magnesium alloy is the simultaneous formation of size and orientation gradients, which is called as dual gradient microstructure throughout this manuscript. Since the mechanical properties of magnesium alloys strongly depend on texture, the orientation gradient would lead to the transition of deformation mechanism from layer to layer. Together with the contribution of size gradient, the dual gradient microstructure might enable the magnesium alloy to be stretched with optimized mechanical properties.

The formation of size gradient during SMAT has been well investigated for large amount of fcc, bcc and hcp metals, which is mainly attributed from twinning and dislocation mechanisms. According to microstructure observations, the twinning mechanism have been activated through whole thickness during SMAT processing (Fig. 23.7 to Fig. 23.9), which plays an important role in refining the coarse-grained microstructure. Along with the formation and evolution of dislocation boundaries (Fig. 23.9), the size gradient microstructure was finally produced in magnesium alloy.

Meanwhile, as evidenced in Fig. 23.3, the orientation gradient was formed during SMAT processing. In comparison with the initial texture (Fig. 23.1b), the c-axes in center area have a change in orientation of about 90° from in-plane direction towards normal direction, as demonstrated by PFs in Fig. 23.3d and Fig. 23.7c. Given the prevalent $\{10\bar{1}2\}$ twinning in center area (Fig. 23.7d–e), it is naturally inferred that the change of orientation was resulted directly from the extensive twinning. That is because the extensive twinning will lead to a rotation of c-axes around $\langle 11\bar{2}0 \rangle$ for 86.3° [18], which is in alignment with the abrupt change of orientation after SMAT processing. Meanwhile, it should be noted that the initial texture in as-received sample is also favorable to extensive twinning, since the peening during SMAT process could generate a compressive stress vertical to $\langle 0001 \rangle$ direction, which is evidenced to be the easy mode to activate extensive twinning in magnesium alloys. Therefore, the extensive twinning prevalent in this study is the primary reason resulting in the strong (0001) texture with c-axes normal to plane. Afterwards, with the increasing of strain and stress amplitude near treated surface, the dislocation multiplication and interaction as well as subsequent dislocation boundaries would weaken the (0001) texture gradually, and thus results in the orientation gradient along thickness.

23.4.2 Influence of Grain Size on Deformation Mechanism

Towards the understanding of deformation mechanisms in this dual gradient microstructure, it is necessary to figure out the influence of grain size on the plastic deformation mechanisms. With regards to dislocation mechanism, the small grains are supposed to be less anisotropic in plastic deformation since the CRSS of prismatic or pyramidal slip is close to the same level of that for basal slip [13, 33]. It is still interesting and important to ask whether there is a critical size under which the non-basal slips are prevalent. For nanosized particle or tensile sample with single crystal structure, it was found that such critical size for magnesium is about 100 nm [13, 33]. However, in polycrystalline AZ31 magnesium alloy, Koike et al. have observed non-basal slips activated in grains with average diameter of 6 μm [4]. As supposed by the authors, compatibility across grain boundaries would raise the local stress and hence reduce the yield anisotropy substantially, leading to the activation of pyramidal slips to accommodate the possible incompatibilities. Therefore, the critical size for triggering non-basal slip depends on the chemical composition, local stress state and so on. Anyway, the refined microstructure of magnesium alloys is definitely beneficial to non-basal slips, which therefore improves the ability of uniform deformation.

In respect to twinning mechanism, it also depends on grain size for magnesium alloys. Broadly speaking, the extensive twinning is always activated in larger grains [14, 34]. No twinning was shown during micropillar compression for single crystal magnesium with diameters of several microns [35]. With the further decreasing of size, as indicated by pillar diameter, deformation twinning would reappear like what have been well observed in fcc metals. However, the critical size under which deformation twinning was observed is always less than 200 nm, and the critical stress for twinning nucleation is higher than 1GPa [16, 36]. Therefore, given that the most fractions of grains in dual gradient microstructure of this study is greater than that value (Fig. 23.2a), the deformation twinning during tensile deformation would be negligible, and thus the deformation mechanism is mostly related to dislocation slips.

Accordingly, for the dual gradient microstructure in present investigation, the plastic deformation is dominated by dislocation slip rather than deformation twinning. The grain refinement along thickness could play an important role in decreasing the yield anisotropy during tensile deformation, and hence enhance the propensity for pyramidal slips to accommodate deformation along c-axes.

23.4.3 Influence of Orientation and Its Gradient on Deformation Mechanism

Owing to the low-symmetric crystallographic structure for hcp metals, their deformation mechanisms activated during tensile deformation and the subsequent mechanical properties are closely dependent on the loading direction. In general, when stretched along or compressed vertical to c-axes, extensive twinning is the easiest mode to be activated. When deformed with loading inclined to c-axes, the basal $\langle a \rangle$ slip will be activated firstly. Additionally, the prismatic $\langle a \rangle$ slip will be triggered if the tensile loading has a direction vertical to c-axes [18].

In current study, the treated surface and center area differ clearly in texture, where c-axes tend to be tilted to normal direction near treated surface, but show a trend parallel with normal direction at center area. Therefore, in terms of plastic deformation mechanism, the gradient layer with more tilted c-axes would be more favorable to basal $\langle a \rangle$ slip, while the prismatic $\langle a \rangle$ slip will dominate the deformation mechanism in center core due to vertical c-axes to loading direction. The difference in dominant slip modes between gradient layer and center core has been evidenced through the investigation of pole figures (Fig. 23.10). The vertical contraction of contour towards center point in (0001) pole figure of gradient layer is attributed from crystal rotations along TD (Fig. 23.10b), which reveals the basal $\langle a \rangle$ or pyramidal $\langle c + a \rangle$ slip with Taylor axis of $\langle 10\bar{1}0 \rangle$. In contrast, the rotation around ND in $(10\bar{1}0)$ pole figure of core sample was observed (Fig. 23.10d), demonstrating the prismatic $\langle a \rangle$ slip with Taylor axis of $\langle 0001 \rangle$. Therefore, the dominant slip modes are exactly changed along thickness.

Meanwhile, the different dominant mechanisms between gradient layer and core due to orientation are also evidenced by the Hall-Petch relationship. According to the available data [37–42], as shown in Fig. 23.13, when grain size is greater than 1 μm, the lower slope of basal slip mechanism manifests that this deformation mechanism has lower grain size sensitivity with respect to yield strength [39]. The CG samples and the core of gradient structure have different deformation mechanisms due to different textures. While the grain size is less than 1 μm, the linear relationship begins to deviate and shows a lower slope, which has been ascribed to the transition of dominant deformation mechanism from prismatic to basal slip [42]. Therefore, it manifests that the plastic deformation mechanisms of different layers in gradient microstructure are changed with depth due to different orientations.

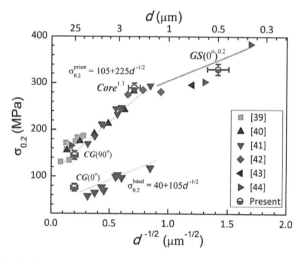

Figure 23.13 The yield stresses of different magnesium alloys plotted against grain size showing Hall-Petch relationship and corresponding deformation mechanisms.

For gradient microstructure, which could be viewed as a pile of layers with gradually changed size and orientation, the difference in deformation modes across layers will generate internal stress in material to remain continuous during plastic deformation. Similar to the grain boundary, as supposed by Koike et al., the compatible stress across layers will significantly improve the activation of non-basal slips [4]. Therefore, for current gradient microstructure (Fig. 23.3), the difference in orientation along thickness will definitely generate internal stresses to accommodate such anomalously dominant slip modes across layers, and hence, in agreement with the investigations of stress relaxation (Fig. 23.6), the pyramidal slip is activated.

23.4.4 Coupling between Size Gradient and Orientation Gradient

As demonstrated that size and orientation gradients exist together in magnesium alloy, it is interesting to ask how the two gradients affect to or couple with each other. The size-gradient microstructure has been evidenced that its strain hardening can be enhanced through the effects of strain gradient and HDI stress [18, 43]. Both the strain gradient and HDI stress effect are relatively insensitive to materials in respect to the triggering principle, because the strain gradient and subsequent stress-state change are caused by the propensity to instability for layers near to surface, while the HDI stress is resulted from the heterogeneous nature in strength along thickness. Therefore, such beneficial effects are still work in magnesium alloy with size-gradient structure.

Besides the above effects, since the microstructural size of magnesium alloy was greatly refined along the entirety of thickness, the yield stress anisotropy would be significantly reduced. Accordingly, the size-gradient microstructure has another effect that is to improve the probability in activating pyramidal slip. As compared with the basal $\langle a \rangle$ and prismatic $\langle a \rangle$ slips, the pyramidal slip is the only one that could provide accommodation for deformation along c-axis.

The pyramidal slip could not be guaranteed by size-gradient itself, since the most grains are still larger than the critical size to trigger isotropic deformation [13]. However, the orientation gradient could generate the internal stresses across layers along thickness, which can further reduce the stress anisotropy to trigger pyramidal slip. Under the collective effects, pyramidal slips can be smoothly activated and thus both dislocation density and interaction are enhanced. The pyramidal slip has been predicted by stress relaxation investigation (Fig. 23.6), further directly observed by TEM (Fig. 23.12). The analysis based on the pole figures has shown that the pyramidal slip might even occurs at the entirety of thickness, since the contraction feature remains to the center area (Fig. 23.11).

23.5 Conclusions

A dual gradient microstructure was obtained in AZ31 magnesium alloy through SMAT processing, where the microstructural size and crystallographic orientation change gradually along thickness. The deformation mechanisms and corresponding influence to mechanical properties are investigated. The main conclusions are drawn as follows:

(1) The size and orientation gradients were generated together in magnesium alloy with thickness of 1.7 mm. Attributed from the nucleation and interaction of $\{10\bar{1}2\}$ extensive twins and dislocations, the average microstructural size decreases from 2.5 μm at center area to about 200 nm near treated surface, while the c-axes gradually tilt from being parallel with normal direction towards inclined with it.

(2) The yield stress of dual gradient microstructure is increased by 4 times as compared with its as-received sample, and 6.5% uniform elongation is retained at the same time. Meanwhile, the diffusive necking appears in dual gradient sample, leading to 5% post-uniform elongation, which is always negligible in magnesium alloys. Furthermore, the ductility of dual gradient microstructure is higher than that of its constituents, i.e., the gradient layer and corresponding core.

(3) The size refinement through the entirety of thickness significantly reduces the yield stress anisotropy for pyramidal slip, and hence improves its propensity to be activated during tensile deformation. The pyramidal slip will then provide a component vector along c-axis, accommodating the incompatibility from basal and prismatic $\langle a \rangle$ slips, and thus make the tensile deformation more uniformly.

(4) The dominant deformation modes change gradually from basal to prismatic slip from treated surface to center area due to orientation gradient. This orientation gradient will raise internal stresses across layers along thickness, which could further lower down the yield stress anisotropy for pyramidal slip besides the contribution of size refinement. The pyramidal slip has been activated along the entirety of thickness. Together with strain gradient and HDI stress effects, the dual gradient microstructure shows a well combination between strength and ductility.

Acknowledgements

The authors would like to appreciate in advance for the financial support by the National Natural Science Foundation of China (NSFC) (Grant nos. 51301187, 11572328, 11222224 and 11472286) and the 973 Programs (Grant nos. 2012CB932203, 2012CB937500 and 6138504).

References

1. Williams JC, Starke Jr EA, *Acta Mater*, 51 (2003) 5775–5799.
2. Koch CC, *Nanostructured Materials: Processing, Properties and Applications*, William Andrew, 2006.
3. Nastasi M, Parkin DM, Gleiter H, *Mechanical Properties and Deformation Behavior of Materials having Ultra-Fine Microstructures*, Springer Science & Business Media, 2012.
4. Koike J, Kobayashi T, Mukai T, Watanabe H, Suzuki M, Maruyama K, Higashi K, *Acta Mater*, **51** (2003) 2055–2065.
5. Agnew SR, Duygulu Ö, *Int J Plast*, **21** (2005) 1161–1193.
6. Máthis K, Nyilas K, Axt A, Dragomir-Cernatescu I, Ungár T, Lukáč P, *Acta Mater*, **52** (2004) 2889–2894.
7. Al-Samman T, Li X, Chowdhury SG, *Mater Sci Eng A*, **527** (2010) 3450–3463.
8. Li B, Ma E, *Phys Rev Lett*, **103** (2009) 035503.
9. Proust G, Tomé CN, Jain A, Agnew SR, *Int J Plast*, **25** (2009) 861–880.
10. Barnett M, *Mater Sci Eng A*, **464** (2007) 8–16.
11. Ando S, Tonda H, *Mater Trans JIM*, **41** (2000) 1188–1191.

12. Al-Samman T, *Acta Mater*, **57** (2009) 2229–2242.
13. Yu Q, Qi L, Mishra RK, Li J, Minor AM, *Proc Natl Acad Sci USA*, **110** (2013) 13289–13293.
14. Sharon JA, Zhang Y, Mompiou F, Legros M, Hemker KJ, *Scr Mater*, **75** (2014) 10–13.
15. Kang F, Liu JQ, Wang JT, Zhao X, *Scr Mater*, **61** (2009) 844–847.
16. Wei Y, Li Y, Zhu L, Liu Y, Lei X, Wang G, Wu Y, Mi Z, Liu J, Wang H, *Nat Commun*, **5** (2014).
17. Fang T, Li W, Tao N, Lu K, *Science*, **331** (2011) 1587–1590.
18. Wu X, Jiang P, Chen L, Yuan F, Zhu YT, *Proc Natl Acad Sci USA*, **111** (2014) 7197–7201.
19. Wu X, Jiang P, Chen L, Zhang J, Yuan F, Zhu Y, *Mater Res Lett*, **2** (2014) 185–191.
20. Moering J, Ma X, Malkin J, Yang M, Zhu Y, Mathaudhu S, *Scr Mater*, **122** (2016) 106–109.
21. Wu X, Yang M, Yuan F, Wu G, Wei Y, Huang X, Zhu Y, *Proc Natl Acad Sci USA*, **112** (2015) 14501–14505.
22. Barnett MR, Jacob S, Gerard BF, Mullins JG, *Scr Mater*, **59** (2008) 1035–1038.
23. Wu XL, Yang MX, Yuan FP, Chen L, Zhu YT, *Acta Mater*, **112** (2016) 337–346.
24. Ma E, Wang Y, Lu Q, Sui M, Lu L, Lu K, *Appl Phys Lett*, **85** (2004) 4932–4934.
25. Kocks U, Mecking H, *Prog Mater Sci*, **48** (2003) 171–273.
26. Wang YM, Hamza AV, Ma E, *Appl Phys Lett*, **86** (2005) 241917.
27. Chen L, Yuan F, Jiang P, Xie J, Wu X, *Mater Sci Eng A*, **618** (2014) 563–571.
28. Caillard D, Martin J-L, *Thermally Activated Mechanisms in Crystal Plasticity*, Elsevier, 2003.
29. Wu Z, Curtin W, *Nature*, **526** (2015) 62–67.
30. Mecking H, Kocks UF, *Acta Mater*, **29** (1981) 1865–1875.
31. Lu K, Lu J, *Mater Sci Eng A*, **375** (2004) 38–45.
32. Sánchez-Martín R, Pérez-Prado M, Segurado J, Molina-Aldareguia J, *Acta Mater*, **93** (2015) 114–128.
33. Hutchinson WB, Barnett MR, *Scr Mater*, **63** (2010) 737–740.
34. Tsai MS, Chang CP, *Mater Sci Technol*, **29** (2013) 759–763.

35. Lilleodden E, *Scr Mater*, **62** (2010) 532–535.
36. Ye J, Mishra RK, Sachdev AK, Minor AM, *Scr Mater*, **64** (2011) 292–295.
37. Jain A, Duygulu O, Brown DW, Tomé CN, Agnew SR, *Mater Sci Eng A*, **486** (2008) 545–555.
38. del Valle JA, Carreño F, Ruano OA, *Acta Mater*, **54** (2006) 4247–4259.
39. Yuan W, Panigrahi SK, Su JQ, Mishra RS, *Scr Mater*, **65** (2011) 994–997.
40. Bhargava G, Yuan W, Webb S, Mishra RS, *Metall Mater Trans A*, **41** (2010) 13–17.
41. Yuan W, Mishra RS, Carlson B, Mishra RK, Verma R, Kubic R, *Scr Mater*, **64** (2011) 580–583.
42. Razavi SM, Foley DC, Karaman I, Hartwig KT, Duygulu O, Kecskes LJ, Mathaudhu SN, Hammond VH, *Scr Mater*, **67** (2012) 439–442.
43. Yang M, Pan Y, Yuan F, Zhu Y, Wu X, *Mater Res Lett*, (2016) 1–7.

Chapter 24

The Evolution of Strain Gradient and Anisotropy in Gradient-Structured Metal

Xiangde Bian,[a,b] **Fuping Yuan,**[a,b] **Xiaolei Wu,**[a,b] **and Yuntian Zhu**[c,d]

[a]*State Key Laboratory of Nonlinear Mechanics, Institute of Mechanics, Chinese Academy of Science, Beijing 100190, China*
[b]*School of Engineering Science, University of Chinese Academy of Sciences, Beijing 100190, China*
[c]*Nano Structural Materials Center, School of Materials Science and Engineering, Nanjing University of Science and Technology, Nanjing, Jiangsu 210094, China*
[d]*Department of Materials Science and Engineering, North Carolina State University, Raleigh, NC 27695, USA*
fpyuan@lnm.imech.ac.cn, bianxiangde@imech.ac.cn

Gradient structured metals have been reported to possess superior mechanical properties, which were attributed to their mechanical heterogeneity. Here we report in-situ observation of evolutions of strain gradient and anisotropy during tensile testing of a gradient

---

Reprinted from *Metall. Mater. Trans. A*, **48**(9), 3951–3960, 2017.

structured metal. Strain gradients and anisotropy in the lateral directions were observed to increase with increasing applied tensile strain. In addition, the equivalent Poisson's ratio showed gradient, which evolved with applied strain. The gradient structure produced higher deformation anisotropy than coarse-grained homogeneous structure, and the anisotropy increased with increasing tensile strain. The strain gradient and anisotropy resulted in strong heterodeformation induced (HDI) hardening, large strain gradients, and high density of geometrically necessary dislocations, which helped with increasing the ductility.

24.1 Introduction

It has been a challenge to produce both high strength and high ductility in metals and alloys [1–5]. The strength can be easily enhanced by well-known strategies such as grain refinement, work hardening, solution strengthening, second-phase particle strengthening, deformation twins, etc. [6–11]. However, high strength is usually accompanied with the sacrifice of ductility in homogeneous metals. For example, ultrafine grained (UFG) or nanostructured (NS) metals obtained by severe plastic deformation (SPD) can have strengths an order of magnitude higher than those of their coarse-grained (CG) counterparts. However, they usually show limited strain hardening and very low (near zero) uniform elongation [6, 7, 12, 13].

Recently, several promising strategies for achieving simultaneous high strength and high ductility have been proposed by tailoring microstructures through heterogeneous and/or hierarchical structures [2–5, 14–29]. Among them, the gradient structure, where the grain size or the substructure size changes gradually along the depth [5, 20–26], has great potential in engineering applications due to their superior combinations of strength and ductility. The tensile properties and underlying deformation mechanisms of gradient structured metals have been reported recently [5, 20, 21, 23–26]. For example, gradient structured Cu was reported 100% stronger than CG Cu, while retaining its ductility [20], which was attributed to the growth of nanocrystalline layer due to the low thermal stability of Cu. In gradient structured metals with stable microstructures the high ductility was attributed to the presence of strain gradient together with stress-state change, which promotes dislocation interactions, and generation of geometrically necessary dislocations (GNDs) [21, 24, 30].

In addition to forest dislocation hardening, hetero-deformation induced (HDI) strengthening and HDI work hardening have also been reported to play an important role in the deformation of heterostructures, such as the gradient structure and heterostructured lamella structure [19, 24, 28]. For the gradient structure, the onset of plastic deformation occurs in soft CG core first, and then propagate gradually to the hard surface layer with increasing applied tensile strain [5, 20]. This progressive yielding of gradient layers results in gradient plastic strain in the of the gradient structured sample, as revealed recently by a crystal-plasticity finite element modeling [25]. The plastic strain gradient needs to be accommodated by geometrically necessary dislocations, which in turn produces significant hetero-deformation induced (HDI) [19, 24, 28, 31–35]. This can simultaneously increase strength as well as ductility.

The gradient structures can be regarded as consisting of numerous thin layers with systematically varying mechanical behaviors. Therefore, mechanical incompatibilities exist between neighboring layers during tensile testing [5, 21, 24, 25], which produces strain gradient and enhanced mechanical properties. However, how the strain gradient and anisotropy are generated and developed during tensile loading is poorly understood, because of the lack of experimental study. It is of critical importance to investigate the evolution of strain gradient and anisotropy in gradient structured materials to better understand the scientific principle behind their observed unique tensile behavior and superior mechanical properties.

In this study, quasi-static tensile testing was coupled with in-situ 3D strain mapping on lateral sample surfaces using digital image correlation (DIC) to investigate the evolutions of strain gradient and anisotropy. Loading-unloading-reloading (LUR) tests were conducted to study the HDI stress evolution. It was revealed for the first time that strain gradient and anisotropy increased with increasing applied strain, indicating that HDI work hardening associated with increasing GND density.

24.2 Materials and Experimental Procedures

24.2.1 Materials

A TWIP steel was used in this work to study the evolutions of strain gradient and anisotropy in the gradient structured sample under tensile load. The steel has a chemical composition of 0.6C, 23Mn, 0.035Nb, 0.035Ti, and the balance of Fe (all in wt %). The initial material was first melted in an induction furnace under the protection of Ar atmosphere and then cast into an ingot with dimensions of 750 × 270 × 170 mm. The ingot was homogenized at 1423 K (1150°C) for 2 h, and then hot-forged between 1423 K (1150°C) and 1173 K (900°C) to form slabs with a thickness of 20 mm. Gradient structures were produced by surface mechanical attrition treatment (SMAT). The disks for SMAT were cut from the slab by wire saw into dimensions of 70 × 50 × 3 mm. These disks were fully austenized at 973 K (700°C) for 1 h followed by immediate quenching in water. This produced samples with a coarse-grained (CG) microstructure with an average grain size of about 15 μm (Fig. 24.1d). Both sides of the sample plates were SMAT-processed by high velocity steel balls (4 mm in diameter) for 30 min. More information on SMAT can be found elsewhere [36].

24.2.2 Microstructural Characterization

Back-scattered electron (BSE) imaging, electron backscattered diffraction (EBSD) and transmission electron microscopy (TEM) were used to study the microstructures and their gradients. The samples for BSE and EBSD were first grinded to 2000 grit with sandpapers, and then polished with a 0.05 μm SiO_2 aqueous suspension, followed by electro-polishing in a solution of 10% $HCLO_4$ and 90% alcohol at 22 V and 253 K (−20°C). Thin disks with thickness of about 300 μm were prepared for TEM observations, and then mechanically ground to about 50 μm, followed by a twin-jet polishing using a solution of 5% perchloric acid and 95% ethanol at 22 V and 253 K (−20°C). X-ray diffraction (XRD) measurements were performed on polished samples to obtain the phase and texture information using a Philips Xpert X-ray diffractometer with Cu Kα radiation. The distribution of microhardness along the sample depth was obtained on the polished cross-section using a Vickers diamond indenter under a load of 10 g for 15 s dwell time. For each depth, five groups of measurements were made.

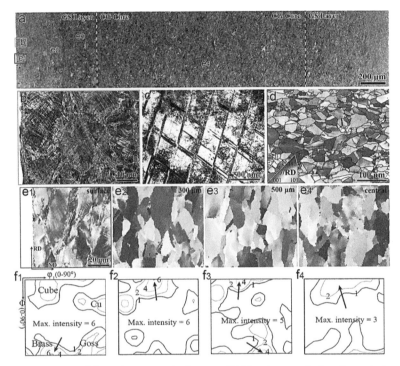

Figure 24.1 (a) Back-scattered electron (BSE) micrograph for the whole gradient structure; (b) BSE micrograph at the treated surface; (c) TEM image at the depth of 50 µm from the treated surface; (d) EBSD (IPF) image for the untreated CG sample; (e) EBSD (IPF) images at various depths for the gradient structure: (e1) surface, (e2) 300 µm, (e3) 500 µm, (e4) center; (f) Texture and intensity at various depths for the gradient structure: (f1) surface, (f2) 300 µm, (f3) 500 µm, (f4) center.

24.2.3 Quasi-Static Uniaxial Tensile Tests Coupled with Digital Image Correlation

Dog-bone samples, with a thickness of about 3 mm, a gauge length of 15 mm and a width of 4 mm, were used for quasi-static tensile tests and loading-unloading-reloading (LUR) tests. These tensile tests and LUR tests were conducted at a strain rate of $5 \times 10^{-4}\ \mathrm{s}^{-1}$ at room temperature under displacement control using an Instron

5565 testing machine. The tensile, thickness and width directions are designated as the y, z, and x directions, respectively, as shown in Fig. 24.2c. Strain contours were measured using digital image correlation (DIC) during tensile tests. A commercial software, ARAMIS®, was used to analyze the DIC data. Initial high-contrast stochastic spot patterns on both the treated surface (xy plane) and the lateral surface (yz plane) were created, in which a background of flexible, adhesive and matte white paint was sprayed first, then a fine layer of black paint spots (with size of about 10 μm) was sprayed onto the white background. The evolution of the spot patterns was recorded using two 1.2 MPx digital CCD cameras at a rate of 1 frame per second to calculate the contours for the in-plane strain and the out-plane height profile for both the treated surface and the lateral surface. The facet size for the strain calculation using DIC method was 50 μm.

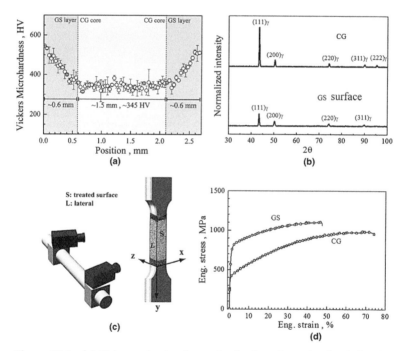

Figure 24.2 (a) Vickers microhardness distribution along the depth for the gradient structure; (b) XRD curves for the untreated CG sample and the surface after SMAT; (c) Illustration of experimental set-up for the tensile tests coupled with DIC; (d) Engineering stress–strain curves for the CG and the gradient structure.

24.3 Experimental Results and Discussions

24.3.1 Microstructural Characterization and Tensile Properties

Figure 24.1 shows the microstructures for the gradient structure after SMAT were characterized. Figure 24.1a shows the BSE image for the whole gradient structure. The close-up BSE image of the treated surface is displayed in Fig. 24.1b. Figure 24.1c shows the TEM image at the depth of 50 μm from the treated surface. The EBSD (IPF) images, textures and intensities at various depths for the gradient structure are shown in Figs. 24.1e1–e4 and 24.1f1–f4, respectively. It is observed that the grain itself is not refined and the grain size is similar at various depths (about 15 μm), while substantial substructures with much refined size in the interior of grains are formed on the treated surface. These substructures are identified as single deformation twins or multiple deformation twins from the TEM image (Fig. 24.1c clearly shows multiple twins from two twinning systems). Twin boundary spacings are less than 300 nm. High density of dislocations can also be found near twin boundaries. Such microstructure refinement by deformation twins and dislocation tangles result in the reduction of dislocation mean free path in the grain interior, rendering an obvious hardness increase at the treated surface. Previous research [16] also revealed that hierarchical deformation twins can produce higher strength/hardness than a single level of primary deformation twins. Moreover, the gradient structure shows weak texture, as indicated from the texture intensity maps (the maximum intensity is less than 6) at various depths in Figs. 24.1e1–e4. Thus the texture effect on the deformation anisotropy is not considered in the present work.

Figure 24.2a shows the distribution of microhardness along the depth from the sample surface after SMAT for 30 min. The 1.5-mm CG core is sandwiched between two 0.6-mm gradient layers. As shown, the microhardness decreases almost linearly from about 550 Hv at the surface to about 345 Hv at the center. This microhardness gradient is caused by the gradient twinned structure. Figure 24.2b displays the XRD curves from the untreated CG sample and the SMAT-processed sample surface, which clearly indicates that there is no phase transformation during the SMAT process. The engineering stress–strain curves for the untreated CG sample and the gradient structured sample are displayed in Fig. 24.2d. The CG sample has a yield strength of about 400 MPa and a uniform elongation of about 65%. As shown, the gradient structured sample achieved a superior combination of strength and ductility: its yield strength is about 2 times that of the CG sample and its uniform elongation still retained about 70% of that of the CG sample. Our previous paper [21] attributed these outstanding tensile properties to the extra strain hardening by the strain gradient and the stress state change due to the mechanical incompatibility. In order to provide better understanding for the detailed mechanisms underlying the observed tensile properties for the gradient structured sample, how the strain heterogeneity and gradient are generated and developed in the gradient structure will be delineated in the following sections.

24.3.2 Strain Contours and Strain Distributions along the Depth

As mentioned earlier, the in-plane strain contours and the contours for the out-plane height profile for both the treated surface and the lateral surface were measured by the 3D DIC techniques. The distribution of strain ε_x on the lateral surface was calculated using the following equation: $\varepsilon_x = \varepsilon_x^{surface} + 2\dfrac{\Delta H}{b_0}$, where $\varepsilon_x^{surface}$ is the average strain along the x direction measured from the treated surface, ΔH is the height difference (in x direction) between each depth and the surface layer measured from the lateral surface, and b_0 is the width of the samples.

Strain measurements for the untreated CG sample are shown in Fig. 24.3. Contours of ε_z and contours of height profiles for the lateral surface at various tensile strains are displayed in Figs. 24.3a and b, respectively. As indicated from the contours, the deformation was not uniform microscopically even in the homogenous CG structure, this is due to the variation in grain size and orientation in the CG sample. Taking the average value along the y direction for the contours in Fig. 24.3a, ε_z distributions along the depth at varying tensile strains are shown in Fig. 24.3c. Similarly, height profile distributions along the depth at various tensile strains can be obtained based on the contours in Fig. 24.3b, and are displayed in

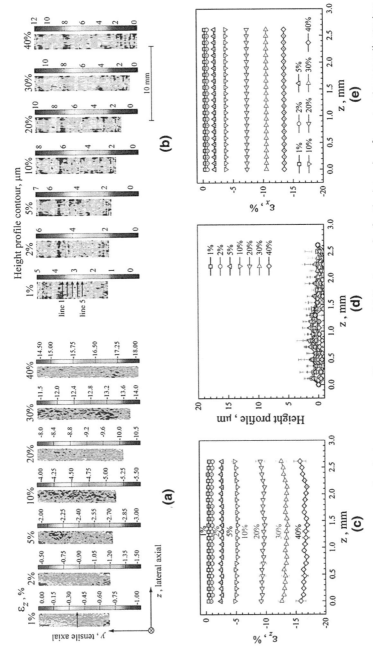

Figure 24.3 Strain measurements for the homogeneous CG structure. (a) Contours of ε_z for the lateral surface at various tensile strains; (b) contours of height profiles for the lateral surface at various tensile strains; (c) ε_z distributions along the depth at various tensile strains; (d) height profile distributions along the depth at various tensile strains; (e) ε_x distributions along the depth at various tensile strains.

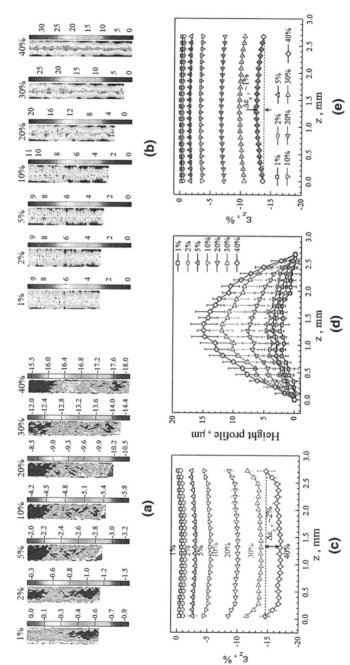

Figure 24.4 Strain measurements for the gradient structure. (a) Contours of ε_z for the gradient structure. (a) Contours of height profiles for the lateral surface at various tensile strains; (b) contours of ε_z for the lateral surface at various tensile strains; (c) ε_z distributions along the depth at various tensile strains; (d) height profile distributions along the depth at various tensile strains; (e) ε_x distributions along the depth at various tensile strains.

Fig. 24.3d. Then the ε_x distributions along the depth at varying tensile strains can be calculated by the aforementioned equation and are shown in Fig. 24.3e. Although microscopic heterogeneity is observed in the strain contours, relatively homogeneous deformation along the depth is observed for the CG structure when taking the average value along the y direction for each layer. As indicated by Figs. 24.3c and e, the lateral deformations (ε_x and ε_z) are nearly homogeneous along the depth for the untreated CG sample, this means that no deformation incompatibility and mechanical interactions exist among different layers at various depths for the homogeneous CG structure.

Figure 24.4 presents the evolution of strains in the gradient structured sample using the same analysis procedure as for the Fig. 24.3. Again, Figs. 24.4a and b show the contours of ε_z and the height profiles of the lateral surface at various tensile strains, respectively. Figures 24.4c–e display ε_z, height profile, and ε_x distributions along the thickness at various tensile strains. As shown in Figs. 24.4c and e, both ε_x and ε_z have gradients along the depth. The strain in the width direction, ε_x, is higher at the treated surface than the center part, while the strain in the thickness direction, ε_z, shows the opposite trend.

24.3.3 Evolutions of Strain Gradient and Anisotropy

The evolution of differences in lateral strains at the various depths from that at the central part, $\Delta\varepsilon_x$ and $\Delta\varepsilon_z$, are shown in Figs. 24.5a and b, respectively. Figures 24.5c and d show the normalized lateral strains ($\varepsilon_x/\varepsilon_y$ and $\varepsilon_z/\varepsilon_y$) as a function of the applied tensile strain at various depths for the gradient structured samples. The corresponding curves for the untreated CG sample are also included in Figs. 24.5c and d for comparison. As indicated in Figs. 24.5c and d, the lateral strain ε_x at various depths for the gradient structured sample is always smaller than that for the untreated CG sample, while the lateral strain ε_z at various depths for the gradient structure is always larger than that for the untreated CG sample. These results indicate that the gradient structure would cause higher anisotropy at the two lateral directions than the homogeneous CG structure. This phenomenon will be discussed further later.

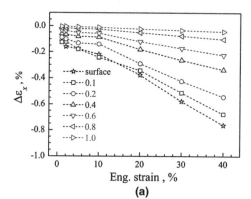

(a)

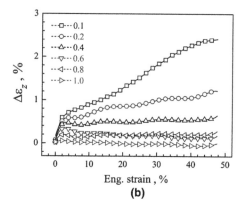

(b)

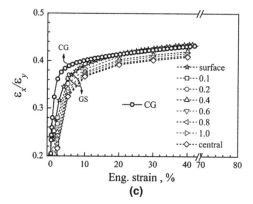

(c)

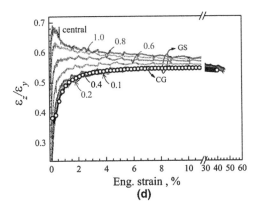

Figure 24.5 (a) $\Delta\varepsilon_x$, the difference of ε_x at the various depths and ε_x at the central part as a function of applied tensile strain for the gradient structured sample; (b) $\Delta\varepsilon_z$, the difference of ε_z at the various depths and ε_x at the central part as a function of applied tensile strain for the gradient structured sample; (c) $\varepsilon_x/\varepsilon_y$ as a function of applied tensile strain at various depths for the gradient structured sample; (d) $\varepsilon_z/\varepsilon_y$ as a function of applied tensile strain at various depths for the gradient structured sample.

Figure 24.5 reveals that strain gradient starts at the beginning of the tensile deformation in the gradient structured sample. The mechanical incompatibilities along the lateral directions should cause constraints and interactions among different layers, and this will lead to lateral stresses in both lateral directions, resulting in a stress state change from the uniaxial stress state to the multiaxial stress state [21]. As shown in Figs. 24.5a and b, the absolute magnitudes of the strain gradients along the depth $\Delta\varepsilon_x$ and $\Delta\varepsilon_z$ increase with increasing applied tensile strain. This means that the constraints and interactions between different layers became stronger and stronger with tensile loading. These observations can be attributed to the gradient microstructures and the corresponding gradient mechanical properties along the depth. These strain gradients are in contrast to the well-known strain-gradient plasticity induced by imposing a non-uniform deformation such as bending [37], torsion [38] and indentation [39]. However, both situations can produce long-range strain gradients, which are accommodated by the GNDs [40].

For materials with anisotropic properties, the strains for two lateral directions are generally different (y is the tensile direction):

$\varepsilon_x \neq \varepsilon_z$. It should be noted that the ε_x, ε_y, and ε_z in this chapter are all true strains. In order to compare the lateral strains of the gradient structured sample with those of isotropic materials, an equivalent lateral strain and an equivalent apparent Poisson's ratio are defined as following for anisotropic materials:

$$\bar{\varepsilon}_{lateral} = \ln\frac{\sqrt{A}}{\sqrt{A_0}} = \frac{1}{2}\ln\frac{A}{A_0} = \frac{1}{2}\ln\left(\frac{l_x}{l_{x0}}\frac{l_z}{l_{z0}}\right) = \frac{1}{2}(\varepsilon_x + \varepsilon_z) \quad (24.1)$$

$$\bar{v} = -\frac{\Delta\bar{\varepsilon}_{lateral}}{\Delta\varepsilon_y} = -\frac{1/2(\Delta\varepsilon_x + \Delta\varepsilon_z)}{\Delta\varepsilon_y} = \frac{1}{2}(v_x + v_z) \quad (24.2)$$

As mentioned earlier, the lateral strain ε_x (width direction) is higher at the treated surface than in the center part, while the lateral strain ε_z (thickness direction) shows the opposite trend. These observations make us wonder whether the equivalent lateral strain $\bar{\varepsilon}_{lateral}$ has gradient along the depth. Based on Eq. (24.1), the evolution of $\Delta\bar{\varepsilon}_{lateral} = \bar{\varepsilon}_{various\ depths} - \bar{\varepsilon}_{centeral}$ at the various depths as a function of applied tensile strain for the gradient structured sample are displayed in Fig. 24.6, which shows that the equivalent lateral strain also has significant gradient along the depth. Moreover, this gradient for the equivalent lateral strain also starts from the beginning of the tensile deformation, and increases with increasing applied tensile strain, especially at the areas close the treated surface.

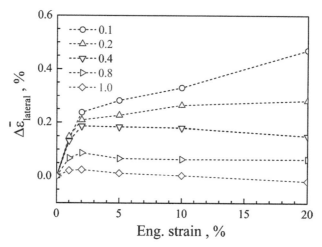

Figure 24.6 The evolution of $\Delta\bar{\varepsilon}_{lateral} = \bar{\varepsilon}_{various\ depths} - \bar{\varepsilon}_{centeral}$ at various depths as a function of applied tensile strain for the gradient structure.

It is well known that the apparent Poisson's ratio remains essentially constant at the elastic range, then gradually increases at the elasto-plastic transition stage, finally approaches, but seldom actually reaches, the values of 1/2 (which is tantamount to the assumption of constant volume) [41]. Variation of the apparent Poisson's ratio with increasing applied tensile strain affects the evolution of cross-sectional area, and this in turn has direct influence on plastic tensile instability. For "perfect" homogeneous materials, the evolution of the apparent Poisson's ratio during the elasto-plastic transition can be described as [41]:

$$\upsilon = \frac{1}{2} - \left(\frac{1}{2} - \upsilon_e\right)\left(\frac{E_S}{E}\right) \qquad (24.3)$$

where υ_e is the Poisson's ratio for elastic stage, E is the elastic modulus and E_S is the tangent modulus at the related strain, i.e. the slope of the stress–strain curve.

In order to better understand the deformation physics of the gradient structure, especially at the elasto-plastic transition stage, the evolution of v_x and v_z at various depths for the gradient structured sample, and the data for the untreated homogeneous CG structure, are provided in Figs. 24.7a and b, respectively. As shown, there is a strong anisotropy for the two lateral apparent Poisson's ratios. In other words, v_x and v_z evolves very differently with increasing applied strain.

Disregarding the anisotropy, the equivalent apparent Poisson's ratio $\left(\bar{\upsilon} = \frac{1}{2}(\upsilon_x + \upsilon_z)\right)$ at various depths for the gradient structured sample and for the untreated homogeneous CG sample is displayed in Fig. 24.7c, which shows a strong gradient along the depth for the equivalent Poisson's ratio in the gradient structured sample. This indicates that the strain gradient occurred during the entire tensile loading, which is consistent with our earlier report [21]. At the plastic deformation stage, the equivalent apparent Poisson's ratio of the gradient structured sample is slightly higher than those of the untreated homogeneous CG sample. This indicates that the reduction of the cross-section area is faster in the gradient structure than the untreated homogeneous CG structure at the plastic deformation stage. This observation is in good agreement with the fact that the strain hardening is weaker and the necking instability tendency is higher in the gradient structured sample than in the untreated homogeneous CG sample.

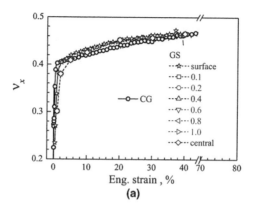

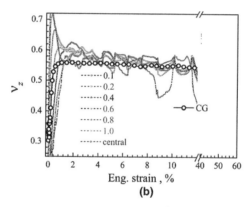

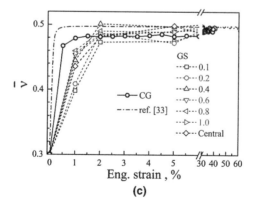

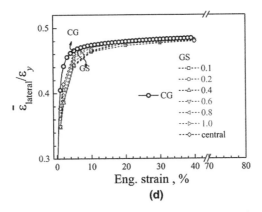

Figure 24.7 (a) v_x as a function of applied tensile strain at various depths for the gradient structured sample; (b) v_z as a function of applied tensile strain at various depths for the gradient structured sample; (c) the equivalent Poisson's ratio $\left(\bar{v} = \frac{1}{2}(v_x + v_z)\right)$ as a function of applied tensile strain at various depths for the gradient structured sample; (d) $\bar{\varepsilon}_{lateral}/\varepsilon_y$ as a function of applied tensile strain at various depths for the gradient structured sample. The theoretical prediction from $v = \frac{1}{2} - \left(\frac{1}{2} - v_e\right)\left(\frac{E_s}{E}\right)$ for the homogeneous CG structure is also provided in (c) for comparison.

In order to reveal the accumulated gradient for the equivalent lateral strain and its evolution at the small strain range, the normalized equivalent lateral strain ($\bar{\varepsilon}_{lateral}/\varepsilon_y$) at various depths as a function of applied tensile strain is shown in Fig. 24.7d. It is interesting to see from Figs. 24.7c and d that the duration of the elasto-plastic transition stage is short for the homogeneous CG sample, which agrees well with the theoretical prediction [41]. While the duration of elasto-plastic transition stage is much longer in the gradient structure. In our previous work [35], a dual-phase steel with heterostructure also exhibited a long and explicit elasto-plastic transition, which is mainly attributed to the load and strain partitioning between the two phases. The observed extraordinary strain-hardening rate in the dual-phase steel has been attributed to the high HDI hardening from the strain incompatibility between the two phases, which produced strain gradient near the phase boundaries [35]. In the gradient structure, strain gradient also exists between different layers, thus the mechanical incompatibility resulted in longer duration of elasto-plastic transition, providing strong HDI hardening, particularly at this stage [24].

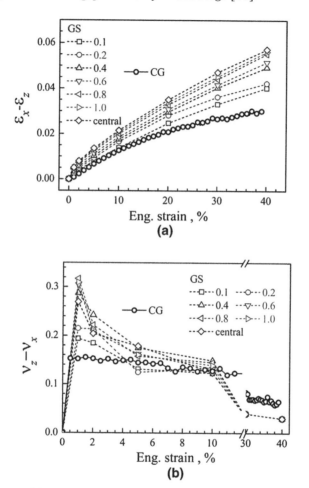

Figure 24.8 (a) ε_x-ε_z and (b) v_z-v_x as a function of applied tensile strain at various depths for the gradient structured and CG samples.

As mentioned earlier, deformation anisotropy in the two lateral directions is induced during the tensile loading in the gradient structure. In order to clearly illustrate this, ε_x-ε_z curves as a function of applied tensile strain at various depths are presented as the accumulated deformation anisotropy in Fig. 24.8a, while v_z-v_x curves are displayed as the instantaneous deformation anisotropy in

Fig. 24.8b. For comparison, the curves for the homogeneous CG sample are also provided in Fig. 24.8. As shown, the deformation anisotropy is higher in the gradient structure than in the homogeneous CG structure from the beginning of the tensile deformation and lasts all the way to the large applied tensile strain. The instantaneous deformation anisotropy is found to decrease while the accumulated deformation anisotropy is observed to increase with increasing applied tensile strain. The high anisotropy in lateral deformation could intensify the interaction and the constraint between different layers at various depths, resulting in higher lateral stresses to promote the operation of additional slip systems and consequently leads to more efficient dislocation storage [21].

24.3.4 HDI Hardening

The gradient structures can be considered consisting of numerous thin layers with varying flow behaviors, which produces mechanical incompatibilities between neighboring layers. This in turn produces HDI stress, which is a long-range stress field that hinder dislocation slip to accommodate the applied strain. As a result, higher flow stress is needed to overcome this stress field and HDI hardening is induced during the plastic deformation. This HDI hardening mechanism have been observed in our recent works on a Ti lamella structure [19], an IF steel gradient structure [24], and a copper/bronze laminate structure [28]. Figure 24.9a shows the true stress–strain curves for LUR tests on both the CG sample and the gradient structured sample, and the inset displays the close-up view on the details of the hysteresis loops. The HDI stress is the stress that drives the mobile dislocations to move reversely to produce unloading yield (σ_u is the unloading yield stress) during the unloading process ($\sigma_b = \sigma_u + \sigma_f$). During the reloading process, the applied stress needs to overcome the frictional stress (σ_f) and the HDI stress to drive the dislocation forward at the reloading yield point (σ_r is the reloading yield stress, $\sigma_r = \sigma_b + \sigma_f$). Thus, the HDI stress can be estimated by the average of the unloading yield stress and the reloading yield stress ($\sigma_b = (\sigma_u + \sigma_r)/2$) [24]. The HDI stress evolutions along with tensile loading (see Fig. 24.9b) can be calculated from the unloading-reloading curves at varying tensile strains using an equation proposed in our recent work [24]. It is found that the resulted HDI stress is much higher in the gradient structure than that in the CG structure, and the HDI hardening is more pronounced in the gradient structure, especially at the early stage of tensile deformation. This high HDI hardening rate at the early stage of tensile deformation for the gradient structure can be attributed to the elasto-plastic transition at this stage, at which the mechanical incompatibility is largest since the hard surface layers are still under elastic deformation while the soft center layer is already under plastic deformation. In the co-deformation stage, all layers are under plastic deformation and HDI hardening rate should become smaller for the gradient structure.

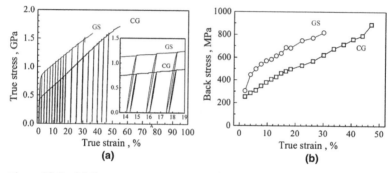

Figure 24.9 (a) True stress–strain curves of LUR tests for the CG structure and the gradient structure; (b) the calculated HDI stress at corresponding tensile strain levels.

It is proposed in the previous work [42, 43] that the HDI hardening in the CG structure can be understood by the development of internal stresses between the newly formed "hard" nanotwin lamellae and the "soft" untwined matrix during the tensile loading. While the mechanical incompatibility between different layers with varying flow behaviors in the gradient structure should contributes significantly to the HDI hardening, besides the HDI hardening originated from the interaction between the "soft" matrix and newly formed "hard" twins in the CG grains at the center of the gradient structure. Consequently, more GNDs accumulate at the boundaries of different layers, resulting in a higher observed HDI stress and a more pronounced HDI hardening in the gradient structure. Based on the current findings, the density of these GNDs due to the strain gradients increased with increasing applied strain to delay necking and to increase the ductility for the gradient structured sample.

24.4 Concluding Remarks

For the first time, the evolutions of the strain gradient and deformation anisotropy in a gradient structured sample during tensile loading were revealed by a series of quasi-static tensile tests, coupled with in-situ 3D strain contour measurements. The main findings are summarized below:

(1) Unlike the homogeneous deformation in the untreated CG sample, strain gradients in both lateral directions started to develop at the beginning of the tensile deformation and increased with increasing applied tensile strain, resulting in increasing GNDs to sustain the uniform deformation for the gradient structured sample.

(2) Deformation anisotropy is much higher in the gradient structure than in the homogeneous CG structure. The accumulated deformation anisotropy is observed to increase with increasing applied tensile strain. This could intensify the interaction and the constraint between different layers at various depths, resulting in higher lateral stresses to promote the operation of additional slip systems to help with dislocation storage.

(3) More pronounced HDI hardening is observed in the gradient structured samples than in the CG structured samples, due to the plastic incompatibility between different layers with varying mechanical behaviors in the gradient structure. The high HDI hardening can suppress necking to improve ductility in the gradient structured sample.

Acknowledgements

This work was financially supported by NSFC under grant Nos. 11472286, 11572328 and 11672313, the 973 Program of China under grant Nos. 2012CB932203 and 2012CB937500, and the Strategic Priority Research Program of the Chinese Academy of Sciences under grant No. XDB22040503. Y.Z. was supported by the US Army Research Office (W911 NF-12-1-0009), the US National Science Foundation (DMT-1104667), and the Nanjing University of Science and Technology. This chapter is dedicated to Prof. Koch's

TMS symposium 2017 (Mechanical Behavior of Advanced Materials). This symposium will honor the outstanding contributions of Prof. Carl C. Koch to many fields in materials science in the last 50 years and celebrate his 80th birthday.

References

1. Y.T. Zhu and X.Z. Liao: *Nat. Mater.*, 2004, vol. 3, pp. 351–352.
2. Y.M. Wang, M.W. Chen, F.H. Zhou, and E. Ma: *Nature*, 2002, vol. 419, pp. 912–915.
3. Y.H. Zhao, Liao X.Z., Cheng S., Ma E., and Y.T. Zhu: *Adv. Mater.*, 2006, vol. 18, pp. 2280–2283.
4. K. Lu, L. Lu, and S. Suresh: *Science*, 2009, vol. 324, pp. 349–352.
5. K. Lu: *Science*, 2014, vol. 345, pp. 1455–1456.
6. R.Z. Valiev: *Nat. Mater.*, 2004, vol. 3, pp. 511–516.
7. M.A. Meyers, A. Mishra, and D.J. Benson: *Prog. Mater. Sci.*, 2006, vol. 51, pp. 427–556.
8. M. Suzuki, T. Kimura, J. Koike, and K. Maruyama: *Scr. Mater.*, 2003, vol. 48, pp. 997–1002.
9. S. Cheng, Y.H. Zhao, Y.T. Zhu, and E. Ma: *Acta Mater.*, 2007, vol. 55, pp. 5822–5832.
10. J. da Costa Teixeira, L. Bourgeois, C.W. Sinclair, and C.R. Hutchinson: *Acta Mater.*, 2009, vol. 57, pp. 6075–6089.
11. Y.T. Zhu, X.Z. Liao, and X.L. Wu: *Prog. Mater. Sci.*, 2012, vol. 57, pp. 1–62.
12. R.Z. Valiev, Y. Estrin, Z. Horita, T.G. Langdon, M.J. Zehetbauer, and Y.T. Zhu: *JOM*, 2016, vol. 68, pp. 1216–1226.
13. K.M. Youssef, R.O. Scattergood, K.L. Murty, J.A. Horton, and C.C. Koch: *Appl. Phys. Lett.*, 2005, vol. 87, pp. 54–421.
14. L. Lu, X. Chen, X. Huang, and K. Lu: *Science*, 2009, vol. 323, pp. 607–610.
15. X.Y. Li, Y.J. Wei, L. Lu, K. Lu, and H.J. Gao: *Nature*, 2010, vol. 464, pp. 877–880.
16. L.L. Zhu, S.X. Qu, X. Guo, and J. Lu: *J Mech. Phys. Solids.*, 2015, vol. 76, pp. 162–179.
17. P.V. Liddicoat, X.Z. Liao, Y.H. Zhao, Y.T. Zhu, M.Y. Murashkin, E.J. Lavernia, R.Z. Valiev, and S.P. Ringer: *Nat. Commun.*, 2010, vol. 1, p. 63.
18. G. Liu, G.J. Zhang, F. Jiang, X.D. Ding, Y.J. Sun, J. Sun, and E. Ma: *Nat. Mater.*, 2013, vol. 12, pp. 344–350.

19. X.L. Wu, M.X. Yang, F.P. Yuan, G.L. Wu, Y.J. Wei, X.X. Huang, and Y.T. Zhu: *Proc. Natl. Acad. Sci. U.S.A.*, 2015, vol. 112, pp. 14501–14505.
20. T.H. Fang, W.L. Li, N.R. Tao, and K. Lu: *Science*, 2011, vol. 331, pp. 1587–1590.
21. X.L. Wu, P. Jiang, L. Chen, F.P. Yuan, and Y.T. Zhu: *Proc. Natl. Acad. Sci. U.S.A.*, 2014, vol. 111, pp. 7197–7201.
22. S. Suresh: *Science*, 2001, pp. 2447–2451.
23. Y.J. Wei, Y. Li, L. Zhu, Y. Liu, X. Lei, G. Wang, Y. Wu, Z. Mi, J. Liu, H. Wang, and H.J. Gao: *Nat. Commun.*, 2014, vol. 5, p. 3580.
24. M.X. Yang, Y. Pan, F.P. Yuan, Y.T. Zhu, and X.L. Wu: *Mater. Res. Lett.*, 2016, vol. 4, pp. 141–151.
25. Z. Zeng, X.Y. Li, D.S. Xu, L. Lu, H.J. Gao, and T. Zhu: *Extre. Mech. Lett.*, 2016, vol. 8, pp. 213–219.
26. X.L. Wu, M.X. Yang, F.P. Yuan, L. Chen, and Y.T. Zhu: *Acta Mater.*, 2016, vol. 112, pp. 337–346.
27. X.L. Ma, C.X. Huang, W.Z. Xu, H. Zhou, X.L. Wu, and Y.T. Zhu: *Scr. Mater.*, 2015, vol. 103, pp. 57–60.
28. X.L. Ma, C.X. Huang, J. Moering, M. Ruppert, H.W. Höppel, M. Göken, J. Narayan, and Y.T. Zhu: *Acta Mater.*, 2016, vol. pp. 43–52.
29. R. Yuan, I.J. Beyerlein, and C.Z. Zhou: *Mater. Res. Let.*, 2016, (http://dx.doi.org/10.1080/21663831.2016.1255264).
30. X.L. Wu, P. Jiang, L. Chen, and Y.T. Zhu: *Mater. Res. Lett.*, 2014, vol. 2, pp. 185–191.
31. Y. Xiang and J.J. Vlassak: *Acta Mater.*, 2006, vol. 54, pp. 5449–5460.
32. J. Rajagopalan, J.H. Han, and M.T.A. Saif: *Scr. Mater.*, 2008, vol. 59, pp. 734–737.
33. S.W. Lee, A.T. Jennings, and J.R. Greer: *Acta Mater.*, 2013, vol. 61, pp. 1872–1885.
34. X. Feaugas: *Acta Mater.*, 1999, vol. 47, pp. 3617–3632.
35. M.X. Yang, F.P. Yuan, Q.G. Xie, Y.D. Wang, E. Ma, and X.L. Wu: *Acta Mater.*, 2016, vol. 109, pp. 213–222.
36. K. Lu and J. Lu: *J Mater. Sci. Technol.*, 1999, vol. 15, pp. 193–197.
37. A.G. Evans and J.W. Hutchinson: *Acta Mater.*, 2009, vol. 57, pp. 1675–1688.
38. N.A. Fleck, G.M. Muller, M.F. Ashby, and J.W. Hutchinson: *Acta Metall. Mater.*, 1994, vol. 42, pp. 475–487.
39. W.D. Nix and H.J. Gao: *J. Mech. Phys. Solids*, 1998, vol. 46, pp. 411–425.

40. M.F. Ashby: *Philos. Mag.*, 1970, vol. 21, pp. 399–424.
41. C.W. Bert, E.J. Mills, and W.S. Hyler: *J. Basic Eng.*, 1967, vol. 89, pp. 35–39.
42. O. Bouaziz: *Scr. Mater.*, 2012, vol. 66, pp. 982–985.
43. J.G. Sevillano and F.D.L. Cuevas: *Scr. Mater.*, 2012, vol. 66, pp. 978–981.

Chapter 25

Influence of Gradient Structure Volume Fraction on the Mechanical Properties of Pure Copper

Xincheng Yang,[a] Xiaolong Ma,[b] Jordan Moering,[b] Hao Zhou,[b] Wei Wang,[a] Yulan Gong,[a] Jingmei Tao,[a] Yuntian Zhu,[b,c] and Xinkun Zhu[a]

[a]*Faculty of Materials Science and Engineering, Kunming University of Science and Technology, Kunming, Yunnan 650093, China*
[b]*Department of Materials Science & Engineering, North Carolina State University, Raleigh, North Carolina 27695, USA*
[c]*School of Materials Science and Engineering, Nanjing University of Science and Technology, Nanjing 210094, China*
ytzhu@ncsu.edu, xk_zhu@hotmail.com

This chapter reports the influence of gradient structure volume fraction on the tensile mechanical behaviors of pure copper processed by surface mechanical attrition treatment at cryogenic

---

Reprinted from *Mater. Sci. Eng. A*, **645**, 280–285, 2015.

Heterostructured Materials: Novel Materials with Unprecedented Mechanical Properties
Edited by Xiaolei Wu and Yuntian Zhu
Text Copyright © 2015 Elsevier B.V.
Layout Copyright © 2022 Jenny Stanford Publishing Pte. Ltd.
ISBN 978-981-4877-10-7 (Hardcover), 978-1-003-15307-8 (eBook)
www.jennystanford.com

temperature. Superior combinations of tensile strength and ductility are observed in a certain volume fraction, in which strain hardening uprising after yielding is also observed. The gradient structure produces a synergetic strengthening and extra work hardening. These findings suggest the existence of an optimum volume fraction of gradient structure for the best mechanical properties.

25.1 Introduction

Making materials both strong and ductile has been an important issue for materials science [1–3]. Severe plastic deformation (SPD) techniques have been extensively investigated to produce ultrafine-grained or nanocrystalline materials over several decades. Grain refinement in bulk metals can produce very high strength but disappointingly limited ductility.

Recently, gradient structures (GS) have been introduced into some metals and produced excellent strength and ductility [7–10, 29]. GS in materials exhibits a macroscopic gradual change in microstructure from surface to a depth. Gradient structures have been evolved over millions of years in nature and contributed to biological optimization against severe natural environments [4–6]. However, the investigation of the relationship between microstructures and mechanical properties, from the atomic to the macro level and their interactions, hasn't garnered scientific attention until quite recently [11]. Further, there are limited studies analyzing the influence of the volume fraction of the GS layer on the tensile behaviors.

In this study, we produced GS layer at the two top most surfaces in commercial purity copper by surface mechanical attrition treatment (SMAT) at cryogenic temperature. The dynamic recovery and local recrystallization were greatly suppressed during cryogenic deformation [12]. In a certain volume fraction range, a synergetic strengthening effect and strain hardening uprising were observed.

25.2 Experimental

Copper samples were first prepared by rolling commercially-pure copper (99.995 wt%) and then annealing in vacuum at 873 K for 2 h to obtain homogeneous coarse grains. Four different sample thicknesses (2, 3, 4, 5 mm) were chosen for this study to systematically vary the volume fraction of GS layers. These samples were polished to a mirror finish before SMAT treatment.

Details of the SMAT process have been described in a previous work [17]. Briefly, 180 stainless steel balls with 8 mm in diameter were placed at the bottom of a cylinder-shaped chamber and vibrated with a frequency of 50 Hz. Both sample sides were processed at cryogenic temperature.

After the SMAT process, dog-bone-shaped tensile specimens with a gauge length of 15 mm and width of 5 mm were cut using wire electric discharge machining. Tensile tests were performed on a Shimazu Universal Tester at room temperature under a strain rate of $5.0 \times 10^{-4}\,s^{-1}$.

The hardness profiles of 3 mm samples along the depth were measured using a Vikers microhardness tester with a load of 10 g and duration of 15 s. Final results were determined by averaging the values of 8 indentation measurements.

The cross-sections of the samples treated by SMAT were polished and etched for microstructure observation. The samples were etched in a solution containing $FeCl_3$ (5 g), HCl (50 mL) and H_2O (100 mL) for 30 s. The cross-sectional microstructure was analyzed using optical microscopy (OM, Olympus BX51M) and scanning electron microscopy (SEM, Hitachi S 3500 N).

The samples prepared for in-situ tensile tests under SEM observation were prepared in a procedure similar to the preparation for dog-bone-shaped tensile specimens. In order to facilitate SEM observation, the middle of these samples was machined narrower with a radius of 30 mm to make strain concentrate in this area (Fig. 25.1).

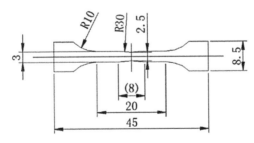

Figure 25.1 The sample geometry for in-situ SEM observation.

25.3 Results

25.3.1 Microstructure

SEM observation revealed grain size gradient from the top most surface to the CG matrix (Fig. 25.2a). The average grain sizes are below 100 nm in the top 10-µm thick layer (NG layer), and increase to about 1 µm in the depth span of 10–100 µm (UFG layer). At depth greater than 100 µm are typical deformation structures in coarse grains. The deformed CG layer is characterized with dislocation tangles or dislocation cells. For convenience, we define the whole gradient structure (GS) layer to include the NG layer, the UFC layer and the deformed CG layer. The thickness of whole GS layer is about 200 µm. Formation of this gradient structure from NG to CG can be understood by a gradient strain and strain rate during the SMAT processing at cryogenic temperature (CT-SMAT). The topmost surface layer was subjected to the largest strains and strain rates [20].

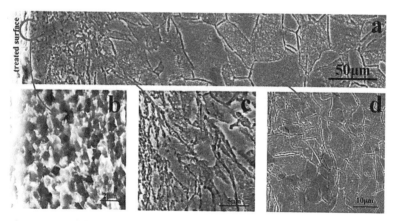

Figure 25.2 (a) An SEM cross-sectional image of a sample processed by SMAT at cryogenic temperature. (b–d) Higher-magnification SEM micrographs at different depths as indicated in (a).

25.3.2 Vickers Hardness

As shown in Fig. 25.3a, the hardness increases from about 0.65 GPa in the CG matrix to about 1.21 GPa in the top treated surface layer. After

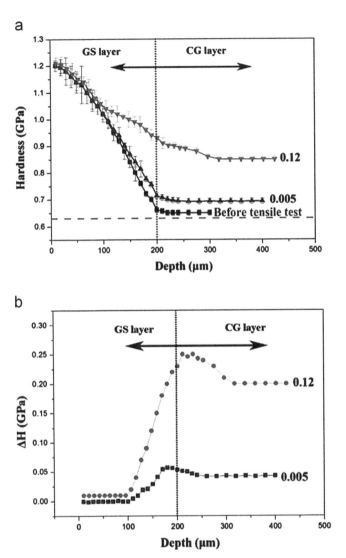

Figure 25.3 (a) Vickers hardness along the depth before tensile test and after varying tensile strains (0.005 and 0.12). (b) Variation of Vickers hardness (ΔH) along the depth after varying tensile strains.

tensile testing, the hardness increases with increasing tensile strain. The strain hardening in the NG and UFG layers is limited. Figure 25.3b shows the hardness increments (ΔH) profile along depth, indicating a ΔH peak. ΔH of CG/GS boundary is much higher than that of the CG matrix, and the peak moves to the CG layer at higher tensile strain, which is consistent with previous observations in GS IF-steel [10]. If the GS layer and CG matrix deforms alone, the large grain sizes of CG matrix will have higher capability of accumulating dislocations to produce higher strain hardening than GS layer. The existence of the ΔH peak near the GS/CG boundary with higher strain hardening than the CG matrix provides direct evidence of extra strain hardening at the boundaries between the virtual necking fine-grained layer and the stable larger grained central layer [10]. The migration of the ΔH peak indicates that the boundaries migrated dynamically.

25.3.3 Mechanical Behaviors

Figure 25.4a shows the engineering stress-strain curves of the CT-SMAT samples with different GS layer volume fraction and the annealed CG samples. Each curve is the average of three tensile tests. The CG samples exhibited yield strength (0.2% offset) of about 53 ± 5 MPa and a uniform elongation of 36 ± 1%. Mean while, the yield strength of the CT-SMAT samples are 160 ± 3 MPa, 190 ± 6 MPa, 220 ± 1 MPa and 235 ± 5 MPa and the uniform elongation are 29 ± 2%, 25 ± 2%, 14 ± 1% and 3 ± 0.3% for the GS volume fractions (VF_{GS}) of 0.08, 0.1, 0.13 and 0.2, respectively.

Figure 25.4b shows that the tensile stress–strain curves systematically vary with VF_{GS}. In general, the curves obtained in this study can be categorized into the following four different characteristic types. Type 1 ($VF_{GS} < 0.08$): The curve shows continuous strain-hardening similar to that in pure CG copper. Type 2 ($0.08 < VF_{GS} < 0.1$): The curve exhibits a transition of tensile stress, in which tensile stress increases slowly first after yielding, then increases steeply for a small strain and finally follows the traditional strain-hardening behavior. Type 3 ($0.1 < VF_{GS} < 0.2$): The curve exhibits a yield peak followed by limited strain hardening. Type 4 ($VF_{GS} > 0.2$): The curve shows a clear yield peak followed by direct necking.

Corresponding strain hardening rate (Θ) curves and true stress-strain curves are shown in Fig. 25.4c, which reveal the transition more clearly. When $VF_{GS} > 0.1$ the strain hardening rate shows an up-turn in Θ curves, which is similar to a previous report in IF-steel with GS [18]. In contrast, curves from pure CG and the

VF$_{GS}$ = 0.08 samples do not show any up-turn, suggesting that the unique behavior is related to the volume fraction of GS layer. No Θ up-turn was observed in SMGT pure copper [28], which also includes a minor fraction of GS layer.

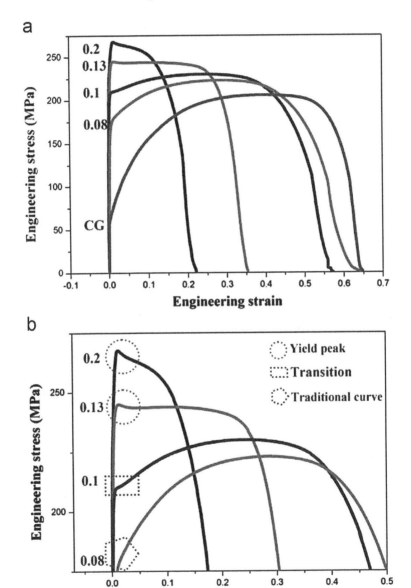

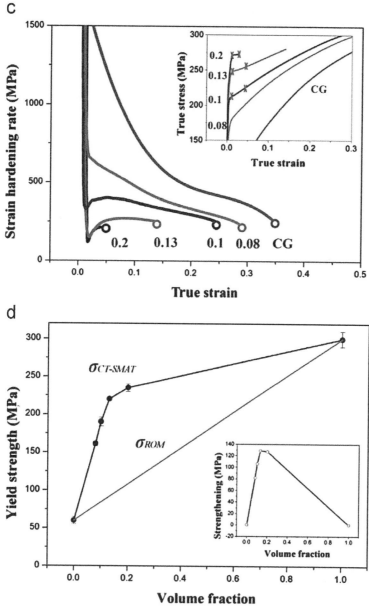

Figure 25.4 (a) Engineering stress-strain curves of samples with different GS layer volume fractions (marked on the curves), and the CG sample. To observe the transition more clearly, a local magnification was given in (b). (c) Strain hardening rate ($\theta = d\sigma/d\varepsilon$) curves. Inset shows corresponding true stress-strain

curves, with the transient response marked by red crosses. (d) The measured yield strength of the CT-SMAT samples ($\sigma_{CT\text{-}SMAT}$) and the calculated yield strength using the ROM (σ_{ROM}) vs GS volume fraction. Inset shows the extra strengthening ($\Delta\sigma = \sigma_{CT\text{-}SMAT} - \sigma_{ROM}$) as a function of the GS volume fraction.

Figure 25.4d shows the strengthening of the GS layer. The strength of the whole sample is much higher than the volume weighed sum of the GS layers and the CG matrix, indicating an synergetic strengthening exist in CT-SMAT samples. The 0 volume fraction represents the CG sample, while the 1 volume fraction of the GS layer represents the data point from the standalone GS layer. The thickness of GS layer is about 200 μm. At such a thin thickness the rough surface and the size effect will influence the experimental accuracy [19]. So an empirical relationship $H_v = 3\sigma_y$ is applied to calculate the average yield strength of standalone GS layer, where H_v is the average hardness of GS layer and σ_y is the ultimate strength of GS layer (standalone GS layer has negligible work hardening, so the ultimate strength approximately equal to the yield strength). The average hardness of the GS layer is obtained from Fig. 25.3, and the estimated yield strength value is 300 MPa. σ_{ROM} is calculated, using the rule of mixture (ROM): $\sigma_{ROM} = VF_{GS}\sigma_{GS} + VF_{CG}\sigma_{CG}$, where VF_{GS} and VF_{CG} are the volume fraction of the GS layers and the CG layers, respectively. σ_{GS} and σ_{CG} are the yield strength at 0.2% strain of GS layer and CG sample, respectively.

The strengthening as a function of the GS volume fraction is shown in the inset of Fig. 25.4d, it can be seen that the strengthening is more significant in the samples with $VF_{GS} = 0.2$ and $VF_{GS} = 0.13$, indicating the existence of an optimum volume fraction for the largest additional strengthening. But ductility is also dramatically decreased, as shown in Fig. 25.4a. The strength-ductility combinations of SMAT samples with different volume fraction will be discussed in section 25.4.

25.3.4 In-situ SEM Observation

As shown in Fig. 25.5, when the stress increased to about 200 MPa, slip bands began to develop in the CG layer while none was observed in the GS layer. This means the CG central was already deforming plastically while whole sample still appeared to deform largely elastically according to the stress strain curve. The plastic deformation of the CG layer would lead to strain hardening, which should have contributed to the additional strengthening shown in Fig. 25.4d.

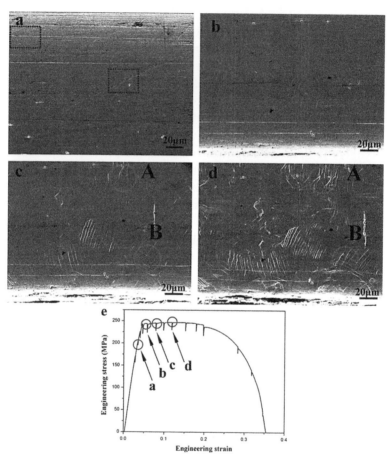

Figure 25.5 An in-situ SEM observation of VF = 0.13 sample (the fluctuation of curve is due to unloading-reloading). (a) Slip bands formed in the CG layer under an engineering stress of 200 MPa. (b) Slip bands developed in GS layer under an engineering stress of 240 MPa (soon after yielding). (c) New slip bands formed at a strain of 0.08. (d) The sample was full of slip bands, but few slip bands were observed in NG layer at a strain of 0.12 (before necking). (e) Engineering stress-strain curve of in-situ tensile test, where the four points correspond to the above SEM images (the curve is a little different with the curves in Fig. 25.2a because of the difference in the sample size).

When the stress in the sample reached about 240 MPa, slip bands began to appear in both the CG and GS layers. Figure 25.5b shows slip bands in the GS layer, which means that the GS layer began to yield when the stress reached 240 MPa. It's important to note that the stress in the GS layers was much higher than the applied average stress because of stress redistribution (detailed description is given in section 25.4.1).

With increasing stress, new slip systems were activated. The lines in individual grains were approximately parallel, but different slip directions could be seen in different grains. Several interesting phenomena can be observed from these slip bands. First, an array of parallel slip bands cut through another array of slip bands advancing in a different direction (marked by B in Fig. 25.5c,d). Second, some slip band bending and branching (marked by A in Fig. 25.5c,d) can be seen. The above observations indicate that in this stage the samples maintained plastic deformation through nucleation and development of slip bands, and dislocations from different slip bands interacted with each other.

As shown in Fig. 25.5d, there were some new slip bands activated. However, an important factor to be noted is that few slip bands are found in the NG layer. This occurred in the CT-SMAT sample at a large strain (0.13) indicating that plastic deformation in the NG layer did not result from the activation of prolific dislocation sources that could continuously emit dislocations. This is consistent with early reports that grain boundaries become primary dislocation sources and sinks in nanocrystalline metals [30–32]. Since grain boundaries emitted dislocations more randomly, no large dislocation steps was observed on the sample surface.

A slip band consist of many dislocations steps and each dislocation step is produced by slip of many dislocations with the same Burgers vector on the same slip plane, we can validate the sequence of plastic deformation in CG and GS layer. The above observation indicates that the CG layer started plastic deformation first. This is the basis for the following discussion on deformation mechanism.

25.4 Discussion

25.4.1 Synergetic Strengthening and Extra Strain Hardening

Figure 25.6a is a 3D schematic of the samples under uniaxial tensile stress. Inelastic-plastic transition stage, the CG matrix deforms plastically first, while the GS layer still deforms elastically, forming plastic-elastic boundary. This boundary migrates into the GS layer toward the sample surface as layers with smaller grains start to deform plastically. Dislocations gliding in grains adjacent to the boundary are blocked at and accumulate near the boundary, which results in enhanced strain hardening and contribute to the observed synergetic strengthening. These points are with previous observations [10, 18, 33].

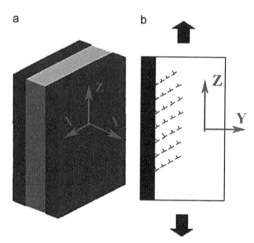

Figure 25.6 (a) A 3D schematic of the samples under uniaxial tensile stress. (b) Schematic illustration of dislocation pileup at the plastic/elastic boundary.

A complex stress-state will be developed during the tensile test. The Poisson's ratio in the plastically deformed region is close to 0.5 to maintain a constant volume, while the Poisson's ratio in the elastically deformed region is close to 0.3 [21], under the condition of $\varepsilon_{zCS} = \varepsilon_{zGS}$, where ε_{xCG} is larger than ε_{xGS}. Due to the continuity of materials, strain has to be continuous at the boundary, causing a strain gradient. When the deformation of CG matrix are constrained by the GS layer, the constraint is realized in the form of lateral stress in the CG layer, which is consistent with the simulation results in Ref. [18].

In plastic deformation stage, the standalone NG layer starts necking at very low tensile strains. However, in CT-SMAT samples the instability of NG layer is constrained by the neighboring stable layer (UFG layer or deformed CG layer). In order to keep material continuity, the strain gradient is produced near the boundary between the unstable surface layers and the stable central core, which also produce a stress gradient. These gradients result in two consequences: (1) The strain gradient near the boundary will lead to the accumulation of geometrically necessary dislocations[10, 22–24]. (2) The stress gradient near the boundary in the Z and X direction can significantly increase the flow stress [10, 18, 25–27]. In addition, the GS structure also converts the applied uniaxial stress into a two-dimensional stress state.

25.4.2 Optimum GS Volume Fraction for Extra Strain Hardening

Figure 25.4c shows that strain hardening rate uprising after yielding only occurred in sample with GS volume fraction of 0.1–0.2. These are in the volume fraction range in which the highest extra strengthening is observed (Fig. 25.4d). These observations indicate that there exist an optimum GS volume fraction that produces the highest extra strengthening and extra strain hardening. It is our hypothesis that this optimum GS volume fraction is largely determined by the maximum mechanical incompatibility [10] between the different layers in the sample. Specifically, during the elastic-plastic deformation stage, the plastically deforming central layer will be under tensile stress laterally (*X*-direction in Fig. 25.6a), while the elastically deforming layers will be under tensile stress laterally [10, 18]. The maximum stress magnitude is hypothesized to produce the highest extra strengthening. During the plastic deformation stage, the virtual fine-grained surface layers will be under compressive stress laterally, while the stable central layer will be under tensile stress laterally. The larger stress magnitude should produce higher extra strain hardening. This hypothesis needs to be verified in future studies.

25.4.3 Strength-Ductility Combinations

The strength and ductility of SMAT samples are compared to other literature values in Fig. 25.7 [28, 34–39]. As shown, the strength and ductility of conventional techniques are mainly distributed in the light blue region, but the strength and ductility of SMAT samples with $VF_{GS} = 0.1$ and $VF_{GS} = 0.13$ are out of this region, indicated a outstanding of strength-ductility combinations that are out of reach of homogeneous samples.

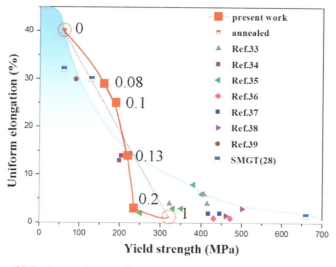

Figure 25.7 Comparing tensile properties of pure copper processed by a variety of processes and SMAT.

The red curve in Fig. 25.7 is the strength-ductility line of SMAT samples with different VF_{GS}. When the VF_{GS} is in the range of 0 to 0.1, the strength-ductility line is arch shaped. When the VF_{GS} is in the range of 0.1 to 1, the strength-ductility line is inverse-arch shaped. The changing of strength-ductility line is closely related with VF_{GS}, which means the VF_{GS} have great influence on strength-ductility combinations. It is evident from Fig. 25.7 that samples have superior combination in the range of 0.08 to 0.1. Interestingly, the $VF_{GS} = 0.1$ is the "θ-up-turn" critical value too, which suggests a possible link between them.

25.5 Conclusions

A gradient structure is generated at top most surfaces in the pure copper by CT-SMAT process, and can be optimized to produce a superior combination of tensile strength and ductility. There exists an optimum gradient structure volume fraction for the highest extra strain hardening and extra strengthening.

Acknowledgements

This work is supported by the National Natural Science Foundation of China (Grant No.50874056 and 51361017). Y.T.Z. is funded by the US Army Research Office (W911 NF-12-1-0009) and the China 1000Plan program.

References

1. Zhu YT, Liao XZ, *Nat Mater.* **3** (2004) 351.
2. Valiev RZ, Alexandrov IV, Zhu YT, Lowe TC, *J Mater Res.* **17** (2002) 5.
3. Jian WW, Cheng GM, Xu WZ, Yuan H, Tsai MH, Wang QD, Koch CC, Zhu YT, Mathaudhu SN, *Mater Res Lett.* **1** (2013) 61.
4. Suresh S, *Science.* **292** (2001) 2447.
5. Gao HJ, Ji BH, Jager IL, Arzt E, Fratzl P, *Proc Natl Acad Sci USA.* **100** (2003) 5597.
6. Ray AK, Das SK, Mondal S, Ramachandrarao P, *J Mater Sci.* **39** (2004) 1055.
7. Fang TH, Li WL, Tao NR, Lu K, *Science.* **331** (2011) 1587.
8. Chen AY, Li DF, Zhang JB, Song HW, Lu J, *Scr Mater.* **59** (2008) 579–582.
9. Hughes DA, Hansen N, *Phys Rev Lett.* **87** (2001) 135503.
10. Wu XL, Jiang P, Chen L, Yuan FP, Zhu YT, *Proc Natl Acad Sci USA.* **111** (2014) 7197.
11. Launey ME, Munch E, Alsem DH, Barth HB, Saiz E, Tomsia AP, et al. *Acta Mater.* **57** (2009) 2919.
12. Darling KA, Tschopp MA, Roberts AJ, Ligda JP, Kecskes LJ, *Scr Mater.* **69** (2013) 461.
13. Johnston WG, Gilman JJ, *J. Appl. Phys.* **30** (1959) 129.
14. Wyrzykowski JW, Grabski MW, *Mater Sci Eng.* **56** (1982) 197.

15. Tsuji N, Ito Y, Saito Y, Minamino Y, *Scr Mater.* **47** (2002) 893.
16. Hayes JS, Keyte R, Prangnell PB, *Mater Sci Tech.* **16** (2000) 1259.
17. Valiev RZ, Islamgaliev RK, Alexandrov IV, *Prog Mater Sci.* **45** (2000) 103.
18. Wu XL, Jiang P, Chen L, Zhang JF, Yuan FP, Zhu YT, *Mater Res Lett,* **2** (2014) 185.
19. Yang L, Lu L, *Scr Mater.* **69** (2013) 242.
20. Wang K, Tao NR, Liu G, Lu J, Lu K, *Acta Mater.* **54** (2006) 5281.
21. Beat CW, Mills EJ, Hyler WS, *J Fluids Eng.* **89** (1967) 35.
22. Ashby MF, *Philos Mag.* **21** (1970) 300.
23. Gao HJ, Huang Y, Nix WD, Hutchinson JW, *Theory J Mech Phys Solids.* **47** (1999) 1239.
24. Gao HJ, Huang YG, *Scr Mater.* **48** (2003) 113.
25. Chaudhar P, Scatterg RO, *Acta Metall.* **14** (1966) 685.
26. Hirth JP, *Philos Mag.* **86** (2006) 3959.
27. Chakravarthy SS, Curtin WA, *Proc Natl Acad Sci USA.* **108** (2011) 15716.
28. Fang TH, Li WL, Tao NR, Lu K, *Science.* **331** (2011) 1587–1590.
29. Wei YJ, Li YQ, Zhu LC, Liu Y, et al., *Nat Commun.* **5** (2014) 3580.
30. Zhu YT, Liao XZ, Wu XL, *Prog Mater Sci.* **57** (2012) 1-62.
31. Liao XZ, Zhou F, Lavernia EJ, Srinivasan SG, et al., *Appl Phys Lett.* **83** (2003) 632–634.
32. Zhu YT, Wu XL, Liao XL, Narayan J, Mathaudhu SN, Kecskes LJ, *Appl Phys Lett.* **95** (2009) 031909.
33. Lugo N, Llorca N, Cabrera JM, Horita Z, *Mater Sci Eng A.* **447** (2008) 366–371.
34. Ranjbar Bahadori Sh, Dehghani K, Bakhshandeh F, *Mater Sci Eng A.* **583** (2013) 36–42.
35. K. Edalati, T. Fujioka, Z. Horita, *Mater Sci Eng A.* **497** (2008) 168–173.
36. Guo J, Wang K and Lu L, *J Mater Sci Technol.* **22** (2006) 789–792.
37. San XY, Liang XG, Cheng LP, Li CJ, Zhu XK, *Mater Des.* **35** (2012) 480–483.
38. Hanazakia K, Shigeiri N, Tsuji N, *Mater Sci Eng A.* **527** (2010) 5699–5707.
39. Lu L, Wang LB, Ding BZ, Lu K, *J Mater Res.* **15** (2000) 270–273.

Chapter 26

The Role of Shear Strain on Texture and Microstructural Gradients in Low Carbon Steel Processed by Surface Mechanical Attrition Treatment

Jordan Moering,[a] Xiaolong Ma,[a] Guizhen Chen,[b]
Pifeng Miao,[b] Guozhong Li,[b] Gang Qian,[b]
Suveen Mathaudhu,[c] and Yuntian Zhu[a,d]

[a]*Department of Materials Science and Engineering,
North Carolina State University, Raleigh, NC 27695, USA*
[b]*Jiangyin Xingcheng Special Steel Works Co., Ltd., Jianyin, Jiangsu, China*
[c]*Department of Mechanical Engineering, University of California,
Riverside, CA, USA*
[d]*School of Materials Science and Engineering, Nanjing University of Science
and Technology, Nanjing, China*
ytzhu@ncsu.edu

In this study, the shear strain at various depths of a low carbon steel processed by Surface Mechanical Attrition Treatment (SMAT)

Reprinted from *Scr. Mater.*, **108**, 100–103, 2015.

Heterostructured Materials: Novel Materials with Unprecedented Mechanical Properties
Edited by Xiaolei Wu and Yuntian Zhu
Text Copyright © 2015 Acta Materialia Inc.
Layout Copyright © 2022 Jenny Stanford Publishing Pte. Ltd.
ISBN 978-981-4877-10-7 (Hardcover), 978-1-003-15307-8 (eBook)
www.jennystanford.com

was measured using deformed carbide bands as internal strain markers. The shear strain gradient is found to strongly correlate with the gradients of texture, microstructure and hardness. The microhardness increases approximately linearly with shear strain. In the top surface, the average ferrite grain size is reduced to 60 nm with a strong {110}//SMAT surface texture.

Gradient structures are known to produce a wide variety of interesting properties including improved wear resistance and fatigue life, and extraordinary mechanical properties [1–6]. In recent years, SMAT has gained attention for its ability to generate gradient structured materials through grain refinement of the surface layer to the nanometer scale [1, 7–10]. This technique is ideal for systematic investigations of gradient structures due to the gradients in strain, strain rate, hardness, grain size, and hardening mechanisms throughout the deformed layer. In order to refine grain sizes to the nanometer scale, large strains and strain rates need to be applied [11–13]. It is well known that the shear component of the applied strain is directly correlated with dislocation slip and microstructure evolution. However, quantitatively mapping a single component of the strain tensor is challenging [14–16]. Markers and photographic evidence have been reported to extract the effect of shear strain on microstructure evolution, but they are not suitable for measuring shear strain in SMAT, due to the complexity of the process [17–19].

In this work, cementite bands are used as internal markers to quantify the shear strain at various depths of the surface, which is the first time shear strain has been quantitatively mapped in SMAT-processed samples. Additionally, the texture evolution is systematically characterized, which has rarely been studied in SMAT-processed structures but profoundly affects mechanical behavior [20–23]. The observations from this work elucidate the effect of shear strain on the development of texture gradient, microstructure gradient, and microhardness gradient in the SMAT-processed samples.

Normalized steel plates with a composition of 0.14% C, 0.33% Si, 1.44% Mn, 0.08% Cr, 0.03% Ni, and balance Fe was used for this study. Samples were cut along the rolling direction so that the SMAT treatment would take place normal to the rolling direction. The pearlite was agglomerated into bands normal to the SMAT surface as shown in Fig. 26.1. The SMAT process was carried out using a

SPEX 8000M Mixer/Mill by replacing the lid of the vial with ¼" thick plates of the sample to be treated. Samples were polished to 1200 grit, sealed in ambient atmosphere, and processed with three ½" 440C steel balls for 120 min. Profilometry revealed that the surface was roughened to an Ra of 8.8 µm. After treatment, cross-sectional samples were Ni-plated to protect the surface microstructure from edge rounding when polishing, and were imaged using a JEOL 6010LA Scanning Electron Microscope (SEM) at 20kV.

Experimentally measuring a single component of the strain tensor is not a trivial matter. Attempts to discern the shear component of the tensor in severe plastic deformation processes have been undertaken primarily by simulation or calculation [11, 14, 16, 18]. Experimentally, an imbedded pin method has been used to measure shear in accumulated roll bonding, where the interface of the pin can be used to map the shear strain, but this is not suitable for SMAT structures [19]. In normalized low carbon steels, clear bands of pearlite along the rolling direction (α-iron + Fe_3C) and ferrite (α-iron) naturally act as markers in the microstructure. Since the plates were cut normal to the rolling direction, these bands were perpendicular to the SMAT surface, as shown in Fig. 26.1. Because the shot impacts from SMAT repeatedly induce plastic flow in the surface, the top deformed layer is theorized to undergo high shear strains [1, 17]. Therefore, simply mapping the slope of the deformed pearlite bands could yield the accumulated shear strain induced at various depths of the SMAT surface. First, a grid with 100 µm blocks was overlayed onto the micrograph seen in Fig. 26.1a. Then, the average slope was measured in each of these regions to calculate the average shear strain at various depths. In this way, the shear strain plotted at 50 µm, 150 µm, 250 µm, etc. represents the average shear strain at each 100 µm interval. The result indicates that the shear strain increases exponentially near the surface, which is visually apparent in the cross-sectional SEM micrograph (Fig. 26.1a). A simple exponential fit ($R^2 = 0.97$) was applied to the data in order to estimate the shear strain at discrete depths from the surface. Extrapolation of this data was used to estimate the shear strain at depths less than 100 µm but it is not clear how big the error is from such an extrapolation (Fig. 26.2). Surprisingly, the average measured shear strain at a depth of 50 µm is 90, and the extrapolated shear strain in the top 10 µm is 119, which is in the realm of shear strains measured in accumulated roll bonding and chip processing, as well as high pressure torsion [14, 18, 19]. Note that in the very top surface, e.g. at layer thickness close to the roughness, the current strain measurement may significantly underestimate the shear strain.

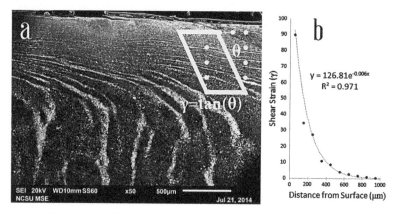

Figure 26.1 (a) An SEM image of the as-received SMAT surface. The inset shows how the slope of the cementite bands was measured in order to calculate shear strain γ. (b) The resulting data calculated at various depths showing a clear exponential increase in shear strain approaching the surface of the SMAT gradient.

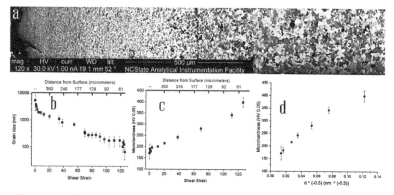

Figure 26.2 (a) A cross sectional FIB image of the SMAT gradient extending to depths greater than 500 μm, (b) the strong linear dependence of microhardness on shear strain, (c) the dependence of grain size on shear strain, and (d) the microhardness approximately follows the Hall–Petch relationship.

Once the shear strain was calculated, grain size and microhardness measurements could be plotted to determine their relationship. Five hardness measurements were averaged at each depth with a Mitutoyo Microhardenss Testing Machine Model HM-11 with a Vickers diamond indenter at a load of 0.05N. Grain size measurements were performed using the line intercept method from micrographs collected from the dual Beam FEI Quanta 3D FEG, the JEOL 2010 F Transmission Electron Microscope. Figure 26.2 shows these relationships and a corresponding FIB micrograph of the gradient structure at the surface. Both hardness and grain size show a strong dependency on the shear strain, and the Hall–Petch plot shows slight deviation from the ideal linear trend.

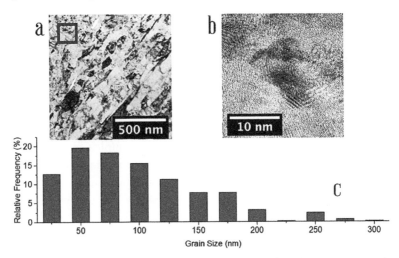

Figure 26.3 (a) A representative image of the top surface shows regions with both nanocrystalline (<100 nm) grains and coarser grains intermixed. (b) High resolution TEM of the nanocrystalline region shows a single grain ~10 nm in diameter. This microstructure has been reported in steels subject to high shear strains and indicates partial decomposition of Fe_3C. (c) Grain size distribution at the top surface.

At the top surface, the grain size is dramatically reduced to 60 nm, as seen in Fig. 26.3. TEM samples confirmed that the grain size at the top surface was skewed, with some regions containing grains less than 10 nm, while other regions had grains 100 nm in diameter. Figure 26.3 shows the distribution of grain size at the top 10 μm of the SMATed surface. In carbon steels, nanocrystallization has been reported in regions subject to very high shear strains, and were first discovered in railroad tracks [14]. These regions were called "white etching layers", which consist of fragmented Fe_3C and even complete Fe_3C dissolution that leads to supersaturation of carbon in nanocrystalline α-iron [6, 15, 24]. These reports are consistent with this observation here.

Because SMAT is a complex deformation scheme consisting of compressive and shear strains at various strain rates, mapping the texture allows for a simple investigation on the underlying deformation schemes at various depths from the surface [1, 23]. Samples were prepared for EBSD imaging by conventional polishing followed by ion milling in a Fischione Ion Mill (Model-1060) at 5kV and 5° tilt for 45 min. An Oxford EBSD detector installed in the dual Beam FEI Quanta 3D FEG was used for collecting images. Figure 26.4 shows an overview of the microstructural and textural development along the depth. The as-received material consists primarily of high-angle grain boundaries (>15°) and displays no strong texture. After SMAT, at 200 μm below the surface, no strong texture can be determined, but the grain size has been reduced and clear subgrain boundaries (>2°) can be seen within large grains. At 100 μm below the surface, there is a clear transition to a complex texture with {110} and some {111} planes//SMAT surface, and low-angle grain boundaries have evolved from the subgrain boundaries. At 50 μm below the surface, although some residual {111}//SMAT surface texture can still be seen, the texture mostly transitioned to {110}//SMAT surface, which is a well-known texture for highly sheared α-iron [21, 23]. At the top surface, the {110}//SMAT surface texture is further strengthened, and most grain boundaries have been converted to high angle. As can be seen in Fig. 26.4, the development of the texture of {110}//SMAT surface is preceded by diminishing {111} components from the depth of 100 μm to the surface.

Texture develops when preferred crystallographic orientations align with applied stress. Slip systems tend to align with the shear direction to maximize the resolve shear stress [1, 22]. In BCC α-iron, {110} are the slip planes, and <111> are the slip directions, so the {110} <1$\bar{1}$1> slip systems are preferred. When shear is induced during plastic deformation, the resolved shear stress is maximized when the {110} lie parallel to the shear direction, like a deck of cards. This is why the <110> α wire texture is commonly found in

HPT, wire drawing, and other deformation modes that induce high shear strains [18, 13, 20].

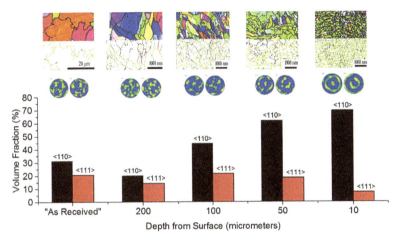

Figure 26.4 Development of microstructure and texture at varying depths from the surface. The percentage of grains indexed normal to the SMAT surface were calculated from EBSD maps and show the relative frequency of the {111}, {110} and planes in ferrite at various depths. Grain boundary maps indicate high angle (>15°) grain boundaries in green, low angle (<15°) grain boundaries in red, and subgrain (>2°) boundaries in black. Pole figures are projected normal to the SMAT surface, indicating a strong {110}//SMAT surface texture developed in the top 50 μm.

The SMAT has also induced nanocrystalline regions with grain sizes less than 10 nm as seen in Fig. 26.3, suggesting the formation of the <110> α wire texture is accompanied with the dissolution of Fe_3C.

The {110}//the surface texture was developed through dislocation slip and generation of geometrically necessary dislocations. At the same time, dislocation accumulation and interactions formed subgrain boundaries, which were eventually transformed to high angle grain boundaries, as is consistent with previous reports [1, 6–10, 23]. These defect structures serve not only to accommodate strain but also to orient the crystallographic directions towards the highest shear directions. The texture, however, is strongest only in the top 100 μm, while grain refinement and shear strain accumulation are also prevalent at depths >350 μm. In these regions, the lack of a strong shear direction may be the reason for the weakness of the texture. Cementite thinning and fragmentation may also play a role. It is clear, however, that the texture forms over a gradient, and is most strongly formed at the highest shear strains with the most severe grain refinement.

Because the gradient texture will locally affect deformation mechanics, it will likely play a role in the global deformation of the sample. For example, the texture gradient will lead to a deformation gradient, which causes macroscopic mechanical incompatibility that has been shown to increase mechanical strength and ductility simultaneously in similar systems [2–4]. Further investigation on the affects of gradient textures and their global response to mechanical stress could provide insight to the exciting developments in gradient structured materials.

In summary, SMAT imparted very large shear strain near the sample surface, which decreases along the depth, which resulted in gradients in hardness, grain size and texture. Fe_3C dissolution occurred near the surface, which helped with the reduction of grain size to 10 nm in local areas. {110}//SMAT texture was produced over a depth of 100 μm, below which the texture is complex. Further investigation on the effects of gradient textures and their global response to mechanical stress is needed to provide insight to the exciting developments in gradient structured materials.

Acknowledgments

This work is funded by the Jiangyin Xingchecng Special Steel Works Co., Ltd, Jiangxu, China. The authors acknowledge the use of the Analytical Instrumentation Facility (AIF) at North Carolina State University, which is supported by the State of North Carolina and the National Science Foundation. YTZ also acknowledges the support of the Support of the China Thousand Talents Plan.

References

1. Lu K, Lu J. *Mater. Sci. Eng. A* 2004;375–377:38.
2. Fang TH, Li WL, Tao NR, Lu K. *Science* 2011;331:1587.
3. Wu XL, Jiang P, Chen L, Yuan FP, Zhu YTT. *Proc. Natl. Acad. Sci. USA* 2014;111:7197.

4. Wu XL, JIang P, Chen L, Zhang JF, Yuan FP, Zhu YT. *Mater. Res. Lett.* 2014;2:185.
5. Lu K. *Science* 2014;345:1455.
6. Li D, Chen HN, Xu H. *Appl. Surf. Sci.* 2009;255:3811.
7. Tao NR, Wang ZB, Tong WP, Sui ML, Lu J, Lu K. *Acta Mater.* 2002;50:4603.
8. Zhang HW, Hei ZK, Liu G, Lu J, Lu K. *Acta Mater.* 2003;51:1871.
9. Wu X, Tao N, Hong Y, Liu G, Xu B, Lu J, Lu K. *Acta Mater.* 2005;53:681.
10. Zhou L, Liu G, Ma XL, Lu K. *Acta Mater.* 2008;56:78.
11. Wang K, Tao NR, Liu G, Lu J, Lu K. *Acta Mater.* 2006;54:5281.
12. Villegas JC, Shaw LL. *Acta Mater.* 2009;57:5782.
13. Dai K, Shaw L. *Mater. Sci. Eng. A* 2007;463:46.
14. Ivanisenko Y, Lojkowski W, Valiev RZ, Fecht HJ. *Acta Mater.* 2003;51:5555.
15. Ramesh A, Melkote SN, Allard LF, Riester L, Watkins TR. *Mater. Sci. Eng. A* 2005;390:88.
16. Murty SVSN, Torizuka S, Nagai K, Koseki N, Kogo Y. *Scr. Mater.* 2005;52:713.
17. Ahn JH, Kwon D. *J. Mater. Res.* 2001;16:3170.
18. Swarninathan S, Shankar MR, Lee S, Hwang J, King AH, Kezar RF, Rao BC, Brown TL, Chandrasekar S, Compton WD, Trumble KP. *Mater. Sci. Eng. A* 2005;410:358.
19. Kamikawa N, Sakai T, Tsuji N. *Acta Mater.* 2007;55:5873.
20. Blonde R, Chan HL, Allain-Bonasso N, Bolle B, Grosdidier T, Lu JA. *J. Alloys Compd.* 2010;504:S410.
21. Chen WY, Tong WP, He CS, Zhao X, Zuo L. *Mater. Sci. Forum* 2012;706–709:2663.
22. Wei Q, Ramesh KT, Kecskes LJ, Mathaudhu SN, Hartwig KT. *Mater. Sci. Forum* 2008;579:75.
23. Barrett CS, Massalski TB. *Structure of Metals*. New York: McGraw-Hill, 1966.
24. Langford G. *Metall. Trans. A* 1977;8:861.

Chapter 27

Bauschinger Effect and Hetero-Deformation Induced (HDI) Stress in Gradient Cu-Ge Alloy

Xianzhi Hu,[a] Shenbao Jin,[b] Hao Zhou,[c] Zhe Yin,[a] Jian Yang,[a] Yulan Gong,[a] Yuntian Zhu,[b] Gang Sha,[b] and Xinkun Zhu[a]

[a]*Faculty of Materials Science and Engineering, Kunming University of Science and Technology, Kunming, Yunnan 650093, China*
[b]*School of Materials Science and Engineering, Nanjing University of Science and Technology, Nanjing, Jiangsu 210094, China*
[c]*The Institute of Microstructures and Properties of Advanced Materials in Beijing University of Technology, Beijing 100124, China*
gang.sha@njust.edu.cn, xk_zhu@hotmail.com

By introducing a gradient structure through surface mechanical attrition treatment (SMAT), the yield strength of Cu-5.7 wt% Ge samples was significantly improved. Unloading-reloading test showed an unusual Bauschinger effect in these GS samples. The hetero-deformation induced (HDI) stresses caused by accumulated

---

Reprinted from *Metall. Mater. Trans. A*, **48A**, 3943–3950, 2017.

Heterostructured Materials: Novel Materials with Unprecedented Mechanical Properties
Edited by Xiaolei Wu and Yuntian Zhu
Text Copyright © 2017 The Minerals, Metals & Materials Society and ASM International
Layout Copyright © 2022 Jenny Stanford Publishing Pte. Ltd.
ISBN 978-981-4877-10-7 (Hardcover), 978-1-003-15307-8 (eBook)
www.jennystanford.com

geometrically necessary dislocations (GNDs) on the GS/CG border increased with increasing strain. The GNDs are mainly distributed in the gradient structured layer measured by electron back scatter diffraction (EBSD), and the density of GNDs increases with the SMAT time. The effect of HDI stress increases with increasing SMAT processing time due to the increase in strain gradient. The pronounced Bauschinger effect in GS sample can improve the resistance to forward plastic flow and finally contribute to the high strength of GS samples.

27.1 Introduction

Metals with gradient structure (GS) can achieve high strength and good ductility simultaneously. Compared to the bulk metals with coarse grains, gradient structured materials exhibit a substantially different deformation behavior and mechanical properties [1–5]. This has drawn more and more attention in recent years. Various mechanisms, including mechanically driven grain growth [3], strain-gradient, multiaxial stresses state and extra strain hardening [6], synergetic strengthening [7, 8], have been proposed to elucidate the intrinsic effects associated with strain and stress gradient. Wu [9] found an unusual Bauschinger effect (BE) in heterostructure lamella (HL) structured Ti, which is responsible for its high strength and good ductility.

A distinct Bauschinger effect, which was considered to significantly contribute to the high strength, has been found in freestanding Cu thin films [10–13], conformally passivated copper nanopillars [14], and heterostructure lamella (HL) structure Ti [6, 15]. The Bauschinger effect is generally attributed to long-range internal stresses caused by dislocation interactions and pile-ups at barriers or the boundary of soft and hard phases [16]. An in-depth study of Bauschinger effect may provide insight into the strengthening mechanism and plasticity theories of materials. However, Bauschinger effect in gradient structured metals is barely reported.

In this study, through cyclic (unloading-reloading) tensile test, we systematically investigated the Bauschinger effect in gradient Cu-Ge alloy. The results give a direct evidence of a pronounced Bauschinger effect observed in GS Cu-Ge alloy compared to their CG counterparts, and provide a new insight to further understanding the inherent property of gradient structure.

27.2 Experiment

Cu-5.7 wt% Ge alloy was produced by induction vacuum melting and rolled to 3 mm thickness, then cut into a 100 × 100 mm^2 plates. These plates were first annealed at 1073 K for 4.5 h in vacuum to obtain homogeneous coarse grains (CG). Some annealed plates were processed by surface mechanical attrition treatment (SMAT) at cryogenic temperature on both sides to obtain gradient structure. SMAT process was described in detail previously [8]. The SMAT duration was 5 min, 10 min and 30 min, respectively. All tensile specimens were dog-bone-shaped with a gauge length of 15 mm, width of 5 mm and thickness of 3 ± 0.1 mm. Tensile tests were conducted at room temperature with a quasi-static strain rate of 5.0 × 10^{-4} s^{-1} on a Shimadzu Universal Tester. An extensometer of 12.5 mm gauge length with full range capacity of 5 mm were used to measure the displacement. Unloading-reloading tests were carried out at 15 varying applied strains. At each applied strain, the specimen was first unloaded with an unloading rate of 500 N·min^{-1} to 20 N, then reloaded with strain rate of 5.0 × 10^{-4} s^{-1} to the next applied strain. Microhardness measurements were carried out on a HX-1 Vickers hardness testing machine, which is performed on the cross-sectional of the samples with a load of 50 g and a loading time of 15 s. Neighboring indentations were separated by a distance of 20 μm and the total test depth of approximately 300 μm. EBSD mapping were carried out on the GS samples (SMAT for 5, 30 min) for comparison, and the channel 5 software was used to analyze the original EBSD data.

27.3 Results and Discussion

Figure 27.1a shows the unloading and reloading stress–strain curves of annealed (CG) and SMAT (GS) samples. Figure 27.1b shows enlarged hysteresis loops taken at true strain of ~0.165 from (a). As shown, the hysteresis loop becomes larger as the SMAT treating time increases, indicating a stronger Bauschinger effect in GS sample. The annealed sample with homogeneous coarse grains, by contract, has a very small hysteresis. The tensile engineering stress–strain curves are given in Fig. 27.1c, which demonstrate that the yield strength of GS samples was enhanced dramatically compared to their CG counterpart. The yield strength (0.2% offset) of different GS samples (SMAT for 5, 15, 30 min) is 183 ± 9 MPa, 267 ± 10 MPa, and 328 ± 7 MPa, respectively, which is more than four times that of the CG samples (48 ± 5MPa). Figure 27.1d shows the microhardness profiles of Cu-5.7 wt% Ge samples. The average microhardness of the CG samples is ~815 MPa, the thickness of GS layer apparently increased with increasing SMAT time.

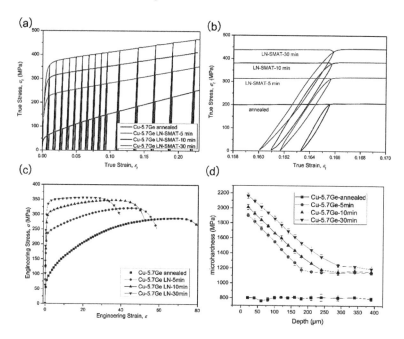

Figure 27.1 (a) The unloading and reloading stress–strain curves of annealed (CG) and SMAT (GS) samples. (b) Typical hysteresis loops taken from (a). (c) The engineering stress–strain curves of Cu-5.7 wt% Ge samples. (d) The microhardness profiles as a function of the depth.

For GS samples under tensile loading, the soft coarse grains in the core begin deformation first and will be restrained by the harder GS layer. The mechanical incompatibility of GS and CG layers produces a large strain gradient which need to be accommodated by geometrically necessary dislocations (GNDs) to maintain the continuity [17]. The GNDs pile ups near the GS/CG border and the GS layer produces a strong long-range internal HDI stress that can enhance the yield strength. In addition, there is a strong HDI hardening, i.e., HDI stress increases with applied strain, which will enhance the ductility.

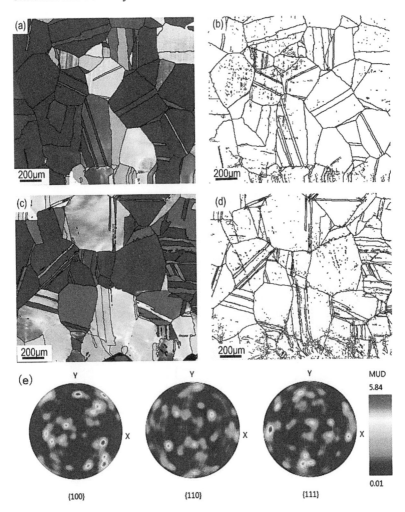

Figure 27.2 EBSD images of GS samples: (a) orientation map of the Cu-5.7Ge LN-SMAT-5 min sample. (b) Boundary misorientation map of the Cu-5.7Ge LN-SMAT-5 min sample. (c) Orientation map of the Cu-5.7Ge LN-SMAT-30 min sample. (d) Boundary misorientation map of the Cu-5.7Ge LN-SMAT-30 min sample. (e) {100}, {110}, {111} pole figures of the Cu-5.7Ge LN-SMAT-30 min sample.

The increment of the yield strength is due to the formation of dislocations rather than texture [20, 21]. The strength contribution via the contribution of the total dislocation density, the flow stress of the Cu-5.7Ge alloy can be expressed as [22, 23]

$$\sigma_f = \sigma_o + M\alpha Gb\sqrt{\rho_{Dis}} \quad (27.1)$$

where σ_f is the flow stress, σ_o is the friction stress (σ_0 = 20 MPa) [24], b is the Burgers vector (b = 0.256 nm), G is the shear modulus (G = 45,000 MPa), α is a constant considered to be 0.24 [25], M is the Taylor factor which is taken to be 3.0, and ρ_{Dis} is the total dislocation density. As we can see, the Cu-5.7Ge LN-SMAT-30 min sample have a higher dislocation density compared to Cu-5.7Ge LN-SMAT-5 min sample in boundary misorientation map, which indicates that flow stress increases as the SMAT treating time increases.

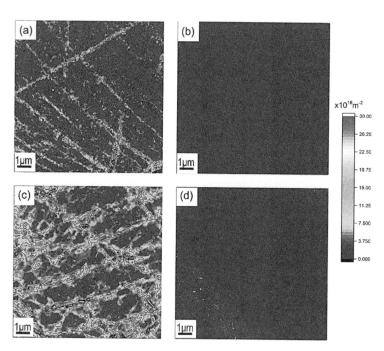

Figure 27.3 GND density map of the GS samples: (a) the surface of Cu-5.7Ge LN-SMAT-5 min sample. (b) The center of Cu-5.7Ge LN-SMAT-5 min sample. (c) The surface of Cu-5.7Ge LN-SMAT-5 min sample. (d) The center of Cu-5.7Ge LN-SMAT-5 min sample.

Figure 27.3 shows GND density distribution map measured by EBSD. The scan size is 10 μm × 10 μm in each scan slice, step size is 50 nm, and the grain boundary misorientation of GND density is drawn from 0 to 2°. The kernel average misorientation (KAM) can be obtained directly from EBSD data, which can be used to measure the local misorientations [19]. GND density can be estimated by the strain gradient model research by Gao and Kubin [26]

$$\rho_{GND} = \frac{2\theta}{\mu b} \qquad (27.2)$$

where μ is the unit length (step size), b is the Burgers vector and θ is the local misorientations. Figure 27.4. shows the histogram of GND density distribution calculated from the EBSD results in Fig. 27.3. Interestingly, Fig. 27.3 also shows that the GNDs are mostly concentrated in shear bands, which is not surprising because the shear bands are where the dislocation slip and pile up occurs. As can be seen in Fig. 27.3. and Fig. 27.4, GND densities are significantly higher in surface than center for GS samples. It can be concluded that higher GND densities are found near the GS/CG border than away from the border as the SMAT time increases, which indicates larger strain gradient.

The long-range internal stress, i.e., HDI stress, can be used to quantify the Bauschinger effect through unloading-reloading cyclic tensile test. Figure 27.5a is the schematic of hysteresis loops for characterizing the Bauschinger effect. The HDI stress σ_b, which is mainly ascribed to the long-range interactions of mobile dislocations, can be calculated by [16]:

$$\sigma_b = \sigma_f - \sigma_{eff} \qquad (27.3)$$

where σ_f is the flow stress, and the σ_{eff} is the effective stress associated with the short-range interactions like forest dislocation hardening.

$$\sigma_{eff} = \frac{\sigma_f - \sigma_u}{2} + \frac{\sigma^*}{2} \qquad (27.4)$$

where σ_u is the reverse yield stress during unloading, and σ^* is the thermal part of the flow stress, as defined in Fig. 27.5a. Recently, Yang et al. [27] derived the following new equation to calculate the HDI stress basing on the assumption that dislocation pileup is reversible during the unloading-reloading process:

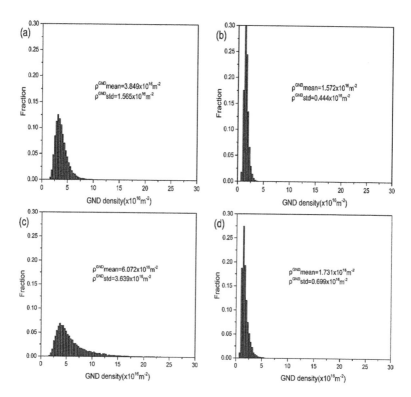

Figure 27.4 Histogram shows GND density distribution calculated from results in Fig. 27.3. The mean and standard deviation of the GND density are labeled in each histogram.

$$\sigma_b = \frac{\sigma_r + \sigma_u}{2} \qquad (27.5)$$

The advantage of Eq. 27.5 is that the calculated HDI stress values are less scattered than those calculated using other equations [27].

In the model, the reloading modulus Er (the slope of segment EF) is assumed equal to unloading Young's modulus Eu (the slope of segment BC) since the defect structures are assumed to remain unchanged/reversible during the unloading–reloading test. In this chapter, we use 5% slope reduction of the Young's modulus to calculate the HDI stress, as detailed in previous work proposed by Yang et al. [27]. The modulus reduction has a clear physical meaning: 5% modulus reduction can be considered as 5 vol % of the sample is deforming plastically. Figure 27.5b shows the measured hysteresis loop of Cu-5.7Ge LN-SMAT-10 min sample at true strain of 13.98% with defined σ_u and σ_r. The calculated HDI stresses of both GS and CG samples are presented in Fig. 27.5d.

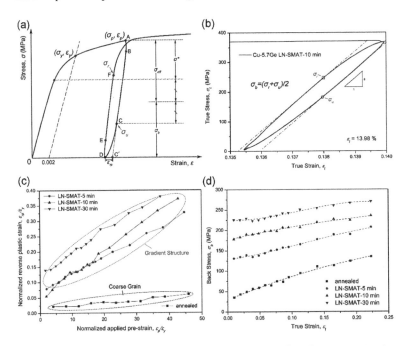

Figure 27.5 (a) The schematic of hysteresis loops for characterizing the Bauschinger effect. (b) A measured hysteresis loop of Cu-5.7Ge LN-SMAT-10 min sample at true strain of 13.98% with defined σ_u and σ_r. (c) The normalized reverse plastic strain varies as a function of the normalized applied pre-strain. (d) The evolution of HDI stresses with increasing applied strain.

As illustrated in Fig. 27.5d, the HDI stress increases with applied true strain. As the strain increases, more and more dislocations interact and pile-up GNDs at grain boundaries. The dislocation density increases and the dislocations motion on the slip plane will be hampered. Since Bauschinger effect is generally associated with the dislocation density, it is reasonable to say that increasing dislocation density and dislocation pile-ups consequently affect the Bauschinger behavior and produce a higher HDI stress.

Figure 27.5d shows that the HDI stress is significantly increased with increasing SMAT treating time. The HDI stress in CG sample is ~35 MPa at true strain of 0.01. After SMAT treating for 30 min, it was increased by ~5.5 times to 225 MPa. This high HDI stress is resulted from the gradient structure. A thicker GS layer with larger strength gradient produces higher strain gradient during tensile test and consequently more GNDs and higher HDI stresses. Owing to the microstructural heterogeneity of GS sample, a multiaxial stress state is also developed during the tensile test, which enhances strain hardening [17].

From Fig. 27.5d shows that the HDI stress hardening increased a little slower with applied tensile strain in GS samples processed by SMAT for longer time. This may be attributed to the severe deformation during the SMAT, which can lead to more dislocation generation in the sample. The high dislocation density in GS sample processed for longer time makes it difficult to accumulate more GNDs during the tensile test, because there is less room for dislocation to pile up and accumulate.

Figure 27.5c shows that the normalized reverse plastic strain varies as a function of the normalized applied pre-strain. The difference between GS and CG samples is obvious. The reverse plastic strain ε_{rp} increased almost linearly with applied strain. All of the samples possess a non-vanishing reverse plastic strain; however, the gradient structure dramatically amplifies this effect by a factor of ~7. It is evident that the GS samples exhibit a quicker raise in reverse plastic strain compared to the CG sample, indicating a much stronger Bauschinger effect. Homogeneous CG sample shows a small Bauschinger effect. This indicates that the Bauschinger effect and the HDI stress are closely related. Stronger HDI stress correlates with stronger Bauschinger effect. This can be understood by the fact that both are created by the pileup of GNDs.

Figure 27.6 shows the HDI stress σ_b, flow stress σ_f, and their ratio (σ_b/σ_f) in CG and GS samples near the yield point (at a strain of 1%). The σ_b/σ_f increases with increasing structural gradient (SMAT processing time), in other words, the contribution of HDI stress on the flow stress increases, which demonstrates that the high HDI stress is caused by gradient structure and play an important role on improving the strength.

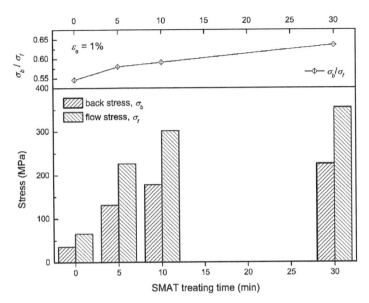

Figure 27.6 HDI stress and flow stress of CG and GS samples near the yield point (at strain of 1%).

The work hardening rate ($\Theta = d\sigma/d\varepsilon$) as a function of true strain is shown in Fig. 27.7. The variation of work hardening rate is characterized with a sharp decline in stage I, followed by a slight increase or tends to a constant in stage II, and then a gradual decline just before the necking in stage III. Homogeneous CG sample shows higher work hardening rate compared to the GS samples and the work hardening rate of the GS samples decline gradually with increasing SMAT treating time. As shown in Fig. 27.7, the three stage is similar to previous experimental observation [28] and [29] studies, which is reported that it is associated with the dislocation density increase and HDI stress evolution during tensile test.

According to the Kocks–Meacking model [30] (K–M model), which is used to characterize the evolution of dislocation density and explain the behavior of work hardening rate. Dislocation density evolution is derived from generation and annihilation rates of dislocation [31], and can be expressed as:

$$\frac{d\rho}{d\gamma} = \left(\frac{d\rho}{d\gamma}\right)^{+} + \left(\frac{d\rho}{d\gamma}\right)^{-} \tag{27.6}$$

where ρ is dislocation density and γ is shear strain, the mechanism of dynamic recovery process is dislocation annihilation.

Stage I starts after yielding of the material, occurs only at lower plastic strain, which does not have enough dislocation storage. A change of strain path was reported in GS interstitial-free (IF) steel, which leads to an initial reduction of dislocation density and strain hardening rate. However, this effect is not obvious in the current study. As shown in Fig. 27.7, the curves of GS samples (SMAT for 5, 30 min) remains a constant in stage II. The CG sample has higher increase in work hardening as compared to the GS samples, despite of the higher HDI hardening in the GS samples. This means that the rate of dislocation accumulation is much greater in the CG sample than in the GS samples.

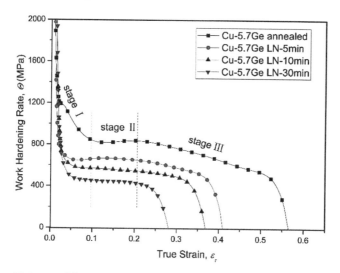

Figure 27.7 Work hardening rate vs. true strain.

It can be seen from the Fig. 27.3d that the values of HDI hardening, which is ~45, ~38, ~25 and ~25 for CG sample and GS samples (SMAT for 5, 10, 30 min), respectively. The above results show that the work hardening rate remain unchanged and increase when HDI hardening is approximately ~25 and more than ~25. The above description is corresponding to the capacity of HDI hardening, which indicates that the high ductility can be attributed to the high HDI hardening in stage II. The strain hardening gradual decline in stage III.

27.4 Summary

In this chapter, we have analyzed and measured HDI stress and the Bauschinger effect in gradient structured Cu-Ge sample. The GS samples exhibit a significant Bauschinger effect and high HDI stresses. The pronounced HDI stress in GS sample can be attributed to the large strain gradient which introduces lots of geometrically necessary dislocations (GNDs). Higher GND densities are found near the GS/CG border as the SMAT time increases. The effect of HDI stress increases with increasing SMAT processing time due to the increasing structural gradient. The HDI stress increases the yield strength and the HDI hardening helps with retaining ductility.

Acknowledgements

The authors would like to acknowledge financial supports by the National Natural Science Foundation of China (NSFC) under Grant No. 51561015 and No. 51664033, and the introduction of talent fund project of Kunming University of Science and Technology (KKSY201407100). YTZ would like to acknowledge the support of the US Army Research Office under the Grant Nos. W911NF-09-1-0427.

References

1. X.L. Wu, M.X. Yang, F.P. Yuan, L. Chen, Y.T. Zhu, Combining gradient structure and TRIP effect to produce austenite stainless steel with high strength and ductility, *Acta. Mater.* **112** (2016) 337–346.
2. Q. Wang, Y. Yang, H. Jiang, C.T. Liu, H.H. Ruan, J. Lu, Superior tensile ductility in bulk metallic glass with gradient amorphous structure, *Sci. Rep.* **4** (2014) 4757.
3. T.H. Fang, N.R. Tao, K. Lu, Tension-induced softening and hardening in gradient nanograined surface layer in copper, *Scr. Mater.* **77** (2014) 17–20.
4. Y. Wei, Y. Li, L. Zhu, Y. Liu, X. Lei, G. Wang, Y. Wu, Z. Mi, J. Liu, H. Wang, H. Gao, Evading the strength–ductility trade-off dilemma in steel through gradient hierarchical nanotwins, *Nat. Commun.* **5** (2014) 3580.
5. K.M. Youssef, R.O. Scattergood, K.L. Murty, J.A. Horton, C.C. Koch, Ultrahigh strength and high ductility of bulk nanocrystalline copper, *Appl. Phys. Lett.* **87** (2005) 091904-091904-091903.

6. X. Wu, P. Jiang, L. Chen, F. Yuan, Y.T. Zhu, Extraordinary strain hardening by gradient structure, *Proc. Natl. Acad. Sci. U.S.A.* **111** (2014) 7197–7201.

7. X.L. Wu, P. Jiang, L. Chen, J.F. Zhang, F.P. Yuan, Y.T. Zhu, Synergetic strengthening by gradient structure, *Mater. Res. Lett.* **2** (2014) 185–191.

8. K. Lu, J. Lu, Nanostructured surface layer on metallic materials induced by surface mechanical attrition treatment, *Mater. Sci. Eng. A* **375–377** (2004) 38–45.

9. X. Wu, M. Yang, F. Yuan, G. Wu, Y. Wei, X. Huang, Y. Zhu, Heterogeneous lamella structure unites ultrafine-grain strength with coarse-grain ductility, *Proc. Natl. Acad. Sci. U.S.A.* **112** (2015) 14501–14505.

10. Y. Xiang, J.J. Vlassak, Bauschinger effect in thin metal films, *Scr. Mater.* **53** (2005) 177–182.

11. Y. Xiang, J.J. Vlassak, Bauschinger and size effects in thin-film plasticity, *Acta. Mater.* **54** (2006) 5449–5460.

12. J. Rajagopalan, J.H. Han, M.T.A. Saif, Bauschinger effect in unpassivated freestanding nanoscale metal films, *Scr. Mater.* **59** (2008) 734–737.

13. J. Rajagopalan, C. Rentenberger, H. Peter Karnthaler, G. Dehm, M.T.A. Saif, In situ TEM study of microplasticity and Bauschinger effect in nanocrystalline metals, *Acta. Mater.* **58** (2010) 4772–4782.

14. S.-W. Lee, A.T. Jennings, J.R. Greer, Emergence of enhanced strengths and Bauschinger effect in conformally passivated copper nanopillars as revealed by dislocation dynamics, *Acta. Mater.* **61** (2013) 1872–1885.

15. D. Zhu, H. Zhang, D.Y. Li, Influence of nanotwin boundary on the Bauschinger's effect in Cu: a molecular dynamics simulation study, *MMTA* **44** (2013) 4207–4217.

16. X. Feaugas, On the origin of the tensile flow stress in the stainless steel AISI 316L at 300 K: back stress and effective stress, *Acta. Mater.* **47** (1999) 3617–3632.

17. X.L. Wu, P. Jiang, L. Chen, F.P. Yuan, Y.T.T. Zhu, Extraordinary strain hardening by gradient structure, *Proc. Natl. Acad. Sci. U.S.A.* **111** (2014) 7197–7201.

18. X.H. An, S. Qu, S.D. Wu, Z.F. Zhang, Effects of stacking fault energy on the thermal stability and mechanical properties of nanostructured Cu–Al alloys during thermal annealing, *J. Mater. Res.* **26** (2011) 407–415.

19. X. Ma, C. Huang, J. Moering, M. Ruppert, H.W. Höppel, M. Göken, J. Narayan, Y. Zhu, Mechanical properties of copper/bronze laminates: role of interfaces, *Acta. Mater.* **116** (2016) 43–52.
20. A. Kundu, D.P. Field, Influence of plastic deformation heterogeneity on development of geometrically necessary dislocation density in dual phase steel, *Mater. Sci. Eng. A* **667** (2016) 435–443.
21. X.C. Liu, H.W. Zhang, K. Lu, Strain-induced ultrahard and ultrastable nanolaminated structure in nickel, *Science* **342** (2013) 337–340.
22. L. Lu, R. Schwaiger, Z.W. Shan, M. Dao, K. Lu, S. Suresh, Nano-sized twins induce high rate sensitivity of flow stress in pure copper, *Acta. Mater.* **53** (2005) 2169–2179.
23. O. Bouaziz, N. Guelton, Modelling of TWIP effect on work-hardening, *Mater. Sci. Eng. A* **319** (2001) 246–249.
24. N. Hansen, Hall–Petch relation and boundary strengthening, *Scr. Mater.* **51** (2004) 801–806.
25. F. Yan, H.W. Zhang, N.R. Tao, K. Lu, Quantifying the microstructures of pure Cu subjected to dynamic plastic deformation at cryogenic temperature, *J. Mater. Sci. Technol.* **27** (2011) 673–679.
26. M. Calcagnotto, D. Ponge, E. Demir, D. Raabe, Orientation gradients and geometrically necessary dislocations in ultrafine grained dual-phase steels studied by 2D and 3D EBSD, *Mater. Sci. Eng. A* **527** (2010) 2738–2746.
27. M. Yang, Y. Pan, F. Yuan, Y. Zhu, X. Wu, Back stress strengthening and strain hardening in gradient structure, *Mater. Res. Lett.* (2016) 1–7.
28. A. Rohatgi, K.S. Vecchio, G.T. Gray, The influence of stacking fault energy on the mechanical behavior of Cu and Cu-Al alloys: deformation twinning, work hardening, and dynamic recovery, *MMTA* **32** (2001) 135–145.
29. F. Hamdi, S. Asgari, Influence of stacking fault energy and short-range ordering on dynamic recovery and work hardening behavior of copper alloys, *Scr. Mater.* **62** (2010) 693–696.
30. U.F. Kocks, H. Mecking, Physics and phenomenology of strain hardening: the FCC case, *Prog. Mater. Sci.* **48** (2003) 171–273.
31. A. Vinogradov, Mechanical properties of ultrafine-grained metals: new challenges and perspectives, *Adv. Eng. Mater.* **17** (2015) 1710–1722.

Chapter 28

Gradient Structured Copper Induced by Rotationally Accelerated Shot Peening

X. Wang,[a] Y. S. Li,[a] Q. Zhang,[b] Y. H. Zhao,[a] and Y. T. Zhu[a,c]

[a]*Nano Structural Materials Center, School of Materials Science and Engineering, Nanjing University of Science and Technology, Nanjing 210094, China*
[b]*School of Mechanical Engineering, Nanjing University of Science and Technology, Nanjing 210094, China*
[c]*Department of Materials Science and Engineering, North Carolina State University, Raleigh, NC 27695, USA*
liyusheng@njust.edu.cn

A new technology, rotationally accelerated shot peening (RASP), was developed to prepare gradient structured materials. By using centrifugal acceleration principle and large balls, the RASP technology can produce much higher impact energy when compared to conventional shot peening. As a demonstration, the RASP-process refined the surface layer in copper to an average grain size of ~85 nm. The grains are increasingly larger along the depth, forming

Reprinted from *J. Mater. Sci. Technol.*, **33**, 758–761, 2017.

a 800 μm thick gradient-structured layer and consequently a microhardness gradient. The difference between the RASP technology and other techniques in preparing gradient structured materials is discussed in the chapter. The RASP technology is conducive to large-scale industrial processing of gradient materials.

28.1 Introduction

Gradient structures with excellent mechanical properties have been found in many biological systems such as bamboo and seashells [1]. Several techniques have been developed to prepare gradient materials such as wire brushing [2], ultrasonic shot peening (USP) [3], surface mechanical attrition treatment (SMAT) [4], surface nanocrystallization and hardening (SNH) [5] and surface mechanical grinding treatment (SMGT) [6]. Gradient structures with grain sizes varying systematically from nanometers to micrometers induced by these techniques showed many superior properties over conventional homogeneously-structured materials, such as low nitriding temperature [7], excellent tribological properties [8], great fatigue life [9] and good combination of high strength and ductility [10–12]. However, there are some shortcomings in the industrial application of these technologies. For example, the thickness of gradient layer formed by wire brushing is usually limited. The energies of the balls in USP, SMAT and SNH are controlled by changing the vibration frequency but the velocity of balls cannot be easily controlled. The emerging SMGT technology is quite promising but it is only suitable for cylindrical samples presently. In a summary, development of new technologies is still a challenge to produce gradient structured materials.

Centrifugal force has been widely used in daily life and advanced technology development. For example, it has been used in washing machines, simultaneous densification of intermetallic [13] or even for measurement of xylem cavitation [14]. In this work, based on the centrifugal acceleration principle, a new technology, RASP has been developed to produce gradient structures on metallic materials, and pure copper was chosen as a model material to be processed by the RASP technology.

28.2 Experimental

The experimental setup of the RASP process is illustrated in Fig. 28.1. A RASP equipment is mainly consisted of power generator, sample room, steel ball recycle system, dust elimination system and control system. Steel balls enter into the rotational system, accelerated to high speed by the centrifugal force, and fly off to impact the surface of a metal sample. The diameter of balls can be chosen in the range of 1 to 8 mm. The velocity of balls is adjusted by controlling the rotational speed and can reach a maximum of 80 m/s. Steel balls are recycled during the whole RASP process and all balls return to a storage tank when the RASP process is finished. Any dust or debris produced during RASP process is collected in a dust elimination system. Treated sample is clamped and put in the sample room and can be rotated 360° around the fixed axis. The rotation speed can be adjusted to improve the treatment uniformity and coverage. The whole RASP process is automatically completed once treatment parameters are set.

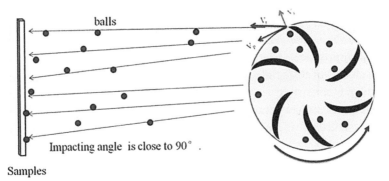

Figure 28.1 A schematic illustration of the RASP process.

A commercial 99.97% pure copper plate ($90 \times 50 \times 10$ mm^3) was selected as the model material for processing by RASP. The initial grain size was about 100 μm. The sample was treated in air at room temperature with 3-mm diameter GGr15 steel balls at a velocity of about 25 m/s for 30 min. The sample was fixed without rotation. An optical microscope (ZEISS AXIO CSM700) was used to observe the microstructures of the treated Cu. The microhardness profile from the surface to interior in the as-prepared Cu was measured using a HMV-705 tester and the spacing between two indentions is greater than the three times value of indention to avoid the overlap of hardening zones of adjacent indents. Electron Back-Scattered Diffraction (EBSD) observation was carried in a Quant 250 FEG scanning electron microscope (SEM). The step size was 50 nm and the scanning voltage was 20 KV. Cross-sectional microstructures of the treated sample were characterized by using a TECNAI G2 20 LaB6 transmission electron microscope (TEM) at the voltage of 200 KV. The TEM samples were cut by electric spark from the cross section and mechanically polished to ~40 μm thick foils. The final thinning was accomplished by ion milling.

28.3 Results and Discussion

Figure 28.2 shows an optical cross-sectional image of the RASP-processed Cu. It is apparent that there is a microstructural gradient from the treated surface to the interior, although it is very difficult to differentiate individual grains in the top 100 μm layer, which implies a heavy plastic deformation-induced grain refinement. The initial equiaxed coarse grains become elongated, which can be attributed to compressive strain induced by impact of steel balls [15]. The thickness of the gradient layer was estimated to be about 800 μm, which is consistent with the result of hardness measurement shown in Fig. 28.3. Every hardness data point is an averaged value from 7 measurements and the error bar is also plotted. The average hardness in ~10 μm depth from the surface is about 144 Hv, which is much higher than that of coarse-grained (CG) sample, which is about 90 Hv as indicated by the red star. As shown in Fig. 28.3, the hardness decreased gradually with distance from the surface, and finally became close to that of CG counterpart (90 Hv) at the depth of 800 μm.

In order to better understand the microstructures in the top surface layer, EBSD observation was performed on the top surface layer of the RASP-processed Cu sample. Fig. 28.4 shows the variation of grain size and orientation along the depth. The black layer corresponds to the layer of nanograins where high strain concentration and fine grains makes it impossible to index individual grains correctly [16]. The fraction of grains that can be indexed increased with the distance from the surface. Some ultrafine grains were observed in the deeper region. As shown, the grains at larger depth are elongated with the long axis parallel to the sample surface. As discussed earlier, this was caused by the compressive stress and strain imparted by the colliding steel balls during the RASP process.

Figure 28.2 Optical micrograph of the cross section of a RASP-processed Cu sample.

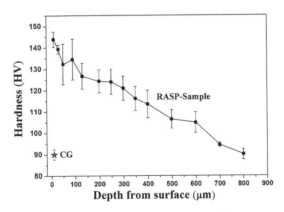

Figure 28.3 The hardness gradient along the depth of a RASP-processed Cu sample.

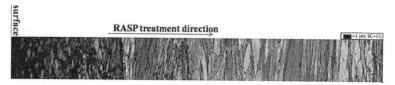

Figure 28.4 An EBSD map along the depth of a RASP-processed Cu sample.

Figures 28.5a and b are representative cross-sectional TEM bright and dark field micrographs from the top surface of the RASP-processed Cu sample, where many nanometer-sized grains can be observed and the crystallographic orientations of these nanometer-sized grains were randomly distributed, as indicated by the corresponding selected area electro diffraction (SAED) pattern (Fig. 28.5a, inset). The statistical distribution of grain sizes and aspect ratios are shown in Figs. 28.5c and d, which were obtained by counting ~200 grains. The grain size distribution can be described by a normal logarithmic distribution with an average of 85 nm in transverse grain size and 136 nm in longitudinal axis.

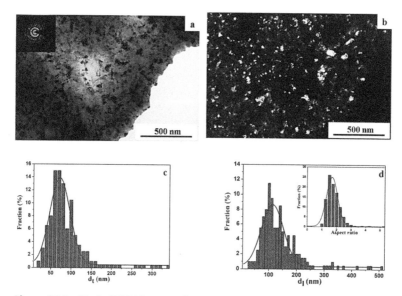

Figure 28.5 Typical TEM images of nanometer-sized grains in the top surface layer: (a) the bright field image and SAED pattern (inset); (b) the dark field image; (c) the histograms of transerve grain sizes (d_t) and (d) longitudal grain sizes (d_l). The inset in (d) is the aspect ratio (d_t/d_l) distribution.

The microstructures at different depths were characterized and shown in Fig. 28.6. From the nano grains in the surface layer (Fig. 28.5) to the ultrafine grains at the depth of 50 μm and the mixture structures of elongated grains and ultrafine grains at 130 μm (Fig. 28.6a, b), dislocation band structures were formed at the depth of 200 μm (Fig. 28.6c), and finally dislocation cell structures at the depth of 300 μm (Fig. 28.6d), as can be clearly seen that a grain/cell size gradient was formed in the surface of Cu samples induced by RASP treatment, which is in good agreement with the micro-hardness gradient shown in Fig. 28.3, implying that the hardness improvement in RASP-processed Cu mainly comes from the grain refinement and dislocation accumulation. Dislocation cells and walls were formed in the early stage of deformation corresponding to the deeper region (300 μm). Dislocation cell boundaries evolved into low-angle grain boundaries and eventually into high-angle boundaries with increasing strain, which was dominated by dislocation activities [17, 18]. The grain refinement and microstructural evolution in Cu during RASP will be further investigated in the future work.

Figure 28.6 Microstrures at the different depth from the surface: (a) ~50 μm, (b) ~130 μm, (c) ~200 μm, and (d) ~300 μm.

The refinement of microstructure is attributed to the severe plastic deformation induced by consecutive impacts of high velocity balls on the sample surface with gradient distribution of strain and strain rate [17]. The RASP technique developed here is different from the conventional shot peening by using much larger balls. It is well known that the kinetic energy of the balls depends on their size and velocity, and Table 28.1 gives a summary of the process parameters and kinetic energy of balls in shot peening, USP, SMAT, SNH and RASP, respectively. It cab be clearly seen that although in conventional shot peening the ball speed can reach as high as 150 m/s, the much larger balls in RASP can possess 100 times higher impact energy as compared to conventional shot peening [19], which give RASP the capability to refine copper grains to smaller sizes and much greater depths than conventional shot peening. The thickness of the gradient layer (~800 µm) in RASP-processed Cu is similar to what is produced by SMAT [17], as compared in Table 28.1, because the kinetic energy of balls is in the same order between the present RASP process and the SMAT process in literature [17]. It can be also seen from Table 28.1 that the kinetic energy of balls in RASP can be varied in a much wider range than other techniques, which is very important because metals with different hardness may need different impact kinetic energy to produce optimum gradient structure and acceptable surface roughness. The influence of RASP processing parameters on the surface roughness and gradient microstructures of other metallic materials are in progress [26].

Table 28.1 Comparison of process parameters between RASP and other techniques

Technique	Diameter of balls or shots (mm)	Impact velocity (m/s)	Density of balls (g/cm^3)	Kinetic energy of balls (J)
Shot Peening [20]	0.25–1	20–150	7.74	3.2×10^{-5}–0.12
USP [21, 22]	3–7.5	<20	7.98	0.06–0.9
SMAT [23, 24]	3–8	2–5	7.98	6×10^{-4}–0.07
SNH [25]	5–7.9	5	14.5	0.03–0.12
RASP	1–8	5–80	7.8	1.3×10^{-4}–17

The RASP technology developed here is conducive to large-scale industrial processing for the following reasons:

Firstly, the RASP technology has lower energy waste and higher processing efficiency. Relatively larger energy waste was caused by the collision between balls and chamber walls in USP [3], SMAT [4] and SNH [5] as a result of random impacts provided by the vibrator. Moreover, the velocity provied by high-frequency vibration is reatively low (<20 m/s) [27], which implies the treatment efficiency is not very high. Compared to these techniques, relatively lower energy consumption of RASP is realized since energy is mainly provided to generate collisons between balls and workpieces because of relativley low scatter angle. Higher impact velocity of RASP process can improve the energy effeicency.

Secondly, there is no shape restriction for RASP samples. By 360° rotation, both work pieces with complex irregular shapes can be simutaneously treated by the RASP technology.

Thirdly, the RASP technology is similar to conventionl shot peening technology, which means that the RASP technology can be easily incorporated into a modified conventional shot peening and there is no need to completely rebuilt the production line. Furthermore, RASP technique has a high degree of automation as discussed previously in section 28.2, which meets the requirements in the large-scale industry application.

It is found that the grain size in the top surface layer is a little larger than what was reproted in copper processed by SMAT [17, 28], where the grain size in topmost surface can be refined to as small as 10 nm. The large grain size in the RASP processed Cu coulf be caused by the recrystallization during RASP processing [6]. Surface heating occurs due to the deformation-induced heat generation within the sample, ball-to-ball impact, and frictional heating during the repeated imapct process [29]. The surface temperature increase has significant effect on the recrustallization and growth of nano-grains [30, 31].

The RASP technique provides a new approach to produce gradient structures in metallic materials. Combining the existing advantages of RASP and the experience from traditional shot peening industries, this new technique, is expected to be easily scaled-up and adapted to industrial production and applications.

28.4 Conclusion

In a summary, a novel technology, RASP, was developed and demonstrated to effectively produce surface nanocrystallization and gradient structure in a bulk copper sample. The obtained results can be summarized below:

(1) RASP technology has higher impacting energy and can be adjusted in a wide energy range.
(2) The total thickness of gradient structure is ~800 μm in RASP processed Cu. A grain/cell size gradient was observed, from micron-sized dislocation cell, evolved to dislocation band structures, ultra-fine grain and finally averagely 85 nm sized grain in the top surface.
(3) The hardness of top surface is about 144 Hv, which is much higher than that in the initial coarse-grained sample, a hardness gradient was observed along with distance from the surface.
(4) RASP processing is expected to be easily scaled-up, which makes it promising for industrial applications.

Acknowledgements

Financial supports from the National Natural Science Foundation of China (Grant No. 51301092) and Pangu Foundation are acknowledged.

References

1. R.O. Ritchie, *Nat. Mater.* **10** (2011) 817–822.
2. M. Sato, N. Tsuji, Y. Minamino, Y. Koizumi, *Sci. Technol. Adv. Mater.* **5** (2004) 145–152.
3. G. Liu, J. Lu, K. Lu, *Mater. Sci. Eng. A* **286** (2000) 91–95.
4. K. Lu, J. Lu, *J. Mater. Sci. Technol.* **15** (1999) 193–197.
5. K. Dai, J. Villegas, L. Shaw, *Scr. Mater.* **52** (2005) 259–263.
6. W.L. Li, N.R. Tao, K. Lu, *Scr. Mater.* **59** (2008) 546–549.
7. W.P. Tong, N.R. Tao, Z.B. Wang, J. Lu, K. Lu, *Science*, **299** (2003) 686–688.

8. Y.S. Zhang, Z. Han, K. Wang, K. Lu, *Wear* **260** (2006) 942–948.
9. T. Roland, D. Retraint, K. Lu, J. Lu, *Scr. Mater.* **54** (2006) 1949–1954.
10. T.H. Fang, W.L. Li, N.R. Tao, K. Lu, *Science* **331** (2011) 1587–1590.
11. X. Wu, P. Jiang, L. Chen, F. Yuan, Y.T. Zhu, *Proc. Natl. Acad. Sci. U.S.A.* **111** (2014) 7197–7201.
12. X.L. Wu, P. Jiang, L. Chen, J.F. Zhang, F.P. Yuan, Y.T. Zhu, *Mater. Res. Lett.* **2** (2014) 185–191.
13. R. Seshadri, *Mater. Manuf. Process.* **17** (2002) 501–518.
14. N.N. Alder, W.T. Pockman, J.S. Sperry, S. Nuismer, *J. Exp. Bot.* **48** (1997) 665–674.
15. Y.S. Li, N.R. Tao, K. Lu, *Acta Mater.* **56** (2008) 230–241.
16. Y. Samih, B. Beausir, B. Bolle, T. Grosdidier, *Mater. Charact.* 83 (2013) 129–138.
17. K. Wang, N.R. Tao, G. Liu, J. Lu, K. Lu, *Acta Mater.* **54** (2006) 5281–5291.
18. J. Huang, Y. Zhu, H. Jiang, T. Lowe, *Acta Mater.* **49** (2001) 1497–1505.
19. K. Dai, L. Shaw, *Mater. Sci. Eng. A* **463** (2007) 46–53.
20. Y.F. Al-Obaid, *Mech. Mater.* **19** (1995) 251–260.
21. N.R. Tao, M.L. Sui, J. Lu, K. Lua, *Nanostruct. Mater.* **11** (1999) 433–440.
22. X. Wu, N. Tao, Y. Hong, B. Xu, J. Lu, K. Lu, *Acta Mater.* **50** (2002) 2075–2084.
23. N. Tao, Z. Wang, W. Tong, M. Sui, J. Lu, K. Lu, *Acta Mater.* **50** (2002) 4603–4616.
24. K.Y. Zhu, A. Vassel, F. Brisset, K. Lu, J. Lu, *Acta Mater.* **52** (2004) 4101–4110.
25. K. Dai, J. Villegas, Z. Stone, L. Shaw, *Acta Mater.* **52** (2004) 5771–5782.
26. X. Wang, Y.S. Li, to be published.
27. K. Dai, L. Shaw, *Mater. Sci. Eng. A* **463** (2007) 46–53.
28. J. Guo, K. Wang, L.U. Lei, *J. Mater. Sci. Technol.* **22** (2006) 789–792.
29. K.A. Darling, M.A. Tschopp, A.J. Roberts, J.P. Ligda, L.J. Kecskes, *Scr. Mater.* **69** (2013) 461–464.
30. M. Ames, J. Markmann, R. Karos, A. Michels, A. Tschöpe, R. Birringer, *Acta Mater.* **56** (2008) 4255–4266.
31. G. Sharma, J. Varshney, A.C. Bidaye, J.K. Chakravartty, *Mater. Sci. Eng. A* **539** (2012) 324–329.

Chapter 29

Microstructure Evolution and Mechanical Properties of 5052 Alloy with Gradient Structures

Yusheng Li,[a] Lingzhen Li,[a] Jinfeng Nie,[a] Yang Cao,[a] Yonghao Zhao,[a] and Yuntian Zhu[a,b]

[a]*Nano Structural Materials Center, School of Materials Science and Engineering, Nanjing University of Science and Technology, Nanjing 210094, China*
[b]*Department of Materials Science and Engineering, North Carolina State University, Raleigh, NC 27695, USA.*
liyusheng@njust.edu.cn

In this chapter, we report on the microstructure evolution and mechanical properties of 5052 Al alloy processed by rotationally accelerated shot peening (RASP). Thick deformation layer of ~2 mm was formed after the RASP process. Nano-sized grains, equiaxed subgrains and elongated subgrains were observed along the depth in the deformation layer. Dislocation accumulation and dynamic recrystallization were found primarily responsible for the grain

---

Reprinted from *J. Mater. Res.*, **32**, 4443–4451, 2017.

Heterostructured Materials: Novel Materials with Unprecedented Mechanical Properties
Edited by Xiaolei Wu and Yuntian Zhu
Text Copyright © 2017 Materials Research Society
Layout Copyright © 2022 Jenny Stanford Publishing Pte. Ltd.
ISBN 978-981-4877-10-7 (Hardcover), 978-1-003-15307-8 (eBook)
www.jennystanford.com

refinement process. An obvious microhardness gradient and a good combination of strength and ductility were found in RASP-processed gradient sample. The superior properties imparted by the gradient structure are expected to expand the application of the 5052 Al alloy as a structural material.

29.1 Introduction

5052 Al alloy is a typical 5xxx Al-Mg alloy, and has been widely used due to its high specific strength, excellent corrosion resistance and good formability. However, its relatively low strength limits its application [1]. Severe plastic deformation (SPD) techniques [2–4], such as accumulative roll bonding (ARB) [5, 6], equal-channel angular pressing (ECAP) [7–11], and high pressure torsion (HPT), etc. [12, 13], have been extensively used to process metals, including Al and Al alloys by introducing ultrafine grains (UFG, grain size below 1000 nm) [14]. The strength of these SPD-processed materials can be improved significantly through grain refinement and other mechanisms, including nanotwins, stack faults, non-equilibrium grain boundaries, etc. [15, 16]. For the 5052 Al alloy, microstructures formed during the SPD processes [11, 17–23] mainly consist of elongated and equiaxed UFGs, dislocation cells and dislocation tangles. Nano-grains were formed in HPT-processed 5052 Al alloy and were attributed to the high strain induced dislocation activities [12].

Low ductility is a bottleneck for the application of SPD-processed UFG materials, which is caused by insufficient ability to strain harden in the deformed metals, causing the onset of early necking [24, 25]. Great efforts have been devoted to increasing the ductility of UFG materials [26]. Very recently, gradient materials, that is, a spatial gradient in grain size of a metal from nanostructures in the top surface to the coarse-grains in the center, have attracted extensive attention due to their unique microstructures and mechanical properties, especially a good strength-ductility combination [27–30]. Widely used techniques for preparing gradient materials include surface mechanical attrition treatment (SMAT) [31, 32], surface mechanical grinding treatment (SMGT) [33, 34] and newly developed rotationally accelerated shot peening (RASP) [35]. The

RASP is a variant of SMAT, but has a greater range of shot diameter and its impact velocity can be easily adjusted according to the materials processing need.

The effect of high strain rate deformation on the microstructure evolution of 5052 Al alloy is unclear, and the mechanical properties of 5052 Al alloy with gradient structures has not been investigated. Therefore, the objective of present work is to study the microstructure evolution and grain refinement mechanism in 5052 Al alloy subjected to RASP treatment, and the mechanical properties of the resultant bulk 5052 Al alloy with gradient structures will also be studied.

29.2 Experimental

Commercial 5052 Al alloy sheets with a chemical composition of 2.5Mg, 0.4Fe, 0.3Cr, 0.25Si, 0.10Mn, 0.10Cu, 0.10Zn, 0.2 others, and the balance Al in wt% were used as raw material for processing by RASP. The 5052 Al alloy sheets with a dimension of 120 × 90 × 4 mm^3 (4 mm in thickness) were homogenized at 823 K for 30 min and furnace cooled to obtain coarse grained (CG) structure prior to RASP treatment. The details of the RASP set-up and processing have been described previously [35]. In brief, 2 mm-diameter steel balls were accelerated to high speed by the centrifugal force and impact the metal surface. During the RASP process, the 5052 Al alloy samples were rotated at 15 r/min, thereby both sides of the samples were processed, forming two GS layers sandwiching a CG center layer. In order to obtain 5052 Al alloy sheets with different microstructures and mechanical properties, several RASP treatment durations and ball velocities were chosen. A 5052 Al alloy sample subjected to RASP treatment with a ball velocity of 50m/s and duration time of 5 minutes (referred to as 50 m/s-5 min hereafter) was chosen as a model material for microstructure characterization and mechanical testing.

An optical microscope (OM, Axio Vert A1) was used to observe the holistic microstructure of the deformed 5052 Al alloy. For OM observations, the samples were prepared by mechanical grinding, polishing, followed by anodic coating. Anodic oxidation was carried out in a solution of 40% fluorine boric acid at the voltage of 18 V for 4 to 5 min. The detailed microstructures and quantitative microstructure statistics of the samples were characterized by means of transmission electron microscopy (TEM), performed on a FEI Tecnai 20 TEM at an operation voltage of 200 kV.

Cross-sectional thin foils for TEM observations were prepared in the following steps: (i) careful mechanical grinding foils to less than 40 μm in thickness; (ii) punching the ~40 μm into semicircles with a diameter of 3 mm with the treated surface in the middle; (iii) sticking the two semicircles on a brass ring with a diameter of 3 mm using AB glue, with the RASP treated sides lie in the center of the ring; (iv) 12 hours later, ion thinning to perforation at room temperature with an accelerating voltage of 3 kV. Electron backscatter diffraction (EBSD) observation was performed on a Quant 250 FEG scanning electron microscope (SEM), and the samples for EBSD analysis were obtained by mechanical grinding and polishing followed by electro-polishing at room temperature. EBSD analysis was carried out with 12 kV applied voltage and a scanning step size of 0.5 μm.

Microhardness measurements were carried out on a Shimadzu microhardness tester with a load of 25 g and a loading time of 15 s. The hardness value obtained was averaged from at least 20 indentations for each sample. Uniaxial tensile tests were performed on a Walter + bai LFM20 KN tensile test machine with a strain rate of $6.7 \times 10^{-4} s^{-1}$ at room temperature. A contactless laser extensometer was used to measure the sample strain upon loading. The gauge section of the dog-bone-shaped tensile specimens was 25 mm in length and 2 mm in width. More than four tensile tests were performed on each sample.

29.3 Results and Discussion

29.3.1 OM/EBSD Observations

Figure 29.1a and b show cross-sectional OM micrographs of the annealed and 50 m/s-5 min RASP-processed 5052 Al alloy samples, respectively. Equiaxed grains with an average size of 70 μm were observed in the annealed sample (Fig. 29.1a). Apparently, the equiaxed coarse grains were deformed into flat grains perpendicular to the RASP direction (Fig. 29.1b), where the left and right sides are the outermost layers of the RASP-processed sample (the black arrows refer to the RASP direction). Note that the RASP deformation penetrated through the whole 4 mm-thickness of the sample, which indicates high deformation energy of RASP processing. The surface of the RASP-processed sample is rough, together with the obvious contrast difference, showing the severe local inhomogeneity of the deformation structures.

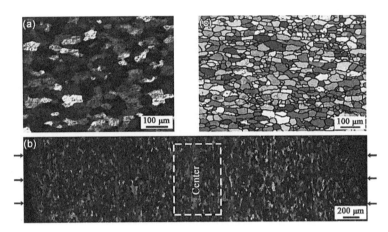

Figure 29.1 OM micrographs of (a) annealed; (b) 50 m/s-5 min RASP-processed 5052 Al alloy and (c) an EBSD image of the center microstructures enlarged from (b).

Figure 29.1c shows a typical EBSD image of the microstructures in the center region marked with a white rectangle in Fig. 29.1b. It can be seen that the grains are elongated perpendicularly to the RASP direction, and the mean grain size is 25 µm, which is much finer than ~70 µm in the CG sample, indicating an obvious grain refinement even in the center region. Moreover, color and contrast variation caused by grain orientation variations exist in the grain interiors, indicating the formation of subgrains and dislocation cells formed by dislocation activities that may include interaction, accumulation, annihilation, and recovery.

29.3.2 TEM Characterization

RASP-processed samples at various depths were characterized using cross-sectional TEM observations to reveal the microstructural characteristics at different levels of strain and strain rate, and thereby to uncover the underlying grain refinement mechanism. Two different grain refinement mechanisms appear to exist: in the deformation layer (>40 μm depth), grain refinement was primarily achieved by dislocation activities; while in the top layer (<40 μm depth), dynamic recrystallization (DRX) played a primary role. In the following sections, the two grain refinement mechanisms will be described in detail.

29.3.2.1 Grain refinement via dislocation activities (depth >40 μm)

In the depth range of 200–2000 μm (center), the microstructures mainly consist of dislocation debris and dislocation entanglement. Figures 29.2a–c show the typical TEM micrographs obtained from the region of 200 μm, 100 μm and 60 μm below the surface, respectively. High density of dislocation was found in 200-μm depth and dislocation entanglement were formed near a grain boundary, as shown in Fig. 29.2a. In the depth of 100 μm (Fig. 29.2b), grain boundaries of an elongated grain (marked by white triangles) can be seen. In addition to dislocation entanglements and dislocation walls, there are small subgrains present at the elongated grain boundary, which could be formed along the grain boundary during the RASP process. The internal contrast of subgrains formed near the grain boundaries is not uniform, which is caused by internal defects, mainly dislocations. For instance, the boundary shared between subgrain A and the elongated grain is sharp while the boundary formed with dislocation walls is wavy and not well delineated (marked by white circles), which is a typical grain boundary morphology caused by SPD [36–38]. High density of dislocation was found in subgrain B whose boundary is poorly defined. It is believed that with increasing deformation strain, more dislocations will accumulate at the boundaries of subgrains A and B, and high-angle grain boundaries will be developed [2, 9]. Typical TEM micrographs taken from the region of 60 μm below the treated surface (Fig. 29.2c) mainly consist of elongated subgrains and dislocation cells. Grain boundaries of subgrains are not obvious because of large deformation. The grain boundaries of the original large grains gradually become curved and fuzzy (marked by white triangles). There are some fine subgrains aggregation areas as shown by the white dotted circles.

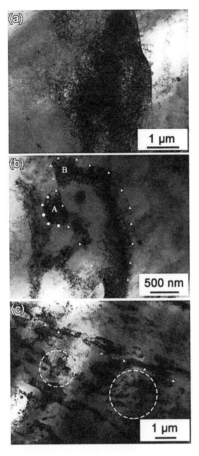

Figure 29.2 Typical TEM micrographs obtained from the region of (a) 200 μm; (b) 100 μm; (c) 60 μm below the treated surface.

Figure 29.3 shows typical TEM images at 40 μm from the surface in a RASP-processed sample. The features of Fig. 29.3a and b are distinctly different, showing the inhomogeneous nature of the high strain rate RASP deformation. Microstructures in Fig. 29.3a consist of many small subgrains with poorly defined boundaries and size of the subgrains is in the range of several hundred nanometers, i.e., UFG range. Dislocation density is high both at interior and at grain boundaries. Selected area electron diffraction (SAED, 600 nm in diameter) pattern shows elongated spots, which indicates polycrystals with small misorientations between these subgrains.

In Fig. 29.3b, the microstructure shows typical dislocation morphology with high density of dislocation walls, dislocation cells and dislocation entanglements. With increasing strain, some big dislocation cells evolve into subgrains with smaller dislocation cells formed in their interior, such as dislocation cells A, B, C. The dislocation cells is a low energy dislocation structure typically formed during the deformation of metals with medium to high stacking fault energy. For the subgrain that generate dislocation cells A, B and C, the dislocations accumulate further with increasing deformation strain, and the dislocations slip to form new dislocation cells in order to reduce the total energy. Dislocation cells typically have low dislocation density in their center areas. This phenomenon is similar to what was observed in different-speed-rolling-processed 5052 Al alloy [17]. Unlike the dislocation cell A, the dislocation densities are relatively high both in the interior and dislocation walls for cells B and C. The subgrains marked by white arrows in Fig. 29.3b have a certain regularity and maybe formed along some grain boundary, which agrees with the phenomenon that dislocation cells and subgrains are preferentially formed at grain boundaries.

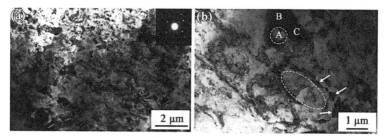

Figure 29.3 Typical TEM micrographs obtained from the region of 40 μm below the treated surface.

29.3.2.2 Grain refinement via DRX (depth <40 μm)

Figure 29.4a–c shows typical TEM micrographs of the RASP-processed 5052 Al alloy sample at the depth of 20 μm, 10 μm and 0 μm (topmost) from the surface, respectively. It can be seen that with increasing deformation strain, transverse boundaries within elongated subgrains were formed, resulting in the reduction of subgrains aspect ratio. It has been proved that the smaller aspect ratio in UFG will affect mechanical properties of materials by inhibiting the formation of dislocation cells during tensile deformation [39]. Similar to that of 5052 Al alloy processed by ECAP [11], the microstructures at 20 μm depth mainly consist of elongated/equiaxed subgrains, and the dislocation density is lower than that of depth >40 μm, implying dynamic recovery that is probably due to the temperature rise induced by the RASP treatment. Parallel, long black and gray bands known as banded-contrast image (BCI) were also observed in the Fig. 29.4a. It was believed that this phenomenon was formed from strain fields of high dislocation density [40]. For dislocations that approximately parallel to each other or even with the same Burgers vector, it will generate BCI when they orient to a certain angle to their strain fields. The elongated subgrains in other reports always have high dislocation density within well-defined boundaries [11, 19, 41].

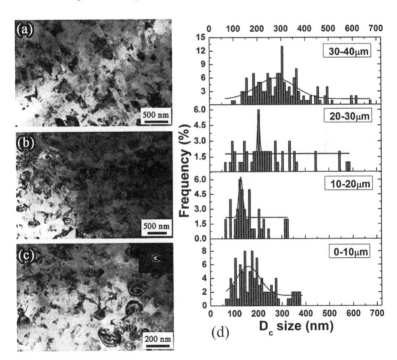

Figure 29.4 Typical TEM micrographs obtained from the region of (a) 20 μm; (b) 10 μm below the treated surface, and (c) topmost surface; (d) gives the grain size distribution of different region as indicated.

The microstructures at 10 μm depth show randomly oriented UFGs (Fig. 29.4b). The grain boundaries are distinguishable but not very sharp and have a strong spread in terms of thickness extinction contours, implying large internal stress and lattice distortions. Figure 29.4c gives the microstructures at the topmost layer, all of which are basically equiaxed grains with sharp grain boundaries. The inserted SAED pattern taken from an area with a diameter of 600 nm contains well-defined rings demonstrating the formation of fine grains with random misorientations.

Systematic TEM observations showed that the critical sizes of subgrain/grain (transverse size statistics from lots of TEM images and refers to Dc) vary with the depth from the surface. Figure 29.4d gives the critical size of the depth range of 30–40 μm, 20–30 μm, 10–20 μm and 0–10 μm. The grain size in these different depth ranges is nearly normally distributed. In the depth range of 30–40 μm, the average grain size is ~300 nm with very few grains have size less than 100 nm. With decreasing depth, the grain size distribution curves move leftward, and the average grain size in 10–20 μm depth region is the smallest (~160 nm). Some percent of grains with size less than 100 nm was found in the range of top 0–30 μm, which indicates that RASP treatment is effective for the grain refinement of 5052 Al alloy.

It is difficult for pure Al and Al alloy to be refined to nano-regime because of their high SFE and low melting temperature. In the top 40 μm depth from surface, dynamic recrystallization (DRX) is believed to play an important role. It is well known that DRX occurs in materials subjected to severe deformation above a specific temperature. In the RASP process, shots with a large kinetic energy hit the surface of the sample sheets. In addition to very high plastic strain, an obvious temperature rise was also expected. The temperature rise in the RASP samples can be calculated using the following equation:

$$\Delta|T| = \frac{\beta}{\rho C} \int_{\varepsilon_1}^{\varepsilon_2} \sigma d\varepsilon$$

in which $\beta = 0.9$ (assuming 90% of work of deformation was converted to heat), ρ is sample density, and C is the specific heat capacity. The estimated transient temperature within shear bands in Cu during plastic deformation at ambient temperature is about 500 K, and such a high thermal pulse might induce grain coarsening or dynamic recovery/recrystallization in pure Cu [42]. For high-strain-rate cryogenically deformed Cu, nano-grains were found in adiabatic shear band, which was attributed to DRX process [43]. A similar phenomenon was also reported in SMAT-processed Mg alloy, where nano-sized grains of about 30 nm were formed via the DRX mechanism [44]. In the present work, heat dissipation is very difficult since the strain rate of RASP is very high [35] and the deformation is approximately adiabatic. In addition, inhomogeneous deformation makes the temperature distribution localized. Moreover, Al alloy has a low melting point and therefore a low recrystallization temperature. All of these factors favors DRX in the top surface of the RASP processed 5052 Al alloy. As evidence, Fig. 29.5 shows a "clean" nanometer grain (white arrow) formed in a highly deformed subgrain with high density of dislocation arrays. This "clean" nanometer grain implies that it is strain free even though there is high strain in the surrounding area. The only possible way to generate such clean nanoscale grains within a highly deformed area is DRX.

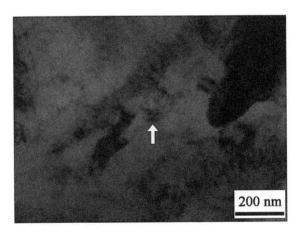

Figure 29.5 Strain-free nanometer grain formed in a highly deformed subgrain along the boundary through DRX.

The effect of solute elements such as Mg on the grain refinement of 5052 Al alloy cannot be neglected. A minimum grain size of 150 nm in pure Al and 20–50 nm in the A356 Al–Si alloy processed by SMAT were observed because Si particles can stabilize grains [45]. The average subgrain size in HPT Al-Mg alloys decreased considerably from 120 to 55 nm as the Mg content increased from 0.5 to 4.1 wt%, which was primarily attributed to the stabilization of Mg solute on grain size [12]. Mg in 5052 Al alloy may play the same role and contribute to the nano-grain formation in the surface of the sample, and this is beyond the scope of this work and will be further investigated.

29.3.3 Mechanical Properties

Figure 29.6 shows variation of microhardness along the depth from the treated surface in the RASP 5052 Al alloy samples with different RASP processing conditions, and the depth in 2000 μm corresponding to the "center" location in Fig. 29.1b. Three features can be clearly seen from the figure: First, microhardness gradient was observed for all of the RASP-processed samples. In the top surface layer, hardness in the 50 m/s-5 min RASP-processed samples reach as high as 116 Hv, which is about twice that of its CG counterpart. In all RASP-processed samples, the microhardness decreases with increasing depth from the surface. Second, microhardness becomes higher with increasing RASP duration and/or ball velocity. For instance, the microhardness in the 50 m/s-5 min RASP-processed sample shows the highest hardness almost in the whole hardness curve, which means increasing RASP duration and ball velocity can increase strain hardening effect. However, small cracks were occasionally found in the 50 m/s-5 min RASP-processed samples. Hence, no RASP processing was performed using ball velocity higher than 50 m/s. Third, all of the RASP-processed samples show an obvious microhardness improvement even in the center over that of the CG sample, indicating that the RASP deformation penetrated the whole thickness of Al alloy samples, which is consistent with OM observation in Fig. 29.1b.

Figure 29.7 shows the tensile engineering strength-strain curves of the RASP-processed 5052 Al alloy samples with deformation conditions at 20 m/s-5 min, 30 m/s-5 min, 40 m/s-5 min and 50 m/s-5 min. Enhanced yield strength were found in each deformed sample. Both yield strength and ultimate tensile strength of the 50 m/s-5 min deformed samples are higher than those deformed at 20 m/s-5 min, 30 m/s-5 min and 40 m/s-5 min, consistent with the microhardness measurements in Fig. 29.6. Meanwhile, these deformed samples with gradient structures show a good ductility, therefore, a good strength-ductility combination is realized. For instance, The yield strength of 50 m/s-5 min RASP-ed 5052 Al alloy is 168 MPa, which is more than twice that of CG samples (70 MPa) while the uniform elongation keeps a decent 12.6%.

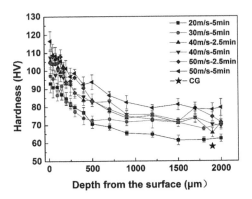

Figure 29.6 Variation of microhardness distribution in RASP-ed 5052 Al alloy as a function of depth from the surface.

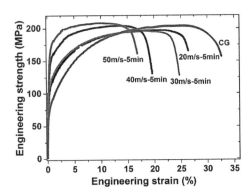

Figure 29.7 Typical tensile engineering strength-strain curves of the RASP-processed 5052 Al alloy samples with different deformation conditions as indicated, in comparison with that of the CG samples.

A set of tensile property data of 5052 Al alloy processed by rolling, ECAP, ARB and differential speed rolling (DSR) were extracted from literature [11, 17, 18, 20, 22, 41], and the yield strength (YS)–uniform elongation (ε_u) data were plotted in Fig. 29.8. The mechanical property data of the RASP-processed 5052 Al alloy in the present work were also plotted in the figure. It can be seen that the YS and ε_u of rolling, DSR, ECAP and ARB samples are mainly distributed in the high strength-low ductility (<10%) region (dark rectangle area), that is, the YS of the samples induced by these methods were improved at the expense of ductility. On the contrary, mechanical properties of RASP-processed 5052 Al alloy is mainly distributed in the middle strength-high ductility region (blue ellipse area), which means that the YS of the sample was obviously improved compared with that of CG, while retained decent ductility. Good strength-ductility synergy have been found in gradient structured Cu, which was attributed the strain hardening in deformed CG and grain-boundary migration-induced softening in nano-grains in the gradient structure [27]. A good combination of strength and ductility was found in gradient structured interstitial-free (IF) steel, and its good mechanical properties came from extraordinary strain hardening introduced by hetero-deformation induced (HDI) strengthening and HDI work hardening due to incompatible deformation along the gradient depth [28, 46]. It was also found recently that the gradient structured IF steel has very high HDI strengthening and HDI work hardening to improve the strength and ductility [47]. It is likely that the HDI stress also played a major role in producing the superior mechanical properties in the gradient 5052 Al alloy.

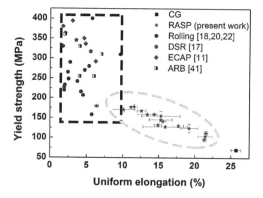

Figure 29.8 A plot of yield strength vs uniform elongation for 5052 Al alloy samples, data were extracted from the tensile curves in literatures and the RASP processed samples in the present work.

29.4 Conclusions

Microstructure evolution and grain refinement in 5052 Al alloy induced by RASP were investigated systematically. The effect of RASP treatment on the mechanical properties of 5052 Al alloy sample is also studied. Major conclusions are:

1. A deformation layer of ~2 mm was formed by RASP treatment, indicating the effectiveness and high energy of RASP process.
2. Different dislocation configurations, elongated subgrains, equiaxed subgrains and nano-sized grains were found along the depth from the sample surface. Dynamic recrystallization was primarily responsible for the nano-grain formation.
3. Microhardness gradient was observed along depth from the surface. The microhardness increased with increasing RASP ball velocity and treatment duration. The yield strength was obviously improved while retaining decent ductility. The good combination of strength and ductility is expected to help with the application of the 5052 Al alloy as a structural material.

Acknowledgments

Financial supports from the National Natural Science Foundation of China (Grant Nos. 51301092, 51501092, and 51601094), Nanjing University of Science and Technology (Grant No. AE89991), Pangu Foundation, and the Jiangsu Key Laboratory of Advanced Micro&Nano Materials and Technology are acknowledged.

References

1. X.Y. Liu, P.P. Ohotnicky, J.B. Adams, C.L. Rohrer, and R.W. Hyland: Anisotropic surface segregation in Al-Mg alloys. *Surf. Sci.* **373**, 357 (1997).
2. R.Z. Valiev, R.K. Islamgaliev, and I.V. Alexandrov: Bulk nanostructured materials from severe plastic deformation. *Prog. Mater. Sci.* **45**, 103 (2000).
3. R.Z. Valiev, I.V. Alexandrov, Y.T. Zhu, and T.C. Lowe: Paradox of strength and ductility in metals processed by severe plastic deformation. *J. Mater. Res.* **17**, 5 (2002).

4. R.Z. Valiev, Y. Estrin, Z. Horita, T.G. Langdon, M.J. Zehetbauer, and Y.T. Zhu: Fundamentals of superior properties in bulk nanoSPD materials. *Mater. Res. Lett.* **4**, 1 (2016).

5. N. Tsuji, Y. Saito, H. Utsunomiya, and S. Tanigawa: Ultra-fine grained bulk steel produced by accumulative roll-bonding (ARB) process. *Scr. Mater.* **40**, 795 (1999).

6. H. Pirgazi, A. Akbarzadeh, R. Petrov, and L. Kestens: Microstructure evolution and mechanical properties of AA1100 aluminum sheet processed by accumulative roll bonding. *Mater. Sci. Eng., A.* **497**, 132 (2008).

7. Y.T. Zhu and T.C. Lowe: Observations and issues on mechanisms of grain refinement during ECAP process. *Mater. Sci. Eng., A.* **291**, 46 (2000).

8. M. Furukawa, Z. Horita, M. Nemoto, and T.G. Langdon: Review: Processing of metals by equal-channel angular pressing. *J. Mater. Sci.* **36**, 2835 (2001).

9. M. Kawasakia, Z. Horitab, T.G. Langdon: Microstructural evolution in high purity aluminum processed by ECAP. *Acta Mater.* **524**, 143 (2009).

10. M. Zha, Y.J. Li, R.H. Mathiesen, R, and H.J. Roven: Microstructure evolution and mechanical behavior of a binary Al–7Mg alloy processed by equal-channel angular pressing. *Acta Mater.* **84**, 42 (2015).

11. T.L. Tsai, P.L. Sun, P.W. Kao, and C.P. Chang: Microstructure and tensile properties of a commercial 5052 aluminum alloy processed by equal channel angular extrusion. *Mater. Sci. Eng., A.* **342**, 144 (2003).

12. M.P. Liu, H.J. Roven, X.T. Liu, M. Murashkin, R.Z. Valiev, T. Ungar, and L. Balogh: Grain refinement in nanostructured Al–Mg alloys subjected to high pressure torsion. *J. Mater. Sci.* **45**, 4659 (2010).

13. H. Jiang, Y.T. Zhu, D.P. Butt, I.V. Alexandrov, and T.C. Lowe: Microstructural evolution, microhardness and thermal stability of HPT-processed Cu. *Mater. Sci. Eng., A.* **290**, 128 (2000).

14. R.Z. Valiev, Y. Estrin, Z. Horita, T.G. Langdon, M.J. Zehetbauer, and Y.T. Zhu: Producing bulk ultrafine-grained materials by severe plastic deformation. *JOM.* **58**, 33 (2006).

15. Y.T. Zhu, X.Z. Liao, and X.L. Wu: Deformation twinning in nanocrystalline materials. *Prog. Mater. Sci.* **57**, 1 (2012).

16. W.W. Jian, G.M. Cheng, W.Z. Xu, H. Yuan, M.H. Tsai, Q.D. Wang, C.C. Koch, Y.T. Zhu, and S.N. Mathaudhu: Ultrastrong Mg alloy via nano-spaced stacking faults. *Mater. Res. Lett.* **2**, 61 (2013).

17. Loorentz, and Y.G. Ko: Effect of differential speed rolling strain on microstructure and mechanical properties of nanostructured 5052 Al alloy. *J. Alloy. Compd.* **586**, S205 (2014).
18. U.G. Gang, S.H. Lee, and W. Jono: The evolution of microstructure and mechanical properties of a 5052 aluminium alloy by the application of cryogenic rolling and warm rolling. *Mater. Trans.* **50**, 82 (2009).
19. Y.B. Lee, D.H. Shin, and W.J. Nam: Effect of deformation temperature on the formation of ultrafine grains in the 5052 Al alloy. *Met. Mater. Int.* **10**, 407 (2004).
20. B. Wang, X.H. Chen, F.S. Pan, J.J. Mao, and Y. Fang: Effects of cold rolling and heat treatment on microstructure and mechanical properties of AA 5052 aluminum alloy. *Trans. Nonferrous Met. Soc.* **25**, 2481 (2015).
21. K.C. Sekhar, R. Narayanasamy, and K. Velmanirajan: Experimental investigations on microstructure and formability of cryorolled AA 5052 sheets. *Mater. Des.* **53**, 1064 (2014).
22. J.T. Shi; L.G. Hou, C.Q. Ma, J.R. Zuo; H. Cui, L.Z. Zhuang, J.S. Zhang: Mechanical properties and microstructures of 5052 Al alloy processed by asymmetric cryorolling. *Mater. Sci. Forum* **850**, 823 (2016).
23. Y.C. Chen, Y.Y. Huang, C.P. Chang, and P.W. Kao: The effect of extrusion temperature on the development of deformation microstructures in 5052 aluminum alloy processed by equal channel angular extrusion. *Acta Mater.* **51**, 2005 (2003).
24. Y.T. Zhu, and X.Z. Liao: Nanostructured metals: retaining ductility. *Nat. Mater.* **3**, 351 (2004).
25. M.A. Meyers, A. Mishra, and D.J. Benson: Mechanical properties of nanocrystalline materials. *Prog. Mater. Sci.* **51**, 427 (2006).
26. K. Lu: Stabilizing nanostructures in metals using grain and twin boundary architectures. *Nat. Rev. Mater.* **1**, 16019 (2016).
27. T.H. Fang, W.L. Li, N.R. Tao, and K. Lu: Revealing extraordinary intrinsic tensile plasticity in gradient nano-grained copper. *Science* **331**, 1587 (2011).
28. X.L. Wu, P. Jiang, L. Chen, F.P. Yuan, and Y.T. Zhu: Extraordinary strain hardening by gradient structure. *Proc. Natl. Acad. Sci. U.S.A.* **111**, 7197 (2014).
29. E. Ma, and T. Zhu: Towards strength–ductility synergy through the design of heterogeneous nanostructures in metals. *Mater. Today.* (2017). Available at: http://dx.doi.org/10.1016/j.mattod.2017.02.003.
30. X.L. Wu, P. Jiang, L. Chen, and Y.T. Zhu: Synergetic strengthening by gradient structure. *Mater. Res. Lett.* **2**, 185 (2014).

31. K. Lu, and J. Lu: Surface nanocrystallization (SNC) of metallic materials-presentation of the concept behind a new approach. *J. Mater. Sci. Technol.* **15**, 193 (1999).

32. K. Lu, and J. Lu: Nanostructured surface layer on metallic materials induced by SMAT. *Mater. Sci. Eng., A.* **375–377**, 38 (2004).

33. W.L. Li, N.R. Tao, and K. Lu: Fabrication of a gradient nano-microstructured surface layer on bulk copper by means of a surface mechanical grinding treatment. *Scr. Mater.* **59**, 546 (2008).

34. X.C. Liu, H.W. Zhang, and Lu K: Strain-induced ultrahard and ultrastable nanolaminated structure in nickel. *Science* **342**, 337 (2014).

35. X. Wang, Y.S. Li, Q. Zhang, Y.H. Zhao, and Y.T. Zhu: Gradient structured copper by rotationally accelerated shot peening. *J. Mater. Sci. Technol.* (2017). Available at: https://doi.org/10.1016/j.jmst.2016.11.006.

36. Z. Horita, D.J. Smith, M. Nemoto, R.Z. Valiev, and T.G. Langdon: Observations of grain boundary structure in submicrometer-grained Cu and Ni using high-resolution electron microscopy. *J. Mater. Res.* **13**, 446 (1998).

37. K. Oh-ishi, Z. Horita, D.J. Smith, and T.G. Langdon: Grain boundary structure in Al–Mg and Al–Mg–Sc alloys after equal-channel angular pressing. *J. Mater. Res.* **16**, 583 (2001).

38. J.Y. Huang, Y.T. Zhu, H.G. Jiang, and T.C. Lowe: Microstructures and dislocation configurations in nanostructured Cu processed by repetitive corrugation and straightening. *Acta Mater.* **49**, 1497 (2001).

39. K.T. Park, and D.H. Shin: Microstructural interpretation of negligible strain-hardening behavior of submicrometer-grained low-carbon steel during tensile deformation. *Metall. Mater. Trans. A.* **33**, 705 (2002).

40. Y.T. Zhu, J.Y. Huang, J. Gubicza, T. Ungar, Y.M. Wang, E. Ma, and R.Z. Valiev: Nanostructures in Ti processed by severe plastic deformation. *J. Mater. Res.* **18**, 1908 (2003).

41. H.R. Song, Y.S. Kim, and W.J. Nam: Mechanical properties of ultrafine grained 5052 Al alloy produced by accumulative roll-bonding and cryogenic rolling. *Met. Mater. Int.* **12**, 7 (2006).

42. A. Mishra, B.K. Kad, F. Gregori, and M.A. Meyers: Microstructural evolution in copper subjected to severe plastic deformation: experiments and analysis. *Acta Mater.* **55**, 13 (2007).

43. Y.S. Li, N.R. Tao, and K. Lu: Microstructural evolution and nanostructure formation in copper during dynamic plastic deformation at cryogenic temperatures. *Acta Mater.* **56**, 230 (2008).

44. H.Q. Sun, Y.N. Shi, M.X. Zhang, and K. Lu: Plastic strain-induced grain refinement in the nanometer scale in a Mg alloy. *Acta Mater.* **55**, 975 (2007).
45. H.W. Chang, P.M. Kelly, Y.N. Shi, and M.X. Zhang: Effect of eutectic Si on surface nanocrystallization of Al–Si alloys by surface mechanical attrition treatment. *Mater. Sci. Eng., A.* **530**, 304 (2011).
46. X.L. Wu, P. Jiang, L. Chen, and Y.T. Zhu: Synergetic strengthening by gradient structure. *Mater. Res. Lett.* **2**, 185 (2014).
47. M.X. Yang, Y. Pan, F.P. Yuan, Y.T. Zhu, and X.L. Wu: Back stress strengthening and strain hardening in gradient structure. *Mater. Res. Lett.* **4**, 145 (2016).

Chapter 30

Quantifying the Synergetic Strengthening in Gradient Material

Y. F. Wang,[a] C. X. Huang,[a] M. S. Wang,[a] Y. S. Li,[b] and Y. T. Zhu[b,c]

[a]*School of Aeronautics and Astronautics, Sichuan University, Chengdu 610065, China*
[b]*Nano and Heterogeneous Materials Center, School of Materials Science and Engineering, Nanjing University of Science and Technology, Nanjing 210094, China*
[c]*Department of Materials Science and Engineering, North Carolina State University, Raleigh, NC 27695, USA*
chxhuang@scu.edu.cn, liyusheng@njust.edu.cn

Synergetic strengthening in heterostructures is a new strengthening mechanism for metals. Here, a simple procedure based on the relationship between hardness increment and yield strength increment of corresponding homogeneous counterparts is proposed to quantitatively predict the synergetic strengthening effect in gradient-structured Cu-30wt%Zn. The synergetic strengthening among incompatible zones accounts for >33% of yield strength. Gradient structures with higher volume fraction of gradient zones

---

Reprinted from *Scr. Mater.*, **150**, 22–25, 2018.

Heterostructured Materials: Novel Materials with Unprecedented Mechanical Properties
Edited by Xiaolei Wu and Yuntian Zhu
Text Copyright © 2018 Acta Materialia Inc.
Layout Copyright © 2022 Jenny Stanford Publishing Pte. Ltd.
ISBN 978-981-4877-10-7 (Hardcover), 978-1-003-15307-8 (eBook)
www.jennystanford.com

exhibits higher synergetic strengthening. These results provide a new method for evaluating synergetic strengthening in heterostructured materials.

Synergetic strengthening induced by the coordinated deformation between mechanically incompatible zones of heterostructured (HS) materials is a new strengthening mechanism for materials, by which the strength can be significantly enhanced while retaining good ductility [1–7]. To achieve superior combination of strength and ductility, material scientists have synthesized various heterogeneous structures in the past decades, including the gradient structure [1, 3–5, 8–12], lamella/layered structure [2, 13–15], multimodal structure [16, 17], etc. [18]. The excellent mechanical responses of these HS materials indicate a promising way to fabricate advanced materials with high performance. However, in most of previous works the improvement of yield strength was just simply attributed to conventional Hall-Petch type strengthening mechanisms such as reduced grain size and high dislocation density in the harder zones [8, 17, 19]. The fundamental consideration of the strengthening effects of mechanical heterogeneity was largely neglected until an extra strength was revealed in gradient IF steel by Wu et al. [1].

Generally, the synergetic strengthening in HS materials is estimated by the difference between the measured strength and the linear summation of properties of standalone components, i.e. the predictions from the volume fraction-based simple rule-of-mixture [1, 13]. However, it is experimentally very difficult, if not impossible, to peel off all homogeneous components from integrated structure and measure their individual mechanical properties [1, 3, 17]. It is reasonable to believe that the extra strengthening observed by Wu et al. [1] in gradient IF steel was much smaller than the real synergetic strengthening of the whole sample, because they simply divided the gradient sample into a sandwich-like structure, i.e. a coarse-grained (CG) core and two gradient surface layers. Yang and Moring et al. [4, 5, 11] only qualitatively investigated the synergetic strengthening in gradient materials from the points of complex stress state and hetero-deformation induced (HDI) stress.

Here, we propose a simple method to quantitatively calculate the synergetic strengthening in gradient structures. The relationship between structural gradient and synergetic strengthening effect is comparatively analyzed in two types of gradient structures with different volume fractions of gradient surface layer.

A brass (Cu-30wt%Zn) plate with a thickness of 3.6 mm was annealed at 600°C for 2 h and used as the baseline metal. In order to fabricate a thicker gradient layer, some samples were subjected to multiple-pass friction stir processing (FSP) firstly under flowing cold water, in which process an unthreaded pin in diameter of 3 mm and length of 1 mm was used [20]. Thereafter, a technique of rotationally accelerated shot peening (RASP) was conducted on both sides of all samples to produce the gradient surface layer [21]. The as-fabricated gradient materials were labeled as G_{RASP} and $G_{FSP+RASP}$, respectively. The gradient microstructures were characterized by transmission electron microscopy (TEM) and scanning electron microscopy (SEM) equipped with electron back-scattered diffraction (EBSD) detector. Dog-bone shaped tensile specimens with a gauge dimension of 20 × 4 × 3.6 mm^3 were machined from the gradient plates. The Vickers micro-hardness was measured on the cross-section using a load of 25 g for 15 s.

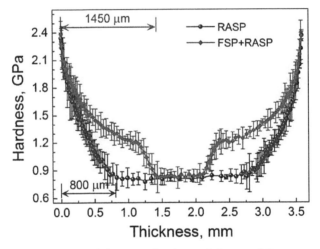

Figure 30.1 Variation of the micro-hardness of G_{RASP} and $G_{FSP+RASP}$ samples along the thickness direction. Every data point was averaged from 4 indents.

Figure 30.1 shows the micro-hardness profile across the whole thickness of gradient samples. The hardness reaches as high as 2.4 GPa in the topmost layers of both G_{RASP} and $G_{FSP+RASP}$ samples, which is ~3 times that of the CG core. Such a high mechanical incompatibility between surface and core layers is expected to produce great strain inhomogeneity during straining [3, 13]. According to the hardness profiles, the thickness of gradient layer in the G_{RASP} and $G_{FSP+RASP}$ samples was measured as about 800 μm and 1450 μm, respectively. Note that the latter is much thicker than that achieved by standalone surface treatment techniques [21, 22]. Here, the FSP process produces a UFG surface layer thicker than 1 mm and UFG-CG transition layer of ~450 mm [20]. After subsequent surface mechanical treatment by RASP to add a top nanocrystalline layer and enhance the mechanical gradient, stronger synergetic strengthening is expected than that of conventional gradient material with only a thin gradient surface layer.

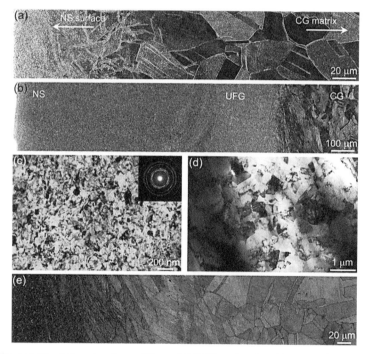

Figure 30.2 Microstructure of the as-RASP (G_{RASP}) and as-FSP+RASP ($G_{FSP+RASP}$) processed gradient materials. (a) and (b) are SEM morphologies showing the cross-sectional gradient surface layer of the G_{RASP} and $G_{FSP+RASP}$ samples, respectively. (c) A typical TEM observation showing the nanostructure at the depth of ~25 μm. (d) A TEM image showing the ultrafine grains at the depth of ~950 μm in the $G_{FSP+RASP}$ material. (d) An EBSD image showing the UFG-CG transitional microstructure in the $G_{FSP+RASP}$ material.

The cross-sectional structure of the gradient surface layer of G_{RASP} material is shown in Fig. 30.2a. A gradient variation of grain size from nano-grains in the topmost layer to equiaxed CG in matrix can be obviously observed. Figure 30.2b shows a much thicker fine-grained layer with a thickness of ~1 mm in the $G_{FSP+RASP}$ sample. Figure 30.2c is a typical bright-field TEM image and corresponding selected area electron diffraction pattern at the depth of ~25 μm below the treated surface in G_{RASP} material, showing well-developed nanostructures with random orientation. A similar microstructural observation was conducted in the top surface of the $G_{FSP+RASP}$ materials as well, and there was no obvious difference from that of the G_{RASP} sample. The $G_{FSP+RASP}$ sample still exhibits a UFG layer at the depth of ~950 μm that was produced by FSP, as seen in Figs. 30.2b and d. Figure 30.2e is an EBSD map showing the UFG-CG transitional zone at the depth from 950 μm to 1450 μm in the $G_{FSP+RASP}$ sample.

Figure 30.3 shows that both the G_{RASP} and $G_{FSP+RASP}$ samples exhibit superior combination of strength and ductility. The yield strength (σ_y) of the G_{RASP} and $G_{FSP+RASP}$ samples are measured as 285 MPa and 422 MPa, respectively, which are about 3–4 times of pure CG sample (103 MPa).

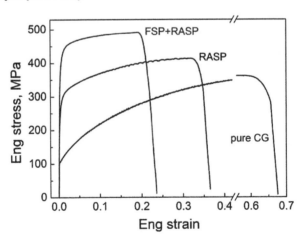

Figure 30.3 Tensile engineering stress-strain curves of the pure CG, G_{RASP} and $G_{FSP+RASP}$ samples.

There often exists a quantitative relationship between microhardness (Hv) and tensile strength for homogeneous-structured material. For example, the widely used empirical formula [23]:

$$\sigma_y = Hv/3, \qquad (30.1)$$

was constructed for materials without strain hardening. Although there is still much debate about the fitting parameters in the relationship between hardness and strength with regarding to different materials, the experimental data of either work-hardening or brittle materials usually can be well fitted by a linear equation [24]:

$$\sigma = k * Hv + y, \qquad (30.2)$$

where k is the ratio that may deviate from $1/3$, and y is the intercept constant. Busby et al. [25] studied this relationship in austenitic and ferritic steels using a large quantity of experimental data, and summarized a more general relationship as

$$\Delta\sigma_y = k * \Delta Hv, \qquad (30.3)$$

where ΔHv and $\Delta\sigma_y$ are the change in hardness and yield strength, respectively.

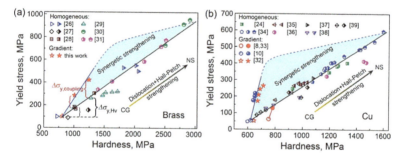

Figure 30.4 The relationship between yield strength and hardness of (a) brass and (b) pure Cu. The data include the experimental results of homogeneous annealed CG (open), work-hardening CG (- center), UFG (half right) and NS (half up) as well as gradient (solid) materials. The light blue shadow region shows possible synergetic strengthening of gradient structure.

In Fig. 30.4a, the tensile yield strength was plotted against the corresponding hardness of bulk homogeneous brass in different grain states, including annealed CG [26, 27], work-hardening CG [28, 29], recrystallized UFG [27, 28], deformed UFG and NS synthesized by severely plastic deformation [26, 30, 31]. Interestingly, these data can be well fitted using Eq. (30.3) (the black line) as

$$\Delta\sigma_y = 0.39 * \Delta Hv. \tag{30.4}$$

Here, it should be noted that the increase in yield strength (or hardness) is mainly due to the grain size refinement and/or dislocation storage. However, the result of studied G_{RASP} and $G_{FSP+RASP}$ samples deviates obviously from this fitting formula and has a higher k value of about 0.83, which implies a high extra yield strength resulted from the synergetic strengthening in gradient structures.

We propose the following procedure to quantify the synergetic strengthening of gradient structure. In a gradient sample, the direct contribution to the strength from both small grain size and enhanced dislocation density ($\Delta\sigma_{y,Hv}$) can be modified (based on Eqs. (30.3) and (30.4)) as

$$\Delta\sigma_{y,Hv} = 0.39 * \overline{\Delta Hv}, \tag{30.5}$$

where $\overline{\Delta Hv}$ represents the hardness increment of whole gradient sample and can be calculated from the hardness profiles (Fig. 30.2) using the simple rule-of-mixture [13]

$$\overline{\Delta Hv} = \frac{\sum V_i H_i}{\sum V_i} - H_{CG}, \tag{30.6}$$

where V_i and H_i are the volume fraction and the hardness of component i, respectively. H_{CG} is the hardness of pure CG sample. It should be noted that the using of rule-of-mixture is based on two assumptions: (1) the gradient material is integrated by i thin layers, and (2) there is no mechanical interaction between neighboring layers during indentation testing. Therefore, the synergetic strengthening induced by coordinated deformation between mechanically incompatible components ($\Delta\sigma_{y,coupling}$) can be expressed as

$$\Delta\sigma_{y,synergetic} = \Delta\sigma_{y,gradient} - \Delta\sigma_{y,Hv}, \tag{30.7}$$

and

$$\Delta\sigma_{y,gradient} = \sigma_{y,gradient} - \sigma_{y,CG}, \tag{30.8}$$

where $\sigma_{y,gradient}$ and $\sigma_{y,CG}$ are the measured tensile yield strength of gradient and pure CG samples, respectively.

The hardness, yield strength and their corresponding increments of both G_{RASP} and $G_{FSP+RASP}$ samples calculated by the above procedure are listed in Table 30.1. Surprisingly, the $\Delta\sigma_{y,synergetic}$ in gradient materials accounts for >33% of the yield strength. In addition, the $\Delta\sigma_{y,\text{synergetic}}$ of $G_{FSP+RASP}$ is much higher than that of G_{RASP}, suggesting that the synergetic strengthening can be enhanced by improving the volume fraction of gradient zones [1, 6].

Table 30.1 Hardness, yield strength and the synergetic strengthening induced by different strengthening mechanisms in as-processed gradient materials

Material	\multicolumn{7}{c}{Hardness & yield strength (MPa)}						
	Hv	ΔHv	σ_y	$\Delta\sigma_{y,\text{gradient}}$	$\Delta\sigma_{y,Hv}$	$\Delta\sigma_{y,\text{synergetic}}$	$\Delta\sigma_{y,\text{synergetic}}/\sigma_y$
CG	805	—	102.5	—	—	—	—
G_{RASP}	1007	202	285.1	182.6	78.8	103.8	36.4%
$G_{FSP+RASP}$	1243	438	421.9	319.4	170.9	148.5	33.9%

To verify the reliability of above procedure, a similar analysis was conducted on homogenous and gradient Cu by collecting reported data [8, 10, 24, 32–39]. Figure 30.4b shows that the hardness increment and yield strength increment of homogeneous Cu in different grain states obey the linear relationship very well. However, the increasing rate of yield stress induced by synergetic strengthening in gradient samples is much higher than that contributed by conventional mechanisms. Moreover, it seems that the synergetic strengthening in gradient Cu is more pronounced than that in brass, indicating that the material properties such as stacking fault energy may also affect the coupling intensity.

According to the TEM observation and stress relaxation test conducted in gradient and lamella structures [2, 3], the physical origins of synergetic strengthening are the HDI stress development from geometrically necessary dislocations [6], unusual dislocation generation, interaction and accumulation behaviors caused by complex stress states. The big difference in elastic limit between NS layer and CG core permits a long elastic-plastic transition stage for gradient materials, during which the plastic shrinking of inner soft layer is constrained by a lateral tensile stress from neighboring elastic layers [2, 3, 8, 12]. As a result, the dislocation movement in soft layer is suppressed, which contributes directly to high yield strength. More importantly, following the migration step of elastic/

plastic boundary during yielding, large quantity of geometrically necessary dislocations (GNDs) could accumulate over the whole gradient layer to accommodate the strain gradient in both the tensile and thickness directions [2, 3, 6, 14, 40]. This process leads to a fast increase of long-range HDI stress, which can significantly hinder dislocation slip, i.e. produce a non-local strengthening [4]. In addition, the gradient distribution of stress in gradient structures can strengthen material as well by acting on the dislocation sources and obstacles [41].

A thicker gradient layer provides a large space for accumulating GNDs and thus contributes to a higher synergetic strengthening. Therefore, other heterogeneous configurations, such as lamellar structure [2, 6, 14], might be an optimal design because of the high density of boundaries for the accumulation of geometrically necessary dislocations. In addition to the thickness of gradient layer, there are many other structural parameters, such as the level of mechanical incompatibility, the volume fraction and arrangement of heterogeneous components, that could affect the intensity of interactions among the heterogeneous zones. Therefore, more studies are needed to optimize the structural design for maximum synergetic effect.

In summary, the extra yield strength induced by synergetic strengthening in gradient-structured brass is quantitatively deduced basing on the relationship between hardness increment and yield strength increment of corresponding homogeneous materials. The synergetic strengthening accounts for >33% of total yield stress. It is revealed that the gradient structure with higher volume fraction of gradient layers exhibits higher synergetic strengthening. This preliminary work provides a quantitative demonstration for the synergetic strengthening of structural heterogeneity, and the proposed derivation procedure can be used for other HS materials as well.

Acknowledgements

This work was supported by the National Natural Science Foundation of China (Nos. 11672195, 51741106 and 51301092) and Sichuan Youth Science and Technology Foundation (2016JQ0047).

References

1. X.L. Wu, P. Jiang, L. Chen, J.F. Zhang, F.P. Yuan, Y.T. Zhu, *Mater. Res. Lett.* **2** (2014) 185–191.
2. X.L. Wu, M.X. Yang, F.P. Yuan, G.L. Wu, Y.J. Wei, X.X. Huang, Y.T. Zhu, *Proc. Natl. Acad. Sci. U.S.A.* **112** (2015) 14501–14505.
3. X.L. Wu, P. Jiang, L. Chen, F.P. Yuan, Y.T. Zhu, *Proc. Natl. Acad. Sci. U.S.A.* **111** (2014) 7197–7201.
4. M.X. Yang, Y. P, F.P. Yuan, Y.T. Zhu, X.L. Wu, *Mater. Res. Lett.* **4** (2016) 145–151.
5. J. Moering, X.L. Ma, J. Malkin, M.X. Yang, Y.T. Zhu, S. Mathaudhu, *Scr. Mater.* **122** (2016) 106–109.
6. X.L. Wu, Y.T. Zhu, *Mater. Res. Lett.* **5** (2017) 527–532.
7. E. Ma, T. Zhu, *Mater. Today* **20** (2017) 323–331.
8. T.H. Fang, K. Lu, *Science* **331** (2011) 1578–1590.
9. K. Lu, *Science* **345** (2014) 1455–1456.
10. Z. Yin, X.C. Yang, X.L. Ma, J. Moering, J. Yang, Y.L. Gong, Y.T. Zhu, X.K. Zhu, *Mater. Des.* **105** (2016) 89–95.
11. X.L. Wu, M.X. Yang, F.P. Yuan, L. Chen, Y.T. Zhu, *Acta Mater.* **112** (2016) 337–346.
12. Y.J. Wei, Y.Q. Li, L.C. Zhu, Y. Liu, X.Q Lei, G. Wang, Y.X. Wu, Z.L. Mi, J.B. Liu, H.T. Wang, H.J. Gao, *Nat. Commun.* **5** (2014) 1–8.
13. X.L. Ma, C.X. Huang, W.Z. Xu, H. Zhou, X.L. Wu, Y.T. Zhu, *Scr. Mater.* **103** (2015) 57–60.
14. X.L. Ma, C.X. Huang, J. Moering, M. Ruppert, H. Höppel, M. Göken, J. Narayan, Y.T. Zhu, *Acta Mater.* **116** (2016) 43–52.
15. M. Huang, G.H. Fan, L. Geng, G.J. Cao, Y. Du, H. Wu, T.T. Zhang, H.J. Kang, T.M. Wang, G.H. Du, H.L. Xie, *Sci. Rep.* **6** (2016) 38461.
16. C. Sawangrat, S. Kato, D. Orlov, K. Ameyama, *J. Mater. Sci.* **49** (2014) 6579–6585.
17. S.K.Vajpai, M. Ota, T. Watanabe, R. Maeda, T. Sekiquchi, T. Kusaka, K. Ameyama, *Mater. Trans. A* **46** (2015) 903–914.
18. M. Calcagnotto, Y. Adachi, D. Ponge, D. Raabe, *Acta Mater.* **59** (2011) 658–670.
19. S. Bahl, S. Suwas, T. Ungàr, K. Chatterjee, *Acta Mater.* **122** (2017) 138–151.
20. P. Xue, B.L. Xiao, Z.Y. Ma, *J. Mater. Sci. Technol.* **29** (2103) 1111–1115.

21. X. Wang, Y.S. Li, Q. Zhang, Y.H. Zhao, Y.T. Zhu, *J. Mater. Sci. Technol.* **33** (2017) 758–761.
22. K. Wang, N.R. Tao, G. Liu, J. Lu, K. Lu, *Acta Mater.* **54** (2006) 5281–5291.
23. D. Tabor, *Hardness of Metals*, Clarendon Press, Oxford, 1951.
24. P. Zhang, S.X. Li, Z.F. Zhang, *Mater. Sci. Eng. A* **529** (2011) 62–73.
25. J.T. Busby, M.C. Hash, G.S. Was, *J. Nucl. Mater.* **336** (2005) 267–278.
26. M. Afrasiab, G. Faraji, V. Tavakkoli, M.M. Mashhadi, K. Dehghani, *Trans. Indian. Inst. Met.* **68** (2015) 873–879.
27. A. Heidarzadeh, T. Saeid, V. Klemm, *Mater. Charact.* 119 (2016) 84–91.
28. N. Xu, R. Ueji, H. Fujii, *Mater. Sci. Eng. A* 610 (2014) 132–138.
29. Y.W. Mai, B. Cotterell, *J. Mater. Sci.* **15** (1980) 2296–2306.
30. H. Bahmanpour, K.M. Youssef, J. Horky, D. Setman, M. A. Atwater, M.J. Zehetbauer, R. O. Scattergood, C. C. Koch, *Acta Mater.* **60** (2012) 3340–3349.
31. Suryadi, A.P. Kusuma, A. Suhadi, D. Priadi, E.S. Siradj, *Int. J. Technol.* **1** (2017) 58–65.
32. X.C. Yang, X.L. Ma, J. Moering, H. Zhou, W. Wang, Y.L. Gong, J.M. Tao, Y.T. Zhu, X.K. Zhu, *Mater. Sci. Eng. A* **645** (2015) 280–285.
33. T.H. Fang, N.R. Tao, K. Lu, *Scr. Mater.* **77** (2014) 17–20.
34. Y. Zhang, N.R. Tao, K. Lu, *Acta Mater.* **56** (2008) 2429–2440.
35. A. Habibi, M. Ketabchi, M. Eskandarzadeh, *J. Mater. Sci. Technol.* **211** (2011) 1085–1090.
36. N. Lugo, N. Llorca, J.M. Cabrera, Z. Horita, *Mater. Sci. Eng. A* **477** (2008) 366–371.
37. P. Xue, B.L. Xiao, Z.Y. Ma, *J. Mater. Sci. Technol.* **29** (2013) 1111–1115.
38. N. Xu, R. Ueji, Y. Morisada, H. Fujii, *Mater. Des.* **56** (2014) 20–25.
39. Y.F. Sun, H. Fujii, *Mater. Sci. Eng. A* **527** (2010) 6879–6886.
40. Z. Zeng, X.Y. Li, D.S. Xu, L. Lu, H.J. Gao, T. Zhu, *Extreme Mech. Lett.* **8** (2016) 213–219.
41. S.S. Chakravarthy, W.A. Curtin, *Proc. Natl. Acad. Sci. U.S.A.* **108** (2011) 15716–15720.

Chapter 31

Achieving Gradient Martensite Structure and Enhanced Mechanical Properties in a Metastable β Titanium Alloy

Xinkai Ma,[a,b] Fuguo Li,[a] Zhankun Sun,[a] Junhua Hou,[a] Xiaotian Fang,[b] Yuntian Zhu,[b] and Carl C. Koch[b]

[a]*State Key Laboratory of Solidification Processing, School of Materials Science and Engineering, Northwestern Polytechnical University, Xi'an 710072, China*
[b]*Department of Materials Science and Engineering, North Carolina State University, Raleigh, NC 27695, USA*
fuguolx@nwpu.edu.cn, ytzhu@ncsu.edu

Gradient materials have been reported to have superior strength-ductility combinations. In this study, gradient α'' martensite were introduced along the radial direction of cylindrical Ti-10V-2Al-3Fe (Ti-1023) samples by torsional straining, which simultaneously improved strength and ductility. The torsional strain gradient produced martensite gradient with increasing density and

---

Reprinted from *Metall. Mater. Trans. A*, **50A**, 2126–2138, 2019.

Heterostructured Materials: Novel Materials with Unprecedented Mechanical Properties
Edited by Xiaolei Wu and Yuntian Zhu
Text Copyright © 2019 The Minerals, Metals & Materials Society and ASM International
Layout Copyright © 2022 Jenny Stanford Publishing Pte. Ltd.
ISBN 978-981-4877-10-7 (Hardcover), 978-1-003-15307-8 (eBook)
www.jennystanford.com

decreasing thickness from center to surface. α'' martensite had parallel and V-shaped morphology, which not only divided coarse β grains into finer β blocks but also blocked dislocation slip. In addition, dislocation slip in the α'' martensite and β blocks led to grain refinement. The formation of geometrically necessary dislocation (GNDs) and the increasing shear stress required for martensitic transformation contributed to high strain hardening. An optimal gradient distribution exists in torsion-processed samples for the optimal mechanical properties.

31.1 Introduction

Materials with gradient microstructure often demonstrate a better combination of mechanical properties than those of homogeneous materials [1-6]. Such superior mechanical properties are caused by synergetic strengthening and work hardening [7], which is produced due to the mechanical incompatibility among gradient layers during deformation. Geometrically necessary dislocations (GNDs) are generated to coordinate the mechanical incompatibility, which produces hetero-deformation induced (HDI) stress, long-range stress that provides for additional strengthening and work hardening [8]. Gradient microstructures reported so far include gradient grain size [9-13], gradient twin spacing [14, 15], gradient grains with embedded twins [11, 16-18], and gradient texture [19]. The samples in these studies were fabricated using varying approaches developed for making nanostructured materials, which have been studied for a few decades [3]. Challenges remain on how to develop an approach that is conducive to broad industrial applications.

In metastable β titanium alloys with low stacking fault energy (SFE), especially for those with molybdenum equivalent (Mo_{eq}) higher than 8%, stress-induced martensitic transformation (SIMT), which is a transformation from β phase to orthorhombic α'' martensite, could occur under applied stresses [20-22]. Deformation mechanisms such as martensitic transformation (transformation induced plasticity, TRIP), mechanical twinning (twinning induced plasticity, TWIP) and dislocation slip may occur during the plastic deformation of metastable β titanium alloys [23, 24]. Previous studies have reported that TRIP, TWIP, or their combination can result in a wide range of mechanical behaviors, which can be used to balance the strength and ductility of metastable β titanium alloys with homogeneous structures [20, 25–29]. The activation sequence of these deformation mechanisms has also been systematically investigated by in-situ observations [21, 30, 31].

Gradient structures have been recently reported to significantly enhance the mechanical properties of metals and alloys [1, 2, 5–9, 32]. It will be of scientific and practical interest to study the effect of gradient structure on TRIP and TWIP in metastable β titanium alloys. In the present study, the formation mechanism of gradient α'' martensite structure in the Ti-1023 alloy during torsional straining [33–37] and its deformation mechanism in subsequent tensile testing were investigated. The influence of gradient magnitude on the mechanical properties was studied, and it is found that there exists an optimal gradient structure for the best mechanical properties.

31.2 Materials and Methods

31.2.1 Sample Preparation

A hot-forged rod of β-metastable titanium alloy (Ti-1023) was used in this study. Its composition is listed in Table 31.1. The β-transus temperature of the alloy was approximately 795°C [26, 38]. The samples were β solution treated at 920°C for 0.5 h, followed by water quenching [38].

Table 31.1 The chemical composition of as-received Ti-1023 alloys (in wt%)

Main component (wt%)				Impurities (wt%)		
V	Al	Fe	Ti	O	N	C
10.69	3.04	1.81	Balance	0.11	0.01	0.007

31.2.2 Sample Processing and Mechanical Testing

Dog-bone-shaped samples with gauge dimensions of 5 mm in diameter and 28 mm in length were used for torsional processing. The samples were twisted into different angles at a rotational speed of 30 turns min^{-1}. Torsional shear strain γ and equivalent strain ε_e can be described by Eq. (31.1) and Eq. (31.2), respectively, which is based on the investigation of high-pressure torsion [39, 40].

$$\gamma = \frac{2\pi N r}{l} \quad (31.1)$$

$$\varepsilon_e = \frac{\gamma}{\sqrt{3}} \quad (31.2)$$

where r is the distance from the center of the torsion axis, N is the number of rotations, and l is its gauge length.

Vickers microhardness along the radius of torsion-processed samples was measured using an Akashi MVK-E3 testing machine under 500 g load with a holding time of 15 s. Moreover, the hardness measurements at each point were repeated at least five times to guarantee the reproducibility of the data Tensile tests were conducted on an INSTRON 3382 testing machine at ambient temperature with a strain rate of 1×10^{-3} s^{-1}. At least three tensile tests were completed for each sample state.

31.2.3 Microstructure Characterization

X-ray diffraction (XRD, Bruker D8) with a Cu Kα radiation source was conducted to identify the phases in the torsion-processed sample. Macroscopic gradient distribution of α'' martensite along the radius direction was revealed by etching the sample with a modified Kroll's Reagent (5 ml HF + 15 ml HNO$_3$ + 80 ml distilled water, in vol%), and observed on an optical microscope (OM, OLYMPUS/PMG3) [38]. To reveal the effect of torsional strain gradient on the morphology of α'' martensite in mesoscopic scale, three different layers (center, half radius, and surface layer) were characterized using scanning electron microscope (SEM, JSM-6700F FEG). In addition, the tensile fracture surface was observed using SEM. After tensile deformation, transmission electron microscopy (TEM) sample with a thickness of 600 μm was cut using electric discharge machining (EDM), and then manually ground to ~50 μm in thickness, followed by twin-jet electropolishing. TEM characterization was performed on a Tecnai G2 F30 operated at an accelerating voltage of 200 kV.

31.3 Results

31.3.1 Martensitic Transformation during Torsion Processing and Subsequent Tensile Deformation

Ti-1023 alloy rods ($\Phi 5 \times 28$) were processed by twisting to produce gradient martensitic transformation along the radial direction. Figure 31.1 shows XRD patterns obtained from samples before and after torsion processing. It can be found that the initial sample with 0° torsional angle, namely β solution treated sample, consists of a single β phase. When the sample was twisted for 90°, which produced a maximum shear strain of 0.14 on the sample surface, the microstructure was found made up of $\beta + \alpha''$ dual phases. With increasing shear strain, diffraction peak broadening was clearly observed for $\alpha''(111)$, $\alpha''(130)$ and $\alpha''(023)$ peaks, which could be attributed to the refinement of α'' phase and/or an increase of the lattice strain [41]. Meanwhile, strong α'' peaks indicate a higher volume fraction of α'' martensite in the β matrix.

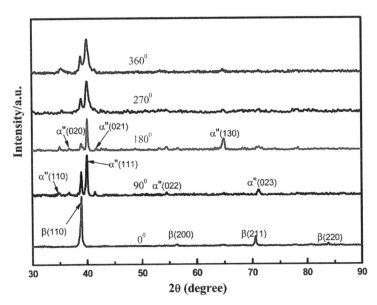

Figure 31.1 XRD patterns of different torsion-processed samples in the Ti-1023 alloy.

Figures 31.2a–d show optical images of the whole cross-section of Ti-1023 samples processed for different torsional angles. The microstructure of the β solution treated sample is β phase with an average grain size of ~216 μm [38]. The dark features in Fig. 31.2 contains stress-induced α'' martensite, which has been confirmed in the XRD data. More detailed examination reveals thick α'' martensite plates that run across the grains of the initial β phase. The grain sizes of the β phase along the radial direction are not changed by the torsional process, while the volume fraction and distribution of α'' martensite vary significantly with increasing shear strain. In addition, the α'' martensite rich region expands to the center of the cross-section with increasing torsional angle. After tensile testing, α'' martensite rich region covers the whole cross-section, as shown in Fig. 31.2e, which indicates the formation of new α'' martensite.

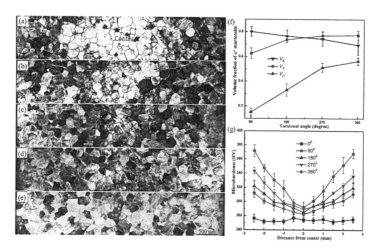

Figure 31.2 Cross-sectional OM images showing a gradient α'' martensite along the radial direction for torsion-processed samples with different torsional angles. (a) 90°, (b) 180°, (c) 270°, (d) 360°, (e) 90° torsion-processed samples after the subsequent tensile test, (f) the volume fraction of α'' martensite as a function of torsional angle, and (g) variation of microhardness in the radial direction.

To evaluate the effect of torsional straining on the martensitic transformation, the variation of α'' volume fraction in the center (V_C), surface (V_S) and the total cross-section (V_T) are plotted in Fig. 31.2f as a function of torsional angle. V_T is calculated from the integrated intensity of all α'' phase and β phase in Fig. 31.1. As shown, V_T

increases with increasing torsional strain and reaches a saturation value of ~77% at 360°. This may be because the martensitic transformation becomes more difficult with increasing torsional strain due to the increasing resistance to martensitic transformation caused by increasing dislocation density and decreasing size of the remaining β phase.

For the torsion-processed sample, strain gradient results in α'' martensite gradient, with increasing volume fraction along the radial direction from the center, namely V_S is higher than V_C in torsion-processed samples. V_C increases from 15% to 56% with the increasing torsional angle. While V_S shows a slight decrease from 80% to 69% with increasing torsional strain. This implies that some α'' martensite transformed back to the β phase with increasing torsional strain. One possible reason is that deformation heating during high-speed torsion deformation resulted in localized temperature rise above M_d [41, 42].

The microhardness variation along the radius of the samples is shown in Fig. 31.2g. The microhardness in the initial sample is a constant of ~276 HV across the total cross section. The microhardness in the torsion-processed sample shows a gradient distribution, decreasing from the surface to the center. This phenomenon is significantly enhanced with the increasing volume fraction of α'' martensite and shear strain.

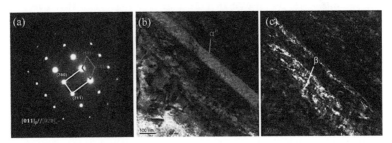

Figure 31.3 (a–c) SAED, BF and DF in 90° torsion-processed sample after tensile deformation to fracture.

At a finer scale, Fig. 31.3 shows the bright field (BF) and dark field (DF) TEM images for 90° torsion-processed samples after tensile deformation, which is made from the center region near the fracture surface. Selected-area electron diffraction (SAED) pattern along the $[011]_\beta$ zone axis in Fig. 31.3a shows a c-centered orthorhombic structure of α'' martensite plates in Fig. 31.3b, as well as strong bcc diffraction spots of β phase in the dark field image in Fig. 31.3c. This demonstrates that α'' martensite has been formed in 90° torsion-processed samples during the subsequent tensile deformation.

31.3.2 Mechanical Behavior of the Torsion-Processed Sample

To study the effect of gradient α'' martensite on the mechanical properties of the Ti-1023 alloy, engineering stress-strain curves under uniaxial tensile test at room temperature are displayed in Fig. 31.4a. There are several salient features in these stress-strain curves. First, typical strain plateau or so-called double yielding phenomenon, which manifests the formation of α'' martensite, has been observed in the initial sample. This is due to the easy activation and continuous martensitic transformation during tensile deformation. The tensile curve of the initial sample has low apparent yield stress, which is the stress that triggers the martensitic transformation. The martensite formed at this stage is called stress-induced martensite [43]. The second yielding on the tensile curve corresponds to the critical stress required to activate dislocation slip [29].

For 90° torsion-processed samples with a ~62% α'' martensite, new α'' martensite was formed in the center of the sample during the tensile tests, as shown in Fig. 31.2 and Fig. 31.3. However, no double yielding phenomenon can be found in its tensile curves, as well as other torsion-processed samples, as shown in Fig. 31.4a. This phenomenon is a result of the higher volume fraction of α'' martensite formed in torsion-processed samples, which largely consume the β grains oriented along the maximum resolved stress and reduce the β zone size. As a result, the presence of α'' martensite in torsion-processed samples can raise the applied stress required for SIMT upon further tensile loading. The higher applied stress required for martensitic transformation led to the disappearance of the first yielding phenomenon at lower stress, which will be discussed in detail later. This indicates that torsional strain significantly affects the deformation behavior of β-solution treated Ti-1023 alloy. The detailed mechanical properties, including triggering stress, ultimate tensile stress and elongation, are listed in Table 31.2.

Table 31.2 Mechanical properties of the torsion-processed Ti-1023 samples

Samples	Equivalent strain ε_e	Triggering stress, σ_T (MPa)	Ultimate tensile strength, (MPa)	Elongation δ	Static toughness U_T (MJ/m³)
Initial	—	301	726	0.242	183
90°	0.08	560	856	0.238	197
180°	0.16	597	870	0.226	206
270°	0.27	765	878	0.205	179
360°	0.32	837	933	0.166	152

It should be noted that increasing torsional strain led to a substantial increase in triggering stress and ultimate tensile stress, as well as a slight decrease in elongation to failure. To estimate the comprehensive mechanical properties, the static toughness U_T is calculated by integrating the engineering stress-stress curves and shown in Table 31.2, which indicates the amount of energy per unit volume absorbed by the samples until rupture. It can be found in Table 31.2 that the static toughness increases until the torsional angle increased to 180°, and then decreased as the torsional angle reached 360°. Therefore, among the five torsion-processed samples, 180° torsion-processed samples show a good combination of strength and ductility.

The work-hardening rate ($\Theta = d\sigma/d\varepsilon$) is plotted in Fig. 31.4b vs. true stain. As shown, the variation of Θ depends on the torsional strain, as well as the α'' martensite gradient. For initial samples, the variation of Θ shows typical three stages, which starts with a decrease (stage I), then a fast increase (stage II) due to SIMT, and finally decreases again (stage III) until fracture. While there are only two stages of Θ in torsion-processed samples, stage I corresponds to the fast decrease, and the absence of stage II is related to the limited nucleation of α'' martensite with higher triggering stress required for SIMT. It should be noted that the torsion-processed samples exhibit a higher work hardening rate than initial samples when the true strain is less than 0.05, then the work hardening rate of torsion-processed samples showed a faster decreasing rate than initial samples until true stain increased to 0.1, finally 90° and 180°

torsion-processed samples showed a slower decreasing rate and maintained a higher work hardening rate until fracture.

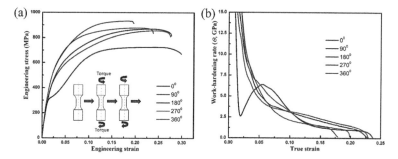

Figure 31.4 Strengthening of the Ti-1023 alloy by torsion processing. (a) Engineering stress-strain curves, (b) work-hardening rate ($\Theta = d\sigma/d\varepsilon$) as a function of true strain.

31.3.3 Gradient Microstructure and Fracture Behavior

The gradient distribution of α'' martensite along the radial direction for Ti-1023 alloys were examined to reveal the plastic deformation mechanism of torsion-processed samples. Figures 31.5a and b present α'' martensite morphology of 90° and 360° torsion-processed samples. As shown, the volume fraction of α'' martensite increases with increasing distance from the center. In 90° torsion-processed samples, there is little α'' martensite at the center. At the half radius, parallel secondary α'' martensite (α_{II}'') formed between primary α'' martensite (α_{I}'') plates, while a large number of α'' martensite plates with complex configuration appear near the surface [44], dividing initial β grains into smaller blocks. In addition, when the torsional angle increased from 90° to 360°, the volume fraction and morphology of α'' martensite changed from the center to the surface. As shown in Fig. 31.5(b), α'' martensite exhibits V-shaped morphology at the center, while series of α'' variants with zig-zag structure were formed at the half radius. The reason for the morphology change is, compared with the center, that higher plastic strain at half radius activates martensitic transformation on different planes. In addition, this zig-zag microstructure disappeared at the surface with the amount of formation of new α'' variants. These

α'' martensite variants with different growth directions can form a twinned α'' martensite (α_T'') structure with increasing strain [45]. Meanwhile, the 360° torsion-processed sample contains finer β blocks subdivided by α'' martensite than the 90° torsion-processed sample, which is due to the high strain/stress occurred on the shear plane. As shown in Fig. 31.5c, a further martensitic transformation occurred during the subsequent tensile deformation, which resulted in very little β phase left after tensile deformation in 360° torsion-processed samples. Meanwhile, more complex arrangements of secondary α'' martensite are also observed on the surface, which could promote the refinement of β blocks.

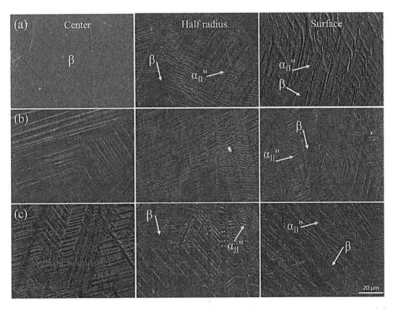

Figure 31.5 Cross-sectional SEM images from different strained regions. (a) 90° torsion-processed sample, (b) 360° torsion-processed sample and (c) 360° torsion-processed sample after tensile deformation to fracture.

The fracture surfaces of the 90° and 360° torsion-processed samples after a tensile deformation are characterized to analyze the effect of gradient α'' martensite on fracture behavior, as shown in Figs. 31.6 and 31.7. Low-magnification images in Figs. 31.6a and 31.7a show that ductile fracture dominates the fracture procedure, though quasi- cleavage features occurred at the edge of samples, which correspond with the gradient distribution of α'' martensite. For 90° torsion-processed samples, Fig. 31.6b shows uniform dimples in the center with an average diameter of 13 μm, but some tearing ridges and micro-voids are shown in Figs. 31.6e and f. In addition, a large shear plane with shallow dimples is shown in Figs. 31.6c and d, which can be regarded as the transition of two fracture modes. As tensile strain increases, these voids would coalesce and grow into microcracks, expanding from the surface to the center.

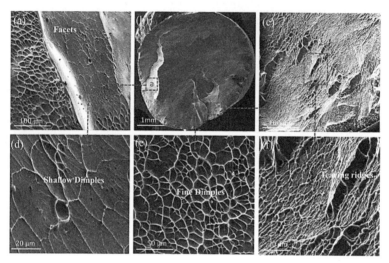

Figure 31.6 Fractographs of 90° torsion-processed sample after tensile deformation revealing a mixed fracture mode: (a), (c) and (e) are the high magnification images of the fracture in the edge and center, respectively, (b) Low magnification image of the whole fracture surface, (d) and (f) are the further blowup of (a) and (c), respectively.

In comparison, although the fracture mechanism of 360° torsion-processed samples is like that of the 90° torsion-processed samples, several distinctive features should be noticed. As shown in Figs. 31.7c and d, tearing ridges of 360° torsion-processed sample indicate lower plasticity, which is typical quasi-cleavage features. In addition, Figs. 31.7e and f show the elongated dimples and shear faces, which are due to the higher shear deformation at the edge. Moreover, the center of 360° torsion-processed sample (Fig. 31.7b) contains a great number of shallow dimples with an average diameter of ~8 μm, caused by the increased content of α'' martensite, which is much smaller than that of 90° torsion-processed sample. These shallow dimples imply lower plasticity in 360° torsion-processed sample. The reduction of plasticity with increasing torsional angle is a result of brittle fracture caused by a higher volume fraction of α'' martensite.

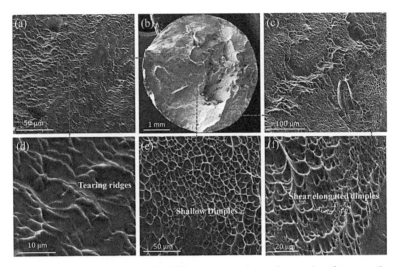

Figure 31.7 Fractographs of 360° torsion-processed sample after tensile deformation revealing a mixed fracture mode: (a), (c) and (e) are the high magnification images of the fracture in the edge and center, respectively, (b) low magnification image of the whole fracture surface, (d) and (f) are the further blowup of (a) and (c), respectively.

31.4 Discussion

31.4.1 Nucleation Mechanism of Gradient α'' Martensite

At a small shear strain during the torsion process, primary α'' martensite plates nucleated from β grain boundaries, which is a stress-induced martensitic transformation. Subsequently, most of α'' martensite nucleate freely to grow across β grains, which accommodates the increasing applied shear strain. As shown in Fig. 31.8, in the 360° torsion-processed sample α'' martensite has divided initial β grains into α''/β lamellae at a nano-scale. Besides, the spacing between α'' martensite lamellas also varies with the thickness of α'' martensite plate. Figure 31.8a shows parallel primary α'' martensite plates with a thickness of ~113 nm and a spacing of ~290 nm in the β matrix. With further shear straining during torsion process, a group of parallel secondary α'' martensite plates and bright (β) plates with a twinning relationship emerged between the primary α'' martensite palate as shown in Fig. 31.8b, which indicates that twining may be responsible for martensitic transformation [46–48]. These newly formed secondary α'' martensite plates have a thickness of ~36 nm and a length of ~180 nm, dividing β blocks into smaller zones, which correspond to the feature shown in Fig. 31.5b. Furthermore, new variants of α'' martensite are required due to the decreasing space for martensitic transformation in partially consumed β grains [49]. Therefore, two variants of α'' martensite in a V-shaped morphology are found in Fig. 31.8c and confirmed with $\{110\}$ type twin relationship, which has the orientation of $(110)[1\bar{1}2]\alpha''1//(\bar{1}10)[1\bar{1}2]\alpha''2$ in Fig. 31.8d [28, 50].

It is noted that primary α'' martensite plates nucleated and grew firstly, then other variants grow subsequently with increasing strain. This led to the formation of a great number of parallel twinned α'' martensite plates in Figs. 31.5b and c. The mutual effect of primary α'' martensite, secondary α'' martensite and twinned α'' martensite will repeatedly divide β phase into smaller β blocks, while the size of α'' martensite will also decrease. Smaller martensite plates would need higher stress to be activated. These typical α'' martensite structures with a parallel arrangement and V-shaped morphology in the torsion-processed sample show a complex nucleation mechanism.

Figure 31.9a schematically illustrates the formation process of forming the gradient α'' martensite structure. It should be noted that the gradient α'' martensite have two distinctive features. The volume fraction of α'' martensite decreases from the surface to the center, while the thickness and spacing of α'' martensite plates increases. On the surface of torsion-processed samples, parallel thin nano-α'' martensite with a thickness of ~25 nm and a spacing of 28 nm are formed, which are also confirmed in Figs. 31.9b and c. Comparing Figs. 31.9b with d reveals that the gradient shear strain resulted in a gradient volume fraction of α'' martensite. Meanwhile, the thickness of α'' martensite increases with the decrease of distance to the center, which is caused by the increasing space left inside β grains for martensitic transformation. Therefore, thick primary α'' martensite plates with a thickness of ~10 µm and spacing of ~22 µm have been formed at the center of torsion-processed samples, as shown in Fig. 31.2.

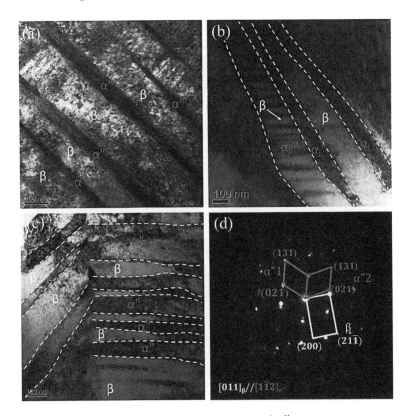

Figure 31.8 Three typical TEM bright field images of α'' martensite structures in 360° torsion-processed sample: (a) primary α'' martensite (α_1''), (b) secondary martensite α'' (α_{II}''), (c) V-shaped α'' martensite and corresponding SAED pattern (d) along the $[011]_\beta$ axis.

During the subsequent tensile deformation, the presence of gradient α'' martensite and limited finer β blocks would increase the critical stress required to initiate new martensitic transformation. At the low-density α'' martensite zone, the martensitic transformation could take place repeatedly in the rest of β blocks until the trigger stress of SIMT on local β phase became higher than the critical resolved shear stress of slip [41], which is corresponding with the second yield point of the stress-strain curve in Fig. 31.4a. As the applied stress required by SIMT increase to the second yield point, there is smaller volume fraction of α'' martensitic forming in the torsion-processed samples during tensile loading. Thus, its tensile curves show the absence of the double-yield phenomenon. Finally, when the volume fraction of α'' martensite increases to a saturated value, dislocation slip becomes a predominant plastic deformation mode [42, 51].

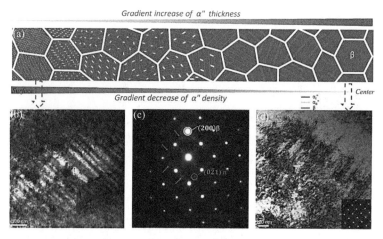

Figure 31.9 (a) A schematic of the cross-sectional microstructure in torsion-processed sample, (b) a TEM bright-field image taken from the surface of 360° torsion-processed sample showing nano-α'' martensite structure, (c) the SAED of (b), and (d) a TEM bright field image taken from the center of 360° torsion-processed sample showing the β matrix with dislocation.

31.4.2 Evolution of β Phase and α'' Martensite

The microstructural evolution of β phase and α'' martensite during tensile deformation was characterized by TEM. As discussed above, the initial β grains were subdivided into high density of submicron-sized β blocks during the torsion process. Upon further straining during tensile deformation, the refined β blocks may be deformed further by dislocation slip and the formation of new α'' martensite. As elucidated in Fig. 31.10, β blocks exhibit a much higher density of dislocation than that of α'' martensite. It is found that the β

block is broken into smaller pieces, which result in the formation of finer rhomboid β blocks. Similarly, fragments of α'' martensite are also revealed in Fig. 31.11. It should be noted that a high density of dislocations in β blocks, especially at the α''/β boundaries, are unlikely to pass through the boundary. This is because α'' martensite has a distorted hexagonal structure with an orthorhombic unit cell [43, 52]. The orthorhombic structure has a low symmetry, resulting in fewer slip systems than the BCC structure of the β phase [20], while the activated dislocations in the β phase have a relatively high symmetry. Therefore, α'' martensite may be more difficult to deform plastically than the β phase, resulting in strain partition between α'' martensite and β phase, which leads to higher flow stress.

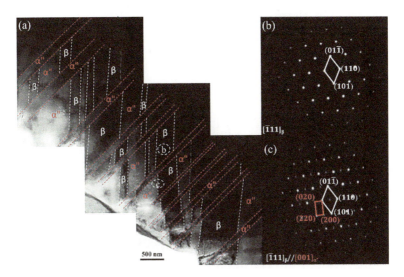

Figure 31.10 (a) Typical TEM images in the center of 360° torsion-processed sample after tensile testing, showing the β phase's shearing along the red dashed lines, (b) the SAED selected in the b area in (a), (c) the SAED selected in the c area in (a).

In general, dislocation slip impeded by α''/β boundaries promotes the deformation induced grain refinement of the β phase and α'' martensite, as schematically illustrated in Fig. 31.12. It can be concluded that stress induced martensitic transformation and dislocation slip are essential for the grain refinement of gradient α'' martensite and β blocks.

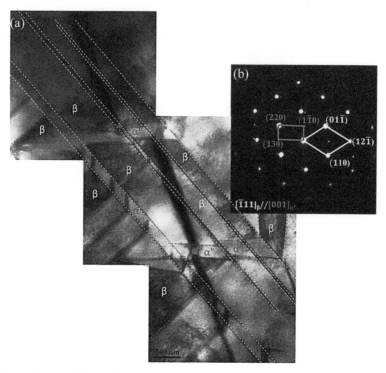

Figure 31.11 (a) Typical TEM images in the center of 360° torsion-processed sample after tensile testing, showing the α'' phase's shearing along the white dashed lines, identified in the SAED in (b).

31.4.3 The Effect of Gradient α'' Martensite Structure on Strain Hardening

Different from common gradient materials, α''/β lamella structure was introduced in torsion-processed samples of Ti-1023 alloy. As shown in Fig. 31.13, large amount of dislocations (marked by white arrows) were blocked by the interlaced α''/β lamella structure in 180° torsion-processed samples, which can significantly increase its strain hardening capability. In addition, the strain hardening is different in α'' martensite and β phase, as well as gradient α'' martensite, as demonstrated by the variation of microhardness in Fig. 31.2g. To accommodate the non-uniform deformation at α''/β boundaries and the mechanical incompatibility between the surface and the center of torsion-processed samples along radius direction, GNDs would be generated. Moreover, GNDs with the same burgers vector would be pinned in β grains, which will form high HDI stress [2, 8], inducing extra strain hardening and strengthening. Therefore, HDI hardening is believed to have contributed to the excellent combination of strength and ductility in torsion-processed samples.

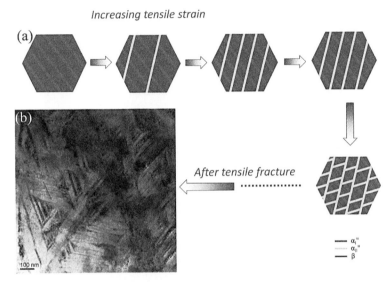

Figure 31.12 (a) A schematic of the grain refinement process in the torsion-processed sample during tensile testing (b) TEM image of abundant nano-α'' structure in the surface of 360° pre-tensioned sample after tensile testing.

Based on current experimental results shown in Table 31.2, a gradient structure of mixed α'' martensite and β phase in the torsion-processed sample, especially 180° torsion-processed samples, significantly enhanced the mechanical properties of Ti-1023 alloy. Compared with the coarse β grains in the initial sample, this optimization in the torsion-processed sample is ascribed to the effective mutual constraint between coarse β grains in the center and nano-lamellae in the surface, as shown in Fig. 31.9. For gradient structure with thin and dense α''/β lamellas in the center, as shown in Figs. 31.2c and d, plastic deformation of torsion-processed samples is dominated by the mechanical behavior of α'' martensite with limited dislocation movement, which led to a poor strain hardening. Therefore, the strain localization of α'' martensite rich zone, which could not be prevented by the coarse β grains, resulted in a quick failure after the yield point. For gradient structure with relatively thick β blocks and large α''/β interfacial spaces in the center, as shown in Fig. 31.2a, plastic deformation in the coarse β grains could not be constrained by α'' martensite. Therefore, a critical size and volume fraction of α'' martensite are necessary to produce an excellent combination of strength and ductility. In our future work, we will further optimize the gradient distribution of α'' martensite and β phase.

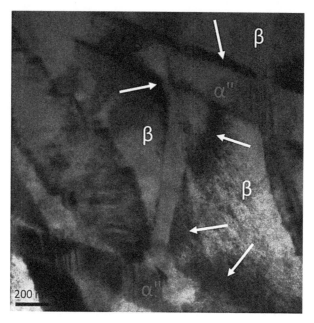

Figure 31.13 The interaction between dislocation motion and α''/β lamella structure in 180° torsion-processed samples.

31.5 Conclusions

The β solution-treated Ti-1023 samples were twisted at room temperature, resulting in the formation of various gradient magnitude in α'' martensite (i.e., volume fraction varies from ~15% to ~80% and thickness varies from ~25 nm to ~22 μm) and

β blocks. The formation mechanism of gradient structure and the effect of gradient magnitude on the microstructural evolution and mechanical properties of Ti-1023 samples were systematically investigated. The main conclusions can be summarized as follows:

(1) The deformation mechanism of Ti-1023 alloys during torsional and tensile tests was revealed by multiscale structural characterization. Stress induced martensitic transformation, {110} type twining occurred at different strain levels.

(2) The nucleation of gradient α'' martensite due to gradient shear strain exhibits parallel morphology (primary α'' martensite and secondary α'' martensite) and V-shaped morphology of two variants with twining relationships. α'' martensite not only divides coarse β grains into finer β blocks repeatedly but also effectively blocks dislocation slip.

(3) The presence of gradient α'' martensite and finer β blocks in the torsion-processed sample significantly increased the stress required for martensitic transformation. As a result, the lower yield point is absent in the torsion-processed sample during the tensile loading.

(4) The shearing in α'' martensite and β block accelerated the process of deformation induced grain refinement. Combining with the mutual interaction between dislocation slips and α''/β boundaries, the GNDs, generating by gradient α'' martensite, played an essential role in the enhancement of strength and ductility, especially in the 180° torsion-processed samples.

Acknowledgments

This work was supported by the China Scholarship Council (No. 201706290055, award to Xinkai Ma for two-year study abroad at the North Carolina State University); the National Natural Science Foundation of China (Grant Nos. 51275414, 51605387); the Fundamental Research Funds for the Central Universities with Grant No. 3102015BJ (II) ZS007; the Research Fund of the State Key Laboratory of Solidification Processing (NWPU) China (Grant No. 130-QP-2015); and the Seed Foundation of Innovation and Creation for Graduate Students in Northwestern Polytechnical University (No. Z2018076).

References

1. K. Lu, *Science*, 2014, vol. 345, pp. 1455–1456.
2. X. Wu, P. Jiang, L. Chen, F. Yuan, Y.T. Zhu, *Proc Natl Acad Sci USA*, 2014, vol. 111, pp. 7197–7201.
3. R. Thevamaran, O. Lawal, S. Yazdi, S.-J. Jeon, J.-H. Lee, E.L. Thomas, *Science*, 2016, vol. 354, pp. 312–316.
4. I. Ovid'ko, R. Valiev, Y. Zhu, *Prog Mater Sci*, 2018, vol. 94, pp. 462–540.
5. X. Bian, F. Yuan, Y. Zhu, X. Wu, *Mater Res Lett*, 2017, vol. 5, pp. 501–507.
6. X. Wu, Y. Zhu, *Mater Res Lett*, 2017, vol. 5, pp. 527–532.
7. X. Wu, P. Jiang, L. Chen, J. Zhang, F. Yuan, Y. Zhu, *Mater Res Lett*, 2014, vol. 2, pp. 185–191.
8. M. Yang, Y. Pan, F. Yuan, Y. Zhu, X. Wu, *Mater Res Lett*, 2016, vol. 4, pp. 145–151.
9. A. Chen, J. Liu, H. Wang, J. Lu, Y.M. Wang, *Mater Sci Eng A*, 2016, vol. 667, pp. 179–188.
10. X. Liu, H. Zhang, K. Lu, *Science*, 2013, vol. 342, pp. 337–340.
11. H. Wang, N. Tao, K. Lu, *Scr Mater*, 2013, vol. 68, pp. 22–27.
12. Z. Ma, J. Liu, G. Wang, H. Wang, Y. Wei, H. Gao, *Sci Rep*, 2016, vol. 6, p. 22156.
13. A. Chen, H. Ruan, J. Wang, H. Chan, Q. Wang, Q. Li, J. Lu, *Acta Mater*, 2011, vol. 59, pp. 3697–3709.
14. Y. Wei, Y. Li, L. Zhu, Y. Liu, X. Lei, G. Wang, Y. Wu, Z. Mi, J. Liu, H. Wang, *Nat Commun*, 2014, vol. 5, p. 3580.
15. L. Lu, X. Chen, X. Huang, K. Lu, *Science*, 2009, vol. 323, pp. 607–610.
16. H. Wang, N. Tao, K. Lu, *Acta Mater*, 2012, vol. 60, pp. 4027–4040.
17. C. Shao, P. Zhang, Y. Zhu, Z. Zhang, Y. Tian, Z. Zhang, *Acta Mater*, 2017, vol. 145, pp. 413–428.
18. Y. Kulkarni, R.J. Asaro, D. Farkas, *Scr Mater*, 2009, vol. 60, pp. 532–535.
19. J. Moering, X. Ma, G. Chen, P. Miao, G. Li, G. Qian, S. Mathaudhu, Y. Zhu, *Scr Mater*, 2015, vol. 108, pp. 100–103.
20. A. Bhattacharjee, V. Varma, S. Kamat, A. Gogia, S. Bhargava, *Metall Mater Trans A*, 2006, vol. 37, pp. 1423–1433.
21. A. Zafari, K. Xia, *Mater Sci Eng A*, 2018, vol. 724, pp. 75–79.
22. D. Qin, Y. Lu, D. Guo, L. Zheng, Q. Liu, L. Zhou, *Mater Sci Eng A*, 2013, vol. 587, pp. 100–109.

23. Z. Wyatt, S. Ankem, *J Mater Sci*, 2010, vol. 45, pp. 5022–5031.
24. W. Xu, K. Kim, J. Das, M. Calin, J. Eckert, *Scr Mater*, 2006, vol. 54, pp. 1943–1948.
25. A. Paradkar, S. Kamat, A. Gogia, B. Kashyap, *Metall Mater Trans A*, 2008, vol. 39, pp. 551–558.
26. C. Li, X. Wu, J. Chen, S. van der Zwaag, *Mater Sci Eng A*, 2011, vol. 528, pp. 5854–5860.
27. F. Sun, J. Zhang, M. Marteleur, C. Brozek, E. Rauch, M. Veron, P. Vermaut, P. Jacques, F. Prima, *Scr Mater*, 2015, vol. 94, pp. 17–20
28. S. Sadeghpour, S. Abbasi, M. Morakabati, A. Kisko, L. Karjalainen, D. Porter, *Scr Mater*, 2018, vol. 145, pp. 104–108.
29. X. Ma, F. Li, J. Cao, J. Li, Z. Sun, G. Zhu, S. Zhou, *Mater Sci Eng A*, 2018, vol. 710, pp. 1–9.
30. P. Barriobero-Vila, J. Gussone, K. Kelm, J. Haubrich, A. Stark, N. Schell, G. Requena, *Mater Sci Eng A*, 2018, vol. 717, pp. 134–143.
31. T. Yao, K. Du, H. Wang, Z. Huang, C. Li, L. Li, Y. Hao, R. Yang, H. Ye, *Acta Mater*, 2017, vol. 133, pp. 21–29.
32. T. Fang, W. Li, N. Tao, K. Lu, *Science*, 2011, vol. 331 p. 1200177.
33. X. Ma, F. Li, J. Cao, J. Li, H. Chen, C. Zhao, *Mater Des*, 2017, vol. 114, pp. 271–281.
34. N. Guo, X. Li, M. Xiao, R. Xin, L. Chai, B. Song, H. Yu, L. Li, *Adv Eng Mater*, 2016, vol. 18, pp. 1738–1746.
35. C. Wang, F. Li, J. Li, J. Dong, F. Xue, *Mater Sci Eng A*, 2014, vol. 598, pp. 7–14.
36. L. Sun, K. Muszka, B. Wynne, E. Palmiere, Palmiere, *Acta Mater*, 2014, vol. 66, pp. 132–149.
37. N. Guo, Z. Zhang, Q. Dong, H. Yu, B. Song, L. Chai, C. Liu, Z. Yao, M.R. Daymond, *Mater Des*, 2018, vol. 143, pp. 150–159.
38. X. Ma, F. Li, J. Cao, Z. Sun, Q. Wan, J. Li, Z. Yuan, *J Alloy Compd*, 2017, vol. 703, pp. 298–308.
39. Y. Estrin, A. Vinogradov, *Acta Mater*, 2013, vol. 61, pp. 782–817.
40. R. Pippan, F. Wetscher, M. Hafok, A. Vorhauer, I. Sabirov, *Adv Eng Mater*, 2006, vol. 8, pp. 1046–1056.
41. W. Chen, Q. Sun, L. Xiao, J. Sun, *Metall Mater Trans A*, 2012, vol. 43, pp. 316–326.
42. W. Xu, X. Wu, M. Calin, M. Stoica, J. Eckert, K. Xia, *Scr Mater*, 2009, vol. 60, pp. 1012–1015.

43. T. Duerig, J. Albrecht, D. Richter, P. Fischer, *Acta Metall*, 1982, vol. 30, pp. 2161-2172.
44. Y. Chai, H. Kim, H. Hosoda, S. Miyazaki, *Acta Mater*, 2009, vol. 57, pp. 4054-4064.
45. M. Kružík, A. Mielke, T. Roubíček, *Meccanica*, 2005, vol. 40, pp. 389-418.
46. A. Zafari, X. Wei, W. Xu, K. Xia, *Acta Mater*, 2015, vol. 97, pp. 146-155.
47. W. Xu, K. Kim, J. Das, M. Calin, B. Rellinghaus, J. Eckert, *Appl Phys Lett*, 2006, vol. 89, p. 031906.
48. S. Sadeghpour, S. Abbasi, M. Morakabati, *J Alloy Compd*, 2015, vol. 650, pp. 22-29.
49. T. Grosdidier, M.-J. Philippe, *Mater Sci Eng A*, 2000, vol. 291, pp. 218-223.
50. D. Ping, Y. Yamabe-Mitarai, C. Cui, F. Yin, M. Choudhry, *Appl Phys Lett*, 2008, vol. 93, p. 151911.
51. C. Wei, Y. Shanshan, L. Ruolei, S. Qiaoyan, X. Lin, S. Jun, *Rare Metal Mat Eng*, 2015, vol. 44, pp. 1601-1606.
52. Y. Ren, F. Wang, S. Wang, C. Tan, X. Yu, J. Jiang, H. Cai, *Mater Sci Eng A*, 2013, vol. 562, pp. 137-143.

Part IV
Heterogeneous Grain Structure

Chapter 32

Dynamically Reinforced Heterogeneous Grain Structure Prolongs Ductility in a Medium-Entropy Alloy with Gigapascal Yield Strength

Muxin Yang,[a,*] **Dingshun Yan,**[a,*] **Fuping Yuan,**[a,b]
Ping Jiang,[a] **Evan Ma,**[c] **and Xiaolei Wu**[a,b]

[a]*State Key Laboratory of Nonlinear Mechanics, Chinese Academy of Sciences, Beijing 100190, China*
[b]*School of Engineering Science, University of Chinese Academy of Sciences, Beijing 100049, China*
[c]*Department of Materials Science and Engineering, Johns Hopkins University, Baltimore, MD 21218, USA*
xlwu@imech.ac.cn, ema@jhu.edu

Ductility, i.e., uniform strain achievable in uniaxial tension, diminishes for materials with very high yield strength. Even for the CrCoNi medium-entropy alloy (MEA), which has a simple face-centered-

---

*These authors contributed equally to this work.

Reprinted from *Proc. Natl. Acad. Sci. U.S.A.*, **115**(28), 7224–7229, 2018.

Heterostructured Materials: Novel Materials with Unprecedented Mechanical Properties
Edited by Xiaolei Wu and Yuntian Zhu
Text Copyright © 2018 the Author(s)
Layout Copyright © 2022 Jenny Stanford Publishing Pte. Ltd.
ISBN 978-981-4877-10-7 (Hardcover), 978-1-003-15307-8 (eBook)
www.jennystanford.com

cubic (FCC) structure that would bode well for high ductility, the fine grains processed to achieve gigapascal strength exhaust the strain hardening ability such that after yielding the uniform tensile strain is as low as <~2%. Here we purposely deploy in this MEA a three-level heterogeneous grain structure (HGS) with grain sizes spanning the nanometer to micrometer range, imparting a high yield strength well in excess of 1 GPa. This heterogeneity results from this alloy's low stacking fault energy (SFE), which facilitates corner twins in recrystallization and stores deformation twins and stacking faults during tensile straining. After yielding the elasto-plastic transition through load transfer and strain partitioning among grains of different sizes leads to an up-turn of the strain hardening rate, and upon further tensile straining at room temperature corner twins evolve into nanograins. This dynamically reinforced HGS leads to a sustainable strain hardening rate, a record-wide hysteresis loop in load-unload-reload stress-strain curve and hence high hetero-deformation induced (HDI) stresses, and consequently a uniform tensile strain of 22%. As such, this HGS achieves in a single-phase FCC alloy a strength-ductility combination that would normally require heterostructures such as in dual-phase steels.

32.1 Introduction

The equimolar multicomponent alloys, recently dubbed as the medium-entropy alloys (MEAs) (1–6) and high-entropy alloys (HEAs) (7–25), are emerging as an interesting class of structural materials. The single-phase CrCoNi and FeCrMnCoNi alloys of the face-centered cubic (FCC) structure, in particular, have exhibited a combination of high ultimate tensile strength, ductility, and fracture toughness (1, 2, 5, 9, 10). However, their high ductility is for micrometer-grained (MG) microstructure, for which the yield strength ($\sigma_{0.2}$) is relatively low (below ~400 MPa at room temperature). When $\sigma_{0.2}$ approaches 1 GPa, achieved for example by refining the grain size, the common dilemma of strength–ductility trade-off rears its ugly head (3, 7): the stress–strain curve peaks immediately after yielding, exhibiting a drastic loss of ductility. The crux of the problem lies in the inadequate strain hardening, which can no longer keep up with the high flow stresses to circumvent plastic instability (localization of strain via necking, as predicted by the Consideré criterion (26)). This is in fact the norm for conventional metals when strengthened by either reducing their grain size or by cold working (27–29). For instance, $\sigma_{0.2}$ of a nanograined (NG) metal can be increased to many times that of its coarse-grained counterpart. However, such strengthening comes at the expense of strain hardening, because the intragranular dislocation storage upon straining, a potent mechanism for strain hardening, almost disappears (27). This is also true in ultrafine-grained (UFG) MEAs (3). Even though stacking faults are stored in large numbers due to the well-known low stacking fault energy (SFE) of these FCC HEAs/MEAs (1, 2, 8, 12), twinning became difficult in UFG grains, and at room temperature strain hardening remained inadequate to sustain uniform elongation after the MEA yields at >1 GPa (3). In general, plasticity-enhancing mechanisms in low SFE alloys such as phase transformation (TRIP) (10, 30–33) and twinning (TWIP) (1, 2, 34–41) all lose their potency in UFG and NG structures (34, 35, 42, 43).

Here we advocate a strategy generally applicable for single-phase low-SFE alloys, to provide adequate strain hardening to delay necking. The CrCoNi MEA is a good model to work with, for setting an example for such alloys. The key idea underlying our microstructural design is as follows. We intentionally push for a highly HGS via partial recrystallization annealing following conventional cold rolling of CrCoNi (see Section 32.3). This HGS, spanning NG, UFG to MG, provides a >1 GPa yield strength similar to that of uniform UFG MEA (3). But the more impactful consequence is that the HGS imparts load partitioning to promote internal stress hardening that produces an up-turn in strain hardening rate, which is further aided by dynamically generated twinned NGs at the grain corners and grain boundaries along with tensile straining. These revived twins and NGs, not previously induced in large numbers at room temperature in uniform grains due to required high stresses, strengthen the alloy and provide strain hardening to enable a large uniform tensile elongation at gigapascal flow stresses after yielding at >1 GPa.

32.2 Manuscript Text

Figure 32.1a shows the high-angle GBs (HAGB, defined as those with misorientation ≥15°) in the CrCoNi MEA after cold rolling and subsequent annealing at 600°C for 1 h (see Section 32.3), observed using electron back scatter diffraction (EBSD) imaging with a resolution of 40 nm. The partial recrystallization resulted in a grain structure that is characterized by three levels of grains with obviously different sizes, MGs, UFGs and NGs. The MGs (white) have a mean grain size of 2.3 µm, the intermediate UFGs (blue) range from 250 nm to 1 µm, and the smallest NGs are less than 250 nm (mostly in red, marked in terms of the magnitude of the Schmid factors for clear recognition from UFGs and MGs). The mean grain size of this HGS is 330 nm. The number percentage of UFGs is 47%, and that of NGs 43%. Interestingly, almost all NGs locate at the GBs and triple junctions (TJs) of UFGs, see an example in the close-up view in the inset. Such a recrystallization microstructure is expected for low SFE metals.

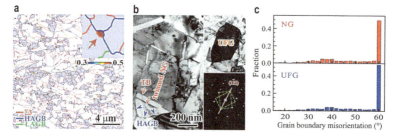

Figure 32.1 HGS in CrCoNi MEA. (a) EBSD grain boundary image showing three grades of grain size. MG (white): micrometer-sized grains; UFG (light blue): ultra-fine grains with submicron grain sizes; NGs (red): nanograins with grain sizes below 250 nm. NGs are colored based on the magnitude of the Schmid factor (see scale bar). Inset: close-up of a twinned NG (indicated by an arrow) nucleated at the triple junction of UFGs. (b) TEM image showing a twinned NG at the lower part of a UFG. Upper-right inset: the entire UFG in (011) zone axis. Lower-right inset: selected area electron diffraction pattern showing twin orientation relationship between the nanograin and parent UFG. (c) Distribution of grain boundary orientation for both UFGs and NGs in (a). This HGS was produced via partial recrystallization annealing at 600°C for 1 h following cold rolling of CrCoNi MEA to 95% thickness reduction.

Figure 32.1b is a TEM image showing clearly a twinned NG locating at the lower part of its parent grain, apparently nucleated at the corner of the HAGB. Such a twinned grain is thus often called the corner twin (44, 45). In fact, this is a well-known mechanism of discontinuous nucleation for recrystallization, especially in metals of low SFE (44, 45). There are many annealing twins in MGs (Fig. 32.1a), as marked using red segments. Based on our EBSD analysis, both the NGs and UFGs are mostly twinned grains, judged based on twinning orientation relationship with their parent grains: out of the HAGB (blue segment) with its parent grain, a significant portion is very often a Σ3 twin boundary (TB, the red segment). Among the total NG or UFG grains, the number percentage of twinned ones is as high as 72% and 43%, respectively. This dominance of twin orientation is obvious in Fig. 32.1c.

Figure 32.2a displays the tensile engineering stress–strain (σ–ε) curve of the HGS MEA tested at 298 K. The alloy exhibits a $\sigma_{0.2}$ as high as 1100 MPa, together with a tensile strain to failure over 30%. Note that the uniform elongation (E_u) is as large as 22%. The σ–ε curve preformed at 77 K is also shown, which indicates a simultaneous increase of $\sigma_{0.2}$, Θ and E_u with respect to room-temperature values. Figure 32.2 also compares the tensile response and strain hardening behavior with three other types of (much more homogeneous) grain structures. The first is the same CoCrNi MEA after high-pressure torsion and subsequent full recrystallization annealing (3), with a mean grain size (199 nm) and yield strength not very different from the our HGS (330 nm). Figure 32.2b shows the true flow stress σ_f (minus $\sigma_{0.2}$ for a close-up view) versus true strain. The 199-nm-MEA has only a ~2% E_u (curve 1). To increase E_u to 20% (to become comparable to our HGS, curve 2), Ref. 3 had to increase the MEA grain size to restore strain hardening, and as a consequence dropped $\sigma_{0.2}$ by ~330 MPa. In contrast, the HGS shows a transient hardening at strains ranging from 4% to ~8%, typical of strain hardening response in the initial stage in heterostructures due to hetero-deformation induced (HDI) hardening (46–49). Figure 32.2c shows the normalized strain hardening rate, $\Theta = \frac{1}{\sigma_f} \cdot (\partial \sigma_f / \partial \varepsilon)$. Uniquely, the HGS shows an up-turn of Θ within the range of the transient and more importantly, Θ is always higher than those of curves 1 and 2 throughout the entire tensile deformation. In contrast, the

199 nm-MEA (curve 1) plunges straightly to the onset of plastic instability predicted by the Considéré criterion (Θ = 1), very early on at a plastic strain of <~2%. The Θ of curve 2 also shows a monotonic drop with strain, even though the stress level is ~330 MPa lower (3).

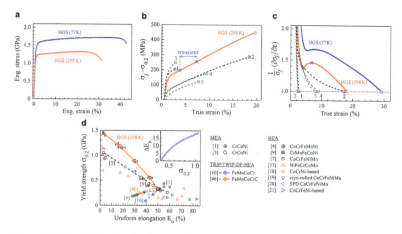

Figure 32.2 Strain hardening and strength–ductility combination in HGS. (a) Tensile engineering stress–strain (σ–ε) curve of HGS CrCoNi MEA at both 298 K and 77 K after cold rolling and recrystallization annealing at 873 K. (b) Flow true stress (minus $\sigma_{0.2}$) versus true strain curves. Note the presence of a transient hardening stage between the two inflexion points (marked by ×) in the HGS curve. Five curves for much more homogenous grained structures are shown for comparison. Curves 1 and 2: 199 nm and 286 nm CoCrNi MEA (910 MPa and 775 MPa) (3). Number in bracket: $\sigma_{0.2}$ (similarly hereinafter). Curves 3 and 4: 316L stainless steel (SS) (975 MPa and 785 MPa, respectively) (50). Curve 5: ultrafine-grained (UFG) Ni (775 MPa) (see Fig. 32.S1 for details). (c) Normalized strain hardening rate (Θ) versus true strain curves. Note the upturn of Θ between two inflexion points (×) in HGS MEA. (d) Combination of yield strength ($\sigma_{0.2}$) and uniform elongation (E_u), in comparison with previous single-phase FCC MEAs and HEAs. Note the curvature of the $\sigma_{0.2}$-E_u relationship (red), which has a slope considerably higher than normally found. Inset shows the higher $\sigma_{0.2}$, the higher the advantage in increased E_u.

Next, we compare our HGS with two uniformly-grained structures in 316L stainless steel (SS) of low SFE (50). Much like the comparison with more uniform grained MEAs above, here again the SS having a $\sigma_{0.2}$ similar to HGS shows a very low E_u (curve 3 in Fig. 32.2b) and a plunging Θ (Fig. 32.2c), and lowering $\sigma_{0.2}$ by ~330 MPa only increased E_u to ~10% (curve 4) (50), with a monotonic decreasing Θ (Fig. 32.2c).

Yet another comparison is made in Fig. 32.2, with a conventional Ni alloy, one that does not have low SFE. This Ni has a mean grain size of 500 nm (see Fig. 32.S1 for details), not too far away from the average in our HGS. Its yield strength is not too much lower either (Fig. 32.2b). However, as seen in Fig. 32.2c, its strain hardening rate is not sustainable (for its load/unload/reload tensile curve, see later). All these comparisons clearly demonstrate the advantage of HGS over those of much more homogeneous grains, regardless of their SFE, even when at a much-lowered $\sigma_{0.2}$. This highlights the crucial role played by our intentionally maximized HGS.

The resultant strength-ductility combination ($\sigma_{0.2}$ and E_u) is displayed in another format in Fig. 32.2d. First, comparing our HGS (red solid line) with the more uniform grained MEA (3) (dashed black line), E_u is much increased at any given $\sigma_{0.2}$. In fact, the higher the $\sigma_{0.2}$, the larger the relative increase in E_u (this difference is highlighted in the inset in Fig. 32.2d). This advantage of HGS is especially pronounced at high strengths towards 1 GPa. The second point to note in Fig. 32.2d is that, although previous reports attempted to use phase transformation (10) or a combination of phase transformation and deformation twinning (51) to evade the trade-off between $\sigma_{0.2}$ and E_u, they offer high E_u usually at low $\sigma_{0.2}$ (<~400 MPa). More detailed comparisons are made in Fig. 32.S2a and S2b.

The loading-unloading-reloading (LUR) tensile tests were conducted to unveil the underlying strain hardening mechanism and defect behaviors (Fig. 32.3a). Each LUR curve exhibits a hysteresis loop (Fig. 32.3b). Of special note is that the reverse (compressive) plastic flow sets in (at 0.2% offset strain) upon elastic unloading, even though the applied stress is still in tension (Fig. 32.3b). This is an unambiguous sign that unequivocally demonstrates the generation of high internal HDI stresses (47–49, 52). Several hysteresis loops in other metals (47, 53) known for strong HDI hardening are also included in the plot for comparison. We have also compared with two homogeneous microstructures. One is UFG-Ni prepared via equal channel angular pressing followed by annealing, as a representative of conventional alloys with uniform small grain size. The other is the coarse-grained (CG) CoCrNi MEA after recrystallization annealing at 1150°C for 1 h (see details in Fig. 32.S3). Obviously, the width of hysteresis loop in both cases is nowhere close to our HGS (see both

Figs. 32.S1f and 32.S3d). The larger the loop width (characterized by residual plastic strain, ε_{rp}, following elastic unloading), the stronger the HDI hardening (47, 52). In fact, the loop widths in the HGS MEA, as plotted in the inset in Fig. 32.3b, are by far the largest ever for any known alloy. Such actual experimental evidence is not available in the literature on HGSs, such as Refs. 54–56, where it was only argued that such behavior may be possible. More directly, we observe that the increase in internal HDI stresses (see Fig. 32.S4 for the method of HDI stress calculations) is the decisive contributor to flow stress after yielding (Fig. 32.3c), accounting for over 80% of the total stress elevation observed during strain hardening, from yield stress to ultimate tensile stress (UTS). We thus conclude that HDI hardening takes a leading role in strain hardening accompanying the tensile elongation. The microstructural origin of the large HDI stress is the dynamically reinforced heterostructure, to be further elaborated in the next section.

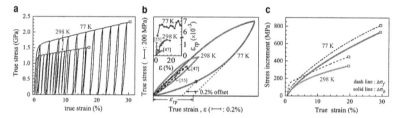

Figure 32.3 Extraordinary HDI hardening in HGS. (a) Tensile true stress-true strain curves during loading-unloading-reloading tests at 298 K and 77 K. (b) Hysteresis loops at the maximum uniform strain. Note inverse (compressive) yielding at 0.2% offset strain following elastic unloading even when the applied stress state is tensile. Hysteresis loops from other representative materials (47, 53) are also included for comparison. Also see comparison with narrow loops observed for both the UFG Ni in Fig. 32.S1f and coarse-grained CoCrNi MEA in Fig. 32.S3d. (c) Comparison of HDI hardening with the total strain hardening (measured flow stress minus that at yielding) in HGS MEA tested at 298 K and 77 K.

The HGS MEA is in fact further refined and reinforced throughout the tensile deformation. To see this, we note that the number density of NGs (ρ_{NG}, grains µm^{-2}) significantly increases from 4.7 (before tensile test in Fig. 32.1a) to 21 (after test at 298 K in Fig. 32.4a), and up to 31 grains µm^{-2} (at 77 K in Fig. 32.4b) by statistical analysis of EBSD observations. The evolution of ρ_{NG} with tensile strain is further shown in Fig. 32.4c. The corresponding area fraction of NGs rises from 2.0% (before), to 5.5% (298 K), and to 8.0% (77 K). Correspondingly, the average grain size of NGs decreases from 180 nm (before) to 144 nm (tested at 298 K), and further down to 140 nm (77 K), as seen in Fig. 32.4c. This clearly indicates the dynamic generation of smaller NGs during tensile deformation. EBSD statistical analysis indicates that the areas of UFGs and MGs decrease, with their number densities largely unchanged, as they provide the locations where NGs form.

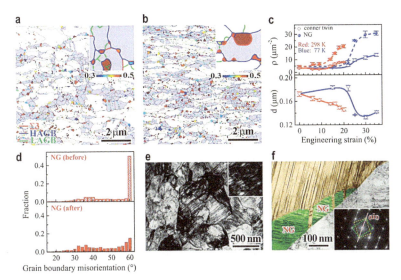

Figure 32.4 Dynamic generation of nanograins at GBs during tensile straining. (a) and (b) EBSD grain boundary maps after tensile test at 298 K and 77 K, respectively. Insets show the generation of NGs mainly at GBs of UFGs. Scale bar: magnitude of the Schmid factor. (c) Upper panel: evolution of number density (ρ) of both NGs and corner twins in tensile tests at 298 K and 77 K. Lower panel: grain size (d) change of NGs during tensile deformation. (d) Distribution of grain boundary orientation in NGs before and after tensile test at 298 K. (e) TEM image showing the dynamically generated NGs at 298 K. Inset shows stacking faults of high density. (f) TEM image showing the formation of twinned NGs at GB at 298 K. Inset is the diffraction pattern showing twinning orientation relationship between the twinned NG and parent grain.

The NGs show several interesting features. First, almost all NGs nucleate at the GBs where stresses are higher especially at GBs of UFGs (see insets in Figs. 32.4a and b). Second, in terms of statistic

EBSD analysis as shown in Fig. 32.4c, ρ of corner twins (with 60°C misorientation in Fig. 32.4d) increases from 3.4 (before) to 8.0 (298 K) and up to 13.6 (77 K) μm^{-2}, respectively, although the TBs evolve with tensile straining and gradually lose the Σ3 character as more and more stacking faults form inside the twinned grains (42). Third, ρ_{NG} begins to increase at a strain of 5% during tensile deformation at 298 K, well consistent with that where the up-turn of Θ occurs (Fig. 32.2c). Namely, the generation of NGs corresponds to the up-turn of Θ. This is because stresses are high at HAGBs of UFGs as a result of load partitioning during tensile straining, facilitating the evolution from Σ3 twin boundaries to HAGBs. With further increased strains, both the corner twins and NGs increase their ρ in the later stage of tensile strains, Fig. 32.4c. In other words, the larger (MG) grains would carry more and more plastic strain as tensile deformation continues. At larger tensile strains stress partitioning towards UFGs has increased the driving stress to sufficiently high levels for twin nucleation at the GBs of UFGs. New twinned NGs are therefore generated, through the emission of twinning partial dislocations (42) at these high-stress GB sites, increasing fast with strain (see Fig. 32.4d). The GBs of these NGs are effective high-angle barriers that impede the slip of dislocations. This refinement also makes the grain structure even more heterogeneous, perpetuating HDI hardening during subsequent deformation.

The dynamic generation of NGs in our HGS is further confirmed by TEM observations. Fig. 32.S5 shows clearly that the NGs indeed form during tensile straining, by monitoring and comparing the microstructural evolution after the cold rolling, recrystallization annealing, and tensile deformation, respectively. After tensile deformation, an original grain is often subdivided, due to the formation of NGs along the GB (Fig. 32.4e). Stacking faults (SFs) are also visible in most grains (inset). Figure 32.4f shows several smaller NGs (green) that form at the GB of a parent UFG (yellow) during tensile deformation. They show twinning relationship with respect to the parent grain. Note here that it is the partially recrystallized HGS, in lieu of uniform-sized grains, that allows the new (twinned) NGs to emerge in large numbers on the fly, with associated strain hardening. A fully recrystallized grain structure would primarily store stacking faults, and not be as conducive to load partitioning and dynamic nucleation of many (twinned) NGs at GBs and TJs via dynamic recrystallization. Also note that the high efficacy of dynamic refinement is also due to the low SFE, which facilitates the formation of corner and necklace twins that evolve into new NGs.

We next explain the evolution of Θ during tensile deformation, from the standpoint of the beneficial effect of the HGS on instigating the high internal HDI stress. Compared with uniform grains, the HGS adds an advantage in that plastic deformation is non-homogeneous due to the inhomogeneous microstructure, inducing strain gradient and large internal stresses (46, 57, 58, 59). In contrast to typical monotonic drop in Θ in the uniformly-grained structures (curves 1 to 5 in Fig. 32.2c), the decreasing Θ of the HGS shows an up-turn, typical of discontinuous yielding in elasto-plastic transition, which in the HGS encompasses "grain to grain yielding": upon tensile straining, the softer grains begin to yield first. Due to the constraints by the still elastic small grains, dislocations in larger grains are piled up and blocked at grain boundaries. The excess geometrically necessary dislocations (GNDs) will be generated due to strain incompatibility, to accommodate the strain gradient (46, 57, 58, 59). These GNDs may interact with mobile dislocations to increase the density of mobile dislocations in varying grains due to dislocation interaction and entanglement. Meanwhile, the MGs preferentially carry plastic strain, relative to the stronger small grains. This is accompanied by stress partitioning: the UFGs/NGs bear an increasing fraction of the applied stress, whilst MGs carry proportionally less stress. This generates HDI stresses, contributing to the HDI hardening during the tensile deformation (47, 48, 52, 53), see Fig. 32.3c. In other words, HDI hardening is associated with strain partitioning, i.e. non-homogeneous plastic strain: a steep strain gradient is established at the grain boundary between grains of very different sizes (47, 57). Due to the constraints by the still elastic small grains, dislocations in larger grains are piled up and blocked at grain boundaries. The long-range HDI stresses are significant as a result (Fig. 32.3), in addition to the hardening from forest dislocations in the grains, such that Θ exhibits an up-turn, see Fig. 32.2c.

One note to add is that the deformation-generated NGs are themselves deformable and hardenable, via the many stacking faults generated inside them (see TEM image in Figs. 32.4e and f). But our emphasis here is that the dynamically generated NGs promote a flow response analogous to that in a multi-phase (such as steels) or composite alloy. Specifically, the HGS is dynamically maintained and always plastically non-homogeneous (when the overall sample is undergoing homogeneous plastic deformation). As demonstrated by the unprecedented magnitude of the long-range HDI stresses shown in Figs. 32.3b and c, the strain inhomogeneity in the self-reinforcing HGS creates unusually large inhomogeneous internal stresses to cause extra hardening (47, 48, 53, 57, 58). The strain hardening rate is therefore sustained (Figs. 32.2c) to prolong uniform elongation to large strains.

Finally, the strength-ductility combination offered by the HGS is much better than those reported before for all the other single-phase FCC HEAs/MEAs (Fig. 32.2d) and TWIP steels (Fig. 32.S2a): we now have extended uniform ductility to 22% even when yield strength is pushed to above 1 GPa. The previous strength-ductility synergy was reported only for those with yield strength <400 MPa (see Fig. 32.2d), where the strain hardening capability is much easier to come by, as the room for defect storage has not been exhausted. There are cases, such as duplex steels or TWIP steels with V and Cr additions (32), which have strength-ductility comparable to our HGS, as shown in Fig. 32.S2b. But that figure is for alloys containing more than one phase as heterogeneities. There multiple mechanisms operate simultaneously for work hardening, including the TRIP effect, precipitates (VC nano-precipitation at GBs, Mo_2C etc.), TWIP, dislocations, etc. In particular, with complex composition (C, Mn, Al, Mo, and V etc.), dual-phases form to provide heterogeneity. In contrast, in MEAs/HEAs (Fig. 32.2d) and TWIP steels (Fig. 32.S2a), only one FCC phase is present, with no TRIP and precipitates in action. In other words, our HGS shows that a single-phase FCC alloy can now be made with a multi-level grain structure heterogeneity to reach properties achievable in a multi-(dual) phase heterostructured steel.

In summary, we advocate a strategy to achieve simultaneous high yield strength and high ductility, generally applicable to low SFE metals and alloys, for which the CrCoNi MEA alloy serves as a model to demonstrate. The key ingredient is to start out with a gigapascal yield strength from a purposely and extraordinarily heterogeneous grain structure that spans the nano-to-micro range. The HGS imparts an additional long-range hardening effect from HDI stresses, and is dynamically reinforced through the in situ production of new nanograins at GBs throughout tensile straining. In other words, the partially recrystallized starting grain structure facilitates non-homogeneous plastic deformation and the associated strain gradient hardening. Also importantly, the HGS evolves towards even more heterogeneous during tensile deformation, as more twins, faults and NGs with high-angel GBs form on the fly due to the low SFE, rather than just dislocations that annihilate easily. This preserves steep strain gradient across neighboring grains and hence large internal HDI stresses (see wide hysteresis loops) to sustain strain hardening over a range of plastic strains. We also showed that these effects would diminish in MEAs after extended annealing towards uniform grains, or in materials with high SFE where full dislocations dominate and could not accumulate efficiently due to recovery and annihilation into GBs. The dynamic HGS supplements the well-known effects of phase transformation (transformation-induced plasticity, TRIP) (10) and twinning (twinning induced plasticity, TWIP) (1, 9, 51), and are particularly useful for boosting strain hardening and uniform tensile ductility in low-SFE materials with high yield strength.

32.3 Materials and Methods

32.3.1 Material Fabrication and Sample Preparation

In the present study, an equimolar CrCoNi medium-entropy alloy (MEA) was fabricated. The purity of each raw elemental material was higher than 99.9 weight percent (wt.%). A bulk ingot, weighting ~3 kg, was produced by electromagnetic levitation melting in a high-purity argon atmosphere and cast to an ingot with a diameter of 120 mm and height of 100 mm. The ingot was then re-melted five times in order to improve the homogeneity of both the chemical composition and microstructure. The actual chemical composition was measured to be 30.77Cr-33.81Co-35.34Ni (wt.%), i.e. 32.46Cr-33.48Co-34.06Ni (at. %). The ingot was homogenized at 1473 K for 12 h, hot-forged in between 1323 K and 1173 K into slabs of 10 mm in thickness, and finally cold rolled into sheets 0.5 mm thick. No cracks were observed on the surfaces of the cold-rolled sheets. The sheets were subsequently annealed at varying temperatures ranging from 473 K to 1473 K for 1 h followed by water quenching. The data presented in this paper are for annealing at 873 K for 1 h, which produced a partially recrystallized microstructure.

32.3.2 Mechanical Property Testing

The flat dog-bone-shaped tensile specimens, with a gauge length of 15 mm and width of 4 mm, were cut from the annealed sheets with their longitudinal axes parallel to the rolling direction. All specimens were mechanically polished prior to tensile tests in order to remove surface irregularities and to ensure a more accurate determination of the cross-sectional area. The quasi-static uniaxial tensile tests were carried out using an MTS Landmark testing machine operating at a strain rate of 5×10^{-4} s^{-1}. The tensile tests were conducted at room temperature (298 K) and liquid-nitrogen (77 K), respectively. The loading-unloading-reloading (LUR) tensile tests were conducted to characterize the HDI stresses. The methods for calculating the HDI stresses from LUR tensile test curves can be found in a previous paper (48). The conditions for LUR tests were the same as those for monotonic tests. All tensile tests were conducted using a 10 mm gauge extensometer to monitor the engineering strain. Five samples were tested for each condition to verify the reproducibility.

32.3.3 Microstructural Characterization

Scanning electron microscopy (SEM, JEOL JSM-7001F) was used to characterize the cross-sectional microstructural evolution of the annealed CrCoNi MEA before and after tensile tests. EBSD measurements were performed with a DigiView camera and the TSL OIM data-collection software (http://www.edax.com/Products/ EBSD/ OIM-Data-Collection-EBSD-SEM.aspx), at an EBSD step size of 40 nm. The grain boundary was defined by misorientation angle larger than 15°. The nano-grains are colored based on the magnitude of the Schmid factor (60), in the un-deformed and deformed microstructure, in order to partially reflect the orientation change, and in turn the deformation level, of nano-grains during tensile deformation. Local grain misorientation information is provided through the grain reference orientation distribution (GROD) (61) in Fig. 32.S6, to help track recrystallized grains in the HGS.

A high-spatial resolution analytical electron microscope (HRTEM, JEM 2010F) operating at 200 kV was used for examination of the microstructural features after tensile testing at designated strains. Thin foils for TEM observations were cut from the gauge sections of the tensile samples, ion thinned to about 80 μm thick and finally thinned by a twin-jet polishing facility using a solution of 5% perchloric acid and 95% ethanol at −25°C.

Acknowledgements

X.L.W. and F.P.Y. were funded by the National Key R&D Program of China 2017YFA0204402, the Strategic Priority Research Program of the Chinese Academy of Sciences (Grant No. XDB22040503), the Natural Science Foundation of China (Grant Nos. 11572328 and 11472286). E.M. was supported by U.S.-DOE-BES, Division of Materials Sciences and Engineering, under Contract No. DE-FG02-16ER46056.

References

1. Gludovatz B, Hohenwarter A, Thurston KVS, Bei H, Wu Z, George EP, Ritchie RO (2016). Exceptional damage-tolerance of a medium-entropy alloy CrCoNi at cryogenic temperatures. *Nat Commun* 7:10602.
2. Zhang ZJ, Sheng HW, Wang ZJ, Gludovatz B, Zhang Z, George EP, Yu Q, Mao SX, Ritchie RO (2017). Dislocation mechanisms and 3D twin architectures generate exceptional strength-ductility-toughness combination in CrCoNi medium-entropy alloy. *Nat Commun* 8:14390.
3. Yoshida S, Bhattacharjee T, Bai Y, Tsuji N (2017). Friction stress and Hall-Petch relationship in CoCrNi equi-atomic medium entropy alloy processed by severe plastic deformation and subsequent annealing. *Scr Mater* 134:33–36.
4. Wu Z, Bei H, Pharr GM, George EP (2014). Temperature dependence of the mechanical properties of equiatomic solid solution alloys with face-centered cubic crystal structures. *Acta Mater* 81:428–441.
5. Laplanche G, Kostka A, Reinhart C, Hunfeld J, Eggeler G, George EP (2017). Reasons for the superior mechanical properties of medium-entropy CrCoNi compared to high-entropy CrMnFeCoNi. *Acta Mater* 128:292–303.

6. Miao J, Slone CE, Smith TM, Niu C, Bei H, Ghazisaeidi M, Pharr GM, Mills MJ (2017). The evolution of the deformation substructure in a Ni-Co-Cr equiatomic solid solution alloy. *Acta Mater* 132:35–48.
7. Shahmir H, He JY, Lu ZP, Kawasaki M, Langdon TG (2016). Effect of annealing on mechanical properties of a nanocrystalline CoCrFeNiMn high-entropy alloy processed by high-pressure torsion. *Mater Sci Eng A* 676:294–303.
8. Yeh JW, Chen SK, Lin SJ, Gan JY, Chin TS, Shun TT, Tsau CH, Chang SY (2004). Nanostructured high-entropy alloys with multiple principal elements: novel alloy design concepts and outcomes. *Adv Eng Mater* 6(5):299–303.
9. Gludovatz B, Hohenwarter A, Catoor D, Chang EH, George EP, Ritchie RO (2014). A fracture-resistant high-entropy alloy for cryogenic applications. *Science* 345:1153–1158.
10. Li ZM, Pradeep KG, Deng Y, Raabe D, Tasan CC (2016). Metastable high-entropy dual-phase alloys overcome the strength-ductility trade-off. *Nature* 534(7606):227–230.
11. Zou Y, Ma H, Spolenak R (2015). Ultrastrong ductile and stable high-entropy alloys at small scales. *Nat Commun* 6:7748.
12. Zhang ZJ, Mao MM, Wang J, Gludovatz B, Zhang Z, Mao SX, George EP, Yu Q, Ritchie RO (2015). Nanoscale origins of the damage tolerance of the high-entropy alloy CrMnFeCoNi. *Nat Commun* 6:10143.
13. Chen BR, Yeh AC, Yeh JW (2016). Effect of one-step recrystallization on the grain boundary evolution of CoCrFeMnNi high entropy alloy and its subsystems. *Sci Rep* 6:22306.
14. He JY, Wang H, Huang HL, Xu XD, Chen MW, Wu Y, Liu XJ, Nieh TG, An K, Lu ZP (2016). A precipitation-hardened high-entropy alloy with outstanding tensile properties. *Acta Mater* 102:187–196.
15. Li DY, Li C, Feng T, Zhang Y, Sha G, Lewandowski JJ, Liaw PK, Zhang Y (2017). High-entropy $Al_{0.3}$CoCrFeNi alloy fibers with high tensile strength and ductility at ambient and cryogenic temperatures. *Acta Mater* 123:285–294.
16. Ma Y, Yuan FP, Yang MX, Jiang P, Ma E, Wu XL (2018). Dynamic shear deformation of a CrCoNi medium-entropy alloy with heterogeneous grain structures. *Acta Mater* 148:407–418.
17. Zaddach AJ, Scattergood RO, Koch CC (2015). Tensile properties of low-stacking fault energy high-entropy alloys. *Mater Sci Eng A* 636:373–378.

18. Wu Z, Gao Y, Bei H (2016). Thermal activation mechanisms and Labusch-type strengthening analysis for a family of high-entropy and equiatomic solid-solution alloys. *Acta Mater* 120:108–119.
19. Stepanov N, Tikhonovsky M, Yurchenko N, Zyabkin D, Klimova M, Zherebtsov S, Salishchev G (2015). Effect of cryo-deformation on structure and properties of CoCrFeNiMn high-entropy alloy. *Intermetallics* 59:8–17.
20. Schuh B, Mendez-Martin F, Voelker B, George EP, Clemens H, Pippan R, Hohenwarter A (2015). Mechanical properties, microstructure and thermal stability of a nanocrystalline CoCrFeMnNi high-entropy alloy after severe plastic deformation. *Acta Mater* 96:258–268.
21. Salishchev GA, Tikhonovsky MA, Shaysultanov DG, Stepanov ND, Kuznetsov AV, Kolodiy IV, Senkov ON (2014). Effect of Mn and V on structure and mechanical properties of high-entropy alloys based on CoCrFeNi system. *J Alloys Compd* 591:11–21.
22. Li ZM, Tasan CC, Pradeep KG, Raabe D (2017). A TRIP-assisted dual-phase high-entropy alloy: grain size and phase fraction effects on deformation behavior. *Acta Mater* 131:323–335.
23. Li ZM, Körmann K, Grabowski B, Neugebauer J, Raabe D (2017). Ab initio assisted design of quinary dual-phase high-entropy alloys with transformation-induced plasticity. *Acta Mater* 136:262–270.
24. Wang MM, Li ZM, Raabe D (2018). In-situ SEM observation of phase transformation and twinning mechanisms in an interstitial high-entropy alloy. *Acta Mater* 147:236–246.
25. Smith TM, Hooshmand MS, Esser BD, Otto F, McComb DW, George EP, Ghazisaeidi M, Mills MJ (2016). Atomic-scale characterization and modeling of 60 degrees dislocations in a high-entropy alloy. *Acta Mater* 110:352–363.
26. Meyers MA, Chawla KK (2009). *Mechanical Behavior of Materials*. Cambridge: Cambridge Univ Press, UK.
27. Wang YM, Chen MW, Zhou FH, Ma E (2002). High tensile ductility in a nanostructured metal. *Nature* 419(6910):912–915.
28. Lu L, Shen YF, Chen XH, Qian LH, Lu K (2004). Ultrahigh strength and high electrical conductivity in copper. *Science* 304(5669):422–426.
29. Zhu YT, Liao XZ (2004). Nanostructured metals: retaining ductility. *Nat Mater* 3(6):351–352.
30. Grassel O, Kruger L, Frommeyer G, Meyer LW (2000). High strength Fe-Mn-(Al, Si) TRIP/TWIP steels development-properties-application. *Int J Plast* 16(10–11):1391–1409.

31. Wu XL, Yang MX, Yuan FP, Chen L, Zhu YT (2016). Combining gradient structure and TRIP effect to produce austenite stainless steel with high strength and ductility. *Acta Mater* 112:337–346.

32. Sohn SS, Song H, Jo MC, Song T, Kim HS, Lee S (2017). Novel 1.5 GPa-strength with 50%-ductility by transformation-induced plasticity of non-recrystallized austenite in duplex steels. *Sci Rep* 7(1):1255.

33. Herrera C, Ponge D, Raabe D (2011). Design of a novel Mn-based 1GPa duplex stainless TRIP steel with 60% ductility by a reduction of austenite stability. *Acta Mater* 59(11):4653–4664.

34. Chen XH, Lu J, Lu L, Lu K (2005). Tensile properties of a nanocrystalline 316L austenitic stainless steel. *Scr Mater* 52(10):1039–1044.

35. Chen AY, Li DF, Zhang JB, Song HW, Lu J (2008). Make nanostructured metal exceptionally tough by introducing non-localized fracture behaviors. *Scr Mater* 59(6):579–582.

36. Dini G, Najafizadeh A, Ueji R, Monir-Vaghefi SM (2010). Improved tensile properties of partially recrystallized submicron grained TWIP steel. *Mater Lett* 64(1):15–18.

37. Niendorf T, Rüsing CJ, Frehn A, Maier HJ (2013). The deformation behavior of functionally graded TWIP steel under monotonic loading at ambient temperature. *Mater Res Lett* 1(2):96–101.

38. Sohn SS, Song H, Kwak JH, Lee S (2017). Dramatic improvement of strain hardening and ductility to 95% in highly-deformable high-strength duplex lightweight steels. *Sci Rep* 7(1):1927.

39. Chen S, Rana R, Haldar A, Ray RK (2017). Current state of Fe-Mn-Al-C low density steels. *Prog Mater Sci* 89:345–391.

40. Wang MM, CC Tasan, D Ponge, Dippel AC, Raabe D (2015). Nanolaminate transformation-induced plasticity–twinning-induced plasticity steel with dynamic strain partitioning and enhanced damage resistance. *Acta Mater* 85:216–228.

41. Gutierrez-Urrutia I, Raabe D (2011). Dislocation and twin substructure evolution during strain hardening of an Fe–22wt.% Mn–0.6wt.% C TWIP steel observed by electron channeling contrast imaging. *Acta Mater* 59(16):6449–6462.

42. Zhu YT, Liao XZ, Wu XL (2012). Deformation twinning in nanocrystalline materials. *Prog Mater Sci* 57(1):1–62.

43. Wu XL, Zhu YT (2008). Inverse grain-size effect on twinning in nanocrystalline Ni. *Phys Rev Lett* 101(2):025503.

44. Sakai T, Belyakov A, Kaibyshev R, Miura H, Jonas JJ (2014). Dynamic and post-dynamic recrystallization under hot, cold and severe plastic deformation conditions. *Prog Mater Sci* 60:130–207.
45. Ponge D, Gottstein G (1998). Necklace formation during dynamic recrystallization: mechanisms and impact on flow behavior. *Acta Mater* 46(1):69–80.
46. Wu XL, Jiang P, Chen L, Yuan FP, Zhu YT (2014). Extraordinary strain hardening by gradient structure. *Proc Natl Acad Sci USA* 111(20):7197–7201.
47. Wu XL, Yang MX, Yuan FP, Wu Guilin, Wei YJ, Huang XX, Zhu YT (2015). Heterogeneous lamella structure unites ultrafine-grain strength with coarse-grain ductility. *Proc Natl Acad Sci USA* 112(47):14501–14505.
48. Yang MX, Pan Y, Yuan FP, Zhu YT, Wu XL (2016). Back stress strengthening and strain hardening in gradient structure. *Mater Res Lett* 4(3):145–151.
49. Wu XL, Jiang P, Chen L, Zhang JF, Yuan FP, Zhu YT (2014). Synergetic strengthening by gradient structure. *Mater Res Lett* 2(4):185–191.
50. Odnobokova M, Belyakov A, Kaibyshev R (2015). Effect of severe cold or warm deformation on microstructure evolution and tensile behavior of a 316L stainless steel. *Adv Eng Mater* 17(12):1812–1820.
51. Li ZM, Tasan CC, Springer H, Gault B, Raabe D (2017). Interstitial atoms enable joint twinning and transformation induced plasticity in strong and ductile high-entropy alloys. *Sci Rep* 7:40704.
52. Sinclair CW, Saada G, Embury JD (2006). Role of internal stresses in co-deformed two-phase materials. *Philos Mag* 86(25–26):4081–4098.
53. Xiang Y, Vlassak JJ (2006). Bauschinger and size effects in thin-film plasticity. *Acta Mater* 54(20):5449–5460.
54. Wu XL, Zhu YT (2017). heterogeneous materials: a new class of materials with unprecedented mechanical properties. *Mater Res Lett* 5(8):527–532.
55. Liu XL, Yuan FP, Zhu YT, Wu XL (2018). Extraordinary Bauschinger effect in gradient structured copper. *Scr Mater* 150:57–60.
56. Ma E, Zhu T (2017). Towards strength–ductility synergy through the design of heterogeneous nanostructures in metals. *Mater Today* 20(6):323–331.
57. Ashby MF (1970). The deformation of plastically non-homogeneous materials. *Philos Mag* 21:399–424.

58. Gao H, Huang Y, Nix WD, Hutchinson JW (1999). Mechanism-based strain gradient plasticity-I. Theory. *J Mech Phys Solids* 47(6):1239–1263.

59. Huang CX, Wang YF, Ma XL, Yin S, Höppel HW, M. Göken M, Wu XL, Gao HJ, Zhu YT (2018). Interface affected zone for optimal strength and ductility in heterogeneous laminate. *Mater Today,* https://doi.org/10.1016/j.mattod.2018.03.006, in press.

60. Blochwitz C, Brechbuehl J, Tirschler W (1996). Analysis of activated slip systems in fatigued nickel polycrystals using the EBSD-technique in the scanning electron microscope. *Mater Sci Eng A* 210:42–47.

61. Gutierrez-Urrutia I, Archie F, Raabe D, Yan FK, Tao NR, Lu K (2016). Plastic accommodation at homophase interfaces between nanotwinned and recrystallized grains in an austenitic duplex-microstructured steel. *Sci Technol Adv Mater* 17(1):29–36.

Supplementary Information Text

32.S1 Fabrication of UFG-Ni by ECAP and Annealing

Commercial pure Ni (99.9% wt%) rods with a diameter of 20 mm were annealed at 980°C for 5 h to form a solid solution with an initial grain size of approximately 40 μm. The equal channel angular pressing (ECAP) processing was carried out using a die with an intersecting channel angle of 90° and an outer arc angle of 45°. This die configuration imposes an effective strain of approximately one per one ECAP pass. The sample was processed for six passes by route Bc in which the work piece was rotated 90 along its longitudinal axis after the first pass. The ECAP-processed UFG Ni rod was then annealed at 580°C for 20 min to obtain the homogeneous grain structure. Finally, thin tensile samples were cut for tensile load/unload/reload tests in an orientation parallel to the pressing direction.

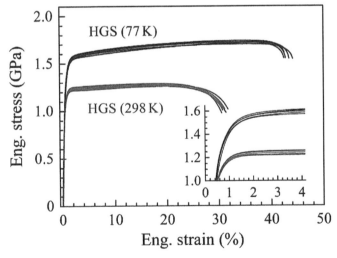

Figure 32.S1 Tensile engineering stress-strain curves of HGS CrCoNi MEA. Five tests were conducted at 298 K and 77 K, respectively. All samples were cut from the same plate after cold rolling and recrystallization annealing at 873 K. Inset: close-up showing yield strength within a narrow range of ~50 MPa, indicating reproducible properties and microstructural stability.

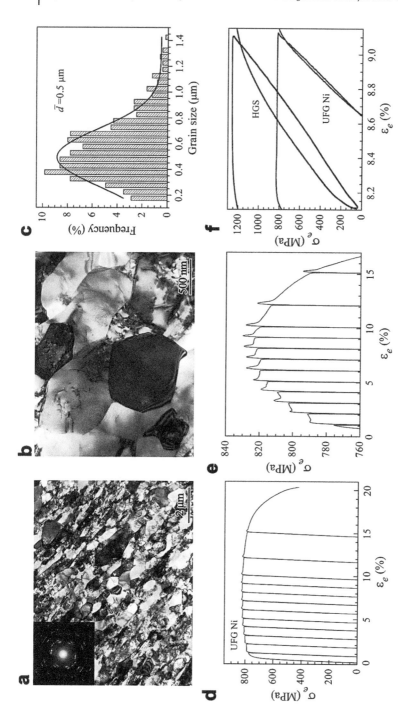

Figure 32.S2 Homogeneous-grained UFG Ni and negligible hysteresis loop width during unload/reload tensile test. (a) and (b) Low- and high-magnification TEM micrographs showing the ultra-fine grained (UFG) Ni after 6-pass ECAP via Bc route and annealing. (c) Grain size distribution. Mean grain size: 500 nm. (d) Load/unload/reload tensile curve. (e) Enlarged view. (f) Hysteresis loop at uniform elongation. The wide loop observed in HGS at 298 K is also shown for comparison.

Supplementary Information Text | 607

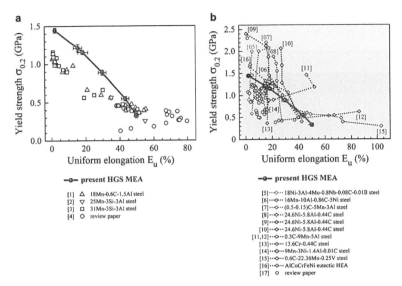

Figure 32.S3 Combination of YS and E_U. (a) TWIP effect as the main work hardening mechanism in single FCC phase alloys. (b) TRIP and multiple hardening mechanisms, usually in dual-phase steels.

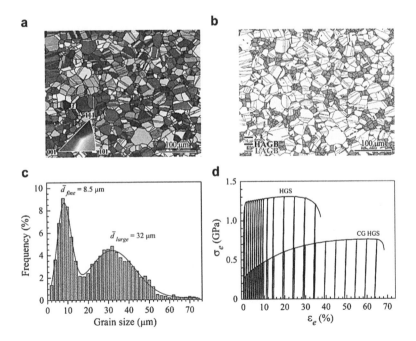

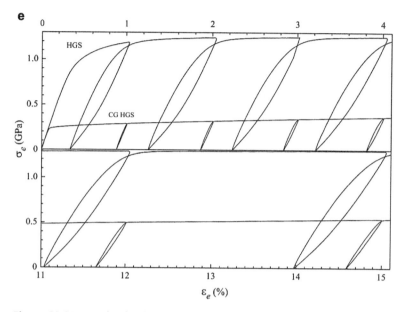

Figure 32.S4 Load-unloading-reloading (LUR) tensile test in coarse-grained CoCrNi MEA after full recrystallization annealing at 1100°C for 1 h. (a) EBSD grain boundary map. (b) Grain size distribution. (c) Tensile LUR stress-strain curves for HGS MEA and CG MEA. (d) Enlarged view for comparison of hysteresis loops at varying strains. The loop width is much narrower compared with HGS. This comparison shows that heterogeneity is the key factor responsible for wide loops, in this low-SFE alloy.

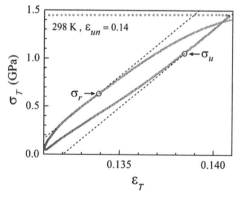

Figure 32.S5 Calculation of internal HDI stress in terms of hysteresis loop. See Ref. 48 for detailed derivation. Ref. 48 derived an equation to calculate the HDI stress (σ_b) from the tensile unloading-reloading hysteresis loop, i.e., $\sigma_b = \frac{1}{2} \cdot (\sigma_u + \sigma_r)$, in which σ_u and σ_r represent the unload and reload yielding stress, respectively.

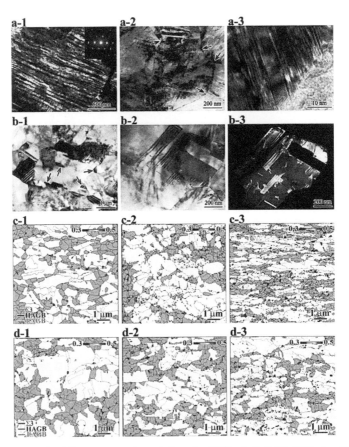

Figure 32.S6 Microstructural evolution in HGS of CoCrNi MEA by TEM, HRTEM, and EBSD observations. (a-raw) Microstructure after cold rolling. (a-1) Lamellae structure. (a-2) Deformed grains with deformation twins and dislocation tangles in their interiors. Note the absence of both corner twins and nanograins at grain boundaries (GBs) and triple junctions (TJs) as indicated by arrows. (a-3) Emission of nano-twins and stacking faults from GB. Note that the NGs are not detected near the GBs in the microstructure after cold rolling. (b-raw) HGS after recrystallization at 600°C for 1 h. (b-1) Typical morphology of UFGs with corner twins in their interiors. Note the corner twins (labeled by arrows) usually at GBs. The dislocation debris and stacking faults (SFs) are also frequently seen. (b-2) and (b-3) Bright- and dark-field TEM images showing the SFs and annealed twins inside large grains where the dislocation density is low. (c-raw) Generation of the NGs in HGS tested at 77 K at varying tensile strains of 14% (c-1), 22% (c-2), and 35% (E_U) (c-3), respectively. It is worthy to note that an obvious increase in number density of NGs begins at tensile strain of 22% (also see Fig. 32.4c for statistical data). (d-row) Generation of the NGs in HGS tested at 298 K at varying tensile strains of 6% (d-1), 14% (d-2), and 20% (E_U) (d-3), respectively. It is worthy to note that an obvious increase in number density of the NGs begins at tensile strain of 6% (also see Fig. 32.4c for statistical data).

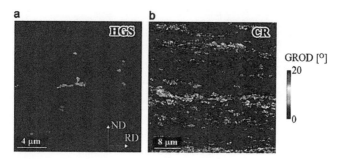

Figure 32.S7 Grain reference orientation deviation (GROD) mapping image. (a) HGS after recrystallization annealing at 600°C for 1 h (i.e. prior to tensile deformation). (b) Cold rolling deformation state.

References

1. Kang S, Jung YS, Jun JH, Lee YK (2010). Effects of recrystallization annealing temperature on carbide precipitation, microstructure, and mechanical properties in Fe-18Mn-0.6C-1.5Al TWIP steel. *Mater Sci Eng A* 527(3):745-751.

2. Mi ZL, Tang D, Jiang HT, Dai YJ, Li SS (2009). Effects of annealing temperature on the microstructure and properties of the 25Mn-3Si-3Al TWIP steel. *Int J Min Metall Mater* 16(2):154-158.

3. Dini G, Najafizadeh A, Ueji R, Monir-Vaghefi SM (2010). Improved tensile properties of partially recrystallized submicron grained TWIP steel. *Mater Lett* 64(1):15-18.

4. Bouaziz O, Allain S, Scott CP, Cugy P, Barbier D (2011). High manganese austenitic twinning induced plasticity steels: a review of the microstructure properties relationships. *Curr Opin Solid State Mater Sci* 15(4):141-168.

5. Jiang S, Wang H, Wu Y, Liu X, Chen H, Yao M, Gault B, Ponge D, Raabe D, Hirata A, Chen M, Wang YD, Lu ZP (2017). Ultrastrong steel via minimal lattice misfit and high-density nanoprecipitation. *Nature* 544(7651):460.

6. Kim SH, Kim H, Kim NJ (2015). Brittle intermetallic compound makes ultrastrong low-density steel with large ductility. *Nature* 518(7537):77-79.

7. He BB, Hu B, Yen HW, Cheng GJ, Wang ZK, Luo HW, Huang MX (2017). High dislocation density-induced large ductility in deformed and partitioned steels. *Science* 357(6355):1029-1032.

8. Furuta T, Kuramoto S, Ohsuna T, Oh-Ishi K, Horibuchi K (2015). Die-hard plastic deformation behavior in an ultrahigh-strength Fe-Ni-Al-C alloy. *Scr Mater* 101:87-90.
9. Edalati K, Furuta T, Daio T, Kuramoto S, Horita Z (2015). High strength and high uniform ductility in a severely deformed iron alloy by lattice softening and multimodal-structure formation. *Mater Res Lett* 3(4):197-202.
10. Ma Y, Yang MX, Jiang P, Yuan FP, Wu XL(2017). Plastic deformation mechanisms in a severely deformed Fe-Ni-Al-C alloy with superior tensile properties. *Sci Rep* 7(1):15619.
11. Sohn SS, Song H, Jo MC, Song T, Kim HS, Lee S (2017). Novel 1.5 GPa-strength with 50%-ductility by transformation-induced plasticity of non-recrystallized austenite in duplex steels. *Sci Rep* 7(1):1255.
12. Sohn SS, Song H, Kwak JH, Lee S (2017). Dramatic improvement of strain hardening and ductility to 95% in highly-deformable high-strength duplex lightweight steels. *Sci Rep* 7(1):1927.
13. Yuan L, Ponge D, Wittig J, Choi P, Jiménez JA, Raabe D (2012). Nanoscale austenite reversion through partitioning, segregation and kinetic freezing: example of a ductile 2 GPa Fe-Cr-C steel. *Acta Mater* 60(6-7):2790-2804.
14. Wang MM, Tasan CC, Ponge D, Raabe D (2016). Spectral TRIP enables ductile 1.1 GPa martensite. *Acta Mater* 111:262-272.
15. Bhattacharjee T, Wani IS, Sheikh S, Clark IT, Okawa T, Guo S, Bhattacharjee PP, Tsuji N (2018). Simultaneous strength-ductility enhancement of a nano-lamellar AlCoCrFeNi$_{2.1}$ eutectic high entropy alloy by cryo-rolling and annealing. *Sci Rep* 8(1):3276.
16. Niendorf T, Rüsing CJ, Frehn A, Maier HJ (2013). The deformation behavior of functionally graded TWIP steel under monotonic loading at ambient temperature. *Mater Res Lett* 1(2):96-101.
17. Sakai T, Belyakov A, Kaibyshev R, Miura H, Jonas JJ (2014). Dynamic and post-dynamic recrystallization under hot, cold and severe plastic deformation conditions. *Prog Mater Sci* 60:130-207.

Chapter 33

Dynamic Shear Deformation of a CrCoNi Medium-Entropy Alloy with Heterogeneous Grain Structures

Yan Ma,[a,b] Fuping Yuan,[a,b] Muxin Yang,[a] Ping Jiang,[a] Evan Ma,[c] and Xiaolei Wu[a,b]

[a]*State Key Laboratory of Nonlinear Mechanics, Institute of Mechanics, Chinese Academy of Sciences, No. 15, West Road, North 4th Ring, Beijing 100190, China*
[b]*School of Engineering Science, University of Chinese Academy of Sciences, Beijing 100190, China*
[c]*Department of Materials Science and Engineering, Johns Hopkins University, 3400 North Charles Street, Baltimore, MD 21218, USA*
fpyuan@lnm.imech.ac.cn

Single-phase CrCoNi medium-entropy alloys (MEA) are emerging recently as an interesting class of metallic materials, but the dynamic response of this MEA at high strain rates remains unknown. Here we have produced this MEA with various heterostructures, using

---

Reprinted from *Acta Mater.*, **148**, 407–418, 2018.

Heterostructured Materials: Novel Materials with Unprecedented Mechanical Properties
Edited by Xiaolei Wu and Yuntian Zhu
Text Copyright © 2018 Acta Materialia Inc.
Layout Copyright © 2022 Jenny Stanford Publishing Pte. Ltd.
ISBN 978-981-4877-10-7 (Hardcover), 978-1-003-15307-8 (eBook)
www.jennystanford.com

cold rolling followed by annealing at various temperatures. The high-strain-rate response of the MEAs was characterized using hat-shaped specimens in Hopkinson-bar experiments. A combination of high dynamic shear yield strength and large uniform dynamic shear strain was observed, exceeding all other metals and alloys reported so far. Even better dynamic shear properties was revealed when the experiments were conducted at cryogenic temperature. The strong strain hardening under dynamic shear loading can be attributed to the dynamic grain refinement and deformation twinning that accompany the homogeneous shear deformation. When compared to room temperature, the efficiency of grain refinement was found to be enhanced at cryogenic temperature, with a higher density of multiple twins, stacking faults, Lomer-Cottrell locks, and hcp phase via phase transformation inside the grains, which could be responsible for the better dynamic shear properties under cryogenic environment.

33.1 Introduction

High-entropy alloys (HEA) [1–14] and medium-entropy alloys (MEA) [15–20], containing multiple elements typically with equal molar fraction, have emerged recently as an interesting class of materials. This class of alloys can be solid solutions with a simple crystal structure, such as fcc, and the random distribution of multiple elements can reduce energy penalty for the formation of stacking fault (SF), rendering low stacking fault energy (SFE). Interestingly, the mechanical properties of fcc FeCrMnCoNi HEA and fcc CrCoNi MEA are better at cryogenic temperatures than at room-temperature, with a simultaneous enhancement in strength, ductility and fracture toughness, due to a transition of deformation mechanism from dislocation slip to deformation twinning [7, 15].

However, HEA and MEA with coarse grains (CG) have a weakness in their relatively low room temperature yield strength. Thus severe plastic deformation (SPD) such as cold rolling (CR) has been applied to obtain increased yield strength in HEA and MEA [1, 3, 5, 9]. Post-deformation annealing produces a multi-modal grain size distribution. Such a heterogeneous grain structure is known to result in a better combination of strength and ductility [21–30]. In our recent work [26, 27], stress/strain partitioning and strain gradients have been found to play an important role during the tensile deformation in heterostructures, resulting in hetero-deformation induced (HDI) stresses that enhance extra strain hardening and uniform elongation. This strategy can also be used for metals and alloys with ultra-low SFE, such as HEAs and MEAs.

The un-dissipated plastic work leads to appreciable temperature rise such that the flow behaviors under dynamic loading are totally different from those under quasi-static conditions, and a clear understanding of the fundamental deformation mechanisms in metals and alloys exposed to high strain rate is critical to the designs of advanced crash relevant and impact-tolerant structures [31–35]. The thermal softening due to the adiabatic heating can facilitate strain localization, such as adiabatic shear band (ASB) [35–37], which is a precursor of the final failure. Thus, the microstructure evolution before and after the formation of ASB is critical for understanding the mechanical properties and the corresponding deformation mechanisms of metals and alloys subjected to impact loading [38–42]. The dynamic shear behaviors in homogeneous metal and alloys with CG, ultra-fine grains (UFG) or NG have been investigated extensively in the literature [35–40]. The CG metals are observed to have low yield strength, high critical strain for onset of ASB and high energy absorption under dynamic loading, while the UFG or NG metals are found to show high yield strength, low uniform shear strain and low energy absorption under dynamic loading. The normal strength-ductility trade-off under quasi-static conditions is found to still exist under dynamic shear loading for homogeneous microstructure. In this regards, the hetero microstructure has the potential to achieve a better combination of strength-ductility even under high strain rate deformation, as suggested in our recent chapter [42, 43].

In the present study, we have designed heterogeneous grain structure through CR followed by annealing, and investigated mechanical properties upon impact loading. Moreover, how the hierarchical grain structure dynamically evolves upon the impact loading and how the microstructure evolution affects the overall dynamic shear properties are critical for the applications of this class of MEAs. In this work, a series of Hopkinson-bar experiments using hat-shaped specimens have been conducted on the MEA with various microstructures to investigate the dynamic shear properties under impact shear loading. Extraordinary dynamic shear properties have been observed, especially at cryogenic temperatures. The underlying deformation mechanisms are discussed using information from microstructural examinations at various controlled shear displacements.

33.2 Materials and Experimental Procedures

The equimolar CrCoNi MEA was produced via electromagnetic levitation melting in a high-purity argon atmosphere, and cast to ingots with a diameter of 120 mm and a height of 100 mm. The actual chemical composition of the ingots was determined to be 35.19Ni–35.67Cr–26.20Co (wt %). The ingots were first homogenized at 1473 K for 12 h, and then hot-forged into slabs with the thickness of 10 mm, at temperatures in between 1323 K and 1173 K. The slabs were then cold rolled into sheets with the final thickness of about 2.7 mm. The CR sheets were subsequently annealed, each at a different temperature, in the range of 873 to 1073 K, for 1 h and immediately water quenched to obtain various heterostructures.

The hat-shaped plate specimens have been machined from the sheets using a wire saw, with the impact direction perpendicular to the rolling direction. All specimens were mechanically polished prior to dynamic shear test in order to remove surface irregularities. The set-up for Hopkinson-bar experiments with hat-shaped plate specimens is displayed in Fig. 33.1a, and the schematics of the hat-shaped specimens and the specimen holders are shown in Fig. 33.1b. The cylindrical high strength maraging steel specimen holder was employed for two purposes: (1) it ensures pure shear deformation by constraining the lateral expansion of two legs for the hat-shaped specimens; (2) the shear displacements for interrupted "frozen" experiments are controlled by changing the height of the specimen holders. The hat-shaped set-up can facilitate the formation of ASB in the narrow concentrated shear zone, and the experimental details for the dynamic shear experiments by Hopkinson-bar technique can be found elsewhere [35, 36, 39–44]. The concentrated shear zone is marked by two white lines in Fig. 33.1, and the shear zone has a width of 0.2 mm and a height of 2 mm. The angle between the shear direction and the impact direction is only about 5°, thus the deformation mode can be considered as nearly pure dynamic shear, and the measured total load can approximately be considered as the shear load [44]. Cryogenic temperature experiments (77 K) were conducted by immerging the specimen into liquid nitrogen for 10 min before dynamic shear tests. The impact velocities of the striker bar for the dynamic shear experiments were about 25 m/s and the nominal shear strain rates were as high as 8×10^4/s.

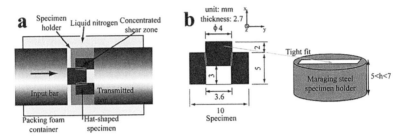

Figure 33.1 (a) Hat-shaped specimen set-up in Hopkinson bar experiment; (b) Illustrations and dimensions of the hat-shaped specimen and the specimen holder.

Scanning electron microscope (SEM), electron backscattered diffraction (EBSD), transmission electron microscope (TEM) and high resolution electron microscope (HREM) have been used to characterize the microstructures before and after dynamic shear testing. The EBSD step size was 30 nm. Thin foils for TEM observations were cut from both the homogeneous shear zone away from the ASB and within the ASB zone, and then mechanically ground to about 50 µm thick and finally thinned by a twin-jet polishing facility using a solution of 5% perchloric acid and 95% ethanol at −25°C. The detailed procedures of sample preparation for SEM, EBSD can be found elsewhere [27, 45].

33.3 Results and Discussions

33.3.1 Microstructural Characterization Before Dynamic Shear Tests

EBSD maps (Inverse pole figure, IPF) for the microstructures annealed at 1073, 973, 923 and 898 K are shown in Figs. 33.2a, b, c and, respectively. The corresponding grain size distributions, plotted in terms of the area fraction, are shown in Fig. 33.2e. In Fig. 33.2e, only high-angle grain boundaries (HAGBs, with misorientation >15°) are considered and annealing twins are excluded when calculating the grain size. Moreover, Fig. 33.2f shows that the misorientation angle distributions have a strong preference for 60°, indicating a large number of annealing twins. As expected, heterogeneous grain structures are observed for the samples annealed at various temperatures: all microstructures are composed of hierarchical grains with grain size on multiple levels. In other words, all microstructures are multimodal in grain size distribution. For example, the grain size ranges from submicron to about 27 μm for the microstructure annealed at 1073 K and the average grain size is 4.70 μm (using the linear intercept method). The grain size ranges from submicron to about 16 μm for the microstructure annealed at 973 K and the average grain size is 2.82 μm. The grain size ranges from submicron to about 15 μm for the microstructure annealed at 923 K and the average grain size is 2.63 μm. The microstructure annealed at 898 K has the grain size ranging from submicron to about 6 μm and the average grain size is 0.75 μm. Moreover, the smallest grains are observed to have the tendency to distribute at the grain boundaries (GBs) or triple-junctions of the large grains.

TEM images for the two microstructures annealed at 1073 and 898 K are shown in Fig. 33.3. As shown in the TEM bright-field image for the microstructure annealed at 1073 K (Fig. 33.3a), sharp GBs are well-known for recrystallized grains, and straight twin boundaries (TB) are clearly observed for the annealing twins (marked by green arrows). The inset of Fig. 33.3a is the selected area diffraction (SAD), clearly indicating the twin relationship. The density of dislocation is low in the grain interior due to the high-temperature annealing (1073 K). In contrast, high density of dislocations along with multiple twins (with much smaller twin boundary spacing, TBS, several tens of nm) are found in the grain interiors for the microstructure annealed at 898 K (Fig. 33.3c). The smaller grain size in the microstructure annealed at 898 K (Fig. 33.3b), and the higher density of substructures and dislocations in the grain interior, will result in much higher strength than that annealed at 1073 K.

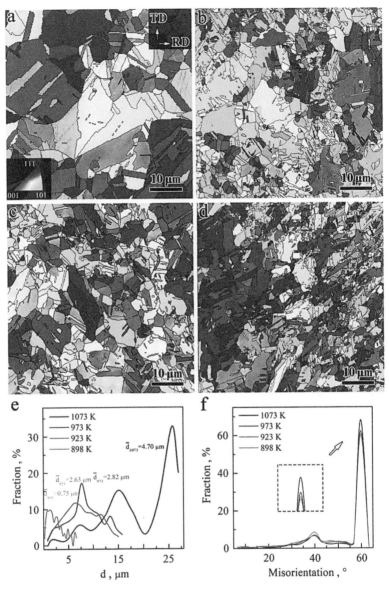

Figure 33.2 (a) IPF map for the microstructure annealed at 1073 K; (b) IPF map for the microstructure annealed at 973 K; (c) IPF map for the microstructure annealed at 923 K; (d) IPF map for the microstructure annealed at 898 K; (e) The corresponding grain size distributions for various microstructures, plotted in terms of the area fraction; (f) The corresponding misorientation angle distributions for various microstructures.

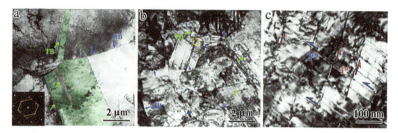

Figure 33.3 (a) TEM bright-field image for the microstructure annealed at 1073 K; (b) TEM bright-field image for the microstructure annealed at 898 K; (c) The corresponding close-up view for the rectangular area in (b).

33.3.2 Dynamic Shear Properties

The dynamic shear tests have been conducted on the CR-samples annealed at different temperatures (873–1073 K), in order to investigate the microstructure effect on the dynamic shear properties. The dynamic experiments have also been conducted at both room temperature (298 K) and cryogenic temperature (77 K) to study the environment temperature effect on the dynamic shear properties.

Figures 33.4a and b display the shear stress-shear displacement curves of the samples annealed at various temperatures for the dynamic shear experiments conducted at room temperature (298 K) and cryogenic temperature (77 K), respectively. It is well known that the point at peak stress can be considered as the initiation point of the ASB according to the widely accepted maximum stress criterion for hat-shaped specimens [35, 41, 44]. Thus, the uniform shear strain can be estimated through dividing the shear displacement by the width of the shear zone before the maximum stress point. The corresponding shear stress-shear strain curves are displayed in Figs. 33.4c and d. The impact shear toughness can be estimated using the area under the shear stress-shear strain curves before the onset of ASB. Then, the uniform shear strain and the impact shear toughness as a function of the dynamic shear yield strength for this MEA, along with other data points for various metal and alloys [39, 41, 42, 44, 46–50], are plotted in Figs. 33.4e and f, respectively. As indicated, this MEA with heterogeneous grain structures has a better combination of dynamic shear properties over all the other metals and alloys reported earlier, offering an excellent candidate for impact-relevant structures. Moreover, this MEA possesses even better dynamic shear properties under cryogenic environment than those at room temperature, rendering it potentially applicable as structural material in the low-temperature environment.

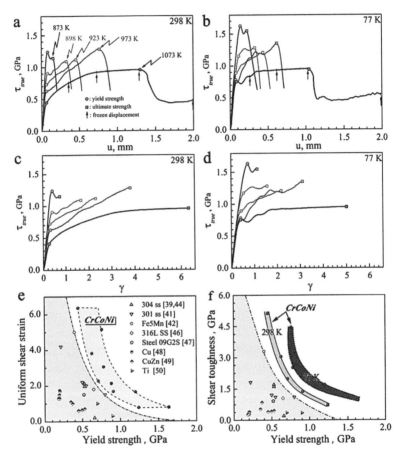

Figure 33.4 Shear stress-shear displacement curves of various microstructures for the experiments conducted: (a) at room temperature; (b) at cryogenic temperature. Shear stress-nominal shear strain curves of various microstructures for the experiments conducted: (c) at room temperature; (d) at cryogenic temperature. (e) Uniform dynamic shear strain vs. dynamic shear yield strength for the present MEA, along with the data for other metals and alloys. (f) Impact shear toughness vs. dynamic shear yield strength for the present MEA, along with the data for other metals and alloys.

It is observed that the uniform shear strain under dynamic shear loading decreases when the annealing temperature decreases. This means that the propensity of flow localization increases with decreasing annealing temperature. The susceptibility to ASB for a metal can be quantitatively expressed as [37, 51]:

$$\chi_{ASB} = (a/m)\min\left\{1, \frac{1}{(n/m) + \sqrt{n/m}}\right\}$$

, where a is the non-dimensional thermal softening parameter which can be calculated by $a = (-\partial\sigma/\partial T)/\rho c$ (T is the temperature, ρ is the mass density, and c is the specific heat of the metal), n is the strain hardening exponent, and m is the strain rate sensitivity. An increased χ_{ASB} quantitatively indicates an increased propensity for flow localization. Thus, lower strain hardening ability and lower strain rate sensitivity will favor the flow localization. The grain/substructure size is smaller when the annealing temperature is lower, as indicated in Fig. 33.2. As we know, the microstructure with smaller grain/substructure size generally has lower strain hardening ability and higher strain rate sensitivity for fcc metals [33, 52]. Lower strain hardening produces smaller uniform shear strain, while higher strain rate sensitivity discourages shear localization under dynamic shear loading based on the aforementioned equation. These two factors will be competing for the samples with smaller grain/substructure size. The strain hardening exponent and the strain rate sensitivity under dynamic loading for various microstructures of this MEA will be investigated later using a series of dynamic compression experiments at various strain rates, thus the quantitative correlation between these two parameters and the critical strain for onset of ASB must await the future work.

33.3.3 Microstructure Evolution during the Homogeneous Dynamic Shear Deformation

In order to understand the deformation mechanisms for the superior dynamic shear properties of this MEA, the experiments were interrupted at controlled shear displacements for the sample annealed at 1073 K. The specimens are "frozen" at these pre-determined shear strain magnitudes for subsequent microstructure observation. The microstructures for the experiments conducted at room temperature and cryogenic temperature are both characterized and compared. For the experiments conducted at room temperature, the specimens are "frozen" at controlled shear displacements of 0.36, 0.73, and 1.28 mm, whereas the experiments are interrupted at shear displacements of 0.26, 0.62, and 1.02 mm for the experiments conducted at cryogenic temperature.

The IPF images at various shear displacements (0.36, 0.73, 1.28 mm) for the experiments conducted at room temperature are shown in Fig. 33.5a, while the IPF images at various shear displacements (0.26, 0.62, 1.02 mm) for the experiments conducted at cryogenic temperature are displayed in Fig. 33.5b. As indicated in Fig. 33.4, the ASB is not formed before these shear displacements (0.36, 0.73, 1.28 mm for room temperature experiments, 0.26, 0.62, 1.02 mm for cryogenic temperature experiments), thus uniform shear deformation can be assumed at these shear displacements, and the EBSD images in Figs. 33.5a and b can represent the microstructures experienced the homogeneous shear deformation. Figs. 33.5c and d display the corresponding misorientation angle distributions at various shear displacements for the experiments conducted at room temperature and cryogenic temperature, respectively. The grain sizes as a function of the homogeneous shear strain for the experiments conducted at both room temperature and cryogenic temperature are plotted in Fig. 33.5e. The grain sizes in Fig. 33.5e are estimated by only counting the GBs with misorientation angle large than 15° (HAGBs) without taking twins into account. As indicated, the grains are observed to be severely elongated along the shear direction and refined with increasing shear displacement for the experiments conducted at both room temperature and cryogenic temperature. This dynamic grain refinement along with homogeneous shear deformation is one of the origins for the strong strain hardening in the dynamic shear tests, as observed in Figs. 33.4c and d. It is interesting to note that the grain refinement is more severe for the experiments conducted at cryogenic temperature than those conducted at room temperature at the same applied shear strain, which could be one of the origins for the better dynamic shear properties under cryogenic environment, as observed in Figs. 33.4e and f. It is well known that the grain refinement can be either mediated by cells with dislocation walls or by strong interactions between dislocations and multiple deformation twins (DTs) [54]. As we know, the efficiency of grain refinement via strong interactions between dislocations and multiple DTs is generally higher than that via dislocation cells [54], which should be the origin of the particularly high efficiency of grain refinement for this MEA under dynamic shear loading due to the ultra-low stacking faults energy (SFE) [18].

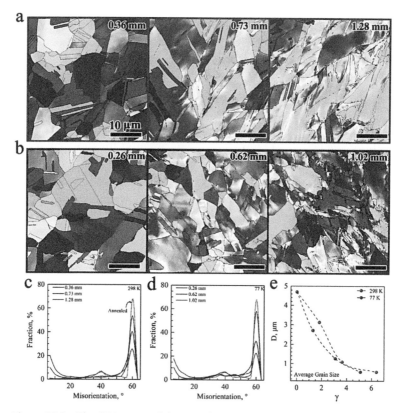

Figure 33.5 The IPF images of the sample annealed at 1073 K at various shear displacements for the experiments conducted: (a) at room temperature; (b) at cryogenic temperature. The corresponding misorientation angle distributions at various shear displacements for the experiments conducted: (c) at room temperature; (d) at cryogenic temperature. (e) The grain size as a function of the homogeneous shear strain for the experiments conducted at both room temperature and cryogenic temperature.

The strain hardening due to this dynamic grain refinement is very similar to the hardening in TWIP steels by DTs (also called to dynamic H-P effect for TWIP steels) [53]. And the strain hardening of grain refinement can be roughly estimated based on the average grain size before and after shear deformation and the coefficient of the Hall-Petch relation: $\Delta\tau = \Delta\sigma/\sqrt{3} = (K_{HP}\bar{d}_{after}^{-1/2} - K_{HP}\bar{d}_{before}^{-1/2})/\sqrt{3}$. The coefficient of the Hall-Petch relation for this MEA has been estimated to be about 500 MPa.µm$^{1/2}$ using our tensile experimental data on the samples with different average grain sizes. Thus, the strain hardening due to the grain refinement during the dynamic shear loading can be roughly estimated to be about 200 MPa for the experiments conducted at both temperatures, which is the main contributor to the total strain hardening observed in the experiments (Figs. 33.4c and d). Based on the misorientation angle distributions, the strong preference for 60° in the undeformed microstructure is weakened during the homogenous shear deformation, while a less obvious preference of 60° is still observed at the shear displacements of 1.02 and 1.28 mm for the dynamic shear experiments at room temperature and cryogenic temperature, respectively. These observations are in contrast with the random misorientations observed within ASB, which is induced by grain refinement through dynamic recrystallization (DRX) at high temperature [35, 40–42]. The lowering fraction of 60° misorientation angle during the shear deformation could be due to the interaction between the dislocations and the coherent TBs.

Deformation twinning has in fact been reported for this CrCoNi MEA with grains ranging from 5 to 50 µm in quasi-static tensile tests and fracture toughness tests [15], while grain refinement with HAGBs was not reported under quasi-static tensile loading for this grain range in this MEA [15]. In order to illustrate the other strain hardening mechanism in the grain interior besides the grain refinement, TEM bright-field images of the homogeneous shear deformation zone (the images are taken away from the ASB zone) after the dynamic shear experiments conducted at room temperature are displayed in Fig. 33.6. As indicated, numerous DTs (including occasional multiple twins, as shown in the inset of Fig. 33.6a) are observed and the TBS is observed to be as small as ~50 nm in the grain interior. A high density of dislocations are also observed in the areas adjacent to the TBs (Fig. 33.6b), indicating strong interactions between the dislocations and the TBs. As summarized in the previous study [55], four possible dislocation-TB reaction processes have been proposed in fcc metals and two of them involve the formation of immobile dislocation locks for strengthening and strain hardening. Thus, besides the dynamically grain refinement, the excellent dynamic shear properties observed for this MEA can also be due to the dislocation accumulation near the TBs for the experiments conducted at room temperature.

Figure 33.6 TEM bright-field images of the homogeneous shear deformation zone for the experiments conducted at room temperature: (a) showing DTs; (b) showing accumulation of high density of dislocation at TBs.

It is well known that an increase in strain rate has similar effect on deformation mechanisms as a decrease in deformation temperature, and vice versa, and the combined effects of strain rate ($\dot{\varepsilon}$) and deformation temperature (T) on the deformation mechanisms can often be represented by a single parameter, namely the Zener-Hollomon parameter (Z) defined as [56]: $Z = \dot{\varepsilon}\exp(Q/RT)$, where R is the gas constant and Q is the related activation energy for deformation. The high strain rate and the low temperature (as the case in the present study) generally represent the extreme deformation conditions, under which some unique deformation mechanisms may occur. In order to illustrate the detailed microstructure mechanisms in the grain interior for the strain hardening besides the grain refinement, bright-field TEM images for the homogeneous dynamic shear deformation zone are also provided for the experiments conducted at cryogenic temperature, as shown in Fig. 33.7. Multiple twinning is regularly observed for the experiments conducted at cryogenic temperature (Fig. 33.7b). Previous research have shown that multiple twinning can induce higher strength and stronger strain hardening than single DTs [18,

57, 58]. The multiple twinning network can present more barriers for dislocation motion to contribute to the higher strength, and the network can also provide adequate pathways for easy glide and cross-slip of dislocations to accommodate significant plastic deformation for stronger strain hardening [18, 57, 58]. Besides the high density of multiple DTs, a high density of SFs are also observed in the experiments conducted at cryogenic temperature. Again, this is rooted in the low SFE of this MEA, which leads to profuse partial dislocations.

Figure 33.7 TEM bright-field images of the homogeneous shear deformation zone for the experiments conducted at cryogenic temperature: (a) showing high density of SFs and phase transformation; (b) showing multiple DTs.

HREM images are also provided in Fig. 33.8 to show the deformation mechanisms in detail at atomic level for the experiments conducted at cryogenic temperature. TBs, SFs and dislocations at TBs can be clearly observed in the grain interior (Fig. 33.8a). Several Lomer-Cottrell (L-C) locks are also formed in the grain interior due to the reaction of leading partials from two different slip planes, and this scenario is seen in the high resolution views in Fig. 33.8b (highlighted by arrows). High density of extended dislocations exist on two 60^0 inter-crossing (111) planes, again due to the low SFE of this MEA [18], which leads to a high probability for the formation of L-C locks. These dissociated dislocations have been proposed to either originate from GBs or from the cross-slip of dislocations in the grain interior [59, 60]. It have been reported that L-C locks can play critical role in the strain hardening of fcc metals [59, 60, 61], which is supported previously by both experiments and molecular dynamics

(MD) simulations. The effectiveness of strain hardening for L-C locks derives from their capability to accumulate dislocations. Four dislocation segments are pinned by each L-C lock, and the length of the L-C lock is usually short, akin to a pinning point. Previously, it has been reported that the elevated strengthening/hardening for L-C lock is similar to Orowan's strengthening/hardening by using an orientation dependent line tension model [61]. Previously, it significant strengthening/hardening can be induced when the distance between locks is small; in other words, high strengthening/hardening is achieved when the density of L-C locks is high in the grain interior. The density of L-C locks is lower than the density of SFs, which can still contribute to the strain hardening for the present experiments conducted at 77 K.

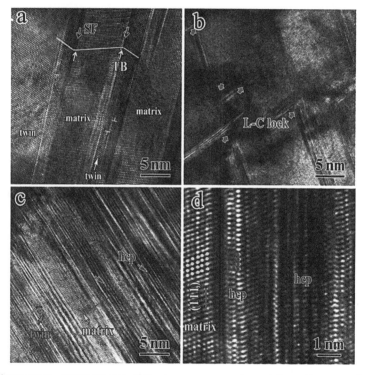

Figure 33.8 HREM images of the homogeneous shear deformation zone for the experiments conducted at cryogenic temperature: (a) showing TBs, SFs and dislocations at TBs; (b) showing L-C locks; (c) showing high density of SFs and phase transformation; (d) the close-up view of the rectangle area in (c) for the details of hcp phase.

Numerous SFs with spacing less than several nm are also seen on multiple {111} planes in the grain interior (Fig. 33.7a, Fig. 33.8c). Although the effects of SFs on the strengthening and strain hardening may not be as strong as TBs [55, 57], the high density of SFs itself can still contribute to the strengthening and the strain hardening according to previous research [62], in which SFs are also considered as barriers to dislocation motion and can accumulate dislocations around them. As indicated in Fig. 33.8c, phase transformation from fcc phase to hcp phase is also occasionally observed due to the successive formation and the overlapping of SFs at every other {111} plane. This phase transformation from fcc phase to hcp phase has also been recently reported in this MEA [63], in the Co-27Cr-5Mo-0.05C HEA [64] and in the Fe-Mn30-Co10-Cr10 HEA [14] at large strain levels. The lamellar structures of hcp phase and the phase boundaries would be effective barriers for further dislocation slip since either <c> or <c+a> dislocations are required for transmission of edge dislocations into the hcp phase, and the activation of these dislocation types typically requires very high critical resolved shear stress. Thus, this strain-induced formation of hcp ε-martensite phase is very similar to the TRIP effect as frequently reported in high-Mn austenitic steels [65], which can result in strong strain hardening for sustaining large uniform elongations. It is well known that the dominant deformation mechanism can be switched between phase transformation and deformation twins, or a mixture of the two, by regulating the SFE through tailoring the Mn content in TRIP/TWIP steels [65]. Our MEA has lower SFE [18] when compared to the equiatomic CrMnFeCoNi HEA, thus a mixture of deformation twins and TRIP effect can be observed in this MEA, in contrast to the deformation twins only in the CrMnFeCoNi HEA [7]. Thus, the better dynamic shear properties for the high-strain-rate dynamic shear experiments conducted at cryogenic temperature should also be partially attributed to higher density of multiple twins, SFs, Lomer-Cottrell locks and hcp phase by phase transformation inside the grains. Since the densities of multiple twins and stacking faults are much higher compared to those of Lomer-Cottrell locks and hcp phase, and the strengthening and strain hardening effects of TBs should be much stronger compared to those of SFs [55, 57], the sequence of importance of the aforementioned four different factors for strain hardening should be as follows: (i) multiple twins, (ii) SFs, (iii, iv) Lomer-Cottrell locks and phase transformation.

According to our two recent review chapters [66, 67], the mechanical properties in heterostructures can be optimized through two factors: (i) high density of zone boundaries (the boundaries between hard zone and soft zone); (ii) maximizing strain partitioning and strain gradients between heterostructured zones, which can be fulfilled by increasing strength difference between zones. Based the current results and the inspiration from these two review chapters [66, 67], the mechanical properties of this MEA could be further optimized by creating a wider range of grain size distribution in the future work, which might be fulfilled by CR under cryogenic temperature and subsequent proper annealing.

33.3.4 Temperature Rise during the Homogeneous Dynamic Shear Deformation

During the dynamic shear process, the shear zone (200 μm width) can be considered as a homogeneous shear deformation zone before the onset of ASB, and the deformation can be approximately considered as an adiabatic process. Thus, the temperature rise during the homogeneous dynamic shear deformation due to the plastic dissipation work within the homogeneous shear zone can be estimated as [39, 41]: $\Delta T = \dfrac{\eta}{\rho C_v} \int \tau d\gamma_p$, where ρ is the mass density, C_v is the heat capacity, η is the Taylor-Quinnery coefficient for plastic work converted to heat (commonly $\eta = 0.85$), γ_p is the plastic shear strain, and τ is the shear stress. For this MEA, ρ is measured to be 11.19 g/cm^3 and C_v can be approximately taken as 540 J/kg [68].

Based on this equation, the temperature rise for all experiments conducted at room temperature and cryogenic temperature can be calculated by utilizing the dynamic shear stress-shear strain curves in Figs. 33.4c and d, and the corresponding temperature rise vs. the homogeneous shear strain curves are displayed in Figs. 33.9a and b. This relates the temperature rises at various shear displacements to the microstructure evolution (the dynamic grain refinement evolution along with the shear deformation, as shown in Fig. 33.5) for the experiments conducted at room temperature and cryogenic temperature. The temperature rises are 126, 344 and

708 K at shear displacements of 0.36, 0.73 and 1.28 mm, respectively, for the experiments conducted at room temperature, whereas the temperature rises are 148, 315 and 579 K at shear displacements of 0.26, 0.62 and 1.02 mm, respectively, for the experiments conducted at cryogenic temperature. It should be noted that the dynamic grain refinement is achieved due to the large shear deformation with assistance of temperature rise at the shear zone. As shown, the temperature rise is lower than 300 K when the homogeneous shear strain is lower than 3. But the temperature rise can be as high as 720 K when the homogeneous shear strain reaches 6. As indicated in Figs. 33.4c and d, strong hardening is observed before the shear strain of 3 (stage I) for the sample annealed at 1073 K, while a plateau of shear flow stress is found after the shear strain of 3 (stage II). This observed trend could be due to the competition between the strong strain hardening from microstructural evolution (grain refinement, deformation twins, SFs, L-C locks, phase transformation) and thermal softening due to adiabatic temperature rise. Stage I is controlled by the strain hardening since the adiabatic temperature rise is low in this stage. The thermal softening can cancel out the strain hardening effect from the microstructural refinement in stage II since the adiabatic temperature rise is as high as 720 K. The cryogenic environment can reduce the actual temperature in the shear zone, thus weakening the thermal softening effect and resulting in better dynamic shear properties for the experiments conducted at cryogenic temperature.

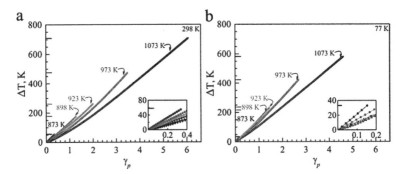

Figure 33.9 The temperature rise vs. the homogeneous shear strain curves for the experiments conducted: (a) at room temperature; (b) at cryogenic temperature.

33.3.5 Characteristics of ASB

Moreover, it is well known that the shear deformation will be concentrated in a thin shear band (generally with width about several to several tens of μm) once ASB is formed, and the shear strain within the ASB can generally be one order of magnitude higher than the nominal homogeneous shear strain, thus temperature rise within the ASB will be even higher. The high shear deformation magnitude and the high temperature rise within the ASB generally result in DRX in ASB. The characteristics of the ASB at the shear displacement of 2.0 mm for the sample annealed at 1073 K are displayed in Fig. 33.10. These images include the SEM images for the ASB zone, the IPF images and the TEM dark-field images for the ASB zone. Fig. 33.10a show the corresponding images for the experiment conducted at room temperature, whereas Fig. 33.10b displays the corresponding images for the experiment conducted at cryogenic temperature. The grain size distributions within ASB are displayed in Fig. 33.10c, and the corresponding misorientation angle distributions of the ASB zone are shown in Fig. 33.10d. Unlike the elongated grains in the homogeneous shear deformation zone, the grains in the ASB are equiaxed, signifying DRX. The grain size is refined to about 300 nm in the ASB for the experiments conducted both at room temperature and cryogenic temperature. The SAD patterns (as shown in the insets of Figs. 33.10a and b for the TEM images) clearly display random misorientation angles (as confirmed by Fig. 33.10d), which is a typical characteristic of DRX grains due to the high temperature rise and the severe plastic strain in the ASB.

According to Figs. 33.10a and b, the widths of the ASB for the experiments conducted at room temperature and cryogenic temperature are 12.5 and 6.4 μm, respectively. The smaller width of the ASB for the experiments at cryogenic temperature indicates that the shear deformation is more concentrated in cryogenic environment. A quantitative description for calculating the width of the ASB has been proposed according to the perturbation analysis of the uniform solutions for the governing equation of the ASB [69]:

$\delta \approx 2\left(\dfrac{kT}{\tau\dot{\gamma}}\right)^{1/2}$, where T is the temperature within the ASB, τ and $\dot{\gamma}$ are the shear stress and the shear strain rate within the ASB, respectively. k is the thermal conductivity. For this MEA, k can be approximately

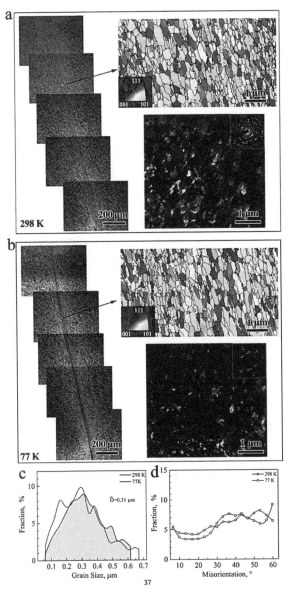

Figure 33.10 The characteristics of the ASB (SEM, EBSD and TEM images) at the shear displacement of 2.0 mm for the sample annealed at 1073 K and for the experiment: (a) conducted at room temperature; (b) conducted at cryogenic temperature. (c) The grain size distributions within ASB for the experiments conducted both at room temperature and cryogenic temperature. (d) The corresponding misorientation angle distributions of ASB zone for the experiments conducted both at room temperature and cryogenic temperature.

taken as 60 W/(K.m) [42, 68]. The shear stress can be taken from the maximum stress of the curves in Figs. 33.4a and b. The actual shear strain rate within the ASB is generally one order of magnitude higher than the nominal shear rate [36, 40], which can be calculated from the strain gage signal of the reflected wave. By comparing the width of ASB (12.5 μm for room temperature, 6.4 μm for cryogenic temperature) and the thickness of the whole shear zone (200 μm), the real shear strain rate within the ASB can be estimated to be about 16 times and 31 times of the nominal shear rate for the room temperature and cryogenic temperature experiments, respectively. As mentioned earlier, the temperature within the ASB is generally much higher than the temperature within the homogeneous shear deformation zone. Thus we take 1500°C (1773 K, close to the melting temperature of this MEA) as the temperature within ASB, for a rough estimate of the ASB width. Based on the aforementioned equation, the widths of ASB have been roughly estimated to be 21 μm and 15 μm for the experiments conducted at room temperature and cryogenic temperature, respectively. This estimate is in reasonable agreement with the experimental observations (12.5 μm for room temperature, 6.4 μm for cryogenic temperature), as Wright [70] has indicated that the agreement of experimental results with the theoretical predictions is usually within a factor of two.

33.4 Concluding Remarks

In the present study, the dynamic shear properties of the CrCoNi MEA have been investigated using a series of Hopkinson-bar experiments and hat-shaped specimens at both room temperature and cryogenic temperature, and the underlying deformation mechanisms have also been revealed by utilizing the "frozen" experiments and subsequent microstructure observations. The main findings are summarized as follows:

(1) Superior dynamic shear properties, in terms of a combination of dynamic shear strength and uniform dynamic shear strain, have been found for this MEA, exceeding all other metals and alloys investigated so far. The dynamic shear properties of this MEA are found to be even better when the experiments were conducted at cryogenic temperature.

(2) The excellent dynamic shear properties can be mainly attributed to the dynamically grain refinement along with the dynamic shear deformation. And the better dynamic shear properties at cryogenic temperature can be attributed to the more efficient grain refinement.

(3) Besides grain refinement, deformation twins and strong interactions between the dislocations and TBs also contribute partly to the strain hardening under dynamic shear loading at room temperature. Higher density of multiple twins, SFs, Lomer-Cottrell locks and phase transformation to hcp phase were observed in the grain interiors for the experiments conducted at cryogenic temperature, which could contribute to the better dynamic shear properties under cryogenic environment.

Acknowledgements

The work was supported by the National Key R&D Program of China [grant number 2017YFA0204402]; the National Natural Science Foundation of China [grant nos. 11472286, 11672313, 11790293, 11572328, 51701228, and 51601204], and the Strategic Priority Research Program of the Chinese Academy of Sciences [grant numbers XDB22040503].

References

1. J.W. Yeh, S.K. Chen, S.J. Lin, J.Y. Gan, T.S Chin, T.T. Shun, C.H. Tsau, S.Y. Chang, Nanostructured high-entropy alloys with multiple principal elements: novel alloy design concepts and outcomes, *Adv. Eng. Mater.* 6 (2004) 299–303.
2. B. Cantor, I.T.H. Chang, P. Knight, A. J. B. Vincent, Microstructural development in equiatomic multicomponent alloys, *Mater. Sci. Eng. A* 375 (2004) 213–218.
3. W.H. Liu, Y. Wu, J.Y. He, T.G. Nieh, Z.P. Lu, Grain growth and the Hall–Petch relationship in a high-entropy FeCrNiCoMn alloy, *Scr. Mater.* 68 (2013) 526–529.
4. A.J. Zaddach, C. Niu, C.C. Koch, D.L Irving, Mechanical properties and stacking fault energies of NiFeCrCoMn high-entropy alloy, *JOM* 65 (2013) 1780–1789.

5. F. Otto, A. Dlouhý, C. Somsen, H. Bei, G. Eggeler, E.P. George, The influences of temperature and microstructure on the tensile properties of a CoCrFeMnNi high-entropy alloy, *Acta Mater.* 61 (2013) 5743–5755.

6. Y.P. Lu, Y. Dong, S. Guo, L. Jiang, H.J. Kang, T.M. Wang, B. Wen, Z.J. Wang, J.C. Jie, Z. Q. Cao, H. H. Ruan, T.J. Li, A promising new class of high-temperature alloys: eutectic high-entropy alloys, *Sci. Rep.* 4 (2014) 6200.

7. B. Gludovatz, A. Hohenwarter, D. Catoor, E.H. Chang, E.P. George, R.O. Ritchie, A fracture-resistant high-entropy alloy for cryogenic applications. *Science* 345 (2014) 1153–1158.

8. Y. Zhang, T.T. Zuo, Z. Tang, M. C. Gao, K.A. Dahmen, P.K. Liaw, Z.P. Lu, Microstructures and properties of high-entropy alloys, *Prog. Mater. Sci.* 61 (2014) 1–93.

9. B. Schuh, F. Mendez-Martin, B. Völker, E.P. George, H. Clemens, R. Pippan, A. Hohenwarter, Mechanical properties, microstructure and thermal stability of a nanocrystalline CoCrFeMnNi high-entropy alloy after severe plastic deformation, *Acta Mater.* 96 (2015) 258–268.

10. Y. Zou, H. Ma, R. Spolenak, Ultrastrong ductile and stable high-entropy alloys at small scales, *Nat. Commun.* 6 (2015) 7748.

11. Y.F. Ye, Q. Wang, J. Lu, C.T. Liu, Y. Yang, High-entropy alloy: challenges and prospects, *Mater. Today* 19 (2015) 349–362.

12. Z. Zhang, M.M. Mao, J. Wang, B. Gludovatz, Z. Zhang, S.X. Mao, E.P. George, Q. Yu, R.O. Ritchie, Nanoscale origins of the damage tolerance of the high-entropy alloy CrMnFeCoNi, *Nat. Commun.* 6 (2015) 10143.

13. J.Y. He, H. Wang, H.L. Huang, X.D. Xu, M.W. Chen, Y. Wu, X.J. Liu, T.G. Nieh, K. An, Z.P. Lu, A precipitation-hardened high-entropy alloy with outstanding tensile properties. *Acta Mater.* 102 (2016) 187–196.

14. Z. Li, K.G. Pradeep, Y. Deng, D. Raabe, C.C. Tasan, Metastable high-entropy dual-phase alloys overcome the strength–ductility trade-off, *Nature* 534 (2016) 227–230.

15. B. Gludovatz, A. Hohenwarter, K.V. Thurston, H. Bei, Z. Wu, E.P. George, R.O. Ritchie, Exceptional damage-tolerance of a medium-entropy alloy CrCoNi at cryogenic temperatures, *Nat. Commun.* 7 (2016) 10602.

16. Z. Wu, H. Bei, F. Otto, G.M. Pharr, E.P. George, Recovery, recrystallization, grain growth and phase stability of a family of FCC-structured multi-component equiatomic solid solution alloys, *Intermetallics* 46 (2014) 131–140.

17. Z. Wu, H. Bei, G.M. Pharr, E.P. George, Temperature dependence of the mechanical properties of equiatomic solid solution alloys with face-centered cubic crystal structures. *Acta Mater.* 81 (2014) 428–441.

18. Z.J. Zhang, H.W. Sheng, Z.J. Wang, B. Gludovatz, Z. Zhang, E.P. George, Q. Yu, S.X. Mao, R.O. Ritchie, Dislocation mechanisms and 3D twin architectures generate exceptional strength-ductility-toughness combination in CrCoNi medium-entropy alloy. *Nat. Commun.* 8 (2017) 14390.
19. Y.L. Zhao, T. Yang, Y. Tong, J. Wang, J.H. Luan, Z.B. Jiao, D. Chen, Y. Yang, A. Hu, C.T. Liu, J.J. Kai, Heterogeneous precipitation behavior and stacking-fault-mediated deformation in a CoCrNi-based medium-entropy alloy, *Acta Mater.* 138 (2017) 72–82.
20. J. Miao, C.E. Slone, T.M. Smith, C. Niu, H. Bei, M. Ghazisaeidi, G.M. Pharr, M.J. Mills, The evolution of the deformation substructure in a Ni-Co-Cr equiatomic solid solution alloy, *Acta Mater.* 132 (2017) 35–48.
21. Y.M. Wang, M.W. Chen, F.H. Zhou, E. Ma, High tensile ductility in a nanostructured metal, *Nature* 419 (2002) 912–915.
22. K. Lu, L. Lu, S. Suresh, Strengthening materials by engineering coherent internal boundaries at the nanoscale. *Science* 324 (2009) 349–352.
23. P.V. Liddicoat, X.Z. Liao, Y.H. Zhao, Y.T. Zhu, M.Y. Murashkin, E.J. Lavernia, R.Z. Valiev, S.P. Ringer, Nanostructural hierarchy increases the strength of aluminium alloys, *Nat. Commun.* 1 (2009) 63.
24. D. Raabe, D. Ponge, O. Dmitrieva, B. Sander, Nanoprecipitate-hardened 1.5 GPa steels with unexpected high ductility, *Scr. Mater.* 60 (2009) 1141–1144.
25. G. Liu, G.J. Zhang, F. Jiang, X.D. Ding, Y.J. Sun, J. Sun, E. Ma, Nanostructured high-strength molybdenum alloys with unprecedented tensile ductility, *Nat. Mater.* 12 (2009) 344–350.
26. X.L. Wu, P. Jiang, L. Chen, F.P. Yuan, Y.T. Zhu, Extraordinary strain hardening by gradient structure, *Proc. Natl. Acad. Sci. USA* 111 (2014) 7197-7201.
27. X.L. Wu, M.X. Yang, F.P. Yuan, G.L. Wu, Y.J. Wei, X.X. Huang, Y.T. Zhu, Heterogeneous lamella structure Unites ultrafine-grain strength with coarsegrain ductility, *Proc. Natl. Acad. Sci. USA* 112 (2015) 14501-14505.
28. X.L. Wu, F.P. Yuan, M.X. Yang, P. Jiang, C.X. Zhang, L. Chen, Y.G. Wei, E. Ma, Nanodomained nickel unite nanocrystal strength with coarse-grain ductility, *Sci. Rep.* 5 (2015) 11728.
29. Y.J. Wei, Y.Q. Li, L.C. Zhu, Y. Liu, X.Q. Lei, G. Wang, Y.X. Wu, Z.L. Mi, J.B. Liu, H.T. Wang, H.J. Gao Evading the strength ductility trade-off dilemma in steel through gradient hierarchical nanotwins, *Nat. Commun.* 5 (2014) 3580.

30. T.H. Fang, W.L. Li, N.R. Tao, K. Lu, Revealing extraordinary intrinsic tensile plasticity in gradient nano-grained copper, *Science* 331 (2011) 1587–1590.

31. C. Zener, J.H. Hollomon, Effect of strain rate upon plastic flow of steel, *J. Appl. Phys.* 15 (1944) 22–32.

32. S. Nemat-Nasser, W.G. Guo, Thermomechanical response of DH-36 structural steel over a wide range of strain rate and temperatures, *Mech. Mater.* 35 (2003) 1023–1047.

33. Q. Wei, Strain rate effects in the ultrafine grain and nanocrystalline regimes influence on some constitutive responses, *J. Mater. Sci.* 42 (2007) 1709–1727.

34. T. Suo, Y.L. Li, F. Zhao, X.L. Fan, W.G. Guo, Compressive behavior and rate-controlling mechanisms of ultrafine grained copper over wide temperature and strain rate ranges, *Mech. Mater.* 61 (2013) 1–10.

35. A. Mishra, M. Martin, N.N. Thadhani, B.K. Kad, E.A. Kenik, M.A. Meyers, High-strain rate response of ultra-fine-grained copper, *Acta Mater.* 56 (2008) 2770–2783.

36. M.A. Meyers, *Dynamic Behavior of Materials*, Wiley-Interscience, New York, USA (1994) pp. 323–326.

37. Q. Wei, L. Kecskes, T. Jiao, K.T. Hartwig, K.T. Ramesh, E. Ma, Adiabatic shear banding in ultrafine-grained Fe processed by severe plastic deformation, *Acta Mater.* 52 (2004) 1859–1869.

38. M.A. Meyers, G. Subhash, B.K. Kad, L. Prasad, Evolution of microstructure and shear-band formation in α-hcp titanium, *Mech. Mater.* 17 (1994) 175–193.

39. Y. Yang, F. Jiang, B.M. Zhou, X.M. Li, H.G. Zheng, Q.M. Zhang, Influence of shock prestraining on the formation of shear localization in 304 stainless steel, *Mater. Sci. Eng. A* 528 (2011) 2787–2794.

40. F.P. Yuan, P. Jiang, X.L. Wu, Annealing effect on the evolution of adiabatic shear band under dynamic shear loading in ultra-fine-grained iron, *Int. J. Impact Eng.* 50 (2012) 1–8.

41. J.X. Xing, F.P. Yuan, X.L. Wu, Enhanced quasi-static and dynamic shear properties by heterogeneous gradient and lamella structures in 301 stainless steels. *Mater. Sci. Eng. A* 680 (2017) 305–316.

42. F.P. Yuan, X.D. Bian, P. Jiang, M.X. Yang, X.L. Wu, Dynamic shear response and evolution mechanisms of adiabatic shear band in an ultrafine-grained austenite-ferrite duplex steel, *Mech. Mater.* 89 (2015) 47–58.

43. X.D. Bian, F.P. Yuan, Y.T. Zhu, X.L. Wu, Gradient structure produces superior dynamic shear properties, *Mater. Res. Lett.* 5 (2017) 501–507.

44. Q. Xue, G.T. Gray III, B.L. Henrie, S.A. Maloy, S.R. Chen, Influence of shock prestraining on the formation of shear localization in 304 stainless steel, *Metall. Mater. Trans. A* 36 (2005) 1471–1486.
45. M.X. Yang, F.P. Yuan, Q.G. Xie, Y.D. Wang, E. Ma, X. L. Wu, Strain hardening in Fe-16Mn-10Al-0.86C-5Ni high specific strength steel, *Acta Mater.* 109 (2016) 213–222.
46. Q. Xue, G.T. Gray III, Development of adiabatic shear bands in annealed 316L stainless steel: Part I. Correlation between evolving microstructure and mechanical behavior. *Metall Mater. Trans. A* 37A (2006) 2435–2446.
47. V.A. Pushkov, A.V. Yurlov, A.M. Podurets, A. Kal'manov, E. Koshatova, Study of adiabatic localized shear in metals by split Hopkinson pressure bar method, *EPJ Web of Conferences* 10 (2010) 00029.
48. V.A. Pushkov, A.V. Yurlov, A.M. Podurets, A.N. Tsibikov, M.I. Tkachenko, A.N. Balandina, Effect of preloading on the formation of adiabatic localized shear in copper. *Combust. Explos. Shock Waves* 49 (2013) 620–624.
49. U. Hofmann, E. El-Magd, Behaviour of Cu-Zn alloys in high speed shear tests and in chip formation processes. *Mater. Sci. Eng. A* 395 (2005) 129–140.
50. Y.B. Gu, V.F. Nesterenko, Dynamic behavior of HIPed Ti-6Al-4V, *Int. J. Impact Eng.* 34 (2007) 771–783.
51. T.W. Wright, *The Physics and Mathematics of Adiabatic Shear Bands*, Cambridge University Press, Cambridge (2002) pp. 176–177.
52. M.A. Meyers, A. Mishra, D.J. Benson, Mechanical properties of nanocrystalline materials. *Prog. Mater. Sci.* 51 (2006) 427–556.
53. O. Bouaziz, Strain-Hardening of twinning-induced plasticity steels, *Scr. Mater.* 66 (2012) 982–985.
54. N.R. Tao, K. Lu, Nanoscale structural refinement via deformation twinning in face-centered cubic metals, *Scr. Mater.* 60 (2009) 1039–1043.
55. L. Lu, Z.S. You, K. Lu, Work hardening of polycrystalline Cu with nanoscale twins, *Scr. Mater.* 66 (2012) 837–842.
56. C. Zener, J.H. Hollomon, Effect of strain rate upon plastic flow of steel, *J. Appl. Phys.* 15 (1944) 22–32.
57. L.L. Zhu, S.X. Qu, X. Guo, J. Lu, Analysis of the twin spacing and grain size effects on mechanical properties in hierarchically nanotwinned face-centered cubic metals based on a mechanism-based plasticity model, *J. Mech. Phys. Solids* 76 (2015) 162–179.

58. L.L. Zhu, H.N. Kou, J. Lu, On the role of hierarchical twins for achieving maximum yield strength in nanotwinned metals, *Appl. Phys. Lett.* 101 (2012) 081906.

59. X.L. Wu, Y.T. Zhu, Y.G. Wei, Q. Wei, Strong strain hardening in nanocrystalline nickel, *Phys. Rev. Lett.* 103 (2009) 205504.

60. V. Yamakov, D. Wolf, S.R. Phillpot, A.K. Mukherjee, H. Gleiter, Dislocation processes in the deformation of nanocrystalline aluminium by molecular-dynamics simulation, *Nat. Mater.* 1 (2002) 45–48.

61. L. Dupuy, M.C. Fivel, A study of dislocation junctions in FCC metals by an orientation dependent line tension model, *Acta Mater.* 50 (2002) 4873–4885.

62. W.W. Jin, G.M. Cheng, W.Z. Xu, H. Yuan, M.H. Tsai, Q.D. Wang, C.C. Koch, Y.T. Zhu, S.N. Mathaudhu, Ultrastrong Mg alloy via nano-spaced stacking faults. *Mater. Res. Lett.* 1 (2013) 61–66.

63. J. Mao, C.E. Slone, T.M. Smith, C. Niu, H. Bei, M. Ghazisaeidi, G.M. Pharr, M.J. Mills, The evolution of the deformation substructure in a Ni-Co-Cr equiatomic solid solution alloy, *Acta Mater.* 132 (2017) 35–48.

64. A. Mani, Salinas-Rodriguez, H.F. Lope, Deformation induced FCC to HCP transformation in a Co-27Cr-5Mo-0.05C alloy, *Mater. Sci. Eng. A* 528 (2011) 3037–3043.

65. O. Grassel, L. Kruger, G. Frommeyer, High strength Fe-Mn-(Al, Si) TRIP/TWIP steels development - properties - application, *Int. J. Plast.* 16 (2000) 1391–1409.

66. X.L. Wu, Y.T. Zhu, Heterogeneous materials: a new class of materials with unprecedented mechanical properties, *Mater. Res. Lett.* 5 (2017) 527–532.

67. E. Ma, T. Zhu, Towards strength-ductility synergy through the design of heterogeneous nanostructures in metals, *Mater. Today* 20 (2017) 323–331.

68. R.D. Pehlke, A. Jeyarajan, H. Wada, *Summary of Thermal Properties for Casting Alloys and Mold Materials*, University of Michigan (1982).

69. Y.L. Bai, B. Dodd, *Adiabatic Shear Localization*, Pergamon Press, New York (1992).

70. T.W. Wright, *The Physics and Mathematics of Adiabatic Shear Bands*, Cambridge University Press, Cambridge (2002) pp. 176–177.

Chapter 34

Superior Strength and Ductility of 316L Stainless Steel with Heterostructured Lamella Structure

Jiansheng Li,[a] Bo Gao,[a] Yang Cao,[a] Yusheng Li,[a]
and Yuntian Zhu[a,b]

[a]*Nano and Heterogeneous Materials Center, School of Materials Science and Engineering, Nanjing University of Science and Technology, Nanjing 210094, China*
[b]*Department of Materials Science and Engineering, North Carolina State University, Raleigh, NC 27695, USA*
liyusheng@njust.edu.cn

Strength and ductility are two of the most important mechanical properties for a metal, but often trade off with each other. Here we report a 316L stainless steel with superior combinations of strength and ductility that can be controlled by fine-tuning its heterostructured lamella structure (HLS). The HLS was produced by 85% cold rolling, which produced lamella coarse grains sandwiched between mixtures of nano-grains and nano-twins. The HLS was fine

---

Reprinted from *J. Mater. Sci.*, **53**, 10442–10456, 2018.

Heterostructured Materials: Novel Materials with Unprecedented Mechanical Properties
Edited by Xiaolei Wu and Yuntian Zhu
Text Copyright © 2018 Springer Science Business Media, LLC, part of Springer Nature
Layout Copyright © 2022 Jenny Stanford Publishing Pte. Ltd.
ISBN 978-981-4877-10-7 (Hardcover), 978-1-003-15307-8 (eBook)
www.jennystanford.com

tuned by annealing at 750°C for 5–25 min, which resulted in varying volume fractions of nano-grains, nano-twins, lamella coarse grains and recrystallized grains. During tensile testing, large amount of geometrically necessary dislocations (GNDs) were generated near the heterostructure boundaries to coordinate the deformation between soft zones and hard zones, which results in high heterodeformation induced (HDI) stress to achieve superior combination of strength and ductility. An optimal high yield strength of ~1 GPa with an elongation-to-failure of ~20% obtained for an optimized HLS sample. Furthermore, the processing technique employed here are conducive to large-scale industrial production at low cost.

34.1 Introduction

316L stainless steel is a widely used engineering material because of its combination of excellent corrosion resistance, great oxidation resistance and good formability [1]. However, the relatively low strength (yield strength of 200–400 MPa in annealed states) limits its application in some critical applications [2]. In recent decades, many technologies such as cold rolling (CR) [3, 4], surface mechanical attribution (SMAT) [5, 6], equal channel angular pressing (ECAP) [7–10] and high pressure torsion (HPT) [11] have been explored to obtain ultrafine-grained/nanocrystalline bulk metals to significantly improve the strength of stainless steels. Unfortunately, strength enhancement by grain refinement is inevitably accompanied with a sacrifice in ductility, which increases the potential for catastrophic failure during service.

Approaches to process the stainless steel for high strength and good ductility are urgently needed. In fact, the mechanical properties of metallic materials are determined by their internal microstructures and external environmental conditions. Many efforts have been reported to optimize microstructure to achieve a good combination of strength and ductility [12–15]. Fang et al. [16] reported that the tensile behavior of Cu can be effectively enhanced by introducing the gradient structures (spatial variation of grain sizes). The gradient nano-grained Cu exhibited one time higher yield strength and a tensile elongation to failure of ~60%. Similar superior strength-ductility combinations were also found by

Wu et al. [17, 18] for the gradient structured IF and 304 stainless steel. The high ductility was attributed to extra strain hardening due to the presence of strain gradient and the change of stress states, which generated geometrically necessary dislocations (GNDs) and promoted the multiplication and interaction of forest dislocations [12, 19].

Another strategy to successfully obtain high strength and good ductility was reported by Wang et al. [20] who developed a thermomechanical processing route to obtain bimodal grains of Cu with micrometer-sized grains randomly embedded among ultrafine grains. They attributed the excellent mechanical property for the action of hard ultrafine grains and numerous GNDs from the ultrafine-coarse grain boundaries (soft/hard boundaries). Ameyama and co-workers [21] further redesigned the bimodal grains to harmonic structure with continuous three-dimensional network of hard ultrafine-grained skeleton filled with islands of soft coarse-grained regions. The high ductility was associated with ultrafine-grained network (soft/hard boundaries). These reports show the advantage of introducing heterostructures in producing good strength and ductility in metals [22].

More recently, heterostructured lamella structures (HLSs) prepared by simple cold rolling and annealing were reported to markedly enhance the strength and ductility of Ti and 301 stainless steel [23, 24]. Unusual high strength was obtained with the assistance of high hetero-deformation induced (HDI) stress, whereas high ductility was attributed to HDI hardening and dislocation hardening [23]. For the 316L austenite stainless steel, a combination of 1.0 GPa tensile strength with an elongation-to-failure of ~27% was achieved in the annealed dynamic plastic deformation (DPD) sample [25]. However, this technique is associated with high cost and limited sample size, which makes it challenging to process materials large enough for real industrial applications. Hirota et al. [26] found the HLS in 316L stainless steel during the process of exploring the recrystallization and grain growth behavior, but the mechanical behaviors were not reported. Belyakov et al. [27] indicated that the 316L stainless steel with HLS possessed enhanced strength and ductility, but its mechanism was not studied. HLS is considered the most effective structure that produces high strength and high ductility [28].

In present work, 316L stainless steel samples with heterostructured lamella structures were prepared by means of conventional cold rolling for a strain of 85% and subsequent annealing. Their structural evolution and mechanical properties are reported herein.

34.2 Experimental

34.2.1 Characterization of the As-Received Sample

The material used in this study was a hot rolled 316L stainless steel sheet with a thickness of 10 mm. Its chemical composition is listed in Table 34.1. Figure 34.1 shows equiaxed austenitic grains and significant amount of annealing twins in its as-received state. The grain size distribution is presented in Fig. 34.1b and the average grain size is around 35 μm.

Table 34.1 Chemical compositions of 316L stainless steel

	Chemical compositions (wt %)											
Material	Cr	Ni	Mo	C	Si	Mn	S	P	Co	Cu	Nb	W	Fe
316L	16.47	10.10	1.97	0.030	0.530	1.42	0.005	0.030	0.244	0.146	0.012	0.031	Bal.

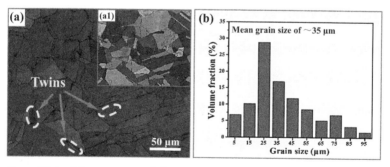

Figure 34.1 Microstructure of the as-received 316L stainless steel: (a) Optical micrograph, and the inset (a1) is the corresponding EBSD orientation map; (b) The volume fraction distribution of the original coarse grains.

34.2.2 Preparation of HLS

The as-received 316L stainless steel sheets were processed by conventional cold rolling at room temperature on a rolling mill with 400-mm-diamter rolls. Both rolls were driven at the same velocity of ~335 mm/s and the mean thickness reduction was ~0.17 mm per pass. The sample sheets were rolled from 10 to 1.5 mm thick after 50 passes with a total rolling reduction of ~85%, and then annealed at 750°C under nitrogen protection for 5, 10, 15, 20 and 25 min for partial recrystallization.

34.2.3 Microstructure Analysis

An automated Bruker-AXS D8 Advance diffractometer with Cu Kα radiation was used to obtain X-ray diffraction (XRD) patterns to study possible phase transformation. 2-Theta ranges from 40° to 100° and the scanning speed was 6°/min. A field emission scanning electron microscope (SEM, Quant 250 FEG) was used to characterize the fractured surface. Electron back-scattered diffraction (EBSD) was carried in a Zeiss Auriga scanning electron microscope (SEM). The step size was 50 nm and scanning voltage was 15 kV. HLSs were characterized on a TECNAI G2 20 LaB6 transmission electron microscope (TEM) at an accelerated voltage of 200 kV. The as-received TEM specimens with a thickness of 0.5 mm were cut from the sample cross-section by wire-electrode cutting and then mechanically grinded down to a thickness of ~50 μm. The final thinning was accomplished by a twin-jet electrochemical polishing in a solution of 8% perchloric acid + 92% ethanol at 50 V (80 mA) and around −10°C.

34.2.4 Mechanical Property Tests

Vickers hardness was measured by a HMV-G hardness tester with a load of 200 g and a holding time of 15 s. Hardness values were obtained by averaging at least 10 indents for each sample. The treated 316L sheets were cut along the rolling direction into dog-bone shaped specimens, with gage length of 20 mm, width of 3 mm, and final polished thickness of 1.5 mm. Uniaxial tensile tests were performed on an electromechanical universal testing machine (LFM-20kN) with a strain rate of $3 \times 10^{-3}\,s^{-1}$ at room temperature. All tests were performed three times under the same condition to guarantee the data consistency.

34.3 Results

34.3.1 Microstructures

34.3.1.1 XRD analysis

XRD data are shown in Fig. 34.2a. The as-received sample is composed of austenite phase (γ, 95.6% in volume) and martensite phase (α', 4.6% in volume). After 85% cold rolling, the volume fraction of martensite phase was increased to 26.4%, due to the strain-induced martensite transformation [29, 30]. Usually, the martensite phase is not stable in 316L stainless steel, especially at high temperature. Roland et al. [31] reported that the volume fraction of martensite reached a minimum value of 5% after annealing at 700°C. In this study, annealing at 750°C effectively eliminated the martensite phase (below 1%) to produce single austenite phase. The volume fractions of martensites are exhibited in Fig. 34.2b and the detailed calculation can be found in previous work [32].

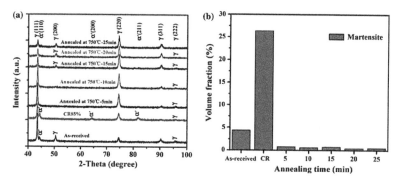

Figure 34.2 (a) XRD data of the as-received and treated 316L stainless steels (γ is austenite phase, α' is martensite phase); (b) The volume fraction of martensite as a function of annealing time.

34.3.1.2 Microstructural evolution characterized by EBSD

Figure 34.1a shows the microstructure of the as-received sample with largely equiaxed austenitic grains and a significant amount of annealing twins. After 85% cold rolling, the original coarse-grained (CG) structure was morphed into HLS, which is composed of the ultrafine grains (UFGs, hard to be identified by EBSD in Fig. 34.3a) and lamella coarse grains (LCG). As shown in Fig. 34.3a1 for its

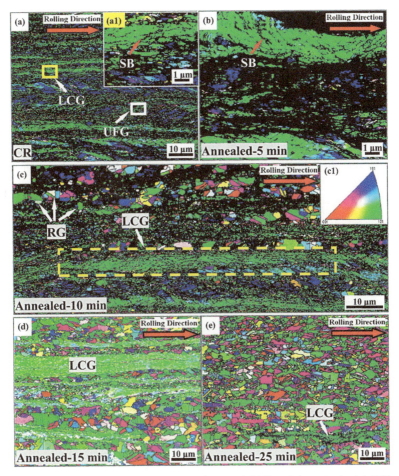

Figure 34.3 EBSD orientation maps of treated 316L stainless steels: (a) 85% cold rolling (CR); (b) CR + annealing at 750°C for 5 min; (c) CR + annealing at 750°C for 10 min, where the inset is inverse pole figure; (d) CR + annealing at 750°C for 15 min; (e) CR + annealing at 750°C for 25 min.

enlarged image (from the yellow framed area in Fig. 34.3a), the lamella coarse grains are sandwiched between UFG zones that mainly consist of elongated subgrains and shear bands (SBs, ~45° to rolling direction). Previous studies [33, 34] have confirmed that the refined grains were formed by shearing of the lamella coarse grains, which will be discussed in section 4.1. After it was annealed at 750°C for 5 min, no obvious change in microstructure was found, compared to that of the deformed sample (Fig. 34.3b). However, when the annealing time was increased to 10 min, a remarkable change in microstructure took place. Figure 34.3c shows some equiaxed recrystallized grains (RGs) in the UFG region and they inclined to form lamellar clusters, while the original lamella coarse grains changed very little. When the annealing time increased to 15 min, the UFG zones mostly transformed into recrystallized grains, forming HLS with soft recrystallized zones and original lamella coarse-grained zones, as shown in Fig. 34.3d. For the sample annealed for 25 min, the recrystallized grain lamellae grew further and lamella coarse-grained zones nearly vanished, which can be clearly seen in Fig. 34.3e.

34.3.1.3 Microstructure observation by TEM

Figure 34.4a is the TEM image of lamella coarse grain in the deformed sample. Numerous dislocation walls induced by shear stress are presented, which resulted in a distorted selected area diffraction pattern (the insert a2 in Fig. 34.4a). It should be emphasized that a distinct shear band with 45° to rolling direction traverses the whole lamella coarse grain. Some previous works [33, 34] attributed the refinement of coarse grains during severe deformation process to the drastic shear effect in the shear bands. The selected area diffraction pattern (the insert a1 in Fig. 34.4a) further confirms that the ultrafine grains are produced within the shear bands. Figure 34.4b shows typical morphologies of the UFG areas, which can be characterized as some nano-twin bundles embedded in the nano-grained matrix. The distributions of their transverse grain sizes are presented in Fig. 34.4c and d, respectively. It reveals that the average transverse size of nano-grains is ~46 nm and the twin/matrix lamellar thickness is ~22 nm.

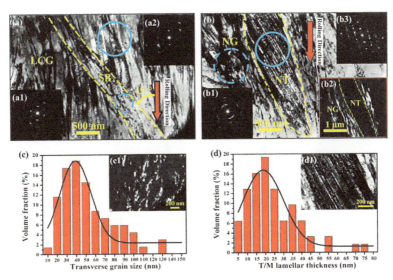

Figure 34.4 Typical cross-sectional TEM image of 85% rolled 316L stainless steel: (a) The TEM image of the lamella coarse grain (LCG), the inset (a1) is selected area diffraction pattern of shear band (SB, circled by blue dashed line), the inset (a2) is selected area diffraction pattern of lamella coarse grain (circled by blue line); (b) The TEM image of nano-twin (NT) bundle surrounded by nano-grains (NGs), the inset (b1) is the selected area diffraction pattern of NGs (circled by blue dashed line), the inset (b2) is the dark-field image of (b), the inset (b3) is selected area diffraction pattern of NT bundle (circled by blue line); (c) and (d) are the distributions of transverse grain sizes of NG and NT, respectively.

Annealing at 750°C for 10 min led to the nucleation and growth of recrystallized grains forming lamellar zones (Fig. 34.5a), which is consistent with the result in Fig. 34.3c. The recrystallized grains have sizes in the range of 0.4-3.6 μm (averagely 0.69 μm) and are equiaxed and dislocation-free. Figure 34.5b shows that an unrecrystallized region consisting of the nano-twins and nano-grains. It is noted that the recrystallization first occurred in regions with high stored energies, such as the regions with nano-grains and nano-twins. Donadille et al. [35] reported that severely deformed regions have high driving force to accelerate the nucleation and growth of recrystallization. As shown in Fig. 34.5c and d, the mean size of nano-twins after annealing increased from ~22 nm (as-CR) to ~50 nm and mean size of nano-grains from ~46 nm (as-CR) to ~89 nm.

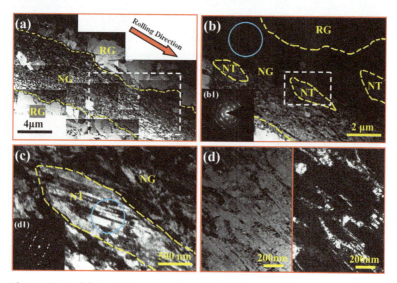

Figure 34.5 (a) Typical cross-sectional TEM image of 316L stainless steel annealed at 750°C for 10 min; (b) The enlarged TEM image of the selected region (the white dashed rectangle) in (a), the inset is the selected area diffraction pattern (circled by blue line) of nano-grains (NGs); (c) The enlarged TEM image of the selected region (the white dashed rectangle) in (b), the inset is the selected area diffraction pattern (circled by blue line) of nano-twins (NTs); (d) The bright-field and dark-field images of NGs.

After annealing at 750°C for 25 min, the microstructure consists of mostly equiaxed recrystallized grains (Fig. 34.6a), in the size range of 1–8 µm, average size of 1.96 µm and volume fraction of >90%. The annealing also produced large quantity of submicro-twins with thickness in the range of 100–800 nm. Figure 34.6b shows small amount of residual lamella coarse-grained zones, which demonstrates the excellent thermal stability of the lamella coarse-grained structure [36]. Figure 34.6c and d further elucidate that these residual lamella coarse grains contain numerous dislocation cells, which might be from the rearrangement of the dislocation walls (Fig. 34.4a) during annealing treatment.

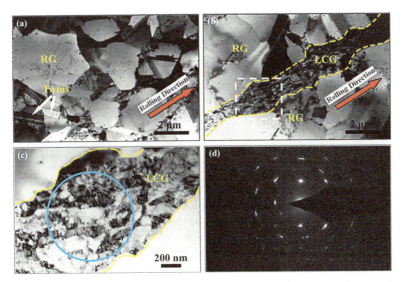

Figure 34.6 (a) Typical cross-sectional TEM image of 316L stainless steel annealed at 750°C for 25 min; (b) TEM image with residual lamella coarse grain (LCG); (c) The enlarged TEM image of the selected region (the white dashed rectangle) in (b); (d) The selected area diffraction pattern (circled by blue line) of lamella coarse grain in (c).

34.3.2 Mechanical Properties

34.3.2.1 Microhardness

Figure 34.7 gives the variation of microhardness of 316L stainless steel with different treatment histories. As shown in the upper right inset, the as-received 316L sample has a low hardness of 177 Hv. After 85% cold rolling, it achieved extremely high hardness of 441 Hv, which exceeds the level of the nanostructured DPD 316L stainless steel [25]. During the annealing at 750°C, the hardness decreased with duration in a typical "S" shape, following the classical Johnson-Mehl-Avrami relation [37]. The microhardness values after 5, 10, 15 and 20 min annealing are 408, 373, 292 and 263 Hv, respectively, and eventually tend to a saturation value of ~250 Hv.

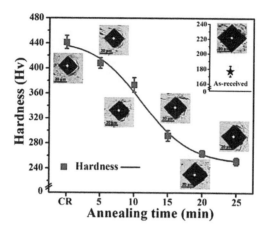

Figure 34.7 Variations of hardness of the 316L stainless steels with annealing time at 750°C, and the inset is the hardness of the as-received 316L sample.

34.3.2.2 Tensile behaviors

Typical engineering stress-strain curves of treated 316L samples are present in Fig. 34.8a and the detailed tensile properties are summarized in Table 34.2. The yield strength of cold-rolled sample is 1421 MPa, which is 3 times higher than that of the as-received sample. However, its low ductility is a roadblock to its practical applications. Figure 34.8b shows that the sample failed in a typical brittle manner without dimples while the typical ductile fracture is found for as-received sample (dimples with the mean sizes of ~5 μm). Meanwhile, an obvious decrease in tensile yield/ultimate strength and an increase in ductility are observed after annealing (Fig. 34.8a and Table 34.2). Annealing for 5 min induced a distinct drop in yield strength (from 1421 to 1108 MPa), accompanied with an increment in elongation-to-failure of ~4%, but without an obvious improvement in uniform elongation (~0.5%). Typical brittle fractography is also observed for the 5-minute-annealed sample (Fig. 34.8b). For the sample annealed for 10 min, the yield strength dropped to ~1 GPa, and the elongation-to-failure increased dramatically to ~20% with an uniform elongation of ~9%. This indicates that the deformed 316L stainless steel annealed at 750°C for 10 min possesses superior combination of strength and ductility. A large number of tiny dimples with the mean sizes of ~1 μm exist on the fracture surface, which is consistent with the good ductility of the 10-minute-annealed sample. As the annealing duration is increased to 15, 20 and 25 min, ductility increased further but strength dropped dramatically (below 670 MPa).

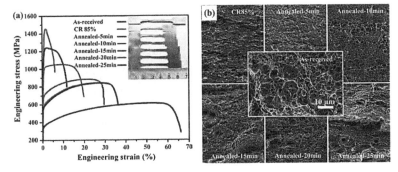

Figure 34.8 (a) Tensile engineering stress-strain curves of as-received and treated 316L stainless steels; (b) Fractographs of the as-received and treated 316L stainless steels.

Table 34.2 Tensile properties of the as-received and treated 316L stainless steels

316L	UTS (MPa)	YS (MPa)	FE (%)	UE (%)
As-received	621 ± 10	356 ± 16	63.1 ± 2.6	53.4 ± 2.0
CR 85%	1451 ± 6	1421 ± 13	5.9 ± 0.7	1.6 ± 0.1
Annealed-5 min	1242 ± 5	1108 ± 112	10.0 ± 1.4	2.1 ± 0.1
Annealed-10 min	1059 ± 5	1000 ± 10	19.4 ± 1.5	9.0 ± 0.5
Annealed-15 min	882 ± 11	662 ± 16	29.2 ± 0.7	21.2 ± 0.3
Annealed-20 min	842 ± 4	603 ± 35	36.3 ± 0.5	26.5 ± 1.5
Annealed-25 min	849 ± 8	572 ± 16	36.1 ± 1.7	27.5 ± 1.3

34.4 Discussion

34.4.1 Formation and Evolution Mechanisms of HLS

Basing on the experimental observations, we propose and illustrate the formation and evolution processes of the HLS in treated 316L stainless steel in Fig. 34.9. The original HLS formed after 85% cold rolling consists of the lamella coarse grains sandwiched between nano-grain/nano-twin zones (Fig. 34.9b). During the cold rolling, equiaxed coarse grains become elongated in the flow direction [38]. Meanwhile many primary micro-twins and subgrains were rapidly formed at early stage of deformation [39, 40], and they will be further refined to form ultrafine twins and grains at higher strain [41]. Further deformation led to the formation of shear bands, which separated the matrix with a pair of sharp boundaries across the lamella coarse grains at 45° to the rolling direction (Fig. 34.3a and Fig. 34.4a) [42]. Figure 34.10 further confirms the formation of nano-grains by cutting twin bundles and deformed austenitic matrix in shear bands, which produced nano-mantensite/autensite grains. Meyers and Xue et al. [33, 43] also reported that nano-twins, elongated ultra-laths and equiaxed ultrafine grains can be commonly created in shear bands due to the formation of secondary nano-twins and shear fracture of the primary micro-twins or the elongated subgrains, which forms the observed nano-twin/nano-grain lamella structure. In fact, deformation is always initiated in some grains with lower Schmid factor. The existence of residual lamella coarse grains may be related to the initial orientation of these coarse grains.

Ultrafine/nano grains in metallic materials are believed metastable. The recrystallization nucleation usually prefers to occur in regions with high stored energy. For annealing at 750°C, since the nano-twins have higher thermal stability (lower stored energy) in comparison with nano-grains [25], recrystallization may occur preferentially in nano-grained regions instead of nano-twin bundles. This hypothesis is confirmed by the observation that the nano-twin bundles largely survived while nano-grains mostly vanished (Fig. 34.11).

As shown in Fig. 34.3c and Fig. 34.5a, recrystallized grains were inclined to form in deformed matrix with lamellar cluster structure, which is consistent with the result of Hiriaki et al. [26]. Yan et al. [25] also reported that shear bands with high stored energy tend to form lamella zone consisting of recrystallized grain clusters. In the un-recrystallized regions, there are slight nano-grain growth and nano-twin thickening with increasing annealing time (Fig. 34.4 and Fig. 34.5). This suggests that some extremely fine nano-grains coalesced through grain boundary migration [44]. Longer annealing duration led to the disappearance of both nano-grains and nano-twin bundles (Fig. 34.6 and Fig. 34.9d). Therefore, subsequent recrystallization may occur in narrow shear bands in the lamella coarse grains, which gradually devours them. The final HLS is composed of micrometer recrystallized grans and the residual lamella coarse grains (Fig. 34.6b). The volume fractions of various microstructures in the treated 316L stainless steels are illustrated in Fig. 34.11. It also indirectly displays the evolution process of the HLS.

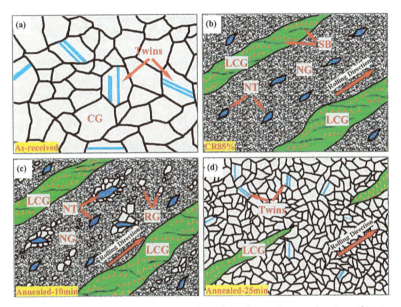

Figure 34.9 A schematic representation of the formation mechanism of HLS.

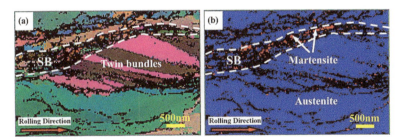

Figure 34.10 (a) EBSD image of the lamella coarse grain in deformed 316L stainless steel; (b) The corresponding phase distribution of (a), where the red phase is martensite and the blue phase is austenite.

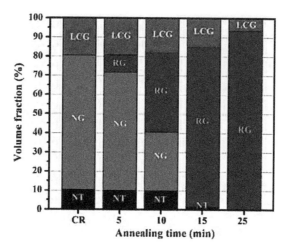

Figure 34.11 Volume fractions of various microstructures in treated 316L stainless steels.

34.4.2 Enhanced Strength and Ductility

The conventional cold rolling with subsequent short annealing can effectively control the formation and evolution of HLS in 316L stainless steel, which produced superior combinations of strength and ductility (Fig. 34.12a) over those reported in previous works for samples processed by CR, hot rolling (HR), SMAT, annealed HR/CR, and comparable to that of processed by DPD and annealing [6, 25, 30–32, 45, 46]. However, DPD can only produce small samples. The current technique has great potential for industrial application with low cost.

The strength of metallic materials is traditionally estimated using the rule of mixture, i.e. the sum of the strength of each structural components weighed by their volume fraction [47]. However, for the HLS structure, the synergistic strengthening caused by the interaction between the soft and hard heterostructured zones, $\Delta\sigma^{Syn}$, should be included [23, 47], which lead to the following equation:

$$\sigma_Y = f^{LCG}\sigma_Y^{LCG} + f^{RG}\sigma_Y^{RG} + f^{NT}\sigma_Y^{NT} + f^{NG}\sigma_Y^{NG} + \Delta\sigma^{Syn} \quad (34.1)$$

where f represents volume fraction, σ_Y the yield strength, LCG the lamella coarse grains, RG the recrystallized grain, NT the nano-twin and NG the nano-grain. It should be noted that the Eq. (34.1) can only be used to estimate the strength. Since zones that yield at higher strength will mean yielding at higher strain, it is logically problematic to simply add the weighted strength of different zones together. For heterostructured materials, the $\Delta\sigma^{Syn}$ contribution could be significant [23, 47].

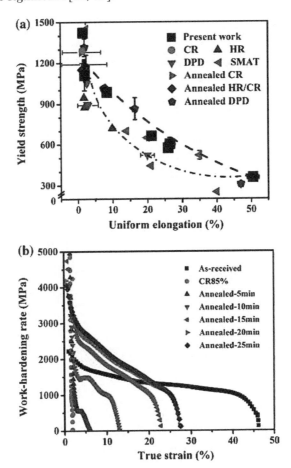

Figure 34.12 (a) Correlations between yield strength and uniform elongation for the 316L stainless steels in present and other works: HR [45], CR [46], DPD [25], SMAT [6, 31, 32], Annealed CR [46], Annealed HR/CR [30], Annealed DPD [25]; (b) Work-hardening rates of the as-received and treated 316L stainless steels in present works.

Twin boundary is a coherent and stable boundary that can strongly obstruct slip transfer of dislocations, which greatly improves the yield strength [15, 48–50]. It is noted that the nano-twin bundles with average T/M lamellar thickness of ~30 nm possess higher yield strength (>2 GPa [25]) than that (~1.45GPa [6]) of nano-grains with a mean size of ~40 nm. As the 316L stainless steel annealed for 10 min, the recrystallized grains nucleated and grew at the cost of nano-grains, which lowers the strength. However, nano-twins were more thermally stable and kept nearly constant volume fraction (Fig. 34.11). This should have made great contribution to the high yield strength.

To estimate the strength using Eq. (34.1), the yield strength of lamella coarse grain in Table 34.3 [6, 25, 51] is calculated with the value of 1003 MPa from CR85% 316L stainless sample. Then the yield strength of lamella coarse grain is assumed as a constant value for 10-minute-annealed 316L sample, which is an overestimation. The yield strength of 10-minute-annealed 316L stainless sample estimated using the first 4 terms is 945 MPa, which is still lower than the observed 1 GPa, despite of the overestimation of the strength of lamella coarse grains and nano-twins. This indicates that the $\Delta\sigma^{Syn}$ contribution is indeed significant here.

Table 34.3 The volume fractions and yield strengths of LCG, RG, NT and NG in CR 85% and 10-minute-annealed 316L stainless steels

316L	LCG	RG	NT	NG	Total
Volume fraction of CR 85% (%)	19.4	0	10.5	70.1	100
Yield strength of CR 85% (MPa)	1003	—	2000 [25]	1450 [6]	1421 (Tested)
Volume fraction of annealed 10 min (%)	18.2	41.1	9.9	30.8	100
Yield strength of annealed 10 min (MPa)	1003*	~520	2000* [25]	~1140 [5,51]	1000 (Tested)

*The data may be higher than the practical value for not taking consideration of the annealing effect.

The excellent ductility for the treated 316L stainless steels is largely determined by the working hardening capability [17, 23, 28,

52]. Work hardening can come from two different contributors: (1) the accumulation of crystalline defects such as dislocations, twins and stacking faults, which makes further deformation harder. This is the primary contributor for conventional homogeneous materials. (2) HDI stress increase with plastic straining. This contributor may become a primary contributor for heterostructured materials [28], such as the heterostructured stainless steel in this study. Figure 34.12b shows the strain hardening with strain for samples annealed for different time periods. As shown, increasing annealing time generally improves the work harden ability, because (1) the recrystallized grains have more room to accumulate dislocations, and (2) HDI hardening may increase when heterostructure is formed.

As mentioned above, HDI strengthening and HDI work hardening should have played an important role in the observed superior mechanical properties. Two kinds of soft/hard zone boundaries exist in the HLSs (Fig. 34.13a). First, the soft recrystallized zones are surrounded by hard nano-grains/nano-twins. During the deformation, there will be strain gradient near their boundaries because of the big difference in their flow stress. The strain gradient will be accommodated by GNDs, which produce HDI strengthening at the yield point and HDI work hardening during subsequent plastic deformation (Fig. 34.13b) [19, 23, 28]. Second, the boundaries between the soft lamella coarse grains and hard nano-grain/nano-twin zones will also produce GNDs and long-range HDI stress although the strength is not as large as the first case.

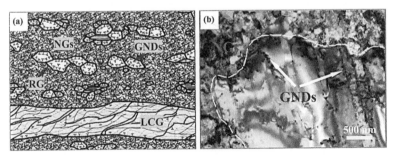

Figure 34.13 (a) A schematic representation of generation of GND induced by various soft/hard boundaries; (b) TEM observation of GNDs in recrystallized grains with an engineering strain of 6.5% in 316L stainless steel annealed at 750°C for 10 min.

34.5 Conclusions

By means of the conventional cold rolling, the heterostructured lamellar structure (HLS) was produced, which is characterized with lamella coarse grains sandwiched between zones consisting of nano-grains and nano-twins in a 316L stainless steel. This original HLS produced high tensile strength (yield strength of ~1421 MPa) but a limited ductility (uniform elongation of ~1.6%). Subsequent annealing annealed at 750°C for 5–25 min led to the evolution of HLS, forming soft zones with recrystallized zones. A superior combination of high yield strength (~1 GPa) and an elongation-to-failure of ~20% was obtained for the sample annealed for 10 min. HDI strengthening and work hardening are believed to have played a significant role in producing the superior mechanical properties. The technique developed here is conducive to large-scale industrial production at low cost.

Acknowledgment

The authors acknowledge the financial support of National Key R&D Program of China (2017YFA0204403), National Natural Science Foundation of China (Grant Nos. 51301092 and 51741106), the Jiangsu Key Laboratory of Advanced Micro & Nano Materials and Technology, and the Pangu Foundation. EBSD & TEM were performed in the Materials Characterization and Research Center of the Nanjing University of Science and Technology. YTZ acknowledge the support of the U.S. Army Research Office (W911 NF-17-1-0350).

References

1. K. H. Lo, C. H. Shek, J. K. L. Lai, Recent developments in stainless steels, *Mater. Sci. Eng. R* 65 (2009) 39–104.
2. W. Martienssen, H. Warlimont, *Springer Handbook of Condensed Matter and Materials Data*, Springer, Berlin, Germany, 2005.
3. A. Belyakov, Y. Kimura, K. Tsuzaki, Microstructure evolution in dual-phase stainless steel during severe deformation, *Acta Mater.* 54 (2006) 2521–2532.

4. M. Odnobokova, A. Belyakov, R. Kaibyshev, Effect of severe cold or warm deformation on microstructure evolution and tensile behavior of a 316L stainless steel, *Adv. Eng. Mater.* 17 (2015) 1812–1820.
5. K. Lu, J. Lu, Nanostructured surface layer on metallic materials induced by SMAT, *Mater. Sci. Eng. A* 375–377 (2004) 38–45.
6. X. H. Chen, J. Lu, L. Lu, K. Lu, Tensile properties of a nanocrystalline 316L austenitic stainless steel, *Scr. Mater.* 52 (2005) 1039–1044.
7. R. Z. Valiev, R. K. Islamgaliev, I. V. Alexandrov, Bulk nanostructured materials from severe plastic deformation. *Prog. Mater. Sci.* 45 (2000) 103–189.
8. H. Ueno, K. Kakihata, Y. Kaneko, S. Hashimoto, A. Vinogradov, Nanostructurization assisted by twinning during equal channel angular pressing of metastable 316L stainless steel, *J. Mater. Sci.* 46 (2011) 4276–4283.
9. C. X. Huang, G. Yang, Y. L. Gao, S. D. Wu, Z. F. Zhang, Influence of processing temperature on the microstructures and tensile properties of 304L stainless steel by ECAP, *Mater. Sci. Eng. A* 485 (2008) 643–650.
10. C.X. Huang, W. P. Hu, Q. Y. Wang, C. Wang, G. Yang, Y. T. Zhu, An ideal ultrafine-grained structure for high strength and high ductility, *Mater. Res. Lett.* 3 (2015) 88–94.
11. J. Gubicza, M. El-Tahawy, Y. Huang, H. Choi, H. Choe, J. L. Labar, T. G. Langdon, Microstructure, phase composition and hardness evolution in 316L stainless steel processed by high-pressure torsion, *Mater. Sci. Eng. A* 657 (2016) 215–223.
12. E. Ma, T. Zhu, Towards strength-ductility synergy through the design of heterogeneous nanostructures in metals, *Mater. Today* 20 (2017) 323–331.
13. Y. T. Zhu, X. Z. Liao, Nanostructured metals: retaining ductility, *Nat. Mater.* 3 (2004) 351–352.
14. K. Lu, Stabilizing nanostructures in metals using grain and twin boundary architectures, *Nat. Rev. Mater.* 1 (2016) 16019.
15. Y. H. Zhao, J. F. Bingert, X. Z. Liao, B. Z. Cui, K. Han, A. V. Sergueeva, A. K. Mukherjee, R. Z. Valiev, T. G. Langdon, Y. T. Zhu, Simultaneously increasing the ductility and strength of ultra-fine-grained pure copper. *Adv. Mater.* 18 (2006) 2949–2953.
16. T. H. Fang, W. L. Li, N. R. Tao, K. Lu, Revealing extraordinary intrinsic tensile plasticity in gradient nano-grained copper, *Science* 331 (2011) 1587–1590.

17. X. L. Wu, P. Jiang, L. Chen, F. P. Yuan, Y. T. Zhu, Extraordinary strain hardening by gradient structure, *Proc. Natl. Acad. Sci. USA* 111 (2014) 7197–7201.
18. X. L. Wu, M. X. Yang, F. P. Yuan, L. Chen, Y. T. Zhu, Combining gradient structure and TRIP effect to produce austenite stainless steel with high strength and ductility, *Acta Mater.* 112 (2016) 337–346.
19. M. X. Yang, Y. Pan, F. P. Yuan, Y. T. Zhu, X. L. Wu, Back stress strengthening and strain hardening in gradient structure, *Mater. Res. Lett.* 4 (2016) 145–151.
20. Y. Wang, M. Chen, F. Zhou, E. Ma, High tensile ductility in a nanostructured metal, *Nature* 419 (2002) 912.
21. T. Sekiguchi, K. Ono, H. Fujiwara, K. Ameyama, New microstructure design for commercially pure titanium with outstanding mechanical properties by mechanical milling and hot roll sintering, *Mater. Trans.* 51 (2010) 39–45.
22. X. L. Ma, C. X. Huang, J. Moering, M. Ruppert, H. W. Hoppel, M. Goken, J. Narayan, Y. T. Zhu, Mechanical properties of copper/bronze laminates: role of interfaces, *Acta Mater.* 116 (2016) 43–52.
23. X. L. Wu, M. X. Yang, F. P. Yuan, G. L. Wu, Y. J. Wei, X. X. Huang, Y. T. Zhu, Heterogeneous lamella structure unites ultrafine-grain strength with coarse-grain ductility, *Proc. Natl. Acad. Sci. USA* 112 (2015) 14501–14505.
24. J. X. Xing, F. P. Yuan, X. L. Wu, Enhanced quasi-static and dynamic shear properties by heterogeneous gradient and lamella structures in 301 stainless steels, *Mater. Sci. Eng. A* 680 (2017) 305–316.
25. F. K. Yan, G. Z. Liu, N. R. Tao, K. Lu, Strength and ductility of 316L austenitic stainless steel strengthened by nano-scale twin bundles, *Acta Mater.* 60 (2012) 1059–1071.
26. N. Hirota, F. X. Yin, T. Inoue, T. Azuma, Recrystallization and grain growth behavior in severe cold-rolling deformed SUS316L steel under an isothermal annealing condition, *ISIJ Int.* 48 (2008) 475–482.
27. A. Belyakov, A. Kipelova, M. Odnobokova, I. Shakhova, R. Kaibyshev, Development of ultrafine grained austenitic stainless steels by large strain deformation and annealing, *Mater. Sci. Forum* 783–786 (2014) 651–656.
28. X. L. Wu, Y. T. Zhu, Heterogeneous materials: a new class of materials with unprecedented mechanical properties, *Mater. Res. Lett.* 5 (2017) 527–532.

29. V. Tsakiris, D. V. Edmonds, Martensite and deformation twinning in austenitic steels, *Mater. Sci. Eng. A* 273–275 (1999) 430–436.
30. M. Eskandari, A. Najafizadeh, A. Kermanpur, Effect of strain-induced martensite on the formation of nanocrystalline 316L stainless steel after cold rolling and annnealing, *Mater. Sci. Eng. A* 519 (2009) 46–50.
31. T. Roland, D. Retraint, K. Lu, J. Lu, Enhanced mechanical behavior of nanocrystallised stainless steel and its thermal stability, *Mater. Sci. Eng. A* 445–446 (2007) 281–288.
32. H. W. Huang, Z. B. Wang, J. Lu, K. Lu, Fatigue behaviors of AISI 316L stainless steel with a gradient nanostructured surface layer, *Acta Mater.* 87 (2015) 150–160.
33. M. A. Meyers, Y. B. Xu, Q. Xue, M. T. Perez-Prado, T. R. McNelley, Microstructural evolution in adiabatic shear localization in stainless steel, *Acta Mater.* 51 (2003) 1307–1325.
34. Y. S. Li, N. R. Tao, K. Lu, Microstructural evolution and nanostructure formation in copper during dynamic plastic deformation at cryogenic temperatures, *Acta Mater.* 56 (2008) 230–241.
35. C. Donadille, R. Valle, P. Dervin, R. Penelle, Development of texture and microstructure during cold-rolling and annealing of FCC alloys: example of an austenitic stainless steel, *Acta Metall.* 37 (1989) 1547–1571.
36. X. C. Liu, H. W. Zhang, K. Lu, Strain-induced ultrahard and ultrastable nanolaminated structure in nickel, *Science* 342 (2013) 337–340.
37. M. A. Arshad, A. K. Maaroufi, Relationship between Johnson-Mehl-Avrami and Sestak-Berggren models in the kinetics of crystallization in amorphous materials, *J. Non-Cryst. Solids* 413 (2015) 53–58.
38. S. G. Chowdhury, S. Das, P. K. De, Cold rolling behaviour and textural evolution in AISI 316L austenitic stainless steel, *Acta Mater.* 53 (2005) 3951–3959.
39. B. R. Kumar, M. Ghosh, Surface and mid-plane texture evolution in austenite phase of cold rolled austenitic stainless steels, *Mater. Sci. Eng. A* 457 (2007) 236–245.
40. M. Odnobokova, A. Kipelova, A. Belyakov, R. Kaibyshev, Microstructure evolution in a 316L stainless steel subjected to multidirectional forging and unidirectional bar rolling, *IOP Conf. Series: Mater. Sci. Eng.* 63 (2014) 012060.
41. Q. Xue, X. Z. Liao, Y. T. Zhu, G. T. Gray III, Formation mechanisms of nanostructures in stainless steel during high-strain-rate severe plastic deformation, *Mater. Sci. Eng. A* 410–411 (2005) 252–256.

42. H. Miura, M. Kobayashi, Y. Todaka, C. Watanable, Y. Aoyagi, N. Sugiura, N. Yoshinaga, Heterogeneous nanostructure developed in heavily cold-rolled stainless steels and the specific mechanical properties, *Scr. Mater.* 133 (2017) 33–36.

43. Q. Xue, E. K. Cerreta, G. T. Gray III, Microstructural characteristics of post-shear localization in cold-rolled 316L stainless steel, *Acta Mater.* 55 (2007) 691–704.

44. J. Hu, Y. N. Shi, X. Sauvage, G. Sha, K. Lu, Grain boundary stability governs hardening and softening in extremely fine nanograined metals, *Science* 355 (2017) 1292–1296.

45. Z. Yanushkevich, A. Lugovskaya, A. Belyakov, R. Kaibyshev, Deformation microstructures and tensile properties of an austenitic stainless steel subjected to multiple warm rolling, *Mater. Sci. Eng. A* 667 (2016) 279–285.

46. I. Ucok, T. Ando, N. J. Grant, Property enhancement in type 316L stainless steel by spray forming, *Mater. Sci. Eng. A* 133 (1991) 284–287.

47. X. L. Wu, P. Jiang, L. Chen, J. F. Zhang, F. P. Yuan, Y. T. Zhu, Synergetic strengthening by gradient structure, *Mater. Res. Lett.* 2 (2014) 185–191

48. Z. S. You, X. Y. Li, L. J. Gui, Q. H. Lu, T. Zhu, H. J. Gao, L. Lu, Plastic anisotropy and associated deformation mechanisms in nanotwinned metals, *Acta Mater.* 61 (2013) 217–227.

49. T. Zhu, H. J. Gao, Plastic deformation mechanism in nanotwinned metals: an insight from molecular dynamics and mechanistic modeling, *Scr. Mater.* 66 (2012) 843–848.

50. Y. T. Zhu, X. Z. Liao, X. L. Wu, Deformation twinning in nanocrystalline materials, *Prog. Mater. Sci.* 57 (2012) 1–62.

51. M. Eskandari, A. Zarei-Hanzaki, H. R. Abedi, An investigation into the room temperature mechanical properties of nanocrystalline austenitic stainless steels, *Mater. Des.* 45 (2013) 674–681.

52. R. Z. Valiev, Y. Estrin, Z. Horita, T. G. Langdon, M. J. Zehetbauer, Y. T. Zhu, Fundamentals of superior properties in bulk nanoSPD materials, *Mater. Res. Lett.* 4 (2016) 1–21.

Part V
Laminate Materials

Chapter 35

Strain Hardening and Ductility in a Coarse-Grain/Nanostructure Laminate Material

X. L. Ma,[a] C. X. Huang,[b] W. Z. Xu,[a] H. Zhou,[a,c] X. L. Wu,[d] and Y. T. Zhu[a,e]

[a]*Department of Materials Science and Engineering, North Carolina State University, Raleigh, NC 27695, USA*
[b]*School of Aeronautics and Astronautics, Sichuan University, Chengdu 610065, China*
[c]*National Engineering Research Center of Light Alloy Net Forming, Shanghai Jiao Tong University, Shanghai 200240, China*
[d]*State Key Laboratory of Nonlinear Mechanics, Institute of Mechanics, Chinese Academy of Sciences, Beijing 100190, China*
[e]*School of Materials Science and Engineering, Nanjing University of Science and Technology, Nanjing 210094, China*
chxhuang@scu.edu.cn, ytzhu@ncsu.edu

A laminate structure with a nanostructured Cu-10Zn layer sandwiched between two coarse-grained Cu layers was produced

---

Reprinted from *Scr. Mater.*, **103**, 57–60, 2015.

Heterostructured Materials: Novel Materials with Unprecedented Mechanical Properties
Edited by Xiaolei Wu and Yuntian Zhu
Text Copyright © 2015 Acta Materialia Inc.
Layout Copyright © 2022 Jenny Stanford Publishing Pte. Ltd.
ISBN 978-981-4877-10-7 (Hardcover), 978-1-003-15307-8 (eBook)
www.jennystanford.com

by high-pressure torsion, rolling and annealing. Sharp interlayer boundaries were developed and remained intact during uniform tensile deformation. Mechanical incompatibility between the different layers during plastic deformation produced high strain hardening, which led to ductility that is higher than prediction by the rule-of-mixture. These observations provide insight into the architectural design and deformation studies of materials with gradient and laminate structures.

Gradient structures (GS) with a grain-size gradient have been recently introduced to structural materials to optimize their mechanical properties with low cost [1–5]. To date, some exceptional combination of enhanced strength and considerable ductility are reported in different material systems with multiscale grain-size structures [6–9], including GS and multi-layered materials. However, the fundamental principles that govern the deformation behaviors of GS are still not fully understood [1, 5]. Elastic/plastic boundary and stable/unstable boundary caused by mechanical incompatibility during deformation have been reported to play a critical role in both strengthening and strain hardening of GS materials [2, 5]. However, such boundaries migrate dynamically during the deformation of GS, which makes it hard to perform quantitative postmortem investigation [1, 2, 10]. In fact, GS can be approximately regarded as the integration of numerous layers and boundaries (laminate structure) [2]. Therefore, there are some similarities in the deformation behaviors of GS structures and laminate structures. For example, both structures have the mechanical incompatibility during the deformation. The advantage of the latter is that its boundaries are stationary [11] and can be easily located after the deformation, making it easier to analyze quantitative mechanics and investigate postmortem microstructures. Therefore, it might be possible to use laminated (or sandwiched) structures to study some fundamentals in deformation behaviors of GS structure.

The fabrication of laminated nanostructured (NS)/coarse-grained (CG) structures with sharp boundaries is also a challenge since interfacial strength and selective grain refinement are required simultaneously [12]. Here, we fabricated a laminate structure with a NS bronze layer sandwiched between two CG copper layers by utilizing two principles: (a) Different grain refinement effectiveness of Cu and bronze during deformation. Bronze has much lower stacking fault energy and deform by twinning while Cu deforms primarily by dislocation slip [13]. As a result, the grain sizes of bronze can be refined much more effectively than those of copper during severe plastic deformation. (b) Different thermal stabilities of Cu and bronze during annealing. Alloying elements are generally effective in pinning grain boundaries and resisting grain growth [14]. Hence, bronze can remain much finer grain size compared to pure copper after proper annealing. Using this novel approach, we were able to produce a laminate structure with sharp and well-bonded NS/CG boundaries. Another objective of this work is to use this NS/CG sandwich to study the effect of grain-size-difference across the boundaries on mechanical behaviors.

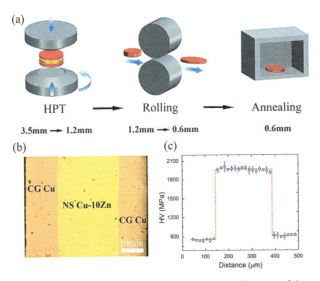

Figure 35.1 (a) Schematic illustration of major procedures to fabricate CG/NS/CG sandwich materials. Deformation history of thickness reduction is also provided below the corresponding step. (b) Optical microscopy observation of as-processed sandwiches with NS Cu-10Zn volume fraction 0.47. (c) Vickers Micro-hardness indentation (with loading 25g) on cross-sectional sample in (b).

Figure 35.1a schematically illustrates the procedure of sample processing. Commercial Cu (99.9 wt.%) and bronze (Cu-10 wt.%Zn) plates were punched into ϕ-10 mm disks and polished to 3 groups of thickness. The total initial thickness of three disks was around 3.5 mm so that sufficient thickness reduction (~83%) after processing can be achieved to form strong interfacial bonding (thickness reduction history as show in Fig. 35.1a). Mechanical polishing and ultrasonic cleaning were carried out before sandwiching the disks together. High pressure torsion (HPT) was applied at room temperature with imposed pressure of 1 GPa for 10 revolutions at 1.5 rpm in order to obtain more homogeneous deformation along the radius direction [15]. Thereafter, as-HPTed sandwiches were rolled to thickness of 0.6 mm from 1.2 mm and then annealed at 240°C for 2 h.

Microstructures near boundaries were characterized by FEI Quanta 3D FEG with Ion Channeling Contrast Microscopy (ICCM), JEOL-2010F Transmission Electron Microscopy (TEM) operated at 200 kV and Electron Dispersive X-ray Spectroscopy (EDS) mapping in FEI Titan 80-300. Dog-bone shaped samples with gauge dimension of $0.6 \times 2 \times 8.4$ mm^3 were cut from the middle of sandwiches and mechanically tested under uniaxial tension at a strain rate of 9×10^{-4} s^{-1}. Scanning electron microscopy (SEM) was used to examine fracture surface and boundary.

Three groups of samples with varying volume fraction of central NS Cu-10Zn layer (Sample A: 0.10, Sample B: 0.22, Sample C: 0.47) were fabricated. Figure 35.1b shows a typical optical micrograph (cross-sectional view) of Sample C. Different color contrast clearly indicates three layers with two sharp boundaries. As shown in Fig. 35.1c, microhardness within each layer is rather homogeneous, while there are abrupt transitions at the boundaries, indicating that the central NS layer has yield strength over twice of that of the CG Cu layers. These hardness levels in each layer do not change much with variation of volume fraction.

Figure 35.2a is a channeling contrast image showing the typical microstructures near the boundary. On the left side is typical CG Cu with grain size of ~4 μm, while on the right side is NS bronze with grain (sub-grain) size of ~100 nm. Magnified image of the NS/CG boundary is shown in Fig. 35.2b, which reveals a void-free transition from NS bronze to CG Cu. The exact Cu/bronze boundary is hard to tell in TEM and can be identified by EDS mapping. Figure 35.2c is a typical high-angle annular dark-field (HAADF) image including a Cu/bronze boundary, whose precise location is unknown. EDS mappings in Fig. 35.2d and e show the elemental distribution of Cu and Zn and resolve the exact boundary (marked by dotted lines) in Fig. 35.2c. Concentrations of Zn are measured as 0.64 wt.% in left side and 10.22 wt.% in right side of Fig. 35.2c, respectively. This agreement of specific composition in Cu and bronze implies no significant bulk diffusion from each side during the processing. Therefore, Fig. 35.2c–e confirm the generation of well-bonded and sharp Cu/bronze boundary.

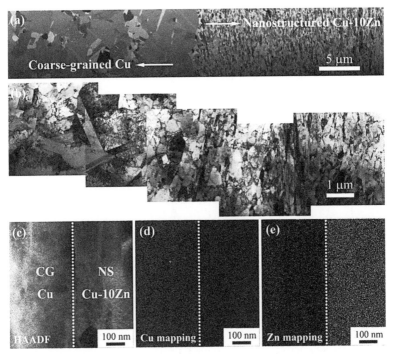

Figure 35.2 (a) ICCM of a typical NS/CG boundary in sandwich (7° tilt of sample while imaging). (b) TEM observation of as-processed boundary. (c) HAADF imaging of an enlarged area around Cu/Cu-10Zn boundary by STEM. EDS mapping of (d) Cu and (e) Zn in corresponding regions in (c).

Figure 35.3a shows the tensile stress-strain curves of the laminated samples with different compositions and those of pure CG Cu and NS Cu-10Zn samples. Pure samples were made from the sandwiches by polishing off other layers. The yield strength (0.2%-strain offset stress) for samples A, B and C were measured as 142 MPa, 201 MPa and 266 MPa, respectively, while their uniform elongation (engineering strain) were measured as 27.1%, 19.6% and 12.3%, respectively. Note that the uniform plastic deformation of NS bronze layer in sandwich is much higher (>12%) than that

(~0.7%) of its pure counterpart. This is because its early necking tendency was constrained by the stable CG Cu layers from both sides via the two boundaries. As shown, the yield strength of the laminate structure increases and tensile ductility decreases with increasing volume fraction of central NS bronze layer.

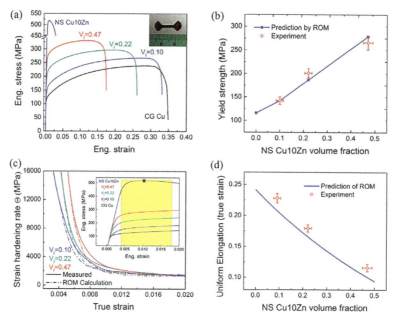

Figure 35.3 (a) Uniaxial tensile engineering strain-stress curves for pure CG Cu and NS Cu-10Zn and sandwiches with various compositions. Inset shows the dimension of tensile test samples 0.6 × 2 × 8.4 mm³. (b) Yield strength versus volume fraction of NS Cu-10Zn from tensile tests and prediction from ROM. (c) Strain hardening curves from corresponding tensile tests and calculation based on ROM. Inset is the magnified tensile curve at low strains where outperformance of strain hardening (yellow shadow) occurs compared to ROM. Black crossover stands for the necking strain level in pure Cu-10Zn. (d) Uniform elongation versus volume fraction of NS Cu-10Zn from tensile tests and prediction from ROM.

In conventional theory of rule of mixture (ROM) for laminated composite structure, yield strength (σ_{ys}), strain hardening ($d\sigma/d\varepsilon$) and uniform elongation (ε_{UE}) (true strain) can be expressed as [2, 16, 17]

$$\sigma_{ys} = \sum_i V_i \sigma'_{i,ys} \qquad (35.1)$$

$$\frac{d\sigma}{d\varepsilon} = \sum V_i \frac{d\sigma_i}{d\varepsilon} \qquad (35.2)$$

$$\varepsilon_{UE} = \frac{\sum V_i \sigma_{i,UE} \varepsilon_{i,UE}}{\sum V_i \sigma_{i,UE}} \qquad (35.3)$$

where V_i is the volume fraction of component i, $\sigma'_{i,ys}$ is the flow stress of component i alone at 0.2% plastic strain of the composite sample [2], σ_i is the true stress of component i, σ and ε represent the true stress and true strain of integrated composite sample. $\sigma_{i,UE}$, $\varepsilon_{i,UE}$ stand for the true stress and true strain of component i at necking point.

It is found that the yield strength of laminate-structured samples agrees reasonably well with the ROM (Eq. 35.1) (shown in Fig. 35.3b). Comparisons of strain hardening and uniform elongation (true strain) between experimental results and theoretical prediction from ROM are shown in Fig. 35.3c and d. When calculating the strain-hardening rate, we assume a constant engineering stress after necking for the pure bronze. This assumption uses the maximum applied stress that a standalone pure bronze layer can sustain before necking according to Considère criterion. Namely, the derived strain-hardening rate is the upper limit based on ROM and would give a conservative extra strain hardening. For Eq. 35.3, it is also assumed that during the uniform elongation, the strain hardening of each component can be expressed by Hollomon law of $\sigma_i = k_i \varepsilon^{n_i}$ [16]. In fact, the Hollomon law has been used on NS materials in compression test [18, 19] despite limit reports of applications for tensile tests due to the plasticity instability. Figure 35.3c shows that the strain-hardening rate of the laminate structure (solid) is higher than that predicted by ROM (dash-dot) at strains around the necking of pure NS bronze layer. This is a limited intermediate strain range as marked by yellow shade in the inset of Fig. 35.3c. The underlying mechanism of this extra strain hardening is discussed later. Consequently, the uniform elongations of laminate-structured sample are measured higher than what is predicted by ROM (Fig. 35.3d).

Our SEM examination revealed that the boundary only failed near the fracture surface, which indicates the boundary did not fail during the tensile tests before the fnal fracture process. Further SEM check of boundary in Fig. 35.4a is a cross-sectional overview of a sample. No inner cracks or failure through the entire uniformly elongated region confirms the concurrent tensile deformation of both CG and NS layers. In addition, the fracture surface shown in Fig. 35.4b reveals a frequently-seen dimple-like feature in NS/CG boundary vertical to tensile direction, which suggests the strong interactions between NS/CG layers at high strain localization. All of above characteristics implies well-bonded boundaries between NS and CG layers.

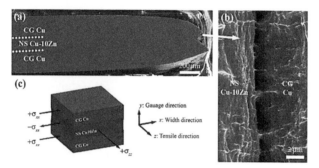

Figure 35.4 (a) Overview of the sandwich piece after tensile fracture. (b) Typical morphology at lateral fracture boundary in (a). (c) Schematic illustration of stress status of the laminates around the strain level where the middle NS layer tends to shrink while outer CG layers stabilize it.

Pure NS bronze has a limited tensile ductility of only 0.7% due to the lack of sufficient strain hardening. However, when sandwiched by CG Cu, its uniform elongation was increased up to over 27% (Fig. 35.3a). Recent modeling work of structures with grain-size difference [21] also revealed enhanced ductility in an otherwise low-ductility layer. CG Cu has considerable strain hardening capability and therefore constrains the NS layer to postpone its necking by preventing the early-emerging tensile instability. Under such constraint, the NS bronze layer should be able to be formed deformed further [5, 22]. In addition, the mutual constraint of the NS bronze layer and the CG Cu layer actually adds extra strain hardening, as shown in Fig. 35.3c. A standalone NS bronze layer will start early necking at very low strain, as manifested by fast local lateral shrinkage. When sandwiched by CG Cu, this necking process is quickly stopped by the stable outer layers on both sides. Therefore, we refer to this as "virtual necking" since it could not proceed very far.

The instable middle layer and stable outer layers mutually constrain each other, which converts uniaxial applied tensile stress to bi-axial stresses that is similar to what was reported in gradient structures [5]. As a result, more dislocations will be accumulated near the boundary in order to accommodate the mechanical incompatibility across the boundaries, which produces the observed extra strain hardening. It follows that the strain-hardening rate of our laminate structure can be described by modifying Eq. 35.2 as

$$\frac{d\sigma}{d\varepsilon} = \sum V_i \frac{d\sigma_i}{d\varepsilon} + \Delta\Theta \qquad (35.4)$$

where $\Delta\Theta$ is the extra strain hardening in addition to what is predicted by conventional ROM. This observation of extra hardening is consistent with the recent discoveries in GS IF-steel [5].

The extra strain hardening observed here is much lower than what is reported in gradient structured IF steel. because this boundary is stationary, which doesn't migrate with straining. Once the dislocation accumulation is saturated to accommodate mechanical incompatibility around the boundary, no more hardening is accessible and the strain behavior will be just similar to ROM in the following deformation. This is believed the most reason for the appearance of limited intermediate strain range in Fig. 35.3c. Indeed, whether boundary is stationary or dynamic during deformation is also the big difference between sandwich structure here and the GS IF steel reported earlier. The stationary boundary becomes ineffective in dislocation accumulation once saturation near the boundaries is reached. In other words, dislocations are accumulating only locally around the fixed boundaries instead of in the whole sample. In contrast, the dynamically migrating boundary in the GS allows the accumulation of dislocations over its entire migrating path in the sample [5]. This can give rise to more strain hardening and higher ductility and possibly accounts for the unique up-turn hardening behavior in earlier report [5]. More systematic investigations are needed to further clarify this issue. Let us the old text and make some minor changes.

In summary, HPT followed by rolling and annealing was used to produce laminated materials by sandwiching a NS layer between CG layers. Uniaxial tensile test is performed and reveals an extra strain hardening that lead to larger uniform elongation than what is predicted by conventional ROM. The boundary is well bonded to maintain the overall uniform deformation for the whole sample. This preliminary work indicates a critical role played by boundaries in the mechanical behaviors in laminate and gradient structured materials.

We are grateful for financial support of 973 programs (grants 2012CB932203, 2012CB937500, and 6138504), National Natural Science Foundation of China (grants 11172187, 11002151, 11072243, 51301187 and 50571110), the Pangu Foundation, the Program for New Century Excellent Talents in University (NCET-12-0372) and the US Army Research Office (grants W911NF-09-1-0427 and W911QX-08-C-0083). We also acknowledge the use of Analytical Instrumentation Facility (AIF) at North Carolina State University, which is supported by the State of North Carolina and the National Science Foundation.

References

1. T.H. Fang, W.L. Li, N.R. Tao, K. Lu, *Science* 331 (2011) 1587–1590.
2. X.L. Wu, P. Jiang, L. Chen, J.F. Zhang, F.P. Yuan, Y.T. Zhu, *Mater. Res. Lett.* (2014) 1–7.
3. X.L. Lu, Q.H. Lu, Y. Li, L. Lu, *Sci. Rep.* 3 (2013).
4. A. Jérusalem, W. Dickson, M.J. Pérez-Martín, M. Dao, J. Lu, F. Gálvez, *Scr. Mater.* 69 (2013) 773–776.
5. X. Wu, P. Jiang, L. Chen, F. Yuan, Y.T. Zhu, Proc. *Natl. Acad. Sci. U. S. A.* 111 (2014) 7197–7201.
6. K. Lu, *Science* 345 (2014) 1455–1456.
7. D.K. Yang, P. Cizek, D. Fabijanic, J.T. Wang, P.D. Hodgson, *Acta Mater.* 61 (2013) 2840–2852.
8. Y. Wei, Y. Li, L. Zhu, Y. Liu, X. Lei, G. Wang, Y. Wu, Z. Mi, J. Liu, H. Wang, H. Gao, *Nat. Commun.* 5 (2014).
9. A.-Y. Chen, D.-F. Li, J.-B. Zhang, H.-W. Song, J. Lu, *Scr. Mater.* 59 (2008) 579–582.
10. S. Suresh, *Science* 292 (2001) 2447–2451.
11. I.J. Beyerlein, J.R. Mayeur, S. Zheng, N.A. Mara, J. Wang, A. Misra, *Proc. Natl. Acad. Sci. U. S. A.* 111 (2014) 4386–4390.
12. X. Li, G. Zu, M. Ding, Y. Mu, P. Wang, *Mater. Sci. Eng. A* 529 (2011) 485–491.

13. Y.H. Zhao, Y.T. Zhu, X.Z. Liao, Z. Horita, T.G. Langdon, *Appl. Phys. Lett.* 89 (2006) 121906.
14. K. Barmak, A. Gungor, C. Cabral, J.M.E. Harper, *J. Appl. Phys.* 94 (2003) 1605–1616.
15. X.H. An, S.D. Wu, Z.F. Zhang, R.B. Figueiredo, N. Gao, T.G. Langdon, *Scr. Mater.* 63 (2010) 560–563.
16. S.L. Semiatin, H.R. Piehler, *Metall. Trans. A* 10 (1979) 85–96.
17. D. Nyung Lee, Y. Keun Kim, *J. Mater. Sci.* 23 (1988) 1436–1442.
18. D. Jia, Y.M. Wang, K.T. Ramesh, E. Ma, Y.T. Zhu, R.Z. Valiev, *Appl. Phys. Lett.* 79 (2001) 611–613.
19. Y.M. Wang, J.Y. Huang, T. Jiao, Y.T. Zhu, A.V. Hamza, *J. Mater. Sci.*, 42 (2007) 1751–1756.
20. X. Guo, J. Tao, W. Wang, H. Li, C. Wang, *Mater. Des.*, 49 (2013) 116–122.
21. L. Jianjun, A.K. Soh, *Model. Simul. Mater. Sci. Eng.* 20 (2012) 085002.
22. X.L. Wu, Y.T. Zhu, Y.G. Wei, Q. Wei, *Phys. Rev. Lett.* 103 (2009) 205504.

Chapter 36

Effect of Strain Rate on Mechanical Properties of Cu/Ni Multilayered Composites Processed by Electrodeposition

Zhengrong Fu,[a] Zheng Zhang,[a] Lifang Meng,[a] Baipo Shu,[a] Yuntian Zhu,[b,c] and Xinkun Zhu[a]

[a]*Faculty of Materials Science and Engineering,*
Kunming University of Science and Technology,
Kunming 650093, China
[b]*Department of Materials Science and Engineering,*
North Carolina State University, Raleigh, NC 27695, USA
[c]*School of Materials Science and Engineering,*
Nanjing University of Science and Technology,
Nanjing 210094, China
ytzhu@ncsu.edu, xk_zhu@hotmail.com

Mechanical properties of Cu/Ni multilayered composites processed by electrodeposition were investigated by tensile tests at different

---

Reprinted from *Mater. Sci. Eng. A*, **726**, 154–159, 2018.

Heterostructured Materials: Novel Materials with Unprecedented Mechanical Properties
Edited by Xiaolei Wu and Yuntian Zhu
Text Copyright © 2018 Elsevier B.V.
Layout Copyright © 2022 Jenny Stanford Publishing Pte. Ltd.
ISBN 978-981-4877-10-7 (Hardcover), 978-1-003-15307-8 (eBook)
www.jennystanford.com

strain rates in the range of 5×10^{-5} to 5×10^{-2} s^{-1} at room temperature. With increasing strain rates, the strength and ductility of Cu/Ni multilayered composites increased simultaneously, while their strain rate sensitivity also increased, which is very different from the constituent pure Cu and Ni. The hetero-deformation induced (HDI) stress caused by the Cu/Ni layer boundaries also increased with strain rate. Strong HDI work hardening is observed, which is the main reason for the observed good ductility.

36.1 Introduction

Improving mechanical properties of conventional engineering materials were inspired by the development of the laminated structures developed in biological materials (1). Therefore, metallic laminated composites have attracted the attention of more and more researchers in recent years (2–4). Many investigations have been carried out on mechanical properties of metallic laminated composites by controlling the constituent layers and boundaries, which revealed a combination of high strength with good ductility (5–9). The high performance of metallic laminated composites was believed to result from the fact that the hard layers can contribute to strength, while the soft layer can improve the uniform elongation by delaying the necking behavior due to the layer boundary effects (3, 10–12). Metallic multilayered composites have been reported prepared by several methods including electron beam evaporation (13), bonding deposition (14, 15) and electrodeposition (16, 17). High strength has been observed (18, 19), whereas the ductility of multilayered composites turned out to be still needed to be improved (20, 21).

Generally speaking, strain rate may significantly affect mechanical properties of metals. Face-centered cubic (fcc) metals usually display higher tensile strength, but lower ductility with increasing strain rates (22–28). However, an electrodeposited nanocrystalline (NC) Cu has been reported to demonstrate an increase in ductility with increasing strain rates (29). Subsequently, the same phenomenon for ductility was found in NC-Ni (30), ultrafine-grained (UFG) Cu(31, 32) and Cu/Ni laminated composites (33, 34). Nevertheless, the mechanism of this abnormal tendency is not well explained.

In this chapter, we investigated the tensile properties of the electrodeposited Cu/Ni multilayered composites with different deposition time and strain rates. Longer deposition time results in thicker individual layers. Effects of deposition time and strain rates on tensile strength and ductility of Cu/Ni multilayered composites will be discussed.

36.2 Experimental Procedure

Cold-rolled Cu sheets 2 mm in thickness were fully annealed at 650°C for 2 h in order to decrease initial dislocation density and to increase good plasticity. Then the Cu sheets were mechanically polished with sandpapers from coarse to fine grades. The polished sheet was used as the substrate to produce Cu/Ni multilayered composites by electrodeposition at a temperature of 50°C. Cu layers were electrodeposited in an electrolyte solution containing 200 g. L^{-1} CuSO$_4$.5H$_2$O, 10 g. L^{-1} H$_3$BO$_4$, 0.08 g. L^{-1} HCl and 0.05 g. L^{-1} H$_2$SO$_4$ in deionized water and the PH value slightly lower than 3, while the Ni layers deposition solution was composed of 350 g. L^{-1} Ni(SO$_3$NH$_2$)$_2$.4H$_2$O, 10 g. L^{-1} NiCl$_2$.6H$_2$O and 30 g. L^{-1} H$_3$BO$_3$ in deionized water and the PH value slightly higher than 3. The Cu and Ni layers were alternately deposited, and the time of deposition was 10 min (10 min-sample) and 30 min (30 min-sample) for per layer. The deposition process lasted 4 h. The first layer of deposition was Ni. The polished stainless steel sheet was used as the substrate to produce the individual pure Cu and Ni by electrodeposition under the same conditions and then removed. The prepared samples with a thickness of 500 μm used for baseline study.

Dog-bone shape tensile specimens of pure Cu, Ni, and Cu/Ni multilayered composites were fabricated by wire-electrode cutting. The samples are 15 mm in gage length and 5 mm in width and their surfaces were polished to mirror finish. Uniaxial tensile tests were performed using a SHIMADZU Universal Tester, with different strain rates ($\dot{\varepsilon}$) ranging from 5×10^{-5} to 5×10^{-2} s^{-1} at room temperature. Jump tests were carried out over the strain rate ranging from 5×10^{-5} to 2×10^{-2} s^{-1} and the loading-unloading-reloading (LUR) tests were conducted at the strain rates 5×10^{-4} s^{-1} and 5×10^{-2} s^{-1}. To ensure the repeatability of the stress-strain curves, at least three tests were carried out under each testing condition. The cross-sectional microstructure of the Cu/Ni multilayered composites with the deposition time of 10 min and 30 min per layer were performed on an optic microscope (OM, Leica DM 5000). The microstructure and grain orientation of Cu and Ni layers were observed on a field emission scanning electron microscope (FE-SEM, NOVA Nano SEM 450) equipped with an electron backscatter diffraction (EBSD) detector for image acquisition. The grain size and distribution of the Ni and Cu layers were estimated using EBSD images using the software (Channel-5), which was used to measure the grain area and subsequently estimate the grain size.

36.3 Results

The OM cross-sectional micrograph of the deposited Cu/Ni multilayered composites for 10 min-samples and 30 min-samples are shown in Fig. 36.1a and b, respectively. The Cu and Ni layers for the 10 min-sample have a mean thickness of 5.34 μm and 6.15 μm, while for the 30 min-sample they are 13.11 μm and 12.84 μm, respectively. The total thickness is about 137 μm for 10 min-sample and 104 μm for 30 min-sample. The thickness ratio of Cu and Ni layers is 0.87 for 10 min-sample and 1.02 for 30 min- sample. The results of EBSD of Cu and Ni layers are shown in Fig. 36.1c and d, respectively. Figure 36.1e and f present distribution of the grain size of Cu and Ni layers, which exhibit that the average grain size is 1.26 μm and 1.47 μm, respectively.

Figures 36.2a and b display a series of typical uniaxial tensile engineering stress-strain responses of the 10 min-samples and the 30 min-samples at $\dot{\varepsilon}$ = 5×10^{-5}–5×10^{-2} s^{-1}, respectively. As the strain rate ($\dot{\varepsilon}$) increased from 5×10^{-5} to 5×10^{-2} s^{-1}, the yield strength and ultimate tensile strength for both the 10 min-sample and the 30 min-sample increased gradually, while the uniform elongation also dramatically increased, as shown in Figs. 36.2c and d. As can be also seen from Figs. 36.2c and d, the yield strength for the 10 min-sample is higher than that for the 30 min-sample, and the ultimate tensile strength follows the same trend. However, the uniform elongation for the 10 min-sample is only slightly lower than that of the 30 min-sample.

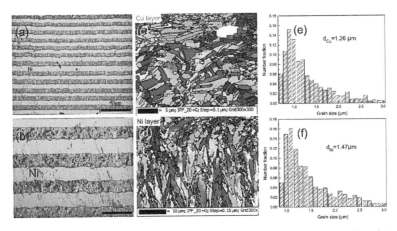

Figure 36.1 The cross-sectional characterization of microstructures of Cu/Ni multilayered composites. OM in (a) 10 min-sample and (b) 30 min-sample; Results of the EBSD of (c) Cu layer and (d) Ni layer; Statistic distribution of grain size of (e) Cu and (f) Ni layers.

In addition, the normalized strain hardening rate ($\Theta = (1/\sigma)(\partial\sigma/\partial\varepsilon)_{\dot{\varepsilon}}$, where σ is the true stress and the ε is the true strain) decreased quickly with increasing ε_t at different strain rates for all samples (Figs. 36.2e and 36.2f). It is clear that Θ of the 10 min-sample increased with increasing $\dot{\varepsilon}$ at the same strain, especially at $\varepsilon_t > 0.02$, so does the that Θ of the 30 min-sample. Furthermore, the Θ of the 30 min-sample is slightly higher than the Θ of the 10 min-sample at the same strain and strain rate, which is why the 10 min-sample has slightly lower ductility. The necking will appear at $\Theta \leq 1$. In other words, the onset of necking for the 10 min-sample is corresponding to $\varepsilon_t = 0.192$ at $\dot{\varepsilon} = 5 \times 10^{-5}$, and $\varepsilon_t = 0.263$ at $\dot{\varepsilon} = 5 \times 10^{-2} s^{-1}$; and for the 30 min-sample is in accordance with $\varepsilon_t = 0.177$ at $\dot{\varepsilon} = 5 \times 10^{-5}$, and $\varepsilon_t = 0.297$ at $\dot{\varepsilon} = 5 \times 10^{-2} s^{-1}$. It is generally known that the strain hardening ability is determined by the competition between the dynamic recovery and storage of dislocations (35), and is important for improving the uniform elongation (namely tensile deformation capacity). It is evident from (e) and (f) in Fig. 36.2 that the work hardening rate rises with increasing strain rate for both the 10 min-sample and the 30 min-sample. This is the best way to delay the onset of necking for the Cu/Ni multilayered composites.

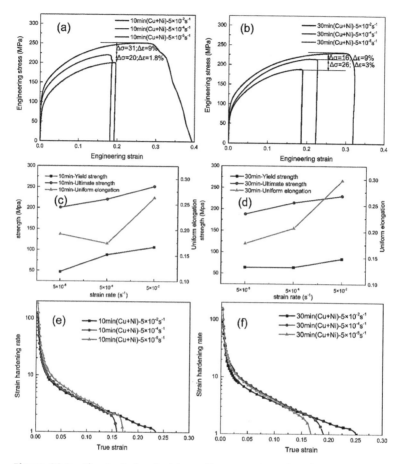

Figure 36.2 The typical uniaxial tensile properties of Cu/Ni multilayered composites: engineering stress-strain curves of (a) 10 min-sample and (b) 30 min-sample, yield strength, ultimate tensile strength and uniform elongation vs. strain rate of (c) 10 min-sample and (d) 30 min-sample and the normalized strain hardening rate dependence on true strain of (e) 10 min-sample and (f) 30 min-sample.

It is well recognized that the mechanical properties of multilayered composites are greatly influenced by the boundaries between Cu/Ni layers. In order to investigate the effect of strain rate on mechanical properties of the layered structures with a high density of boundaries, we conducted the tensile tests of pure Cu and Ni under different strain rates (Fig. 36.3a), respectively. It can be seen that the ultimate tensile strength increments of pure Ni is

87 MPa but no uniform elongation increment with increasing strain rate from 5×10^{-4} to 5×10^{-2} s^{-1}; while for pure Cu, the ultimate tensile strength is lower but the uniform elongation increased with increasing $\dot{\varepsilon}$. The ultimate tensile strength and the uniform elongation increments are 23 MPa and 4%, respectively.

To investigate the effect of Cu/Ni boundaries on mechanical properties, Zhang et al. (33) assumed that there is a stress gradient near the boundaries. According to the ultimate tensile strength of pure Cu and Ni, the schematic illustration of stress gradient effect of Cu/Ni multilayered composites are shown in Fig. 36.3b. We can see that the stress gradient near Cu/Ni layer boundaries increases with increasing strain rate. The existence of stress gradient will lead to the plastic strain gradient near the Cu/Ni boundaries, which needs to be accommodated by the geometrically necessary dislocations, which in turn produces HDI-stress (36, 37).

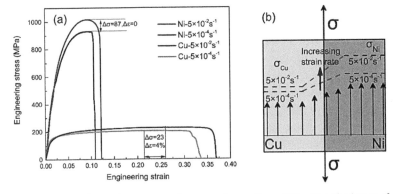

Figure 36.3 (a) Tensile stress-strain curve of pure Cu and Ni with a thickness of about 500 μm prepared at the same conditions as the multilayered composites at strain rates at 5×10^{-4} s^{-1} and 5×10^{-2} s^{-1}. (b) Schematic illustration of stress gradient effect of Cu/Ni multilayered composites.

The HDI-stress, a long-range stress caused by geometrically necessary dislocations (37) can simultaneously increase strength as well as retain uniform elongation (38). The Cu/Ni multilayered composite is a type of heterostructured material according to the definition by Wu and Zhu (37), and for heterostructured materials, the strengthening of HDI-stress can be much higher than that of the statistically stored dislocations (3, 37).

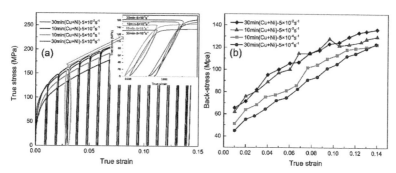

Figure 36.4 Bauschinger effect and HDI-stress of Cu/Ni multilayered composites. (a) LUR stress-strain curves of 10 min-sample and 30 min-sample at the strain rates 5×10^{-4} s^{-1} and 5×10^{-2} s^{-1} and the magnified view of third hysteresis loop (inset). (b) The HDI-stress dependence on the true strain of Cu/Ni multilayered composites. The quick increase in HDI-stress with increasing strain indicates a strong HDI work hardening.

We conducted loading-unloading-reloading (LUR) testing of Cu/Ni multilayered composites at the strain rates 5×10^{-4} s^{-1} and 5×10^{-2} s^{-1} (Fig. 36.4a) to measure the HDI-stress using a procedure developed by Yang et al. (39). Also, we investigated the HDI-stress dependence on strain rate and the relationship between the HDI-stress and the flow stress. Interestingly, it is evident from the inset in Fig. 36.4a that the Cu/Ni multilayered composites display the strong Bauschinger effect. Stronger Bauschinger effect is linked to higher HDI-stress (40). The HDI-stress dependence on the true strain of Cu/Ni multilayered composites at the strain rates 5×10^{-4} s^{-1} and 5×10^{-2} s^{-1} is shown in Fig. 36.4b. It can be seen that the HDI-stress of Cu/Ni multilayered composites increased with increasing true strain from 1% to 14% and strain rate.

In addition to HDI-stress, another mechanical property that might be related to the boundary of Cu/Ni layers is the strain rate sensitivity (m). To measure the m on strength and uniform elongation of the Cu/Ni multilayered composites, we carried out strain rate jump tests on pure Cu, Ni and Cu/Ni multilayered composites over the strain rate ranging from 5×10^{-5} to 2×10^{-2} s^{-1}. Figure 36.5a shows the jump tests of Cu/Ni multilayered composites. The m value for each jump test was calculated using the following formula (41)

$$m \equiv \left\{\frac{\partial \ln \sigma}{\partial \ln \dot{\varepsilon}}\right\}_{T,\varepsilon} \approx \left\{\frac{\ln(\sigma_2/\sigma_1)}{\ln(\dot{\varepsilon}_2/\dot{\varepsilon}_1)}\right\}_{T,\varepsilon}, \qquad (36.1)$$

where T represents the absolute temperature; σ_1 and σ_2 are the true stress before and after the strain rate jump immediately at the same strain and room temperature, respectively; $\dot{\varepsilon}_1$ and $\dot{\varepsilon}_2$ represent the strain rate prior to and after the strain rate-change test, respectively.

The m value of pure Cu, Ni and Cu/Ni multilayered composites at varying strain rates are shown in Fig. 36.5b. The reported m value of 21 nm/21 nm Cu/Ni multilayer thin films (34) is also shown in Fig. 36.5b. It is worth noting that for the pure Ni the m is 0.014 when the strain rate is 5×10^{-5} s^{-1} and decreases to 0.007 at 2×10^{-2} s^{-1}; the m value changed very little for pure Cu with an average value of m 0.009. However, the m value increased from 0.004 to 0.017 when the strain rate increased from 5×10^{-5} to 2×10^{-2} s^{-1} for the Cu/Ni multilayered composites. In other words, the strain sensitivity of the Cu/Ni layered composites behaves very differently from that of its pure Cu and Ni constituents and increased quickly with increasing strain rate.

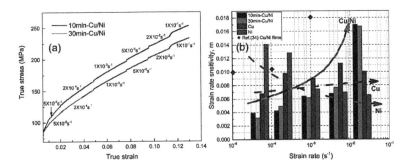

Figure 36.5 (a) The tensile true stress-strain curves obtained in strain rate jump tests with strain rate ranging from 5×10^{-5} to 2×10^{-2} s^{-1} for the Cu/Ni multilayered composites at room temperature. (b) Strain rate sensitivity vs strain rate plot of pure Cu, pure Ni, Cu/Ni multilayered composites in our work and 21 nm/21 nm Cu/Ni multilayer thin films from the Ref. (34).

36.4 Discussions

In terms of strength of Cu/Ni multilayered composites, according to the formula $\sigma \propto \dot{\varepsilon}^m$ (29) (σ, $\dot{\varepsilon}$, and m represent flow stress, strain rate and strain rate sensitivity, respectively), the flow stress increases with increasing $\dot{\varepsilon}$ and m. And the strain rate sensitivity

(m) of both the 10 min-sample and the 30 min-sample are shown in Fig. 36.5b, where m value rises with increasing $\dot{\varepsilon}$. On the other hand, according to the relation between the strain rate and the dislocation motion $\dot{\varepsilon} = b\rho\bar{\upsilon}$ (41) (b represents the magnitude of Burgers vector; ρ is the density of mobile dislocations and υ is the average dislocation velocity related to resolved shear stress τ) and the fact that when some dislocations begin to move, the stress needed to move the dislocations can drop and the average dislocation velocity can decrease (41), the density of mobile dislocations increased with increasing $\dot{\varepsilon}$, which can generate a great number of dislocation interactions and pile-up and lead to a greater strain hardening. This is beneficial to enhancing the ductility.

In addition, as can be seen from Fig. 36.4b, the HDI-stress of Cu/Ni multilayered composites increased with increasing strain rate, which is mostly due to the higher geometrically necessary dislocations near the Cu/Ni layer boundaries at higher strain rate associated with HDI-stress. Furthermore, HDI-stress is related to plastic strain gradient, and the plastic strain is accommodated by geometrically necessary dislocations (3, 42). Consequently, the pile-up of a great amount of geometrically necessary dislocations creates strain gradient as well as stress gradient. This is consistent with the assumption of Zhang et al. (33). This stress gradient will be increased with increasing $\dot{\varepsilon}$ (33), which is also evidenced in Fig. 36.3. In other words, the higher stress gradient means the higher pile-up of geometrically necessary dislocations of Cu/Ni multilayered composites at higher strain rate, which produces higher HDI-stress. And higher HDI-stress will enhance the strength of Cu/Ni multilayered composites. Also, a larger strain gradient of Cu/Ni layer boundary at higher strain rate can promote the HDI work hardening during the tensile tests, which consequently helps with achieving a higher uniform elongation (ductility).

Furthermore, the strain rate sensitivity (m) could have an effect on uniform elongation. Based on Hart's criterion for necking (43):

$$\frac{1}{\sigma}\left(\frac{\partial \sigma}{\partial \varepsilon}\right)_{\dot{\varepsilon}} - 1 + m \leq 0, \tag{36.2}$$

where the first term and m represent the normalized strain hardening rate and the strain rate sensitivity, respectively. From Hart's criterion, for rate-sensitive materials, the m is an important factor to effectively delay the onset of necking and to improve tensile ductility (41). For grain boundary deformation mechanisms including grain boundary sliding and Coble creep, m could be 0.5–1 (44), i.e. much higher than what we observed. Therefore, over the strain rate ranging from 5×10^{-5} to 2×10^{-2} s^{-1}, the deformation mechanism of present pure Cu, Ni, and Cu/Ni multilayered composites is based on dislocation-based plasticity.

Some investigations of strain rate sensitivity for multilayered films have been reported. Cao et al. (45, 46) found the m value for Ti/Ni multilayered films remains largely constant with increasing strain rate. Wang et al. (47) believed that the m value decreased with increasing strain rate for the Cu/Ta multilayered films because a faster loading strain rates (LSR) enabled more boundary activities per unit time and easier boundary transmission of dislocations at the Cu/Ta incoherent boundaries, resulting in less pile-up at the boundary and lower HDI-stress. However, it is evident from Fig. 36.4b that the m value for Cu/Ni multilayered composites increased with increasing strain rate, which is similar to the trend of 21 nm/21 nm Cu/Ni multilayer thin films (34). This is not observed in pure Cu, Ni, the electrodeposition NC Ni (25) and other multilayered films (45–47). Carpenter et al. (34) thought the increase in m implies that the strong evolution of dislocation interaction is not observed in nanocrystalline metals, even including nano-twinned grains.

Does the m value of present Cu/Ni multilayered composites, whose grain size is different from thin films, increase for the similar reason? For the dislocation-based mechanism, the m should increase when the dislocation density is increased in a metal with a given grain size. But the increase of m with dislocation density is relatively small for large grain size (22, 48). It is known that in fcc metals, the dynamic recovery of dislocations could be restricted with increasing the strain rate (31). This means the higher storage of dislocations at higher strain rate. However, it is puzzling that the m value increased for Cu/Ni multilayered composites, which is opposite to pure Cu and Ni.

It is well known that the strain rate sensitivity (m) is inversely proportional to the activation volume (V^*). As discussed above, there is a great amount of geometrically necessary dislocations with very small spacing near the Cu/Ni layer boundaries, which can act as concentrated obstacles to the dislocation movement (49). This leads to a small activation volume near the Cu/Ni layer boundaries. Besides, the geometrically necessary dislocations increase with increasing strain rate and eventually increases strain rate sensitivity. In other words, the effect of Cu/Ni layer boundaries on m value of Cu/Ni multilayered composites is relatively small at the low strain rate and relatively large at the high strain rate. Because the m increases with increasing $\dot{\varepsilon}$, it can help with improving uniform elongation as shown in Fig. 36.2a and b.

36.5 Conclusions

The Cu/Ni multilayered composites synthesized by electrodeposition displayed simultaneously increased in strength and ductility with increasing tensile rates from 5×10^{-5} to 5×10^{-2} s^{-1} for both the 10 min-samples and the 30 min-samples. The increase of strength is mainly due to the HDI-stress of Cu/Ni multilayered composites caused by the Cu/Ni layer boundary, which increases with increasing strain rate. And the increase of uniform elongation can be attributed to the increase of strain rate sensitivity and the HDI work hardening by the Cu/Ni layer boundaries, which help delay the onset of necking during tensile tests.

Acknowledgments

The authors would like to acknowledge financial support by the National Natural Science Foundation of China (NSFC) under Grant No. 51561015, No. 51664033 and No. 51501078, and the introduction of talent fund project of Kunming University of Science and Technology (KKSY201407100). YTZ acknowledges the support of the US Army Research Office (W911NF-17-1-0350).

References

1. Wang J, Misra A. An overview of interface-dominated deformation mechanisms in metallic multilayers. *Current Opinion in Solid State and Materials Science*. 2011;15(1):20–28.

2. Carreño F, Chao J, Pozuelo M, Ruano OA. Microstructure and fracture properties of an ultrahigh carbon steel–mild steel laminated composite. *Scripta Materialia*. 2003;48(8):1135–1140.
3. Wu X, Yang M, Yuan F, Wu G, Wei Y, Huang X, et al. Heterogeneous lamella structure unites ultrafine-grain strength with coarse-grain ductility. *Proceedings of the National Academy of Sciences of the United States of America*. 2015;112(47):14501–14505.
4. Zhang B, Kou Y, Xia YY, Zhang X. Modulation of strength and plasticity of multiscale Ni/Cu laminated composites. *Materials Science and Engineering: A*. 2015;636:216–220.
5. Ma XL, Huang CX, Xu WZ, Zhou H, Wu XL, Zhu YT. Strain hardening and ductility in a coarse-grain/nanostructure laminate material. *Scripta Materialia*. 2015;103:57–60.
6. Misra A, Hirth JP, Hoagland RG. Length-scale-dependent deformation mechanisms in incoherent metallic multilayered composites. *Acta Materialia*. 2005;53(18):4817–4824.
7. Li YP, Zhang GP, Wang W, Tan J, Zhu SJ. On interface strengthening ability in metallic multilayers. *Scripta Materialia*. 2007;57(2):117–120.
8. Inoue J, Nambu S, Ishimoto Y, Koseki T. Fracture elongation of brittle/ductile multilayered steel composites with a strong interface. *Scripta Materialia*. 2008;59(10):1055–1058.
9. Koseki T, Inoue J, Nambu S. Development of multilayer steels for improved combinations of high strength and high ductility. *Materials Transactions*. 2014;55(2):227–237.
10. Liu HS, Zhang B, Zhang GP. Delaying premature local necking of high-strength Cu: a potential way to enhance plasticity. *Scripta Materialia*. 2011;64(1):13–16.
11. Ma X, Huang C, Moering J, Ruppert M, Höppel HW, Göken M, et al. Mechanical properties of copper/bronze laminates: role of interfaces. *Acta Materialia*. 2016;116:43–52.
12. Seok M-Y, Lee J-A, Lee D-H, Ramamurty U, Nambu S, Koseki T, et al. Decoupling the contributions of constituent layers to the strength and ductility of a multi-layered steel. *Acta Materialia*. 2016;121:164–172.
13. Huang HB, Spaepen F. Tensile testing of free-standing Cu, Ag and Al thin films and Ag/Cu multilayers. *Acta Materialia*. 2000;48(12):3261–3269.
14. Carpenter JS, Vogel SC, LeDonne JE, Hammon DL, Beyerlein IJ, Mara NA. Bulk texture evolution of Cu–Nb nanolamellar composites during accumulative roll bonding. *Acta Materialia*. 2012;60(4):1576–1586.

15. Tayyebi M, Eghbali B. Study on the microstructure and mechanical properties of multilayer Cu/Ni composite processed by accumulative roll bonding. *Materials Science and Engineering: A*. 2013;559:759-764.

16. Ren F, Zhao S, Li W, Tian B, Yin L, Volinsky AA. Theoretical explanation of Ag/Cu and Cu/Ni nanoscale multilayers softening. *Materials Letters*. 2011;65(1):119-121.

17. Qian Y, Tan J, Liu Q, Yu H, Xing R, Yang H. Preparation, microstructure and sliding-wear characteristics of brush plated copper–nickel multilayer films. *Surface and Coatings Technology*. 2011;205(15):3909-3915.

18. Ghosh SK, Limaye PK, Bhattacharya S, Soni NL, Grover AK. Effect of Ni sublayer thickness on sliding wear characteristics of electrodeposited Ni/Cu multilayer coatings. *Surface and Coatings Technology*. 2007;201(16-17):7441-7448.

19. Carpenter JS, Misra A, Anderson PM. Achieving maximum hardness in semi-coherent multilayer thin films with unequal layer thickness. *Acta Materialia*. 2012;60(6-7):2625-2636.

20. Mara NA, Bhattacharyya D, Hoagland RG, Misra A. Tensile behavior of 40 nm Cu/Nb nanoscale multilayers. *Scripta Materialia*. 2008;58(10):874-877.

21. Li YP, Zhang GP. On plasticity and fracture of nanostructured Cu/X (X=Au, Cr) multilayers: The effects of length scale and interface/boundary. *Acta Materialia*. 2010;58(11):3877-3887.

22. Cheng S, Ma E, Wang Y, Kecskes L, Youssef K, Koch C, et al. Tensile properties of in situ consolidated nanocrystalline Cu. *Acta Materialia*. 2005;53(5):1521-1533.

23. Gu C, Lian J, Jiang Z, Jiang Q. Enhanced tensile ductility in an electrodeposited nanocrystalline Ni. *Scripta Materialia*. 2006;54(4):579-584.

24. Chen J, Lu L, Lu K. Hardness and strain rate sensitivity of nanocrystalline Cu. *Scripta Materialia*. 2006;54(11):1913-1918.

25. Guduru RK, Murty KL, Youssef KM, Scattergood RO, Koch CC. Mechanical behavior of nanocrystalline copper. *Materials Science and Engineering: A*. 2007;463(1-2):14-21.

26. Shen X, Lian J, Jiang Z, Jiang Q. High strength and high ductility of electrodeposited nanocrystalline Ni with a broad grain size distribution. *Materials Science and Engineering: A*. 2008;487(1-2):410-416.

27. Kvačkaj T, Kováčová A, Kvačkaj M, Pokorný I, Kočiško R, Donič T. Influence of strain rate on ultimate tensile stress of coarse-grained and ultrafine-grained copper. *Materials Letters*. 2010;64(21):2344-2346.

28. Dalla Torre F, Van Swygenhoven H, Victoria M. Nanocrystalline electrodeposited Ni: microstructure and tensile properties. *Acta Materialia*. 2002;50(15):3957–3970.
29. Lu L, Li SX, Lu K. An abnormal strain rate effect on tensile behavior in nanocrystalline copper. *Scripta Materialia*. 2001;45(10):1163–1169.
30. Schwaiger R, Moser B, Dao M, Chollacoop N, Suresh S. Some critical experiments on the strain-rate sensitivity of nanocrystalline nickel. *Acta Materialia*. 2003;51(17):5159–5172.
31. Suo T, Xie K, Li YL, Zhao F, Deng Q. Tensile ductility of ultra-fine grained copper at high strain rate. *Advanced Materials Research*. 2010;160–162:260–266.
32. Zhang H, Jiang Z, Lian J, Jiang Q. Strain rate dependence of tensile ductility in an electrodeposited Cu with ultrafine grain size. *Materials Science and Engineering: A*. 2008;479(1–2):136–141.
33. Tan HF, Zhang B, Luo XM, Sun XD, Zhang GP. Strain rate dependent tensile plasticity of ultrafine-grained Cu/Ni laminated composites. *Materials Science and Engineering: A*. 2014;609:318–322.
34. Carpenter JS, Misra A, Uchic MD, Anderson PM. Strain rate sensitivity and activation volume of Cu/Ni metallic multilayer thin films measured via micropillar compression. *Applied Physics Letters*. 2012;101(5):051901.
35. Estrin Y, Mecking H. A unified phenomenological description of work hardening and creep based on one-parameter models. *Acta Metallurgica*. 1984;32(1):57–70.
36. Chakravarthy SS, Curtin WA. Stress-gradient plasticity. *Proceedings of the National Academy of Sciences of the United States of America*. 2011;108(38):15716–15720.
37. Wu X, Zhu Y. Heterogeneous materials: a new class of materials with unprecedented mechanical properties. *Materials Research Letters*. 2017;5(8):527–532.
38. Bian X, Yuan F, Wu X, Zhu Y. The evolution of strain gradient and anisotropy in gradient-structured metal. *Metallurgical and Materials Transactions A*. 2017;48(9):3951–3960.
39. Yang M, Pan Y, Yuan F, Zhu Y, Wu X. Back-stress strengthening and strain hardening in gradient structure. *Materials Research Letters*. 2016;4(3):145–151.
40. Liu X, Yuan F, Zhu Y, Wu X. Extraordinary Bauschinger effect in gradient structured copper. *Scripta Materialia*. 2018;150:57–60.
41. Dieter GE. *Mechanical Metallurgy*. McGraw-Hill. 1986:pp.200,95–301.

42. H. Gao YH, W. D. Nix, J. W. Hutchinson. Mechanism-based strain gradient plasticity—I. Theory. *Journal of the Mechanics and Physics of Solids*. 1999;47:1239–1263.

43. Hart EW. Theory of the tensile test. *Acta Metallurgica*. 1967;15(2):351–355.

44. Mohamed FA, Li Y. Creep and superplasticity in nanocrystalline materials: current understanding and future prospects. *Materials Science and Engineering: A*. 2001;298(1-2):1–15.

45. Shi J, Cao ZH, Zheng JG. Size dependent strain rate sensitivity transition from positive to negative in Ti/Ni multilayers. *Materials Science and Engineering: A*. 2017;680:210–213.

46. Shi J, Wei MZ, Ma YJ, Xu LJ, Cao ZH, Meng XK. Length scale dependent alloying and strain-rate sensitivity of Ti/Ni multilayers. *Materials Science and Engineering: A*. 2015;648:31–36.

47. Zhou Q, Li JJ, Wang F, Huang P, Xu KW, Lu TJ. Strain rate sensitivity of Cu/Ta multilayered films: Comparison between grain boundary and heterophase interface. *Scripta Materialia*. 2016;111:123–126.

48. Wei Q, Cheng S, Ramesh KT, Ma E. Effect of nanocrystalline and ultrafine grain sizes on the strain rate sensitivity and activation volume: fcc versus bcc metals. *Materials Science and Engineering: A*. 2004;381(1-2):71–79.

49. Wang YM, Hamza AV, Ma E. Activation volume and density of mobile dislocations in plastically deforming nanocrystalline Ni. *Applied Physics Letters*. 2005;86(24):241917.

Part VI
Dual-Phase Structure

Chapter 37

Simultaneous Improvement of Tensile Strength and Ductility in Micro-Duplex Structure Consisting of Austenite and Ferrite

L. Chen, F. P. Yuan, P. Jiang, J. J. Xie, and X. L. Wu

State Key Laboratory of Nonlinear Mechanics, Institute of Mechanics, Chinese Academy of Science, Beijing 100190, China
xlwu@imech.ac.cn

A micro-duplex structure consisting of austenite and ferrite was produced by equal channel angular pressing and subsequent intercritical annealing. As compared to coarse-grained (CG) counterpart, the strength and ductility of micro-duplex samples are enhanced simultaneously due to smaller grain sizes in both phases and more uniformly distributed austenite in ferrite matrix. The average yield stress and uniform elongation are increased to 540 MPa and 0.3 as compared to 403 MPa and 0.26 of its CG counterpart. The

---

Reprinted from *Mater. Sci. Eng. A*, **618**, 563–571, 2014.

Heterostructured Materials: Novel Materials with Unprecedented Mechanical Properties
Edited by Xiaolei Wu and Yuntian Zhu
Text Copyright © 2014 Elsevier B.V.
Layout Copyright © 2022 Jenny Stanford Publishing Pte. Ltd.
ISBN 978-981-4877-10-7 (Hardcover), 978-1-003-15307-8 (eBook)
www.jennystanford.com

Hall–Petch coefficients of austenite and ferrite grain boundaries were quantitatively measured as 224.9 and 428.9 MPa·µm$^{1/2}$. In addition, a Hall–Petch type coefficient was used to describe the ability of phase boundary to obstruct dislocation motion, which was measured as 309.7 MPa·µm$^{1/2}$. Furthermore, the surface-to-volume ratio of phase boundary in micro-duplex structure was estimated to be 1.17×10^6 m^{-1}, which is increased by an order of magnitude as compared to 1.2×10^5 m^{-1} of its CG counterpart. Based on the strain gradient theory, a model was proposed to describe the effect of surface-to-volume ratio of phase boundary on strain hardening rate, which shows a good agreement with the experimental results.

37.1 Introduction

Extensive investigations over the past few decades have demonstrated that the nanostructured (NS) materials have poor ductility although the strength is significantly increased as compared to their CG counterparts [1–4]. The low tensile ductility of NS materials is mainly attributed to their low strain hardening ability because the conventional dislocation mechanism is suppressed by the extremely small grains [1, 5]. However, some researches during past decade also exhibit that the well-designed microstructures could achieve high strain hardening ability, including the introduction of a gradient or bimodal grain size distribution [6, 7], the preexisting nano-scale growth twins [2], dispersion of nano-precipitates [8, 9], transformation and twinning induced plasticity [10, 11], and a mixture of two or multiple phases with varying size scales and properties [12].

Many natural and man-made materials consist of dual or multiple phases, which make them exhibit much better strength-ductility synergy than those single phase materials [13]. To elucidate the relationship between microscopic mechanical behaviors and bulk mechanical properties of dual and even multiple phase materials, more and more advanced in-situ experiments and computer simulations were conducted over past two decades [14–18]. In terms of micro-mechanics, three conclusions could be addressed from those investigations. Firstly, the soft phase always tends to yield earlier than the hard one, leading to an inhomogeneous distribution of plastic strain across phase boundary even under uniaxial tensile test. Thus, secondly, the plastic strain gradient occurs between two dissimilar phases when plastic deformation happens. Thirdly, with different plastic strain accommodated by the hard and soft phases, the applied load born by hard phase is greater than that by the soft, resulting in inhomogeneous stress partitioning between two phases.

The above plastic deformation features are believed to be responsible for the optimized mechanical properties of dual and multiple phase materials. During plastic deformation, the hard phase is relatively elastically deformed and bears most of the applied load, while the soft one provides strain hardening ability and accommodates most of the plastic deformation. Thus high strength and good ductility could be achieved simultaneously. Moreover, the plastic strain gradient across phase boundary requires the generation of geometrically necessary dislocations (GNDs), which would lead to an extra strength over the rule-of-mixture (ROM) prediction. In addition, with grain size decreasing to nanometer range, the lattice dislocations could glide on the phase boundaries and penetrate them into adjacent phases [19–22]. The phase boundary therefore play a similar role for NS dual phase materials as the twin boundary does for nano-twinned metals, thus reducing the strain localization and enhancing the interaction between dislocations during plastic deformation [2].

Since the interaction between two phases results from the strength difference in essence, it is reasonable to deduce that those features of inhomogeneous stress and strain distribution also occur in NS dual phase materials if strength difference still exists. This interaction has a potential to improve the dislocation storage ability due to the generation of GNDs, especially considering the increased surface-to-volume ratio of phase boundary. In the case of Cu/M (M = Nb, Ta, Fe, etc.) NS composites, the bulk strengths tested by experiments are exceeding of the ROM predictions [23]. To describe this strengthening behavior qualitatively, an additional term of yield stress was introduced by considering the interplay between the phases [24, 25].

Although promising in mechanical properties [26–30], the contributions of individual phase to the overall strength and strain hardening are difficult to analyze quantitatively, which are important in establishing the microscopic mechanical models and tailoring the macroscopic mechanical properties. These naturally raise two fundamental questions: How strong does the phase boundary impede the dislocation motion as compared to the grain boundaries of individual phases? And how much contribution does the phase boundary have to the overall strain hardening ability of duplex microstructure?

In current study, the influences of phase boundary on strength and strain hardening rate are investigated. The micro-duplex samples were fabricated by equal channel angular pressing (ECAP) and subsequent intercritical annealing. Then the strengthening abilities of grain and phase boundaries were quantitatively analyzed. Moreover, the strain gradient theory was used to describe the effect of phase boundary on strain hardening behavior of micro-duplex structure.

37.2 Experimental Procedures

The UNS S32304 duplex stainless steel (DSS) was used in this investigation, with chemical compositions (wt.%) of 0.02C, 0.5Si, 1.2Mn, 23.5Cr, 4.0Ni, 0.4Mo, 0.13N, 0.024P, 0.002S, and balanced Fe.

The as-received billets of 10 mm in diameter were annealed at 1373 K for 2 h, followed by oil quenching in vacuum of about 10^{-4} Pa, in order to obtain the CG samples with nearly 50:50 phase balance between austenite and ferrite.

To fabricate the micro-duplex samples, the as-received billets of 10 mm in diameter were firstly solutionized at 1623 K for 2 hours to form single ferrite microstructure (in vacuum of about 10^{-4} Pa and followed by oil quenching). ECAP technique was then used to refine the grain size of ferrite *via* a split die with two channels intersecting at inner angle of 90° and outer angle of 30° [31, 32]. The ECAP was conducted at ambient temperature for 1 pass since further pressing is hugely difficult. At last, the ECAPed samples were intercritically annealed at 1173 K for different time to generate micro-duplex structure (in vacuum of about 10^{-4} Pa and followed by water quenching).

The dog-bone shaped tensile specimens were designed with rectangular cross-section of 2 × 1 mm^2 and gauge length of 8 mm and machined by electrical discharging along extrusion direction on Y plane [32, 33]. Tensile tests were conducted using an Instron

8871 test machine at room temperature with strain rate of 5 × 10^{-4} s^{-1}. At least three times of tensile testing were conducted for each microstructure.

An Olympus PMG3 optical microscope (OM) was used to examine the microstructures and measure the phase fractions. The chemical etchant used for OM observation consists of 30 g $K_3Fe(CN)_6$, 10 g KOH and 100 ml H_2O. Remaining the solution temperature at 353 K, the OM sample was immersed into it for 3 min.

X-ray diffraction (XRD) was taken to investigate the effect of tensile deformation on volume fraction of individual phases by using Rigaku D/max 2400 X-ray diffractometer with Cu K_α radiation, and a step size of 0.02°.

The micro-duplex structure before and after tensile tests were investigated by electron back-scattered diffraction (EBSD) using a field emission gun scanning electron microscope. Specimens for EBSD investigation were prepared on Y plane by standard mechanical grinding and polishing procedures. In the final step, samples were electro-polished using a solution of 95% ethyl alcohol and 5% perchloric acid ($HClO_4$) at 253 K with voltage of 38 V. The EBSD scans were carried out at 15 kV in the center of the gauge section at a step size of 100 nm. The raw data were post-analyzed using TSL OIM software, and the average misorientation of a given point relative to its neighbors was calculated using an orientation gradient kernel method. In this study, the kernel average misorientation (KAM) was calculated up to the second neighbor shell with a maximum misorientation angle of 2°.

37.3 Results

37.3.1 Mechanical Property

Uniaxial tensile tests were conducted to investigate the mechanical properties of different dual phase microstructures. The engineering stress–strain curves are shown in Fig. 37.1a. The yield stress of micro-duplex samples annealed for 10 min at 1173 K is increased to 540 MPa, as compared to 403 MPa of CG DSS. More importantly, the uniform elongation is increased simultaneously, which is 0.3 as compared to 0.26 of CG DSS.

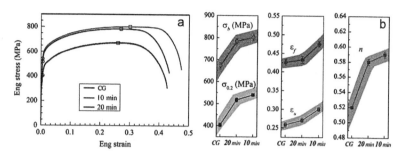

Figure 37.1 (a) Tensile engineering stress vs. strain curves; (b) Simultaneous increase of mechanical properties such as yield stress ($\sigma_{0.2}$), ultimate tensile strength (σ_b), uniform elongation (ε_u), elongation to failure (ε_f) and strain hardening exponent (n).

The corresponding mechanical properties, including the yield stress, ultimate tensile stress, uniform elongation, elongation to failure and strain hardening exponent are shown in Fig. 37.1b. The strain hardening exponents were calculated using the Ludwik equation [34]. The micro-duplex sample annealed for 20 min appear small weak mechanical properties as compared to that annealed for 10 min, manifesting better mechanical performance for finer-grained dual phase microstructure.

37.3.2 Microstructure Observation

The mechanical properties of DSS depend closely on the volume fraction and phase morphology of austenite, as well as the grain sizes of austenite and ferrite. The dual phase microstructure could be adjusted effectively by the combination of plastic deformation and intercritical annealing. For example, at relatively high annealing temperature, the deformed nonequilibrium ferrite would recrystallize first and followed by precipitation of austenite along grain boundaries of ferrite. But when annealed at relatively low temperature, the austenite would precipitate first from deformed ferrite along its (sub-) grain boundaries, and therefore suppress the recrystallization of ferrite. These two different thermodynamic processes lead to different dual phase microstructures, especially in phase morphology and volume fraction of austenite [35]. Therefore, the ECAPed samples with single ferrite phase were annealed at different temperatures to obtain a dual phase microstructure with finer grains and more uniform morphology to achieve better mechanical properties.

Figure 37.2a shows the microstructure of CG DSS, in which the austenite islands embedded in ferrite matrix are about 20 μm in transversal size. The ECAPed samples were intercritically annealed at 1173 K and 1273 K for same 30 min. Figure 37.2b shows the microstructure annealed at 1173 K, in which the austenite islands precipitate along the previous shear flow lines produced by ECAP. When annealed at 1273 K, as shown in Fig. 37.2c, nearly equiaxed austenite islands are exhibited. Based on Figs. 37.2b and c, finer austenite and ferrite could be obtained if annealed at 1173 K. Because at this temperature, the austenite precipitate first and suppress the recrystallization of ferrite, retaining the small size of ferrite. Figure 37.2d shows the transversal size distributions of austenite and ferrite exhibited in Fig. 37.2b, and both the average sizes are typically less than 2 μm.

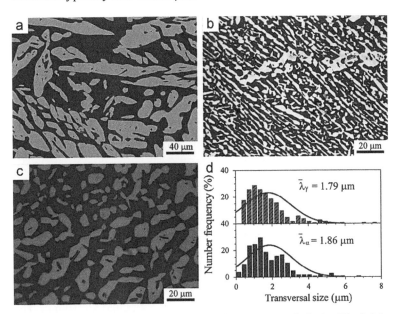

Figure 37.2 Optical micrographs of DSS with various morphologies (The bright islands are austenite). (a) Microstructure of CG; (b) Banded morphology in sample annealed at 1173 K for 30 min after ECAP; (c) Near equiaxed morphology in sample annealed at 1273 K for 30 min after ECAP; (d) Distributions of transverse spacing for austenite and ferrite in (b).

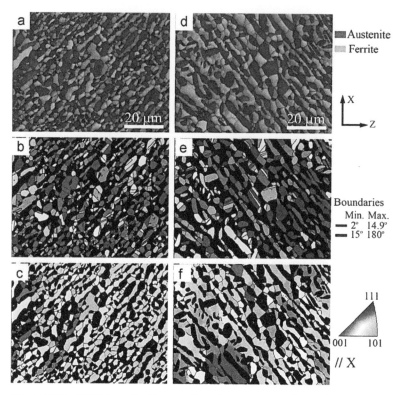

Figure 37.3 EBSD investigation on the microstructure of samples annealed for different time. For sample annealed at 1173 K for 10 min after ECAP, (a) showing the map of IQ plus phase distribution; (b) and (c) showing the IPF plus grain boundary distribution maps of austenite and ferrite, respectively; (d), (e) and (f) exhibiting corresponding maps to (a), (b) and (c) for sample annealed for 20 min.

The micro-duplex structure depends on not only the intercritical annealing temperature, but also the annealing time. Keeping the temperature at 1173 K, two groups of ECAPed samples were intercritically annealed for 10 and 20 min respectively. Figure 37.3a displays the micrograph of image quality (IQ) overlapped with inverse pole figure (IPF) for the specimen annealed for 10 min. The volume fraction of austenite is 48%. In corresponding to Fig. 37.3a, b and c are the IPF and grain boundary distribution maps of austenite and ferrite. Figure 37.3d shows the IQ micrograph overlapped with phase map for the sample annealed for 20 min. The volume fraction of austenite is 49%. Figure 37.3e and f are the corresponding IPF with grain boundary distribution maps of austenite and ferrite in Fig. 37.3d. Based on the comparison of two sets of micrographs, it could be concluded that both the austenite and ferrite are coarsened with annealing time increasing. Therefore, a short time annealing is beneficial to keep fine-grained dual phase microstructure.

The statistical distributions of grain sizes of austenite and ferrite in Fig. 37.3a and d are shown in Fig. 37.4a. The grain diameters are determined by assuming spherical grains in shape for both phases. The average grain size of austenite is smaller than that of ferrite in both annealing conditions. It can be seen in Fig. 37.4a that grain size of ferrite is far greater than that of austenite when annealed for 20 min, which may be caused by the coalescence of ferrite. For comparison, the corresponding distributions of intercept length (with 69 horizontal lines) in two phases are depicted in Fig. 37.4b. The average intercept lengths of austenite grains are also smaller than that of ferrite grains in both conditions.

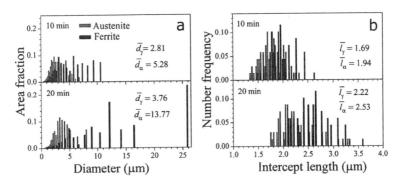

Figure 37.4 (a) Statistical distributions of grain diameter of micro-duplex sample based on EBSD investigation with an assumption of spherical grains; (b) Distributions of the intercept length of grains.

Austenite is always metastable and could transform to martensite during tensile deformation. But in DSS used in this investigation, austenite is very stable and there is no detectable martensite transformation upon tensile test. Figure 37.5 shows the XRD spectra of two micro-duplex samples before and after tensile deformation. The relative intensity of all the main characteristic peaks are nearly not changed with tensile strain. It manifests that the austenite could keep stable during tensile deformation. Our previous research found that the austenite of this DSS are relatively stable even during ECAP, and less than 5vol.% of austenite was transformed after 4 passes pressing at room temperature [30].

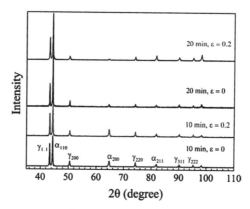

Figure 37.5 XRD spectra of two micro-duplex samples before and after tensile deformation.

Inhomogeneous plastic strain distribution would occur during tensile deformation as the austenite and ferrite have different strengths due to their different yield stresses and strain hardening rates. Thus plastic strain gradient arises across phase boundary and GNDs are generated to meet the requirement of strain gradient. The kernel average misorientation (KAM) map, calculated from local orientation gradients, are used to reflect the GNDs density. Figure 37.6a and b display the KAM maps of austenite experienced 0 and 0.2 tensile strains, and the corresponding maps of ferrite are shown in Fig. 37.6c and d. Before tensile deformation, the KAM maps of Fig. 37.6a and c indicate that the some GNDs have existed in both austenite and ferrite. This is probably produced by plastic deformation at oil quenching stage due to different thermal expansion coefficient between two phases. After 0.2 tensile strain, the KAM maps of Fig. 37.6b and d display higher values, indicating the generation of new GNDs during tensile deformation.

The change of KAM values during tensile test is further expressed by their statistical distributions. As shown in Fig. 37.7, the peak positions of austenite and ferrite move right after 0.2 tensile straining, clearly manifesting the formation of GNDs. In addition, the austenite have slightly higher peak value than that of the ferrite, indicating a little more storage of GNDs in austenite.

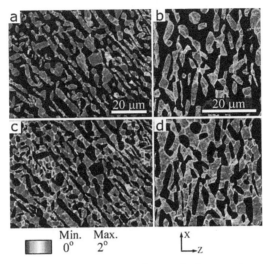

Figure 37.6 Compatible deformation of micro-duplex sample annealed for 10 min. (a) KAM maps of austenite before tensile deformation; (b) KAM maps of austenite after 0.2 tensile strain; (c) and (d) showing the corresponding maps of ferrite to (a) and (b).

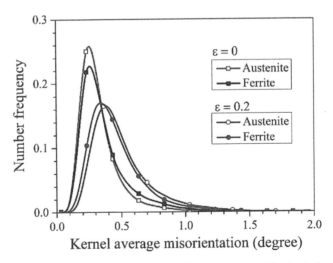

Figure 37.7 Distributions of KAM values for austenite and ferrite before and after tensile deformation in sample annealed for 10 min.

37.4 Discussion

37.4.1 Strengthening Mechanism of Dual Phase Microstructure

In micro-duplex structure, a large number of phase boundaries were introduced, and thus the mechanical properties may largely depend on the phase boundaries rather than grain boundaries. It is therefore necessary to investigate the effects of phase interaction on the mechanical behaviors of dual phase materials. Phase boundary could be a strong barrier to obstruct the dislocation motion [25]. Considering the Hall–Petch type relationship, the strengthening coefficient of phase boundary in DSS is measured quantitatively, and then the contributions of phase and grain boundaries on the overall strength are separated.

The dependence of yield stress on grain size for individual phase could be expressed by classic Hall–Petch relations:

$$\sigma_{y\gamma} = \sigma_{0\gamma} + k_\gamma d_\gamma^{-1/2}, \quad (37.1a)$$

$$\sigma_{y\alpha} = \sigma_{0\alpha} + k_\alpha d_\alpha^{-1/2}. \quad (37.1b)$$

The first terms on the right hands of Eqs. (37.1a) and (37.1b) represent the contribution of lattice friction and solid solution of chemical elements. Fan et al. [36, 37] had extended the Hall–Petch relationship to describe the size dependence of yield stress of dual phase materials. Within this method, as shown in Fig. 37.8, the dual phase microstructure with any grain size, grain shape and phase distribution could be transformed into an idealized microstructure, which consists of two single phase parts subdivided by only grain boundaries and one dual phase part including phase boundaries alone. Therefore, the yield stress of dual phase microstructure could be expressed as:

$$\sigma_{yC} = (\sigma_{0\gamma} + k_\gamma d_\gamma^{-1/2}) \cdot q_\gamma + (\sigma_{0\alpha} + k_\alpha d_\alpha^{-1/2}) \cdot q_\alpha + (\sigma_{0\alpha\gamma} + k_{\alpha\gamma} \bar{d}_{\alpha\gamma}^{-1/2}) \cdot q_{\alpha\gamma},$$

$$(37.2)$$

where q_γ, q_α and $q_{\alpha\gamma}$ are volume fractions of γ, α and α-γ parts in the idealized microstructure; $\bar{d}_{\alpha\gamma}$ is a volume-fraction-weighted grain size in α-γ part. In order to calculate q_γ, q_α, $q_{\alpha\gamma}$ and $\bar{d}_{\alpha\gamma}$, two parameters of contiguities (C_γ and C_α) and separations (S_γ and S_α) are introduced and can be measured experimentally [36, 37].

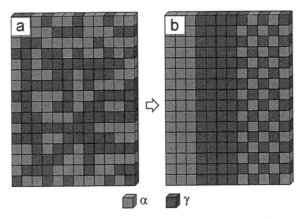

Figure 37.8 Schematic illustration of the topological transformation [36]. (a) The dual phase microstructure with randomly distributed phases; (b) The transformed body consisting of three regular parts.

$$C_\gamma = \frac{2N_L^{\gamma\gamma}}{2N_L^{\gamma\gamma} + N_L^{\alpha\gamma}}, \qquad (37.3a)$$

$$C_\alpha = \frac{2N_L^{\alpha\alpha}}{2N_L^{\alpha\alpha} + N_L^{\alpha\gamma}}, \qquad (37.3b)$$

$$S_\gamma = \frac{N_L^{\alpha\gamma}}{2N_L^{\gamma\gamma} + N_L^{\alpha\gamma}}, \qquad (37.3c)$$

$$S_\alpha = \frac{N_L^{\alpha\gamma}}{2N_L^{\alpha\alpha} + N_L^{\alpha\gamma}}, \qquad (37.3d)$$

where $N_L^{\gamma\gamma}$, $N_L^{\alpha\alpha}$ and $N_L^{\alpha\gamma}$ are intercept numbers of γ/γ, α/α and α/γ boundaries with a random line of unit length on a polished plane. Then, q_γ, q_α, $q_{\alpha\gamma}$ and $\bar{d}_{\alpha\gamma}$ in Eq. (37.2) could be calculated by:

$$q_\gamma = C_\gamma f_\gamma, \qquad (37.4a)$$

$$q_\alpha = C_\alpha f_\alpha, \qquad (37.4b)$$

$$q_{\alpha\gamma} = S_\gamma f_\gamma + S_\alpha f_\alpha, \qquad (37.4c)$$

$$\bar{d}_{\alpha\gamma} = \frac{S_\gamma f_\gamma d_\gamma + S_\alpha f_\alpha d_\alpha}{q_{\alpha\gamma}}, \quad (37.4d)$$

where f_γ and f_α are volume fractions of γ and α phases. The yield stress of dual phase microstructure can also be expressed as:

$$\sigma_{yC} = \sigma_{0C} + k_C \bar{d}^{-1/2}, \quad (37.5)$$

where $\bar{d}(=d_\gamma f_\gamma + d_\alpha f_\alpha)$ is average grain size; σ_{0C} and k_C are the overall friction stress and Hall–Petch coefficient, which could be determined experimentally. Based on Eqs. (37.2) and (37.5), $\sigma_{0\alpha\gamma}$ could be derived as:

$$\sigma_{0\alpha\gamma} = \frac{\sigma_{0C} - q_\gamma \sigma_{0\gamma} - q_\alpha \sigma_{0\alpha}}{q_{\alpha\gamma}}. \quad (37.6)$$

Some of the microstructural parameters, such as d_γ, d_α, $\bar{d}_{\alpha\gamma}$, q_γ, q_α and $q_{\alpha\gamma}$, could be experimentally determined and mathematically calculated by Eq. (37.4). The intercepts ($\sigma_{0\gamma}$, $\sigma_{0\alpha}$) and slope parameters (k_γ, k_α) were expected to be obtained from the samples with single γ and α phase, respectively. However, it is difficult to fabricate a single phase alloy which possesses the same chemical compositions as those in the dual phase materials. Hirota et al. [38] measured the $\sigma_{0\gamma}$ and $\sigma_{0\alpha}$ of another DSS using micro-hardness testing. Here we show a simple method to obtain $\sigma_{0\gamma}$, $\sigma_{0\alpha}$, k_γ and k_α by Pickerings equation [39] and multiple linear regression analysis.

The alloying elements have strong effects on the intercepts of both phases. Sieurin et al. [40] had deduced the hardening coefficients of chemical elements by linear regression, which was shown as $\sigma_{0\gamma}(\text{MPa}) = 700 f_{Cr} + 2000 f_{Mn} + 3300 f_{Si} + 290 f_{Ni} + 770\sqrt{f_N}$. Where the symbols f denote the weight fractions of chemical elements (e.g., $f_{Cr} = \frac{\text{wt.}_{Cr}}{\text{wt.}_{total}}$), so they are dimensionless parameters. Therefore, all the coefficients on the right side of this equation are in units of MPa. Since the concentrations of alloying elements are not the same in the two phases α and γ, the real chemical compositions in two phases (as shown in Table 37.1) are obtained by considering of the equilibrium distribution factors suggested by Charles [41]. Thus, the $\sigma_{0\gamma}$ in UNS S32304 is calculated through multiplying the experimental $\sigma_{0\gamma}$ of Fe-25Cr-7Ni alloy by a ratio considering the difference in chemical concentration. The measured value of $\sigma_{0\gamma}$ in Fe-25Cr-7Ni alloy is 345 MPa [38]. The sum of contributions of all elements to $\sigma_{0\gamma}$ is 283.7 MPa in Fe-25Cr-7Ni and 246.4 MPa in UNS S32304 respectively. Thus the ratio is equal to 246.4/283.7, and $\sigma_{0\gamma}$ of UNS S32304 can be estimated as: 246.4/283.7 × 345 MPa = 299.6 MPa.

Table 37.1 Chemical compositions (in wt.%) of γ- and α-phase in UNS S32304, which calculated using the equilibrium distribution factor investigated by Charles [41]

	C	Si	Mn	Cr	Ni	Mo	N	P
γ	0.02	0.45	1.26	22.02	5.00	0.31	0.24	0.01
α	0.02	0.55	1.14	24.93	3.00	0.49	0.02	0.03

The σ_{0C} and k_C could be calculated through multiple linear regression by considering five different dual phase microstructures listed in Table 37.2, which gives σ_{0C} = 314.8 MPa and k_C = 333.4 MPa·μm$^{1/2}$. The overall friction stress of ferrite in DSS could not be derived as that in austenite because the function is not simply linear any more due to high concentration of Cr [42, 43]. Since σ_{0C} could be expressed as $\sigma_{0C} = \sigma_{0\gamma}f_\gamma + \sigma_{0\alpha}f_\alpha$ and the volume fraction of austenite is about 0.45, $\sigma_{0\alpha}$ can be obtained as 326.9 MPa.

Table 37.2 Summary of the obtained yield stresses, volume fractions, average grain diameters and topological parameters

Sample	σ_{yC} (MPa)	f_γ	f_α	$d\gamma$ (μm)	d_α (μm)	q_γ	q_α	$q_{\alpha\gamma}$	$\bar{d}_{\alpha\gamma}$ (μm)	$\bar{d}$ (μm)
CG [48]	403	0.45±0.01	0.55	17.6	19.5	0.13	0.13	0.74	18.68	18.65
	457	0.39±0.01	0.61	4.81	5.72	0.11	0.13	0.76	5.38	5.37
Micro-	540	0.48±0.03	0.52	1.69	1.94	0.15	0.13	0.72	1.83	1.82
duplex	517	0.49±0.04	0.51	2.22	2.53	0.18	0.18	0.64	2.38	2.38
NS [30]	1100	0.45±0.02	0.55	0.11	0.24	0.44	0.54	0.02	0.20	0.18

The friction stresses of austenite and ferrite have been obtained now, retaining only three unknown parameters in Eq. (37.2): k_γ, k_α and $k_{\alpha\gamma}$. However, there are five independent equations based on the data shown in Table 37.2, and thus the k_γ, k_α and $k_{\alpha\gamma}$ could be calculated by multiple linear regression. The results are displayed in Table 37.3, from which the phase boundary appears a moderate coefficient (307.9 MPa·μm$^{1/2}$) as compared to those of grain boundaries of austenite (224.9 MPa·μm$^{1/2}$) and ferrite (428.9 MPa·μm$^{1/2}$). The coefficient of phase boundary is in agreement with the investigation of Fan el al. (345 MPa·μm$^{1/2}$) [36, 44].

Table 37.3 Hall–Petch coefficients of duplex stainless steel obtained by multiple linear regressions. The statistical parameters are R^2 = 0.999 and p = 10^{-3} (Unit: MPa·μm$^{1/2}$)

Alloys	k_γ	k_α	$k_{\alpha\gamma}$	Refs.
UNS S32304	224.9	428.9	309.7	This study
Fe-Cr-Ni	458	281	345	[36]
SUS 316L	164	—	—	[38]
Fe-25Cr-7Ni-0.15N	239	954	—	[38]

During plastic deformation, the dislocation transmission across phase boundary could occur in DSS when the slip planes in austenite and ferrite are nearly parallel [45]. It is the possible reason for the moderate coefficient of phase boundary. In addition, contrary to the investigation of Fan et al. [36] but in accordance with the result of Sieurin et al. [40], the ferrite in present study has a larger coefficient than that of the austenite. Since the coefficient depends closely on the chemical compositions, the segregation of solute atoms at grain boundaries may raise the coefficient of ferrite.

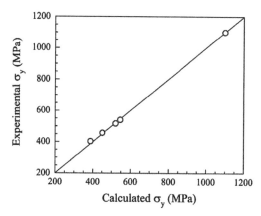

Figure 37.9 Comparison of the yield stresses between theoretical predictions and experimental results for different DSS samples.

As all the parameters in Eq. (37.2) have been obtained, the yield stresses of dual phase materials with different grain size can be calculated. Figure 37.9 exhibits the calculated and experimental values of yield stress, and the result indicates a good agreement between them.

37.4.2 Effect of Phase Interaction on Strain Hardening Rate

One characteristic for the plastic deformation of dual phase materials is the plastic strain gradient across phase boundaries induced by the inhomogeneous distribution of plastic strain. According to the plastic strain gradient theory, it requires the generation of GNDs to make the deformation compatible. There is no obvious difference in physics between the GNDs and traditional lattice dislocations, but the GNDs in essence is an extra storage of dislocations due to geometrical necessity. It is the reason why extra strength occur when plastic strain gradient exist. Thus, the strain hardening rate of dual phase material would be influenced by this extra strength since the strain gradient exists all through the tensile deformation.

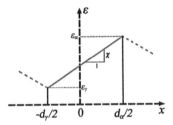

Figure 37.10 Schematic illustration of the gradient plastic strain in dual phase microstructure. For simplicity, the plastic strain is varied linearly with distance, and therefore generate a constant absolute value of strain gradient in dual phase part.

The gradient plastic strain in dual phase microstructure is schematically illustrated in Fig. 37.10. Assuming linear variation of plastic strain with distance, the absolute value of strain gradient in space therefore remain constant in dual phase part shown in Fig. 37.8. Based on the strain gradient theory, the flow stress of dual phase material can be expressed as:

$$\sigma' = q_\gamma \sigma_\gamma + q_\alpha \sigma_\alpha + q_{\alpha\gamma} \sigma'_{\alpha\gamma}, \tag{37.7}$$

where $\sigma'_{\alpha\gamma}$ denotes the flow stress of dual phase part in Fig. 37.8 which has incorporated the effect of strain gradient. Taking into account the works of Nix and Gao [46], the $\sigma'_{\alpha\gamma}$ could be expressed as:

$$\frac{\sigma'_{\alpha\gamma}}{\sigma_{\alpha\gamma}} = \sqrt{1+l\chi}, \tag{37.8}$$

where l is an characteristic material length scale, and χ is strain gradient. Substituting Eq. (37.8) into Eq. (37.7), and incorporating a stress partitioning coefficient $p_{\alpha\gamma}$, one can easily get:

$$\frac{\sigma'}{\sigma} = 1 + q_{\alpha\gamma} p_{\alpha\gamma} \left\{ [1+l\chi]^{1/2} - 1 \right\}, \tag{37.9}$$

where $p_{\alpha\gamma}$ is equal to $\sigma_{\alpha\gamma}/\sigma$, and the characteristic material length scale is expressed as:

$$l = b \left(\frac{u}{q_{\alpha\gamma} \sigma} \right)^2, \tag{37.10}$$

where b is the magnitude of Burgers vector, and u is the shear modulus. Based on Eq. (37.9), it is clear that the more the volume fraction of dual phase part and the larger the strain gradient are, the higher the strength becomes.

In order to analyze the strain hardening behavior, Mecking-Kocks theory was always used and exhibited a linear equation with respect to flow stress [47]:

$$\Theta = \Theta_0 - K(\sigma - \sigma_0), \tag{37.11}$$

where $\Theta = \dfrac{d\sigma}{d\varepsilon}$, Θ_0 is an athermal hardening rate, and K denotes the rate of dynamic recovery. Taking derivatives of Eq. (37.9) and (37.10) with respect to strain, combining the differential equations, and assuming nearly constant $p_{\alpha\gamma}$, the strain hardening rate of dual phase material which incorporates the strain gradient effect was obtained (detailed derivation process is shown in Appendix section):

$$\Theta' = \Theta'_0 - K(1 - \zeta S_V)(\sigma - \sigma_0), \tag{37.12}$$

where S_V is surface-to-volume ratio of phase boundary, equaling to $3q_{\alpha\gamma}/\overline{d}_{\alpha\gamma}$ when the grains in dual phase part are considered as spherical.

Based on Eq. (37.12), the dynamic recovery rate of dual phase material depends closely on the surface-to-volume ratio of phase boundary. The higher the ratio is, the lower the dynamic recovery rate becomes, and *vice versa*.

The values of S_V, $K(1 - \zeta S_V)$ and Θ'_0 are listed in Table 37.4. K and ζ are obtained as 4.26 and 0.03 m by linear regression method. Therefore, the theoretical $K(1 - \zeta S_V)$ of three different duplex structures can be calculated, and the results demonstrated that the calculated $K(1 - \zeta S_V)$ is in good agreement with experimental results.

Table 37.4 Summary of surface-to-volume ratios (S_V) of phase boundary in different microstructures, as well as the experimental (Exp.) and theoretical (Theor.) values of dynamic recovery rate

Sample	S_V (×10^5 m^{-1})	Θ'_0 (MPa)	$K(1 - \zeta S_V)$ (Exp.)	$K(1 - \zeta S_V)$ (Theor.)
CG	1.2	2802.7 ± 26.1	4.31 ± 0.4	4.11
Micro-duplex	11.7	2468.3 ± 76.7	2.80 ± 0.6	2.76
	8.1	2685.4 ± 87.2	3.44 ± 0.9	3.21

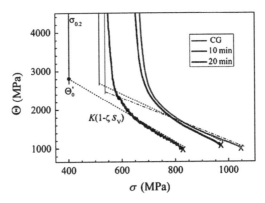

Figure 37.11 The strain hardening rate *vs.* true stress curves upon tensile deformation of CG and micro-duplex samples.

Figure 37.11 show the comparison of theoretical Θ–σ data with those obtained from experiments at linear decreasing stage. The results show that theoretical predictions have only little overestimations on the dynamic recovery rate. In general, the model as expressed by Eq. (37.12) could well describe the influence of phase interaction on strain hardening behavior of duplex stainless steel.

37.5 Conclusion

Micro-duplex structured samples consisting of austenite and ferrite (with average grain sizes near 2 μm in both phases) were fabricated by ECAP and subsequent thermal annealing. As compared to their coarse-grained counterpart, the micro-duplex structures have finer grain sizes in both phases and higher surface-to-volume ratios of phase boundary. The yield stress and uniform elongation of micro-duplex samples are increased simultaneously. The strengthening mechanism is quantitatively investigated, and the influence of surface-to-volume ratio of phase boundary on strain hardening behavior is analyzed based on strain gradient theory. The main conclusions are summarized as follows:

1. The average grain sizes of austenite and ferrite are 1.69 and 1.94 μm in micro-duplex structure annealed at 1173 K for 10 min, while the yield stress and uniform elongation increase to 540 MPa and 0.3 as compared to 403 MPa and 0.26 of their coarse-grained counterpart.
2. The α/γ boundary has an intermediate ability to impede the motion of dislocations as compared to γ/γ and α/α grain boundaries. A Hall–Petch type coefficient is used to quantitatively describe this strengthening ability, which are measured as 309.7, 224.9 and 428.9 MPa·μm$^{1/2}$ for α/γ, γ/γ and α/α boundaries, respectively.
3. The surface-to-volume ratio of phase boundary in micro-duplex structure is estimated to be 1.17×10^6 m^{-1}, which is increased by an order of magnitude as compared to 1.2×10^5 m^{-1} of its coarse-grained counterpart.
4. The tensile test show that the dynamic recovery rate is decreased significantly in micro-duplex structure. Based

on strain gradient theory, a simple model was proposed to describe the influence of surface-to-volume ratio of phase boundary on strain hardening rate, which show good agreement with experimental results.

Acknowledgments

The authors would like to thank the financial supports by 973 Program of China (Nos. 2012CB932203 and 2012CB937500) and NSFC (Grants Nos. 51301187, 11222224, 11021262 and 11072243).

Appendix A

Since the flow stress of dual phase microstructure incorporating the effect of plastic strain gradient has been expressed as Eq. (37.9), the strain hardening rate can be obtained by taking the derivative of both sides with respect to strain:

$$\frac{d\sigma'}{d\varepsilon} = \left\{1 + q_{\alpha\gamma}p_{\alpha\gamma}\left[(1+l\chi)^{1/2} - 1\right]\right\} \cdot \frac{d\sigma}{d\varepsilon}$$
$$+ \frac{q_{\alpha\gamma}p_{\alpha\gamma}}{2(1+l\chi)^{1/2}}\left(\chi\frac{dl}{d\varepsilon} + l\frac{d\chi}{d\varepsilon}\right) \cdot \sigma \quad (37.A1)$$

In the equation above, the characteristic material length scale l and strain gradient χ are strain dependent. The variation of $p_{\alpha\gamma}$ with plastic strain was neglected for simplicity. Taking the Eq. (37.10) into account, the derivative of l over ε is equal to:

$$\frac{dl}{d\varepsilon} = -\frac{2l}{p_{\alpha\gamma}^2}\frac{1}{\sigma}\frac{d\sigma}{d\varepsilon} \quad (37.A2)$$

Substituting Eq. (37.A2) into Eq. (37.A1), Eq. (37.A1) can be rewritten as:

$$\frac{d\sigma'}{d\varepsilon} = \left\{1 + q_{\alpha\gamma}p_{\alpha\gamma}\left[(1+l\chi)^{1/2} - 1\right] - \frac{q_{\alpha\gamma}}{p_{\alpha\gamma}}\frac{l\chi}{(1+l\chi)^{1/2}}\right\} \cdot \frac{d\sigma}{d\varepsilon}$$
$$+ \frac{lq_{\alpha\gamma}}{2p_{\alpha\gamma}(1+l\chi)^{1/2}}\frac{d\chi}{d\varepsilon} \cdot \sigma \quad (37.A3)$$

Substituting Eq. (37.11) into Eq. (37.A3), one can get:

$$\frac{d\sigma'}{d\varepsilon} = (\Theta_0 + K\sigma_0)\left\{1 - q_{\alpha\gamma}p_{\alpha\gamma}\left[1 + \frac{l\chi(1-p_{\alpha\gamma}^2)-p_{\alpha\gamma}^2}{(1+l\chi)^{1/2}p_{\alpha\gamma}^2}\right]\right\}$$
$$- \left\{K\left\{1 - q_{\alpha\gamma}p_{\alpha\gamma}\left[1 + \frac{l\chi(1-p_{\alpha\gamma}^2)-p_{\alpha\gamma}^2}{(1+l\chi)^{1/2}p_{\alpha\gamma}^2}\right]\right\} - \frac{q_{\alpha\gamma}}{2p_{\alpha\gamma}}\frac{l}{(1+l\chi)^{1/2}}\frac{d\chi}{d\varepsilon}\right\} \cdot \sigma$$

(37.A4)

If the difference of plastic strain between two phases are relatively constant, Eq. (37.A4) can be simplified as:

$$\frac{d\sigma'}{d\varepsilon} = \underbrace{\Theta_0\left\{1 - q_{\alpha\gamma}p_{\alpha\gamma}\left[1 + \frac{l\chi(1-p_{\alpha\gamma}^2)-p_{\alpha\gamma}^2}{(1+l\chi)^{1/2}p_{\alpha\gamma}^2}\right]\right\}}_{\Theta_0'}$$
$$\underbrace{- K\left\{1 - q_{\alpha\gamma}p_{\alpha\gamma}\left[1 + \frac{l\chi(1-p_{\alpha\gamma}^2)-p_{\alpha\gamma}^2}{(1+l\chi)^{1/2}p_{\alpha\gamma}^2}\right]\right\}}_{K'} \cdot (\sigma - \sigma_0) \quad (37.A5)$$

Therefore, Eq. (37.A5) and Eq. (37.11) are identical in form. Since the surface-to-volume ratio S_V of phase boundary is equal to $3q_{\alpha\gamma}/\bar{d}_{\alpha\gamma}$ when assume the grains are spherical in dual phase part, Eq. (37.A5) can be further rewritten as:

$$\frac{d\sigma'}{d\varepsilon} = \Theta_0' - K(1-\zeta S_V)(\sigma - \sigma_0) \quad (37.A6)$$

where ζ is equal to $\dfrac{\bar{d}_{\alpha\gamma}}{3}\left[1 + \dfrac{l\chi(1-p_{\alpha\gamma}^2)-p_{\alpha\gamma}^2}{(1+l\chi)^{1/2}p_{\alpha\gamma}^2}\right]$; Θ_0' is equal to $\Theta_0(1-\zeta S_V)$.

References

1. Fang TH, Li WL, Tao NR, Lu K. *Science* 2011;331:1587.
2. Lu K, Lu L, Suresh S. *Science* 2009;324:349.
3. Wang YM, Ma E. *Acta Mater* 2004;52:1699.

4. Zhu YT, Liao XZ. *Nat Mater* 2004;3:351.
5. Wang YM, Ma E, Chen MW. *Appl Phys Lett* 2002;80:2395.
6. Wu XL, Jiang P, Chen L, Yuan FP, Zhu YTT. *Proc Natl Acad Sci* 2014;111:7197.
7. Wang YM, Chen MW, Zhou FH, Ma E. *Nature* 2002;419:912.
8. Zhao YH, Liao XZ, Cheng S, Ma E, Zhu YT. *Adv Mater* 2006;18:2280.
9. Liu G, Zhang GJ, Jiang F, Ding XD, Sun YJ, Sun J, Ma E. *Nat Mater* 2013;12:344.
10. Wu X, Tao N, Hong Y, Lu J, Lu K. *Scr Mater* 2005;52:547.
11. Ma YQ, Jin JE, Lee YK. *Scr Mater* 2005;52:1311.
12. Koch CC, Scattergood RO, Murty KL. *JOM* 2007;59:66.
13. Militzer M. *Science* 2002;298:975.
14. Jia N, Lin Peng R, Wang YD, Johansson S, Liaw PK. *Acta Mater* 2008;56:782.
15. Harjo S, Tomota Y, Lukáš P, Neov D, Vrána M, Mikula P, Ono M. *Acta Mater* 2001;49:2471.
16. Dakhlaoui R, Braham C, Baczmański A. *Mater Sci Eng A* 2007;444:6.
17. Jia N, Lin Peng R, Chai GC, Johansson S, Wang YD. *Mater Sci Eng A* 2008;491:425.
18. Woo W, Em VT, Kim EY, Han SH, Han YS, Choi SH. *Acta Mater* 2012;60:6972.
19. Wang J, Hoagland RG, Hirth JP, Misra A. *Acta Mater* 2008;56:5685.
20. Wang J, Hoagland RG, Liu XY, Misra A. *Acta Mater* 2011;59:3164.
21. Wang J, Misra A, Hoagland RG, Hirth JP. *Acta Mater* 2012;60:1503.
22. Sinclair CW, Embury JD, Weatherly GC. *Mater Sci Eng A* 1999;272:90.
23. He L, Allard LF, Ma E. *Scr Mater* 2000;42:517.
24. Ankem S, Margolin H. *Metall Trans A* 1983;14:500.
25. Ankem S, Margolin H, Greene CA, Neuberger BW, Oberson PG. *Prog Mater Sci* 2006;51:632.
26. Shin DH, Park KT. *Mater Sci Eng A* 2005;410:299.
27. Son YI, Lee YK, Park K-T, Lee CS, Shin DH. *Acta Mater* 2005;53:3125.
28. Song R, Ponge D, Raabe D, Speer JG, Madock DK. *Mater Sci Eng A* 2006;441:1.
29. Tian YZ, Wu SD, Zhang ZF, Figueiredo RB, Gao N, Langdon TG. *Acta Mater* 2011;59:2783.

30. Chen L, Yuan FP, Jiang P, Wu XL. *Mater Sci Eng A* 2012;551:154.
31. Langdon TG, Furukawa M, Nemoto M, Horita Z. *JOM* 2000;52:30.
32. Valiev RZ, Islamgaliev RK, Alexandrov IV. *Prog Mater Sci* 2000;45:103.
33. Iwahashi Y, Horita Z, Nemoto M, Langdon TG. *Acta Mater.* 1998;46:3317.
34. Kleemola H, Nieminen M. *Metall Trans* 1974;5:1863.
35. Maki T, Furuhara T, Tsuzaki K. *ISIJ Int* 2001;41:571.
36. Fan Z, Tsakiropoulos P, Smith P, Miodownik A. *Philos Mag A* 1993;67:515.
37. Fan Z, Tsakiropoulos P, Miodownik AP. *Mater Sci Technol* 1992;8:922.
38. Hirota N, Yin FX, Azuma T, Inoue T. *Sci Technol Adv Mater* 2010;11.
39. Pickering FB, Cahn RW. *Constitution and Properties of Steels*. Weinheim: VCH Verlasgesellschaft GmbH, 1992.
40. Sieurin H, Zander J, Sandstrom R. *Mater Sci Eng A* 2006;415:66.
41. Charles J. Duplex Stainless Steels '94. In: Gooch TG, ed. Glasgow: The Welding Institute, 1994.
42. Gutierrez I, Altuna MA. *Acta Mater* 2008;56:4682.
43. Rodriguez R, Gutiérrez I. *Mater Sci Forum* 2003;426:4525.
44. Tomota Y, Kuroki K, Mori T, Tamura I. *Mater Sci Eng* 1976;24:85.
45. Verhaeghe B, Louchet F, Doisneau-Cottignies B, Bréchet Y, Massoud J. *Philos Mag A* 1997;76:1079.
46. Nix WD, Gao H. *J Mech Phys Solids* 1998;46:411.
47. Kocks UF, Mecking H. *Prog Mater Sci* 2003;48:171.
48. Chen L, Wu XL. *Mater Sci Forum* 2011;682:123.

Chapter 38

Strain Hardening in Fe–16Mn–10Al–0.86C–5Ni High Specific Strength Steel

M. X. Yang,[a] F. P. Yuan,[a] Q. G. Xie,[b] Y. D. Wang,[b] E. Ma,[c] and X. L. Wu[a]

[a]*State Key Laboratory of Nonlinear Mechanics, Institute of Mechanics, Chinese Academy of Sciences, 15 Beisihuan West Road, Beijing 100190, China*
[b]*School of Materials Science and Engineering, Beijing Institute of Technology, 30 Xueyuan Road, Beijing 100081, China*
[c]*Department of Materials Science and Engineering, Johns Hopkins University, 3400 North Charles Street, Baltimore, MD 21218, USA*
xlwu@imech.ac.cn

We report a detailed study of the strain hardening behavior of a Fe–16Mn–10Al–0.86C–5Ni (weight percent) high specific strength (i.e. yield strength-to-mass density ratio) steel (HSSS) during uniaxial tensile deformation. The dual-phase (γ-austenite and B2 intermetallic compound) HSSS possess high yield strength of 1.2–1.4 GPa and uniform elongation of 18–34%. The tensile deformation

---

Reprinted from *Acta Mater.*, **109**, 213–222, 2016.

Heterostructured Materials: Novel Materials with Unprecedented Mechanical Properties
Edited by Xiaolei Wu and Yuntian Zhu
Text Copyright © 2016 Acta Materialia Inc.
Layout Copyright © 2022 Jenny Stanford Publishing Pte. Ltd.
ISBN 978-981-4877-10-7 (Hardcover), 978-1-003-15307-8 (eBook)
www.jennystanford.com

of the HSSS exhibits an initial yield-peak, followed by a transient characterized by an up-turn of the strain hardening rate. Using synchrotron based high-energy in situ X-ray diffraction, the evolution of lattice strains in both the γ and B2 phases was monitored, which has disclosed an explicit elasto-plastic transition through load transfer and strain partitioning between the two phases followed by co-deformation. The unloading-reloading tests revealed the Bauschinger effect: during unloading yield in γ occurs even when the applied load is still in tension. The extraordinary strain hardening rate can be attributed to the high hetero-deformation induced (HDI) stresses that arise from the strain incompatibility caused by the microstructural heterogeneity in the HSSS.

38.1 Introduction

Steels are the strongest ductile bulk materials currently available [1–7]. High-strength and supra-ductile steels are chief cornerstones for the next generation of energy and industrial technologies applied in various engineering sectors such as automotive, aviation, aerospace, power, transport, building and construction. They have been developed based on several design principles; typical categories include the transformation-induced plasticity (TRIP) steels [8, 9], twinning-induced plasticity (TWIP) steels [10, 11], dual-phase (DP) steels [12, 13], nano-structured steels [14, 15], and even hypereutectoid steel wires with ultrahigh (6.35 GPa) tensile strength [16].

Recently, the high-aluminium low-density steels have been actively studied for the purpose of increasing the specific strength (i.e. yield strength-to-mass density ratio) [4, 5, 17–27]. These low-density steels, mainly based on an Fe–Al–Mn–C alloy system containing high contents of Mn (16–28 wt.%), Al (3–12 wt.%) and C (0.7–1.2 wt.%), consist of face-centred cubic (fcc) austenite matrix and body-centred cubic (bcc) ferrite matrix and finely dispersed nanometre-sized κ-carbides of $(Fe, Mn)_3$ AlC type (the so-called TRIPLEX steel) [18, 19]. Recently, Kim et al. [27] showed a new variation: their Fe–16Mn–10Al–0.86C–5Ni high specific strength steel (HSSS) has a hard FeAl-type (B2) intermetallic compound as the strengthening second phase, and the alloying of Ni catalyzes the precipitation of B2 particles in the fcc matrix. The combination of specific strength and elongation is outstanding for this HSSS, when compared with other high-specific-strength alloys [27].

In developing various high strength steels aforementioned, a primary issue is the strain hardening capability. Strain hardening is a prerequisite for large uniform ductility. However, the mechanism of strain hardening remains an open issue for most high strength steels [5, 22, 27], because they deform very heterogeneously due to their inhomogeneous microstructures. Even for an initial single-phase alloy, for instance TWIP and TRIP steels, deformation twins and martensite make the strain hardening behavior complex [20, 22]. Bhadeshia [28] pointed out that in TRIP steels it is unlikely that the large tensile elongation is predominantly caused by the transformation from austenite into martensite alone. The martensite colonies act as strong inclusions, akin to reinforcing components in a composite, and should have also played a role in strain hardening [29, 30]. A similar conclusion has been reached for TWIP steels where the twinning strain itself makes a significant though small contribution to the total elongation [31].

Indeed, plastic deformation in most high strength steels, much like in composites, is characterized by pronounced plastic heterogeneity between the constituent phases, as well as among grains with different orientations and mechanical responses towards an externally applied load. This causes complex internal stresses, which develop because of intra- and inter-granular variations of plastic strain. The load redistribution and strain partitioning resulting from the microstructural heterogeneity enable a high capacity of strain hardening, affecting the large ductility. The high internal stresses have, in fact been, reported to contribute to strain hardening and large ductility in TRIP steels [32], TWIP steels [33, 34, 35], nano-composites [36, 37], and dual-phase alloys [38, 39].

The development of internal stresses during deformation of an inhomogeneous micro- structure with yield stress mismatch has been well described before [13, 40, 41]: upon tensile loading, plastic yield starts in the soft phase, and the applied load will be transferred from the soft phase to the hard one that is still in elastic state. Thus, internal stresses will build up at the phase boundaries. Upon unloading, the macroscopic stress remains higher than the stress in the soft phase until it reaches the unloaded state, where the soft phase is subjected to an (elastic) compression stress (a tensile stress in the hard phase) [42, 43]. If the two-phase alloy is subsequently subjected to compressive loading, it initially behaves elastically until the soft phase enters the plastic regime in compression, a situation that will take place at a much lower absolute stress compared to the tensile loading case because of the initial compression of the soft phase. A consequence is an asymmetry in the forward (tensile) and reverse (compressive) yield stresses. Such a phenomenon is known as the Bauschinger effect [35, 38, 44]. Recently, the use of diffraction techniques has supplanted this macroscopic description of internal stresses by the measurements of lattice strains in individual phases [45, 46]. The unload-reload tests [36, 38, 47] are also used for the study of internal stresses in thin films or composite wires where compression cannot be applied.

In this chapter, we analyze the strain hardening in the Fe–16Mn–10Al–0.86C–5Ni HSSS composed of an γ-austenite matrix containing the B2 FeAl second phase. Based on in situ high energy X-ray diffraction data, the lattice strain evolution in both phases has been monitored and then used to correlate with the mechanical responses such as the stress and strain partitioning, the elasto-plastic transition and co-deformation, and the HDI strain hardening. Different from Kim et al. [27], who treated this HSSS as a case of precipitation strengthening with brittle and non-deformable B2 FeAl, here we show that this steel is better understood as a dual-phase microstructure, with plastic behaviors much like a composite. In particular, the B2 phase is deformable, with significant strain hardening capability.

38.2 Materials and Experimental Procedures

38.2.1 Materials

Similar to the procedures in Ref. [27], an Fe–16Mn–10Al–0.86C–5Ni (wt.%) HSSS was produced using arc melting in a high frequency induction furnace under pure argon atmosphere, and then cast to a cylindrical ingot with a diameter of 130 mm and length of 200 mm. The chemical composition of the ingot was Fe–16.4Mn–0.86C–9.9Al–4.8Ni–0.008P– 0.004S (wt.%). The ingot was homogenized at 1180°C for 2 h, forged in between 1150°C and 900°C into slabs with a thickness of 14 mm, and hot-rolled with a starting temperature of 1050°C into strips with a thickness 7.3 mm. The hot-rolled striped were finally cold rolled to sheets with the final thickness of 1.5 mm. No cracks were detected on the surfaces of the cold-rolled sheets. The final annealing of the cold-rolled sheets was conducted at 900°C for 2–15 min followed immediately water quenched.

38.2.2 Mechanical Property Tests

The dog-bone-shaped plate tensile specimens, with a gauge length of 18 mm and width of 4 mm, were cut from the annealed sheets with longitudinal axes parallel to the rolling direction. All specimens were mechanically polished prior to tensile tests in order to remove surface irregularities and to guarantee an accurate determination of the cross-sectional area. The quasi-static, uniaxial tensile tests were carried out using an MTS Landmark testing machine operating at a strain rate of 5×10^{-4} s^{-1} at room temperature. The tensile load-unload-reload (LUR) tests were conducted. The condition for LUR tests was the same as that of monotonic tensile test. All tensile tests were conducted using a 10 mm gauge extensor- meter to monitor the strain. The resolutions for stress and strain measurements were 1.0×10^{-2} MPa and 1.0×10^{-5}, respectively. Vickers micro-hardness indentations (25 g load) were made at the ends of the gauge section to check the error in strain measurements by means of an optical traveling microscope. The tensile testing was performed five times on average for each condition to verify the reproducibility of the monotonic and cyclic tensile stress-strain curves.

38.2.3 Synchrotron Based High Energy X-Ray Diffraction

In-situ high energy X-ray diffraction measurements were carried out on the beam-line 11-ID-C, at the Advanced Photon Source (APS), Argonne National Laboratory, USA. The experimental set-up was detailed in Ref. [46]. Dimensions of the tensile specimen in the gauge part were 10 mm (length) × 3 mm (width) × 0.5 mm (thickness). During tensile loading, a monochromatic X-ray beam with energy 105.1 keV (λ = 0.11798 Å) and beam size of 500 μm (height) × 500 μm (width) was used. The 2-D detector was placed 2 m behind the tensile sample to collect the intensity data of the diffraction rings. Crystallographic planes were determined from the diffraction patterns and the lattice strains were calculated from the change of the measured inter-planar spacing. The in situ tensile testing of three times was performed to verify the reproducibility.

38.2.4 Microstructural Characterization

Optical microscopy (OM, Olympus BX51) and scanning electron microscopy (SEM, JSM-7001F) were used to characterize the microstructure of the HSSS before and after tensile tests. The major and minor axis lengths of the γ grains, and lamellar and granular B2 grains were measured by the linear intercept method following the procedures given in ASTM E1382, and at least 600 target objects were measured for each corresponding statistical distribution.

A transmission electron microscope (TEM, JEM-2100) and a high-spatial resolution analytical electron microscope (HRTEM, FEI Tecnai G20) both operated at 200 kV for examinations of the typical microstructural features in γ and B2 grains. Thin foils for TEM observations were cut from the gauge sections of tensile samples, mechanically ground to about 50 μm thick and finally thinned by a twin-jet polishing facility using a solution of 5% perchloric acid and 95% ethanol at −20°C.

In addition, the microstructural features of B2 precipitates were also examined by electron back-scattered diffraction (EBSD) using a high-resolution field emission Cambridge S-360 SEM equipped with a fully automatic Oxford Instruments Aztec 2.0 EBSD system (Channel 5 Software). During the EBSD acquisition, a scanning area of 10×14 μm^2 was chosen and a scanning step of 0.03–0.08 μm was performed. Due to spatial resolution of the EBSD system, the collected Kikuchi patterns can be obtained automatically at a minimum step of 0.02 μm and correspondingly misorientations less than 2° cannot be identified. The longitudinal sections of samples for EBSD examinations were mechanically polished carefully followed by electro-polishing using an electrolyte of 90 vol.% acetic acid and 10 vol.% perchloric acid with a voltage of 40–45 V at about −40°C in a Struers LectroPol-5 facility.

38.3 Experimental Results

38.3.1 Microstructural Characterization

Figure 38.1a is an optical microscope image of the longitudinal section of the hot forged sample. Two phases are visible. One is the equi-axed grains of recrystallized fcc γ-austenite, while the other is the thick lamellar B2 phase parallel to the rolling direction. After cold rolling with a rolling strain of 80%, the γ grains change from granular to elongated shape, as shown in Fig. 38.1b. The B2 phase exhibits a much reduced thickness, indicative of its capability for plastic deformation. After annealing at 900°C for 2 min, as shown in Fig. 38.1c, a large number of granular B2 phase precipitates from γ. Figure 38.1d shows the high resolution electron back-scattered diffraction (EBSD) image. It is seen that B2 is much inclined to precipitate at both the grain boundaries and triple junctions of γ grains, instead of their interiors. This is further evidenced by an enlarged EBSD image as shown in Fig. 38.1e, where γ grains (upper) and B2 phase (lower) are marked in color. Little precipitation

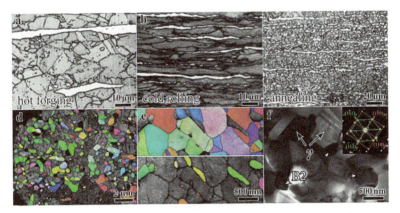

Figure 38.1 Microstructural characterization of HSSS. (a–c) Optical images of longitudinal sections of the HSSS samples after hot forging, further cold rolling, and finally annealing at 900°C for 2 min, respectively. (d) High resolution electron back-scattered diffraction (EBSD) image from (c). B2 grains are colored. (e) Enlarged EBSD images. Colored grains are γ (upper half of the image) and B2 (lower half). (f) Transmission electron microscope image showing γ and B2 dual-phase. Arrows indicate γ and triangles indicate B2. Inset in (f) is the indexed selected area electron diffraction pattern with $[011]_\gamma$ and $[001]_{B2}$ zone axes, respectively.

is visible inside the interiors of the γ grains. Figure 38.1f is a transmission electron microscopic (TEM) image showing both the γ and B2 grains. They are nearly free of dislocations due to annealing at high temperatures. Annealing twins are often seen in the γ grains. The B2 grains are seldom observed inside γ.

The above microstructural evolution is similar to that reported by Kim et al. [27], except for B2 precipitation. Kim et al. observed the B2 phase in three different morphologies, namely stringer bands parallel to the rolling direction (type 1), fine B2 particles of sizes 200 nm to 1 μm (type 2) at phase boundaries, and finer particles of sizes 50 to 300 nm inside γ grains (type 3). These finer B2 grains precipitate along shear bands in the non-recrystallized coarse γ grains. The volume fraction of B2 in these different types was not reported.

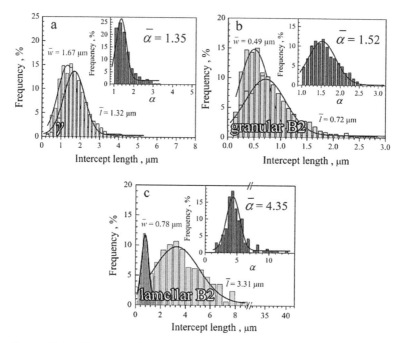

Figure 38.2 Histograms showing the statistics of the grain size distribution in γ(a), granular (b), and lamellar B2 (c). $\bar{l}$ and $\bar{w}$ indicate the average spacing in axial and transverse direction, respectively. Aspect ratio (α) is defined as $\bar{l}/\bar{w}$.

Figures 38.2a–c show the statistical distribution of the grain sizes in the γ phase, and in granular and lamellar B2, respectively, after annealing. Several hundreds of grains were counted, in extensive EBSD and TEM observations. It is seen that the γ grains are nearly equi-axed with the mean length ($\bar{l}$) and mean width ($\bar{w}$) of 1.67 and 1.32 μm, respectively, and an aspect ratio (α) of 1.35. α of the granular and lamellar B2 is larger, at 1.52 and 4.35, respectively. It is interesting to note that both $\bar{l}$ and $\bar{w}$ are less than 1 μm for granular B2, and $\bar{w}$ is also less 1 μm for lamellar B2.

The volume fraction of granular and lamellar B2 is 7.5% and 14.5%, respectively. Hence, it is more reasonable to consider the present HSSS as a microstructure composed of two phases, with B2 as a micrometer-sized co-existing phase rather than nano-precipitates in a precipitation-hardened γ matrix.

Figure 38.3 is the XRD spectrum of the HSSS after annealing at 900°C for 2 min. The diffraction peaks of fcc γ-austenite and bcc B2 are clearly identified.

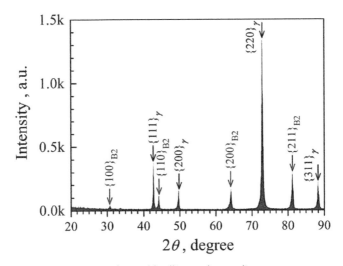

Figure 38.3 XRD spectra after cold rolling and annealing.

38.3.2 Tensile Properties

Figure 38.4 shows the tensile properties of the HSSS. Figure 38.4a displays the engineering stress versus strain (σ–ε) curves. The as-rolled sample yields at a strength of nearly 2 GPa but soon necks (curve A). After annealing at various temperatures, the tensile uniform elongation increases but with a decrease of strength, as can be seen in the σ–ε curves from B to D. The combination of both the strength and ductility is similar to those reported in Ref. [27], which are also included (dotted lines) in Fig. 38.4a for comparison. Note that the rolling strain is 80% in the present case, which is larger than the 66% in Ref. [27]. As a summary of the mechanical properties measured so far for high specific strength steels, Figs. 38.4c and d compare the tensile properties of our HSSS with that of Kim et al. [27], as well as with the TRIPLEX steel.

Interestingly, all the σ–ε curves in Fig. 38.4a show non-continuous yielding. A yield-peak appears first, followed by a transient deformation stage, seen as a concave segment on the σ–ε curve over a range of strains. This is distinctly different from the continuous yielding (at strain rate of 10^{-3} s^{-1}) in the same HSSS

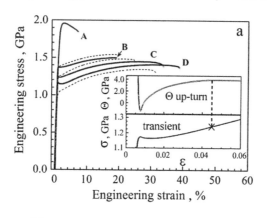

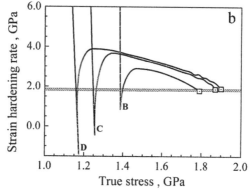

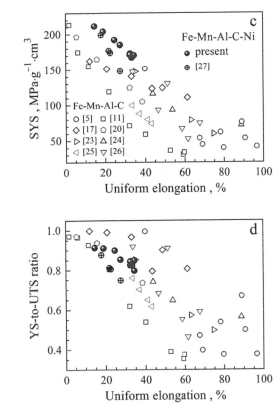

Figure 38.4 (a) Tensile engineering stress-strain ($\sigma - \varepsilon$) curves at a strain rate of 5×10^{-4} s^{-1}. A: cold rolling with strain 80%, B and C: annealing at 800 and 900°C for 2 min, D: annealing at 900°C for 15 min. Inset shows a close-up of both true $\sigma - \varepsilon$ curve and strain hardening rate (Θ) vs true strain curve of sample D. Note that the Θ maximum corresponds to the inflection point marked by a 'x'. (b) Θ vs true stress curves. Square indicates the ultimate tensile strength. (c) Specific yield strength (SYS, i.e. yield strength-to-mass density ratio) versus uniform elongation. (d) Yield stress-ultimate tensile stress ratio (YS-to-UTS ratio) versus uniform elongation. The present HSSS is compared with the Fe-Mn-Al-C-Ni based HSSS reported by Kim et al. (Ref. [27]) and the conventional Fe-Mn-Al-C based steels [5, 11, 17, 20, 23–36].

reported previously by Kim et al. [27]. Figure 38.4b gives the strain hardening rate ($\Theta = \dfrac{\partial \sigma}{\partial \varepsilon}$) versus true stress curves. The hardening rate Θ evolves in a similar trend. That is, Θ drops rapidly first, even to below zero in curves C and D, followed by an up-turn to reach its maximum. In particular, as seen in the inset of Fig. 38.4a, the Θ maximum corresponds to the inflection point marked by 'x'. In other words, Θ recovers to its maximum when the transient ends. Such a transient was widely observed before [48–51], indicative of the Bauschinger effect (BE). It usually appears for alloys of multi-phases with varying yield stresses, over a strain regime corresponding to heterogeneous elasto-plastic deformation [36, 38]. Similar behavior of Θ has also been observed recently in gradient structure [52, 53] and heterostructured lamella structure [41].

38.3.3 Strain Hardening due to HDI Stress

To probe the mechanism of strain hardening and especially the origin of the Θ up-turn, LUR tests (Fig. 38.5a) were conducted. Two typical features are seen in unload-reload cycles. One is the hysteresis loop (Fig. 38.5b), while the other is the unload yield effect, i.e. yield-drop $\Delta\sigma_y$ (inset in Fig. 38.5b), upon each reload. These are schematically depicted in Fig. 38.5c. These hysteresis loops are evidence of the BE. Interestingly, the hysteresis loop appears even from the onset of the transient (upper panel in Fig. 38.4b). Moreover, this BE appears to be strong: the reverse plastic flow (σ_{rev}) starts even when the overall applied stress is still in tension during unloading. The BE could be described by the reverse plastic strain (ε_{rp}), Fig. 38.5c, which increases as plastic strains increase (Fig. 38.5d).

The most common explanation for the hysteresis loops in interrupted tensile tests is based on the HDI stresses resulting from inhomogeneous plastic deformation [36, 38, 54, 55]. Here, the macroscopic stress is separated into the internal stress (σ_{HDI}) and effective stress (σ_{eff}) [56]. The former is generally associated with a long-range stress on mobile dislocations and the latter is the stress required for a dislocation to overcome local obstacles. σ_{HDI} can be calculated as [56],

$$\sigma_{HDI} = \frac{(\sigma_{flow} + \sigma_{rev})}{2} - \frac{\sigma^*}{2} \tag{38.1}$$

where σ_{flow} is the flow stress upon unloading and σ^* is the thermal part of the flow stress. These parameters are defined in Fig. 38.5c. The values of σ_{rev} are monitored, as shown in Fig. 38.5e, at three selected strain offsets, with increasing deviation from the initial linear segment on each loop.

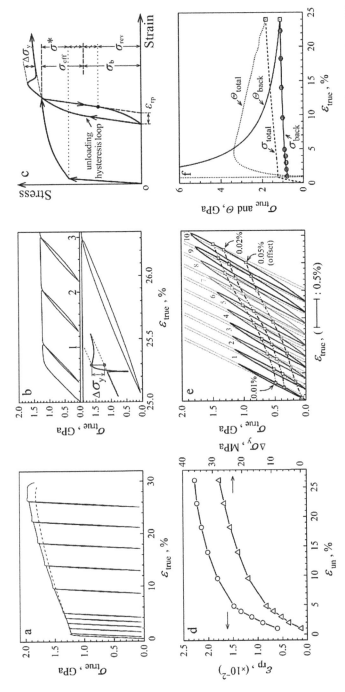

Figure 38.5 HDI hardening. (a) Tensile LUR true $\sigma - \varepsilon$ curve for sample D. The monotonic $\sigma - \varepsilon$ curve is also shown for comparison. (b) Close-up of hysteresis loops of the first three in a (upper) and the last loop (lower). Inset shows the unloading yielding effect ($\Delta\sigma_y$). (c) Schematic of two typical features in each unloading-reloading cycle, i.e. hysteresis loop and $\Delta\sigma_y$. (d) Reverse plastic strain (ε_{rp}) and $\Delta\sigma_y$ versus unloading strains (ε_{un}). (e) Loop assembly showing the unloading yield stress (σ_{un}) marked by square by three strain offsets, namely 0.01%, 0.02%, and 0.05%. (f) HDI stress and hardening rate versus true strain.

Figure 38.5f shows the σ_{HDI} and the hardening rate due to HDI stresses, θ_{HDI} as the tensile strain increases, based on σ_{rev} calculated at 0.02% offset. With increasing strain, σ_{HDI} increases, while θ_{HDI} decreases. Interestingly, θ_{HDI} is initially much higher than Θ_{total} but shows a steep drop during the transient. This indicates that σ_{HDI} is responsible for the Θ up-turn. Meanwhile, when selecting σ_{rev} with different strain offset (Fig. 38.5e), the similar trend appears in θ_{HDI}, even though the σ_{HDI} values differ significantly.

38.3.4 Load Transfer and Strain Partitioning

During the plastic deformation, the softer γ grains are easier to deform than the harder B2 grains. This causes plastic strain partitioning where the soft grains carry higher plastic strains. As seen in Figs. 38.6a–c, the present HSSS is a dual-phase material. That is, the load transfer and strain redistribution will occur due to the highly heterostructure. In situ high energy X-ray diffraction measurements have been employed to provide atomic-level strains to help understand the underlying mechanism.

38.3.4.1 Load transfer revealed by in situ diffraction measurements

Figure 38.6a shows both the true $\sigma - \varepsilon$ and $\Theta - \varepsilon$ curves, measured in situ together with the high-energy X-ray diffraction measurements of the lattice strain. The stress-strain behavior and the Θ up-turn are similar to those observed in Fig. 38.4a. The stress partitioning during tensile loading was reflected from the lattice strain evolution of both phases. Figure 38.6b is the lattice strain as a function of the applied tensile strain in the axial (loading) and transverse direction, respectively, in both the γ and B2 phases. The lattice strains were calculated using $(d^{hkl} - d_0^{hkl})/d_0^{hkl}$, where d_0^{hkl} is the d-spacing of the (hkl) plane at zero applied stress. As the applied strain increases, both the ε_γ^{111} and ε_{B2}^{110} in axial direction roughly follow a three-stage evolution. In stage I, both the ε_γ^{111} and ε_{B2}^{110} coincide well and increase linearly. In stage II, ε_γ^{111} deviates from the straight line at the strain of 0.06, while ε_{B2}^{110} still keeps rising linearly. A close-up is shown in Fig. 38.6c. As the tensile strain increases further to ~0.07, a rapid drop of ε_{B2}^{110} appears. In stage III, both ε_γ^{111} and ε_{B2}^{110} begin to rise once again, but with obviously different slopes.

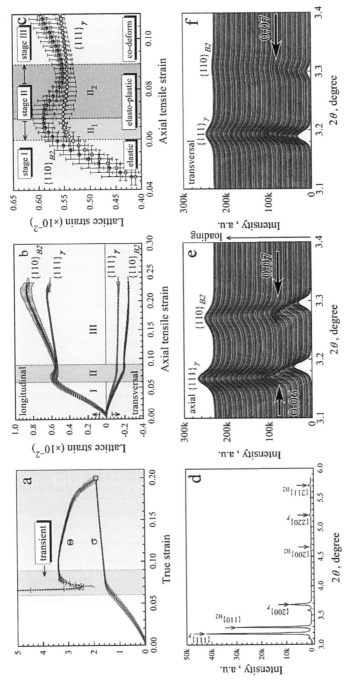

Figure 38.6 Stress and strain partitioning from in situ X-ray diffraction measurements. (a) $\sigma - \varepsilon$ and $\Theta - \varepsilon$ curves. Note the Θ up-turn. (b) Lattice strains in axial (loading) and transverse direction, respectively, in both γ and B2 versus applied tensile strain. Note the three-stage evolution of lattice strains. (c) Close-up of the three stages in (b) in axial direction. Also note stage II consists of II_1 where γ begins to yield while B2 stays elastic, and II_2 where B2 starts to yield. (d) High-energy X-ray diffraction spectra. (e and f) Diffraction patterns in axial and transverse directions, respectively. The applied tensile strain increases from bottom to top.

Figure 38.6d shows the high energy X-ray diffraction pattern. As seen in Fig. 38.6e, both $\{111\}_\gamma$ and $\{110\}_{B2}$ diffraction peaks maintain constant intensity initially, while their positions shift to smaller 2ϑ along the axial direction; an opposite trend appears in transverse direction (see Fig. 38.6f). For the axial strains over the range studied, the intensities of all the diffraction peaks keep almost constant. A sudden increase in the full width at half-maximum (FWHM) is visible for $\{111\}_\gamma$ and $\{110\}_{B2}$, respectively, at a strain of 0.06 and 0.07. At the same time, it is interesting to observe that the $\{110\}_{B2}$ peak starts to diminish in the transverse direction. In addition, both $\{200\}_{B2}$ and $\{211\}_{B2}$ peaks lower their heights gradually with plastic deformation and disappear eventually, similar to what has happened to $\{110\}_{B2}$.

38.3.4.2 Strain partitioning from aspect ratio measurements

The strain partitioning is such that γ bears a large amount of plastic strains. This is measured by the aspect ratio changes before and after tensile tests, monitored also in sample D. The aspect ratio information before tensile testing was in Figs. 38.2a to c, for γ, granular B2, and lamellar B2, respectively. Figures 38.7a–c show the SEM images of the longitudinal section after tensile testing to a true strain of 25%. Most of the initially equi-axed γ grains (Fig. 38.2c) now become strongly elongated (gray contrast) along the tensile direction as seen in the inset of Fig. 38.7a. The mean spacing $\bar{l}$ and $\bar{w}$, and aspect ratio α in the tensile direction are shown in Figs. 38.7d–f. The mean true strain in each phase can be calculated as $\bar{\varepsilon} = (2/3) \cdot \ln(\alpha_{\text{after}}/\alpha_{\text{before}})$. The derivation process of the formulas and calculation method is fully described in Ref. [41]. The strain experienced by γ, granular and lamellar B2, respectively, is about 27%, 7%, and 1%, respectively, at the total applied strain of ~26%.

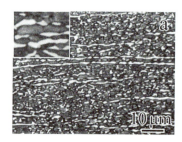

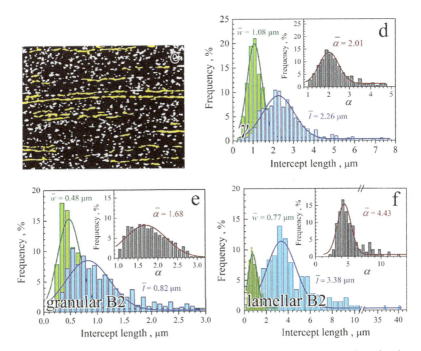

Figure 38.7 Aspect ratio measurements after tensile testing in sample D. (a–c) SEM images after tensile tests in γ (light-grey) and granular (red) and lamellar (yellow) B2. (d–f) Corresponding histograms of aspect ratios. In (b) and (c), the B2 particles and lamellae were colored by means of manual analysis based on the SEM image.

38.4 Discussion

38.4.1 Plastic Deformation in HSSS

Figure 38.6 shows three stages of the lattice strain evolution with applied axial strains in the γ and B2 phases. In stage I, both γ and B2 phases deform elastically, with a linear increase of both ε_γ^{111} and ε_{B2}^{110}. In stage II, as shown in Fig. 38.6c, ε_γ^{111} is the first to deviate from linearity at a strain of 0.06, while ε_{B2}^{110} continues to rise linearly. This indicates the onset of yielding in γ, at the microscopic level, due to its lower yield stress than that of B2 [39]. Exactly from this very moment, the stress shifts from γ to the still elastic B2. With further deformation, the load shared by γ will decrease. This leads to a drop in ε_γ^{111} and a continued rise in ε_{B2}^{110}, soon after this micro-yielding. This stage is marked as II$_1$ in Fig. 38.6c, reflecting the "grain to grain yielding" in γ due to the varying Schmid factor and strain hardening of individual grains [39, 57]. As the strain increases to 0.07, a rapid drop of ε_{B2}^{110} appears, which signals the onset of the stage II$_2$. This drop of ε_{B2}^{110} indicates declined stresses that are delivered from γ to B2 through phase boundaries. That is, the B2 phase starts to yield. The fact that ε_{B2}^{110} still rises to large strains rules out the possibility of failure at the γ/B2 boundaries. With a severe stress concentration built up at the phase boundaries between γ and B2 during stage II$_1$, the B2 also begins to deform to provide necessary accommodation between the two phases. Hence, both strains and stresses will relax to a large degree at the γ/B2 phase boundaries. As a result, ε_{B2}^{110} drops rapidly. As shown in Fig. 38.6e, an increase in FWHM of the $\{111\}_\gamma$ and $\{110\}_{B2}$ diffraction peak at axial strain of 0.06 and 0.07 is due to dislocation generation in the crystal, indicative of the onset of micro-yielding in γ and B2, respectively. Hence, stage II indicates the elasto-plastic transition, as the γ phase deforms plastically first (II$_1$) followed by the B2 phase (II$_2$).

In stage III, a rise in both ε_γ^{111} and ε_{B2}^{110} is visible once again at strains larger than 0.088. This indicates the onset of the co-deformation in both γ and B2. Interestingly, the slope ($\partial\varepsilon_{B2}^{110}/\partial\varepsilon_{axial}$) for lattice strains to rise in B2 is much larger than the $\partial\varepsilon_\gamma^{111}/\partial\varepsilon_{axial}$. This indicates that the load transfer occurs once again during the co-deformation of two phases. As a result, strain partitioning occurs between the γ and B2 phases, as is further evidenced by the aspect ratio measurements shown in Fig. 38.7. Moreover, an increase in ε_{B2}^{110} with axial strains is significant, indicating the strong and continued strain hardening in B2 all the way to final fracture. In addition, an increase in both ε_γ^{111} and ε_{B2}^{110} also indicates the recovery of hardening capacity in both γ and B2 in this stage. Both γ and B2 strain harden and participate in the load sharing, even though B2 carries the most part of applied load. As the axial strain increases further, the intensities of all the diffraction peaks stay almost constant, as seen in Fig. 38.6e. This indicates the co-deformation of both the γ and B2 phases.

As seen in Fig. 38.6f, the $\{110\}_{B2}$, as well as $\{200\}_{B2}$ and $\{211\}_{B2}$ peaks, begin to disappear with strains beyond the inflection point during the tensile testing. This indicates crystal re-orientation, e.g. rotation of the grains, in the B2 phase for the purpose of accommodating strain.

38.4.2 Strain Hardening

By comparing Figs. 38.6a and c, the onset of B2 yielding corresponds to the rapid drop of Θ, while the Θ starts its up-turn at the end of this yielding. This means that the stress–strain curve in the transient before the inflection point corresponds to the elasto-plastic transition. The interesting Θ evolution (Figs. 38.4b and 38.6a) originates, therefore, from the dual-phase nature of plastic deformation in HSSS. Under tensile loading, γ will plastically deform first. However, due to the constraint by the still elastic B2, dislocations in γ are piled up and blocked at phase boundaries. Geometrically necessary dislocations (GNDs) will be generated at phase boundaries due to the strain incompatibility of the two phases [40, 58]. This produces the internal stresses to make it difficult for dislocations to slip in γ grains until B2 grains start to yield at a larger global strain. The long-range HDI stresses develop as a result. Additionally, the intra-granular internal stresses also occur, which is due to the dislocations inside the γ grains, along the way with the "grain to grain yielding" due to the misfit plastic strains among neighboring γ grains [59]. This is the reason of a very high θ_{HDI} at the onset of the elasto-plastic transition as shown in Fig. 38.5f. Therefore, the high strain hardening rate is originated from both the HDI hardening and forest dislocation hardening in γ. Plastic incompatibilities that result from microstructural heterogeneity induce large internal stresses. Also, after the whole B2 grains yield, Θ is sustained at a level corresponding to the HDI stresses resulting from the strain incompatibility, together with dislocation multiplication in both γ and B2.

Note that the B2 grains are plastically deformable and can store dislocations, as is shown in Fig. 38.8 by TEM observations. During the co-deformation of both the γ and B2 phases (stage III), as seen in Figs. 38.6b and c, the lattice strain in B2, ε_{B2}^{110}, increases with the axial strain, ε_{axial}. But the slope $\partial\varepsilon_{B2}^{110}/\partial\varepsilon_{axial}$ is smaller than that of elastic stage I. If B2 is elastic in stage III, its slope should remain unchanged. This indicates that plastic deformation does occur in B2, which is further supported by the aspect ratio measurements (Fig. 38.7). The average macroscopic plastic strain in the granular and lamellar B2 is 10%, and 2%, respectively. On the other hand, the slope for lattice strains to rise in B2 is much larger than that in γ. This indicates even stronger strain hardening in B2 than that in γ. The load transfer and HDI stress is probably responsible for this strong strain hardening in B2. Moreover, the observation of a high density of dislocations after tensile testing lends support to the assertion that B2 can sustain strain hardening during tensile deformation.

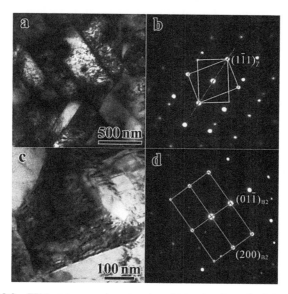

Figure 38.8 TEM micrographs showing high density of dislocations in both γ (a) and B2 (c) grains at tensile strain of 25% in sample D. (b and d) Corresponding selected area diffraction patterns with $[011]_\gamma$ and $[011]_{B2}$ zone axes, respectively.

In addition, the strains should be continuous at the phase boundaries between γ and B2, with a strain gradient near the phase boundaries [60, 61]. As a result, the GNDs will be generated to accommodate the strain gradient [52], contributing to the dislocation pile-up near phase boundaries that elevates the HDI stresses [37, 40, 62]. In other words, the HDI stress associated with strain partitioning should have contributed to the observed high strain hardening rate. Moreover, γ grains are observed to be stretched along their length direction by as much as 100 pct on average (by comparison of $\bar{l}$ before and after tensile testing in Figs. 38.2a and 38.7d); this plastic deformation is expected to increase the dislocation density inside γ grains interior.

38.4.3 Unloading Yield Effect

As seen in Fig. 38.4a, there is a yield-drop in the initial yielding. The inset in Fig. 38.5b and d also indicate that the new yield stress upon each reloading is also higher than the flow stress at the point of unloading. These two observations may have the same underlying mechanism. Such a behavior resembles the yield-drop phenomenon in carbon steels [63].

The yield-drop phenomenon in an interrupted tensile test was first observed in a single crystal Al-4.5% Cu alloy [64]. A more systematic study on commercial aluminium alloys 2024, 7075and 6061 was reported subsequently by Nieh and Nix [65] who used the term "unloading yield effect" to describe this phenomenon. We shall also use this latter term as it avoids being ambiguous with the distinct yield point phenomenon observed in carbon steels. Nieh and Nix [65] proposed a qualitative model based on the healing of sheared coherent precipitates for the appearance of the unloading yield effect in Al alloys. According to their model, the precipitates sheared by moving dislocations are quickly healed by diffusion processes when plastic deformation is paused by unloading. However, this mechanism does not seem to be applicable if a sample with little precipitates still exhibits unloading yield effect.

The unloading yield effect may be understood as follows. Once unloaded, the B2 phase becomes elastic. Upon reloading, firstly, the B2 stays elastic while γ begins to deform plastically. The yield peak appears once again upon reloading due to the load transfer. As the tensile strain increases, the mobile dislocation density decreases in γ and a higher stress is needed for the γ to yield, leading to an increased $\Delta \sigma_y$. Secondly, once the B2 yields, rapid relaxation of elastic stresses and strains on the γ/B2 boundaries causes the stress-drop.

38.5 Conclusions

We have analyzed the strain hardening process in the Fe–16Mn–10Al–0.86C–5Ni high specific strength steel. As the steel has a heterostructure composed of a γ-austenite matrix containing the B2 FeAl second phase, our treatment is from the perspective of a dual-phase that are both deformable with significant strain hardening capability. In situ high-energy X-ray diffraction revealed the lattice strain evolution in both phases. The softer phase deformed first, shedding load to harder regions, eventually causing the latter to deform. The load transfer and strain partitioning unraveled from the in situ data indicates an elasto-plastic transition, with grain-to-grain yielding until the yielding of all grains, in the order of softer γ and harder B2 and finally co-deformation. As such, the atomic-scale information shed light on the origin of the high HDI stresses that have been measured, of the pronounced Bauschinger effect, as well as of the rapid yield drop and subsequent up-turn of the strain hardening rate in the transient. The HDI hardening is believed to play a crucial role for the high strain hardening rate that has sustained the large elongation in the HSSS.

Acknowledgments

This work was financially supported by the National Natural Science Foundation of China (NSFC) under grant nos. 11572328, 11072243, 11222224, 11472286, and 11021262, the 973 Programs under grant nos. 2012CB932203, 2012CB937500, and 6138504. E.M. was supported at the Johns Hopkins University by U.S.-DOE-BES, Division of Materials Sciences and Engineering, under Contract No. DE-FG02-09ER46056.

References

1. M. Militzer, A synchrotron look at steel, *Science* 298 (2002) 975–976.
2. Y. Kimura, T. Inoue, F. Yin, K. Tsuzaki, Inverse temperature dependence of toughness in an ultrafine grain-structure steel, *Science* 320 (2008) 1057–1060.

3. F. G. Caballero, H. K. D. H. Bhadeshia, Very strong bainite, *Curr. Opin. Solid State Mater. Sci.* 8 (2004) 251–257.
4. D. Raabe, H. Springer, I. Gutierrez-Urrutia, F. Roters, M. Bausch, J. B. Seol, K. Koyama, P. -P. Choi, K. Tsuzaki, Alloy design, combinatorial synthesis, and microstructure-property relations for low-density Fe-Mn-Al-C austenitic steels, *JOM* 66 (2014) 1845–1856.
5. D. Raabe, C. C. Tasan, H. Springer, M. Bausch, From high-entropy alloys to high-entropy steels, *Steel Res. Int.* 86 (2015) 1127–1138.
6. O. Bouaziz, S. Allain, C. P. Scott, P. Cugy, D. Barbier, High manganese austenitic twinning induced plasticity steels: a review of the microstructure properties relationships, *Curr. Opin. Solid State Mater. Sci.* 15 (2011) 141–168.
7. T. Furuta, S. Kuramoto, T. Ohsuna, K. Oh-ishi, K. Horibuchi, Die-hard plastic deformation behavior in an ultrahigh-strength Fe-Ni-Al-C alloy, *Scr. Mater.* 101 (2015) 87–90.
8. G. Frommeyer, U. Brüx, P. Neumann, Supra-ductile and high-strength manganese-TRIP/TWIP steels for high energy absorption purposes, *ISIJ Int.* 43 (2003) 438–446.
9. D. Raabe, D. Ponge, O. Dmitrieva, B. Sander, Nanoprecipitate-hardened 1.5 GPa steels with unexpected high ductility, *Scr. Mater.* 60 (2009) 1141–1144.
10. O. Bouaziz, S. Allain, C. Scott, Effect of grain and twin boundaries on the hardening mechanisms of twinning-induced plasticity steels, *Scr. Mater.* 58 (2008) 484–487.
11. S. Kang, Y. -S Jung, J. -H Jun, Y. -K. Lee, Effects of recrystallization annealing temperature on carbide precipitation, microstructure, and mechanical properties in Fe-18Mn-0.6C-1.5Al TWIP steel, *Mater. Sci. Eng. A* 527 (2010) 745–751.
12. Y. I. Son, Y. K. Lee, K. T. Park, C. S. Lee, D. H. Shin, Ultrafine grained ferrite-martensite dual phase steels fabricated via equal channel angular pressing: microstructure and tensile properties, *Acta Mater.* 53 (2005) 3125–3134.
13. M. Calcagnotto, Y. Adachi, D. Ponge, D. Raabe, Deformation and fracture mechanisms in fine-and ultrafine-grained ferrite/martensite dual-phase steels and the effect of aging, *Acta Mater.* 59 (2011) 658–670.
14. H. K. D. H. Bhadeshia, The first bulk nanostructured metal, *Sci. Tech. Adv. Mater.* 14 (2013) 014202.

15. R. Ueji, N. Tsuji, Y. Minamino, Y. Koizumi, Ultragrain refinement of plain low carbon steel by cold-rolling and annealing of martensite, *Acta Mater.* 50 (2002) 4177–4189.
16. Y. Li, D. Raabe, M. Herbig, P. P. Choi, S. Goto, A. Kostka, H. Yarita, C. Borchers, R. Kirchheim, Segregation stabilizes nanocrystalline bulk steel with near theoretical strength, *Phys. Rev. Lett.* 113 (2014) 106104.
17. G. Frommeyer, U. Bruex, Microstructures and mechanical properties of high-strength Fe-Mn-Al-C light-weight TRIPLEX steels, *Steel Res. Int.* 77 (2006) 627–633.
18. R. Rana, C. Lahaye, R. K. Ray, Overview of lightweight ferrous materials: strategies and promises, *JOM* 66 (2014) 1734–1746.
19. H. Kim, D. W. Suh, N. J. Kim, Fe-Al-Mn-C lightweight structural alloys: a review on the microstructures and mechanical properties, *Sci. Tech. Adv. Mater.* 14 (2013) 014205.
20. S. S. Sohn, H. Song, B. C. Suh, J. H. Kwak, B. J. Lee, N. J. Kim, S. Lee, Novel ultra-high-strength (ferrite + austenite) duplex lightweight steels achieved by fine dislocation substructures (Taylor lattices), grain refinement, and partial recrystallization, *Acta Mater.* 96 (2015) 301–310.
21. C. L. Lin, C. G. Chao, J. Y. Juang, J. M. Yang, T. F. Liu, Deformation mechanisms in ultrahigh-strength and high-ductility nanostructured FeMnAlC alloy, *J. Alloys Compd.* 586 (2014) 616–620.
22. I. Gutierrez-Urrutia, D. Raabe, Multistage strain hardening through dislocation substructure and twinning in a high strength and ductile weight-reduced Fe-Mn-Al-C steel, *Acta Mater.* 60 (2012) 5791–5802.
23. I. Gutierrez-Urrutia, D. Raabe, Influence of Al content and precipitation state on the mechanical behavior of austenitic high-Mn low-density steels, *Scr. Mater.* 68 (2013) 343–347.
24. J. D. Yoo, S. W. Hwang, K. T. Park, Factors influencing the tensile behavior of a Fe-28Mn-9Al-0.8 C steel, *Mater. Sci. Eng. A* 508 (2009) 234–240.
25. S. W. Hwang, J. H. Ji, E. G. Lee, K. T. Park, Tensile deformation of a duplex Fe-20Mn-9Al-0.6 C steel having the reduced specific weight, *Mater. Sci. Eng. A* 528 (2011) 5196–5203.
26. Z. Q. Wu, H. Ding, X. H. An, D. Han, X. Z. Liao, Influence of Al content on the strain-hardening behavior of aged low density Fe-Mn-Al-C steels with high Al content, *Mater. Sci. Eng. A* 639 (2015) 187–191.
27. S. H. Kim, H. Kim, N. J. Kim, Brittle intermetallic compound makes ultrastrong low-density steel with large ductility, *Nature* 518 (2015) 77–79.

28. H. K. D. H. Bhadeshia, TRIP-assisted steels ?, *ISIJ Int.* 42 (2002) 1059–1060.
29. C. C. Tasan, M. Diehl, D. Yan, C. Zambaldi, P. Shanthraj, F. Roters, D. Raabe, Integrated experimental-simulation analysis of stress and strain partitioning in multiphase alloys, *Acta Mater.* 81 (2014) 386–400.
30. H. Ghassemi-Armaki, P. Chen, S. Bhat, S. Sadagopan, S. Kumar, A. Bower, Microscale-calibrated modeling of the deformation response of low-carbon martensite, *Acta Mater.* 61 (2013) 3640–3652.
31. B. Qin, H. K. D. H. Bhadeshia, Plastic strain due to twinning in austenitic TWIP steels, *Mater. Sci. Tech.* 24 (2008) 969–973.
32. S. Cheng, X. L. Wang, Z. Feng, B. Clausen, H. Choo, P. K. Liaw, Probing the characteristic deformation behaviors of transformation-induced plasticity steels, *Metall. Mater. Trans. A* 39A (2008) 3105–3112.
33. O. Bouaziz, S. Allain, C. Scott, Effect of grain and twin boundaries on the hardening mechanisms of twinning-induced plasticity steels, *Scr. Mater.* 58 (2008) 484–487.
34. J. G. Sevillano. An alternative model for the strain hardening of FCC alloys that twin, validated for twinning-induced plasticity steel, *Scr. Mater.* 60 (2009) 336–339.
35. A. A. Saleh, E. V. Pereloma, B. Clausen, D. W. Brown, C. N. Tomé, A. A. Gazder, On the evolution and modelling of lattice strains during the cyclic loading of TWIP steel, *Acta Mater.* 61 (2013) 5247–5262.
36. L. Thilly, S. Van Petegem, P. O. Renault, F. Lecouturier, V. Vidal, B. Schmitt, H. Van Swygenhoven, A new criterion for elasto-plastic transition in nanomaterials: application to size and composite effects on Cu-Nb nanocomposite wires, *Acta Mater.* 57 (2009) 3157–3169.
37. Y. Xiang, J. J. Vlassak, Bauschinger effect in thin metal films, *Scr. Mater.* 53 (2005): 177–182.
38. C. W. Sinclair, G. Saada, J. D. Embury, Role of internal stresses in co-deformed two-phase materials, *Phil. Mag.* 86 (2006) 4081–4098.
39. S. Harjo, Y. Tomota, D. Neov, P. Lukas, M. Vrana, P. Mikula, Bauschinger effect in alpha-gamma dual phase alloys studied by in situ neutron diffraction, *ISIJ int.* 42 (2002) 551–557.
40. M. F. Ashby, The deformation of plastically non-homogeneous materials, *Phil. Mag.* 21 (1970) 399–424.
41. X. L. Wu, M. X. Yang, F. P. Yuan, G. L. Wu, Y. J. Wei, X. X. Huang, Y. T. Zhu, Heterogeneous lamella structure unites ultrafine-grain strength with

coarse-grain ductility, *Proc. Natl. Acad. Sci. USA* 112 (2015) 14501–14505.

42. J. Rajagopalan, J. H. Han, M. T. A. Saif, Bauschinger effect in unpassivated freestanding nanoscale metal films, *Scr. Mater.* 59 (2008) 734–737.

43. J. Rajagopalan, C. Rentenberger, H. P. Karnthaler, G. Dehm, M. T. A. Saif, In situ TEM study of microplasticity and Bauschinger effect in nanocrystalline metals, *Acta Mater.* 58 (2010) 4772–4782.

44. J. Llorca, A. Needleman, S. Suresh, The Bauschinger effect in whisker-reinforced metal-matrix composites, *Scr. Metall. Mater.* 24 (1990) 1203–1208.

45. Z. H. Cong, N. Jia, X. Sun, Y. Ren, J. Almer, Y. D. Wang, Stress and strain partitioning of ferrite and martensite during deformation, *Metall. Mater. Trans. A* 40A (2009) 1383–1387.

46. S. Cheng, Y. D. Wang, H. Choo, J. Almer, Y. Lee, P. K. Liaw, An assessment of the contributing factors to the superior properties of a nanostructured steel using in situ high-energy X-ray diffraction, *Acta Mater.* 58 (2010) 2419–2429.

47. Z. L. Wang, P. Zheng, Z. H. Nie, Y. Ren, Y. D. Wang, P. Müllner, D. C. Dunand, Superelasticity by reversible variants reorientation in a Ni-Mn-Ga microwire with bamboo grains, *Acta Mater.* 99 (2015) 373–381.

48. T. Hasegawa, T. Yakou, U. F. Kocks, Forward and reverse rearrangements of dislocations in tangled walls, *Mater. Sci. Eng.* 81 (1986) 189–199.

49. M. G. Stout, A. D. Rollett, Large-strain Bauschinger effects in fcc metals and alloys, *Metall. Trans. A* 21A (1990) 3201–3213.

50. A. Deschamps, B. Decreus, F. De Geuser, T. Dorin, M. Weyland, The influence of precipitation on plastic deformation of Al-Cu-Li alloys, *Acta Mater.* 61 (2013) 4010–4021.

51. G. Fribourg, Y. Bréchet, A. Deschamps, A. Simar, Microstructure-based modelling of isotropic and kinematic strain hardening in a precipitation-hardened aluminium alloy, *Acta Mater.* 59 (2011) 3621–3635.

52. X. L. Wu, P. Jiang, L. Chen, F. P. Yuan, Y. T. Zhu, Extraordinary strain hardening by gradient structure, *Proc. Natl. Acad. Sci. USA* 111 (2014) 7197–7201.

53. X. C. Yang, X. L. Ma, J. Moering, H. Zhou, W. Wang, Y. L. Gong, J. M. Tao, Y. T. Zhu, X. K. Zhu, Influence of gradient structure volume fraction on the mechanical properties of pure copper, *Mater. Sci. Eng. A* 645 (2015) 280–285.

54. G. D. Moan, J. D. Embury, A study of the Bauschinger effect in Al-Cu alloys, *Acta Metall.* 27 (1979) 903–914.

55. H. Mughrabi, Dislocation wall and cell structures and long-range internal stresses in deformed metal crystals, *Acta Metall.* 31 (1983) 1367–1379.

56. X. Feaugas, On the origin of the tensile flow stress in the stainless steel AISI 316L at 300 K: back stress and effective stress, *Acta Mater.* 47 (1999) 3617–3632.

57. Y. Tomota, P. Lukas, S. Harjo, J. H. Park, N. Tsuchida, D. Neov, In situ neutron diffraction study of IF and ultra-low carbon steels upon tensile deformation, *Acta Mater.* 51 (2003) 819–830.

58. H. Mughrabi, On the role of strain gradients and long-range internal stresses in the composite model of crystal plasticity, *Mater. Sci. Eng. A* 317 (2001) 171–180.

59. B. Chen, J. N. Hu, Y. Q. Wang, S. Y. Zhang, S. Van Petegem, A. C. F. Cocks, D. J. Smith, P. E. J. Flewitt, Role of the misfit stress between grains in the Bauschinger effect for a polycrystalline material, *Acta Mater.* 85 (2015) 229–242.

60. X. L. Wu, P. Jiang, L. Chen, J. F. Zhang, F. P. Yuan, Y. T. Zhu, Synergetic strengthening by gradient structure, *Mater. Res. Lett.* 2 (2014) 185–191.

61. J. J. Li, S. H. Chen, X. L. Wu, A. K. Soh, A physical model revealing strong strain hardening in nano-grained metals induced by grain size gradient structure, *Mater. Sci. Eng. A* 620 (2015) 16–21.

62. A. T. Jennings, C. Gross, F. Greer, Z. H. Aitken, S. W. Lee, C. R. Weinberger, J. R. Greer, Higher compressive strengths and the Bauschinger effect in conformally passivated copper nanopillars, *Acta Mater.* 60 (2012) 3444–3455.

63. N. Tsuji, Y. Ito, Y. Saito, Y. Minamino, Strength and ductility of ultrafine grained aluminum and iron produced by ARB and annealing, *Scr. Mater.* 47 (2002) 893–899.

64. G. Greetham, R. W. K. Honeycombe, The deformation of single crystals of aluminium 4.5-percent copper alloy, *J. Inst. Met.* 89 (1960) 13–21.

65. T. G. Nieh, W. D. Nix, Unloading yield effects in aluminum alloys, *Metall. Trans. A*, 17A (1986) 121–126.

Chapter 39

Deformation Mechanisms for Superplastic Behaviors in a Dual-Phase High Specific Strength Steel with Ultrafine Grains

Wei Wang,[a,b] Muxin Yang,[a] Dingshun Yan,[a] Ping Jiang,[a] Fuping Yuan,[a,b] and Xiaolei Wu[a,b]

[a]*State Key Laboratory of Nonlinear Mechanics, Institute of Mechanics, Chinese Academy of Science, No. 15, North 4th Ring, West Road, Beijing 100190, China*
[b]*School of Engineering Science, University of Chinese Academy of Sciences, Beijing 100190, China*
fpyuan@lnm.imech.ac.cn

The superplastic behaviors of a high specific strength steel (HSSS) with dual-phase microstructure and ultrafine grains have been investigated under a temperature range of 873–973 K and at a wide strain rate range of 10^{-4}–10^{-1}/s. The ultrafine grained HSSS exhibits excellent superplastic properties. The microstructure observations

---

Reprinted from *Mater. Sci. Eng. A*, **702**, 133–141, 2017.

Heterostructured Materials: Novel Materials with Unprecedented Mechanical Properties
Edited by Xiaolei Wu and Yuntian Zhu
Text Copyright © 2017 Elsevier B.V.
Layout Copyright © 2022 Jenny Stanford Publishing Pte. Ltd.
ISBN 978-981-4877-10-7 (Hardcover), 978-1-003-15307-8 (eBook)
www.jennystanford.com

at interrupted strains for tests under temperature of 973 K and at strain rate of 10^{-3}/s have provided evidences of different mechanisms for two stages. At the first stage (strain range from 0% to 400%), the superplastic flow is attributed to the diffusional transformation from fcc austenite phase to intermetallic compound B2 phase coupled with grain boundary sliding. While intragranular dislocation activities should be the dominant mechanism for the second stage (strain range from 400% to 629%) due to the increased realistic strain rate by diffusive necking. The grain sizes of both phases are observed to be relatively stable and remain always sub-micron level during the high temperature tensile deformation, facilitating the superplastic flow.

39.1 Introduction

Metals and alloys generally fail at relatively small elongations (<100%) when subjected to uniaxial tensile loading at room temperature. High elongation to fracture, i.e., superplasticity is possible under certain conditions [1–14]. It is now well established that two fundamental conditions should be satisfied in order to achieve superplastic deformation in metals and alloys: (1) High homologous temperature (generally with a testing homologous temperature of $T/T_m > 0.5$, where T_m is the absolute melting temperature) is needed; (2) A relatively small grain size is required since grain boundary sliding (GBS) is an important deformation mechanism during superplastic flow and GB density increases with decreasing grain size. Previous studies have also indicated that ultrafine grained (UFG) metals processed by severe plastic deformation (SPD) can achieve excellent superplastic properties in bulk materials [15, 16]. However, these two requirements are generally incompatible in pure metals and solid solution alloys due to the easy grain growth at high temperatures. Thus, the UFG metals are better to either consist of dual-phase or contain finely dispersed second phase, which can inhibit grain growth.

Stronger, tougher and lighter steels are always desirable in various industry or defense applications. Such expectations have been realized in recent decades by low-density steels, which are mainly based on Fe-Al-Mn-C alloy system and are so called TRIPLEX steels. These TRIPLEX steels generally consist of fcc austenite, bcc ferrite and finely dispersed nanometer-sized κ-carbides with (Fe, Mn) AlC type [17–23]. More recently, Kim et al. [24] has developed a high specific strength steel (HSSS) with composition of Fe-16Mn-10Al-0.86C-5Ni (wt %), which consists of both fcc austenite phase and intermetallic compound B2 phase. This HSSS shows excellent combination of specific strength and elongation when compared to the other low-density metals and alloys. In our previous research [25, 26], it has been shown that this UFG HSSS should be better understood as a dual-phase microstructure since the B2 phase is deformable. Thus, the stress/strain partitioning between the constituent phases and the hetero-deformation induced (HDI) strain hardening should play important roles during the plastic deformation for this HSSS.

Superplasticity is of both academic and industrial interests because it enables the production of complex parts [27]. Superplastic forming may be more widely used if the superplasticity can be achieved under lower deformation temperatures and at higher deformation strain rates [1, 28, 29]. The UFG HSSS (Fe-16Mn-10Al-0.86C-5Ni, wt %) has the potential for the application in automobile industry due to its excellent mechanical properties. This UFG HSSS also has potential for easy superplastic forming in the industry due to its dual-phase microstructure and small grain size. However, the superplastic behaviors and the corresponding deformation mechanisms of this HSSS under high temperature tensile deformation are still unknown. In this regard, the UFG HSSS was produced first by cold rolling and subsequent short-time annealing. Then, the superplastic flow behaviors of this UFG HSSS were studied by tensile tests under a temperature range of 600–700°C and at a wide strain rate range of 10^{-4}–10^{-1}/s, and the corresponding mechanisms for superplasticity are investigated by a series of interrupted tests at varying strains and the subsequent microstructure observations.

39.2 Materials and Experimental Procedures

The details for the preparation of the HSSS can be found in our previous papers [25, 26]. The hot-rolled plates with a thickness of

7.3 mm were cold rolled into a final thickness of 1.5 mm, and then were annealed at 1173 K for 15 min followed immediately by water quenching. The tensile specimens for the superplastic tests have a gauge section of 10 × 4 × 1.5 mm^3, and the tensile direction is parallel to the rolling direction. All surfaces of the specimens were carefully polished to remove any irregularities. High-temperature uniaxial tensile tests were performed on an MTS Landmark testing machine with temperatures from 873 to 973 K and engineering strain rates of 10^{-4} s^{-1}-10^{-1} s^{-1}. In order to ensure uniform temperature distributions, the specimens were held for 5 min at the given temperatures before tensile testing.

Followed by cold rolling and annealing, the microstructures of the HSSS were characterized by electron back-scattered diffraction (EBSD) and transmission electron microscopy (TEM). The details of the sample preparations and the operation procedures for obtaining EBSD and TEM images can be found in our previous papers [25]. The "frozen" microstructures from the interrupted tests (extracted from the center part of the gauge section) were also revealed by EBSD to investigate the deformation mechanisms of the superplastic flow in this HSSS. During the EBSD acquisition, a scanning area of 40 × 40 μm^2 was chosen and a scanning step of 0.07 μm was used. Grain boundaries (GBs) are defined by misorientations larger than 15°. Texture is calculated using rank 16 harmonic series expansion with 5 degree Gaussian smoothing. The favored slip system is deduced by sorting the Schimd factor of potential slip systems at each EBSD nodes. Kernel average misorientation (KAM) is calculated against all neighbors within 280 nm distance (the misorientation angles larger than 2° are excluded) [30].

39.3 Results and Discussions

The microstructures of the specimens prior to high-temperature tensile tests are shown in Fig. 39.1. In the EBSD phase image (Fig. 39.1a), two phases are clearly visible, in which one is the fcc γ-austenite phase with equiaxed grains and the other is the B2 phase (FeAl intermetallic compound) with both granular and lamellar grains. In the TEM image (Fig. 39.1b), the B2 phase is observed to be much inclined to precipitate at either GBs or triple junctions of γ-austenite matrix, instead of the γ grain interiors. Annealing twins are often seen in the γ-austenite grains. The corresponding indexed selected area diffraction patterns for both phases are also displayed in the inset of Fig. 39.1b. The average grain size is about 0.934 μm for the fcc γ-austenite phase while is about 0.475 μm for the B2 phase prior to high-temperature tensile tests. The area fraction of γ-austenite is about 82% while the area of B2 phase is about 18% for the untested sample.

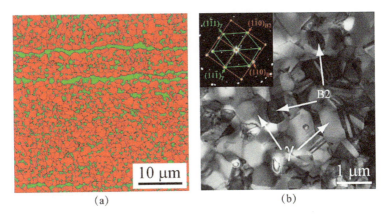

Figure 39.1 (a) EBSD phase distribution for the untested sample. (b) TEM image for the untested sample. In Fig. 39.1a, the red color is for the fcc austenite phase while the blue color is for the B2 phase. The same color coding is used for the following figures with phase distribution. The inset of Fig. 39.1b shows the indexed selected area diffraction pattern for the TEM image with an electron beam closely parallel to both the $[011]_\gamma$ and $[001]_{B2}$ zone axes.

The engineering stress-strain curves under different elevated temperatures (873, 923, 973 K) for the HSSS are shown in Fig. 39.2a, b and c, respectively. After elastic deformation, the flow stress reaches the peak stress rapidly. With further tensile deformation, the flow stress decreases slowly, and then stays at a steady state (with a very small stress level) for a long strain interval before the final fracture at low strain rates. However, the flow stress decreases directly to the final fracture rapidly at high strain rates. The undeformed and the final fracture shapes for the sample under temperature of 973 K and at strain rate of 10^{-3}/s are also shown in the inset of Fig. 39.2c. Obviously, the flow behaviors are highly dependent on the deformation conditions. The flow softening rate after peak stress is observed to decrease monotonically with increasing temperature and decreasing strain rate.

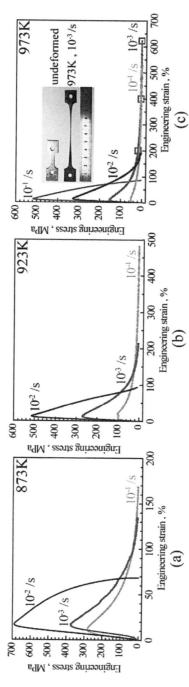

Figure 39.2 The tensile properties of the HSSS at a wide range of strain rates and under different elevated temperatures: (a) 873 K; (b) 923 K; (c) 973 K. Several interrupted tests at varying strains (200%, 400%, 629%, marked as squares in the Fig. 39.2c) have been conducted under temperature of 973 K and at strain rate of 10^{-3}/s.

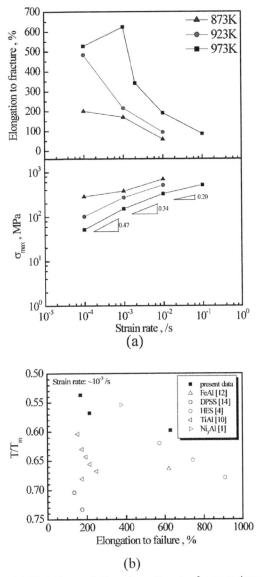

Figure 39.3 (a) Variations of the elongation to fracture (upper) and the maximum flow stress (lower) as a function of the imposed strain rate for the HSSS tested under different elevated temperatures. (b) The normalized deformation temperature vs. the elongation to fracture for the HSSS, along with the data for the other steels and the intermetallic compounds [1, 4, 10, 12, 14]. In Fig. 39.3a, the strain rate sensitivities (SRS, m) are also given for the experiments conducted at 973 K.

Figure 39.3a shows the elongation to failure and the maximum flow stress as a function of strain rate. The logarithmic coordinates are used for the strain rate and the maximum flow stress. The peak flow stress is observed to increase monotonically with decreasing temperature and increasing strain rate. These trends could be due to the following aspects: (1) The lower deformation temperature might reduce the effects of dynamic softening due to the decreasing thermal activation process; (2) The higher strain rate might decelerate the dislocation annihilation thus increase the dislocation density. The elongation to fracture increases with increasing temperature. At lower temperatures (873, 923 K), the elongation to fracture decreases monotonically with increasing strain rate. However, the elongation to fracture increases first and then decreases with increasing strain rate at higher temperature (973 K). At the same strain rate (10^{-3}/s), the normalized deformation temperature (T/T_m, where T_m is the melting temperature) is plotted against the elongation to fracture in Fig. 39.3b for the HSSS, along with the data for the other steels and the intermetallic compounds [1, 4, 10, 12, 14]. As observed, the HSSS displays excellent superplastic properties when compared to the other steels and the intermetallic compounds. For example, when compared with the duplex stainless steel (DPSS) steel and the TiAl intermetallic compound, the normalized deformation temperature for this HSSS is much lower when the elongations to fracture are similar (~200%). Moreover, when compared with the hyper-eutectoid steel (HES), the Martensitic steel and the FeAl intermetallic compound, the normalized deformation temperature for this HSSS is also lower when the elongations to fracture are similar (~600%). The HSSS shows the highest superplasticity (629%) under temperature of 973 K and at strain rate of 10^{-3}/s, thus several interrupted tests at varying strains (200%, 400%, 629%) have also been conducted in order to investigate the corresponding deformation mechanisms for the superplastic behaviors.

The EBSD phase distributions at varying times under static annealing (under temperature of 973 K and without deformation) are shown in Figs. 4a–c. While, Figs. 4d–f show the EBSD phase distributions at varying times under high temperature tensile deformation (under temperature of 973 K and strain rate of 10^{-3}/s). It should be noted that the time intervals for EBSD images are exactly the same for both static annealing condition and high temperature tensile deformation condition. The evolutions for volume fraction of B2 phase as a function of time are shown in Fig. 39.4g for both conditions. It is observed that the B2 phase with lamellar shape has a tendency to turn into granular shape under high temperature tensile deformation, while the lamellar B2 precipitates are reserved under static annealing condition. Moreover, although the phase transformation from fcc austenite phase to B2 phase occurred during the static annealing, the transformation was found to be enhanced during high temperature tensile deformation under the same temperature.

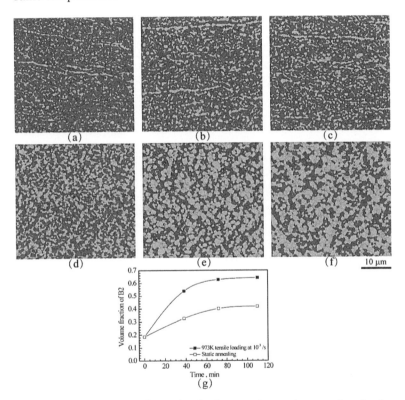

Figure 39.4 The EBSD phase distributions under static annealing (under temperature of 973 K) at varying times: (a) 38 min; (b) 72 min; (c) 110 min. The EBSD phase distributions under high temperature tensile deformation (under temperature of 973 K and strain rate of 10^{-3}/s) at varying times: (d) 38 min; (e) 72 min; (f) 110 min. (g) The evolutions of volume fraction of B2 phase as a function of time for both conditions.

The phase transformation was observed to take place randomly, and the average grain size of the B2 phase is plotted as a function of time for both static annealing and high temperature tensile deformation in Fig. 39.5a, which clearly follows the well-known $t^{-1/2}$ diffusion rule. A classical equation can be used to calculate the flow activation energy in the HSSS as following:

$$\dot{\varepsilon} = A\sigma^{1/m} \exp\left(\frac{-Q}{RT}\right) \tag{39.1}$$

where, $\dot{\varepsilon}$ is the strain rate, σ is the maximum flow stress, m is SRS, Q is the activation energy, T is the deformation temperature, A is a material constant. The logarithmic value of the maximum flow stress as a function of $10000/T$ is plotted in Fig. 39.5b for this HSSS. Thus, the activation energy during the superplastic flow behavior of this HSSS can be estimated as:

$$Q = \frac{R}{m} \frac{\partial \ln \sigma}{\partial \left(\frac{1}{T}\right)} \tag{39.2}$$

Taking the average value for SRS (m) in the temperature range of 923–973 K for this HSSS, the activation energy Q can be estimated to be about 216 kJ/mol, which is very similar to the value for solute element diffusion in the literature [31]. Moreover, this value of Q for the HSSS is much lower than the creep activation energy of 370 kJ/mol reported for binary Fe-28Al [32] and 450 kJ/mol reported for binary Fe-40Al [33]. The lower flow activation energy for this HSSS indicates that the superplastic deformation process in this HSSS should be controlled by sub-boundary and grain-boundary diffusion instead of by lattice diffusion [34].

As we know, the well-known Avrami equation can be used to describe how solids transform from one phase to another phase at constant temperature, the transformation characteristic exponent n is usually an indicator whether or not the transformation is diffusion controlled or diffusionless, and is also an indicator of what type of diffusion controlled growth [31, 35]. The Avrami equation can be described as follows:

$$V'_{B_2} = 1 - \exp(-kt^n) \tag{39.3}$$

$$\ln\ln\left(\frac{1}{1-V'_{B_2}}\right) = k + n(\ln t) \tag{39.4}$$

where $V'_{B_2} = V_{B_2}/V_{B_2,saturation}$ ($V_{B_2,saturation}$ is the volume fraction of transformed B2 phase after complete transformation) is the normalized volume fraction of transformed B2 phase, t is the time and k is a constant.

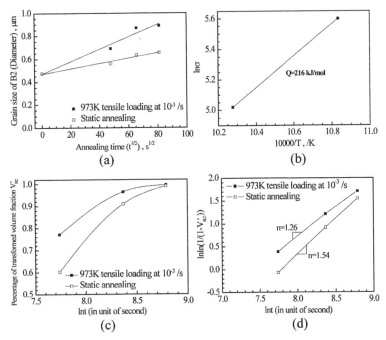

Figure 39.5 (a) The average grain size of the B2 phase as a function of time for both static annealing and high temperature tensile deformation, following the well-known $t^{-1/2}$ diffusion rule. (b) The logarithmic value of the maximum flow stress as a function of 10000/T for this HSSS. (c) The normalized volume fraction of transformed B2 phase as a function of time for both conditions. (d) The $\ln\ln\left(\dfrac{1}{1-V'_{B_2}}\right)$ vs. ln t curves for both conditions.

The normalized volume fraction of transformed B2 phase as a function of time for both static annealing and high temperature tensile deformation are plotted in Fig. 39.5c. Then the $\ln\ln\left(\dfrac{1}{1-V'_{B_2}}\right)$ vs. ln t curves for both conditions are shown in Fig. 39.5d. The transformation characteristic exponent n for both conditions can be extracted from Fig. 39.5c as the slopes of the two curves. The obtained n is 1.54 for static annealing while is 1.26 for high temperature tensile deformation. These values are relatively small, suggesting that the transformation is indeed diffusion controlled for the B2 phase growth along GBs or triple junctions [31, 35]. These values also suggest that the diffusion controlled growth is all shapes growing from small dimensions for static annealing (n value is about 1.5), while the diffusion controlled growth is growth of particles with appreciable initial volume for high temperature tensile deformation (n value is between 1 and 1.5) [35]. These suggestions are also consistent with the EBSD observations in Fig. 39.4. Based on the above discussions, the B2 phase growth is shown to be diffusion controlled along defects (sub-boundaries, GBs). Thus, the strain-enhanced transformation mentioned earlier should be due to the kinetic effect, since the defect concentrations (dislocation wall, subboundary, etc.) would increase with increasing strain in the superplastically loaded specimens, resulting in higher diffusion-controlled phase transformation.

The evolutions of grain morphology, misorientation and size (Inverse Pole Figure, IPF) under high temperature tensile deformation (under temperature of 973 K and strain rate of 10^{-3}/s) for both fcc austenite phase and B2 phase are displayed in Fig. 39.6 and Fig. 39.7, respectively. B2 phase in Fig. 39.6 and fcc austenite phase in Fig. 39.7 are blacked out for clarity. In each figure, texture is also provided as an inset and the tensile direction is also indicated by arrows. At the first stage (from undeformed state to 400% strain), it is can be clearly seen that the maximum texture intensity decreases from 1.905 to 1.157 for fcc austenite phase, and decreases from 2.389 to 1.126 for B2 phase. High temperature deformation at the first stage results in the weakening and randomization of the initial textures. It is well known that the weakening of texture and intensity during deformation are closely related to GBS and grain rotation [36–39]. As indicated in Fig. 39.7, <101> texture is much weakened for B2 phase when the tensile strain is from 0% to 200%, which could also be attributed to the texture weakening by transformation besides by GBS and grain rotation. The GBS might also be partly accommodated by the deformation-enhanced phase transformation from fcc austenite phase to B2 phase, similar to

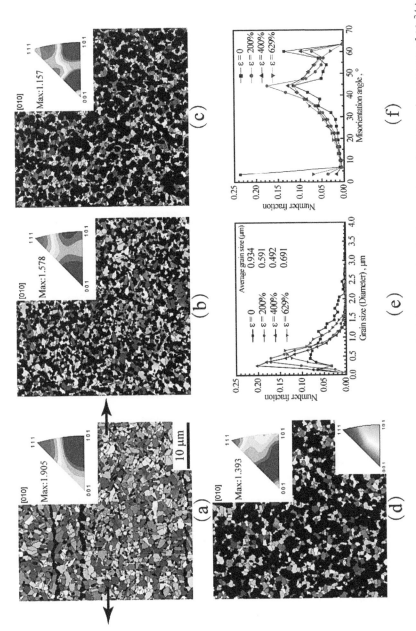

Figure 39.6 The EBSD IPF images under high temperature tensile deformation (under temperature of 973 K and strain rate of 10^{-3}/s) for fcc austenite phase: (a) $\varepsilon = 0$; (b) $\varepsilon = 200\%$; (c) $\varepsilon = 400\%$; (d) $\varepsilon = 629\%$. (e) The corresponding grain size distributions at varying tensile strains. (f) The corresponding misorientation angle distributions at varying tensile strains.

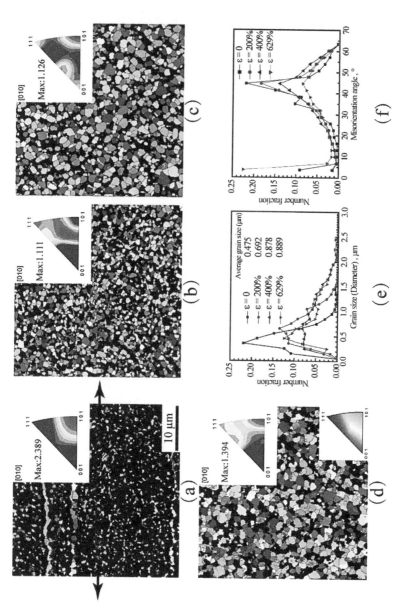

Figure 39.7 The EBSD IPF images under high temperature tensile deformation (under temperature of 973 K and strain rate of 10^{-3}/s) for B2 phase: (a) $\varepsilon = 0$; (b) $\varepsilon = 200\%$; (c) $\varepsilon = 400\%$; (d) $\varepsilon = 629\%$. (e) The corresponding grain size distributions at varying tensile strains. (f) The corresponding misorientation angle distributions at varying tensile strains.

the idea in the previous research [31]. The misorientation angle distributions at varying tensile strains for fcc austenite phase and B2 phase are shown in Fig. 39.6f and Fig. 39.7f, respectively. It is observed that the fraction of very low angle grain boundaries (LAGBs) with misorientation angle <5° decreases at the first stage for both phases. These changes indicate that the very LAGBs have been transited and migrated into high angle grain boundaries (HAGBs with misorientation angle >15°), and this transition could be attributed to dynamic recrystallization (DRX) by generation of numerous new grains with HAGBs [37].

The grain size distributions at varying tensile strains for fcc austenite phase and B2 phase are shown in Fig. 39.6e and Fig. 39.7e, respectively. It is clearly observed that the grain size of fcc austenite phase monotonically decreases while the grain size of B2 phase monotonically increases at the first stage. The grain size reduction for fcc austenite phase can be attributed to the DRX and the phase transformation inside the fcc austenite grains. While the grain growth for B2 phase can be attributed to the nucleation and the propagation of phase transformation process. Although slight grain refinement/growth is observed for both phases, the grain sizes of both phases are observed to always remain sub-micron level during the high temperature deformation. As indicated in previous research [1], UFG metals processed by SPD usually can achieve better superplastic behavior than conventional CG metals. Thus, this relatively stable UFG grains in both phase also help to achieve excellent superplasticity in this HSSS.

B2 phase is harder than fcc austenite phase, thus the benefits of the phase transformation in this HSSS are apparent to accommodate the plastic strain under high temperature. This effect is similar to the effect of transformation induced plasticity (TRIP) for austenite steels [40, 41], where stress induced martensite can accommodate the local strain concentration as a medium. The only difference between TRIP steels and the HSSS is the transformation is diffusion controlled for the present case. The strain-enhanced transformation can provide two advantages contributing to large elongation during the high temperature deformation: Firstly, the transformation strain itself can prevent early void formation by accommodating the large stress during GBS; Secondly, the B2 phase can strengthen the strain concentration region and thus help to maintain large homogeneous deformation by preventing early necking formation [41]. Thus, as a summary, the superplastic behaviors at the first stage can be attributed to the diffusional transformation from fcc austenite phase to B2 phase coupled with GBS.

While at the second stage (from 400% to 629%), it is observed that the maximum texture intensity slightly rises again from 1.157 to 1.393 for fcc austenite phase, and increases from 1.126 to 1.394 for B2 phase. It is also observed that the fraction of very LAGBs with misorientation angle <5° rises again at the second stage for both phases. In order to illustrate the deformation mechanisms for the second stage, the KAM images at varying tensile strains for fcc austenite phase and B2 phase are shown in Fig. 39.8 and Fig. 39.9, respectively. The magnitude of KAM generally represents the dislocation density (especially for geometrically necessary dislocations, GNDs) in the grain interior. Then, the KAM distributions at varying tensile strains for both phases are summarized in Fig. 39.8e and Fig. 39.9e. It is observed that the average KAM decreases from 0.768 to 0.663 for fcc austenite phase and reduces from 0.786 to 0.513 for B2 phase at the first stage, which is consistent with the reduction of the fraction of very LAGBs and DRX as mentioned earlier. While the average KAM rises again for both phases at the second stage, which will result in an increase of the fraction of very LAGBs as shown in Fig. 39.6f and Fig. 39.7f. This increase for KAM also indicates the increase of dislocation density, thus indicates the significant dislocation activities in the grain interior at the second stage.

It is indicated from previous research [15] that the superplastic behaviors for metals and alloys at a given temperature and varying strain rates can be categorized into three well defined regions having different values for the SRS (m): In region I at low strain rates, SRS is generally low (~0.2) and the behavior is controlled by impurity effects; In region II over a range of intermediate strain rates, SRS is generally high (~0.5) and the flow behavior can be attributed to GBS; In region III at high strain rates, SRS is low again (~0.2) and the metals and alloys deform by the glide and climb of dislocations within the grains (dislocation creep). In Fig. 39.3a, it is observed that the SRS is about 0.47 at the strain rate range of 10^{-4} to 10^{-3}/s, is about 0.34 at the strain rate range of 10^{-3} to 10^{-2}/s, and is about 0.20 at the strain rate range of 10^{-2} to 10^{-1}/s at a given temperature

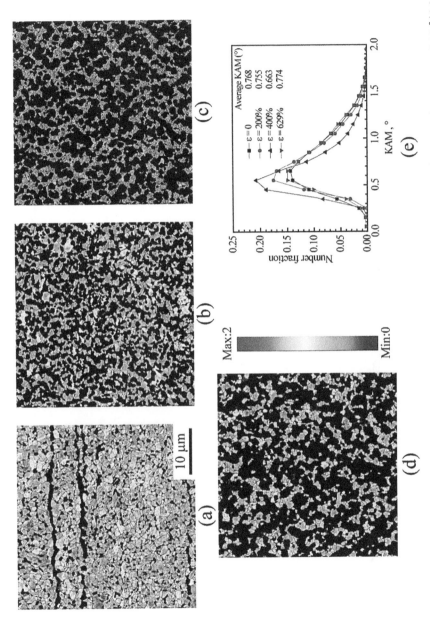

Figure 39.8 The KAM images under high temperature tensile deformation (under temperature of 973 K and strain rate of 10^{-3}/s) for fcc austenite phase: (a) $\varepsilon = 0$; (b) $\varepsilon = 200\%$; (c) $\varepsilon = 400\%$; (d) $\varepsilon = 629\%$. (e) The corresponding KAM distributions at varying tensile strains.

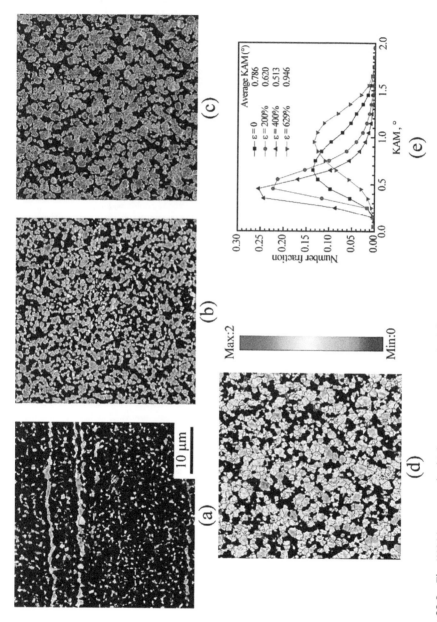

Figure 39.9 The KAM images under high temperature tensile deformation (under temperature of 973 K and strain rate of 10^{-3}/s) for B2 phase: (a) $\varepsilon = 0$; (b) $\varepsilon = 200\%$; (c) $\varepsilon = 400\%$; (d) $\varepsilon = 629\%$. (e) The corresponding KAM distributions at varying tensile strains.

of 973 K. Thus, regions II and III are observed in strain rate range of the present experiments, region I should be in the even lower strain rate range. As indicated from the inset of Fig. 39.2c, the final fracture shape indicates a diffusive necking at the center part of the gauge section, thus the realistic strain rate at the center part of the gauge section at the second stage should be much higher than the initial strain rate. Thus, the observed increase for KAM and the resultant intragranular slips at the second stage should be attributed to this increase in the realistic strain rate by diffusive necking (this suddenly increase in strain rate can turn the initial region II into the current region III).

The SEM images of final fracture surface for the high temperature deformation under temperature of 973 K and strain rate of 10^{-3}/s are shown in Fig. 39.10. The fracture is observed to be ductile with dimples. These dimples appear to be formed by pulling out the B2 particle from the fcc austenite matrix or fracturing the B2 particle during the high temperature deformation. Thus, the higher density of the boundaries, the more energy needed to be consumed for the final fracture. Thus, the fraction of boundaries should be considered as a controlling factor for the superplasticity. As a result, the superplasticity of this HSSS could be further enhanced by increasing the density of boundaries, i.e., refining the sizes of both phases. Refining grain sizes can promote GBS for better superplasticity by an increase of GB density on one hand, and can raise the resistance to final fracture by an increase of phase boundary density on the other hand.

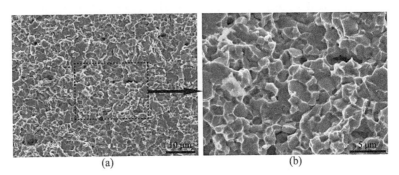

Figure 39.10 (a) The SEM images of final fracture surface for the high temperature deformation under temperature of 973 K and strain rate of 10^{-3}/s; (b) The corresponding close-up view for the rectangular area in Fig. 39.10a.

39.4 Conclusions

In the present study, the superplastic behaviors of a HSSS with dual-phase microstructure and ultrafine grains have been investigated. The main findings are summarized as follows:

(1) The UFG HSSS exhibits excellent superplastic properties. The microstructure observations at interrupted strains for tests under temperature of 973 K and at strain rate of 10^{-3}/s have provided evidences of different mechanisms for two stages.

(2) At the first stage (strain range from 0% to 400%), the superplastic flow can be attributed to the strain-enhanced diffusional transformation from fcc austenite phase to B2 phase coupled with GBS. The average size of the B2 precipitates as a function of time follows the well-known $t^{-1/2}$ diffusion rule, the activation energy is about 216 kJ/mol and the transformation characteristic exponent n is close to 1.26 under high temperature tensile deformation, suggesting that the transformation is indeed diffusion controlled along defects. The grain sizes of both phases are observed to be relatively stable and remain always sub-micron level during the superplastic deformation, facilitating the superplastic flow.

(3) At the second stage (strain range from 400% to 629%), the diffusive necking results in an increase in the realistic strain rate and a transition from region II to region III, thus the glide and climb of dislocations in the grain interior should be the dominant mechanism for this stage. The current results should provide insights for better understanding of the superplastic behaviors and for the part forming of this HSSS in automobile industry.

Acknowledgements

This work was supported by NSFC [Grant nos. 11572328, 11472286, and 11672313]; and the Strategic Priority Research Program of the Chinese Academy of Sciences [Grant no. XDB22040503].

References

1. S.X. McFadden, R.S. Mishra, R.Z. Valiev, A.P. Zhilyaev, A.K. Mukherjee, Low-temperature superplasticity in nanostructured nickel and metal alloys, *Nature* 398 (1999) 684–686.
2. G. Rai, N.J. Grant, On the measurements of superplasticity in an Al-Cu alloy, *Metall. Trans. A* 6A (1975) 385–390.
3. M. Mabuchi, K. Ameyama, H. Iwasaki, K. Higashi, Low temperature superplasticity of AZ91 magnesium alloy with non-equilibrium grain boundaries, *Acta Mater.* 7 (1999) 2047–2057.
4. G. Frommeyer, J.A. Jiménez, Structural superplasticity at higher strain rates of hypereutectoid Fe-5.5Al-1Sn-1Cr-1.3C steel, *Metall. Mater. Trans. A* 36A (2005) 295–300.
5. H. Zhang, B. Bai, D. Raabe, Superplastic martensitic Mn-Si-Cr-C steel with 900% elongation, *Acta Mater.* 59 (2011) 5787–5802.
6. G. Frommeyer, W. Kowalski, R. Rablbauer, Structural superplasticity in a fine-grained eutectic intermetallic NiAl-Cr alloy, *Metall. Mater. Trans. A* 37A (2006) 3511–3517.
7. L. Lu, M.L. Sui, K. Lu, Superplastic extensibility of nanocrystalline copper at room temperature, *Science* 287 (2000) 1463–1466.
8. T.G. Langdon, The mechanical properties of superplastic materials, *Metall. Trans. A* 13A (1982) 689–701.
9. E.M. Taleff, M. Nagao, K. Higashi, O.D. Sherby, High-strain-rate superplastocoty in ultrahigh-carbon steel containing 10 wt.% Al(UHCS-10Al), *Scr. Mater.* 34 (1996) 1919–1923.
10. R.M. Imayev, O.A. Kaibyshev, G.A. Salishchev, Mechanical behaviour of fine grained TiAl intermetallic compound-I. Superplasticity, *Acta Metall. Mater.* 40 (1992) 581–587.
11. G. Frommeyer, C. Derder, J.A. Jiménez, Superplasticity of $Fe_3Al(Cr)$, *Mater. Sci. Technol.* 18 (2002) 981–986.
12. D.L. Lin, A.D. Shan, D.Q. Li, M.W. Chen, Y. Liu, Superplastic behavior of large grained iron alluminides, *Mater. Sci. Forum* 243–245 (1997) 619–630.
13. D.L. Lin, D.Q. Li, Y. Liu, Superplasticity in large-grained FeAl based intermetallic alloys, *Intermetallics* 6 (1998) 243–256.
14. J.A. Jiménez, G. Frommeyer, M. Carsí, O.A. Ruano, Superplastic properties of a δ/γ stainless steel, *Mater. Sci. Eng. A* A307 (2001) 134–142.

15. M. Kawasaki, T.G. Langdon, Principles of superplasticity in ultrafine-grained materials, *J. Mater. Sci.* 42 (2007) 1782–1796.
16. T.G. Langdon, Seventy-five years of superplasticity: historic developments and new opportunities, *J. Mater. Sci.* 44 (2009) 5998–6010.
17. G. Frommeyer, U. Bruex, Microstructures and mechanical properties of high-strength Fe-Mn-Al-C light-weight TRIPLEX steels, *Steel Res. Int.* 77 (2006) 627–633.
18. J.D. Yoo, S.W. Hwang, K.-T. Park, Factors influencing the tensile behavior of a Fe-28Mn-9Al-0.8C steel, *Mater. Sci. Eng. A* 508 (2009) 234–240.
19. S.W. Hwang, J.H. Ji, E.G. Lee, K.-T. Park, Tensile deformation of a duplex Fe-20Mn-9Al-0.6C steel having the reduced specific weight, *Mater. Sci. Eng. A* 528 (2011) 5196–5203.
20. I. Gutierrez-Urrutia, D. Raabe, Influence of Al content and precipitation state on the mechanical behavior of austenitic high-Mn low-density steels, *Scr. Mater.* 68 (2013) 343–347.
21. D. Raabe, H. Springer, I. Gutierrez-Urrutia, F. Roters, M. Bausch, J.-B. Seol, M. Koyama, P.-P. Choi, K. Tsuzaki, Alloy design, combinatorial synthesis, and microstructure–property relations for low-density Fe-Mn-Al-C austenitic steels, *JOM* 66 (2014) 1845–1856.
22. R. Rana, C. Lahaye, and R.K. Ray, Overview of lightweight ferrous materials strategies and promises, *JOM* 66 (2014) 1734–1746.
23. S.S. Sohn, H. Song, B.C. Suh, J.H. Kwak, B.J. Lee, N.J. Kim, S. Lee, Novel ultra-high-strength (ferrite + austenite) duplex lightweight steels achieved by fine dislocation substructures (Taylor lattices), grain refinement, and partial recrystallization, *Acta Mater.* 96 (2015) 301–310.
24. S.H. Kim, H. Kim, N.J. Kim, Brittle intermetallic compound makes ultrastrong low-density steel with large ductility, *Nature* 518 (2015) 77–79.
25. M.X. Yang, F.P. Yuan, Q.G. Xie, Y.D. Wang, E. Ma, X.L. Wu, Strain hardening in Fe-16Mn-10Al-0.86C-5Ni high specific strength steel, *Acta Mater.* 109 (2016) 213–222.
26. W. Wang, H.S. Zhang, M.X. Yang, P. Jiang, F.P. Yuan, X.L. Wu, Shock and spall behaviors of a high specific strength steel: effects of impact stress and microstructure, *J. Appl. Phys.* 121 (2017) 135901.
27. O.D. Sherby, Advances in superplasticity and in superplastic materials, *ISIJ Int.* 29 (1989) 698–716.

28. W.J. Kim, K. Higashi, J.K. Kim, High strain rate superplastic behaviour of powder-metallurgy processed 7475Al+0.7Zr alloy, *Mater. Sci. Eng. A* 260 (1999) 170–177.
29. Z.Y. Ma, R.S. Mishra, M.W. Mahoney, Superplastic deformation behaviour of friction stir processed 7075Al alloy, *Acta Mater.* 50 (2002) 4419–4430.
30. M. Calcagnotto, D. Ponge, E. Demir, D.Raabe, Orientation gradients and geometrically necessary dislocations in ultrafine grained dual-phase steels studied by 2D and 3D EBSD, *Mater. Sci. Eng. A* 527 (2010) 2738–2746
31. K.L. Yang, J.C. Huang, Y.N. Wang, Phase transformation in the β phase of super α_2 Ti$_3$Al base alloys during static annealing and superplastic deformation at 700–100°C, *Acta Mater.* 51 (2003) 2577–2594.
32. C.G. McKamey, P.J. Maziasz, J.W. Jones, Effect of addition of molybdenum or niobium on creep-rupture properties of Fe$_3$Al, *J. Mater. Res.* 7 (1992) 2089–2106.
33. J.D. Whittenberger, The influence of grain size and composition on slow plastic flow in FeAl between 1100 and 1400 K, *Mater. Sci. Eng.* 77 (1986) 103–113.
34. D. Lin, T.L. Lin, A. Shan, M. Chen, Superplasticity in large-grained Fe$_3$Al alloys, *Intermetallics* 4 (1996) 489–496.
35. J.W. Christian, *The Theory of Transformations in Metals and Alloys*, Oxford, NY: Pergamon Press, 1975, pp. 538–546.
36. Y. Huang, Evolution of microstructure and texture during hot deformation of a commercially processed supral100, *J. Mater. Sci. Technol.* 28 (2012) 531–536.
37. Z.G. Liu, P.J. Li, L.T. Xiong, T.Y. Liu, L.J. He, High-temperature tensile deformation behavior and microstructure evolution of Ti55 titanium alloy, *Mater. Sci. Eng. A* 680 (2017) 259–269.
38. H. Watanabe, K. Kurimoto, T. Uesugi, Y. Takigawa, K. Higashi, Isotropic superplastic flow in textured magnesium alloy, *Mater. Sci. Eng. A* 558 (2012) 656–662.
39. L. Cheng, J.S. Li, X.Y. Xue, B. Tang, H.C. Kou, E. Bouzy, Superplastic deformation mechanisms of high Nb containing TiAl alloy with ($\alpha_2+\gamma$) microstructure, *Intermetallics* 75 (2016) 62–71.
40. O. Grässel, L. Krüger, G. Frommeyer, L.W. Meter, High strength Fe-Mn-(Al, Si) TRIP/TWIP steels development-properties-application, *Int. J. Plast.* 16 (2000) 1391–1409.

41. X. Ren, M. Hagiwara, Displacive precursor phenomena in Ti-22Al-27Nb intermetallic compound prior to diffusional transformation, *Acta Mater.* 49 (2001) 3971–3980.

Chapter 40

Plastic Deformation Mechanisms in a Severely Deformed Fe-Ni-Al-C Alloy with Superior Tensile Properties

Yan Ma,[a,b] Muxin Yang,[a] Ping Jiang,[a] Fuping Yuan,[a,b] and Xiaolei Wu[a,b]

[a]*State Key Laboratory of Nonlinear Mechanics, Institute of Mechanics, Chinese Academy of Sciences, Beijing 100190, China*
[b]*School of Engineering Science, University of Chinese Academy of Sciences, Beijing 100190, China*
fpyuan@lnm.imech.ac.cn

Nanostructured metals have high strength while they usually exhibit limited uniform elongation. A yield strength of approximately 2.1 GPa and a uniform elongation of about 26% were achieved in a severely deformed Fe-24.8%Ni-6.0%Al-0.38%C alloy in the present work. The plastic deformation mechanisms for the coarse-grained (CG) sample and the cold-rolled (CR) samples of this alloy were investigated by a series of mechanical tests and microstructure

---

Reprinted from *Sci. Rep.*, **7**, 15619, 2017.

Heterostructured Materials: Novel Materials with Unprecedented Mechanical Properties
Edited by Xiaolei Wu and Yuntian Zhu
Text Copyright © 2017 The Author(s)
Layout Copyright © 2022 Jenny Stanford Publishing Pte. Ltd.
ISBN 978-981-4877-10-7 (Hardcover), 978-1-003-15307-8 (eBook)
www.jennystanford.com

characterizations before and after tensile tests. No obvious phase transformation was observed during the tensile deformation for the CG sample, and the plastic deformation was found to be mainly accommodated by deformation twins and dislocation. While significant phase transformation occurs for the CR samples due to the facts that the deformed grains by CR are insufficient to sustain the tensile deformation themselves and the flow stress for the CR samples is high enough to activate the martensite transformation. The amount of phase transformation increases with increasing thickness reduction of CR, resulting in excellent ductility in the severely deformed alloy. The hetero-deformation induced (HDI) hardening was found to play a more important role in the CR samples than in the CG sample due to the dynamically reinforced heterostructure by phase transformation.

40.1 Introduction

Grain refinement has been extensively utilized to strengthen metals and alloys [1–5]. Bulk ultrafine-grained (UFG) and nanostructured (NS) metals can be many times stronger when compared to their conventional coarse-grained (CG) counterparts [1–5], but with poor strain hardening and limited ductility. Previous studies have shown that several strategies can be employed to obtain both high strength and good ductility in metals and alloys, such as gradient structures [6–8], heterostructured lamella structure [9, 10], nanotwins with coherent twin boundaries (CTB) [11, 12], nano-precipitates [13, 14], bimodal/multimodal structure [15, 16], twinning-induced plasticity (TWIP) [17, 18], transformation-induced plasticity (TRIP) [19, 20], and lattice softening by composition control [21–25]. Recently, a new design to lattice-softened alloys with both ultrahigh strength and high ductility has been proposed [26, 27], in which C is added as stabilizer for the austenite phase, Al is added as stabilizer for the bcc phase and the final chemical composition is Fe-24.8%Ni-6.0%Al-0.38%C (in wt.%). The high strength has been attributed to the grain refinement by severe plastic deformation (SPD), the high strain hardening and the high ductility have been attributed to the appropriate choice of chemical composition for the lattice softening and the multimodal-structure formation [27].

The mechanism of strain hardening for most ultrahigh strength steels is still not fully clear since their plastic deformation is inhomogeneous due to the heterostructure [28, 29]. Even for an alloy with initial single austenite phase and homogeneous grain size, martensite transformation or deformation twins would make the plastic deformation inhomogeneous and the strain hardening behavior complex [17, 19]. In our recent work [10], the high strain hardening and the high ductility in a commercial pure Ti with heterostructured lamella structure were attributed to the HDI hardening associated with the plastic incompatibility between the lamellae with different grain sizes. In our more recent paper [30], the extraordinary strain hardening rate in a dual-phase high specific strength steel (HSSS) can be attributed to the high HDI stresses that arise from load transfer and strain partitioning between the two phases with different mechanical properties. The HDI hardening due to the high internal stresses have in fact been reported in TWIP steels [31–33], TRIP steels [34], and dual-phase alloys [30, 35, 36] to account for the strong strain hardening and the large ductility.

Indeed, the plastic deformation in the Fe-24.8Ni-6.0Al-0.38C (in wt.%) alloy with heterostructure should be similar to that in composites, and can be characterized by the strong HDI hardening due to the load redistribution and the strain partitioning between the constituent phases. In this regard, we analyze the mechanisms of plastic deformation and the strain hardening due to the HDI stress in the Fe-24.8Ni-6.0Al-0.38C (in wt.%) alloy consisting of γ austenite phase and α' martensite phase in this chapter. The initial CG materials were prepared by melting under Ar atmosphere and solution treatment at 1373 K for 24 h following quenched in water. Then different heterostructures with varying yield strength were introduced by cold rolling (CR) with different thickness reduction. A series of load-unload-reload (LUR) tests have been conducted to investigate the HDI stress evolutions during the tensile deformation for various microstructures, and then the mechanisms of HDI hardening for various microstructures have been carefully analyzed. Generally, martensite transformation can be considered as a stress-induced process based on the thermodynamic action with a local threshold stress during the transformation [17, 37]. The transformation behaviors should be different for various microstructures with different flow stresses, then the quantitative analysis for the martensite transformation of various microstructures has been conducted by a series of X-ray diffraction (XRD) measurements before and after tensile tests. The detailed microstructure evolutions have also been obtained by electron backscattered diffraction (EBSD) and transmission electron microscopy (TEM) before and after tensile tests to further clarify the plastic deformation mechanisms.

40.2 Results

40.2.1 Microstructures before Tensile Tests and Quasi-Static Uniaxial Tensile Properties

The microstructure evolution during CR need be characterized first. Figure 40.1 shows the microstructures before the tensile tests for the solution treated sample and the CR samples with different thickness reductions. The optical microscope (OM) and EBSD (inverse pole figure, IPF) images for the solution treated sample are shown in Figs. 40.1a and b respectively. The TEM images for the solution treated sample are shown in Figs. 40.1c and d. As shown, the solution treated sample displays a dual-phase microstructure, composed of an austenite phase matrix (γ phase) with an average grain size of 12.4 μm and the second martensite phase (α' phase) with an average grain size of 3.6 μm. It is observed that the α' martensite grains are much inclined to precipitate at either the triple junctions or the grain boundaries of γ austenite grains. The area fraction is about 87% for the γ phase while is about 13% for α' phase in the solution treated sample. As indicted in the TEM images for the solution treated sample, the grain interiors within both the γ and α' grains are relatively clean due to the solution treatment at high temperature although a few dislocations can be seen.

Figures 40.1e and f display the OM and EBSD (IPF) images for the CR 50% sample. After CR with thickness reduction of 50%, the area fraction for the α' phase is slightly increased to about 17% due to the martensite transformation during cold working. The plastic deformation is mainly carried by the heavily deformed structure in the soft γ grains and partially carried by the dislocation behaviors in the hard α' grains during CR. As indicated, an inhomogeneous

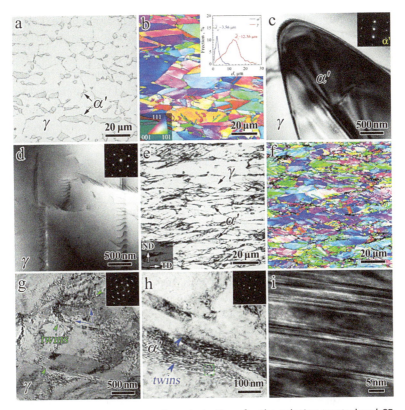

Figure 40.1 Microstructure characterizations for the solution treated and CR samples before tensile testing. (a,b) OM and EBSD (IPF) images for the solution treated sample, respectively; (c,d) TEM images for the solution treated sample; (e,f) OM and EBSD (IPF) images for the CR 50% sample, respectively; (g,h) TEM images for the CR 90% samples; (i) HREM image for the corresponding rectangle area in (h).

microstructure is observed for the CR 50% sample. A mixture of two areas, namely an area with large grain size of about 10 μm and an area with small grain size of about 1–3 μm or even sub-micron, is observed and this observation is consistent with that reported in previous research [27]. This indicates that the CR samples possess a hierarchical microstructure with both ultrafine grains (UFG) and coarse grains (CG) in both γ and α′ phases. After CR with thickness reduction of 90%, the microstructure can hardly be revealed by EBSD due to the even more heavily deformed structure. Thus, the detailed results for the severely deformed microstructure in the soft γ grains and the deformed microstructure in the hard α′ grains after CR with thickness reduction of 90% are displayed by TEM and high resolution electron microscope (HREM) images in Figs. 40.1g–i. Lamellae with thickness of about 50~100 nm are observed in the γ grains, and these lamellae are indentified to be deformation twins (DTs, marked by green arrows) with the exact twinning orientation relationship for fcc metals. As marked by blue arrows, multiple twins formed at two {111} planes are found in the grain interior of γ grains. High density of dislocation are also seen in the heavily deformed γ grains. Very high density of lamellar structures are observed in the hard α′ grains (as shown in Fig. 40.1h), and these structures are identified to be nanotwins (with thickness of a few nm) by the indexed selected area diffraction (SAD), which is confirmed by the HREM image of Fig. 40.1i. These microstructures (DTs and high density of dislocations) with nanoscale can result in the strengthening associated with CR.

The quasi-static uniaxial tensile properties for the solution treated sample and the CR samples are displayed in Fig. 40.2. Figures 40.2a and b show the engineering stress-strain curves and the true stress-strain curves, respectively. The yield points are marked by circles and the points for ultimate strength (UTS) are marked by squares in Figs. 40.2a and b. Then, the Holloman's equation, $\sigma = \sigma_0 + K\varepsilon^n$ (where σ_0 is the yield stress, K is the strength coefficient, n is the strain hardening exponent), is used to fit the true stress-strain curves and the strain hardening exponent is plotted as a function of thickness reduction of CR in Fig. 40.2c. As indicated, the CG sample has a round continuous yielding and strain hardening behaviors, the yield strength is about 400 MPa and the uniform elongation is about 42%. While three-stages are observed before necking in the engineering stress-strain curves for the CR samples: a yield drop is followed by a stress plateau stage and a strain hardening stage. The strain duration for the stress plateau stage is observed to increase while the strain duration for the strain hardening stage is seen to decrease with the increasing thickness reduction for the CR samples. It is interesting to note that the yield strength is dramatically increased while the uniform elongation is still relatively reserved after CR. For example, the yield strength is elevated to about 2.1 GPa and the uniform elongation still remains about 26% after CR with thickness reduction of 90%. Compared to the CG sample, the strain hardening exponent is also observed to increase slightly after CR and be similar for all CR samples. The slope in the true stress-strain curves is observed to be similar for all samples, indicating that the CR process does not obviously reduce the strain hardening ability. As a summary, Fig. 40.2d plots yield strength versus uniform elongation curves for the present data and for other high-strength advanced steels [29, 38–42]. The other conventional metals and alloys clearly show a banana curve for the trade-off between strength and ductility, while the present data exhibit ultrahigh strength and large ductility with clear deviation from the other high strength metals and alloys. Specially, when the thickness reduction is increased from 50% to 90% for the CR samples, the yield strength is observed to be significantly elevated while the uniform elongation is found to be almost unchanged.

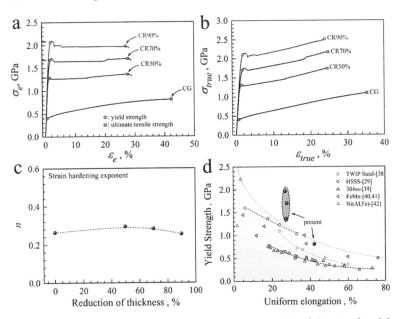

Figure 40.2 Tensile properties for the solution treated and CR samples. (a) Engineering stress-strain curves; (b) True stress-strain curves; (c) Strain hardening exponent versus the thickness reduction of CR; (d) Yield strength versus uniform elongation for the present data and other high-strength advanced steels.

40.2.2 Deformation Mechanisms during Tensile Deformation for Solution Treated and CR Samples

In order to reveal the deformation mechanisms for the solution treated and CR samples, the behaviors of martensite transformation during CR and subsequent tensile deformation need be quantitatively characterized. Thus, the XRD spectra in the solution treated sample and the CR samples with different thickness reductions are given in Fig. 40.3a, and the XRD spectra after tensile testing for all samples are shown in Fig. 40.3b. Based on these XRD spectra, the volume fraction of α' martensite phase can be calculated using the same equations and methods as in our previous paper [43]. The volume fractions of α' phase for the solution treated and CR samples before and after tensile tests are plotted as a function of thickness reduction of CR in Fig. 40.3c. The EBSD phase distributions for the solution treated and CR 50% samples before and after tensile testing are also shown in the Supplementary Information (Fig. S1). The volume fraction of α' phase can also be counted from these EBSD phase images (Fig. S1), which is consistent with the observations from the XRD data (Fig. 40.3c).

As indicated, almost no phase transformation occurs during the tensile deformation for the solution treated sample. This observation indicates two aspects: (i) The soft γCG, even the hard α' grains, can deform plastically by deformation twins or dislocation behaviors without martensite transformation on one hand; (ii) It also shows that the austenite phase in this alloy is relatively stable and the flow stress for the CG sample is too low to reach the threshold stress and activate the martensite transformation on the other hand.

In order to illustrate the deformation mechanisms for the solution treated sample, the uniaxial stress-relaxation results for the solution treated sample are displayed in Fig. S2, and the TEM images before and after tensile tests for the solution treated sample are shown in Fig. 40.4. The evolution of mobile dislocations can be examined through these repeated uniaxial stress-relaxation tests to reveal the strain hardening mechanism of the solution treated sample. Figure S2a displays the engineering stress versus engineering strain curve for the stress-relaxation test on the solution treated sample. The detailed data analysis for the repeated stress-relaxation tests

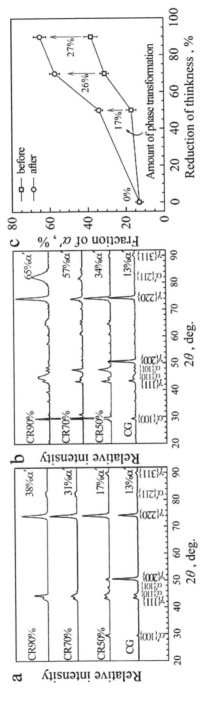

Figure 40.3 Martensite transformation during CR and subsequent tensile deformation. (a) The evolution of XRD spectra during CR; (b) The evolution of XRD spectra during subsequent tensile deformation; (c) The volume fractions of α' phase as a function of CR thickness reduction for the solution treated and CR samples before and after tensile tests.

can be found in our previous paper [7, 44]. The calculated data are shown in Figs. S2b–d. Figure S2b shows physical activation volume as a function of engineering strain for the solution treated sample, Fig. S2c displays the exhaustion curves of mobile dislocation with respect to time at various preset strains for the solution treated sample, while Fig. S2d exhibits the retained density of mobile dislocation at the end of each relaxation against engineering strain. It is observed that the physical activation volume decreases while the mobile dislocation density increases with increasing engineering strain for the solution treated sample. As we know, the physical activation volume is proportional to the size of barrier for dislocations and the mean free path between barriers. Therefore, the mean free path generally decreases with increasing strain due to the increase of dislocation density and the multiplication of dislocations, thus resulting in reduction of physical activation volume. The increasing mobile dislocation density during the tensile loading for the solution treated sample indicates that dislocation behaviors in both phase should be a dominant deformation mechanism for the solution treated sample during tensile deformation, given that no obvious phase transformation is observed (as shown in Fig. 40.3c).

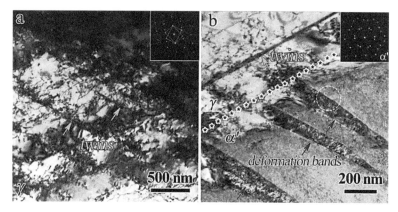

Figure 40.4 TEM observations for the solution treated sample after tensile deformation. (a) DTs and dislocations in the γ austentite grain; (b) Deformation bands and dislocations in the α' martensite grain. The phase boundary between the γ austentite grain and the α' martensite grain is marked by five-pointed stars.

As shown in Figs. 40.1c and d, the dislocation density is low and the grain interior is relatively clean for both phases in the solution treated sample before tensile testing. While, high density of DTs (marked by green arrows, indicated by the indexed SAD in Fig. 40.4a) with a very small average twin boundary spacing (TBS) of several hundreds of nm are formed inside the soft γ grains after tensile deformation for the solution treated sample (Figs. 40.4a and b), and a few of deform bands with high density of dislocations are observed in the grain interior of α' phase (Fig. 40.4b). The inset of Fig. 40.4b shows the indexed SAD for the area marked by white circle in the α' phase, which clearly indicates the deformed structure by dislocations for bcc α' phase. As observed in Fig. 40.4a, high density of dislocations near twin boundaries (TBs) are also observed in the grain interior of γ phase and TBs of DTs are no longer coherent due to the interactions between the dislocations and the TBs, which is consistent with the earlier results from the measurement of mobile dislocation density by stress relaxation tests. Previous research [11, 12, 45] suggested that TBs of DTs are effective obstacles to the motion of dislocations and can accumulate the pile-up of dislocations near TBs, which could provide great resistance to the plastic deformation and strong strain hardening for sustaining uniform elongation.

As indicated, the volume fractions of α' phase is slightly increased from 13% to 17% after CR with thickness reduction of 50%, and then increased to 31% after CR with thickness reduction of 70%, finally increased to 38% after CR with thickness reduction of 90%. These observations indicate that the martensite transformation also contributes to the strengthening during CR besides the microstructure refinement by UFG and DTs, and the increasing dislocation density inside grains. While, as indicated from results of both XRD data and EBSD images, significant amount phase transformation occurs during the tensile loading for CR samples, which is contrast to the CG sample (no phase transformation is observed during tensile deformation). For examples, the volume fractions of α' phase is increased from 17% to 34% for CR 50% sample, from 31% to 57% for CR 70% sample, and from 38% to 65% for CR 90% sample during tensile deformation. The amounts of phase transformation during tensile testing are 17%, 26% and 27% for CR 50% sample, CR 70%

sample and CR 90% sample, respectively (as shown in Fig. 40.3c). It is interesting to note that the amount of phase transformation increases with increasing thickness reduction of CR, which can result in strong strain hardening for sustaining excellent ductility in the severely deformed alloy even the deformed grains by CR are insufficient to sustain the tensile deformation themselves. These observations indicate two things: (i) The plastic deformation ability by the soft γ CG or the hard α' grains themselves almost exhausts during CR, and the plastic deformation during subsequent tensile loading need be accommodated by martensite transformation; (ii) The flow stress for the CR samples is high enough to reach the threshold stress and activate the martensite transformation.

The TEM images after tensile tests for the CR 90% sample are shown in Fig. 40.5. As indicated from Fig. 40.5a, high density of dislocations are accumulated in the γ grain around the α' martensite nano-precipitate, which could result in strong strain hardening [5, 13, 14]. As indicated from Fig. 40.5b, nanograins are formed in both phases according to the indexed SAD, even higher density of dislocations are observed within both γ grains and α' grains after tensile testing when compared to those before the tensile deformation for the CR 90% sample. Thus, the strong hardening for the CR samples during the tensile deformation could be attributed to two aspects: (i) The obvious martensite transformation can contribute to the strain hardening and accommodate the plastic strain since the martensite phase is harder than the austenite phase. This stress-induced transformation can provide the transformation strain itself to prevent early void formation at the phase boundaries on one hand, and can strengthen the strain concentration region and thus help to maintain large homogeneous tensile deformation by preventing early necking formation on the other hand [19, 46]. (ii) Followed CR, the continuous interactions between dislocations and TBs during the tensile deformation could involve the formation of nanograins and immobile dislocation locks for strengthening and hardening [45], and the accumulation of dislocations around the α' martensite nano-precipitates could also result in strengthening and hardening [5, 13, 14].

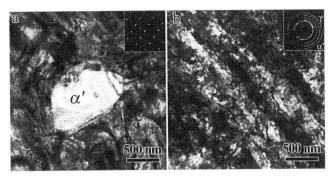

Figure 40.5 TEM observations for the CR 90% sample after tensile deformation. (a) Bright-field image showing accumulation of dislocations in the γ austentite grain around the α' martensite nano-precipitate; (b) Bright-field image and the indexed SAD showing formation of nanograins and high density of dislocations in both phases.

40.3 Discussions

As indicated in our previous paper [30], the high strain hardening and the excellent ductility in the HSSS with dual-phase microstructure can be attributed to the high HDI stresses that arise from plastic deformation incompatibility between the two phases with different mechanical properties. The HDI hardening has been found to also play an important role in the heterostructured lamella structure Ti with both UFG and CG lamellae [10]. Thus, the HDI hardening should also play an important role in the current alloy with dual-phase microstructure. The HDI hardening might contribute more to the strain hardening and the ductility in the CR samples than the solution treated sample due to the fact that the initial microstructure for the CR samples is hierarchical with both UFG and CG in both phases and this heterostructure is dynamically reinforced due to the significant martensite transformation during tensile deformation for the CR samples.

Figures S3a,b shows Vickers micro-hardness 2D contours for the CG sample and the CR 70% sample before and after tensile tests. Then the corresponding micro-hardness distributions are plotted in Figs. S3c,d. As indicated, the average hardness is increased from 420 to 462 Hv for the solution treated sample, and from 550 to 663 Hv for the CR 50% sample after tensile deformation. It is interesting to note that the hardness increment is larger for the CR 50% sample than for the solution treated sample, which could be due to the martensite transformation for the CR 50% sample. Another interesting thing to note is that a wide distribution of micro-hardness is observed for both the solution treated sample and the CR samples, which indicates that the microstructures are highly heterogeneous with different mechanical properties at different areas for both the solution treated sample and the CR samples. These heterostructures with different yield strengths/hardness at different areas should result in strong stress/strain partitioning at different areas. For the solution treated sample, this strain partitioning generally happens between two phases (i.e., the softer γ grains accommodate more plastic strain and achieve more strain hardening than the harder α' grains during tensile loading). This strain partitioning can be measured for both phases by the aspect ratio changes before and after tensile tests in the solution treated sample since no phase transformation occurs. It is observed that the softer γ grains are more severely elongated along the tensile direction (horizontal direction) than the harder α' grains, as indicated in Figs. S1a and b. These results can qualitatively indicate that the strong strain partitioning indeed happens between two phase with different mechanical properties. Thus, this plastic incompatibility due to the load transfer and the strain partitioning between two phases is the main origin for the HDI hardening in the solution-treated sample.

In order to understand the deformation mechanism of the stress plateau stage observed for the CR samples, additional in-situ digital image correlation (DIC) experiment along with tensile testing was also conducted for the CR 70% sample. The evolution of strain contours for the gauge section along with tensile strain is shown in Fig. S4. It can be clearly seen that the stress plateau stage for the CR 70% sample is due to the nucleation and propagation of the deformation band (Lüders band). This discontinuous yielding followed by large Lüders strain is similar to the results observed for a deformed and partitioned high strength steels in the recent work [47], and this phenomenon can help to sustain large ductility.

The true stress-strain curves for LUR tests on the solution treated sample, the CR 50% sample and the CR 90% sample are shown in Fig. 40.6a, and the inset displays the close-up view for the yield drop phenomenon in the reloading curve of the CR samples. According to our previous paper [30], this unloading yield effect could be understood as follows: once unloaded, the hard α' grains become elastic. Upon reloading, the hard α' grains stays elastic while the soft γ grains begins to deform plastically. Upon reloading, the yield peak appears due to the load transfer between two phases. Then, rapid relaxation of elastic stresses and strains at the phase boundaries causes the stress drop once the hard α' grains yields upon reloading. As shown in Fig. 40.6b, the unloading yield effect is much obvious in the CR samples than in the solution treated sample due to the facts that the volume fraction of hard α' phase is higher and is significantly increased during tensile deformation for the CR samples.

The close-up views of typical hysteresis loops for the solution treated sample and the CR samples are shown in Fig. 40.6c, and the HDI stress can be estimated by the average of the unloading yield stress and the reloading yield stress ($\sigma_{back} = (\sigma_u + \sigma_r)/2$), which was proposed in our previous paper [48]. The effective stress σ_{eff}, which contributes to the forest dislocation hardening, can be calculated by detracting the HDI stress from the total flow stress. Then the evolutions of both the HDI stress and the effective stress along with tensile strain can be obtained from the unloading-reloading curves at varying tensile strains, and are shown in Fig. 40.6d. Moreover, the evolution of $\sigma_{back}/\sigma_{total}$ along with tensile strain is shown in Fig. 40.6e for the solution treated sample and the CR samples. The magnitudes of $\sigma_{back}/\sigma_{total}$ for all samples increase with increasing strain, indicating the high HDI hardening for all samples. $\sigma_{back}/\sigma_{total}$ can represent the contribution of HDI hardening to the overall strain hardening, thus the contribution of HDI hardening is observed to be much larger for the CR samples than for the solution treated sample. The HDI hardening can be only originated from the long internal stress between the soft γ phase and the hard α' phase for the solution treated sample. While, the HDI hardening is accommodated by the pile-up and accumulation of geometrically necessary dislocations at the phase boundaries or at the boundaries between UFG and CG for the CR samples, which is caused by the long-range internal stress among the γ CG, the γ UFG and the hard α' grains. Thus, stronger HDI hardening in the CR samples could be attributed to the much more complex interplay among the γ CG, the γ UFG and the hard α'

grains and the dynamically reinforced heterostructure in the CR samples due to the significant phase transformation.

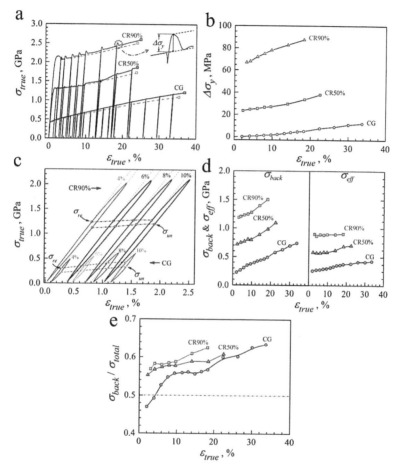

Figure 40.6 Bauschinger effect and HDI hardening for the solution treated and CR samples. (a) The true stress-strain curves for LUR tests; (b) $\Delta\sigma_y$ as a function of true tensile strain due to unloading yield effect; (c) The close-up views of typical hysteresis loops; (d) The evolutions of σ_{back} and σ_{eff} along with tensile strain; (e) The evolution of $\sigma_{back}/\sigma_{total}$ along with tensile strain.

In summary, the plastic deformation mechanisms of the solution treated sample and the CR samples in a Fe-24.8%Ni-6.0%Al-0.38%C alloy have been investigated in the present work, and the main findings are summarized as follows. The yield strength is dramatically elevated after CR without significant reduction of the uniform ductility. The present results exhibit ultrahigh strength and large ductility with clear deviation from the trade-off banana curve for the other high strength metal and alloys. The austentite in this alloy is relatively stable and no obvious phase transformation occurs during the tensile loading of the solution treated sample. Deformation twins and dislocation behaviors in the CG are the main plastic deformation carriers during the tensile deformation for the solution treated sample. Obvious phase transformation was observed during the tensile loading for the CR samples, and the amount of phase transformation was found to increase with increasing thickness reduction of CR, resulting in excellent ductility in the severely deformed alloy. On the one hand, the deformation ability of the CG exhausts during CR and the deformed CG by CR are insufficient to sustain the tensile deformation themselves. On the other hand, the tensile flow stress of the CR samples is high enough to activate the martensite transformation. The ductility in this alloy with dual-phase microstructure can be attributed to the strong HDI hardening due to the load transfer and the strain partitioning between the two phases, and the HDI hardening was observed to play a more important role in the CR samples than in the solution treated sample due to the much more complex interplay among the γ CG, the γUFG and the hard α' grains, and the dynamically reinforced heterostructure by the phase transformation. The present results could provide better understanding for the deformation mechanisms of the Fe-Ni-Al-C alloy and could provide strategies to achieve both ultrahigh strength and excellent ductility in metals and alloys.

40.4 Materials and Experimental Procedures

40.4.1 Materials

The Fe-24.8Ni-6.0Al-0.38C (in wt.%) alloy was first melted in an induction furnace under the protection of Ar atmosphere, and then cast into ingots with dimensions of 750 × 120 × 120 mm^3. The ingots were then hot forged between 1423 and 1223 K into slabs with a thickness of 6 mm. Next, the slabs were solution treated at

1373 K for 24 h in Ar atmosphere and were then quenched in water. After solution treatment, the slabs were cold rolled into sheets with different thickness reductions (50%, 70%, and 90%).

40.4.2 Microstructure Characterizations

OM, EBSD and TEM were used to study the microstructures before and after the tensile tests. The sample surfaces for OM and EBSD were first grinded to 3000 grit with sandpapers, and then polished with a 0.05 μm SiO_2 aqueous suspension, followed by electro-polishing in a solution of 5% $HClO_4$ and 95% alcohol at 37 V and 253 K (−20°C). For TEM observations, thin disks with thickness of about 200 μm were prepared, and then mechanically polished to about 40 μm, followed by a twin-jet polishing using a solution of 5% perchloric acid and 95% ethanol at 25 V and 253 K (−20°C). X-ray diffraction (XRD) measurements were performed on polished samples to obtain the phase transformation information during CR and tensile tests using a Philips Xpert X-ray diffractometer with Cu Kα radiation. The phase volume fractions were estimated from the peak integrated intensities I_{hkl} after background subtraction.

40.4.3 Mechanical Testing

The specimens for quasi-static uniaxial tensile tests and LUR tests have a dog-bone plate shape and a gauge section of $10 \times 4 \times 1.5$ mm^3. The tensile direction is parallel to the rolling direction. These tensile tests and LUR tests were conducted at a strain rate of 5×10^{-4}/s and at room temperature under displacement control using an Instron 5565 testing machine. In order to obtain the evolution of mobile dislocation density during the tensile tests, uniaxial tensile stress-relaxation tests were also performed under strain-control mode at room temperature with a series of preset strains. Upon reaching a designated relaxation strain, the strain was maintained constant while the stress was recorded as a function of time. After the first relaxation over an interval of 90 s, the specimen was reloaded by a strain increment of 0.6% at a strain rate of 5×10^{-4}/s for next relaxation. Four stress relaxations were conducted for each designated strain. During the tensile tests, LUR tests and tensile stress-relaxation tests, an exensometer was used to accurately measure and control the strain. The distributions (2D contours) of Vickers micro-hardness before and after tensile tests were also obtained on the polished sample surfaces using a Vickers diamond indenter under a load of 5 g for 15 s dwell time. The area for microhardness measurement is 140 × 210 µm², and the distance for each indentation is about 5 µm. Strain contours were measured using DIC during tensile tests. A commercial software, ARAMIS®, was utilized to analyze the DIC data. Initial high-contrast stochastic spot patterns on the sample surface were created. The evolution of the spot patterns was recorded using two 1.2 MPx digital CCD cameras at a rate of 1 frame per second. The facet size for the strain calculation using DIC method was 50 µm. The other details for the DIC method can be found in our recent paper [49].

Acknowledgments

This work was financially supported by National Key R&D Program of China [Grant No. 2017YFA0204402]; NSFC [Grant Nos. 11572328, 11472286, and 11672313]; and the Strategic Priority Research Program of the Chinese Academy of Sciences [Grant No. XDB22040503].

Additional Information

Supplementary information accompanies this chapter at http://www.nature.com/scientificreports.

References

1. Valiev, R. Z., Islamgaliev, R. K. & Alexandrov, I. V. Bulk nanostructured materials from severe plastic deformation. *Prog. Mater. Sci.* **45**, 103–189 (2000).
2. Langdon, T. G. Twenty-five years of ultrafine-grained materials: Achieving exceptional properties through grain refinement. *Acta Mater.* **61**, 7035–7059 (2013).
3. Valiev, R. Z. Nanostructuring of metals by severe plastic deformation for advanced properties. *Nat. Mater.* **3**, 511–516 (2004).
4. Zhu, Y. T. & Liao, X. Z. Nanostructured metals: retaining ductility. *Nat. Mater.* **3**, 351–352 (2004).

5. Wu, X. L., Yuan, F. P., Yang, M. X., Jiang, P., Zhang, C. X., Chen, L., Wei, Y. G. & Ma, E. Nanodomained nickel unite nanocrystal strength with coarse-grain ductility. *Sci. Rep.* **5**, 11728 (2015).

6. Fang, T. H., Li, W. L., Tao, N. R. & Lu. K. Revealing extraordinary intrinsic tensile plasticity in gradient nano-grained copper. *Science* **331**, 1587–1590 (2011).

7. Wu, X. L., Jiang, P., Chen, L., Yuan, F.P. & Zhu, Y. T. Extraordinary strain hardening by gradient structure. *Proc. Natl. Acad. Sci. U.S.A.* **111**, 7197–7201 (2014).

8. Wei, Y. J., Li, Y. Q., Zhu, L. C., Liu, Y., Lei, X. Q., Wang, G., Wu, Y. X., Mi, Z. L., Liu, J. B., Wang, H. T. & Gao, H. J. Evading the strength–ductility trade-off dilemma in steel through gradient hierarchical nanotwins. *Nat. Commun.* **5**, 3580 (2014).

9. Ma, X. L., Huang, C. X., Moering, J., Ruppert, M., Höppel, H. W., Göken, M., Narayan, J. & Zhu, Y. T. Mechanical properties of copper/bronze laminates: role of Interfaces. *Acta Mater.* **116**, 43–52 (2016).

10. Wu, X. L., Yang, M. X., Yuan, F. P., Wu, G. L., Wei, Y. J., Huang, X. X. & Zhu, Y. T. Heterogeneous lamella structure unites ultrafine-grain strength with coarse-grain ductility. *Proc. Natl. Acad. Sci. U.S.A.* **112**, 14501–14505 (2015).

11. Lu, L., Chen, X., Huang, X. & Lu, K. Revealing the maximum strength in nanotwinned copper. *Science* **323**, 607–610 (2009).

12. Li, X. Y., Wei, Y. J., Lu, L., Lu, K. & Gao, H. J. Dislocation nucleation governed softening and maximum strength in nano-twinned metals. *Nature* **464**, 877–880 (2010).

13. Liddicoat, P.V., Liao, X.Z., Zhao, Y. H., Zhu, Y. T., Murashkin, M.Y., Lavernia, E. J., Valiev, R. Z. & Ringer, S. P. Nanostructural hierarchy increases the strength of aluminum alloys. *Nat. Commun.* **1**, 63 (2010).

14. Liu, G., Zhang, G. J., Jiang, F., Ding, X. D., Sun, Y. J., Sun, J. & Ma, E. Nanostructured high-strength molybdenum alloys with unprecedented tensile ductility. *Nat. Mater.* **12**, 344–350 (2013).

15. Wang, Y. M., Chen, M. W., Zhou, F. H. & Ma, E. High tensile ductility in a nanostructured metal. *Nature* **419**, 912–915 (2002).

16. Zhao, Y. H., Liao, X. Z., Cheng, S., Ma, E. & Zhu, Y. T. Simultaneously increasing the ductility and strength of nanostructured alloys. *Adv. Mater.* **18**, 2280–2283 (2006).

17. Bouaziz, O., Allain, S., Scott, C. P., Cugy, P. & Barbier, D. High manganese austenitic twinning induced plasticity steels: a review of the

microstructure properties relationships. *Curr. Opin. Solid State Mater. Sci.* **15**, 141–168 (2011).

18. Sevillano, J. G. & Cuevas, F. D. L. Internal stresses and the mechanism of work hardening in twinning-induced plasticity steels, *Scr. Mater.* **66**, 978–981 (2012).

19. Grassel, O., Kruger, L. & Frommeyer, G., High strength Fe-Mn-(Al, Si) TRIP/TWIP steels development - properties - application, *Int. J. Plast.* **16**, 1391–1409 (2000).

20. Fischer, F. D., Reisner, G., Werner, E., Tanaka, K., Cailletaud, G. & Antretter, T. A new view on transformation induced plasticity (TRIP). *Int. J. Plast.* **16**, 723–748 (2000).

21. Saito, T., Furuta, T., Hwang, J. H., et al. Multifunctional alloys obtained via a dislocation-free plastic deformation mechanism. *Science* **300**, 464–467 (2003).

22. Kuramoto, S., Furuta, T., Nagasakko, N. & Horita, Z. Lattice softening for producing ultrahigh strength of iron base nanocrystalline alloy. *Appl. Phys. Lett.* **95**, 211901 (2009).

23. Li, T., Morris, Jr. JW., Nagasako, N., Kuramoto, S. & Chrzan, D. C. "Ideal" engineering alloys. *Phys. Rev. Lett.* **98**, 105503 (2007).

24. Chrzan, D. C., Sherburne, M. P., Hanlumyuang, Y., Li, T. & Morris, Jr. JW. Spreading of dislocation cores in elastically anisotropic body-centered-cubic materials: The case of gum metal. *Phys. Rev. B* **82**, 184202 (2010).

25. Edalati, K., Toh, S., Furuta, T., Kuramoto, S., Watanabe, M. & Horita, Z. Development of ultrahigh strength and high ductility in nanostructured iron alloys with lattice softening and nanotwins. *Scr. Mater.* **67**, 511–514 (2012).

26. Furuta, T., Kuramoto, S., Ohsuna, T., Oh-ishi, K. & Horibuchi, K. Die-hard plastic deformation behavior in an ultrahigh-strength Fe-Ni-Al-C alloy, *Scr. Mater.* **101**, 87–90 (2015).

27. Edalati, K., Furuta, T., Daio, T., Kuramoto, S. & Horita, Z. High strength and high uniform ductility in a severely deformed iron alloy by lattice softening and multimodal-structure formation, *Scr. Mater.* **101**, 87–90 (2015).

28. Gutierrez-Urrutia, I. & Raabe, D. Multistage strain hardening through dislocation substructure and twinning in a high strength and ductile weight-reduced Fe-Mn-Al-C steel, *Acta Mater.* **60**, 5791–5802 (2012).

29. Kim, S. H., Kim, H. & Kim, N. J. Brittle intermetallic compound makes ultrastrong low-density steel with large ductility, *Nature* **518**, 77–79 (2015).

30. Yang, M. X., Yuan, F. P., Xie, Q. G., Wang, Y. D., Ma, E. & Wu, X. L. Strain hardening in Fe-16Mn-10Al-0.86C-5Ni high specific strength steel. *Acta Mater.* **109**, 213–222 (2016).
31. Bouaziz, O. Strain-Hardening of twinning-induced plasticity steels. *Scr. Mater.* **66**, 982–985 (2012).
32. Sevillano, J. G. & Cuevas, F. D. L., Internal stresses and the mechanism of work hardening in twinning-induced plasticity steels. *Scr. Mater.* **66**, 978–981 (2012).
33. Gutierrez-Urrutia, I., Valle, J. A. D., Zaefferer, S. & Raabe, D. Study of internal stresses in a TWIP steel analyzing transient and permanent softening during reverse shear tests. *J. Mater. Sci.* **45**, 6604–6610 (2010).
34. Cheng, S., Wang, X. L., Feng, Z., Clausen, B., Choo, H. & Liaw, P. K. Probing the characteristic deformation behaviors of transformation-induced plasticity steels. *Metall. Mater. Trans. A* **39A**, 3105–3112 (2008).
35. Sinclair, C. W., Saada, G. & Embury, J. D. Role of internal stresses in co-deformed two-phase materials, *Phil. Mag.* **86**, 4081–4098 (2006).
36. Harjo, S., Tomota, Y., Neov, D., Lukas, P., Vrana, M. & Mikula, P. Bauschinger effect in alpha-gamma dual phase alloys studied by in situ neutron diffraction, *ISIJ Int.* **42**, 551–557 (2002).
37. Taleb, L. & Sidoroff, F. A micromechanical modeling of the Greenwood-Johnson mechanism in transformation induced plasticity. *Int. J. Plast.* **19**, 1821–1842 (2003).
38. Schinhammer, M., Pecnik, C. M., Rechberger, F., Hanzi, A. C., Loffler, J. F. & Uggowitzer, P. J. Recrystallization behavior, microstructure evolution and mechanical properties of biodegradable Fe-Mn-C(-Pd)TWIP alloys. *Acta Mater.* **60**, 2746–2756 (2012).
39. Mataya, M. C., Brown, E. L., Riendeau, M. P. Effect of hotworking on structure and strength of type 304 L austenitic stainless steel. *Metall Trans A.* **21**, 1969–1987 (1990).
40. Gutierrez-Urrutia, I. & Raabe, D. Influence of Al content and precipitation state on the mechanical behavior of austenitic high-Mn low-density ateels. *Scr. Mater.* **68**, 343–347 (2013).
41. Raabe, D., Ponge, D., Dmitrieva, O. & Sander, B. Nanoprecipitate-hardened 1.5GPa steels with unexpected high ductility. *Scr. Mater.* **60**, 1141–1144 (2009).
42. Jiang, S. H., Wang, H., Wu, Y., Liu, X. J., Chen, H. H., Yao, M. J., Gault, B., Ponge, D., Raabe, D., Hirata, A., Chen, M. W., Wang, Y. D. & Lu,

Z. P. Ultrastrong steel via minimal Lattice misfit and high-density nanoprecipitation. *Nature* **544**, 460–464 (2017).

43. Wu, X. L., Yang, M. X., Yuan, F. P., Chen, L. & Zhu, Y. T. Combining gradient structure and TRIP effect to produce austenite stainless steel with high strength and ductility. *Acta Mater.* **112**, 337–346 (2015).

44. Chen, L., Yuan, F. P., Jiang, P., Xie, J. J. & Wu, X. L. Mechanical properties and deformation mechanism of Mg-Al-Zn alloy with gradient microstructure in grain size and orientation. *Mater. Sci. Eng. A* **694**, 98–109 (2017).

45. Lu, L., You, Z. S. & Lu, K. Work hardening of polycrystalline Cu with nanoscale twins. *Scr. Mater.* **66**, 837–842 (2012).

46. Wang, W., Yang, M. X., Yan, D. S., Jiang, P., Yuan, F. P. & Wu, X. L. Deformation mechanisms for superplastic behaviors in a dual-phase high specific strength steel with ultrafine grains. *Mater. Sci. Eng. A* **702**, 133–141 (2017).

47. He, B. B., Hu, B., Yen, H. W., Cheng, G. J., Wang, Z. K., Luo, H. W. & Huang, M. X. High dislocation density-induced large ductility in deformed and partitioned steels. *Science* **357**, 1029–1032 (2017).

48. Yang, M. X., Pan, Y., Yuan, F. P., Zhu, Y. T. & Wu, X. L. Back stress strengthening and strain hardening in gradient structure. *Mater. Res. Lett.* **4**, 141–151 (2016).

49. Bian, X. D., Yuan, F. P., Wu, X. L. & Zhu, Y. T. The evolution of strain gradient and anisotropy in gradient-structured metal. *Metall. Mater. Trans. A* **48A**, 3951–3960 (2017).

Chapter 41

Hetero-Deformation Induced (HDI) Strengthening and Strain Hardening in Dual-Phase Steel

X. L. Liu,[a,b,*] Q. Q. Xue,[a,c,*] W. Wang,[a] L. L. Zhou,[a] P. Jiang,[a] H. S. Ma,[a] F. P. Yuan,[a,c] Y. G. Wei,[d] and X. L. Wu[a,c]

[a]*State Key Laboratory of Nonlinear Mechanics, Institute of Mechanics, Chinese Academy of Sciences, Beijing 100190, China*
[b]*School of Mechanical Electronic & Control Engineering, Beijing Jiaotong University, Beijing 100044, China*
[c]*College of Engineering Sciences, University of Chinese Academy of Sciences, Beijing 100049, China*
[d]*College of Engineering, Peking University, Beijing 100871, China*
xlwu@imech.ac.cn

Strain hardening still remains challenging at high strength levels specifically in hetero- microstructures inherently with large mechanical incompatibility among various phases. In this chapter,

---

*These authors contributed equally to this work.

Reprinted from *Materialia*, **7**, 100376, 2019.

Heterostructured Materials: Novel Materials with Unprecedented Mechanical Properties
Edited by Xiaolei Wu and Yuntian Zhu
Text Copyright © 2019 Acta Materialia Inc.
Layout Copyright © 2022 Jenny Stanford Publishing Pte. Ltd.
ISBN 978-981-4877-10-7 (Hardcover), 978-1-003-15307-8 (eBook)
www.jennystanford.com

both tensile deformation and strain hardening were investigated in detail in an as-annealed dual-phase steel, with chemical composition (wt. %) of 0.86 C, 16 Mn, 10 Al, 5 Ni, balance Fe. The dual-phase microstructure consists of both ductile face-centered-cubic γ-austenite and almost non-deformable B2 intermetallic compound. The tensile response exhibited a yield-drop followed by the transient with an up-turn of strain hardening rate due to the initial deficiency of mobile dislocations. The load-unload-reload tensile testing indicated solid evidence of the operation of hetero-deformation induced (HDI) stresses, by the presence of hysteresis loops with large residual plastic strains and repeated occurrence of yield-drop during each unload-reload cycle. The microstructural origin of HDI stresses was ascribed to plastic mismatches near grain/phase boundaries between two phases as evidenced by the change of Schmid factor and KAM values. The production of geometrically necessary dislocations was responsible to both HDI stress and HDI hardening, which accounted for a large proportion of global flow stress and strain hardening. A micro- structure-based model was developed to calculate HDI stress in the dual-phase structure to correlate the evolution of HDI stresses with applied strains. The modeling results were well consistent with the experimental ones. The effect of microstructural parameters was discussed on HDI stresses in the dual-phase structure.

41.1 Introduction

Both strength and ductility are two key mechanical properties of metallic materials. However, ductility, i.e. uniform elongation during tensile testing, becomes disappointedly low as yield-strength increases significantly [1–3]. This is mainly due to the deficiency of strain hardening in strongly grain-refined microstructures where dislocation plasticity almost disappears [4–6]. So far, the strategies of two kinds are developed to enhance strain hardening and therefore, ductility. The first is to promote the storage of defects to recover intra-granular dislocation plasticity for forest hardening, e.g. the nano-twinned grains [7], nano-precipitation [8, 9], and grain refinement [10] etc. The second is to introduce the inter-granular

HDI stress via trans-scale hetero-structuring for the long-range back stress-induced hardening [11, 12].

Recently, great interests arouse in the heterostructures of metallic materials [11–22]. The heterostructures usually feature a trans-scale microarchitecture in terms of grain sizes, such as the gradient structure [13, 14, 21], lamellar structure [12, 17, 18], and heterogeneous grain structure [15, 22]. Therein the nano- and submicron-sized grains of hardly any ductility are in sharp contrast with coarse grains of large ductility. As a result, large mismatch of mechanical responses happens inevitably during plastic deformation [11, 15], such as yield strength, flow stress, and strain hardening. Both the plastic incompatibility and resultant strain gradient thus exist among trans-scale grains. This will trigger the introduction of intergranular back stress which plays a crucial role in work hardening [11–22]. From this, the heterostructure harvests acceptable ductility, even comparable to that in the coarse-grained counterpart [12–22]. The dual-phase steel is, obviously, representative of heterostructures, especially in case of the presence of large mechanical contrasts between two phases [23–25].

The traditional method to measure the back stress is the classical Bauschinger tensile-compressive loading [26, 27]. Recently, a modified load-unload-reload tensile testing is developed [28]. Several physically based-models are proposed to measure the back stress [29–36]. For example, the back stress relies on the number of Orowan loops and the radius of precipitates in the precipitation-hardening alloy [30, 31]. The back stress is twice shear modulus times eigenstrain according to the Eshelby's equivalent inclusion method in particle- or fiber-reinforced composites [37, 38]. However, the microstructure-based model is scarce to calculate directly the HDI stress in various heterostructures.

In this study, both the plastic deformation and HDI strain hardening were investigated during tensile deformation in a dual-phase steel. A high specific strength steel was selected as a model dual-phase structure, where the matrix is ductile γ-austenite, while hard and nearly non-deformable B_2 intermetallic compound serves as the second phase. Large mechanical mismatch makes it an ideal dual-phase microstructure to reveal the formation and evolution of both back stress and HDI hardening during tensile deformation. Moreover, γ-austenite is, actually, an ultrafine-grained microstructure, with much low initial mobile dislocations and low tendency of dislocation propagations. This facilitates to further reveal the crucial role by HDI hardening during the whole tensile deformation, instead of traditional forest dislocation hardening. The latter is usually considered to be responsible for ductility in coarse-grained dual-phase steels of low yield strength.

41.2 Materials and Experimental Methods

The high specific strength dual-phase (DP) steel was produced by arc melting in a high frequency induction furnace under pure argon atmosphere. The chemical composition is 16Mn, 10Al, 5Ni, 0.86C, balance Fe, in weight percent. Figure 41.1 presents the schematic of processing route for the HSSS. First, the cylindrical ingots were cast of a diameter of 130 mm and length of 200 mm and homogenized at 1180°C for 2 h. Then, the homogenized ingots were hot-forged in between 1150°C and 900°C into slabs with a thickness of 14 mm, and hot-rolled with a starting temperature of 1050°C into strips with a thickness 7.3 mm. The hot-rolled striped were finally cold rolled to sheets with the final thickness of 1.5 mm. The final annealing of the cold-rolled sheets was conducted at 900°C for 2–15 min followed immediately by water quenched.

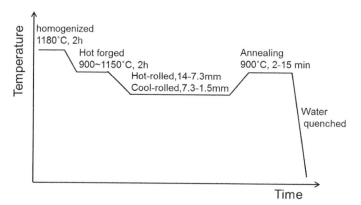

Figure 41.1 Schematic of processing route for the HSSS.

Both the monotonic and interrupted load-unload-reload (LUR) tensile testing were performed by using an MTS Landmark testing machine operating at quasi-static strain rate of 5×10^{-4} s^{-1} and at room temperature. The tensile specimens were cut from the annealed plate with longitudinal axes parallel to rolling direction. The specimens were dog-bone- shaped, with cross-sectional area of 4×1.5 mm^2 and gauge length of 20 mm. During LUR testing, the tensile specimen was stretched at first to certain strain under the strain control mode, then unloaded under stress control mode, and finally reloaded to the next unloading strain under strain control mode. The extensometer was attached to measure tensile strains during the whole tensile deformation. Three tensile tests were conducted for reproductive results.

The microstructural characterization was performed by using electron back-scattered diffraction (EBSD, JSM-7001F) and transmission electron microscopy (TEM, JEM-2100). Both the Schmid factor and Kernel Average Misorientation (KAM) values were measured near the grain boundaries and phase boundaries [39, 40], for the purpose of revealing micro- structural responses with respect to the HDI stress.

41.3 Experimental Results

41.3.1 Microstructural Characterization

Figures 41.2a and b are, respectively, the EBSD inverse pole figure (IPF) and phase map in the DP steel. Two phases are, respectively, the face-centered-cubic (fcc) γ-austenite and body-centered-cubic (bcc) FeAl-type B2 intermetallic compound [23, 24]. The volume fraction of B2 phase is 22.5%. The granular and lamellar B2 phases of two kinds are visible (Fig. 41.2b). The granular B2 grains locate mainly at the grain boundaries (GBs) of γ-grains. As seen from TEM micrograph in Fig. 41.2c, both γ-grains and B2 phase (labeled by arrows) are almost free of dislocations after recrystallization annealing. Figures 41.2d–f summarize the grain size distribution of two phases. The average grain size ($\bar{d}$) is 1.32 μm in γ-grains and 0.78 μm in granular B2. The average intercept of short- and long-axis ($\bar{\lambda}_S$ and $\bar{\lambda}_L$) in lamellar B2 is 0.77 μm and 3.38 μm, respectively.

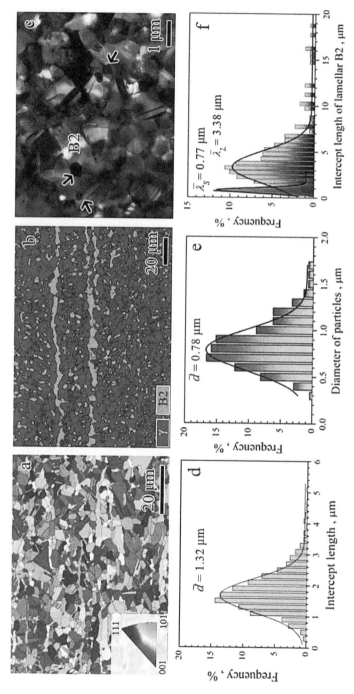

Figure 41.2 Microstructure characterization in as-annealed dual-phase microstructure before tensile testing. (a) EBSD inverse pole figure. (b) EBSD phase map of both γ-austenite and B2 intermetallic compound. Note lamellar and granular morphology of B2 phase of two kinds. (c) TEM bright-field image. Arrows: B2 grains mainly at grain boundaries. (d–f) Size distribution in γ-austenite, granular B2, and lamellar B2. $\bar{d}$: average grain size. $\bar{\lambda}_S$ and $\bar{\lambda}_L$ in (f): average intercept of short and long axis in lamellar B2.

41.3.2 Yield-Point Phenomenon

Figure 41.3a shows the tensile engineering stress-strain ($\sigma_e - \varepsilon_e$) curve. The DP steel shows the superior combination of yield strength ($\sigma_{0.2}$) of 1250 MPa, uniform elongation (E_U) of 27%, and ultimate tensile strength (σ_{UTS}) of 1449 MPa. Of special note is the presence of dis-continuous yield-point phenomenon soon after yielding. The yield-point phenomenon appears, see inset A, in the form of a yield-drop at first and the transient later, i.e. a concave curve with two inflection points labeled by ×. During the transient, the strain hardening rate, θ, exhibits a unique up-turn, see inset B. The yield-point phenomenon is well consistent with that in hetero-structures of other kinds [12, 13]. The yield-drop is ascribed to the deficiency of initial mobile dislocations upon yielding, while the up-turn of θ is due to the newly-generated dislocations [11–13].

The LUR tensile testing was further performed to display the details of evolution of plastic responses during the whole tensile deformation. Figure 41.3b is the LUR $\sigma_{true} - \varepsilon_{true}$ curve. The monotonic curve (dashed line) is also shown for comparison. Interestingly, two curves almost coincide at first but later, the LUR curve goes gradually above monotonic one. This indicates that unceasing load-unload cycles play a special role in strain hardening, leading to an increase in both σ_{UTS} and E_U.

Several distinct features are noteworthy. Firstly, the hysteresis loop appears during each unload-reload cycle, see Fig. 41.3c which shows a loop at unload strain of 25%. Usually, the hysteresis loop is a definitive sign of the presence of HDI stress during tensile deformation [12, 15, 26, 28]. The loop width is characterized by the residual plastic strain, ε_{rp}, upon each unloading (Fig. 41.3c) [15]. It is visible that the larger unload strain, the larger ε_{rp} will be, see Fig. 41.3d. Interestingly, ε_{rp} becomes even larger than 0.2% at unload strains larger than 18%. This indicates the onset of compressive yield flow, defined at an offset strain of 0.2%, even though the sample is still under applied tensile loading [15, 23]. Namely, this is an undoubtedly sign for the generation of HDI stresses [15]. Secondly, it is worthy to note that yield-drop, $\Delta\sigma_y$, recurs every time after each reload, see the close-up view of tensile curve as shown in Fig. 41.3e. Moreover, $\Delta\sigma_y$ increases with tensile straining, see inset. This indicates that the larger tensile strain, the less initial

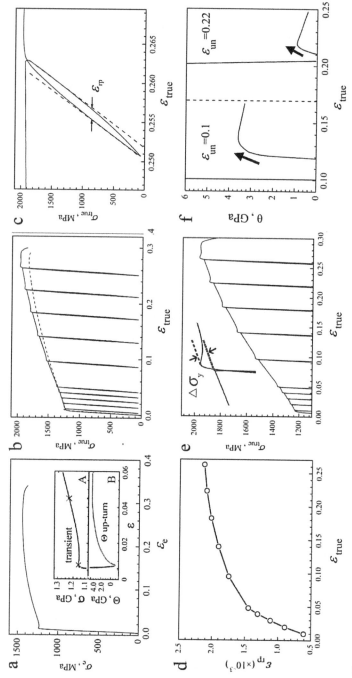

Figure 41.3 Strain hardening in dual-phase steel. (a) Tensile engineering stress-strain ($\sigma_e - \varepsilon_e$) curve. Inset A: close-up view showing yield-drop soon after yielding, followed by transient hardening. Two ×: inflection points. Inset B: up-turn of strain hardening rate (θ). (b) Interrupted load-unload-reload tensile $\sigma_{true} - \varepsilon_{true}$ curve. Dashed line: monotonic tensile $\sigma_{true} - \varepsilon_{true}$ curve. (c) Hysteresis loop at unload strain of 25%. ε_{rp}: residual plastic strain. (d) Change of ε_{rp} unloading strains with ε_e (e) Reoccurrence of yield-drop during LUR testing. Inset; close-up view of yield-drop ($\Delta\sigma_y$) at unload strain (ε_{un}) of 0.25. (f) Strain hardening rate (θ) vs true strain (ε_T) at ε_{un} of 0.1 and 0.22. Arrow: up-turn of θ.

dislocations will be after each unload. In other words, mobile dislocations are insufficient all through tensile deformation. This is inconsistent with large hardening component and stress increment, usually indicative of the operation of dominant forest hardening. Thirdly, Fig. 41.3f shows, respectively, two curves of θ-true strain (ε_T) at unload strain of 0.1 and 0.22 (E_U). Of special note is the upturn of θ which incessantly recurs till co-deformation stage of B2 phase and γ-grains, e.g. at strain of 0.22. It is generally recognized, however, that dynamic recovery, instead of dislocation production, happens in the form of prevailed dislocation cross-slips during co-deformation stage of B2 phase and γ-grains of hardening [26, 29]. It is, therefore, speculated that dislocations associated with the upturn of θ are probably not the traditional forest dislocations.

41.3.3 HDI Stress during Tensile Deformation

Both the HDI stress (σ_b) and HDI hardening rate ((θ_b) are calculated further based on the method developed in Ref. [28]. Figure 41.4a shows the increment of both HDI stress, $\Delta\sigma_b$ (= $\sigma_b - \sigma_b^0$), and flow stress, $\Delta\sigma_f$ (= $\sigma_f - \sigma_{0.2}$), where σ_b^0 is HDI stress at yield stress ($\sigma_{0.2}$) and σ_f is flow stress. It is interesting to note that $\Delta\sigma_b$ accounts for the proportion of nearly 100% of $\Delta\sigma_f$ when plastic strain is smaller than 9% and still for 70% of $\Delta\sigma_f$ even at E_U. Figure 41.4b shows the strain hardening rate versus true strain curves. The unusually high proportion of HDI hardening rate, θ_b, is noted at the initial stage of small strains. This proportion keeps higher than 50% with further straining even with a declined trend.

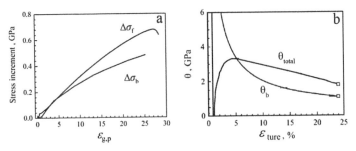

Figure 41.4 HDI stress and HDI hardening rate. (a) Increment of HDI stress ($\Delta\sigma_b$) and flow stress ($\Delta\sigma_f$) vs the applied strain. (b) Strain hardening rate (θ) vs true strain.

41.3.4 Evolution of Schmid Factor and KAM Value

Both the Schmid factor and KAM values were measured after tensile deformation to unveil the microstructural origin of HDI stresses. Figures 41.5a-1 and a-2 are EBSD maps before and after tensile deformation. Only γ-grains are imaged based on the magnitude of Schmid factor. Figures 41.5b-1 and b-2 are corresponding results for B2 phase. An obvious increase is visible in the volume fraction of γ-grains with low Schmid factors after tensile deformation. The statistic result is shown in Fig. 41.5a-3. By contrast, B2 phase only shows the unconspicuous change of Schmid factors in the range of large values, see Figs. 41.5b-1 to b-3. This indicates that during tensile deformation, γ-grains bear the vast majority of plastic strains. In other words, strain partitioning happens in dual phases. This is due to the almost non-deformable B2 phase. As a result, large mechanical mismatches exist between two phases during tensile deformation, which facilitates to produce the HDI stress.

Figure 41.6a-1 is the KAM mapping of dual phases after tensile deformation. The GBs between γ/γ grains and phase boundaries between γ/B2 grains are colored, respectively, in white and black. In general, the large KAM values appear mainly near GBs between γ/γ grains. Figures 41.6a-2 and a-3 are close-up views for two representative examples. Figure 41.6a-2 shows two granular B2 grains. The large KAM values appear in the neighboring γ-grain near phase boundaries. However, KAM values are obviously larger in grain 2 than those in grain 1. This indicates plastic accommodation in granular B2 grains. This is well in line with the change of Schmid factor in B2 as shown in Fig. 41.5b-3. In contrast, the interior of lamellar B2 shows small KAM values, see Fig. 41.6a-3, indicative of nearly absence of plastic deformation. Figures 41.6b-1 and b-2 show the statistic distribution of KAM values based on the measurements on several tens of GBs. Interestingly, the KAM values near γ/B2 phase boundaries are only a little bit smaller than those near γ/γ GBs. This indicates strong strain incompatibility between γ-austenite and B2 phase. Then the GND density information can be obtained using a simple method based on the strain gradient theory, which was proposed by Gao and Kubin [33, 35]: $\rho_{GND} = \theta/lb$, Where ρ_{GND} is the GND density at local points, θ represents the KAM values at local points, l is the unit length for the local points, and b is the Burger's vector for the materials. l and b are constant parameters. Therefore, the larger KAM value is, the larger the GND density.

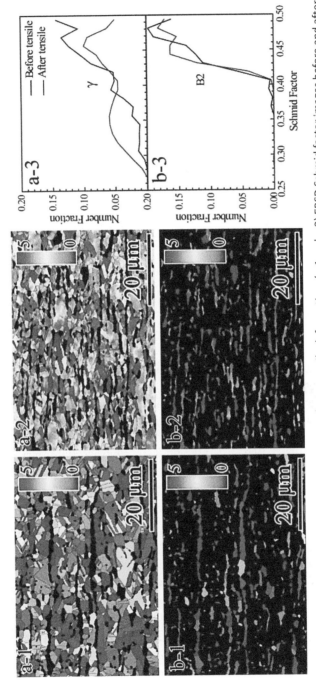

Figure 41.5 Change of Schmid factor in two phases during tensile deformation. (a-1 and a-2) EBSD Schmid factor images before and after tensile testing. Only γ-grains are imaged. Scale bar: Schmid factor range. (a-3) Schmid factor distribution in γ-grains. (b-1–b-3) Corresponding results in B2 phase.

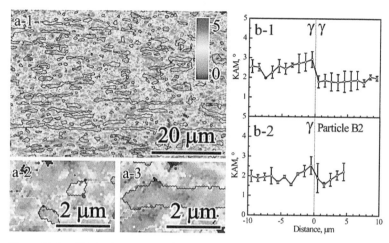

Figure 41.6 KAM values in two phases. (a-1) EBSD image showing KAM value distribution in dual-phase microstructure after tensile testing. Line in black and white: B2/γ phase boundaries and γ/γ grain boundary. Scale bar: KAM value range. (a-2 and a-3) Close-up views showing KAM value distribution in small granular B2 and lamellar B2. (b-1–b-2) Statistic distribution of KAM value distribution at γ/γ and γ/B2 boundaries.

41.4 Modeling HDI Stress in Dual-Phase Microstructure

A microstructure-based model is developed to understand the generation of HDI stress in the dual phase structure. As shown in Fig. 41.2a, The matrix is γ-grains, which integrates granular and lamellar units of B2 phase, and the shape of lamellar and granular B2 is not regular. For modeling purpose we set some ideal conditions. Figure 41.7a is a schematic microstructural model. Here, two dimensional B2 phase is simplified as a circle (Fig. 41.7b) and an ellipse (Fig. 41.7c), respectively. The microstructure, thus, was ideally regarded as consisting of m units of granular B2 and n units of lamellar B2. In this model, B2 grains are treated as inclusions due to their almost non-deformable feature (Fig. 41.5b-3). The equivalent inclusion method is, therefore, developed suitable to measure the HDI stress in the DP steel in terms of the disturbance of inclusion-induced uniform stress [37, 38].

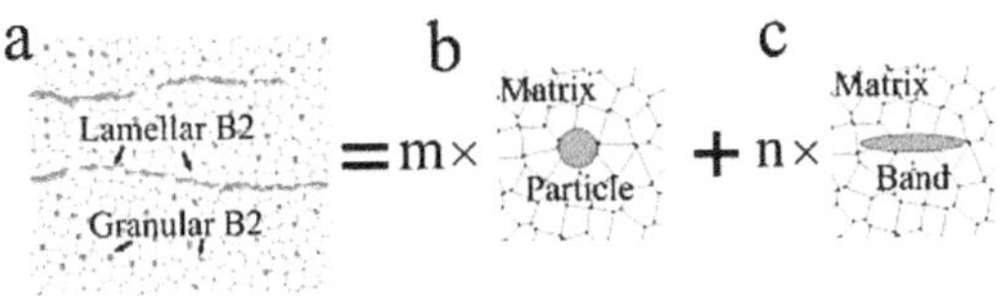

Figure 41.7 Microstructural-based model in dual-phase structure with granular and lamellar hard B2 phase on soft γ-grain matrix. The granular and lamellar B2 are simplified to circle (b) and ellipse (c), respectively.

The HDI stress on an active slip plane under uniform shear is given by [30]:

$$\tau_b = 2\gamma\mu f \varepsilon_p^* \tag{41.1}$$

where μ is shear modulus of matrix, γ is accommodation factor, f_V is volume fraction of hard second phase, ε_p^* is symmetrical plastic shear strain.

The plastic strain due to stress relaxation is [41]:

$$\varepsilon_p^* = 0.7\left(\frac{b\varepsilon_{sp}}{r}\right)^{1/2} \tag{41.2}$$

where r is hard phase radius, b is Burges vector, and ε_{sp} is shear strain.

The back stress in tensile direction is:

$$\sigma_b^* = M\tau_b \tag{41.3}$$

$$\varepsilon_{sp} = M\varepsilon_{l,p}/2 \tag{41.4}$$

where σ_b^* and are $\varepsilon_{l,p}$ local HDI stress and plastic strain in tensile direction, and M is Taylor effective orientation factor.

Combining equations from (41.1) to (41.4), the HDI stress σ_b^* is:

$$\sigma_b^* = M^{2/3}\mu\gamma f\left(\frac{b\varepsilon_{l,p}}{r}\right)^{1/2} \tag{41.5}$$

The global HDI stress is f_V-weighted average of back stress in each unit:

$$\sigma_b^* = \sum M^{2/3}\mu\gamma f\left(\frac{b\varepsilon_{g,p}}{r}\right)^{1/2} \tag{41.6}$$

Table 41.1 lists physical constants to measure the back stress by using Eq. (41.6). f_V is shown in Figs. 41.2e and f, respectively, for granular and lamellar B2 with varying grain sizes. The same modulus is selected for both γ-austenite and B2 phase. The method to determine accommodation factor in both granular and lamellar B2, γ_g and γ_l, is shown in Appendix.

Table 41.1 Physical constants and parameters used in the model [24, 42]

Parameters	Physical meaning	Value
E_i	Modulus of B2	200 GPa
E_m	Modulus of γ-austenite	195 GPa
v	Poisson's ratio	0.24
μ	Shear modulus of γ-austenite	80.6 GPa
M	Taylor factor	3.06
b	Burgers vector	2.5×10^{-10} m

Figures 41.8a and b show, respectively, the increment of back stress ($\Delta\sigma_b$) and HDI strain hardening rate (θ_b) by Eq. (41.6), see red lines. It is visible that the calculating results of the model agree well with experimental ones. Further, the roles are analyzed of several key microstructural parameters on back stress. Figure 41.8c shows the effect of volume fraction of B2 f_V, with a fixed ratio of 1:2 for granular to lamellar B2 based on the microstructure (Fig. 41.2b). It is seen that the larger f_V, the higher HDI stress will be. Figure 41.8d shows the effect of the ratio of volume fraction $f_{g/l}$ of granular to lamellar B2, with a fixed f_V of B2 of 22.5%. It is seen that the more granular B2, the larger back stress will be. Figure 41.8e shows the size effects of B2 phase. The smaller B2 plays a larger role in HDI stress. A rapid decrease in back stress appears when the grain size of B2 phase is less than 1 μm.

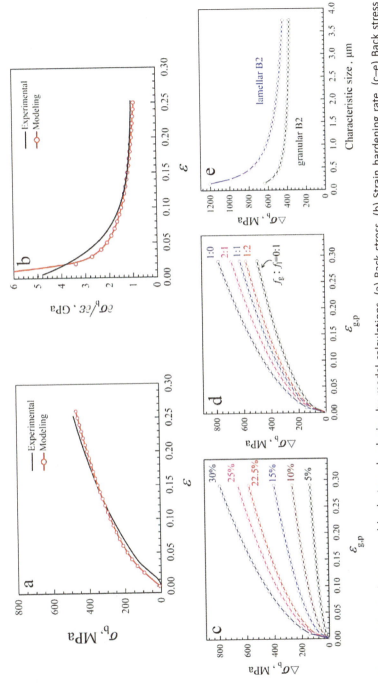

Figure 41.8 Back stress and back stress hardening by model calculations. (a) Back stress. (b) Strain hardening rate. (c–e) Back stress increment vs global plastic strain ($\varepsilon_{g,p}$) and grain size. Number in (c): volume fraction of B2 (f_{B2}) at fixed volume fraction ratio, 2:1, of lamellar to granular B2. Number in (d): volume fraction ratio of granular to lamellar B2, $f_{g,l}$, at fixed total volume fraction, 22.5%, in B2. (e) Internal back stress at $\varepsilon_{g,p}$ = 0.25 with the variation of the characteristic size for lamellar and granular B2.

41.5 Discussions

Soon after yielding, the present dual-phase steel exhibits a yield-drop followed by the transient with an up-turn of strain hardening rate, see insets A and B in Fig. 41.3a, typical of a yield-point phenomenon. The shortage of mobile dislocations appear in most of ultrafine- grained γ-grains after recrystallization (Fig. 41.2c), along with limited sources to activate dislocations [42, 43]. It is, therefore, concluded that the yield-point phenomenon is due to the shortage of mobile dislocations upon yielding. It is then necessary to increase the flow stress to facilitate yielding. Once the dislocation sources get activated, the flow stress will decrease. This is the dislocation source-limiting mechanism as shown in the varying recrystallized ultra-fine-grained microstructures [42, 43]. Furthermore, plastic yielding is, actually, elasto-plastic co-deformation in this DP steel as a result of large difference in yield strength between soft γ-grains and hard B2 phase [11, 24, 28]. The micro-hardness of γ-grains is about 175 Hv, and that of B2 phase with hot-rolled treatment is more than 1235 Hv [45]. The γ-grains begin to yield at first, while B2 grains still maintain elastic. The γ-grains are thus constrained by neighboring B2 grains such that the dislocations in γ-grains are piled up and blocked at phase boundaries. This produces a long-range back stress to make it difficult for dislocation slip inside γ-grains. In other words, the back stress renders an increased flow stress in γ-grains by the time the whole sample yields. The elasto-plastic mismatch will induce the generation of geometrically necessary dislocations (GNDs) near γ/B2 phase boundary for compatible deformation [46, 47]. These GNDs further lead to an extra strengthening effect to enhance yield strength [11–15, 48]. With the continuous production of mobile dislocations and GNDs as well, transient hardening occurs soon after yield-drop (inset A in Fig. 41.3a). The strain hardening rate therefore shows an up-turn (inset B in Fig. 41.3a) during the transient due to the production of GNDs, along with the recovery of intra-granular dislocation plasticity [11, 13, 14].

Of special note is the repeated presence of yield-drop followed by transient as evidenced by LUR testing, see Figs. 41.3d and e. This indicates the persistent shortage of mobile dislocations during tensile deformation. In other words, the traditional dislocation plasticity is not enough to be sustainable to strain hardening during the entire tensile deformation. Note continued presence of plastic mismatch in dual phases, as evidenced by the large change of Schmid factor in γ-grains and just the opposite in B2 phase, see Fig. 41.4. In response to tensile deformation, plastic incompatibility exists steadily at GBs, where strain gradient appears. It is, therefore, concluded that the GNDs are incessantly generated [11, 15]. These GNDs will induce the long-range inter-granular back stress [11]. Hence, the HDI hardening operates. This is actually the result of hetero-deformation due to the microstructural heterogeneity in the DP steel.

Hence, two kinds of mechanisms for strain hardening operate in the DP steel. One is HDI hardening associated with the accumulation of GNDs [46, 47], while the other is forest hardening related to the statistically stored dislocations (SSDs). Strain hardening can be derived as [11, 46]:

$$\frac{d\tau}{d\varepsilon} = \alpha G b \frac{d\sqrt{\rho_{S+\rho_G}}}{d\varepsilon} \quad (41.7)$$

where ε is applied strain, τ is shear flow stress, α is a constant, G is shear modulus, b is the magnitude of Burgers vector, ρ_S and ρ_G are density of SSDs and GNDs, respectively.

In the DP steel, B2 phase, on the whole, only bears a small portion of plastic strains. This is evidenced by hardly any change in Schmid factor distribution. Hence, the larger tensile strains, the larger plastic incompatibility between two phases, and therefore, the higher density of GNDs will be. In other words, the HDI hardening will play a growing role during tensile deformation. The repeated and increased yield-drop during LUR testing lends strong support that HDI hardening dominates. This gets further supported by the HDI stress model. As seen from Figs. 41.8a and b, back stress hardening reigns over the entire tensile deformation. Meanwhile, plastic strain partitioning is noted in this DP steel. The soft γ-grains carry much larger plastic strain, while strong B2 phase bears higher stress. Actually, the partitioning of both strain and stress is an inevitable result of back stress hardening. The strain partitioning gives rise to the presence of strain gradient which makes the continuous production of GNDs for back stress.

A universal microstructure-based model is developed to describe the evolution of back stress in a dual phase microstructure. The back stress predicted by the model is well consistent with that by experimental measurements. In terms of the model, the back stress will increase by both increasing volume fraction in B2 phase and decreasing grain sizes. This is understandable because both lead to an increase in plastic incompatibility and therefore to increase the GNDs.

41.6 Conclusions

The strain hardening behavior is investigated in a high specific strength dual-phase steel, with chemical composition (wt. %) of Fe–16Mn–10Al–0.86C–5Ni. The dual phases are ductile γ-austenite and hardly deformable B2 intermetallic compound. The main conclusions are drawn as follows.

(1) The tensile response exhibits the yield-drop soon after yielding, followed by transient hardening when strain hardening rate exhibits an up-turn. The interrupted load-unload-reload tensile testing shows the persistent presence of both hysteresis loop and yield-drop during each unload-reload cycle. Both the dislocation source-limiting mechanism and HDI hardening are proposed to be responsible for the yield-drop and up-turn as a result of the deficiency of mobile dislocations.

(2) The back stress and back stress-related hardening account for, respectively, a large proportion of flow stress and global strain hardening. The microstructural origin of back stress is plastic mismatch between two phases in response to tensile deformation. The γ-austenite bears almost all plastic strains.

(3) A universal microstructure-based model is proposed to describe the back stress in a dual-phase microstructure. The back stress predicted by the model is well consistent with that experimental result.

Acknowledgments

This work was supported by the National Key R&D Program of China under Grant No. 2017YFA0204402, Natural Science Foundation of China under Grant Nos. 11572328, 11472286, 11802011,

51601204, and 51471039, and Strategic Priority Research Program of CAS under Grant No. XDB22040503.

Appendix

Determination of accommodation factors

According to the Eshelby's theory [37, 38], the accommodation factor is determined by both the inclusion shape and orientation between the inclusion and slip system. For the isotropic granular B2 particles, the orientation between B2 and slip plane is considered to be consistent. The accommodation factor, γ_g, is determined as [41],

$$\gamma_g = \frac{(7-5v)}{15(1-v)} \quad (41.A1)$$

where v is Poisson's ratio.

For the anisotropy lamellar B2, the accommodation factor, γ_l, is a function of the angle between the major axis and slip system and thus, determined as [41],

$$\gamma_l = 1 - 2S \quad (41.A2)$$

where S is determined as,

$$S = \frac{1}{4C}\left(\cos^4\alpha - 2C\cos^2\alpha\cos^2\beta + \cos^4\beta\right) + \frac{1}{8(1-v)}\left(5\cos^2\alpha\cos^2\beta + 3(1-C)\right)$$
$$- \frac{v}{2(1-v)C^2}\left(2(1-C-C^2)\cos^2\alpha\cos^2\beta + (1-C)(\cos^4\alpha + \cos^4\beta)\right)$$

$$\quad (41.A3)$$

$$C = \cos^2\alpha + \cos^2\beta \quad (41.A4)$$

where α is the angle between the major axis and slip direction, β is the angle between major axis and direction normal to slip plane.

The lamellar B2 is parallel to loading direction in the present dual-phase structure. Then, the angle is $\pi/4$ between loading direction and slip direction and is $\pi/4$ also between loading and direction normal to slip plane. The accommodation factor of lamellar B2 γ_l is thus [41],

$$\gamma_l = \frac{11 - 8v}{16(1-v)} \quad (41.A5)$$

References

1. Y. Zhu, X. Liao. Nanostructured metals: retaining ductility, *Nat. Mater.* 3 (2004) 351–352.
2. R. Valiev, I. Alexandrov, Y. Zhu, T. Lowe. Paradox of strength and ductility in metals processed by severe plastic deformation, *J. Mater. Res.* 17 (2002) 5–8.
3. E. Ma. Eight routes to improve the tensile ductility of bulk nanostructured metals and alloys, *JOM* 58 (2006) 49–53.
4. Y. Cao, N. Song, X. Liao, M. Song, Y. Zhu. Structural evolutions of metallic materials progressed by severe plastic deformation, *Mater. Sci. Eng. R* 133 (2018) 1–50.
5. I. Ovid'ko, R. Valiev, Y. Zhu. Review on superior strength and enhanced ductility of metallic nanomaterials, *Prog. Mater. Sci.* 94 (2018) 462–540.
6. Y. Zhu, X. Wu. Ductility and plasticity of nanostructured metals: differences and issues, *Mater. Today Nano* 2 (2018) 15–20.
7. L. Lu, X. Chen, X. Huang, K. Lu. Revealing the maximum strength in nanotwinned copper, *Science* 323 (2009) 607–610.
8. Y. Zhao, T. Topping, J. Bingert, J. Thornton, A. Dangelewicz, Y. Li, W. Liu, Y. Zhu, Y. Zhou, E. Lavernia. High tensile ductility and strength in bulk nanostructured nickel, *Adv. Mater.* 16 (2008) 3028–3033.
9. T. Yang, Y. L. Zhao, Y. Tong, Z. B. Jiao, J. Wei, J. X. Cai, X. D. Han, D. Chen, A. Hu, J. J. Kai, K. Lu, Y. Liu, C. T. Liu. Multicomponent intermetallic nanoparticles and superb mechanical behaviors of complex alloys, *Science* 362 (2018) 933–937.
10. Z. Li, K. G. Pradeep, Y. Deng, D. Raabe, C. C. Tasan. Metastable high-entropy dual-phase alloys overcome the strength–ductility trade-off, *Nature* 534 (2016) 227–230.
11. X. Wu, Y. Zhu. Heterogeneous materials: a new class of materials with unprecedented mechanical properties, *Mater. Res. Lett.* 5 (2017) 527–532.
12. X. Wu, M. X. Yang, F. Yuan, G. Wu, Y. Wei, X. Huang, Y. Zhu. Heterogeneous lamella structure unites ultrafine-grain strength with coarse-grain ductility, *Proc. Natl. Acad. Sci. U.S.A.* 112 (2015) 14501–14505.
13. X. Wu, P. Jiang, L. Chen, F. Yuan, Y. Zhu. Extraordinary strain hardening by gradient structure, *Proc. Natl. Acad. Sci. U.S.A.* 111 (2014) 7197–7201.

14. X. Wu, P. Jiang, L. Chen, J. Zhang, F. Yuan, Y. Zhu. Synergetic strengthening by gradient structure, *Mater. Res. Lett.* 4 (2014) 185–191.
15. M. X. Yang, D. Yan, F. Yuan, P. Jiang, E. Ma, X. Wu. Dynamically reinforced heterogeneous grain structure prolongs ductility in a medium-entropy alloy with gigapascal yield strength, *Proc. Natl. Acad. Sci. U.S.A.* 115 (2018) 7224–7229.
16. B. B. He, B. Hu, H. W. Yen, G. J. Cheng, Z. K. Wang, H. W. Luo, M. X. Huang. High dislocation density induced large ductility in deformed and partitioned steels, *Science* 357 (2017) 1029–1032.
17. J. G. Kim, M. J. Jang, H. K. Park, K. G. Chin, S. Lee, H. S. Kim. Back-stress effect on the mechanical strength of TWIP-IF steels layered sheet, *Met. Mater. Int.* (2019) 1–6.
18. J. G. Kim, S. M. Baek, H. H. Lee, K. G. Chin, S. Lee, H. S. Kim. Suppressed deformation instability in the twinning-induced plasticity steel-cored three-layer steel sheet, *Acta Mater.* 147 (2018) 304–312.
19. H. Wang, Z. You, L. Lu. Kinematic and isotropic strain hardening in copper with highly aligned nanoscale twins, *Mater. Res. Lett.* 6 (2018) 333–338.
20. R. Z. Valiev, Y. Estrin, Z. Horita, T. G. Langdon, M. J. Zehetbauer, Y. T. Zhu. Fundamentals of superior properties in bulk nanoSPD materials, *Mater. Res. Lett.* 4 (2016) 1–21.
21. H. H. Lee, J. I. Yoon, H. K. Park, H. S. Kim. Unique microstructure and simultaneous enhancements of strength and ductility in gradient-microstructured Cu sheet produced by single-roll angular-rolling, *Acta Mater.* 166 (2019) 638–649.
22. S. Shukla, D. Choudhuri, T. Wang, K. Liu, R. Wheeler, S. Williams, B. Gwalani, R. S. Mishra. Hierarchical features infused heterogeneous grain structure for extraordinary strength-ductility synergy, *Mater. Res. Lett.* 6 (2018) 676–682.
23. S. H. Kim, H. Kim, N. J. Kim. Brittle intermetallic compound makes ultrastrong low-density steel with large ductility, *Nature* 518 (2015) 77–79.
24. M. Yang, F. Yuan, Q. Xie, Y. Wang, E. Ma, X. Wu. Strain hardening in Fe-16Mn-10Al-0.86C-5Ni high specific strength steel, *Acta Mater.* 109 (2016) 213–222.
25. W. Wang, M. Yang, D. Yan, P. Jiang, F. Yuan, X. Wu. Deformation mechanisms for superplastic behaviors in a dual-phase high specific strength steel with ultrafine grains, *Mater. Sci. Eng. A* 702 (2017) 133–141.

26. X. Feaugas. On the origin of the tensile flow stress in the stainless steel AISI 316L at 300 K: back stress and effective stress. *Acta Mater.* 13 (1999) 3617–3632.
27. X. Liu, F. Yuan, Y. Zhu, X. Wu. Extraordinary Bauschinger effect in gradient structured copper, *Scr. Mater.* 150 (2018) 57–60.
28. M. Yang, Y. Pan, F. Yuan, Y. Zhu, X. Wu. Back stress strengthening and strain hardening in gradient structure, *Mater. Res. Lett.* 6 (2016) 1–7.
29. J. Gilsevillano. An alternative model for the strain hardening of FCC alloys that twin, validated for twinning-induced plasticity steel, *Scr. Mater.* 60 (2009) 336–339.
30. L. M. Brown, W. M. Stobbs. The work-hardening of copper-silica, *Philos. Mag.* 23 (1971) 1185–1199.
31. L. M. Brown, W. M. Stobbs. The work-hardening of copper-silica v. equilibrium plastic relaxation by secondary dislocations, *Philos. Mag.* 34 (1976) 351–372.
32. L. M. Brown, D. R. Clarke. The work hardening of fibrous composites with particular reference to the copper-tungsten system, *Acta Mater.* 25 (1977) 563–570.
33. K. Tanaka, T. Mori. The hardening of crystals by non-deforming particles and fibres, *Acta Mater.* 18 (1970) 931–941.
34. P. J. Withers, W. M. Stobbs, O. B. Pedersen. The application of the eshelby method of internal stress determination to short fibre metal matrix composites, *Acta Mater.* 37 (1989) 3061–3084.
35. G. Masing. Zur Heyn'schen Theorie der Verfestigung der Metalle durch verborgen elastische Spannungen. in: Harries CD, (ed.). Wissenschaftliche Veröffentlichungen aus dem Siemens-Konzern: III. Band. Springer Berlin Heidelberg, Berlin, Heidelberg, 1923. pp. 231–239.
36. S. Allain, O. Bouaziz, M. Takahashi. Toward a new interpretation of the mechanical behaviour of as-quenched low alloyed martensitic steels, *ISIJ Int.* 52 (2012) 717–722.
37. J. D. Eshelby. The determination of the elastic field of an ellipsoidal inclusion, and related problems, *Proc. R. Soc. Lon. Ser. A* 241 (1957) 376–396.
38. J. D. Eshelby. The elastic field outside an ellipsoidal inclusion, *Proc. R. Soc. Lon. Ser. A* 252 (1959) 561–569.
39. N. Allain-Bonassoa, F. Wagnera, S. Berbennia, D. Field. A study of the heterogeneity of plastic deformation in IF steel by EBSD, *Mater. Sci. Eng. A* 548 (2012) 56–63.

40. I. Gutierrez-Urrutia, F. Archie, D. Raabe, F. Yan, N. Tao, K. Lu. Plastic accommodation at homophase interfaces between nanotwinned and recrystallized grains in an austenitic duplex-microstructured steel, *Sci. Tech. Adv. Mater.* 17 (2016) 29–36.
41. K. Tanaka, T. Mori. The hardening of crystals by non-deforming particles and fibres, *Acta Mater.* 18 (1970) 931–941.
42. D. Chanbi, L. A. Amara, E. Ogam, S. E. Amara, Z. El A. Fellah. Microstructural and mechanical properties of binary Ti-rich Fe–Ti, Al-rich Fe–Al, and Ti–Al alloys, *Materials* 12(3) (2019) 433.
43. X. X. Huang, N. Hansen, N. Tsuji. Hardening by annealing and softening by deformation in nanostructured metals, *Science* 312 (2006) 249–251.
44. N. Kamikawa, X. X. Huang, N. Tsuji, N. Hansen. Strengthening mechanisms in nanostructured high-purity aluminium deformed to high strain and annealed, *Acta Mater.* 57 (2009) 4198–4208.
45. M. Krasnowski, S. Gierlotk, T. Kulik. FeAl-B composites with nanocrystalline matrix produced by consolidation of mechanically alloyed powders, *J. Alloys Compd.* 791 (2019) 75–80.
46. M. F. Ashby. The deformation of plastically non-homogeneous materials, *Philos. Mag.* 21 (1970) 399–424.
47. C. S. Hana, H. J. Gao, Y. G. Huang, W. D. Nix. Mechanism-based strain gradient crystal plasticity—I. Theory, *J. Mech. Phys. Solid* 53 (2005) 1188–1203.
48. C. X. Huang, Y. F. Wang, X. L. Ma, S. Yin, H. W. Höppe, M. Göken, X. L. Wu, H. J. Gao, Y. T. Zhu. Interface affected zone for optimal strength and ductility in heterogeneous laminate, *Mater. Today* 21 (2018) 713–719.

Index

accumulative roll bonding (ARB) 131, 133, 138, 140, 160, 168, 170, 193, 528, 539
adiabatic shear bands (ASB) 311–312, 314–319, 405, 407, 537, 615–617, 620, 622–623, 625, 632–634
alloys 4, 13, 43, 47, 527–532, 534–541, 586–587, 591–592, 596, 614–615, 620–621, 750–751, 773–775, 779–780, 788–789
  magnesium 47, 418–419, 424, 426, 439–440, 442–444
  solid solution 750
applied strain, increasing 5, 62, 121–125, 158, 161, 166, 184, 195, 197, 213, 220, 261–262, 264, 302, 367
ARB, *see* accumulative roll bonding
ASB, *see* adiabatic shear bands
ASB zone 617, 625, 632–633
aspect ratio 82, 88, 269, 520, 535, 728–729, 736–737
austenite 293–294, 299–300, 655, 697–698, 702–703, 705–707, 711–712, 721, 723, 727, 729, 798–802, 806, 810, 814
austenite grains 295, 297, 705, 753, 774–776, 780
austenite phase 646, 752–753, 765, 784
austenitic stainless steel 14, 46
axial direction 216, 734–736
axial strains 228, 736, 738, 740

back-scattered electron (BSE) 423, 433, 435, 452–453
base metal (BM) 206–208
Bauschinger effect 9, 20, 23, 78–79, 87, 361–363, 366–368, 499–500, 505, 507–508, 722, 724, 732
BM, *see* base metal
boundary spacing 132–133, 140, 146, 148–149, 164–167
BSE, *see* back-scattered electron

CG, *see* coarse-grained
coarse-grain ductility 48–49, 73, 86
coarse-grained (CG) 74–77, 90–95, 106–109, 112–114, 193–195, 197–199, 226, 239–240, 288–290, 324–326, 389–390, 500–502, 647–648, 697–698, 715–716
coarse-grained layer 91, 93–95
coarse-grained sample 57, 93, 228, 350
coarse-grained zones 648, 650
coarse grains 289, 293–294, 364, 367, 399, 405, 474, 476, 500–501, 642, 644, 648–649, 651, 654–655, 658–659
cold rolling (CR) 38, 588, 590, 609, 614–615, 641–644, 646–647, 653–654, 656–658, 710–711, 727, 751–752, 773–781, 783–786, 789
core samples 430–431, 435–436, 442
CR, *see* cold rolling
critical resolved shear stress (CRSS) 419, 629

CRSS, *see* critical resolved shear stress
cryogenic temperature 13, 474–476, 501, 614, 616, 620–621, 623–635

DDWs, *see* dense dislocation walls
deformation
  heterogeneous 18, 20, 25, 215, 363
  high strain rate 529, 615
  high temperature 760, 763, 767
  started plastic 126, 303–304, 483
deformation anisotropy 455, 466–467, 469
deformation behavior 5, 13, 49, 150, 328, 420, 500, 566, 668
deformation conditions 538–539, 753
deformation layer 527, 532, 541
deformation mechanisms 372, 374, 383, 385, 417–420, 422, 424, 426, 428, 436, 440–442, 444, 560–561, 626–627, 780
  dominant 372–373, 380, 385, 442, 629, 782
  transition of 149, 439, 614
deformation modes 38–39, 373, 419, 443, 495, 617
deformation temperature 410, 419, 626, 758
  normalized 755–756
deformation twinning 38, 44, 46, 147, 284, 425, 440–441, 591, 614, 625
deformation twins (DTs) 34, 70, 148, 336, 433–434, 450, 455, 623–625, 629–631, 635, 774–775, 778, 780, 782–783, 789
dense dislocation walls (DDWs) 397, 399
density gradient 178, 183–184

depth range 122, 257–258, 532, 536
DIC, *see* digital image correlation
differential speed rolling (DSR) 539–540
digital image correlation (DIC) 161, 169, 174, 193, 204, 206–207, 212, 227–228, 239, 241, 254, 256, 451, 453–454, 786
dislocation activities 40, 61, 195, 205, 218, 231, 521, 531–532, 764
dislocation arrays 163–164, 216, 537
dislocation behaviors 371–372, 385, 419, 435, 776, 780, 782, 789
dislocation cells 55–56, 108, 121–122, 124, 138, 299, 397, 399, 476, 528, 531–532, 534–535, 624, 650
dislocation debris 61, 110, 232, 532, 609
dislocation density 123–125, 166, 232, 367–368, 373, 397, 399, 411, 432, 507, 509–510, 533–535, 689, 764, 782–783
  evolution of mobile 292, 429, 790
  high 124, 138, 270, 508, 535, 548
  total 82, 192, 199–200, 219, 292, 363, 504
dislocation dynamics 87, 177–186
dislocation emissions 21, 148, 163, 179
dislocation entanglements 126, 232, 299, 313, 532, 534
dislocation interactions 78, 99–101, 180, 186, 451, 500, 595, 688–689
dislocation line 184–185, 331, 381

dislocation loop 181–182
dislocation mechanisms 23, 432, 439–440
dislocation models 20–21, 23
dislocation motion 143, 374, 507, 578, 627, 629, 688, 700, 708
dislocation nucleation 45, 372, 384–385
dislocation segments 180, 184–185, 195, 628
dislocation sources 8–9, 21, 42, 78, 80, 148, 166, 179, 182–183, 555, 812
  primary 178–179, 183–184, 483
dislocation storage 61–62, 469, 510, 553
dislocation substructures 122
dislocation tangles 56, 397, 476, 528, 609
dislocation types 380–381, 629
dislocation velocity 69–70
dislocation walls 197, 372, 384–385, 532, 534, 623, 648, 650, 760
  propagation and formation of 372, 383–384
dislocations 178, 507
DP, *see* dual-phase
DPD, *see* dynamic plastic deformation
DSR, *see* differential speed rolling
DSS, *see* duplex stainless steel
DTs, *see* deformation twins
dual-phase (DP) 721–722, 727, 742, 750, 775, 800
dual-phase microstructure 724, 749, 751, 768, 776, 785, 789, 798, 808–809, 811, 814
dual-phase steel 7, 465, 586, 607, 797–800, 802, 804, 806, 808, 810, 812, 814, 816
dual-phase structure 107, 798, 809

duplex stainless steel (DSS) 700, 702–703, 705–706, 708, 710–712, 716, 756
dynamic compression 394, 400, 402–408, 412
dynamic plastic deformation (DPD) 643, 656–657
dynamic shear experiments 315, 616–617, 620, 625
dynamic shear loading 312, 314–315, 614–615, 622, 624–625, 635

EBSD, *see* electron back-scattered diffraction
elastic layer 123–124, 127, 554
elastic-plastic boundaries 123–125
elastic-plastic transition stage 261–262, 265
elasto-plastic transition 463, 466, 468, 586, 595, 724, 738–739, 742
elasto-plastic transition stage 463, 465
electron back-scattered diffraction (EBSD) 76, 134–135, 206, 241, 393, 395, 423–424, 433–434, 452–453, 645, 647, 682–683, 726–727, 752, 776–777
engineering strains 35, 37, 168, 394, 421, 426, 429, 432, 598, 659, 671, 782
engineering stress-strain curves 36, 91, 94, 127, 210, 399, 454, 478, 480, 482, 502, 566, 568, 684, 753, 778–779
exhaustion curves 430, 782

flow stress 92–93, 111, 164–165, 198–199, 319, 329–330, 333, 377, 407, 409–411, 504–505, 686–687, 713–714, 753, 812

extra 209, 219
higher 62, 299, 372, 467, 575
maximum 755-756, 758-759
shear 7, 318, 363, 631, 813
tensile 87, 789
forest dislocation hardening 233, 362-363, 367, 451, 505, 739, 787
forward stress 18, 20, 23-26, 106-107, 110, 124, 127, 192, 197, 246-247, 367
fracture surface 40-41, 565, 569-571, 653, 670, 673-674
Frank-Read sources 159, 161, 178-185, 195, 198
friction stir processing (FSP) 206-208, 549, 551
FSP, *see* friction stir processing
function of distance 8-9, 162, 173, 194, 213

GBs, *see* grain boundaries
geometrically necessary dislocations (GNDs) 6-9, 21, 106-107, 124-125, 146-147, 191-192, 197-199, 216-220, 327-329, 331-333, 345-350, 500, 642-643, 688-690, 812-814
GL, *see* gradient layers
GNDs, *see* geometrically necessary dislocations
GNG, *see* gradient nano-grained
gradient 105-109, 112-115, 253-256, 267-268, 285-290, 301-304, 314-316, 324-326, 340-350, 449-452, 455-456, 459, 461-469, 552-554, 571-573
  structural 54, 121, 548
gradient boundary 203, 205-206, 208, 211, 214-215, 217-220
gradient distribution 132, 194, 522, 555, 565, 568, 570, 578

gradient layers (GL) 56-57, 89-93, 284-287, 289-291, 293-304, 343, 421, 423, 425, 427-429, 432-433, 435, 441-442, 444, 555
gradient magnitude 425, 561, 578-579
gradient materials 12, 120-121, 127, 132, 159, 205, 253, 265, 312, 319, 362, 516, 528, 553-554, 559
  mechanical behavior of 126, 253, 362
  synergetic strengthening in 547-556
gradient microstructure 258, 273, 420-421, 424-428, 430, 432-433, 438, 442-443, 461, 522, 549, 560
  dual 426, 436, 439-441, 444-445
gradient nano-grained (GNG) 44, 62-63, 86, 371-372, 374, 376, 378, 380, 382, 384, 386, 391, 642
gradient plastic strain 451, 713
gradient region 328, 337-338, 340, 342, 348
gradient structure 53-56, 89-101, 105-114, 283-286, 298-305, 311-320, 324-336, 350-352, 450-456, 465-469, 474, 499-501, 527-529, 547-549, 577-579
  nano-layer of 225-234
  strengthening and strain hardening in 50, 105
gradient structured (GS) 54-57, 105-109, 112-115, 121-122, 226-228, 285-290, 301-304, 361-368, 420-421, 429-430, 449-452, 461-469, 500-502, 515-516, 668

gradient surface layers (GSL) 325–326, 340, 342, 345, 548–549, 551
gradient textures 496, 560
gradient zones 213, 547, 554
grain boundaries (GBs) 42–43, 330–331, 345, 380–381, 383–385, 494–495, 532–534, 540, 593–598, 609, 708, 752, 760, 763–764, 801–802, 806
grain boundary migration 252, 269–270, 273, 654
grain boundary sliding 39, 43, 49, 82, 689, 750
grain coalescence 372, 382–383
grain coarsening 252, 269, 536
grain diameter 423, 425, 705
grain growth 63, 106, 136, 138, 232, 312, 373, 383, 500, 750, 763
grain orientations 382–383, 682
grain reference orientation distribution (GROD) 598, 610
grain refinement mechanisms 529, 532
grain rotations 39, 82, 372, 383, 760
grains 374–375, 377–378, 380–385, 434, 518–520, 577–579, 594–595, 625–629, 726–729, 738–742, 778, 783–784, 786–789, 805–808, 812–813
  austenite 782, 785
  larger 266, 338, 385, 424, 440, 595
  parent 589, 593–594
  uniform 587, 595, 597
grain size distribution 68, 90, 122, 139, 241–242, 364, 422, 424, 493, 520, 608, 618, 630, 632–633, 644

grain size gradient 54–56, 66, 106, 371, 373, 378, 385, 391, 405, 411–412, 418, 424, 426, 476, 668
grain sizes 42–43, 54–56, 68, 91, 133, 209, 241, 325, 327–328, 337, 339, 341, 346, 348, 364, 374–375, 391, 395, 411, 442, 450, 455, 493–494, 523, 536, 618, 668
  average transversal 269, 364
  decreasing 43, 373–374, 383, 419, 750, 814
  increasing 106, 252, 285, 362, 427
  reference 333, 338, 340
  smaller average 390, 410–411, 413
  transverse 520, 648–649
  ultrafine 49
grain structures 172, 588–589, 594, 597
grain to grain yielding 595, 738–739
grant numbers 26, 115
GROD, *see* grain reference orientation distribution
GS, *see* gradient structured
GSL, *see* gradient surface layers

HBAR, *see* hetero-boundary-affected region
HDI, *see* hetero-deformation-induced
HEAs, *see* high-entropy alloys
height profiles 59, 210–211, 220, 327, 341–342, 456, 459
hetero-boundary-affected region (HBAR) 26, 132–133, 143, 149–150, 157–164, 166–168, 170, 172, 174, 192–200, 204–205, 207, 211, 215, 217–218

826 | Index

hetero-deformation-induced (HDI) 4, 17–18, 43, 105–114, 132, 158–159, 178, 191–200, 204, 218, 226, 450–451, 499–500, 508–510, 659–660

heterogeneous grain structure (HGS) 338, 340, 342–343, 348–349, 585–592, 594–598, 606, 608–610, 613–615, 618, 620, 799

heterogeneous grain structures 596, 613–614, 618, 620, 799

heterostructured boundary 177–186, 192, 194–196, 198, 200, 205, 247

heterostructured lamella (HL) 4, 7, 11, 19, 24, 73–82, 84, 86, 88, 159, 500, 641, 643–644, 774–775, 785

heterostructured lamella structure (HLS) 4, 11, 19, 74, 159, 451, 641, 643–645, 647–648, 653–655, 659–660, 732, 774–775, 785

heterostructured zone boundaries 192, 245

HGS, see heterogeneous grain structure

high-entropy alloys (HEAs) 586, 590, 614–615, 629

high resolution electron microscope (HREM) 617, 778

high specific strength steel (HSSS) 721–724, 726–730, 732, 734, 736, 738, 740, 742, 749, 751–756, 758–759, 763, 767–768, 775, 799–800

HL, see heterostructured lamella

HL60 76–77, 79, 81

HLS, see heterostructured lamella structure

HSSS, see high specific strength steel

hysteresis loops 20, 79–80, 108, 111, 196, 501, 505, 507, 591–592, 606, 608, 732–733, 798, 803–804, 814

ICCM, see ion channeling contrast microscopy

inversed pole figure (IPF) 422, 424, 432–434, 453, 455, 617, 704, 760, 776–777, 801

ion channeling contrast microscopy (ICCM) 135–136, 160, 168, 670–671

IPF, see inversed pole figure

KAM, see kernel average misorientation

kernel average misorientation (KAM) 141, 505, 701, 706, 752, 764, 767, 801

LAGBs, see low angle grain boundaries

lamella coarse grains (LCG) 641–642, 647–651, 654–656, 658–660

laminate structures 4, 10, 19, 133, 146, 150, 205, 213, 667–669, 672–673, 675

lateral surface 59, 68, 207, 210, 212, 326–327, 329, 331, 333, 341–343, 345, 349–350, 355, 454, 456–459

lattice strains 563, 722, 724, 726, 734–735, 738, 740

layer boundaries 97, 208–209, 211–213, 328, 332

layers
  bronze 136, 138–140, 146–147, 160, 168, 172
  component 215, 324, 327, 330, 333–334, 336, 345, 351, 353
  copper 136, 138–140, 141, 145–147

free-standing 328–329, 353, 355
gradient-structured 91–94
nano-grained 62, 324, 348, 350
nanostructured 56, 91, 147, 226–227, 254, 259
outermost 55, 396–397, 531
plastic 124, 126–127, 302
stable 58, 62, 334, 485
surface 94–96, 120, 122, 124, 265, 271, 273, 375–379, 397, 399, 405, 423, 425, 515, 521
surface gradient 390, 405, 412
thick 364, 476
topmost 255, 258, 270, 338, 342, 345–346, 348, 536, 549, 551
top surface 393–394, 518, 520, 523, 538
unstable surface 58, 485
layer thickness 94, 136, 160, 168, 193, 196, 332, 337, 340, 346, 492
LCG, *see* lamella coarse grains
LD, *see* loading direction
loading direction (LD) 83, 161, 425, 433, 435–437, 441, 815
loading-unloading-reloading (LUR) 79, 84, 196, 451, 453, 591, 598, 608, 681, 686, 725, 775, 801, 803–804, 812–813
load transfer 586, 722, 734, 738, 740–742, 775, 786–787, 789
localized strain zone (LSZ) 226–234
local misorientations 141–142, 505
long-range internal stresses 19, 21, 69, 86, 197, 199, 246, 248, 332, 500, 505, 787
low angle grain boundaries (LAGBs) 295, 763–764
LSZ, *see* localized strain zone
LUR, *see* loading-unloading-reloading

martensite 288, 293–295, 297–301, 559–560, 562–579, 646, 655, 705, 723
martensite grains 299, 776, 782
martensite phase 629, 646, 775, 780, 784
martensite plates 284, 289, 300, 564, 566, 568, 571–573
martensite structures 561, 572–574
martensite transformation 289, 295, 298–305, 705, 774–776, 780–781, 783–786, 789
martensitic steel 756
martensitic transformation 284–285, 293, 297, 303, 560, 563–566, 568–569, 572–573, 579
materials community 18–20, 22–23, 132
maximum stress criterion 36, 315, 318–319, 620
MEA, *see* medium-entropy alloy
measurements, experimental 337, 341, 343, 814
medium-entropy alloy (MEA) 585–587, 597, 613–616, 618, 620–622, 624–630, 632, 634
metallic materials 34–35, 63, 158, 516, 522–523, 613, 642, 654, 656, 798–799
metallic nanomaterials 44
microhardness gradient 424, 455, 490, 516, 521, 538, 541
microhardness measurements 393, 412, 791
microstructural evolution 293, 299, 315–316, 318, 521, 574, 579, 594, 631, 647, 728
microstructural heterogeneity 5, 115, 242, 508, 722–723, 739, 813
microstructural observations 168, 286, 551

microstructure evolution   317, 490, 527, 529, 615, 622, 630, 776
microstructure gradient   254, 256, 268, 490
microstructures
　coarse-grained   66, 90, 439
　deformed   138, 395, 397, 433–434, 598, 777–778
　mobile dislocation density   60–61, 67, 69–70, 292, 429, 432, 741, 782–783, 790
model material   286, 420, 516–517, 529
model predictions   337–338, 341, 343, 345, 347

nanocrystalline materials   13, 44, 474
nano-grained (NGs)   62–63, 121–122, 348, 350, 371–372, 374, 376, 378, 380, 390–391, 587–589, 593–595, 609, 648–650, 656–658
nanostructured gradient surface layer (NGSL)   258, 261
nanostructured materials   11–12, 14, 41, 48, 85, 254, 560
nanostructured metals   4, 12–14, 33–44, 46, 48, 50, 74, 86, 225, 284, 325, 392, 418, 428, 773
　ductility of   34–35, 42–43
nanostructured surface layer   60, 75, 114, 258, 315–316, 318–319
nano-twin (NT)   45, 284, 353, 372, 391, 609, 641–642, 648–650, 654–656, 658–660, 689, 699, 798
NGs, *see* nano-grained
NGSL, *see* nanostructured gradient surface layer
NT, *see* nano-twin

peak stress   620, 753
PFs, *see* pole figure
phase boundaries   698–700, 706, 708, 711–716, 718, 723, 738–741, 782, 784, 787, 798, 801, 806, 808, 812
phase transformation   285, 299, 302–303, 305, 587, 591, 627–631, 635, 757–758, 763, 774, 780, 782–784, 786, 788–789
plastic deformation   77–78, 114–115, 159–160, 165–166, 252–253, 264–265, 301, 303, 433–434, 440–441, 467–468, 482–483, 699, 739–741, 774–776
plastic deformation mechanisms   392, 440–442, 568, 773–774, 776, 778, 780, 782, 784, 786, 788, 790
plastic deformation problem   333, 335–336
plastic deformation stage   26, 463, 485
plastic strain   5–6, 8, 19, 82, 165, 167, 182–183, 243, 594–595, 699, 713, 717–718, 805–806, 809, 813–814
　higher   81–82, 178, 240, 568, 734
　normalized reverse   80, 507–508
　residual   592, 803–804
pole figure (PFs)   138, 140–142, 395, 418, 424–426, 434–437, 439, 441–442, 444, 495, 503, 617, 647, 704, 801–802
pyramidal dislocations   418, 431, 438

RASP, *see* rotationally accelerated shot peening
residual stress   119–124, 126–128, 206, 253

reverse plastic strain 79, 508, 732–733
room temperature 37–38, 42–43, 48, 66–67, 84, 136, 287, 530, 586–587, 620–621, 623–626, 631–635, 645–646, 680–681, 687
rotationally accelerated shot peening (RASP) 254–255, 272, 515–518, 520–524, 527–530, 537–538, 540–541, 549–550

SAED, *see* selected area electron diffraction
sample
  alloy 529–530, 534, 538–541
  edge 97, 569
  initial 563, 565–567, 577
  micro-duplex 697, 700–702, 705–707, 715–716
  size 34, 41, 43, 482
  surface 66, 75, 100, 134, 168, 179–182, 287, 325, 393, 421, 455, 483–484, 519, 522, 790–791
  thickness 59, 77, 91, 96, 100, 210–211, 213, 268, 289–290, 301, 317, 475
  total 340, 346
sample thickness direction 162, 212, 256
sample width direction 211, 244, 256, 260
SBs, *see* shear bands
  dispersed 247, 266, 268–269, 271, 273
selected area electron diffraction (SAED) 294–295, 299, 520, 533, 565, 574–576
severe plastic deformation (SPD) 12–14, 44, 47–48, 85, 232, 390, 450, 474, 521, 528, 532, 614, 669, 750, 774

shear band delocalization 225–234
shear bands (SBs) 226–227, 229, 233, 243–248, 251–254, 259–264, 266–273, 295, 297, 330–331, 505, 508, 648–649, 654, 805
shear deformation 625, 630, 632
shear displacements 314–318, 616, 620, 623–625, 630–633
shear strain 489–496, 510, 565, 571, 631–632, 809
  high 491, 493–495
  uniform 620, 622
shear stress 265, 315, 318, 429, 494, 630, 632, 634, 648
  applied 8, 19, 21, 23, 69, 246
SIMT, *see* stress-induced martensitic transformation
SMAT, *see* surface mechanical attrition treatment
SMGT, *see* surface mechanical grinding treatment
soft zones 5–6, 10–11, 18–19, 21, 23–25, 43, 105–107, 192, 240, 302, 363, 630, 642, 660
SPD, *see* severe plastic deformation
SRS, *see* strain rate sensitivity
  dynamic 410, 412
SS, *see* stainless steel
SSDs, *see* statistically stored dislocations
stainless steel (SS) 45, 186, 284, 304, 590, 641–644, 646–660
statistically stored dislocations (SSDs) 7–8, 43, 218, 363, 368, 685, 813
steel balls 491, 517–518
strain accumulation 198, 243, 245, 252
strain compatibility 97, 372–373, 376–377
strain contours 456, 459, 786, 791

830 | Index

strain curves 289, 701–702
  true 289, 402, 590, 731, 805
  true stress-true 343, 395, 403–404
strain distributions 81, 161, 194, 204, 213–214, 256, 262, 456, 699
strain gradient 5–6, 8–10, 58–59, 61–62, 97–100, 158–159, 178–179, 184, 191–195, 197–198, 211–217, 449–452, 460–466, 468–470, 713–714
  accumulation of 204, 215
  high 162, 368, 393, 397
  increasing 192–193, 197
  intensity of 195, 197–198, 200, 204, 215, 220
  large 4, 268, 450, 502, 511
  macroscopic 54, 100
  maximum 59, 230
  negative 178, 182–183, 194, 200
  plastic 5, 8, 192, 205, 451, 685, 688, 699, 706, 713, 717
strain gradient effects 714
strain gradient evolution 133, 193, 449
strain gradient peaks 97, 211, 213–214
strain gradient plasticity 100, 164, 166, 219
strain gradient theory 142, 698, 700, 713, 716–717, 806
strain gradient zone 211, 214, 216–219
strain hardening 55–57, 105–114, 122–124, 301–302, 362–363, 587, 624–631, 667–676, 721–724, 732, 738–740, 784–787, 797–800, 802–804, 812–814
  ability 399, 402, 405, 586, 683, 699–700, 779
  high 79, 121, 290, 324, 560, 774–775, 785
strain hardening behaviors 126, 206, 210, 290, 299, 341, 391–392, 400, 412, 714, 716, 721, 723, 775, 778
  apparent 390, 399, 402–403, 405, 412–413
strain hardening capability 192, 214, 230, 247, 269, 273, 324, 326, 345, 576, 596, 674, 723–724, 742
strain hardening effects 538, 629, 631
strain hardening exponent 390, 402, 412, 622, 702, 778–779
strain hardening rate 37, 42–44, 55–57, 108–109, 140, 210, 289, 343, 345, 428, 478, 480, 586–587, 673, 675, 698, 717, 804–805, 811–812
strain incompatibility 220, 465, 595, 722, 739
strain increment 67, 168, 287, 423, 790
strain intensity 162, 166, 245, 259, 261–262, 267
strain levels 136, 145, 285, 303, 579, 674
strain localization 35, 226, 229–230, 233, 243, 267, 314, 362, 373, 377, 391, 578, 615, 699
strain measurements 394, 456–458, 492, 725
strain partitioning 6, 10, 81–82, 284–286, 300–303, 315, 317, 722–724, 734–736, 738, 741–742, 775, 786, 789, 806
  dynamic 284–286, 300–302, 305
strain path 87, 215, 232, 510
strain peaks 161–162, 173, 261–262

strain range 60, 284, 286, 298,
    302–303, 465, 750, 768
strain rate 66–67, 84, 287, 312,
    389–395, 399–410, 412–413,
    423, 475–476, 613, 615, 626,
    679–690, 753–758, 760–762,
    764–768
  shear 69, 632, 634
strain rate jump tests 410,
    686–687
strain rate sensitivity (SRS) 37,
    42–45, 389–392, 394, 396,
    398, 400, 402, 404, 408–412,
    622, 686–690, 755, 758, 764
strain softening behaviors 402,
    405
strain stages, early 197, 199–200,
    261
strength differences 4, 699
strength-ductility 324, 346–347,
    586, 596, 615
strength-ductility combinations
    258, 486, 586, 591, 596
  good 324, 528, 539
  superior 204, 258, 312, 559,
    642
strength-ductility synergy 346,
    348
strength increment 547, 554–555
stress 17–26, 78–80, 97–100, 107,
    109–113, 366–368, 419–420,
    425–427, 467, 592–595,
    711–713, 722–725, 732–734,
    798–800, 809–814
  friction 504, 710–711
  local 184–185, 440
  long-range 8, 106–107, 560,
    685, 732
  true 36, 226, 394, 402,
    427–428, 431–432, 590, 673,
    683, 687
stress anisotropy 419–420,
    443–445

stress concentration 21, 23–24,
    107, 195, 205, 245, 247,
    264–265, 271, 273
  high 245, 247–248
stress concentration spot
    179–180
stress concept 20, 23
stress gradient 8, 90, 99–101,
    485, 500, 685, 688
  macroscopic 90, 101, 106
stress hardening 127, 811, 813
stress-induced martensitic
    transformation (SIMT) 560,
    566–567, 571, 573–574
stress relaxation tests 60–61, 292,
    410, 423, 429, 554, 783
stress states 106, 185, 199, 232,
    285, 327, 373, 376, 420, 643
stress-strain curves 41, 96, 168,
    341, 456, 463, 566, 574, 681,
    739
  tensile 49, 108, 139, 209, 671
  true 36, 467–468
strong strain hardening 230, 315,
    317, 327, 614, 623, 629, 631,
    740, 775, 783–784
subsequent tensile deformation
    563, 566, 569, 573, 780–781
superior dynamic shear properties
    311–320, 622, 634
superplastic behaviors 749–752,
    754, 756, 758, 760, 762, 764,
    766, 768
surface mechanical attrition
    treatment (SMAT) 55, 66,
    90–91, 121–122, 313,
    420–421, 425–426, 452,
    454–455, 473–476, 489–491,
    494–496, 501–502, 522–523,
    528–529
surface mechanical grinding
    treatment (SMGT) 325, 376,
    389–390, 392–393, 395, 397,
    399, 412, 479, 516, 528

Index

surface roughness 59, 68, 211, 227, 252, 254–255, 257, 271, 273, 522
surface-to-volume ratio 698, 715, 717–718

TBs, *see* twin boundaries
TD, *see* transversal direction
tensile curves 141, 149, 365, 367, 426–427, 540, 566, 574, 591, 606, 803
tensile deformation 435–436, 441, 565–566, 569–571, 592–595, 597–598, 705–707, 750–751, 756–762, 765–766, 774–775, 782–787, 798–801, 805–807, 812–814
  early stage of 125, 468
  uniform 13, 44, 66, 84, 668
tensile direction 81–82, 161–162, 175, 194, 210, 212–213, 259, 267, 368, 736, 752, 760, 786, 790, 809
tensile ductility 39, 74, 149, 672, 689
tensile samples 57, 68, 75–77, 84–85, 97, 136, 169, 196, 206, 209, 227, 288, 421, 440, 726
tensile strain 56, 67–68, 95–96, 112–113, 141–143, 212–214, 291, 301, 379–380, 382–384, 428, 430, 456–459, 705–707, 786–788
  applied 123, 126, 167, 228, 230, 233, 252–253, 287, 298–301, 459, 461–462, 465–466, 508, 734–735
  function of 298, 377–379, 382–383
  function of applied 298, 461–462, 465–466
  increasing 56, 60, 79–80, 112–113, 125, 142, 162, 166, 212, 253, 450, 477

varying 57, 84, 108–109, 456, 459, 467, 477, 609, 761–766, 787
tensile straining 198, 586–587, 593–595, 597, 706, 803
tensile strength 681–682, 684–685, 697–698, 700, 702, 704, 706, 708, 710, 712, 714, 716, 718, 722, 731
tensile stress 35, 59, 361, 478, 484–485, 567, 723
  lateral 58–59, 303, 376–377, 554
tensile tests 35–36, 41–43, 66–67, 84–85, 287–288, 508–509, 591–593, 598, 672–673, 699–701, 725–726, 776, 780–781, 784–786, 790–791
thin films 20, 185, 500, 687, 689, 724
top surface 57, 126, 291, 395, 399, 490, 492–494, 520, 524, 528, 537, 551
torsion-processed samples 560, 562–579
torsional strain, increasing 565, 567
transversal direction (TD) 425, 433, 435–437, 441
true strain 77, 81–82, 86, 291–292, 394, 399, 407, 409–410, 427–428, 507–508, 510, 672–673, 733, 736, 804–805
  fixed 407, 410
true stress curves 715, 731
true stress-strain curves 165, 198, 262, 290, 292, 399, 407, 410, 478, 778–779, 786, 788
twin boundaries (TBs) 8, 45–46, 158, 192, 246, 253, 313, 425, 455, 589, 594, 618, 625–629, 635, 783–784

UFG, *see* ultra-fine grains
ultimate strength   139–140, 150, 209, 481, 778
ultimate tensile stress (UTS)   566–567, 592, 653, 702, 731, 778
ultrafine-grained materials   12, 85
ultrafine grains   82, 132, 397, 519, 521, 528, 550, 643, 647–648, 749, 768, 777
ultra-fine grains (UFG)   74, 76–78, 205–207, 239–240, 246, 312, 390–392, 405, 587–590, 592–594, 615, 783, 785, 787, 789
ultrasonic shot peening (USP)   516, 522–523
uniaxial stress, applied   54, 58, 82, 97, 371, 385, 485
uniform deformation   37, 108, 229, 290, 292, 298–299, 302, 334, 428, 440, 469, 676
uniform elongation   34–37, 41–43, 142–143, 226, 258–259, 301, 304–305, 341, 589–590, 652, 671–674, 682–686, 701–702, 730–731, 778–779

unique extra strain hardening   54, 57
unloading-reloading process   108, 110, 505
unloading strains   108–109, 733, 801
unload strain   803–805
USP, *see* ultrasonic shot peening
UTS, *see* ultimate tensile stress

volume fraction   91–93, 298–299, 481, 553–555, 563–564, 568, 578, 646, 655–656, 658, 701–702, 710–711, 780–781, 783, 809–810

work hardening rate   34, 166, 509–510, 567, 683

X-ray diffraction (XRD)   121, 287–288, 418, 423, 425–426, 452, 562, 645, 701, 776, 790
XRD, *see* X-ray diffraction

zone boundaries   4–6, 8–10, 19, 21, 23–24, 107, 181, 186, 192–193, 205, 240, 245–248, 265–266, 273